NATURAL HISTORY
UNIVERSAL LIBRARY

U0215810

西方博物学大系

主编：江晓原

THE HERBALL OR GENERALL HISTORIE OF PLANTES

本草要义：改订版

[英] 约翰·杰拉德　著
[英] 托马斯·约翰逊　改订

华东师范大学出版社

图书在版编目（CIP）数据

本草要义：改订版 = The Herball or Generall Historie of Plantes：英文 /（英）约翰·杰拉德著. — 上海：华东师范大学出版社，2018
（寰宇文献）
ISBN 978-7-5675-7716-9

Ⅰ.①本… Ⅱ.①约… ②托… Ⅲ.①草本植物–英文 Ⅳ.①Q949.4

中国版本图书馆CIP数据核字(2018)第096292号

本草要义：改订版
The Herball or Generall Historie of Plantes
（英）约翰·杰拉德著 （英）托马斯·约翰逊改订

特约策划　黄曙辉　徐　辰
责任编辑　庞　坚
特约编辑　许　倩
装帧设计　刘怡霖

出版发行　华东师范大学出版社
社　　址　上海市中山北路3663号　邮编 200062
网　　址　www.ecnupress.com.cn
电　　话　021-60821666　行政传真　021-62572105
客服电话　021-62865537
门市（邮购）电话　021-62869887
地　　址　上海市中山北路3663号华东师范大学校内先锋路口
网　　店　http://hdsdcbs.tmall.com/

印刷者　虎彩印艺股份有限公司
开　　本　16开
印　　张　108.75
版　　次　2018年6月第1版
印　　次　2018年6月第1次
书　　号　ISBN 978-7-5675-7716-9
定　　价　1880.00元（精装全二册）

出版人　王　焰

（如发现本版图书有印订质量问题，请寄回本社客服中心调换或电话021-62865537联系）

总　目

《西方博物学大系》总序

江晓原

　　《西方博物学大系》收录博物学著作超过一百种，时间跨度为15世纪至1919年，作者分布于16个国家，写作语种有英语、法语、拉丁语、德语、弗莱芒语等，涉及对象包括植物、昆虫、软体动物、两栖动物、爬行动物、哺乳动物、鸟类和人类等，西方博物学史上的经典著作大备于此编。

中西方"博物"传统及观念之异同

　　今天中文里的"博物学"一词，学者们认为对应的英语词汇是Natural History，考其本义，在中国传统文化中并无现成对应词汇。在中国传统文化中原有"博物"一词，与"自然史"当然并不精确相同，甚至还有着相当大的区别，但是在"搜集自然界的物品"这种最原始的意义上，两者确实也大有相通之处，故以"博物学"对译Natural History一词，大体仍属可取，而且已被广泛接受。

　　已故科学史前辈刘祖慰教授尝言：古代中国人处理知识，如开中药铺，有数十上百小抽屉，将百药分门别类放入其中，即心安矣。刘教授言此，其辞若有憾焉——认为中国人不致力于寻求世界"所以然之理"，故不如西方之分析传统优越。然而古代中国人这种处理知识的风格，正与西方的博物学相通。

　　与此相对，西方的分析传统致力于探求各种现象和物体之间的相互关系，试图以此解释宇宙运行的原因。自古希腊开始，西方哲人即孜孜不倦建构各种几何模型，欲用以说明宇宙如何运行，其中最典型的代表，即为托勒密（Ptolemy）的宇宙体系。

　　比较两者，差别即在于：古代中国人主要关心外部世界"如何"运行，而以希腊为源头的西方知识传统（西方并非没有别的知识传统，只是未能光大而已）更关心世界"为何"如此运行。在线性发展无限进步的科学主义观念体系中，我们习惯于认为"为何"是在解决了"如何"之后的更高境界，故西方的分析传统比中国的传统更高明。

然而考之古代实际情形，如此简单的优劣结论未必能够成立。例如以天文学言之，古代东西方世界天文学的终极问题是共同的：给定任意地点和时刻，计算出太阳、月亮和五大行星（七政）的位置。古代中国人虽不致力于建立几何模型去解释七政"为何"如此运行，但他们用抽象的周期叠加（古代巴比伦也使用类似方法），同样能在足够高的精度上计算并预报任意给定地点和时刻的七政位置。而通过持续观察天象变化以统计、收集各种天象周期，同样可视之为富有博物学色彩的活动。

还有一点需要注意：虽然我们已经接受了用"博物学"来对译 Natural History，但中国的博物传统，确实和西方的博物学有一个重大差别——即中国的博物传统是可以容纳怪力乱神的，而西方的博物学基本上没有怪力乱神的位置。

古代中国人的博物传统不限于"多识于鸟兽草木之名"。体现此种传统的典型著作，首推晋代张华《博物志》一书。书名"博物"，其义尽显。此书从内容到分类，无不充分体现它作为中国博物传统的代表资格。

《博物志》中内容，大致可分为五类：一、山川地理知识；二、奇禽异兽描述；三、古代神话材料；四、历史人物传说；五、神仙方伎故事。这五大类，完全符合中国文化中的博物传统，深合中国古代博物传统之旨。第一类，其中涉及宇宙学说，甚至还有"地动"思想，故为科学史家所重视。第二类，其中甚至出现了中国古代长期流传的"守宫砂"传说的早期文献：相传守宫砂点在处女胳膊上，永不褪色，只有性交之后才会自动消失。第三类，古代神话传说，其中甚至包括可猜想为现代"连体人"的记载。第四类，各种著名历史人物，比如三位著名刺客的传说，此三名刺客及所刺对象，历史上皆实有其人。第五类，包括各种古代方术传说，比如中国古代房中养生学说，房中术史上的传说人物之一"青牛道士封君达"等等。前两类与西方的博物学较为接近，但每一类都会带怪力乱神色彩。

"所有的科学不是物理学就是集邮"

在许多人心目中，画画花草图案，做做昆虫标本，拍拍植物照片，这类博物学活动，和精密的数理科学，比如天文学、物理学等等，那是无法同日而语的。博物学显得那么的初级、简单，甚至幼稚。这种观念，实际上是将"数理程度"作为唯一的标尺，用来衡量一切知识。但凡能够使用数学工具来描述的，或能够进行物理实验的，那就是"硬"科学。使用的数学工具越高深越复杂，似乎就越"硬"；物理实验设备越庞大，花费的金钱越多，似乎就越"高端"、越"先进"……

这样的观念，当然带着浓厚的"物理学沙文主义"色彩，在很多情况下是不正确的。而实际上，即使我们暂且同意上述"物理学沙文主义"的观念，博物学的"科学地位"也仍然可以保住。作为一个学天体物理专业出身，因而经常徜徉在"物理学沙文主义"幻影之下的人，我很乐意指出这样一个事实：现代天文学家们的研究工作中，仍然有绘制星图，编制星表，以及为此进行的巡天观测等等活动，这些活动和博物学家"寻花问柳"，绘制植物或昆虫图谱，本质上是完全一致的。

这里我们不妨重温物理学家卢瑟福（Ernest Rutherford）的金句："所有的科学不是物理学就是集邮（All science is either physics or stamp collecting）。"卢瑟福的这个金句堪称"物理学沙文主义"的极致，连天文学也没被他放在眼里。不过，按照中国传统的"博物"理念，集邮毫无疑问应该是博物学的一部分——尽管古代并没有邮票。卢瑟福的金句也可以从另一个角度来解读：既然在卢瑟福眼里天文学和博物学都只是"集邮"，那岂不就可以将博物学和天文学相提并论了？

如果我们摆脱了科学主义的语境，则西方模式的优越性将进一步被消解。例如，按照霍金（Stephen Hawking）在《大设计》（*The Grand Design*）中的意见，他所认同的是一种"依赖模型的实在论（model-dependent realism）"，即"不存在与图像或理论无关的实在性概念（There is no picture- or theory-independent concept of reality）"。在这样的认识中，我们以前所坚信的外部世界的客观性，已经不复存在。既然几何模型只不过是对外部世界图像的人为建构，则古代中国人干脆放弃这种建构直奔应用（毕竟在实际应用中我们只需要知道七政"如何"运行），又有何不可？

传说中的"神农尝百草"故事，也可以在类似意义下得到新的解读："尝百草"当然是富有博物学色彩的活动，神农通过这一活动，得知哪些草能够治病，哪些不能，然而在这个传说中，神农显然没有致力于解释"为何"某些草能够治病而另一些则不能，更不会去建立"模型"以说明之。

"帝国科学"的原罪

今日学者有倡言"博物学复兴"者，用意可有多种，诸如缓解压力、亲近自然、保护环境、绿色生活、可持续发展、科学主义解毒剂等等，皆属美善。编印《西方博物学大系》也是意欲为"博物学复兴"添一助力。

然而，对于这些博物学著作，有一点似乎从未见学者指出过，而鄙意以为，当我们披阅把玩欣赏这些著作时，意识到这一点是必须的。

这百余种著作的时间跨度为15世纪至1919年，注意这个时间跨度，正是西方列强"帝国科学"大行其道的时代。遥想当年，帝国的科学家们乘上帝国的军舰——达尔文在皇家海军"小猎犬号"上就是这样的场景之一，前往那些已经成为帝国的殖民地或还未成为殖民地的"未开化"的遥远地方，通常都是踌躇满志、充满优越感的。

作为一个典型的例子，英国学者法拉在（Patricia Fara）《性、植物学与帝国：林奈与班克斯》（*Sex, Botany and Empire, The Story of Carl Linnaeus and Joseph Banks*）一书中讲述了英国植物学家班克斯（Joseph Banks）的故事。1768年8月15日，班克斯告别未婚妻，登上了澳大利亚军舰"奋进号"。此次"奋进号"的远航是受英国海军部和皇家学会资助，目的是前往南太平洋的塔希提岛（Tahiti，法属海外自治领，另一个常见的译名是"大溪地"）观测一次比较罕见的金星凌日。舰长库克（James Cook）是西方殖民史上最著名的舰长之一，多次远航探险，开拓海外殖民地。他还被认为是澳大利亚和夏威夷群岛的"发现"者，如今以他命名的群岛、海峡、山峰等不胜枚举。

当"奋进号"停靠塔希提岛时，班克斯一下就被当地美丽的土著女性迷昏了，他在她们的温柔乡里纵情狂欢，连库克舰长都看不下去了，"道德愤怒情绪偷偷溜进了他的日志当中，他发现自己根本不可能不去批评所见到的滥交行为"，而班克斯纵欲到了"连嫖妓都毫无激情"的地步——这是别人讽刺班克斯的说法，因为对于那时常年航行于茫茫大海上的男性来说，上岸嫖妓通常是一项能够唤起"激情"的活动。

而在"帝国科学"的宏大叙事中，科学家的私德是无关紧要的，人们关注的是科学家做出的科学发现。所以，尽管一面是班克斯在塔希提岛纵欲滥交，一面是他留在故乡的未婚妻正泪眼婆娑地"为远去的心上人绣织背心"，这样典型的"渣男"行径要是放在今天，非被互联网上的口水淹死不可，但是"班克斯很快从他们的分离之苦中走了出来，在外近三年，他活得倒十分滋润"。

法拉不无讽刺地指出了"帝国科学"的实质："班克斯接管了当地的女性和植物，而库克则保护了大英帝国在太平洋上的殖民地。"甚至对班克斯的植物学本身也调侃了一番："即使是植物学方面的科学术语也充满了性指涉。……这个体系主要依靠花朵之中雌雄生殖器官的数量来进行分类。"据说"要保护年轻妇女不受植物学教育的浸染，他们严令禁止各种各样的植物采集探险活动。"这简直就是将植物学看成一种"涉黄"的淫秽色情活动了。

在意识形态强烈影响着我们学术话语的时代，上面的故事通常是这样被描述的：库克舰长的"奋进号"军舰对殖民地和尚未成为殖民地的那些地方的所谓"访问"，其实是殖民者耀武扬威的侵略，搭载着达尔文的"小猎犬号"军舰也是同样行径；班克斯和当地女性的纵欲狂欢，当然是殖民者对土著妇女令人发指的蹂躏；即使是他采集当地植物标本的"科学考察"，也可以视为殖民者"窃取当地经济情报"的罪恶行为。

后来改革开放，上面那种意识形态话语被抛弃了，但似乎又走向了另一个极端，完全忘记或有意回避殖民者和帝国主义这个层面，只歌颂这些军舰上的科学家的伟大发现和成就，例如达尔文随着"小猎犬号"的航行，早已成为一曲祥和优美的科学颂歌。

其实达尔文也未能免俗，他在远航中也乐意与土著女性打打交道，当然他没有像班克斯那样滥情纵欲。在达尔文为"小猎犬号"远航写的《环球游记》中，我们读到："回程途中我们遇到一群黑人姑娘在聚会，……我们笑着看了很久，还给了她们一些钱，这着实令她们欣喜一番，拿着钱尖声大笑起来，很远还能听到那愉悦的笑声。"

有趣的是，在班克斯在塔希提岛纵欲六十多年后，达尔文随着"小猎犬号"也来到了塔希提岛，岛上的土著女性同样引起了达尔文的注意，在《环球游记》中他写道："我对这里妇女的外貌感到有些失望，然而她们却很爱美，把一朵白花或者红花戴在脑后的髮髻上……"接着他以居高临下的笔调描述了当地女性的几种发饰。

用今天的眼光来看，这些在别的民族土地上采集植物动物标本、测量地质水文数据等等的"科学考察"行为，有没有合法性问题？有没有侵犯主权的问题？这些行为得到当地人的同意了吗？当地人知道这些行为的性质和意义吗？他们有知情权吗？……这些问题，在今天的国际交往中，确实都是存在的。

也许有人会为这些帝国科学家辩解说：那时当地土著尚在未开化或半开化状态中，他们哪有"国家主权"的意识啊？他们也没有制止帝国科学家的考察活动啊？但是，这样的辩解是无法成立的。

姑不论当地土著当时究竟有没有试图制止帝国科学家的"科学考察"行为，现在早已不得而知，只要殖民者没有记录下来，我们通常就无法知道。况且殖民者有军舰有枪炮，土著就是想制止也无能为力。正如法拉所描述的："在几个塔希提人被杀之后，一套行之有效的易货贸易体制建立了起来。"

即使土著因为无知而没有制止帝国科学家的"科学考察"行为，这事也很像一个成年人闯进别人的家，难道因为那家只有不懂事的小孩子，闯入者就可以随便打探那的隐私、拿走那家的东西、甚至将那家的房屋土地据为己有吗？事实上，很多情况下殖民者就是这样干的。所以，所谓的"帝国科学"，其实是有着原罪的。

如果沿用上述比喻，现在的局面是，家家户户都不会只有不懂事的孩子了，所以任何外来者要想进行"科学探索"，他也得和这家主人达成共识，得到这家主人的允许才能够进行。即使这种共识的达成依赖于利益的交换，至少也不能单方面强加于人。

博物学在今日中国

博物学在今日中国之复兴，北京大学刘华杰教授提倡之功殊不可没。自刘教授大力提倡之后，各界人士纷纷跟进，仿佛昔日蔡锷在云南起兵反袁之"滇黔首义，薄海同钦，一檄遥传，景从恐后"光景，这当然是和博物学本身特点密切相关的。

无论在西方还是在中国，无论在过去还是在当下，为何博物学在它繁荣时尚的阶段，就会应者云集？深究起来，恐怕和博物学本身的特点有关。博物学没有复杂的理论结构，它的专业训练也相对容易，至少没有天文学、物理学那样的数理"门槛"，所以和一些数理学科相比，博物学可以有更多的自学成才者。这次编印的《西方博物学大系》，卷帙浩繁，蔚为大观，同样说明了这一点。

最后，还有一点明显的差别必须在此处强调指出：用刘华杰教授喜欢的术语来说，《西方博物学大系》所收入的百余种著作，绝大部分属于"一阶"性质的工作，即直接对博物学作出了贡献的著作。事实上，这也是它们被收入《西方博物学大系》的主要理由之一。而在中国国内目前已经相当热的博物学时尚潮流中，绝大部分已经出版的书籍，不是属于"二阶"性质（比如介绍西方的博物学成就），就是文学性的吟风咏月野草闲花。

要寻找中国当代学者在博物学方面的"一阶"著作，如果有之，以笔者之孤陋寡闻，唯有刘华杰教授的《檀岛花事——夏威夷植物日记》三卷，可以当之。这是刘教授在夏威夷群岛实地考察当地植物的成果，不仅属于直接对博物学作出贡献之作，而且至少在形式上将昔日"帝国科学"的逻辑反其道而用之，岂不快哉！

2018 年 6 月 5 日
于上海交通大学
科学史与科学文化研究院

约翰·杰拉德
（1545-1612）

约翰·杰拉德（John Gerarde）生于英国柴郡的楠特威奇，年少时没受过多少教育就去伦敦当学徒，出师后成为草泽医生。他对植物学和园艺颇有兴趣，不但自学成才，还成功在伦敦城内的霍本营造了一座庭园，并和多名冒险家签合同，委托他们从海外收集珍奇的罕见植物。他主张植物学家应多多随船出海，到外国进行实地研究，自己却属于闭门造车的典范，一辈子几乎没离开过自己的园子，只在一艘航行北海的商船上短期受雇当过医官。

1597年，杰拉德出版了自己生涯中最重要的著作《本草要义》，这本书集伦贝特·多东恩斯、马提亚斯·罗贝尔、雅各·狄奥多尔等植物学大家的研究于一身，加上杰拉德自己的植物学知识和文学性极佳的笔法，将西方人所知及未知世界的植物传承故事、栽培方法、分类法等内容娓娓道来，可谓脍炙人口，出版后即风行一时。本书厚达1400页，配以2000幅插绘，在一定程度上打破了迪奥科里斯所著《药物论》一千五百年来的独霸地位，是英国当代植物学研究的奠基之作。

不过，杰拉德本人的植物学专业知识毕竟有限，书中引用及叙述谬误之处颇多。他去世二十年后，坊间风传约翰·帕金森将要推出一本质量更佳的植物学著作（就是后来的《植物学剧场》）。为了竞争，杰拉德的继承人委托著名的药草种子商人兼植物学家托马斯·约翰逊（Thomas Johnson）对《本草要义》进行更正改订。约翰逊改订版《本草要义》于1633年初版，之后多次再版。改订版的专业程度远远优于原初版本，且厚度达到1700页，配图2700余幅，并添加了一篇优雅的序言。

《本草要义：改订版》在之后的二百年间一直是西方实用本草书籍，早期的英国殖民者赶赴北美大陆定居时，本书为生活必备书籍。

本书改订者托马斯·约翰逊却没能像杰拉德那样寿终正寝，在十七世纪中期的英国内战期间，他作为保王党的一员上了战场。因负伤过重，他的草药学知识救不了自己，死在病榻上。

Ceres

Pomona

יהוה

Ecce dedi vobis omnes herbas sementantes semen, quæ sunt. 1.29.

Excideret ne tibi diuini muneris Author
Præsentem monstrat quælibet herba Deum.

THEOPHRASTVS

DIOSCORIDES

THE HERBALL OR GENERALL Historie of Plantes.

Gathered by Iohn Gerarde of London Master in CHIRVRGERIE

Very much Enlarged and Amended by. Thomas Iohnson Citizen and Apothecarye. of LONDON

London Printed by Adam Islip Ioice Norton and Richard Whitakers Anno 1636.

Io: Payne Sculp:

VIRIS

PRVDENTIA, VIRTVTE,
ARTE, RERVMQVE VSV SPECTATISSIMIS,
DIGNISSIMIS,

RICHARDO EDWARDS
RECTORI, SIVE MAGISTRO;

EDWARDO COOKE, LEONARDO
STONE, GVARDIANIS,

CÆTERISQVE CLARISS. SOCIET.
PHARMACEVT. LOND. SOCIIS

HOS SVOS IN EMA-
CVLANDO, AVGENDOQVE
HANC PLANTARVM
HISTORIAM,

LABORES, STVDIORVM BOTANICORVM
SPECIMEN, AMORIS SYMBOLVM
EX ANIMO

D. D.

VESTRÆ, PVBLICÆQVE VTILI-
TATIS STVDIOSISSIMVS

THOM. IOHNSON.

TO THE RIGHT HONORABLE
HIS SINGVLAR GOOD LORD AND
MASTER, SIR WILLIAM CECIL KNIGHT, BARON OF
Burghley, Master of the Court of Wards and Liueries, Chancellor of the
Vniuersitie of Cambridge, Knight of the most noble Order of the Garter,
one of the Lords of her Majesties most honorable Priuy Coun-
cell, and Lord high Treasurer of *England*.

Mong the manifold creatures of God (right Honora-
ble, and my singular good Lord) that haue all in all
ages diuersly entertained many excellent wits, and
drawn them to the contemplation of the diuine wisdome, none haue prouoked mens studies more, or satisfied their desires so much as plants haue done, and
that vpon just and worthy causes : for if delight may
prouoke mens labor, what greater delight is there
than to behold the earth apparelled with plants, as
with a robe of embroidered worke, set with Orient
pearles and garnished with great diuersitie of rare and costly jewels? If this varietie and perfection of colours may affect the eie, it is such in herbs and floures,
that no *Apelles*, no *Zeuxis* euer could by any art expresse the like : if odours or if
taste may worke satisfaction, they are both so soueraigne in plants, and so comfortable that no confection of the Apothecaries can equall their excellent vertue.
But these delights are in the outward senses: the principal delight is in the mind,
singularly enriched with the knowledge of these visible things, setting forth to vs
the inuisible wisdome and admirable workmanship of Almighty God. The delight is great, but the vse greater, and ioyned often with necessitie. In the first ages
of the world they were the ordinary meate of men, and haue continued euer
since of necessary vse both for meates to maintaine life, and for medicine to recouer health. The hidden vertue of them is such, that (as *Pliny* noteth) the very   Plin.l.8,ca.27.
bruit beasts haue found it out: and (which is another vse that he obserues) from   Ibid.l.22,ca.2.
thence the Dyars tooke the beginning of their Art.

   Furthermore, the necessary vse of those fruits of the earth doth plainly appeare
by the great charge and care of almost all men in planting & maintaining of gardens, not as ornaments onely, but as a necessarie prouision also to their houses.
And here beside the fruit, to speake againe in a word of delight, gardens, especia-
ly such as your Honor hath, furnished with many rare Simples, do singularly delight, when in them a man doth behold a flourishing shew of Summer beauties
in the midst of Winters force, and a goodly spring of flours, when abroad a leafe
is not to be seene. Besides these and other causes, there are many examples of
those that haue honoured this science : for to passe by a multitude of the Philoso-
phers, it may please your Honor to call to remembrance that which you know of
some noble Princes, that haue ioyned this study with their most important mat-

Plut. de Difer.
adul. & amic.
Plin. lib.25.
cap.2.

ters of ftate : *Mithridates* the great was famous for his knowledge herein,as *Plu-tarch* noteth. *Euax* alfo King of Arabia,the happy garden of the world for prin-cipall Simples,wrot of this argument,as *Pliny* fheweth. *Dioclefian* likewife,might haue had his praife,had he not drowned all his honour in the bloud of his perfe-cution. To conclude this point, the example of *Solomon* is before the reft, and greater,whofe wifdome and knowledge was fuch, that hee was able to fet out the nature of all plants from the higheft Cedar to the loweft Moffe. But my very good Lord,that which fometime was the ftudy of great Phylofophers and migh-tie Princes,is now neglected,except it be of fome few,whofe fpirit and wifdome hath carried them among other parts of wifdome and counfell,to a care and ftu-die of fpeciall herbes,both for the furnifhing of their gardens,and furtherance of their knowledge:among whom I may iuftly affirme and publifh your Honor to be one,being my felfe one of your feruants,and a long time witneffe thereof: for vnder your Lordfhip I haue ferued,and that way emploied my principall ftudy and almoft all my time,now by the fpace of twenty yeares. To the large and fin-gular furniture of this noble Ifland I haue added from forreine places all the va-rietie of herbes and floures that I might any way obtaine, I haue laboured with the foile to make it fit for plants, and with the plants, that they might delight in the foile, that fo they might liue and profper vnder our clymat, as in their natiue and proper countrey : what my fucceffe hath beene,and what my furniture is, I leaue to the report of they that haue feene your Lordfhips gardens, and the lit-tle plot of myne owne efpeciall care and husbandry. But becaufe gardens are priuat,and many times finding an ignorant or a negligent fucceffor,come foone to ruine,there be that haue follicited me,firft by my pen, and after by the Preffe to make my labors common,and to free them from the danger wherunto a gar-den is fubject: wherein when I was ouercome,and had brought this Hiftory or report of the nature of Plants to a juft volume, and had made it (as the Reader may by comparifon fee)richer than former Herbals,I found it no queftion vnto whom I might dedicate my labors;for confidering your good Lordfhip,I found none of whofe fauour and goodnes I might fooner prefume,feeing I haue found you euer my very good Lord and Mafter.Againe,confidering my duty and your Honors merits, to whom may I better recommend my Labors,than to him vnto whom I owe my felfe, and all that I am able in your feruice or deuotion to per-forme ? Therefore vnder hope of your Honorable and accuftomed fauor I pre-fent this Herball to your Lordfhips protection ; and not as an exquifite Worke (for I know my meanneffe)but as the greateft gift and chiefeft argument of duty that my labour and feruice can affoord:wherof if there be no other fruit,yet this is of fome vfe, that I haue miniftred Matter for Men of riper wits and deeper judgements to polifh,and to adde to my large additions where any thing is de-fectiue, that in time the Worke may be perfect. Thus I humbly take my leaue, befeeching God to grant you yet many daies to liue to his glory,to the fupport of this State vnder her Majeftie our dread Soueraigne,and that with great increafe of honor in this world,and all fulneffe of glory in the world to come.

Your Lordfhips moft humble

and obedient Seruant,

IOHN GERARD.

LANCELOTVS BRVNIVS MEDICVS REGINEVS
IOHANNI GERARDO *Chirurgo peritißimo*
& rei Herbariæ callentißimo S.P.D.

VM fingularum medicinæ partium cognitio atque intelligentia libero homine digna cenfenda eft; tum earum nulla vel antiquitate, vel dignitate, vel vtilitate, vel denique iucunditate cum ftirpium cognitione iure comparari debet. Antiquiffimam eam effe ex eo liquet, quòd quum ceteræ medicinæ partés (ficut reliquæ etiam artes) ab ipfis hominibus (prout eos dura preffit neceffitas) primum excogitatæ & inuentæ fuerunt : fola herbarum arborumque cognitio ante hominem formatum condita, eidemque mòx creato ab ipfo mundi architecto donata videri poteft. Cuius tanta apud antiqua fecula exiftimatio ac dignitas erat, vt & ipfius inuentionem fapientiffimo Deorum Apollini veteres tribuerint, & reges celeberrimi in ftirpium viribus indagandis ftudium laboremque fuum confumere, fummæ fibi apud pofteros laudi honorique futurum cenfuerint. Iam verò plantarum vtilitas, atq; etiam neceffitas, adeò latè patet, vt eius immenfitatem nullius vel acutiffimi hominis animus capere, nedum meus calamus exprimere queat. Stirpium enim complurimæ nobis in cibos, alimentumque cedunt: innumeræ aduerfus morbus remedia fuppeditant : ex alijs domos, naues, inftrumenta tam bellica quam ruftica fabricamus : aliquot etiam earum veftes noftris corporibus fubminiftrant. In quibus fingulis recenfendis diutiùs perfiftere, hominis effet intemperantèr abutentis & otio & literis. Quantas autem & quam varias voluptates ex ftirpium fiue amœnitate oculis capiamus, fiue fragrantia naribus hauriamus, fine fumma in earum conditorem impietate inficiari non poffumus. Adeò vt abfque ftirpium ope & fubfidio vita nobis ne vitalis quidem haberi, debeat.

Quum igitur res plantaria reliquis omnibus medicinæ partibus antiquitate antecedat, dignitate, nulli cedat, vtilitate infuper oblectationeque cæteras longè fuperet, quis futurus eft, adeo, aut infenfatus vt non exploratum habeat, aut ingratus, vt non ingenuè agnofcat, quanta vniuerfis Anglis commoda, quantafque voluptates tuus mi *Gerarde* in ftirpium inueftigatione & cultu labor indefeffus, ftudium inexhauftum, immenfique fumptus hoc de ftirpibus edito libro allaturi funt. Macte itaque ifta tua virtute, iftoque de republica benè merendi ftudio, & quod infigni tua cum laude ingreffus es vertutis gloriæque curriculum, eidem infifte animofè & gnauitèr, neq; à re plantaria promouenda prius defifte, quam eam à te ad vmbilicum jam fermè productam ipfe plenè abfoluas atque perficias. Sic enim & tibi adhuc fuperftiti gloriam paries immortalem, & poft obitum tantam tui nominis celebritatem relinques, vt tuarum laudum opus pofteros noftros nulla vnquam captura fit oblivio. Bene vale. Ex Aula Reginea Weftm. ipfis Cal. Decemb. 1597.

MAT-

MATTHIAS DE LOBEL
IOHANNI GERARDO
felicitatem.

Vum Londinum appuli,in sinu gauisus sum Gerarde amicissime,dum typographo formis excudenda Plantarum collectanea tua commissa vidi, de quibus summas,nulla die perituras laudes Anglia tibi Rei-herbaria familiam vniuersam, medicatricis artis partem, antiquissimum, iucundissimum & vtilissimum studium,retegere cupido,debet. Priscorum enim Theophrasti,Dioscoridis, Plinij, & Galeni scripta, passim toto orbe pervulgata, tanquam fontes ; Neotericorum autem, seu rivulos,Brunselsij, Fuchsij, Tragi, Ruellij,Matthioli,Dodonæi,Turneri, Clusij, Daleschampij, Camerarij,Tabernæmontani, Penæ,nostramque novam methodum & ordinem, à Gramine & notioribus ad Triticea, generatim & speciatim,materno idiomate, Anglicæ genti tuæ Cultissimæ,Reipublicæ voluptabili commodo, recludis ; quò ipsa stimulata, herbarum delitias & hortorum suauissimum & amænissimum cultum amplectetur, maximorum Imperatorum, Regum & Heroum tam priscorum quam nuperorum exemplo. Nec satis hoc tibi fuit ; sed multò magis insuper præstitisti, quòd copiam multarum elegantissimarum plantarum in Anglia sponte nascentium ab alijs hactenus prætermissarum, historiam descripsisti,magno hoc studio captorum vtilitate & oblectamento : Singulas enim regiones peculiares quasdam plantas, quas in alias non facilè reperias, gignere certum. Neque magni tibi fuit hæc inspectione & è viuis Naturæ typis nosse,quippe qui diu herbas indigenas,inquilinas & peregrinas cum nuperrimè solo erumpentes & pululantes, tum adultas, semineque prægnantes, hortulo tuo suburbano aluisti & souisti:Exactum enim cognoscendarum ex figura aut facie superficiaria herbarum studium generatim consistit (Dioscoride teste) in frequenti & assidua, temporis omnis, inspectione. Sed alia est interioris & substantialis formæ plantarum, quæ oculis cerni non potest, solers cognitio ; quam etiam, quantum potes percunctando,seniorum Græcorum medicorum more, aperire conaris. Solebant autem antiqui suorum Medicaminum experimenta,in Reipublica vtilitatem,scriptis tabellis dare,quibus apud Epheseos templi syluaticæ Dianæ parietes vestiebantur. Compertum etiam est Hippocratem discendi cupidum, permultis regionibus peragratis, idem præstitisse, & in methodum commemorabiliorem restituisse & illustrasse. Melius enim est Reipublica quam nostris commodis prospicere.Non est igitur quod huius inuidiosæ procacis ætatis conuiciatores maledici Zoili scripta tua obtrectent : dedisti enim gratis quod potuisti, cætera doctioribus iudicijs relinquens ; exortiuis & exoticis incompartarum penè adhuc virium mangonizatis & lenocinijs allectis Floristarum floribus à Flora Dea meretrice nobili dictis, valetudini & vtilitati potius consulens, quam voluptati,valere jussis.Nonnulli siquidem ex alijs libris herbarum transcriptores rapsodi, ignotis sibi vivis plantis ad medendum maximè necessarijs, assignant incertis, dubijs & supposititijs stirpibus aut simplicibus facultates legitimi simplicis medicamenti,maximo errore & summa periclitatione (vnum enim sæpe simplex compositionem ineptam reddit,pernertit aut depranat) quibus nec tutò nec temerè credendum ; multoque etiam minus multis herbarum experimentis fallacibus,quibus etiamneque nisi notissimis morbis simplicibus, compositis & implicatis, eorundemque sæuissimis symptomatibus,vtendum,ne inoportunus earum vsus sæpius uenenum quam remedium sit. Summo enim ægrotantium dispendio & exercitatissimorum Medicorum tædio p̃ riclitatores procaces, contemptis & neglectis artis institutionibus, Hippocratis & Galeni præceptis,per salutis discrimina & hominum strages medentum tentamenta agunt. Omitto, breuitatis ergô, vulgi opisices,textores sellularios, sordidissimos fabros, interpolatores, circulatores forenses & veteratores scutica dignos, qui professionibus & mechanicis artibus suis fastiditis,scelerato insaniæ lucro, se Medicos Theophrasteos, quem vix vnquam summis labris degustarunt,profteuntur. Non innenistè Syluius in huiusmodi homines inuehit,dum ait,Quam quisque non artem,hanc exerceat vnam,atque excolat,& totus in ea versetur;&c. Et sub finem præfationis rursus ait,Faxit Deus vt quisque quam exercet Artem,pernoscat,& Medicus nihil eorum quæ ad morbos citè & tutò curandos vtilia vel necessaria esse consueuerunt,ignoret. Præualet Medicus vbi Pharmacopœi fides suspecta est, qui ipse simplicia & composita pernoscit;imò quam insamiæ notam imprudens inurit, dum ignarus horum simplicium medicamentorum, tanquam asinus quidam ad omnia Pharmacopœi rogata, auribus motis,velut annuit : quid quod illi sæpe etiam volens Pharmacopæus illudit. Absurdissimus est ac sæpè ridiculus qui medicinam facit,harum rerum ignarus ; & Pharmacopœo ignorantiæ suspectum meritò se reddit. Plura si vis require apud Sylurum, ibidem loci.

Authoris ne-cessaria diligentia in stirpium siue Materiæ Medicæ cognitione commendatur.

Præstigiosas popularium medicastrorū fallacias detegimus & inueteratos depulimus errores.

Initio prologi Pharmac,Præparand.

e Medico

Medico quàm plurima perscrutanda, vt satis superque ad artem medicatricem perdiscendam, annos paucos haudquaquam sufficere, testantur ipsius experientissimi & diuini senis verba vbi inquit ; Ego enim ad finem medicinæ non perueni, etiamsi jám senex sim. Et statim per initia Aphorismorum vitam breuem & artem longam pronunciauit. Quomodo ergo tuto medebuntur multi laruati Medici aut Medicastri tam repente creati, nulla Medicinæ parte, Medicamentorumve facultatibus perspectis ? Hujusmodi adulatores, assentatores, dubitatores, rixatores, periclitatores & Gnathonicos parasistratos histrionibus qui in tragœdijs introducuntur simillimos fecit Hippocrates. Quemadmodum enim illi (inquit) figuram quidem & habitum ac personam eorum quos referunt habent, illi ipsi autem vere non sunt : Sic & medici fama quidem & nomine multi, * re autem & opere valde pauci. *Itaque cum paulo ante Medicinam omnium artium præclarissimam esse dixerit :* Verum propter ignorantiam eorum qui eam exercent, & ob vulgi ruditatem, qui tales pro Medicis judicat & habet, jam eo res deuenisse, vt omnium artium longè vilissima censeatur. At verò hoc peccatum ob hanc potissimum causam committi videtur ; soli namque Medicinæ nulla pœna in rebus-publicis statuta est, præterquam ignominiæ. Ne animam & famam læderit, aut illi insignis ignominia inureretur ob huiusmodi ardua & noxia discrimina, bonus ille & synceruus Dodonæus (quamvis multas herbas ex alijs & Fuchsio transcripserit, cuius methodo vsus est, quemq; inchoauerat, vt ipsemet mihi retulit, vernacula Germanica inferiori lingua vertere) vulgatissimis, notissimis ijsque paucis ex tot herbarum millibus, quinquagenis aut septuagenis herbis quibus vtebatur potius contentus fuit, quam innumeris sibi ignotis periclitari : melius enim omnino medicamento carere, abstinere, & naturæ committere, quam abuti. Vtinam huius nostræ ætatis quamplures anso potiti, medicinam factitantes, eo studio, candore & voto mederentur : Illis id forsitan nequaquam eueneerit, quod Philosophis (Hippocrate defuncto) discipulis suis inexpertis & parum adhuc exercitatis medendo, id est necando (vt memoria traditum est) contingit : quamobrem ars medica Athænis, Roma & per vniuersam Græciam centum & septuaginta annis, interdicta & exul fuit. Merito igitur cautè & tutè agendum : Opiatis & Diagrediatis, Colocynthide, Tithymalis, Esula, Lathyride, Mercurio, Stibio, & similibus molestissimis simplicibus cum cautione vtendum : optimis ducibus & experientissimis senioribus præceptoribus adhærendum, quorum sub vexillis fidissimè & tutissimè rara & præclara, ob barbarie ferè extincta, patrum & auorum remedia, maximo & pristino artis ornamento & proximi vtilitate renouantur, & in vsum reuocantur ; neglectis, spretis, & exclusis Empiricis verbosis, inuidiosis, suspensis, ambagiosis & exitiosis opinionibus, quibus mundus immundus regitur & labitur, qui cum decepi velit, decipiatur : in cuius fallacias per appositè finxit & cecinit olim hos versiculos eruditissimus collega D. Iacobus Paradisius nobilis Gandauensis alludens ad nomen tanti versutissimi herois Nostradami Salonensis Galloprouincia,

Nostra. damus, cum verba damus, quia fallere nostrum ;
Et cum verba damus, nil nisi nostra damus.

Vale. Londini ipsis Calendis Decemb. 1597.

In GERARDI Botanologian
προσφώνημα.

VLtimus ecce Gerardus : at edit an optimus herbæ ?
  Quidni ? non notas sed dedit ille novas.
Ergo ne invideas, videas cum nomen & omen
  αυτοδήκ, mirum est ardua quanta gerit.
Οὗτος ἄνακτα, κρήνατα, πῆ ἐστιν τιλθχμα τ' ἔλθι :
  Sic liber est promus, condus vt hortus erat.
Et ραχίω ἀρθὰν Cælúmque solúmque subegit.
  τῷ γ' ἐρα διαπασό'αι ἡ μῆλαι ἔπι γέρεα.

                  ANTON. HVNTONVS
                  Medicinæ candidatus.

Ad Iohannem Gerardum *Chirurgum Herba-*
riumque peritißimum.

NVlla oculos hominum species magis allicit illa,
  Quam præstante manu duxit generosus Apelles.
Nulla aures animosque magis facundia, quam quæ
Se fusam loquitur Ciceronis ab ore deserti :
Hæc eadem hunc librum commendat causa, *Gerarde,*
Cui pro laude satis tali natum esse parente,
Artifices cui inter dextras pro numine, nomen
Nobilius reliquis herbæ, plantæque magistris.
Illi etenim Europæ succos, Asiæque liquores
Quæque arente solo sitiens parit Africa, tractant:
Tu veterum inventis nova consuis omnia, si qua
Indus vterque dedit nostram forura salutem,
Sive aliunde vehit nostras mercator ad oras,
Hoc ipso vtilius. Quia quæ sunt credita scriptis,
Illa manu expertus medico, & bene dives ab horto
Explorato diu multumque emittis in auras,
Que curent hominum languentia corpora, multi
Præstantesque viri docuere fideliter artem.
Sed si sustuleris plantas quem verba iuvabunt
Sic animo, sic fronte minax. In prælia miles
Prosilit, at stricto cedit victoria ferro
Quæ tibi pro tanto cedit victoria ferro
Præmia persolvet, Myrti laurique coronas ?
Istam novit edax mercedem abolere vetustas,
At tibi pro studio impensisque laboribus istis,
Queis hominum curas sertam tectamque salutem,
Ille opifex rerum, custosque authorque salutis
Æternâ statuit frontem redimire coronâ.

                  G. Launæus Medicus.

In historiam plantarum, Io. Gerardi *civis & Chirurgi Lond.* M. Iacobi
Ihonstonij *Scoti Ballinerisa Regij pagi portionarij Epigramma.*

DEsine quæ vastis pomaria montibus Atlas
  Clauserat (Hesperij munera rara soli)
Auratis folijs auratos desine ramos
  Mirari & ramis pendula poma suis.
Singula cum Domino periere, & Gorgone viso
  In montis riguit vescera versus Atlas.
Alcinoi perijt qui, cedat pensilis hortus,
  Quem celebrat prisci temporis aura fugax:
Vna Gerardini species durabilis horti
  Æterno famæ marmore sculpta manet.
Hic quicquid Zephyrus produxit, quicquid & Eurus,
  Antiquus quicquid & novus orbis habet,

                                         Intulit

Intulit in patriam naturamque exprimit arte.
Sic nullo cedit terra Britanna folo.
Quod magis eſt Graium & Latium concludit in vno
Margine, & Anglorum jam facit ore loqui:
Sic erit æternum hinc vt vivas, horte *Gerardi,*
Cultoris ſtudio nobilitate tui.

In Plantarum hiſtoriam, à ſolertiſſimo viro, Reiq; Herbariæ peritiſſimo,
D. Iohanne Gerardo, Anglice editam Epigramma.

EGregiam certè laudem, decus immortale refertis
Tu, ſocijq; tui, magnum & memorabile nomen
(Illuſtris D E V O R A X) raptoribus orbis I B E R I S
Devictis claſſe Anglorum; Tuque (Dicaſta
Maxime E G E R T O N E) veterem ſuperans Rhadamanthum,
H E R O V M merito ἡμιϑεὸϛ cenſendus in albo.
Nec laus veſtra minor (ſacræ pietatis alumni)
Qui mentes hominum divina paſcitis eſca.
Ornatis Patriam cuncti, nomenq; Britannum
Augetis, vobiſq; viam munitis ad aſtra.
Quin agite, in partem ſaltem permittite honoris
Phœbei veniant Vates, qui pellere gnari
Agmina morborum, humanæ inſidiantia vitæ.
Hujus & ingentes, ſerena fronte labores
A N G L O - D I O S C O R I D I S, Patriæ, veſtræq; ſaluti
Excipite exhauſtos: paulum huc divortite in H O R T O S
Quos C H O R T E I A colit, quos Flora exornat, & omnes
Naiades, & Dryades, Charites, Nymphæq; Britannæ.
Corporibus hic grata ſalus, animiſq; voluptas.
Hic laxate animos: H A B I T A V I T N V M E N I N H O R T I S.

Fran. Hering, Med. D.

Tho. Newtonus, Ceſtreſhyrius, D. *Io. Gerardo,*
Amico non vulgari, S.

POſt tot ab ingenuis conſcripta volumina myſtis,
Herbarum vires qui reſerare docent,
Tu tandem prodis Spartamq; hanc gnaviter ornas,
Dum reliquis palmam præripuiſſe ſtudes.
Nec facis hoc, rutilo vt poſſis ditarier auro,
Nec tibi vt accreſcat grandis acervus opum;
Sed prodeſſe volens, veſtitos gramine colles
Perluſtras, & agros, frondiferumq; nemus.
Indeq; Pæonias (apis inſtar) colligis herbas,
Inq; tuum ſtirpes congeris alueolum.
Mille tibi ſpecies plantarum, milleq; notæ;
Hortulus indicio eſt, quem colis ipſe domi.
Pampineæ vites, redolens cedrus, innuba laurus,
Nota tibi, nota eſt pinguis oliva tibi.
Balſama, narciſſus, rhododaphne, nardus, amomum,
Salvia, dictamnus, galbana nota tibi.
Quid multis? radix, ſtirps, flos, cum cortice ramus,
Spicaq; cum ſiliquis eſt bene nota tibi.
Gratulor ergo tibi, cunctiſq; (*Gerarde*) Britannis,
Namptwicoq; tuo gratulor atq; meo.
Nam Ceſtreſhyrij te ac me genuere parentes,
Tu meliore tamen ſydere natus eras:
Macte animo, pergaſq; precor, cœptumq; laborem
Vrge etiam vlterius. Vivitur ingenio.
Aurum habeant alij, gemmas, nitidoſq; pyropos,
Plantas tu & flores ſcribe *Gerarde.* Vale.

Vere & ex animo tuus, Tho. Newton, *Ilfordenſis*
ᾐπιφᾷ τηϛ·

To

To the well affected Reader and Peruser of this Booke, *St. Bredwell* Physitian, greeting.

Plin.Iun.
in Pan.

 Pen is the Campe of glory and honour for all men, saith the younger Pliny, *not onely men of great birth and dignitie, or men of office endued with publique charge and titles, are seene therein, and haue the garland of praise and preferment waiting to crowne their merits, but euen the common Souldier likewise : so as he whose Name and note was erst obscure, may by egregious acts of valour obtaine a place amongst the Noble. The schoole of Science keepeth semblable proportion : whose amplitude, as not alwaies, nor onely, men of great titles and degrees labor to illustrate ; so whosoeuer doth, may confidently account of, at the least, his name to be immortall. What is he then that will deny his voice of gratious commendation to the Authors of this Booke : To euery one, no doubt, there is due a condigne measure. The first Gatherers out of the Antients, and Augmenters by their owne paines, haue already spred the* Turnerus.
Dodonæus.
Pena.
Lobelius.
Tabernam. *odour of their good names thorow all the lands of learned habitations. Doctor* Priest, *for his translation of so much as* Dodonæus, *hath thereby left a Tombe for his honourable sepulture. Master* Gerard *comming last, but not the least, hath many waies accommodated the whole Worke vnto our English Nation : For this History of Plants, as it is richly replenished by those fine mens Labors laid together, so yet could it full ill haue wanted that new accession he hath made vnto it. Many things hath he nourished in his Garden, and obserued in our English fields, that neuer came into their pens to write of. Againe, the greatest number of these plants, hauing neuer been written of in our English tongue, would haue wanted names for the vulgar sort to call them by : in which defect hee hath been curiously carefull, touching old and new names to make supply. And lest the Reader should too often languish with frustrate desire to finde some plant he readeth, of rare vertue, he spareth not to tell (if himselfe haue seen it in England) in what wood, pasture, or ditch the same may be seene and gathered. Which when I thinke of, and therewithall remember with what chearefull alacritie and resolute attendance he hath many yeares tilled this ground, and now brought forth the fruit of it, whether I should more commend his diligence to attaine this skill, or his large beneuolence in bestowing it on his Countrey, I cannot easily determine. This Booke-birth thus brought forth by* Gerard, *as it is in forme and disposition faire and comely, (euery* Species *being referred to his likeliest* Genus, *of whose stocke it came) so is it accomplished with surpasing varietie, vnto (such spreading growth and strength of euery limme, as that it may seeme some heroicall Impe of illustrious Race, able to draw the eyes and expectation of euery man vnto it. Somewhat rare it will be heere for a Man to moue a question of this nature, and depart againe without some good satisfaction. Manifold will bee the vse both to the Physitian and others : for euery man delighteth in knowledge naturally, which* Laert.l.5.ca.1 *(as* Aristotle *said) is in prosperitie an ornament, in aduersitie a refuge. But this booke aboue many others will sute with the most, because it both plentifully administreth knowledge (which is the food of the minde) and doth it also with a familiar and pleasing taste to euery capacitie. Now as this commoditie is communicated to all, and many shall receiue much fruit thereof, so I wish some may haue* Iuuen.7 Sat. *the minde to returne a benefit againe : that it might not be true in all, that* Iuuenal *saith,* Scire volunt omnes, mercedem soluere nemo, *(id est)* All desire to know, none to yeeld reward. Let men thinke, That the perfection of this knowledge is the high aduancement of the health of man : That perfection is not to be attained but by strong endeauour : neither can strong endeauour be accomplished without free maintenance. This hath not hee, who is forced to labour for his dayly bread : but if he, who from the short houres of his daily and necessarie trauell, stealing as it were some for the publique behoofe, and setting at length these pieces together, can bring forth so comely a garment as this, meet to couer or put away the ignorance of many ; what may be thought he would do, if publique maintenance did free him from that priuate care, and vnite his thoughts to be wholly intent to the generall good. O Reader, if such men as this sticke to rob themselues of such wealth as thou hast, to enrich thee with that sustenance thou wantest, detract not to share out of thine abundance to merit and encourage their paines, that so fluxable riches and permanent sciences may the one be-* Cic.Offic. 1. *come a prop to the other. Although praise and reward, ioyned as companions to fruitfull endeauors, are in part desired of all men that vndertake lossis, labours, or dangers for the publique behoofe, because* Simpl.Comm.
in Epict. *they adde sinues (as it were) vnto Reason, and able her more and more to refine her selfe : yet do they not embrace that honour in respect of it selfe, nor in respect of those that conferred it vpon them, but*

as hauing thereby an argument in themselues, that there is something in them worthy estimation among men: which then doubleth their dilligence to deserue it more abundantly. Admirable and for the imitation of Princes, was that act of Alexander, who setting Aristotle to compile commentaries of the brutt creatures, allowed him for the better performance thereof, certaine thousands of men, in all Asia and Greece, most skilfull obseruers of such things, to giue him information touching all beasts, fishes, soules, serpents and flies. What came of it? A booke written, wherein all learned men in all ages since do exercise themselues principally, for the knowledge of the creatures. Great is the number of those that of their owne priuate haue laboured in the same matter, from his age downe to our present time, which all do not in comparison satisfie vs. Whereas if in those ensuing ages there had risen still new Alexanders, there (certainely) would not haue wanted Aristotles to haue made the euidence of those things an hundred fold more cleered vnto vs, than now they be. Whereby you may perceiue the vnequall effects that follow those vnsutable causes of publicke and priuate maintenances vnto labours and studies. Now that I might not dispaire in this my exhortation, I see examples of this munificence in our age to giue me comfort: Ferdinand the Emperor and Cosmus Medices Prince of Tuscane are herein registred for furthering this science of Plants, in following of it themselues and becomming skilfull therein. which course of theirs could not be holden without the supporting and aduancing of such as were studious to excell in this kinde. Bellonius likewise (whom for honours cause I name) a man of high attempts in naturall science, greatly extolleth his Kings liberalitie, which endued him with free leasure to follow the studie of Plants, seconded also herein by Montmorencie the Constable, the Cardinals Castilion and Lorraine, with Oliuerius the Chancellor, by whose meanes he was enabled to performe those his notable peregrinations in Italy, Africa, and Asia: the sweet fruit whereof, as we haue receiued some taste by his obseruations, so we should plenteously haue beene filled with them, if violent death by most accursed robbers had not cut him off. And as I finde these examples of comfort in forraine nations so we are (I confesse) much to be thankefull to God, for the experience we haue of the like things at home. If (neuerthelesse) vnto that Physicke lecture lately so well erected, men who haue this Worlds goods shall haue hearts also of that spirit, to adde some ingenious labourer in the skill of simples, they shall mightily augment and adorne the whole science of Physicke. But if to that likewise they joyne a third, namely the art of Chymicall preparation; that out of those good creatures which God hath giuen man for his health, pure substances may be procured for those that be sicke, (I feare not to say it, though I see how Momus scorneth) this present generation would purchase more to the perfection of Physicke, than all the generations past since Galens time haue done: that, I say, nothing of this one fruit that would grow thereof, to wit, the discouering and abolishing of these pernitious impostures and sophistications, which mount promising Paracelsians euery where obtrude, through want of a true and constant light among vs to discerne them by. In which behalfe, remembring the mournefull speech of graue Hippocrates; The art of Physicke truly excelleth all arts, howbeit, through the ignorance partly of those that exercise it, and partly of those that iudge rashly of Physitions, it is accounted of all Arts the most inferiour: I say in like manner, the Art of Chymistrie is in it selfe the most noble instrument of naturall knowledges; but through the ignorance and impiety, partly of those that most audaciously professe it without skill, and partly of them that impudently condemne that they know not, it is of all others most basely despised and scornfully rejected. A principall remedy to remoue such contumelious disgrace from these two pure virgins of one stocke and linage, is this that I haue now insinuated, euen by erecting the laboratory of an industrious Chymist, by the sweet garden of flourishing simples. The Physicke Reader by their meanes shall not onely come furnished with authorities of the Antients, and sensibles probabilities for that he teacheth, but with reall demonstrations also in many things, which the reason of man without the light of the furnace would neuer haue reached vnto. I haue vttered my hearts desire, for promoting first the perfection of my profession, and next by necessary consequence, the healthie liues of men. If God open mens hearts to prouide for the former, it cannot be but the happy fruits shall be seene in the later. Let the ingenious learned iudge whether I haue reason on my side: the partiall addicted sect I shun, as men that neuer meane good to posterity.

George

Riʃtotle, a Prince amongʃt the Philoʃophers, writing in his Metaphyʃicks of the nature of mankind, ʃaith, that man is naturally inclined and deʃirous of ʃcience. The which ʃentence doth teach vs, that all creatures (being vertuouʃly giuen) doe ʃtriue to attain to perfection, and draw neere in what they can to the Creator; and this knowledge is one of the principall parts which doth concerne the perfection of vnderʃtanding: for of the ʃame doth follow, that all ʃuch are generally inclined to know the meanes by the which they may conʃerue their life, health, and reputation. And although it be neceʃʃary for man to learn and know all ʃciences, yet neuertheleʃʃe the knowledge of naturall philoʃophie ought to be preferred, as being the moʃt neceʃʃary; and moreouer, it doth bring with it a ʃingular pleaʃure and contentment. The firʃt inuentor of this knowledge was *Chiron* the Centaure, of great renowne, ʃonne to *Saturne* and *Phillyre*: and others ʃay that it was inuented of *Apollo*; & others of *Eʃculape* his ʃon; eʃteeming that ʃo excellent a ʃcience could neuer proceed but from the gods immortall, and that it was impoʃʃible for man to finde out the nature of Plants, if the great worker, which is God, had not firʃt inʃtructed and taught him. For, as *Pliny* ʃaith, if any think that theʃe things haue bin inuented by man, he is vngratefull for the workes of God. The firʃt that we can learne of among the Greeks that haue diligently written of herbes, haue bin *Orpheus, Muʃæus,* and *Heʃiode,* hauing bin taught by the Ægyptians: then *Pythagoras* of great renown for his wiʃdome, which did write bookes of the nature of Plants, and did acknowledge to learne the ʃame from *Apollo* and *Eʃculape. Democrite* alʃo did compoʃe bookes of Plants, hauing firʃt trauelled ouer all Perʃia, Arabia, Ethiopia, and Egypt. Many other excellent ʃpirits haue taken great pleaʃure in this ʃcience, which to accompliʃh haue hazarded their liues in paʃʃing many vuknowne regions, to learne the true knowledge of *Elleborus,* and other Medicaments: of which number were *Hippocrates, Crateua, Ariʃtotle, Theophraʃt, Diocles Cariʃtius, Pamphylus, Montius, Hierophile, Dioʃcorides, Galen, Pliny,* and many others, which I leaue to name, fearing to be too long. And if I may ʃpeake without partiality of the Author of this book, his great paines, his no leʃʃe expenʃes in trauelling far and neere for the attaining of his ʃkill haue bin extraordinary. For he was neuer content with the knowledge of thoʃe ʃimples which grow in thoʃe parts, but vpon his proper coʃt and charges hath had out of all parts of the world all the rare ʃimples which by any means he could attaine vnto, not only to haue them brought, but hath procured by his excellent knowledge to haue them growing in his garden, which as the time of the yeare doth ʃerue may be ʃeene: for there ʃhall you ʃee all manner of ʃtrange trees, herbes, roots, plants, floures, and other ʃuch rare things, that it would make a man wonder, how one of his degree, not hauing the purʃe of a number, could euer accompliʃh the ʃame. I proteʃt vpon my conʃcience, I do not thinke for the knowledge of Plants, that he is inferiour to any: for I did once ʃee him tried with one of the beʃt ʃtrangers that euer came into England, and was accounted in Paris the onely man, being recommended vnto me by that famous man Maʃter *Amb. Pareus,*

Pareus ; and he being here was defirous to goe abroad with fome of our Herbarifts, for the which I was the meane to bring them together, and one whole day we fpent herein, fearching the rareft Simples : but when it came to the triall, my French man did not know one to his foure. What doth this man deferue that hath taken fo much paines for his countrey, in fetting out a Booke, that to this day, neuer any in what language foeuer did the like ? Firft, for correcting their faults in fo many hundred places, being falfly named, miftaken the one for the other ; and then the pictures of a great number of plants now newly cut. If this man had taken this paines in Italy and Germany where *Matthiolus* did write, he fhould haue fped as well as he did : For (faith he) I had fo great a defire euer to finifh my Book, that I neuer regarded any thing in refpect of the publique good, not fo much as to think how I fhould finifh fo great a charge, which I had neuer carried out, but that by Gods ftirring vp of the renowned Emperour *Ferdinando* of famous memorie, and the excellent Princes had not helped mee with great fums of money, fo that the common wealth may fay, That this bleffing doth rather proceed of them than from me. There haue beene alfo other Princes of Almain, which haue been liberal in the preferring of this Book, and the moft excellent Elector of the Empire the Duke of Saxony, which fent me by his Poft much mony toward my charges : the liberalitie of the which, & their magnificence toward me I cannot commend fufficiently. They which followed in their liberalitie were the excellent *Fredericke* Count Palatine of the Rhine, and the excellent *Ioachim* Marques of Brandeburg, which much fupplied my wants : and the like did the reuerend Cardinall and Prince of Trent, and the Excellent Archbifhop of Saltzperg, the Excellent Dukes of Bauare and Cleues, the duke of Megapolencis, Prince of Vandalis, the State Republique of Noremberg, the liberalitie of whom ought to be celebrated for euer : and it doth much reioyce me that I had the help and reward of Emperors, Kings, Electors of the Roman Empire, Archdukes, Cardinalls, Bifhops, Dukes and Princes, for it giueth more credit to our Labors that any thing that can be faid. Thus far *Matthiolus* his owne writing of the liberalitie of Princes towards him. What age do we liue in here that wil fuffer all vertue to go vnrewarded? Mafter *Gerrard* hath taken more pains than euer *Matthiolus* did in his Commentaries, and hath corrected a number of faults that he paffed ouer ; and I dare affirme (in reuerence be it fpoken to that Excellent man) that Mafter *Gerrard* doth know a great number of Simples that were not knowne in his time : and yet I doubt whether he fhall tafte of the liberalitie of either Prince, Duke, Earle, Bifhop, or publique Eftate Let a man excell neuer fo much in any excellent knowledge, neuertheles many times he is not fo much regarded as a Iefter, a Boafter, a Quackfaluer or Mountebanke : for fuch kinde of men can flatter, diffemble, make of trifles great matters, in praifing of this rare fecret, or that excellent fpirit, or this Elixer or Quinteffence ; which when it fhall come to the triall, nothing fhall be found but boafting words.

VALE.

To the courteous and well willing Readers.

Lthough my paines hau not beene ſpent (courteous Reader) in the gracious diſcouerie of golden Mines, nor in the tracing after ſiluer veines, whereby my natiue country might be inriched with ſuch merchandiſe as it hath moſt in requeſt and admiration ; yet hath my labour (I truſt) been otherwiſe profitably imploied, in deſcrying of ſuch a harmeleſſe treaſure of herbes, trees, and plants, as the earth frankely without violence offereth vnto our moſt neceſſary vſes. Harmeleſſe I call them, becauſe they were ſuch delights as man in the perfecteſt ſtate of his innocencie did erſt inioy : and treaſure I may well terme them ſeeing both Kings and Princes haue eſteemed them as Iewels ; ſith wiſe men haue made their whole life as a pilgrimage to attaine to the knowledge of them : by the which they haue gained the hearts of all, and opened the mouthes of many, in commendation of thoſe rare vertues which are contained in theſe terreſtriall creatures. I confeſſe blind Pluto is now adaies more ſought after than quicke ſighted Phœbus : and yet this duſty mettall, or excrement of the earth (which was firſt deeply buried leaſt it ſhould be an cie-ſore to grieue the corrupt heart of man) by forcible entry made into the bowels of the earth, is rather ſnatched at of man to his owne deſtruction, than directly ſent of God to the comfort of this life. And yet behold in the compaſſing of this wordly droſſe, what care, what coſt, what aduentures, what myſticall proofes, and chymicall trials are ſet abroach ; when as notwithſtanding the chiefeſt end is but vncertaine wealth. Contrariwiſe, in the expert knowledge of herbes, what pleaſure ſtill renewed with variety ? what ſmall expence ? what ſecurity ? and yet what an apt and ordinary meanes to conduct man to that moſt deſired benefit of health ? which as I deuoutly wiſh vnto my natiue countrey, and to the carfull nurſing mother of the ſame ; ſo hauing bent my labours to the benefit of ſuch as are ſtudiouſly practiſed in the conſeruation thereof, I thought it a chiefe point of my dutie, thus out of my poore ſtore to offer vp theſe my far fetched experiments, together with mine owne countries vnknowne treaſure, combined in this compendious Herball (not vnprofitable though vnpoliſhed) vnto your wiſe conſtructions and courteous conſiderations. The drift whereof is a ready introduction to that excellent art of Simpling, which is neither ſo baſe nor contemptible as perhaps the Engliſh name may ſeeme to intimate : but ſuch it is, as altogether hath beene a ſtudy for the wiſeſt, an exerciſe for the nobleſt, a paſtime for the beſt. From whence there ſpring floures not onely to adorne the garlands of the Muſes, to decke the boſomes of the beautifall, to paint the gardens of the curious, to garniſh the glorious crownes of Kings ; but alſo ſuch fruit as learned Dioſcorides long trauelled for ; and princely Mithridates reſerued as precious in his owne cloſet : Mithridates I meane, better knowne by his ſoueraigne Mithridate, than by his ſometime ſpeaking two and twenty languages. But what this famous Prince did by tradition, Euax king of the Arabians did deliuer in a diſcourſe written of the vertues of herbes, and dedicated it vnto the Emperor Nero. Every greene Herbariſt can make mention of the herbe Lyſimachia, whoſe vertues were found out by King Lyſimachus, and his vertues no leſſe eterniſed in the ſelfe ſame plant, than the name of Phydias, queintly beaten into the ſhields of Pallas, or the firſt letters of Ajax or Hyacinthus (whether you pleaſe) regiſtred in that beloued floure of Apollo. As for Artemiſia, firſt called ﬡﬡﬡﬡﬡ, whether the title thereof ſprang from κρρρρ, Diana her ſelfe, or from the renowned Queene of Caria, which diſcloſed the vſe thereof vnto poſterity, it ſuruiueth as a monument to reuiue the memories of them both for euer. What ſhould we ſpeake of Gentiana, bearing ſtill the cogniſance of Gentius? or of diuers other herbes taking their denominations of their princely inuertors ? What ſhould I ſay of thoſe royall perſonages, Iuba, Attalus, Climenus, Achilles, Cyrus, Maſyniſſa, Semyramis, Diocleſian ? but onely thus, to ſpeake their Princely loues to Herbariſme, and their euerlaſting honors (which neither old Plinius dead, nor yong Lipſius liuing will permit to die?) Creſcent herbæ, creſcetis amores : creſcent herbæ, creſcetis honores. But had this wonted faculty wanted the authoriſement of ſuch a royall company, King Solomon, excelling all the reſt for wiſdome, of greater royalty than they all (though the Lillies of the field out-braued him) he onely (I ſay) might yeeld hereunto ſufficient countenance and commendation, in that his lofty wiſedome thought no ſcorne to ſtoupe vnto the lowly plants. I liſt not ſeek the

common

common colours of antiquitie, when notwithstanding the world can brag of no more antient Monument than Paradise and the garden of Eden; and the Fruits of the earth may contend for seniority, seeing their Mother was the first Creature that conceiued, and they themselues the first fruit she brought forth. Talke of perfect happinesse or pleasure, and what place was so fit for that as the garden place wherein Adam was set to be the Herbarist? Whither did the Poets hunt for their sincere delights, but into the gardens of Alcinous, of Adonis, and the Orchards of the Hesperides? Where did they dreame that heauen should be, but in the pleasant garden of Elysium? Whither doe all men walke for their honest recreation, but thither where the earth hath most beneficially painted her face with flourishing colours? And what season of the yeare more longed for than the Spring, whose gentle breath inticeth forth the kindly sweets, and makes them yeeld their fragrant smells? Who would therefore looke dangerously vp at Plancts, that might safely looke downe at Plants? And if true bee the prouerb, Quæ supra nos, nihil ad nos; I suppose this new saying cannot be false, Quæ infra nos, ea maximè ad nos. Easie therefore is this treasure to be gained, and yet pretious. The science is nobly supported by wise and kingly Fauorits: the subiect thereof so necessarie and delectable, that nothing can be confected, either delicate for the taste, dainty for smell, pleasant for sight, wholsome for body, conseruatiue or restoratiue for health, but it borroweth the rellish of an herb, the sauor of a flour, the colour of a leafe, the juice of a plant, or the decoction of a root. And such is the treasure that this my Treatise is furnished withall; wherein though myne art be not able to counteruaile Nature in her liuely portraitures, yet haue I counterfeited likenesse for life, shapes and shadowes for substance, being ready with the bad Painter to explaine the imperfections of my pensill with my pen, chusing rather to score vpon my pictures such rude marks as may describe my meaning, than to let the beholder to guesse at random and misse. I haue here therefore set downe not only the names of sundry Plants, but also their natures, their proportions and properties, their affects and effects, their increase and decrease, their flourishing and fading, their distinct varieties and seuerall qualities, as well of those which our owne country yeeldeth, as of others which I haue fetched further, or drawn out by perusing diuers Herbals set forth in other languages: wherein none of my countrymen haue to my knowledge taken any paines, since that excellent Worke of Master Doctor Turner. After which time Master Lyte a worshipfull Gentleman translated Dodonæus out of French into English; and since that, Doctor Priest one of our London Colledge hath (as I heard) translated the last edition of Dodonæus, and meant to publish the same; but being preuented by death, his translation likewise perished. Lastly my selfe, one of the least among many, haue presumed to set forth vnto the view of the World, the first fruits of these myne owne Labours, which if they be such as may content the Reader, I shall thinke my selfe well rewarded; otherwise there is no man to be blamed but my selfe, being a Worke I confesse for greater Clerks to vndertake: yet may my blunt attempt serue as a whetstone to set an edge vpon sharper wits, by whom I wish this my course: Discourse might be both fined and refined. Faults I confesse haue escaped, some by the Printers ouersight, some through defects in my selfe to performe so great a work, and some by means of the greatnesse of the Labour, and that I was constrained to seeke after my liuing, being void of friends to beare some part of the burthen. The rather therefore accept this at my hands (louing Country-men) as a token of my goodwill, and I trust that the best and well minded will not rashly condemne me, although something haue passed worthy reprehension. But as for the slanderer or Envious I passe not for them, but return vpon themselues any thing they shall without cause either murmure in corners, or iangle in secret. Farewell.

From my House in Holborn, within the Suburbs of London, this first of December, 1597.

Thy sincere and vnfeigned Friend,

IOHN GERARD.

TO THE READER.

Courteous READER,

 Here are many things which I thinke needfull to impart vnto thee, both concerning the knowledge of Plants in generall, as also for the better explaining of some things pertinent to this present Historie, which I haue here set forth much amended and inlarged. For the general differences, affections, &c. of Plants, I hold it not now so fitting nor necessary for me to insist vpon them; neither do I intend in any large discourse to set forth their many and great vses and vertues : giue me leaue onely to tell you, That God of his infinite goodnesse and bounty hath by the *medium* of Plants, bestowed almost all food, clothing, and medicine vpon man. And to this off-spring we also owe (for the most part) our houses, shipping, and infinite other things, though some of them *Proteus*-like haue run through diuers shapes, as this paper whereon I write, that first from seed became Flaxe; then after much vexation thread, then cloath, where it was cut and mangled to serue the fashions of the time, but afterward reiected and cast aside; yet vnwilling so to forsake the seruice of man for which God had created it, again it comes (as I may terme it) to the hammer, from whence it takes a more noble form and aptitude to be imployed to sacred, ciuill, forrein, and domesticke vses. I will not speake of the many and various obiects of delight that these present to the sences, nor of sundry other things which I could plentifully in this kinde deliuer; but rather acquaint you from what Fountaines this Knowledge may be drawne, by shewing what Authours haue deliuered to vs the Historie of Plants, and after what manner they haue done it. And this will be a meanes that many controuersies may be the more easily vnderstood by the lesse learned and iuditious Reader.

He whose name we first find vpon record (though doubtlesse some had treated thereof before) that largely writ of Plants, was the wisest of men, euen King *Solomon*, who certainly would not haue medled with this subiect, if he in his wisedome had not known it worthy himselfe, and exceeding fitting : First, for the honour of his Creator, whose gifts and blessings these are : Secondly, for the good of his Subiects, whereof without doubt he in this Work had a speciall regard, for the curing of their diseases and infirmities. But this kingly Worke being lost, I will not insist vpon it, but come to such as are yet extant, of which (following the course of antiquity) that of *Theophrastus* first takes place.

Now *Theophrastus* succeeded *Aristotle* in the gouerment of the Schoole at Athens, about the 114 Olymp. which was some 322 yeares before Christ. He among many other things writ a Historie of Plants in ten bookes, and of the causes of them, eight bookes; of the former ten there are nine come to our hands reasonable perfect; but there now remain but six of the eight of the causes of Plants. Some looking vpon the Catalogue of the bookes of *Theophrastus* his writing, set forth in his life, written by *Diogenes Laertius*, may wonder that they finde no mention of these bookes of Plants, amongst these he reckons vp, and indeed I thought it somewhat strange, and so much the more, because this his life is set forth by *Daniel Heinsius*, before his edition * of *Theophrastus*, and there also no mention neither in the Greeke nor Latine of those Workes. Considering this, I thinking to haue said something thereof, I found the doubt was long since cleared by the learned *Causabone* in his notes vpon " *Laertius*, where *pag.* 331, for Ρ.ρι φυσικᾶς ἱερῶν and φυσικᾶ αἰνῶν, hee wishes you to reade ... ἱερῶν and αἰνῶν. Thus being certaine of the Authour, let mee say somwhat of the Worke, which though by the iniury of time it hath suffered much, yet is it one of the chiefe pieces of Antiquitie, from whence the knowledge of plants is

Solomon.

Theophrastus.

Lugd.Batau. 1613.

Excuf.ab Henr. Steph. 1593.

to

to be drawn. *Theophraſtus* as he followed *Ariſtotle* in the Schoole, ſo alſo in his manner of writing; for according as *Ariſtotle* hath deliuered his *Hiſtoria Animalium*, ſo hath he ſet forth this of Plants, not by writing of each *ſpecies* in particular, but of *their differences and nature, by their parts, affections, generations, and life.* Which how hard a thing it was he tells you in his ſecond chapter, and renders you this reaſon, *Becauſe there is nothing common to all Plants, as the mouth and belly is to other liuing creatures, &c.* Now by this manner of writing you may learne the generall differences and affections of Plants, but cannot come to the particular knowledge of any, without much labor; for you muſt go to many places to gather vp the deſcription of one plant: neither doth he (nor is it neceſſarie for any writing in this manner) make mention of any great number, and of many it may be but once. His Works being in Greeke were tranſlated into Latine by *Theodore Gaza*, who did them but *Græca fide*, for he omitted ſome things, otherwhiles rendred them contrarie to the minde of the Author; but aboue all, he took to himſelfe too much libertie in giuing of names, in imitation of the Greeke, or of his own inuention, when it had been better by much for his Reader to haue had them in the Greeke; as when hee renders Βλαπίριον, *Agitatorium*; Ἡλιότρόπιον, *Solaris, &c.* The learned *Iulius Scaliger* hath ſet forth *Animaduerſiones* vpon theſe Books, wherein he hath both much explained the minde of *Theophraſtus*, and ſhewed the errors of *Gaza*. Some ſince his time haue promiſed to do ſomething to this Authour, as *Daniel Heinſius* and *Spigelius*, but twenty yeares are paſt ſince, and I haue not yet heard of any thing done in this kinde by either of them. Thus much for *Theophraſtus*.

margin: *Theoph. Hiſt. pl. l. 1. cap. 1.*

margin: Σπήσιων δ' τὸ μηδὲν είναι κοινὸν καζεῖς ὁ ποτ' ἰμφανερζοίς, &c.

Let me not paſſe ouer *Ariſtotle* in ſilence, though his bookes writ of this ſubiect were but two, and theſe according to the conjecture of *Iulius Scaliger* (who hath made a large and curious examination of them, haue either periſhed, or come to vs not as they were originally written by *Ariſtotle*, but as they haue been by ſome later man put into Greeke. Amongſt other things *Scaliger* hath theſe concerning thoſe two bookes; *Reor è textrina Theophraſti detracta fila quædam, iiſq; clavos additos, tametſi neque aureos, neque purpureos. Quod ſi protinus autorem tibi dari vis ad Arabum diligentiam propius accedit.* And afterwards thus; *Attribuere viri docti, alius alij, at quidem qui aliorum viderem nihil Planudem autorem facienti malim aſſentiri; extant enim illius alijs in libris ſimilis veſtigia ſemilatinietatis, &c.* Thus much for *Ariſt.* whom as you ſee I haue placed after his Scholler, becauſe there is ſuch doubt of theſe bookes caried about in his name, and for that *Scaliger*, as you ſee, thinks them rather taken out of *Theophraſtus*, than written by his Maſter.

margin: *Ariſtotle.*

The next that orderly followes is *Pedacius Dioſcorides Anazarbeus*, who liued (according to *Suidas*) in the time of *Cleopatra*, which was ſome few yeares before the birth of our Sauior. Now *Suidas* hath confounded * *Dioſcorides Anazarbeus* with *Dioſcorides Phacas*, but by ſome places in *Galen* you may ſee they were different men: for our Anazarbean *Dioſcorides* was of the Emperick Sect, but the other was a follower of *Herophylus* and of the Rational Sect. He writ not only of Plants, but *de tota materia medica*; to which ſtudy he was addicted euen from his childehood, which made him trauell much ground, and leade a military life, the better to accompliſh his ends: and in this hee attained to that perfection, that few or none ſince his time haue attained to. Of the exellencie of his work, which is as it were the foundation and ground-worke of all that hath bin ſince deliuered in this kinde, heare what *Galen*, one of the excellenteſt of Phyſitians, and one who ſpent no ſmal time in this ſtudy, affirmes: *But* (ſaith he) *the Anazarbean* Dioſcorides *in fiue bookes hath written of the neceſſary matter of Medicine, not onely making mention of Herbes, but alſo of Trees, Fruits,* * *Liquors and Iuyces, as alſo of all Minerals, and of the parts of liuing Creatures: and in myne opinion he hath with the greateſt perfection performed this worke of the matter of Medicine; for although many before him haue written well vpon this Subiect, yet none haue writ ſo well of all.* Now *Dioſcorides* follows not the method of *Theoph.* but treats of each herb in particular, firſt giuing the Names, then the deſcription, and then the place where they vſually grow, and laſtly their vertues. Yet of ſome, which as then were as frequently known with them as Sage, Roſemary, an Aſh or oke tree are with vs, he hath omitted the deſcriptions as vnneceſſarie, as indeed at that time when they were ſo vulgarly known they might ſeeme ſo to bee, but now wee know the leaſt of theſe, and haue no certaintie, but ſome probable coniectures to direct vs to the knowledge of them. He was not curious about his words nor method, but plainly and truly deliuered that whereof he had certaine and experimentall knowledge, concerning the deſcription and nature of Plants. But the generall method he obſerued you may finde ſet forth by *Banhine*, in his edition of *Matthiolus*, immediatly after the Preface of the firſt booke; whereto I refer the Curious, being too long for me in this place to inſiſt vpon. His Works that are come to vs are fiue Books *de materia Medica*. One *De letalibus venenis, eorumq; præcautione & curatione.* Another, *De Cane rabido*,

margin: *Dioſcorides.*

margin: Διοπκρίδης Αναζαρζεὺς Ἰατρὸς ὁ Φακᾶς ὅντις οὐμρὲ &c. Suid.

margin: *De ſimpl. med. facult. lib. 6. proem.*

margin: χημα, ἡ ὀπῶι.

rabido, deq; notis quæ morsus ictusve animalium venenum relinquentium sequuntur. A third, *De eorum curatione.* These eight books within these two last centuries of years haue bin translated out of Greeke into Latine, and commented vpon by diuers, as *Hermolaus Barbarus, Iohannes Ruellius, Marcellus Virgilius,&c.* But of these and the rest, as they offer themselues, I shall say somewhat hereafter. There is also another Worke which goes vnder his name, and may well be his. It is ... ..., *siue de facile parabilibus,* diuided into two books, translated and confirmed with the consent of other Greeke Physitions, by the great labor of *Iohn Moibane* a Physition of Auspurge, who liued not to finish it, but left it to be perfected and set forth by *Conrade Gesner.*

Pliny.

The next that takes place is the laborious *Caius Plinius secundus,* who liued in the time of *Vespasian,* and was suffocated by the sulphurious vapours that came from mount Vesuvius, falling at that time on fire; he through ouermuch curiositie to see and finde out the cause thereof, approching too nigh, and this was *Anno Dom.* 79. He read and writ exceeding much, though by the injury of time we haue no more of his than 37 books, *De historia Mundi,* which also haue receiued such wounds, as haue tried the best skill of our Criticks, and yet in my opinion in some places require *Medicas manus.* From the twelfth to the end of the twenty seuenth of these books he treats of plants, more from what he found written in other Authors, than from any certain knowledge of his owne, in many places following the method and giuing the words of *Theophrastus,* and in other places those of *Dioscorides,* though he neuer make mention of the later of them: he also mentions and no question followed many other Authors, whose writings are long since perished. Sometimes he is pretty large, and otherwhiles so briefe, that scarce any thing can from thence be gathered. From the seuenteenth vnto the twenty seuenth he variously handles them; what method you may quickly see by his *Elenchus* contained in his first booke, but in the twenty seuenth he handles those whereof he had made no, or not sufficient mention, after an alphabeticall order, beginning with *Æthiopis, Ageratum, Aloe,&c.* so going on to the rest.

Galen.
Paulus.
Aetius.

I must not passe ouer in silence, neither need I long insist vpon *Galen, Paulus Ægineta,* and *Aetius,* for they haue only alphabetically named Plants and other simple medicines, briefly mentioning their temperature and faculties, without descriptions (some very few, and those briefe ones excepted) and other things pertinent to their historie.

Macer.

The next that present themselues are two counterfeits, who abuse the world vnder feined titles, and their names haue much more antiquitie than the works themselues. The first goes vnder the title of *Æmilius Macer* a famous Poet, of whom *Ovid* makes mention in these verses:

Sæpe suas volucres legit mihi grandior ævo,
Quæq; nocet Serpens, quæ juvat herba Macer.

Pliny also makes mention of this *Macer.* He in his Poems imitated *Nicander;* but this Worke that now is caried about vnder his name, is written in a rude and somwhat barbarous verse, far different from the stile of those times wherein *Macer* liued, and no way in the subiect imitating *Nicander.* It seemes to haue been written about 400 or 500 yeares agoe.

Apuleius.

The other also is an vnknowne Author, to whom the Printers haue giuen the title of *Apuleius Madaurensis,* and some haue beene so absurdly bold of late, as to put it vnto the Works of *Apuleius:* yet the vncurious stile and method of the whole Worke will conuince them of error, if there were no other argument. I haue seen some foure Manuscripts of this Author, and heard of a fifth, and all of them seem to be of good antiquitie: the Figures of them all for the most part haue some resemblance each of other. The first of these I saw some nine yeares agoe, with that worthy louer and storer of Antiquitie, Sr *Robert Cotton;* it was in a faire Saxon hand, and as I remember in the Saxon tongue, but what title it carried I at that time was not curious to obserue. I saw also another after that, which seemed to be of no small standing, but carelesly obserued not the title. But since I being informed by my friend Mr *Goodyer* (as you may finde in the chapter of Saxifrage of the Antients) that his Manuscript, which was very antient, acknowledged no such Author as *Apuleius,* I begun a little to examine some other Manuscripts; so I procured a very faire one of my much honored friend Sr *Theod. Mayern:* In the very beginning of this is writ, *In hoc continentur libri quatuor medicinæ Ypocratis, Platonis Apoliensis vrbis de diuersis Herbis; Sexti Papiri placiti ex Animalibus,&c.* A little after in the same page at the beginning of a table which is of the Vertues, are these words; *In primo libro sunt herbæ descriptæ, quas Apoliensis Plato descripsit,&c.* and thus also he is named in the title of the Epistle or Proem ; but the end of the work is *Explicit liber Platonis de herbis masculinis,&c.* With this in all things agrees that of Mr *Goodyer*, as he hath affirmed to me. Besides these, I found one with Mr *Iohn Tradescant*, which was written in a more ignorant and barbarous time, as one may cojecture by the title, which is thus at the very beginning ; *In nomine domini incipit herbarium Apulei Platonis quod accepit à Scolapio, & Chirone Centauro magistro.* Then followes (as also in the former and in the printed bookes) the tractat ascribed to *Antonius Musa, de herba Betonica.* After that are these words, *Liber medicina Platonis herbaticus explicit.* By this it seems the Author of this Worke either was named, or else called himself *Plato*, a thing not without example in these times. This worke was first printed at Basil, 1528, amongst some other works of physicke, and one *Albanus Torinus* set it forth by the helpe of many Manuscripts, of whose imperfections he much complains, and I think not without cause. After this, *Gabriel Humelbergius* of Rauenspurge in Germany set it forth with a Comment vpon it, who also complaines of the imperfections of his copies, and thinkes the Worke not perfect. Indeed both the editions are faulty in many places ; and by the help of these Manuscripts I haue seene they might be mended (if any thought it worth their labour) in some things, as I obserued in cursorily looking ouer them. One thing I much maruell at, which is, That I finde not this Author mentioned in any Writer of the middle times, as *Platearius, Bartholomaus Anglus, &c.* Now I conjecture this Worke was originally written in Greeke, for these reasons ; first, because it hath the Greeke names in such plenty, and many of them proper, significant, and in the first place. Secondly, some are only named in Greek, as *Hierobulbon, Artemisia Leptophyllos,* and *Artemisia tagantes, Batrachion, Gryas,* (which I iudge rather Greeke than Latine) &c. Besides, in both the written books in very many places amongst the names I finde this word *Ombes,* but diuersly written ; for I conjecture the Greeke names were written in the Greeke character, and *ὁμοίως* amongst them : and then also when the rest of the Worke was translated, which afterwards made the transcribers who vnderstood it not to write it variously, for in the one book it is alwaies written *Amoeos,* and in the other, *Omoeos,* and sometimes *Omeos,* as in the chapter of *Brittanica,* the one hath it thus, *Nomen herbæ istius Britanica, Amocos dicunt eum Damasinium, &c.* The other thus, *Nomen herbæ Brittanica, Omeos Damasinius, &c.* And in the chapter of *Althæa* the one hath it thus, *Nomen hujus herbæ Altea Amocos vocant hanc herbam Moloche, &c.* The other, *Nomen herbæ Ibiscus Omoeos Moloche, &c.* If it be certain which *Philip Ferrarius* affirmes in his *Lexicon Geographicum,* That the City Apoley is Constantinople, then haue I found *Apoliensis vrbs,* of which I can find no mention in any antient or modern Geographer besides ; and then it is more than probable that this was written in Greeke, and it may bee thought differently translated, which occasions such diuersitie in the Copies, as you shall finde in some places. Now I conjecture this Worke was written about some 600 yeares ago.

From the Antients haue sprung all or the greatest part of the knowledge that the middle or later times haue had of Plants ; and all the controuersies that of late time haue so stuffed the books of such as haue writ of this subiect, had their beginning by reason that the carelesnes of the middle times was such, that they knew little but what they transcribed out of these Antients , neuer endeauoring to acquire any perfect knowledge of the things themselues : so that when as learning (after a long winter) began to spring vp again, men began to be somwhat more curious, and by the notes and descriptions in these antient Authors they haue labored to restore this lost knowledge, making inquirie, first, Whether it were knowne by *Theophrastus, Dioscorides,* or any of the Antients, then by what name. But to return to my Authors.

About *An. Dom.* 1100, or a little after liued the Arabians *Avicen, Averrhoes, Mesue, Rhasis* and *Serapio;* most of these writ but briefly of this subiect, neither haue we their Works *The Arabians.* in the Arabicke wherein they were written, but barbarously translated into Latine, and most of these works were by them taken out of the Greeks, especially *Dioscorides* and *Galen;* yet so, as they added somwhat of their own, and otherwhiles confounded other things with those mentioned by the Greeks, because they did not well know the things whereof they writ. *Avicen, Averrhoes,* and *Rhasis* alphabetically and briefly (following the method of *Galen)* giue the names, temperature, and vertues of the chiefest simple medicines. But *Avicen.* *Serapio* after a particular Tract of the Temperature and Qualities of simple medicines in *Averrhoes. Rhasis.* generall, comes to treat of them in particular, and therein followes chiefly *Dioscorides, Galen,* and *Paulus,* and diuers Arabians that went before him. This is the chiefe work in this *Serapio.* kinde of the Arabians which hath come to vs : he himselfe tells vs his method in his preface,

face,which is (when he comes to particulars) firſt of medicines temperat,then of thoſe that are hot and dry in the firſt degree,then thoſe cold and dry in the ſame degree ; after that,thoſe hot and dry in the ſecond degree,&c. and in each of theſe tracts he followeth the order of the Arabick Alphabet.

In or after the times of the Arabians,vntill about the yere 1400,there were diuers ob-ſcure and barbarous writers,who by ſight knew little whereof they writ,but tooke out of the Greeks,Arabians,and one another, all that they writ, giuing commonly rude figures, ſeldome ſetting down any deſcriptions. I will only name the chiefe of them that I haue ſeen,and as neere as I can gueſſe,in that order that one of them ſucceeded another:for the particular times of their liuing is ſomewhat difficult to be found out. One of the antien-teſt of them ſeems to be *Iſidore:* then *Platearius,*whoſe work is alphabetical, and intituled *Circa inſtans.*The next *Matthæus Sylvaticus,*who flouriſhed about the yeare 1319,his work is called *Pandecta.*A little after him was *Bartholomæus Anglus,* whoſe Works (as that of *Iſodore,*and moſt of the reſt of thoſe times)treat of diuers other things beſides Plants, as Beaſts,Birds,Fiſhes,&c. His worke is called *De proprietatibus rerum :* the Authors name was *Bartholmew Glanvill,*who was deſcended of the noble family of the Earles of Suffolk, and he writ this Work in *Edw.*the thirds time,about *An.Dom.*1397. After all theſe,and much like them,is the *Hortus ſanitatis,*whoſe Author I know not. But to leaue theſe ob-ſcure men and their writings,let me reckon ſome of later time,who with much more lear-ning and iudgement haue endeauored to illuſtrate this part of Phyſicke.

Iſidore.
Platearius.
Barthol. Angl.

Hortus ſanitat.

About ſome 200 years ago learning again beginning to flouriſh,diuers began to leaue and loath the confuſed and barbarous writings of the middle times,and to haue recourſe to the Antients , from whence together with puritie of Language they might acquire a more certain knowledge of the things treated of,which was wanting in the other. One of the firſt that tooke pains in this kinde was *Hermolaus Barbarus* Patriarch of Aquileia,who not onely tranſlated *Dioſcorides,*but writ a Commentary vpon him in fiue bookes, which he calls his *Corollarium.* In this Work he hath ſhewed himſelfe both iuditious and lear-ned.

Hermol. Barb.

After him *Marcellus Virgilius* Secretary to the State of Florence,a man of no leſſe lear-ning and iudgement than the former, ſet forth *Dioſcorides* in Greeke and Latine, with a Comment vpon him.

Merc.Virg.

Much about their time alſo *Iohn Ruellius* a French phyſitian,who flouriſhed in the yere 1480,tranſlated *Dioſcorides* into Latine,whoſe tranſlation hath bin the moſt followed of all the reſt. Moreouer,he ſet forth a large Worke, *De natura Stirpium,*diuided into three bookes,wherein he hath accurately gathered all things out of ſundry Writers, eſpecially the Greeks and Latines : for firſt hauing(after the maner of *Theophraſtus)*deliuered ſome common precepts and aduertiſements pertaining to the forme, life,generation,ordering, and other ſuch accidents of Plants , hee then comes to the particular handling of each *Species.*

John Ruellius.

Much about this time the Germanes began to beautifie this ſo neceſſarie part of Phy-ſicke ; and amongſt them *Otho Brunfelſius* a phyſitian of good account, writ of Plants,and was the firſt that gaue the liuely figures of them ; but he treated not in all of aboue 288 plants.He commonly obſerues this method in his particular chapters;firſt the figure(yet he giues not the figures of all he writes of) then the Greeke,Latine,and German names; after that,the deſcription and hiſtorie out of moſt former Authors;then the temperature and vertues ; and laſtly the Authors names that had treated of them. His Worke is in three parts or Tomes, the firſt was printed in *An.*1530. the ſecond in 1531, and the third in 1536.

Otho Brunfel.

. Next after him was *Hieronymus Tragus,* a learned ingenious and honeſt writer,who ſet forth his works in the German tongue, which were afterwards tranſlated into Latine by *David Kiber.* He treats of moſt of the Plants commonly growing in Germany, and I can obſerue no generall method he keeps,but his particular one is commonly this;he firſt gi-ueth the figure with the Latine &high Dutch name;then commonly a good deſcription; after that,the names,then the temperature,and laſtly the vertues, firſt inwardly,then out-wardly vſed.He hath figured ſome 567,and deſcribed ſome 800. his figures are good,and ſo are moſt of the reſt that follow.His works were ſet forth in Latine,*An.*1552.

Hieron.Tragus.

In his time liued *Leonhartus Fuchſius* a German Phyſitian,being alſo a learned and di-ligent writer,but he hath taken many of his deſcriptions as alſo vertues word for word out of the Antients,and to them hath put figures:his generall method is after the Greek Al-phabet,and his particular one thus : Firſt,the names in Greeke and Latine, together oft-times

Leonhar. Fuch.

times with their etymologies, as also the German and French names, then the kinds, after that the form, the place, time, temperature, then the vertues; firſt out of the Antients, as *Dioſcorides, Galen, Pliny, &c.* and ſometimes from the late writers, whom he doth not particulariſe, but expreſſes in generall, *ex recentioribus*. His Worke was ſet forth at Baſil, 1542 in *Fol.* containing 516 figures; alſo they were ſet forth in *Octavo*, the hiſtorie firſt, with, al the figures by themſelues together at the end, with the Latine and high Dutch names.

About this time and a little after flouriſhed *Conrade Geſner*, a Germane phyſitian alſo, who ſet forth diuers things of this nature, but yet liued not to finiſh the great and general Work of plants, which he for many yeres intended, and about which he had taken a great deale of pains, as may be gathered by his Epiſtles. He was a very learned, painfull, honeſt, and iuditious writer, as may appeare by his many & great works, whereof thoſe of Plants was firſt a briefe alphabeticall hiſtory of plants without figures, gathered out of *Dioſcorides, Theophraſtus, Pliny, &c.* with the vertues briefly, and for the moſt part taken out of *Paulus Ægineta*, with their names in Greek and French put in the margent. This was printed at Venice, 1541, in a ſmall form. He ſet forth a catalogue of plants in Latine, Greek, high Dutch and French printed at Zurich, 1542. Alſo another tract *De Lunarijs & noctu lucentibus cum montis fracti, ſive Pilati Lucernatum deſcriptione, Ann.* 1552. in *quarto*. He alſo ſet forth the foure books of *Valerius Cordus* (who died in his time) and his *Sylva obſervationum* in Strausburgh, 1561, in *fol.* and to theſe he added a Catalogue of the Germane gardens, with an appendix and *Corollarium* to *Cordus* his hiſtory. Alſo another treatiſe of his, *de ſtirpium collectione*, was ſet forth at Zurich by *Wolphius, An.* 1587, in *Octavo*.

At the ſame time liued *Adam Lonicerus* a phyſitian of Frankford, whoſe natural hiſtory was there printed, *An.* 1551. and the firſt part thereof is of plants, and foure years after he added anothe r part thereto, treating alſo of plants. I find no general method obſerued by him: but his particular method vſually is this; firſt he giues the figure, then the names in Latine and Dutch, then the temperature, &c. as in *Tragus*, from whom and *Cordus* he borrowes the moſt part of his firſt tome, as he doth the ſecond from *Matth.* & *Amat Luſitanus.*

In his time the Italian phyſitian *Petrus Andreas Matthiolus* ſet forth his Commentaries vpon *Dioſcorides*, firſt in Latine, with 957 large and very faire figures, and then afterwards in Latine at Venice, with the ſame figures, *An.* 1568. After this he ſet forth his epitome in *quarto*, with 921 ſmaller figures. Now theſe his Commentaries are very large, and hee hath in them deliuered the hiſtory of many plants not mentioned by *Dioſcorides* : but he is iuſtly reprehended by ſome, for that he euery where taxes & notes other writers, when he himſelfe runs into many errors, & ſome of them wilfull ones; as when he giues figures framed by his own fancie, as that of *Dracontium majus, Rhabarbarum, &c.* and falſified other ſome in part, the better to make them agree with *Dioſcor.* his deſcription, as when hee pictures *Arbor Iuda* with prickles, and giues it for the true *Acatia* : and hee oſt times giues bare figures without deſcription of his owne, but ſaith it is that deſcribed by *Dioſcorides, Nullis reclamantibus notis*, for which the Authors of the *Adverſaria* much declaim againſt him. It had bin fit for him, or any one that takes ſuch a worke in hand, to haue ſhewed by deſcribing the plant he giues, and conferring it with the deſcription of his Authour, that there is not any one note wanting in the deſcription, vertues, or other particulars which his Author ſets down ; and if he can ſhew that his is ſuch, then will the contrary opinions of all others fall of themſelues, and need no confutation.

Amatus Luſitanus alſo about the ſame time ſet forth Commentaries vpon *Dioſcorides*, adding the names in diuers languages, but without figures, at Strausburgh, *Ann.* 1554, in *Quarto*. He diſſented from *Matthiolus* in many things, whereupon *Matthiolus* writ an Apologie againſt him. He hath performed no great matter in his enarrations vpon *Dioſcorides*, but was an author of the honeſty of *Matthiolus* ; for as the one deceiued the World with counterfeit figures, ſo the other by feined cures to ſtrengthen his opinion, as *Crato* iudged of his *Curationes Medicinales* (another Worke of his) which he thinkes *potius ficta quam facta.*

Rembertus Dodonæus, a Phyſitian born at Mechlin in Brabant, about this time began to write of Plants. Hee firſt ſet forth an hiſtory in Dutch, which by *Cluſius* was turned into French, with ſome additions, *An.* 1560. And this was tranſlated out of French into Engliſh by Mr *Henry Lyte*, and ſet forth with figures, *Ann. Dom.* 1578. and diuers times ſince printed, but without figure. In the yeare 1552, *Dodonæus* ſet forth in Latine his *Frugum hiſtoria*, and within a while after, his *Florum, purgantium, & deleteriorum hiſtoria*. And afterwards he put them all together, his former and thoſe his later Works, and diuided them into 30 books, and ſet them forth with 1305 figures, in *fol, Ann.* 1583. This edition was

also

[marginal notes: conrade Geſnkr. / Leonicerus. / P. And. Matthiolu. / Amatus Luſitanus. / Rremb. Dodonæs]

also tranflated into Englifh, which became the foundation of this prefent Worke, as I fhall fhew hereafter. It hath fince been printed in Latine, with the addition of fome few new figures; and of late in Dutch, *Ann.*1618, with the addition of the fame figures, and moft of thefe in the *Exoticks* of *Clufius*, and great ftore of other additions. His generall method is this; firft he diuides his works into fix Pemptades or fiues: the firft Pemptas or fiue books of thefe contain plants in an alphabetical order, yet fo, as that other plants that haue affinitie with them are comprehended with them, though they fal not into the order of the alphabet. The fecond Pempt. contains *Flores Coronaria, Planta odorata & vmbellifera.* The third is *De Radicibus, Purgantibus herbis, convolvulis, deleterijs ac perniciofis plantis, Filicibus, Mufcis, & Fungis.* The fourth is *De Frumentis, Leguminibus, paluftribus, & aquatilibus.* The fifth, *De Oleribus & Carduis.* The fixt, *De Fruticibus & Arboribus.* The particular method is the fame vfed by our Author.

Peter Pena.   In the yeare 1570, *Peter Pena* and *Matthias Lobel* did here at London fet forth a Worke
Matth. Lobel. intituled *Stirpium Adverfaria nova*; the chief end and intention wherof being to find out the *Materia medica* of the Antients. The generall method is the fame with that of our Author, which is, putting things together as they haue moft refemblance one with another in external forme, beginning with Graffes, Cornes, &c. They giue few figures, but fomtimes refer you to *Fuchfius, Dodon.* & *Matthiolus*; but where the figures war nor giuen by former Authors, then they commonly giue it; yet moft of thefe figures are very fmall and vnperfect, by reafon (as I conjecture) they were taken from dried plants. In this work they infift little vpon the vertues of plants, but fuccinctly handle Controuerfies, and giue their opinions of plants, together with their defcriptions and names, which fomtimes are in all thefe languages, Greek, Latine, French, high & low Dutch, and Englifh; otherwhile in but one or two of them. Some writers for this Work call them *Doctiffimi Angli*; yet neither of them were born here, for *Pena* (as I take it) was a French man, and *Lobel* was born at Ryffele in Flanders, yet liued moft part of his later time in this kingdom, and here alfo ended his daies. In the yeare 1576 he fet forth his Obferuations, and ioyned them with the *Adverfaria*, by them two to make one entire work: for in his Obferuations he giues moft part of the figures and vertues belonging to thofe herbs formerly defcribed only in the *Adverf.* and to thefe alfo addes fome new ones not mentioned in the former worke. After which he fet forth an Herbal in Dutch, wherein he comprehended all thofe plants that were in the two former Works, and added diuers other to them, the Worke containing fome 2116 figures, which were printed afterward in a longifh form, with the Latine names, and references to the Latine & Dutch books. After all thefe, at London, *An.*1605, he again fet forth the *Adverfaria*, together with the fecond part thereof, wherein is contained fome forty figures, being moft of them of Graffes and floures, but the defcriptions were of fome 100 plants, varieties and all. To this he added a Treatife of Balfam, which alfo was fet forth alone in quarto, *An.*1598; and the *Pharmacopaa* of *Rondeletius*, with Annotations vpon it. He intended another great Worke, whofe title fhould haue bin *Stirpium illuftrationes*, but was preuented by death.

Carol. Clufius.   Some fix yeares after the edition of the *Adverfaria, Anno* 1576, that learned diligent and laborious Herbarift *Carol.Clufius* fet forth his Spanifh Obferuations, hauing to this purpofe trauelled ouer a great part of Spaine; and being afterwards called to the Imperiall Court by *Maximilian* the fecond, he viewed Auftria and the adiacent prouinces, and fet forth his there Obferuations, *An.*1583. He alfo tranflated out of Spanifh the Works of *Garcias ab Orta* and *Chriftopher Acofta*, treating of the fimple medicines of the Eaft Indies, and *Nicolas Monardus*, who writ of thofe of the Weft Indies. After this he put into one body both his Spanifh and Pannonick Obferuations, with fome other, and thofe hee comprehends in fix books, intituled *Rariorum Plantarum Hiftoria*: whereto he alfo adds an Appendix, a treatife of Mufhromes, fix Epiftles treating of plants, from *Honorius Bellus*
Honor. Bellus. an Italian phyfitian liuing at Cydonia in Candy; as alfo the defcription of mount Baldus, being a catalogue with the defcription and figures of fome rare and not before written of plants there growing, written by *Iohn Pona* an Apothecarie of Verona. (This De-
Iohn Pona. fcription of *Pona's* was afterwards with fome new defcriptions and thirty fix figures fet forth alone in quarto, *Anno* 1608.) This firft Volume of *Clufius* was printed in Antwerpe, *Anno* 1601, in Folio: and in the yeare 1605 he alfo in Folio fet forth in another volume fix bookes of *Exotickes* containing various matter, as plants, or fome particles of them, as Fruits, Woods, Barks, &c. as alfo the forenamed tranflations of *Garcias, Acofta*, and *Monardus*: Three Tracts befides of the fame *Monardus*; the firft, *De lapide Bezaar, & Herba Scorfonera.* The fecond, *De Ferro & eius facultatibus*: The third, *De Niue & eius commodis.*

To

To thefe he alfo added *Bellonius* his Obfervations or Singularities, and a tract of the fame Author. *De neglecta Stirpium cultura*, both formerly tranflated out of French into Latine by him. He way borne at Atrebas or Arras, the chiefe city of Artois, *Anno* 1526. and died at Leyden, *Anno* 1609. After his death, by *Euerard Vorftius, Peter Paw*, or fome others, were fet forth fome additions and emendations of his former Works, together with his funerall oration made by *Vorftius*, his Epitaph, &c. in Quarto, *Anno* 1611. by the name of his *Curæ pofteriores.*

In the yeare 1583 *Andreas Cafalpinus* an Italian Phyfitian, and Profeffor at Pifa, fet forth an hiftory of Plants, comprehended in fixteene bookes: his Worke is without figures, and he oft times giues the Tufcane names for Latine; wherefore his Worke is the more difficult to be vnderftood, vnleffe it be by fuch as haue beene in Tufcanie, or elfe are already well exercifed in this ftudy. He commonly in his owne words diligently for the moft part defcribes each Plant, and then makes enquirie whether they were knowne by the Antients. Hee feldome fets downe the faculties, vnleffe of fome, to which former Writers haue put downe none. In the firft booke he treats of Plants in generall, according as *Theophraftus* doth: but in the following bookes hee handles them in particular: he maketh the chiefe affinity of Plants to confift in the fimilitude of their feeds and feed veffels. *Andr. Cafalp.*

Ioachimus Camerarius a Phyfition of Noremberg flourifhed about this time: Hee fet forth the Epitome of *Matthiolus*, with fome additions and accurate figures, in Quarto, at Frankfort, 1586. in the end of which Worke (as alfo in that fet forth by *Matthiolus* himfelfe) is *Iter baldi*, or a journey from Verona to mount Baldus, written by *Francis Calceolarius* an Apothecary of Verona. Another Worke of *Camerarius* was his *Hortus Medicus*, being an Alphabeticall enumeration of Plants, wherein is fet forth many things concerning the names, ordering, vertues, &c. of Plants. To this he annexed *Hyrcinia Saxonothuringica Iohannis Thalij*, or an alphabeticall Catalogue written by *Iohn Thalius*, of fuch Plants as grew in Harkwald a part of Germany betweene Saxony and Durengen. This was printed alfo at Frankfort in quarto, *Anno* 1588. *Ioach. Camer. Fr. Calceolarius Ioh. Thalius*

In the yeare 1587. came forth the great Hiftory of Plants printed at Lyons, which is therefore vulgarly termed *Hiftoria Lugdunenfis*: it was begun by *Dalechampius*: but hee dying before the finifhing thereof, one *Iohn Molinæus* fet it forth, but put not his name thereto. It was intended to comprehend all that had written before, and fo it doth, but with a great deale of confufion; which occafioned *Bauhine* to write a treatife of the errors committed therein, in which hee fhewes there are about foure hundred figures twice or thrice ouer. The whole number of the figures in this worke are 2686. This Hiftory is diuided into eighteene bookes, and the Plants in each booke are put together either by the places of their growings, as in Woods, Copfes, mountains, Watery places, &c. or by their externall fhape, as vmbelliferous, bulbous, &c. or by their qualities, as purging, poyfonous, &c. Herein are many places of *Theophraftus* and other antient Writers explained. He commonly in each chapter giues the names, place, forme, vertue, as moft other do. And at the end thereof there is an Appendix containing fome Indian plants, for the moft part out of *Acofta*; as alfo diuers Syrian and Ægiptian plants defcribed by *Reinold Rauolfe* a Phyfition of Ausburgh. *Hift. Lugd. Leon. Rawolf.*

At this time, to wit, *Anno* 1588. *Iacobus Theodorus Tabernamontanus* fet forth an Hiftory of Plants in the Germane tongue, and fome twelue yeares after his Figures being in all 2087. were fet forth in a long forme, with the Latine and high-Dutch names put vnto them; and with thefe fame Figures was this Worke of our Author formerly Printed. *Tabernamont.*

Profper Alpinus a Phyfition of Padua in Italy, in the yeare 1592 fet forth a Treatife of fome Ægyptian Plants, with large yet not very accurate figures: hee there treats of fome 46. plants, and at the end thereof is a Dialogue or Treatife of Balfam. Some fix yeares agone, *Anno* 1627. his Son fet forth two bookes of his Fathers, *De Plantis Exoticis*, with the figures cut in Braffe: this worke contains fome 136 Plants. *Profp. Alpinus.*

Fabius Columna a gentleman of Naples, of the houfe of *Columna* at Rome, *Anno* 1592. fet forth a Treatife called *Phytobafanos*, or an Examination of Plants; for therein he examines and afferts fome plants to be fuch and fuch of the Antients: and in the end of this worke he giues alfo the Hiftory of fome not formerly defcribed Plants. Hee alfo fet forth two other bookes, *De minus cognitis*, or of leffe knowne Plants: the firft of which was Printed at Rome, *Anno* 1606; and the other 1616. He in thefe works, which in all containe little aboue two hundred thirty fix plants, fhewes himfelfe a man of an exquifit iudgment, and very learned and diligent, duely examining and weighing each circumftance in the writings of the Antients. *Fab. Columna*

Cafpar

Casp. Bauhine. Caspar Bauhine, a Physition and Professor of Basil, besides his Anatomicall Works, set forth diuers of Plants. *Anno* 1596 he set forth his *Phytopinax*, or Index of Plants, wherein he followes the best method that any yet found : for according to *Lobels* method (which our Author followed) he begins with Grasses, Rushes, &c. but then he briefely giues the Etymologie of the name in Greeke and Latine, if any such be, and tells you who of the Antients writ thereof, and in what part of their Works : and lastly (which I chiefly commend him for) he giues the *Synonima's* or seuerall names of each Plant giuen by each late Writer, and quoteth the pages. Now there is nothing more troubles such as newly enter into this study, than the diuersitie of names, which sometimes for the same plant are different in each Author ; some of them not knowing that the plant they mention was formerly written of, name it as a new thing ; others knowing it writ of, yet not approuing of the name. In this Worke he went but through some halfe of the history of Plants. After this, *Anno* 1598, he set forth *Matthiolus* his Commentaries vpon *Dioscorides*, adding to them 330 Figures, and the descriptions of fifty new ones not formerly described by any ; together with the *Synonima's* of all such as were described in the Worke. He also *Anno* 1613 set forth *Tabernamontanus* in Dutch, with some addition of history and figures. In *Anno* 1620 he set forth the *Prodromus*, or fore-runner of his *Theatrum Botanicum*, wherein he giues a hundred and forty new figures, and describes some six hundred plants, the most not described by others. After this, *Anno* 1623, he sets forth his *Pinax Theatri Botanici*, whose method is the same with his *Phytopinax*, but the quotatians of the pages in the seuerall Authors are omitted. This is indeed the Index and summe of his great and generall Worke, which should containe about six thousand plants, and was a Worke of forty yeares : but he is dead some nine yeares agone, and yet this his great Worke is not in the Presse, that I can heare of.

Basil Besler. Basil Besler an Apothecary of Noremberg, *Anno* 1613 set forth the garden of the Bishop of Eystet in Bauaria, the figures being very large, and all curiously cut in brasse, and printed vpon the largest paper : he onely giues the *Synonima's* and descriptions, and diuideth the worke first into foure parts, according to the foure seasons of the yeare; and then againe he subdiuides them, each into three, so that they agree with the moneths, putting in each Classis the Plants that flourish at that time.

These are the chiefe and greatest part of those that either in Greeke or Latine (whose Works haue come to our hands) haue deliuered to vs the history of Plants; yet there are some who haue vsed great diligence to helpe forward this knowledge, whose names I wil *Aloys. Anguill.* not passe ouer in silence. The first and antientest of these was *Aloysius Anguillara* a Physition of Padua, and President of the publique Garden there: his opinions of some plants were set forth in Italian at Venice, 1561.

Melchior Guillandinus. Melchior Guillandinus, who succeeded *Anguillara* in the garden at Padua, writ an Apologie against *Matthiolus*, some Epistles of Plants, and a Commentary vpon three chapters of *Pliny, De Papyro.*

Fer. Imperato. Ferantes Imperatus an Apothecary of Naples also set forth a Naturall History diuided into twenty eight bookes, printed at Naples *Anno* 1599. In this there is something of Plants : but I haue not yet seene the opinions of *Anguillara*, nor this Naturall History : yet you shall find frequent mention of both these in most of the forementioned Authors that writ in their time, or since, wherefore I could not omit them.

Will. Turner. Let me now at last looke home, and see who we haue had that haue taken pains in this kinde. The first that I finde worthy of mention is Dr *William Turner*, the first of whose works that I haue seene, was a little booke of the names of herbes, in Greeke, Latine, English, Dutch, and French, &c. Printed at London *Anno* 1548. In the yeare 1551 he set forth his Herbal or History of Plants, where he giues the figures of *Fuchsius*, for the most part : he giues the Names in Latine, Greeke, Dutch, and French: he did not treat of many Plants ; his method was according to the Latine alphabet. He was a man of good iudgment and learning, and well performed wat he tooke in hand.

Hen. Lyte. After this, *Dodonæus* was translated into English by Mr *Lyte*, as I formerly mentioned. And some yeares after, our Author set forth this Worke, whereof I will presently treat, hauing first made mention of a Worke set forth betweene that former Edition, and this I now present you withall.

Ioh. Parkinson. Mr *Iohn Parkinson* an Apothecary of this City (yet liuing and labouring for the common good) in the yeare 1629 set for a Worke by the name of *Paradisus terrestris*, wherein he giues the figures of all such plants as are preserued in gardens, for the beautie of their floures, for vse in meats or sauces, and also an orchard of all trees bearing fruit, and such shrubs as for their raritie or beauty are kept in Orchards and gardens, with the ordering,

planting,

planting and preseruing of all these. In this Worke hee hath not superficially handled these things, but accurately descended to the very varieties in each species: wherefore I haue now and then referred my Reader addicted to these delights, to this worke especially in floures and fruits, wherein I was loth to spend too much time, especially seeing I could adde nothing to what he had done vpon that subiect before. He also there promised another Worke, the which I thinke by this time is fit for the Presse.

Now am I at length come to this present Worke, whereof I know you will expect I should say somewhat; and I will not frustrate your expectation, but labour to satisfie you in all I may, beginning with the Author, then his Worke, what it was, and lastly, what it now is.

For the Author Mr *Iohn Gerard* I can say little, but what you also may gather out of this Worke; which is, he was borne in the yeare 1545. in Cheshire, at Namptwich, from whence he came to this City, and betooke himselfe to Surgery, wherein his endeauours were such, as he therein attained to be a Master of that worthy Profession: he liued some ten yeares after the publishing of this worke, and died about the yeare 1607. His chiefe commendations is, that he out of a propense good will to the publique aduancement of this knowledge, endeauoured to performe therein more then he could well accomplish; which was partly through want of sufficient learning, as (besides that which he himselfe saith of himselfe in the chapter of Water Docke) may be gathered by the translating of diuers places out of the *Aduersaria*; as this for one in the description of * *After Atticus*, *Caules pedales terni aut quaterni*: which is rendred, A stalke foure or fiue foot long. He also by the same defect called burnt Barley, * *Hordeum distichon*; and diuided the titles of honour from the name of the person whereto they did belong, making two names therof, beginning one clause with * *Iulius Alexandrinus* saith, &c. and the next with, *Cæsarius Archiater* saith. He also was very little conuersant in the writings of the Antients, neither, as it may seeme by diuers passages, could hee well distinguish betweene the antient and moderne writers: for he in one place saith, [* Neither by *Dioscorides*, *Fuchsius*, or any other antient writer once remembred.] Diuers such there are, which I had rather passe ouer in silence, than here set downe: neither should I willingly haue touched hereon, but that I haue met with some that haue too much admired him, as the only learned and iudicious writer. But let none blame him for these defects, seeing he was neither wanting in pains nor good will, to performe what hee intended; and there are none so simple but know, that heauie burthens are with most paines vndergone by the weakest men: and although there were many faults in the worke, yet iudge well of the Author; for as a late Writer well saith, *Falli & hallucinari humanum est*; *solitudinem quærat oportet, qui vult cum perfectis viuere. Pensanda vitijs bona cuiusque sunt, & qua maior pars ingenij stetit, ea iudicandum de homine est.*

Now let me acquaint you how this Worke was made vp. *Dodonæus* his Pemptades comming forth *Anno* 1583, were shortly after translated into English by Dr *Priest* a Physition of London, who died either immediately before or after the finishing of this translation. This I had first by the relation of one who knew Dr *Priest* and Mr *Gerard*: and it is apparant by the worke it selfe, which you shall finde to containe the Pemptades of *Dodonæus* translated, so that diuers chapters haue scarce a word more or lesse than what is in him. But I cannot commend my Author for endeauouring to hide this thing from vs, cauilling (though commonly vniustly) with *Dodonæus*, wheresoeuer he names him, making it a thing of heare-say, * that Dr *Priest* translated *Dodonæus*: when in the Epistle of his friend Mr *Bredwell*, prefixed before this Worke, are these words: [The first gatherers out of the Antients, and augmenters by their owne paines, haue already spread the odour of their good names through all the lands of learned habitations: Dr *Priest* for translating so much as *Dodonæus*, hath hereby left a tombe for his honourable sepulture. Mr *Gerard* comming last, but not the least, hath many waies accomodated the whole Worke vnto our English Nation, &c.] But that which may serue to cleare all doubts, if any can be in a thing so manifest, is a place in *Lobels* Annotations vpon *Rondeletius* his *Pharmacopeia*, where pag. 59. he findes fault with *Dodonæus*, for vsing barbarously the word *Seta* for *Sericum* · and with Dr *Priest*, who (saith he) at the charges of Mr *Norton* translated *Dodonæus*, and deceiued by this word *Seta*, committed an absurd errour in translating it a bristle, when as it should haue beene, silke. This place so translated is to be seen in the chapter of the Skarlet Oke, at the letter F. And *Lobel* well knew that it was Dr *Priest* that committed this error, and therefore blames not Mr *Gerard*, to whom hee made shew of friendship, and who was yet liuing: but yet hee couertly gaue vs to vnderstand, that the worke wherein that error was committed, was a translation of *Dodonæus* and that made by

Iohn Gerard.

See the former Edition in the places here mentioned
*p.g.*391.
*p.*65
*p.*147.

*p.*518.

Cun.li.3 to.3. de R.p Hcb.

See his Epistle to the Reader.

Dr *Priest*

Dr *Priest*; and set forth by Mr *Norton*. Now this translation became the ground-worke whereupon Mr *Gerard* built vp this Worke : but that it might not appeare a translation, he changes the generall method of *Dodonæus*, into that of *Lobel*, and therein almost all ouer followes his *Icones* both in method and names, as you may plainly see in the Grasses and *Orchides*. To this translation he also added some plants out of *Clusius*, and other-some out of the *Aduersaria*, and some fourteene of his owne not before mentioned. Now to this history figures were wanting, which also Mr *Norton* procured from Frankfort, being the same wherewith the Works of *Tabernamontanus* were printed in Dutch : but this fell crosse for my Author, who (as it seemes) hauing no great iudgment in them, frequently put one for another : and besides there were many plants in those Authors which hee followed, which were not in *Tabornamontanus*, and diuers in him which they wanted, yet he put them altogether, and one for another ; and oft times by this meanes so confounded all, that none could possibly haue set them right, vnlesse they knew this occasion of these errors. By this meanes, and after this manner was the Worke of my Author made vp, which was printed at the charges of Mr *Norton Anno* 1597.

Now it remaines I acquaint you with what I haue performed in this Edition which is either by mending what was amisse, or by adding such as formerly were wanting : some places I helped by putting out, as the Kindes in the Chapter of Stonecrop, wheare there was but one mentioned. I haue also put out the Kindes in diuers places else where, they were not very necessary, by this meanes to get more roome for things more necessarie : as also diuers figures and descriptions which were put in two or three places, I haue put them out in all but one, yet so, as that I alwaies giue you notice where they were, and of what. Some words or passages are also put out here and there, which I thinke needlesse to mention. Sometimes I mended what was amisse or defectiue, by altering or adding one or more words, as you may frequently obserue, if you compare the former Edition with this, in some few chapters almost in any place. But I thinke I shall best satisfie you if I briefely specifie what is done in each particular, hauing first acquainted you with what my generall intention was: I determined, as well as the shortnesse of my time would giue me leaue, to retaine and set forth whatsoeuer was formerly in the booke described, or figured without descriptions (some varieties that were not necessary excepted) and to these I intended to adde whatsoeuer was figured by *Lobel, Dodonæus*, or *Clusius*, whose figures we made vse of ; and also such plants as grow either wilde, or vsually in the gardens of this Kingdome, which were not mentioned by any of the forenamed Authors ; for I neither thought it fit nor requisite for me, ambitiously to aime at all that *Bauhine* in his *Pinax* reckons vp, or the Exotickes of *Prosper Alpinus* containe, not mentioned in the former. This was my generall intention. Now come I to perticulars, and first of figures : I haue, as I said, made vse of those wherewith the Workes of *Dodonæus, Lobel*, and *Clusius* were formerly printed, which though some of them be not so sightly, yet are they generally as truly exprest, and sometimes more. When figures not agreeable to the descriptions were formerly in any place, I giue you notice thereof with a marke of alteration before the title, as also in the end of the Chapter ; and if they were not formerly in the booke, then I giue you them with a marke of addition. Such as were formerly figured in the booke, though put for other things ; and so hauing no description therein, I haue caused to be new cut and put into their fit places, with descriptions to them, and only a marke of alteration. The next are the descriptions, which I haue in some places lightly amended, without giuing any notice thereof ; but when it is much altered, then giue I you this marke † at the beginning thereof ; but if it were such as that I could not helpe it but by writing a new one, then shall you finde it with this marke ‡ at the beginning and end thereof, as also whatsoeuer is added in the whole booke, either in description or otherwise. The next is the Place, which I haue seldome altered, yet in some places supplied, and in others I haue put doubts, & do suspect othersome to be false, which becaufe I had not yet viewed I left as I found. The Time was a thing of no such moment, for any matter worth mentioning to be performed vpon, wherefore I will not insist vpon it. Names are of great importance, and in them I should haue bin a little more curious if I had had more time, as you may see I at the first haue beene ; but finding it a troublesome worke, I haue only afterwards where I iudged it most needfull insisted vpon it : *Bauhinus* his *Pinax* may supply what you in this kinde finde wanting. In many places of this worke you shall finde large discourses and sometimes controuersies handled by our Author in the names ; these are for the most part out of *Dodonæus*, & some of them were so abbreuiated, and by that meanes confounded, that I thought it not worth my paines to mend them, so I haue put them out in some few places, and referred you to the places in *Dodonæus* out of

which

which they were taken, as in the chapter of Alehoofe: it may be they are not so perfect as they should be in some very few other places, (for I could not compare all) but if you suspect any such thing, haue recourse to that Author, and you shall find full satisfaction.

Now come I to the Temper and Vertues: These commonly were taken forth of the fore-mentioned Author, and here and there out of *Lobels* Observations, and *Camerarius* his *Hortus medicus*. To these he also added some few receipts of his owne: these I haue not altered, but here and there shewed to which they did most properly belong, as also if I found them otherwise than they ought, I noted it; or if in vnfit places, I haue transferred them to the right place, and in diuers things whereof our Author hath been silent, I haue supplied that defect.

For my additions I will here say nothing, but refer you to the immediate insuing Catalogue, which will informe you only what is added in figure, or description, or in both, by which, and these two formerly mentioned markes, you may see what is much altered or added in the Worke; for this marke † put either to figure, or before any clause, shews it to haue bin otherwise put before; or that clause whether it be in description, Place, Time, Names, or Vertues to be much altered. This other marke ‡ put to a figure shewes it not to haue been formerly in the worke, but now added; and put in any other place it shewes all is added vntill you come to another of the same marks. But because it is sometimes omitted, I will therefore giue notice in the *Errata* where it should be put, in those places where I obserue either the former or later of them to be wanting.

Further, I must acquaint you how there were the descriptions of a few plants here and there put in vnfitting places, which made me describe them as new added, as *Saxifraga maior Matthioli, Persicaria siliquosa*, of which in the chapter of *Persicaria* there was an ill description, but a reasonable good one in the chapter of *Astrantia nigra. Papauer spinosum*, was figured and described amongst the Cardui; now all these (as I said) I added as new in the most fitting places: yet found them afterwards described, but put them out all, except the last, whose history I still retained, with a reference to the preceding figure and Historie. Note also, wheresoeuer my Author formerly mentioned *Clusius*, according to his Spanish or Pannonicke Obseruations, I haue made it, according to this Historie, which containes them both with additions.

Also I must certifie you, because I know it is a thing that some will thinke strange, that the number of the pages in this booke do no more exceed that of the former (considering there is such a large accession of matter and figures) the cause hereof is, each page containes diuers lines more than the former, the lines themselues also being longer; and by the omission of descriptions and figures put twice or thrice ouer, and the Kinds, vnnecessarily put in some places, I gained as much as conueniently I could, being desirous that it might be bound together in one volume.

Thus haue I shewed what I haue performed in this Worke, entreating you to take this my Labor in good part; and if there be any defect therein (as needs there must in all humane works) ascribe it in part to my haste and many businesses, and in some places to the want of sufficient information, especially in Exoticke things; and in othersome, to the little conuersation I formerly had with this Author, before such time as (ouercome by the importunitie of some friends, and the generall want of such a Worke) I tooke this taske vpon me. Furthermore I desire, that none would rashly censure me for that which I haue here done; but they that know in what time I did it, and who themselues are able to do as much as I haue here performed; for to such alone I shall giue free liberty, and will be as ready to yeeld further satisfaction if they desire it, concerning any thing I haue here asserted, as I shall be apt to neglect and scorne the censure of the Ignorant and Vnlearned, who I know are still forward to verifie our English prouerbe * | *A fooles bolt is soone shot.*

I must not in silence passe ouer those from whom I haue receiued any fauour or incouragement, whereby I might be the better enabled to performe this Taske. In the first place let me remember the only Assistant I had in this Worke, which was Mr *Iohn Goodyer* of Maple-Durham in Hampshire, from whom I receiued many accurate descriptions, and some other obseruations concerning plants; the which (desirous to giue euery man his due) I haue caused to be so printed, as they may be distinguished from the rest: and thus you shall know them; in the beginning is the name of the plant in Latine in a line by it selfe, and at the end his name is inserted; so that the Reader may easily finde those things that I had from him, and I hope together with me will be thankfull to him, that he would so readily impart them for the further increase of this knowledge.

Mr *George Bowles* of Chisselhurst in Kent must not here be forgot, for by his trauells and industry I haue had knowledg of diuers plants, which were not thought nor formerly knowne to grow wilde in this kingdome, as you fhall finde by diuers places in this book. My louing friends and fellow Trauellers in this ftudy, and of the fame profeſſion, whoſe companie I haue formerly enjoyed in ſearching ouer a great part of Kent, and who are ftill ready to doe the like in other places, are here alſo to be remembred, and that the rather, becauſe this knowledge amongſt vs in this city was almoſt loft, or at leaſt too much neglected, eſpecially by thoſe to whom it did chiefely belong, and who ought to be aſhamed of ignorance, eſpecially in a thing ſo abſolutely neceſſary to their profeſſion. They ſhould indeed know them as workemen do their tooles, that is readily to call them by their names, know where to fetch, and whence to procure the beft of each kinde; and laftly, how to handle them.

Thomas Hickes
Iohn Buggs.
William Broad.
Job Weale.
Leon. Buckner.
Iames Clarke.
Robert Lorkin.

I haue already much exceeded the bounds of an Epiſtle, yet haue omitted many things of which I could further haue informed thee Reader, but I will leaue them vntill ſuch time as I finde a gratefull exceptance; or ſome other occaſion that may againe inuite me to ſet Pen to Paper; which, That it may be for my Countreyes good and Gods glory, ſhall euer be the prayers and
Endeauours of thy Well
Wiſher.

From my houſe on Snow-hill;
October. 22. 1633.

THOMAS **I**OHNSON.

A Catalogue of Additions.

BEcaufe the markes were not fo carefully and right put to thefe Figures, which were not formerly in the booke, I haue thought good to giue you the names of all fuch as are added, either in figure or defcription, or both: together with the booke, Chapter, and number or place they hold in each chapter. *F* ftands for Figure, *D* for Defcription, and where both are added, you fhall find both thefe letters; and where the letter *C* is put, the hiftorie of the whole chapter is added.

BOOKE I.

　　　　　　　　　Chap

Chap.

Ch.507 *Faba vulgaris,*d.
Ch.508.5,6,7,8,9,*Phaseolor.perigrin.var.*9.cum.
    fig.3.d.9.
Ch.509 4.*Lupinus mai.flo.car.*f.d.
Ch.515. 1 *Vicia,*f,
    2 *Vicia max.dumet.*d.
    3 *Vicia syl.flo.alb.*f.d.
    5 *Vicia syl.sue Cracca min.*f.d.
Ch.516.1 *Lathyrus mai.latifol.*f.
    2 *Lath.ang.flo.alb.*f.
    3 *Lath.angust.flo.purp.*f.d.
    4 *Lath.Ægypt.*f.d.
    5 *Lath.ann.sil. Orobi,*f,d.
    6 *Lath syl.flo.lut.*f.
Ch.518.2 *Hedysarum glycyrrhizatum,*f,
    3 *Hedysar.mai.siliquis artic.*f.
    4 *Securidaca min.pal.cærul.*f,
    5 *Secur.min.lut.*f,d,
    6 *Secur.sil.plan.dent.*f,d,
    7 *Hedysar.clyp.*f.
Ch.519.2 *Astragalus syl.*f.d.
Ch.520.3. *Astragalus Matth.*f.
    4 *Astragaloides,*f.
Ch.521,3.*Ornithopodium mai.*f.
    4 *Ornithopod,min.*f.
    5 *Scorpioides leguminosa,*f,d.
C. { Ch.526 1 *Orobus venet,*f,d.
    { 2 *Orobus syl.vernus,*f,d.
    { 3 *Orob.mont.flo.alb.*f,d.
    { 4 *Orob.mon.angust.*f,d.
C. { Ch.527.1 *Ochrus sue Eruilia,*f,d.
    { 2.*Eruum sylv.*f,d.
    { 3 *Aphaca,*f,d.
    { 4 *Legumen mar.long.rad.*d.
Ch.528.3 *Talictrum mai.Hispan.*d.
Ch.531.6*Ruta canin.*f,d.
        LIB. 3.
CH.2.6 *Rosa lut.multipl.*f,d.
    8 *Rosa Cinnam.flo.simpl.*f.
Ch.3.2 *Rosa syl.odor.flo dup.*f.
Ch.4. 2 *Rubus repens fructu casio,*d.
Ch.5.19 *Cistus ann.flo.mac.*f,d.
    20 *Cistus folio sampsuch.*f,d.
Ch.7.*Chamæcistus serpillifol.*f,d.
    8 *Chamæcistus Fris,*f,d.
Ch.7.15 *Cistus Ledon folijs Rosm.*f,d.
Ch.12.*Glycyrrhiza vulg.*f.
Ch.17. *Orobanches triplex var.*f,3.
Ch.20.5 *Genista spinosa humilis,*d.
Ch.25.2 *Tragacantha min,icon,accur.*
    3 *Poterion Lob.*f.
Ch.26. 1 *Acacia Diosc.*f.
Ch.27.2 *Lycium Hisp.*f.
Ch.28.1 *Rhamnus flo.alb.*f.
    *Rhamn.alt.flo.purp.*f,d.
    2 *Rhamnus 2 Cluf.*f,d.
    3 *Rhamnus 3 Clus.*d.
Ch.30.1 *Rhamnus solut.*f.
    2*Rhamn.sol.min.*f,d.
    3 *Rham.sol.pumil.*d.
Ch.34. *Ilicis ramus flor.*f,
Ch.35.*Cerri minoris ram.cum flo.*f,
Ch.37.2 *Galla maior alt.*f,d,
Ch.40.2 *Picea pumila,*f,
Ch.42. 8 *Pinaster Austr.*f.d.
    9 *Pinaster mar.min.*f.d,
Ch.43.2 *Abies mas.*f,
    *Abietis ramus cum julis,*f.

Ch.47.*Taxus glandif,& baccif.*d,
    *Taxus tant.flor.*d,
Ch,48.3 *Iuniperus Alp.min.*f,d,
Ch.49.3 *Cedrus lycia alt.*f,d.
Ch.50.3 *Sabina bacc.alt.*f.d.
Ch.52.3 *Erica mai.flo.alb,*d,
    9 *Erica baccif.procumbens,*f,
    10 *Erica baccif.ten.*d,
    11 *Erica pum,*3.*Dod.*f,d,
    12 *Erica ternis per internalla ramis,*f,d,
    13 *Erica peregrin.Lob.*f,d,
    14 *Erica coris folio* 7 *Cluf.*f,d,
    15 *Erica Coris fol.* 9 *Clus.*f,d,
Ch.54.2 *Vitex lat.serat.folio,*f,d,
Ch.55. 8 *Salix hum.repens,*f,
Ch.61.3 *Syringa Arabica,*f,d,
Ch.71.2 *Myrtus Batica lat.*f,
    3 *Myrt.exot.*f,
    4 *Myrt.fruct.alb.*f,
    5 *Myrtus min.*f,
    6 *Myrt.Batica syl.*f,d,
Ch.73.6 *Vitis Idea fol.subrotund.mai.*d,
Ch.77.2 *Sambucus fructu alb.*f,
Ch.89.*Avellana pum.Byz.*f,d,
Ch.91.3 *Castanea Peru.fruct.*f,d,
Ch.94.5 *Persica flo.pleno,*d,
Ch.98.2 *Mespilus satina alt.*f,d,
    4 *Chamæmespilus,*f,
Ch.113.2 *Alnus hirsut.*f,d,
Ch.116. 1 *Vlmus vulgat.fol.lato scabro,*d,
    2 *Vlmus min.fol.angusto scabro,*f,d,
    3 *Vlmus fol.latis.scab.*f,d,
    4 *Vlmus fol.glab.*d,
Ch.118.1 *Acer mai.*f,
Ch.119.5 *Populus alba folijs minor,*f
Ch.122.2 *Zizipha Cappadocica,*f,
Ch.124.*Guaiacum Patau.angust.*d,
Ch.133.2 *Chamæficus,*f,
Ch.139 *Musa fructus exact.icon,*f,d,
Ch.145. 3 *Balsamum Alp.*f,d,
Ch.146.2 *Molle arboris adulte ramus,*f.
Ch.153.5 *Piper caudatum,*f,
C. { Ch.159.*Fructus Indici & exotic.quorum fig.
    { ad.*26.*descr.* 35,
Ch.162.6 *Muscus Pyxidatus,*f,
    12 *Musc.clauat.fol.Cypr.*d,
    14 *Musc.parv.stell.*f,d,
Ch.164.3 *Lichen mar.rotund.*f,d,
    4 *Quernus mar.var.*f,d,
    5 *Quern mar.secund.*f,d,
    6 *Quern.mar.tertia,*f,d,
    7 *Quern.mar.quarta,*f,d,
    8 *Alga,*f,d,
    9 *Fucus phasganoides & polys.*f,d,
    10 *Fucus spong.nod.*f,d,
    11 *Conferua,*f,d,
Ch.165. 7 *Fucus ferul.*f,d,
    8 *Fucus tenuifol.alt.*f,d,
    9 *Muscus mar.Clus.*f,d,
    10 *Muscus mar.tertius Dod.*f,d,
    11 *Abies mar. Belg.Clus.*f,d
Ch.166.5 *Coraloides albif.*d,
    6 *Coral.rub.*f,d,
    8 *Spong.infundibuli forma.*f,d,
    9 *Spongia ramos.*f,d,
Ch.167.*Fungorum fig.* 14,
The Appendix contains fig. 46.*descrip.*72.

THE.

THE FIRST BOOKE OF
THE HISTORY OF PLANTS.

Containing Graſſes, Ruſhes, Reeds, Corne, Flags, and Bulbous or Onion-rooted Plants.

IN this Hiſtorie of Plants it would be tedious, to vſe by way of introduction any curious diſcourſe vpon the generall diuiſion of Plants, contained in Latine vnder *Arbor, Frutex, Suffrutex, Herba:* or to ſpeake of the differing names of their ſeuerall parts, more in Latine than our vulgar tongue can well expreſſe. Or to go about to teach thee, or rather to beguile thee by the ſmell or taſte, to gheſſe at the temperature of Plants: when as all and euery of theſe in their place ſhall haue their true face and note, whereby thou mayſt both know and vſe them.

In three bookes therefore, as in three gardens, all our Plants are beſtowed; ſorted as neere as might be in kindred and neighbourhood.

The firſt booke hath Graſſes, Ruſhes, Corne, Reeds, Flags, Bulbous or Onion-rooted Plants. The ſecond, moſt ſorts of Herbs vſed for meat, medicine, or ſweet ſmell. The third hath Trees, Shrubs, Buſhes, Fruit-bearing Plants, Roſins, Gummes, Roſes, Heaths, Moſſes, Muſhroms, Corall, and ther ſeuerall kindes.

Each booke hath Chapters, as for each Herb a bed: and euery Plant preſents thee with the Latine and Engliſh name in the title, placed ouer the picture of the Plant.

Then followeth the Kinds, Deſcription, Place, Time, Names, Natures, and Vertues, agreeing with the beſt receiued Opinions.

Laſt of all thou haſt a generall Index, as well in Latine as Engliſh, with a carefull ſupply likewiſe of an *Index Bilinguis*, of Barbarous Names.

And thus hauing giuen thee a generall view of this Garden, now with our friendly Labors wee will accompanie thee and leade thee through a Graſſe-plot, little or nothing of many Herbariſts heretofore touched; and begin with the moſt common or beſt knowne Graſſe, which is called in Latine *Gramen Pratenſe*; and then by little and little conduct thee through moſt pleaſant gardens and other delightfull places, where any Herbe or Plant may be found fit for Meat or Medicine.

CHAP. I.   *Of Medow-Graſſe.*

THere be ſundry and infinite kindes of Graſſes not mentioned by the Antients, either as vnneceſſarie to be ſet downe, or vnknowne to them: only they make mention of ſome few, whoſe wants we meane to ſupply, in ſuch as haue come to our knowledge, referring the reſt to the curious ſearcher of Simples.

¶ *The Deſcription.*

1 COmmon Medow Graſſe hath very ſmall tufts or roots, with thicke hairy threds depending vpon the higheſt turfe, matting and creeping on the ground with a moſt thicke and apparant ſhew of wheaten leaues, lifting vp long thinne jointed and light ſtalks, a foot or a cubit high, growing ſmall and ſharpe at the top, with a looſe eare hanging downward, like the tuft or top of the common Reed.

2   Small Medow Graſſe differeth from the former in the varietie of the ſoile; for as the firſt kinde groweth in medowes, ſo doth this ſmall Graſſe clothe the hilly and more dry grounds vntilled, and barren by nature; a Graſſe more fit for ſheepe than for greater cattell. And becauſe the kindes of Graſſe do differ apparantly in root, tuft, ſtalke, leaſe, ſheath, eare, or creſt, we may aſſure our ſelues that they are endowed with ſeuerall Vertues, formed by the Creator for the vſe of man, although they haue been by a common negligence hidden and vnknowne. And therefore in this our Labor we haue placed each of them in their ſeuerall bed, where the diligent ſearcher of Nature may, if ſo he pleaſe, place his learned obſeruations.

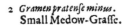

    1 *Gramen pratenſe.*
    Medow-Graſſe.

    2 *Gramen pratenſe minus.*
    Small Medow-Graſſe.

¶ *The Place.*

Common Medow-Graſſe groweth of it ſelfe vnſet or vnſowne, euery where; but the ſmall Medow-Graſſe for the moſt part groweth vpon dry and barren grounds, as partly we haue touched in the Deſcription.

¶ *The Time.*

Concerning the time when Graſſe ſpringeth and ſeedeth, I ſuppoſe there is none ſo ſimple but knoweth it, and that it continueth all the whole yeare, ſeeding in Iune and Iuly. Neither needeth it any propagation or replanting by ſeed or otherwiſe; no not ſo much as the watery Graſſes, but that they recouer themſelues againe, although they haue beene drowned in water all the Winter long, as may appeare in the wilde fennes in Lincolnſhire, and ſuch like places.

¶ *The Names.*

Graſſe is called in Greeke, Ἀγρωϛις: in Latine, *Gramen*; as it is thought, *à gradiendo, quod geniculatis internodijs ſerpat, crebroque nouas ſpargat radices.* For it groweth, goeth, or ſpreadeth it ſelfe vnſet or vnſowne naturally ouer all fields or grounds, cloathing them with a faire and perfect green. It is yearely mowed, in ſome places twice, and in ſome rare places thrice. Then is it dried and withered by the heate of the Sun, with often turning it; and then is it called *Fænum, neſcio an à fænore aut fætu.* In Engliſh, Hay: in French, *Le herbe du praiz.*

¶ *The Nature.*

The roots and ſeeds of Graſſe are of more vſe in Phyſicke than the herbe, and are accounted of all Writers moderately to open obſtructions, and prouoke vrine.

¶ *The*

¶ *The Vertues.*

The decoction of Grasse with the roots of Parsley drunke, helpeth the Dissurie, and prouoketh A vrine.

The roots of Grasse, according to *Galen*, doe glew and consolidate together new and bleeding B wounds.

The juice of Grasse mixed with hony and the pouder of Sothernwood taken in drinke, killeth C wormes in children; but if the childe be yong, or tender of nature, it shall suffice to mix the juice of Grasse and the gall of an Oxe or Bull together, and therewith anoint the childes belly, and lay a clout wet therein vpon the nauell.

Fernelius saith, That Grasse doth helpe the obstruction of the liuer, reines, and kidnies, and the D inflammation of the reins called *Nephritis*.

Hay sodden in water till it be tender, and applied hot to the chaps of beasts that be chap-fallen B through long standing in pound or stable without meat, is a present remedie.

Chap. 2.   *Of Red Dwarfe-Grasse.*

¶ *The Description.*

1   DWarfe-Grasse is one of the least of Grasses. The root consists of many little bulbs, couered with a reddish filme or skin, with very many small hairy and white strings: the tuft or eare is of a reddish colour, and not much differing from the grasse called *Ischæmon*, though the eare be softer, broader, and more beautifull.

† 1   *Gramen minimum rubrum, siue*
    Xerampelinum.
Red Dwarfe-Grasse.

2 *Gramen minimum album.*
White Dwarfe-Grasse.

† 2   This kinde of Grasse hath small hairy roots; the leaues are small and short, as also the stalke, which on the top thereof beares a pannicle not much vnlike the small medow Grasse, but lesse: the colour thereof is sometimes white, and otherwhiles reddish; whence some haue giuen two figures, which I thinking needlesse, haue only retained the later, and for the former giuen the figure of another Grasse, intended by our Author to be comprehended in this Chapter.

A 2   3 Small

3   Small hard Graſſe hath ſmall roots compact of little ſtrings or threds, from which come forth many foure ruſhy leaues of the length of an inch and a halfe : the tuft or eare is compact of many pannicles or very little eares, which to your feeling are very hard or harſh. This Graſſe is vnpleaſant, and no wholeſome food for cattell.

4   Ruſh-Graſſe is a ſmall plant ſome hand-full high, hauing many ſmall ruſhy leaues, tough and pliant, as are the common Ruſhes : wheron grow ſmall ſcaly or chaffie husks, in ſtead of flours: the ſeeds are like thoſe of Ruſhes, but ſmaller : the root is threddy like the former. ‡ There is a varietie of this to be found in bogs, with the ſeeds bigger, and the leaues & whole plant leſſer. ‡

3 *Gramen minus duriuſculum.*
Small hard Graſſe.

4 *Gramen junceum.*
Ruſh-graſſe, or Toad-graſſe.

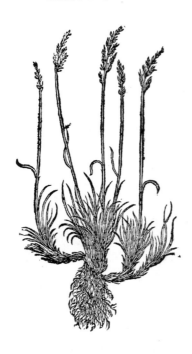

¶ *The Place.*

The Dwarfe-graſſe growes on heathy rough and dry barren grounds, in moſt places of England. ‡ That which I haue giuen you I haue not yet obſerued growing in any part of England. ‡

The white Dwarfe-graſſe is not ſo common as the former, yet doth it grow very plentifully among the Hop gardens in Eſſex, and many other places.

Small Hard-graſſe groweth in moiſt freſh mariſhes and ſuch like places.

Ruſh-graſſe groweth in ſalt mariſhes neere vnto the ſea, where the mariſhes haue beene ouerflowne with ſalt water. ‡ It alſo groweth in many wet woods, lanes, and ſuch like places; as in the lane going by Totenam Court toward Hampſted. The leſſer varietie hereof growes on the bogs vpon Hampſted heath. ‡

¶ *The Time.*

Theſe kindes of Graſſes do grow, floure, and flouriſh, when the common Medow-graſſe doth.

¶ *The Names.*

It ſufficeth what hath been ſaid of the Names in the Deſcription, as well in Engliſh as Latine; only that ſome haue deemed white Dwarfe-graſſe to be called *Xerampelinum.*

Ruſh-graſſe hath been taken for *Holoſteum Matthioli.*

‡ ¶ *The Names in particular.*

1   This I here giue you in the firſt place is *Gramen minimum Xerampelinum* of *Lobel*: it is the

Gramen

Gramen of *Matthiolus*, and *Gramen Bulboſum* of *Daleſchampius*. Our Author did not vnderſtand what *Xerampelinus* ſignified, when as he ſaid the white Dwarfe-graſſe was ſo termed; for the word imports red or murrey, ſuch a colour as the withered leaues of Vines are of.   2.   *Tabern.* calls this, *Gramen panniculatum minus*.   3.   *Lobel* calls this, *Exile Gramen durius*.   4.   This by *Matthiolus* was called *Holoſteum* : by *Thalius*, *Gramen epigonatocaulon* : by *Tabern. Gra. Bufonium*, that is, Toad-graſſe.

¶ The Nature and Vertues.

Theſe kindes of Graſſe do agree as it is thought with the common Medow-graſſe in nature and vertues, notwithſtanding they haue not been vſed in phyſicke as yet, that I can reade of.

† The firſt figure was onely a varietie of the ſecond, according to *Bauhinus*; yet in my iudgement it was the ſame w*th* the third, which is *Gramen minus duri- uſculum.*

CHAP. 3.   Of Corne-Graſſe.

¶ The Deſcription.

1   COrne-graſſe hath many graſſie leaues reſembling thoſe of Rie, or rather Oats, amongſt the which come vp ſlender benty ſtalkes, kneed or jointed like thoſe of Corne; whereupon groweth a faire tuft or pannicle not much vnlike to the feather-like tuft of common Reed, but rounder compact together like vnto Millet. The root is threddy like thoſe of Oats.

1 *Gramen ſegetale.*   2 *Gramen harundinaceum.*
  Corne-graſſe.   Reed-graſſe, or Bent.

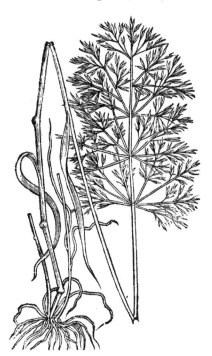 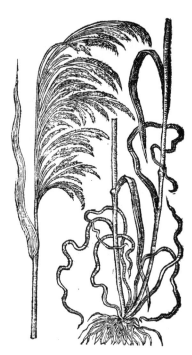

2   Reed-graſſe hath many thin graſſie leaues like the former : the buſhy top with his long feather-like pannicles do reſemble the common Reed, which is lightly ſhaken with the winde, branched vpon a long ſlender Reeden ſtalke, kneed or jointed like corne. The root is ſmall and fibrous.

¶ The Place and Time.

Theſe kindes of Graſſes grow for the moſt part neere hedges, & in fallow fields in moſt places. Their time of ſpringing, flouring, and fading, may be referred to the common Medow-graſſe.

¶ *The Names.*

† The firſt is called in Engliſh, Corne-graſſe. *Lobel* calls this, *Segetum gramen pannicula ſpe-cioſa latiore :* others terme it *Gramen ſegetale* , for that it vſually groweth among corne ; the which I haue not as yet ſeene.

The ſecond is called in Engliſh, Reed-graſſe : of *Lobel* in Latine, *Gramen agrorum latiore, arundi-nacea, & comoſa pannicula,* for that his tuft or pannicles do reſemble the Reed : and *Spica venti agro-rum,* by reaſon of his feather top, which is eaſily ſhaken with the wind. ‡ Some in Engliſh, much agreeing to the Latine name, call theſe Windle-ſtrawes. Now I take this laſt to be the Graſſe with which we in London do vſually adorne our chimneys in Sommer time : and wee commonly call the bundle of it handſomely made vp for our vſe, by the name of Bents. ‡

¶ *The Temperature and Vertues.*

Theſe Graſſes are thought to agree with common Graſſe, as well in Temperature as Vertues, although not vſed in phyſicke.

C<small>HAP</small>. 4.   *Of Millet Graſſe.*

<table>
<tr><td>1 *Gramen Miliaceum.*
Millet Graſſe.</td><td>† 1 *Gramen majus aquaticum.*
Great Water-Graſſe.</td></tr>
</table>

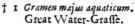

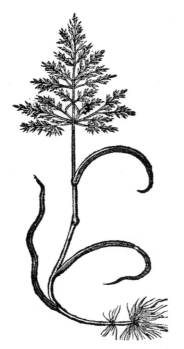

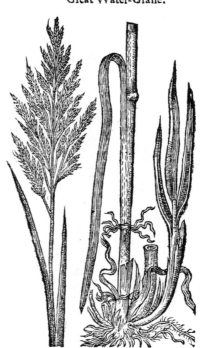

¶ *The Deſcription.*

1   MIllet Graſſe is but a ſlender Graſſe, bearing a tuft or eare like vnto the common Me-dow-graſſe, but conſiſting of ſmall ſeeds or chaffie heads like to Milium or Millet, whereof it tooke the name. The ſtalke or leaues do reſemble the Bent, wherewith countrey people do trim their houſes.

2   The great Water-Graſſe, in root, leafe, tuft, and reeden ſtalke, doth very well reſemble the Graſſe called in Latine, *Gramen ſulcatum,* or *Pictum ;* and by our Engliſh women, Lady-laces, be-cauſe it is ſtript or furrowed with white and greene ſtreakes like ſilke laces, but yet differs from that, that this Water-Graſſe doth get vnto it ſelfe ſome new roots from the middle of the ſtalkes and joints, which the other doth not. ‡ This is a large Graſſe, hauing ſtalkes almoſt as thicke as ones little finger, with the leaues anſwerable vnto them, and a little roughiſh. The tuft is ſome-what like a Reed, but leſſe, and whitiſh coloured. ‡

¶ *The*

¶ *The Place, Time, Names, Nature, and Vertues.*

The former growes in medowes, and about hedges ; and the later is to be found in moſt fennie, and waterie places ; and haue their Vertues and Natures common with the other Graſſes, for any thing that wee can finde in writing. The reaſon of their names may be gathered out of the deſcription.

† This which I giue you in the ſecond place is not of the ſame plant that was figured in the former edition ; for that picture was of *Gramen aquſticum harun-dinaceum panniculatum* of *Tabern.* which hath a running root and large ſpecious pannicle like to a Reed, of a browne colour ; but it is moſt apparant that our Author meant this, and framed his deſcription by looking vpon this figure, eſpecially the later part thereof. The true figure of this was in the ſecond place in the next Chapter.

C H A P. 5.

Of Darnell Graſſe.

¶ *The Deſcription.*

1　Arnell graſſe, or *Gramen Sorghinum*, as *Lobel* very properly termed it, hath a browniſh ſtalke thicke and knotty, ſet with long ſharpe leaues like vnto the common Dogs-graſſe ; at the top whereof groweth a tuft or eare of a grayiſh colour, ſomewhat like *Sorghum*, whereof it tooke his name.

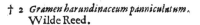

1　*Gramen Sorghinum.*　　　† 2　*Gramen harundinaceum panniculatum.*
　　Darnell Graſſe.　　　　　　　Wilde Reed.

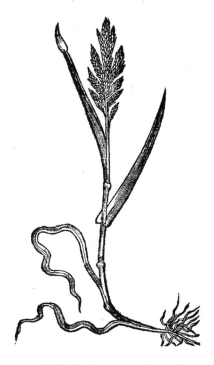

2　Wilde Reed, or *Gramen harundinaceum panniculatum*, called alſo *Calamogroſtis*, is farre bigger than Couch-graſſe or Dogs graſſe, and in ſtalkes and leaues more rough, rugged, and cutting. It is bad food for cattell, though they want, or be very hungry ; and deadly to Sheepe, becauſe that, as the Husbandman ſaith, it is a cauſe of leanneſſe in them, thirſt, and conſumption : it cutteth their

A 4　　　　　　　　　　　　tongue,

‡ 3 *Gramen harundinaceum minus.*
The lesser Reed-grasse.

tongue, straitneth the gullet or throat, and draweth downe bloud into the stomacke or maw ; whereof ensueth inflammation, and death for the most part. And not onely this *Calamogrostis* is hurtfull, but also all other kindes of shearing leaued Reeds, Flagges, Sedge, and the like, which haue as it were edges, and cut on both sides like kniues, as well mens fingers, as cattels mouthes. This herb is in a meane between reed and grasse. The root is white, creeping downewards very deepe. The spike or eare is like vnto the Reed, being soft and cottony, somewhat resembling Panicke.

‡ 3 This in root, stalks, and leaues, is like to the last described, but that they are lesser. The top or head is a long single spike or eare, not seuered or parted into many eares, like the top of the precedent, and by this and the magnitude it may chiefely be distinguished from it. This was in the twelfth place in the sixteenth chapter, vnder the title of *Gramen harundinaceum minus*: and the *Calamogrostis* but now described, was likewise there againe in the eleuenth place. ‡

¶ *The Place.*

The first growes in fields and Orchards almost euerie where.

The other grow in fenny waterie places.

¶ *The Names.*

2   This in Lincolnshire is called Sheere-grasse or Henne ; in other parts of England, Wilde Reed : in Latine, *Calamogrostis*, out of the Greeke *Kalamogrostis*.

As for their natures and vertues, we do not finde any great vse of them worth the relating.

† The figure that was in the second place was of *Gramen maius aquaticum*, being the second of the precedent Chapter. The true figure of this was pag. 11 vnder the title of *Gramen harundinaceum maius*. The third being there also, as I haue touched in the Description.

Chap. 6.   *Of Feather-top, Ferne, or Wood-Grasse.*

¶ *The Description.*

‡ 1   THis might fitly haue been put to those mentioned in the fore-going chapter, but that our Author determined it for this, as may appeare by the mention made of it in the Names, as also by the description hereof, framed from the figure we here giue you. ‡ This Grasse is garnished with chaffie and downie tufts, set vpon a long benty stalk of two cubits high, or somewhat more, naked without any blades or leaues for the most part. His root is tough and hard. ‡ The top is commonly of a red or murrey colour, and the leaues soft and downy. ‡

‡ 2   This, whose figure was formerly by our Author giuen for the last described, though very much different from it, is a very pretty and elegant grasse : it in roots and leaues is not vnlike to the vsuall medow Grasse : the stalke riseth to the height of a foot, and at the top thereof it beareth a beautifull pannicle, whence the French and Spanish Nations call it *Amourettes*, that is, Louely Grasse. This head consists of many little eares, shaped much like those of the ordinarie Quaking Grasse, longer and flatter, being composed of more skales, so that each of them somewhat resembles the leafe of a small Fern ; whence I haue called it Fern grasse. These tops when they are ripe are white, and are gathered where they grow naturally, to beautifie garlands. ‡

3   Wood-grasse hath many small and thready roots compact together in manner of a tuft ; from which spring immediately out of the earth many grassie leaues, among the which are sundrie

benty

‡ 1 *Gramen tomentosum arundinaceum.*
Feather-top, or Woolly Reed-grasse.

2 *Gramen panniculatum elegans.*
Ferne-grasse.

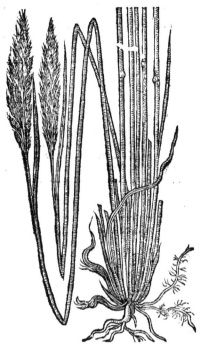

3 *Gramen sylvaticum majus.*
The greater Wood-grasse.

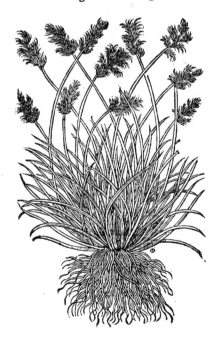

benty stalkes, naked and without leaues or blades like the former, bearing at the top a soft spikie tuft or eare much like vnto a Fox-taile, of a brownish colour.

‡ 4 This in leaues, stalks, roots, manner and place of growing is like the last described : The only difference betweene them is, That this hath much lesse, yet sharper or rougher eares or tufts. The figure and description of this was formerly giuen by our Author in the sixteenth chapter and ninth place, vnder the title of *Gramen sylvaticum minus.* But becaufe the difference between the last described and this is so small, we haue spared the figure, to make roome for others more different and note-worthy. ‡

¶ *The Time and Place.*

1   This kind of Grasse growes in fertil fields and pastures.

2   The second growes in diuers places of Spain and France.

The other two grow in woods.

¶ *The Names.*

1   *Lobel* in Latine calleth this, *Gramen tomentosum & acerosum.* Some haue taken it for the second kinde of *Calamogrostis*; but most commonly

it

it is called *Gramen plumoſum :* and in Engliſh, a Bent or Feather-top graſſe.

  2   *Gramen panniculatum* is called by ſome, *Heragroſtis* in Greeke. *Lobel* calls this, *Gramen panni-culoſum phalaroides.* And it is named in the *Hiſt.Lugd.Gramen filiceum, ſeu polyanthos :* that is, Ferne, or many-floured Graſſe. ‡

  3   *Gramen ſylvaticum,* or as it pleaſeth others, *Gramen nemoroſum,* is called in our tongue, wood Graſſe, or ſhadowie Graſſe.

<h2 align="center">Cʜᴀᴘ. 8.   Of great Fox-taile Graſſe.</h2>

¶ The Deſcription.

  1   THe great Fox-taile Graſſe hath many threddy roots like the common Medow graſſe ; and the ſtalke riſeth immediatly from the root, in faſhion like vnto Barley, with two or three leaues or blades like Otes ; but is nothing rough in handling, but ſoft and downie, and ſomewhat hoary, bearing one eare or tuft on the top, and neuer more ; faſhioned like a Fox-taile, whereof it tooke his name. At the approch of Winter it dieth, and recouereth it ſelfe the next yeare by falling of his ſeed.

<table>
<tr><td align="center">1 Gramen Alopecuroides majus.
Great Fox-taile graſſe.</td><td align="center">† 2 Gramen Alopecuroides minus.
Small Fox-taile graſſe.</td></tr>
</table>

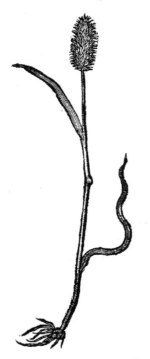 

  2   The leſſer Fox-taile Graſſe hath a tough or hard root compact of many ſmall ſtrings, yeel-ding a ſtrawie ſtalke like the former, though ſomewhat leſſer, with the like top or creſt, but of a whi-tiſh colour.

  3   Great baſtard Fox-taile Graſſe hath a ſtrawie ſtalke or ſtem, which riſeth to the height of a cubit and a halfe, hauing a ſmall root conſiſting of many fibres. His leafe is ſmall and graſſie, and hath on his top one tuft or ſpike, or eare of a hard chaffie ſubſtance, ſome three inches long, com-poſed of longiſh ſeeds, each hauing a little beard or awne.

  4   Small baſtard Fox-taile Graſſe doth reſemble the former, ſauing that this kinde doth not
ſend

send forth such large stalkes and eares as the other, but smaller, and not so close packed together, neither hauing so long beards or awnes.

1 *Gramen Alopecurinum majus.*
Great bastard Fox-taile grasse.

2 *Gramen Alopecurinum minus.*
Small bastard Fox-taile grasse.

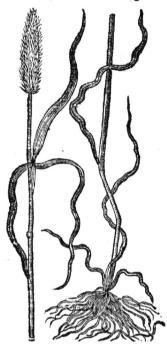

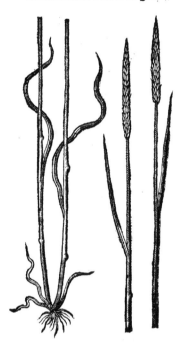

¶ *The Place and Time.*

These wilde bastard Fox-taile Grasses do grow in the moist furrowes of fertile fields, towards the later end of Sommer.

¶ *The Names.*

‡ The first by *Lobel* and *Tabern.* is called *Gramen Phalaroides.* The other, *Lobel* calleth 2 *Gramen Alopecuroides.* 3, *Minus.* 4, *Minus alterum.*

CHAP. 8.   *Of great Cats taile Grasse.*

¶ *The Description.*

1   GReat Cats-taile Grasse hath very small roots compact of very small skinnes or threds, which may easily be taken from the whole root. The stalke riseth vp in the middest, and is somewhat like vnto wilde Barley, kneed and ioynted like corne, of a foot high or thereabout; bearing at the top a handsome round close compact eare resembling the Cats-taile.

2   The small Cats-taile Grasse is like vnto the other, differing chiefely in that it is lesser than it. The root is thicke or cloued like those of Rush Onions or Ciues, with many small strings or hairy threds annexed vnto it.

‡ 3   There is another that growes plentifully in many places about London, the which may fitly be referred to this Classis. The root thereof is a little bulbe, from whence ariseth a stalke some two foot or better high, set at each joint with long grassie leaues, the spike or eare is commonly

Gramen Typhinum minus.
Small Cats taile graſſe.

monly foure or fiue inches long, cloſely
and handſomly made in the faſhion of
the precedent, which in the ſhape it doth
very much reſemble. ‡

¶ *The Place and Time.*

Theſe kindes of Graſſes do grow very
well neere waterie places, as *Gramen Cyperoides* doth, and flouriſh at the ſame
time that all the others do.
‡ The later may be found by the
bridge entring into Chelſey field, as one
goeth from S. *Iames* to little Chelſey. ‡

¶ *The Names.*

The Latines borrow theſe names of the
Greeks, and call it *Gramen Typhinum,* of
Typha, a Cats taile : and it may in Engliſh as well be called round Bent-Graſſe,
or Cats taile Graſſe.
‡ The laſt deſcribed is by *Bauhine,*
who firſt gaue the figure and deſcription
thereof in his *Prodromus, pag.* 10. called
Gramen Typhoides maximū ſpica longiſſima;
that is, The largeſt Fox-tail Graſſe with
a very long eare.

CHAP. 9. *Of Cyperus Graſſe.*

1 *Gramen Cyperoides.*
Cyperus Graſſe.

2 *Gramen Iunceum aquaticum.*
Ruſhy Water-Graſſe.

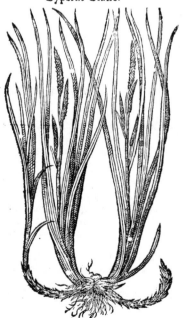

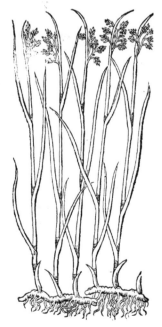

¶ *The Description.*

1　CYperus Grasse hath roots somewhat like Cyperus, whereof it tooke his name. His leaues are long and large like vnto the common Reed : the stalke growes to the height of a cubit in some places, vpon which growe little scaly knobs or eares, spike-fashion, somewhat like vnto Cats taile or Reed-Mace, very chaffie, rough, and rugged.

2　Rushy Water-Grasse hath his roots like the former, with many fibrous strings hanging at them, and it creepes along vpon the vppermost face of the earth, or rather mud, wherein it growes, bearing at each joint one slender benty stalke, set with a few small grassy blades or leaues ; bringing forth at the top in little hoods small feather-like tufts or eares.

¶ *The Place, Time, and Names.*

They grow (as I said) in myrie and muddy grounds, in the same season that others doe. And concerning their names there hath been said enough in their titles.

CHAP. 10.　*Of Water-Grasse.*

1 *Gramen aquaticum.*
Water-Grasse.

2 *Gramen aquaticum spicatum.*
Spiked Water-Grasse.

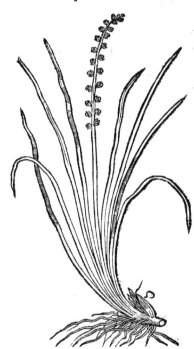

¶ *The Description.*

†　1　WAter-grasse, or as we terme it, Water-Burre-grasse, hath a few long narrow slender and jointed leaues : among which riseth vp a stalke of two foot high, bearing vpon his smal and tender branches many little rough knobs, or brownish sharpe pointed seeds, made vp into cornered heads : his root is small and thready.

‡　The figure of this plant is not well exprest, for it should haue had the leaues made narrower, and joints exprest in them, like as you may see in the *Gramen junceum sylvaticum*, which is the ninth in the sixteenth chapter ; for that and this are so like, that I know no other difference betweene them, but that this hath leaues longer and narrower than that, and the heads smaller and whiter. There is a reasonable good figure of this in the *Hist. Lugd. p.*1001. vnder the name of *Arundo minima.*

2 Spiked Water-Graſſe hath long narrow leaues ; the ſtalke is ſmall, ſingle, and naked, without leaues or blades, bearing alongſt the ſame toward the top an eare or ſpike made of certain ſmal buttons, reſembling the buttony floures of ſea Wormewood. His root is thicke and tough, full of fibres or threds.

¶ _The Place and Time._

They differ not from the former kinds of Graſſes in Place and Time : and their Names are maniſeſt.

¶ _The Nature and Vertues._

Their Nature and Vertues are referred vnto Dogs graſſe, whereof we will ſpeake hereafter.

CHAP. II. _Of Flote-Graſſe._

1 _Gramen Fluviatile._ Flote-Graſſe.

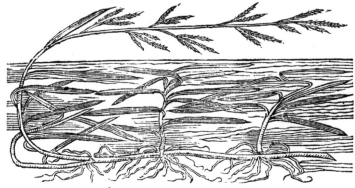

2 _Gramen fluviatile ſpicatum._ Spiked Flote-graſſe.

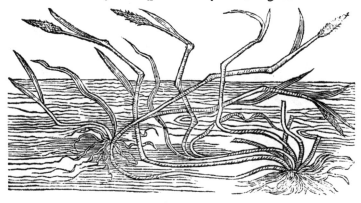

¶ _The Deſcription._

1 FLote-Graſſe hath a long and round root ſomewhat thicke, like vnto Dogs-Graſſe, ſet on euen ioynts with ſmall ſtrings or threds ; from the which riſe vp long and crooked ſtalks, croſſing, winding, and folding one within another, with many flaggie leaues, which horſes eat greedily of. At the top of theſe ſtalks, and ſomwhat lower, there come forth very many little eares of a whitiſh colour, compoſed of two rankes of little chaſſie ſeeds ſet alternatiſy, each of theſe ſmall eares being almoſt an inch in length.

2 Spike Flote-Graſſe, or Spiked Flote-Graſſe beareth at the top of each ſlender creeping ſtalke one ſpiked eare and no more, and the other many, which maketh a difference betwixt them ; otherwiſe they are one like the other. His root is compact, tuſted, and made of many thrummie threds.

¶ _The Place._

The firſt of theſe growes euery where in waters. The ſecond is harder to be found.

¶ _The_

¶ *The Names.*

The firſt is called *Gramen fluviatile*, and alſo *Gramen aquis innatans :* in Engliſh, Flote-graſſe. *Tragus* calls it *Gramen Anatum*, Ducks Graſſe.

The ſecond is called *Gramen-fluviatile ſpicatum*, and *fluviatile album*, by *Tabernamontanus*. Likewiſe in Engliſh it is called Flote-Graſſe, and Floter-Graſſe, becauſe they ſwimme and flote in the water.

CHAP. 12. *Of Kneed-Graſſe.*

¶ *The Deſcription.*

1　KNeed-Graſſe hath ſtreight and vpright ſtrawy ſtalkes, with joints like to the ſtraw of corne, and beareth ſmall graſſy leaues or blades ſpiked at the top like vnto Panicke, with a rough eare of a darke browne colour. His roots are hairy and threddy, and the joints of the ſtraw are very large and conſpicuous.

1 *Gramen geniculatum.*　　　2 *Gramen geniculatum aquaticum.*
Kneed-Graſſe,　　　　　　　Water Kneed-Graſſe.

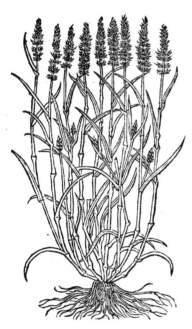

2　Water Kneed-Graſſe hath many long and ſlender ſtemmes, jointed with many knobby and gouty knees like vnto Reed, ſet with broad flaggy leaues ſomwhat ſharpe pointed, bearing at the top a tuft or pannicle, diuided into ſundry ſmall branches of a duskiſh colour. His root is threddie like the other.

¶ *The Place, Time, and Names.*

Theſe Graſſes do grow in fertile moiſt medowes, not differing in time from others. And they are called *Geniculata*, becauſe they haue large joints like as it were knees.

We haue nothing deliuered vs of their Nature and Properties.

CHAP.

Chap. 13. *Of Bearded Panicke-Grasse.*

1 *Gramen Paniceum.*
Bearded Panick-Grasse.

¶ *The Description.*

1 BEarded Panicke grasse hath broad and large leaues like Barly, somwhat hoarie, or of an ouer-worne Russet colour. The stalkes haue two or three joints at the most, and many eares on the top without order ; vpon some stalkes more eares , on others fewer, much like vnto the eare of wilde Panicke, but that this hath many beards or awnes , which the other wants.

2 Small Panicke Grasse, as *Lobel* writeth, in roots, leaues, joints, and stalks, is like the former, sauing that the eare is much lesse, consisting of fewer rowes of seed, contained in small chaffie blackish huskes. This, as the former, hath many eares vpon one stalke.

‡ 3 This small Pannicke Grasse from a threddy root sendeth forth many little stalks, whereof some are one hand-full, other-some little more than an inch high ; and each of these stalks on the top sustains one single eare, in shape very like vnto the eare of wilde Panick, but about halfe the length. The stalks of this are commonly crooked, and set with grassie leaues like the rest of this kinde.

The figure hereof was vnfitly placed by our Author in the sixteenth place in the eighth chapter, vnder the title of *Gramen Cyperoides spicatum.*

2 *Gramen Paniceum parvum.*
Small Panicke-Grasse.

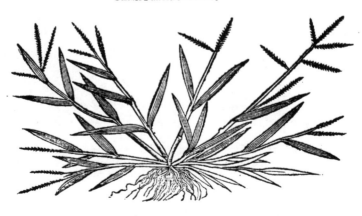

¶ *The Place and Time.*

The first of these two doth grow neere vnto mud walls, or such like places not manured, yet fertile or fruitfull.

The

The fecond groweth in fhallow waterie plafhes of paftures, and at the fame time with others.
‡ I haue not as yet obferued any of thefe growing wilde. ‡

† 3 *Gramen Panici efigie fpica fimplici.*
Single eared Panicke Graffe.

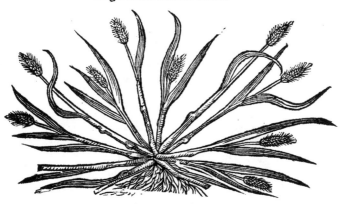

¶ *The Names and Vertues.*
They are called Panicke Graffes, becaufe they are like the Italian corne called Panicke.
Their nature and vertues are not knowne.

CHAP. 14. *Of Hedge-hog Graffe.*

‡ 1 *Gramen paluftre Echinatum.*
Hedge-hog Graffe.

2 *Gramen exile Hirfutum.*
Hairy-graffe.

‡ 3 *Gramen Capitulis globosis.*
Round headed Siluer-grasse-

¶ *The Description.*

1 HEdge-hog Grasse hath long stiffe flaggy leaues with diuers stalkes proceeding from a thicke spreading root ; and at the top of euery stalke grow certaine round and pricking knobs fashioned like an hedge-hog.

† 2 The second is rough and hairie : his roots do spread and creep vnder the mud and mire as Cyperus doth ; and at the top of the stalkes are certaine round soft heads, their colour being browne, intermixed with yellow, so that they looke prettily when as they are in their prime.

‡ 3 This Grasse (whose figure was formerly in the first place in this Chapter) hath a small and fibrous root, from which rise leaues like those of Wheat, but with some long white hairs vpon them like those of the last described : at the tops of the stalks (which are some foot or better high) there grow two or three round heads consisting of soft and white downie threds. These heads are said to shine in the night, and therefore they in Italy call it (according to *Casalpinus*) *Luciola, quia noctu lucet.*

4 To this I may adde another growing also in Italy, and first described by *Fabius Columna.* It hath small creeping joynted roots, out of which come small fibres, and leaues little and very narrow at the first, but those that are vpon the stalks are as long again, incompassing the stalkes, as in Wheat, Dogs-grasse, and the like. These leaues are crested all along, and a little forked at the end : the straw or stalk is very slender, at the top whereof growes a sharpe prickly round head, much after the manner of the last described : each of the seed-vessels, whereof this head consists, ends in a prickly stalke hauing fiue or seuen points, wherof the vppermost that is in the middle is the longest. The seed that is contained in these prickly vessels is little and transparent, like in colour to that of Cow-wheat. The floures (as in others of this kinde) hang trembling vpon yellowish small threds. ‡

¶ *The Place and Time.*

† 1 2 They grow in watery medowes and fields, as you may see in Saint *Georges* fields and such like places.

3 4 Both these grow in diuers mountainous places of Italy ; the later whereof floures in May.

¶ *The Names.*

The first is called Hedge-hog Grasse, and in Latine, *Gramen Echinatum,* by reason of those prickles which are like vnto a hedge-hog.

The second hairy Grasse is called *Gramen exile hirsutum Cyperoides,* because it is small and little, and rough or hairy like a Goat : and *Cyperoides,* because his roots do spring and creepe like the *Cyperus.*

‡ 3 This by *Anguillara* is thought to be *Combretum* of *Pliny*; it is *Gram. lucidum* of *Tabernamontanus* ; and *Gramen hirsutum Capitulo globoso,* of *Bauhine, Pin. pag.* 7.

4 *Fabius Columna* calls this *Gramen montanum Echinatum tribuloides capitatum* : and *Bauhine* nameth it, *Gramen spica subrotunda echinata.* We may call it in English, Round headed Caltrope Grasse.

¶ *The Vertues.*

3 The head of this (which I haue thought good to call Siluer-grasse) is very good to be applied to greene wounds, and effectuall to stay bleeding, *Casalp.* ‡

† It is euident by the name and description, that our Author meant this which we here giue you in the first place; yet his figure was of another Grasse somewhat like the second, which figure and description you may find here exprest in the third place.

Chap. 15.　Of Hairy Wood-Graſſe.

¶ *The Deſcription.*

Hairy Woodgraſſe hath broad rough leaues ſomewhat like the precedent, but much longer, and they proceed from a thready root, which is very thicke, and full of ſtrings, as the common Graſſe, with ſmall ſtalkes riſing vp from the ſame roots; but the top of theſe ſtalkes is diuided into a number of little branches, and on the end of euery one of them ſtandeth a little floure or huske like to the top of *Allium Vrſinum*, or common Ramſons, wherein the ſeed is contained when the floure is fallen.

2 Cyperus Wood-graſſe hath many ſhearie graſſie leaues, proceeding from a root made of many hairy ſtrings or threds: among which there riſeth vp ſundry ſtraight and vpright ſtalkes, on whoſe tops are certaine ſcaly and chaffie huskes, or rather ſpikie blackiſh eares, not much vnlike the catkins or tags which grow on Nut-trees, or Aller trees.

1 *Gramen hirſutum nemoroſum.*
Hairy Wood-graſſe.

2 *Gramen Cyperinum nemoroſum.*
Cyprus Wood-graſſe.

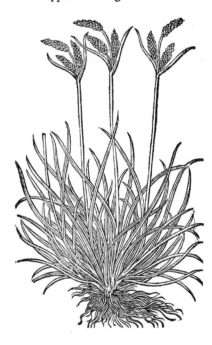

¶ *The Place, Time, and Names.*

Theſe two grow in woods or ſhadowie places, and may in Engliſh be called Wood-graſſe. Their time is common with the reſt.

¶ *Their Nature and Vertues.*

There is nothing to be ſaid of their nature and vertues, being as vnknowne as moſt of the former.

CHAP. 16.   *Of Sea Spike-Graſſe.*

¶ *The Deſcription.*

† 1 SEa Spike-graſſe hath many ſmall hollow round leaues about ſix inches long, riſing
from a buſhie threddy white fibrous root, which are very ſoft and ſmooth in handling. Among theſe leaues there doe ſpring vp many ſmall ruſhie ſtalkes ; alongſt
which are at the firſt diuers ſmall flouring round buttons ; the ſides whereof falling away, the middle part growes into a longiſh ſeed-veſſell ſtanding vpright.

1 *Gramen marinum ſpicatum.*
Sea Spike-graſſe.

2 *Gramen ſpicatum alterum.*
Salt marſh Spike-graſſe.

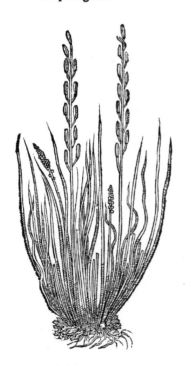   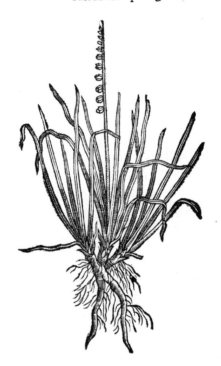

† 2   Salt marſh Spike-graſſe hath a wooddie tough thicke root with ſome ſmall hairy threds
faſtned thereunto ; out of which ariſe long and thicke leaues very like thoſe of the Sea-graſſe we
vulgarly call Thrift. And amongſt theſe leaues grow vp ſlender naked ruſhie ſtalkes which haue
on one ſide ſome ſmall knobs or buttons of a greeniſh colour hanging on them.

3   The third hath many ruſhy leaues tough and hard, of a browne colour, well reſembling Ruſhes : his root is compact of many ſmall tough and long ſtrings. His ſtalke is bare and naked of
leaues vnto the top, on which it hath many ſmall pretty chaffie buttons or heads.

4   The fourth is like the third, ſauing that it is larger ; the ſtalke alſo is thicker and taller than
that of the former, bearing at the top ſuch husks as are in Ruſhes.

5   Great Cypreſſe Graſſe hath diuers long three-ſquare ſtalkes proceeding from a root compact of many long and tough ſtrings or threds. The leaues are long and broad, like vnto the ſedge
called *Carex.* The ſpike or eare of it is like the head of Plantaine, and very prickly, and commonly
of a yellowiſh greene colour.

6   Small Cypreſſe graſſe is like vnto the other in root and leaues, ſauing that it is ſmaller.
His ſtalke is ſmooth and plaine, bearing at the top certaine tufts or pannicles, like to the laſt deſcribed in roughneſſe and colour.

7 The

3 *Gramen junceum marinum.*
Sea Ruſh-graſſe.

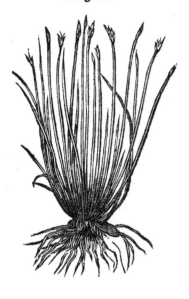

4 *Gramen junceum maritimum.*
Mariſh Ruſh-graſſe.

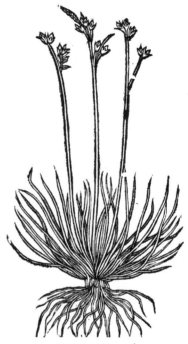

5 *Gramen paluſtre Cyperoides.*
Great Cypreſſe-graſſe.

6 *Gramen Cyperoides parvum,*
Small Cypreſſe-graſſe.

7 *Gramen aquaticum* **Cyperoides** *vulgatius.*
Water Cypreſſe Graſſe.

‡ 8 *Gramen Cyperoides ſpicatum.*
Spike Cypreſſe Graſſe.

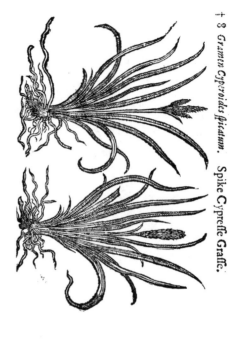

9 *Gramen iunceum ſylvaticum.*
Wood Ruſhy graſſe.

7　The firſt of theſe two kindes hath many crooked and crambling roots of a wooddy ſubſtance, very like vnto the right Cyperus, differing from it onely in ſmell, becauſe the right Cyperus roots haue a fragrant ſmell, and theſe none at all. His leaues are long and broad, rough, ſharpe or cutting at the edges like Sedge. His ſtalke is long, big, and three-ſquare like to Cyperus, and on his top hath a chaffie vmbel or tuft like vnto the true Cyperus,

‡ 8　The ſecond kind hath many broad leaues like vnto thoſe of Gillouers, but of a freſher greene; amongſt the which riſeth vp a ſhort ſtalk ſome handfull or two high, bearing at the top three or foure ſhort eares of a reddiſh murrey colour, and theſe ears grow commonly together at the top of the ſtalk, and not one vnder another. There is alſo another leſſer ſort hereof, with leaues and roots like the former, but the ſtalke is commonly ſhorter, and it hath but one ſingle eare at the top thereof. You haue the figures of both theſe expreſt in the ſame table or piece.

This kinde of Graſſe is the *Gramen ſpicatum folijs Vetonicæ* of Lobel. ‡

9　This hath long tough and hairy ſtrings, growing deepe in the earth like a turfe, which make the root; from which riſe many crooked tough & ruſhy ſtalkes, hauing toward the top ſkaly and chaffie knobs or buttons.

‡ This

‡ This growes some halfe yard high, with round brownish heads, and the leaues are ioynted as you see them expressed in the figure we heare giue you. ‡

¶ *The Place, Time, Names, Nature, and Vertues.*

All the Grasses which we haue described in this Chapter do grow in marish and watery places neere to the sea or other fenny grounds, or by muddy and myrie ditches, at the same time that the others do grow and flourish. Their names are easily gathered of the places they grow in, or by their Descriptions, and are of no vertue or propertie in medicine, or any other necessarie vse as yet knowne.

† Formerly in the eighth place (but very vnfitly) was the figure of *Gramen panici effigie spica simp.* being the third in the thirteenth Chapter. The ninth also is restored to his due place, being the fourth in the sixt chap. The two Reed-grasses that were in the 11 and 12 places are also before in the fift Chap.

CHAP. 17.   *Of Couch-Grasse or Dogs-Grasse.*

1 *Gramen Caninum.*
Couch-grasse or Dogs-grasse.

2 *Gramen Caninum nodosum.*
Knotty Dogs-grasse.

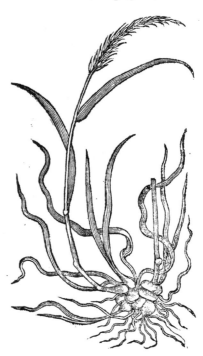

¶ *The Description.*

† 1   THe common or best known dogs grasse or Couch-grasse hath long leaues of a whitish greene colour: the stalke is a cubit and a halfe high, with ioints or knees like wheaten strawes, but these ioints are couered with a little short downe or woollinesse. The plume or tuft is like the reed, but smaller and more chaffie, and of a grayish colour: it creepeth in the ground hither and thither with long white roots, ioynted at certaine distances, hauing a pleasant sweet tast: they are platted or wrapped one within another very intricately, insomuch as where it hapneth in gardens amongst pot-herbes, great labour must be taken before it can be destroyed, each piece being apt to grow, and euery way to dilate it selfe.

† 2 Knotty Dogs-graffe is like vnto the former in ftalke and leafe,but that they are of a dee-per colour,alfo the fpike or eare is greener,and about fome two handfuls long,it much in fhape re-fembles an Ote, yet far fmaller, and is much more difperfed than the figure prefents to you. The roots of this are fomewhat knotty and tuberous,but that is chiefely about the Spring of the yere; for afterwards they become leffe and leffe vntill the end of Sommer. And thefe Bulbes doe grow confufedly together, not retaining any certaine fhape or number.

¶ *The Place.*

1   The firft grows in gardens and arable lands,as an infirmitie or plague of the fields,nothing pleafing to husbandmen; for after the field is plowed,they are conftrained to gather the roots to-gether with harrowes and rakes,and being fo gathered and laid vpon heaps, they fet them on fire left they fhould grow againe.

2   The fecond growes in plowed fields and fuch like places, but not euery where as the other. I haue found of thefe in great plenty,both growing,and plucked vp with harrowes, as before is re-hearfed, in the fields next to Saint *Iames* wall,as ye go to Chelfey,and in the fields as yee goe from the Tower hill in London,to Radcliffe.

¶ *The Time.*

Thefe Graffes feldome come to fhew their eare before Iuly.

¶ *The Names.*

It is called *Gramen Caninum*, or *Sanguinale*, and *Vniola*. The countreymen of Brabant name it 𝕻een: others, 𝕷𝖊𝖉𝖙 𝖌𝖗𝖆𝖋𝖋𝖊: of the Greeks, *ἄγρωστις*: of the Latines by the common name, *Gramen*. It is of fome named *ἄγρωστις*: in Englifh, Couch-graffe, Quitch-graffe,and Dogs-graffe.

Gramen Caninum bulbofum or *nodofum*, is called in Englifh, Knobby or knotty Couch-graffe.

¶ *The Nature.*

The nature of Couch-graffe, efpecially the roots, agreeth with the nature of common Graffe. Although that Couch-graffe be an vnwelcome gueft to fields and gardens, yet his phyficke ver-tues do recompenfe thofe hurts; for it openeth the ftoppings of the liuer and reins,without any manifeft heate.

The learned Phyfitions of the Colledge and Societie of London do hold this bulbous Couch-graffe in temperature agreeing with the common Couch-graffe, but in vertues more effectuall.

¶ *The Vertues.*

A   Couch-graffe healeth green wounds. The decoction of the root is good for the kidnies & blad-der: it prouoketh vrin gently,and driueth forth grauell. *Diofcorides* and *Galen* doe agree, that the root ftamped and laid vpon greene wounds,healeth them fpeedily.

B   The decoction thereof ferueth againft griping paines of the belly, and difficultie of making water.

C   *Marcellus* an old Author maketh mention, Chap.26. That 27 knots of the herbe called *Gramen* or Graffe,boiled in wine till halfe be confumed,preffed forth, ftrained, and giuen to drinke to him that is troubled with the ftranguric, hath fo great vertue, that after the Patient hath once begun to make waterwithout paine,it may not be giuen any more. But it muft be giuen with water only to fuch as haue a feuer. By which words it appeareth,that this knotted Graffe was taken for that which is properly called *Gramen*, or *Agroftis*; and hath been alfo commended againft the Stone and difeafes of the bladder.

The later Phyfitions vfe the roots fomtimes of this, and fomtimes of the other indifferently.

C<small>HAP.</small> 18.   *Of Sea Dogs-Graffe.*

¶ *The Defcription.*

1   T<small>H</small>e Sea Dogs-graffe is very like vnto the other before named: his leaues are long and flender, and very thicke compact together,fet vpon a knotty ftalke fpiked at the top like the former. Alfo the root crambleth and creepeth hither and thither vnder the earth,occupying much ground by reafon of his great encreafe of roots.

‡   This Graffe (whereof *Lobel* gaue the firft figure and defcription,vnder the name of *Gramen geniculatum Caninum marinum*) I conjecture to be that which growes plentifully vpon the banks in the falt marithes by Dartford in Kent,and moft other falt places by the fea; as alfo in many banks and orchards about London, and moft other places farre from the fea. Now *Lobels* figure beeing not good, and the defcription not extant in any of his Latine Workes, I cannot certainly affirme any thing. Yet I thinke it fit to giue you an exact defcription of that I do probably iudge to be it; and

and not onely so, but I iudge it to be the same Grasse that *Bauhinus* in his *Prodromus* hath set forth, *pag.* 17. vnder the name of *Gramen latifolium spica triticea compacta.* This is a very tall Grasse ; for it sends forth a stalke commonly in good ground to the height of a yard and a halfe : the leaues are large, thicke, and greene, almost as big as those of white Wheat ; the which it also very much resembles in the eare, which is vsually some handfull and an halfe long, little spokes standing by course with their flat sides toward the straw. About the beginning of Iuly it is hung with little whitish yellow floures such as Wheat hath. The roots of this are like those of the first described. This sometimes varies in the largenesse of the whole plant, as also in the greatnesse, sparsednesse, and compactednesse of the eare. ‡

1 *Gramen Caninum marinum.*
Sea Dogs-grasse.

2 *Gramen Caninum marinum alterum.*
Sea Couch-grasse.

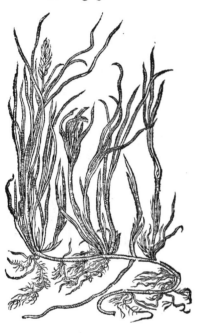

2   The second Sea Dogs-grasse is according vnto *Lobel* somewhat like the former : his rootes are more spreading and longer, dispersing themselues farther vnder the ground than any of the rest. The leaues are like the former, thicke bushed at the top, with a cluster or bush of short thick leaues one folded within another. The stalke and tuft is of a middle kinde betweene *Ischæmon* and the common Couch-grasse.

¶ *The Place, Time, Names, Nature, and Vertues.*

They grow on the sea shore at the same time that others do, and are so called because they grow neere the sea side. Their nature and vertues are to be referred vnto Dogs grasse.

CHAP. 19. *Of vpright Dogs-Grasse.*

¶ *The Description.*

1   VPright Dogges-Grasse, or Quitch-Grasse, by reason of his long spreading ioynted roots is like vnto the former, and hath at euery knot in the root sundry strings of hairie substance, shooting into the ground at euery ioynt as it spreadeth : the stalkes lye creeping, or rise but a little from the ground, and at their tops haue spoky pannicles farre smaller than the

common

common Couch-graſſe. By which notes of difference it may eaſily be diſcerned from the other kindes of Dogs-graſſe.

1 *Gramen Caninum ſupinum.*
Vpright Dogs-graſſe.

2 *Gramen ſtriatum.*
Lady-lace Graſſe.

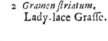

2 Lady-laces hath leaues like to Millet in faſhion, rough or ſharp pointed like to the reed, with many white veins or ribs, and ſiluer ſtreaks running along through the midſt of the leaues, faſhioning the ſame like to laces or ribbons wouen of white and greene ſilke, very beautifull and faire to behold. It groweth to the height of wilde Pannicke, with a ſpoky top not very much vnlike, but more compact, ſoft, white, and chaffie. The root is ſmall and hairy, and white of colour like vnto the medow Graſſe.

¶ *The Place.*

1 Vpright Dogs-graſſe groweth in dunged grounds and fertile fields.

2 Lady-Laces grow naturally in wooddy and hilly places of Sauoy, and anſwers common Graſſe in his time of ſeeding.

It is kept and maintained in our Engliſh gardens rather for pleaſure, than vertue which is yet knowne.

¶ *The Names.*

Lobel calleth the later, *Gramen ſulcatum* and *ſtriatum,* or *Gramen pictum* : in Engliſh, the furrowed Graſſe, the white Chameleon Graſſe, or ſtreaked graſſe; and vſually of our Engliſh women it is called Lady-laces, or painted Graſſe : in French, *Aiguillettes d'armes.*

¶ *The Nature and Vertues.*
The Vertues are referred vnto the Dogs-graſſes.

                                         CHAP.

Chap. 20.   Of Dew-Grasse.

¶ *The Description.*

1   DEw Grasse hath very hard and tough roots long and fibrous ; the stalkes are great, of three or foure cubits high, very rough and hairy, jointed and kneed like the common Reed : the leaues are large and broad like vnto corne. The tuft or eare is diuided into sundry branches, chaffie, and of a purple colour; wherein is contained a seed like *Milium*, wherwith the Germans make pottage and such like meat, as we in England do with Ote-meale ; and it is sent into Middleborow and other townes of the Low-countries, in great quantitie for the same purpose, as *Lobel* hath told me.

2   The second kinde of Dew-grasse or *Ischæmon* is somewhat like the first kinde of Medowgrasse, resembling one the other in leaues and stalks, sauing that the crest or tuft is spread or stretched out abroad like a Cocks foot set downe vpon the ground, whereupon it was called *Galli crus*, by *Apuleius*. These tops are cleere and vpright, of a glistering Purple colour, or rather Violet, and it is diuided into foure or fiue branches, like the former Dew-grasse. The root consists of a great many small fibres.

‡   3   To these may fitly be added another Grasse, which *Clusius* hath iudged to be the medicinall Grasse of the Antients : and *Lobel* refers it to the Dogs grasses, because it hath a root jointed thicke, and creeping like as the Dogs grasses. The stalks are some foot high, round, and of a purplish colour : but the top is very like to that of the last described, of a darke purple colour.

1 *Gramen Mannæ esculentum*   2 *Ischæmon vulgare.*
  Dew-grasse.   Cocks foot Grasse.

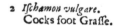

¶ *The Place and Time.*

1   The first groweth naturally in Germanie, Bohemia, and Italy, and in the territories of Goritia and Carinthia, as *Matthiolus* reporteth.

2   The second groweth neere vnto rough bankes of fields, as I haue seene in the hilly bankes neere Greenhithe in Kent. It differeth not in time from those we haue spoken of.

‡   3   This

‡ 3 *Gramen dactiloides radice repente.*
Cocks-foot Grasse with creeping roots.

‡ 3 This groweth plentifully in most parts of Spaine and France: and it is probable, that this was the grasse our Author found neere Greenhithe in Kent.

¶ *The Names.*

1 The Germans call it **Himeltau:** That is to say, *Cæli ros;* whereupon it was called *Gramen Manna:* it seemes to be *Milij sylvestris spurium quoddam genus,* a certaine wilde or bastard kinde of Millet. *Leonicerus* and *Ruellius* name it *Capriola* and *Sanguinaria:* some would haue it to be *Gramen aculeatum Plinij;* but because the description thereof is very short, nothing can be certainly affirmed. But they are far deceiued who thinke it to be *Coronopus,* as some very learned haue set downe: but euerie one in these dayes is able to controll that error. *Lobel* calleth it *Gramen Manna esculentum,* for that in Germany and other parts, as Bohemia and Italy, they vse to eat the same as a kinde of bread-corne, and also make pottage therewith, as we do with Ote-meale; for the which purpose it is there sowne as corne, and sent into the Low-Countries, and there sold by the pound. In English it may be called Manna Grasse or Dew-grasse, but more fitly, Rice-grasse.

2 This is iudged to be *Ischæmon* of *Pliny;* and *Galli Crus* of *Apuleius.*

¶ *The Nature.*

A   These Grasses are astringent and drying, in taste sweet like the common Dogs grasse.

¶ *The Vertues.*

B   *Apuleius* saith, if a plaister be made of this Grasse, Hogs grease, and the Leauen of houshold bread, it cureth the biting of mad dogs.

As in the description I told you, this Plant in his tuft or eare is diuided into sundry branches, some tuft into three, some foure, and so ne fiue clouen parts like Cocks toes. *Apuleius* reporteth, if you take that eare which is diuided only into three parts, it wonderfully helpeth the running or dropping of the eyes, and those that begin to be bleare eyed, being bound about the necke, and so vsed for certaine dayes together, it turneth the humors away from the weake part.

C   ‡ Manna-grasse or Rice-grasse is said to be very good to be put into pultesses, to discusse hard swellings in womens breasts.

D   The Cocks-foot Dogs grasse is very good in all cases as the other Dogs-Grasses are, and equally as effectuall. ‡

‡ Chap. 21.   *Of diuers Cyperus Grasses.*

¶ *The Description.*

‡ 1   THe first of these hath reasonable strong fibrous roots, from whence rise stiffe long and narrow leaues like those of other Cyperus Grasses: the stalkes also (as it is proper to all the plants of this kindred) are three square, bearing at their tops some three brownish eares soft and chaffie like the rest of this kinde, standing vpright, and not hanging downe as some others do.

2   This hath pretty thicke creeping blacke roots, from whence arise three stalks, set with shorter leaues, yet broader than those of the last described; and from the top of the stalke come forth three or foure foot-stalks, whereupon hang longish rough skaly and yellowish heads.

3   The roots of this are blacke, without smell, and somewhat larger than those of the last described:

‡ 1 *Gramen Cyperoides anguſtifolium majus.*
Great narrow leaued Cyperus Graſſe.

‡ 2 *Pſeudocyperus.*
Baſtard Cyperus.

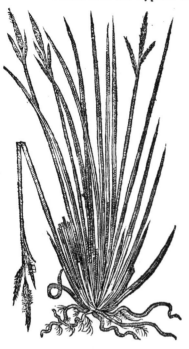

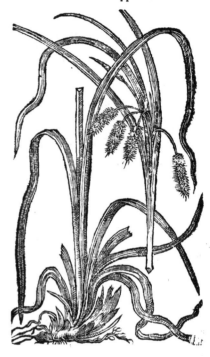

‡ 3 *Cyperus longus inodorus ſylueſtris.*
Long Baſtard Cyperus.

deſcribed : the 3 .ſquare ſtalke alſo is ſome
two cubits high, bearing at the top diſper-
ſedly round ſcaly heads ſomewhat like
thoſe of the wood Ruſh-graſſe : the leaues
are ſomewhat ſharpe and triangular like
thoſe of the other Cyperus.

4   This Cyperus hath creeping blacke
roots, hauing here and there knotty tuberous heads for the moſt part, putting vp
leaues like thoſe of the laſt deſcribed, as
alſo a ſtalke bearing at the top long chaffie
eares like to ſome others of this kinde.

5   This Cyperus-Graſſe hath pretty
thicke fibrous and blacke roots, from
whence ariſeth a ſtalke ſome cubit high,
pretty ſtiffe, triangular, jointed, ſet at each
joynt with a large greene leafe which at
the bottome incompaſſes the ſtalke, which
is omitted in the figure. At the top of the
ſtalke, as in the true Cyperus, come forth
two or three pretty large leaues, betweene
which riſe vp many ſmall foot-ſtalkes very
much branched, and bearing many blacke
ſeeds ſomewhat like Millet or Ruſhes.

¶ *The Place and Time.*
All theſe grow in ditches and waterie
placs,

places,and are to bee found with their heads about the middle of Summer, and ſome of them ſoo-ner.

¶ The Names.

The firſt of theſe by *Lobel* is called *Gramen paluſtre majus*.

2 This by *Geſner, Lobel,*and *Dodonæus* is called *Pſeudocyperus.*

3 *Lobel* names this,*Cyperus longus inodorus ſylueſtris.*

4 He alſo calls this,*Cyperus aquaticus ſeptentrionalis.*

5 This is the *Cyperus graminea miliacea* of *Lobel* and *Pena :* the *Iuncus latus* in the *Hiſtor.Lugdun.* *pag.*988.and the *Pſeudocyperus polycarpos* of *Thalius.*

‡ 4 *Cyperus rotundus inodorus ſylueſtris.*   ‡ 5 *Cyperus gramineus miliaceus.*
Round Baſtard Cyperus.   Millet Cyperus graſſe.

¶ The Temperature and vertue.

None of theſe are made vſe of in Phyſicke; but by their taſte they ſeeme to be of a cold and aſtringent qualitie. ‡

---

‡ C H A P. 22. *Of diuers other Graſſes.*

¶ The Deſcription.

‡ 1 THis Ote or Hauer-graſſe,deſcribed by *Cluſius,*hath ſmall creeping roots:the ſtalkes are ſome cubit high,ſlender,jointed,and ſet with ſhort narrow leaues : at the top of the ſtalke growes the eare,long, ſlender, and bending, compoſed of downie huskes containing a ſeed like to a naked Ote.The ſeed is ripe In Iuly. It growes in the mountainous and ſhadowie woods of Hungarie, Auſtria, and Bohemia Our Authour miſtaking himſelfe in the fi-gure,and as much in the title,gaue the figure of this for Burnt Barley, with this title, *Hordeum Di-ſtichon.* See the former edition,*pag.66.*

2 I cannot omit this elegant Graſſe,found by M^r.*Goodyer* vpon the walls of the ancient city of Wincheſter, and not deſcribed as yet by any that I know of. It hath a fibrous and ſtringy root, from which ariſe leaues long and narrow,which growing old become round as thoſe of *Spartum* or

Mat-

Mat.weed:among theſe graſſie leaues there growes vp a ſlender ſtalke ſome two foot long, ſcarſe
ſtanding vpright, but oft times hanging down the head or top of the eare: it hath ſome two joints,
and at each of theſe a pretty graſſie leafe.   The eare is almoſt a foot in length, compoſed of many
ſmall and ſlender hairy tufts,which when they come to maturitie looke of a grayiſh or whitiſh colour,and doe very well reſemble a Capons taile; whence my friend the firſt obſerued thereof, gaue
it the title of *Gramen* Ἀλεϰϑυϱἑϱᾳ, or Capons-taile graſſe; by which name I receiued the ſeed thereof,
which ſowne, tooke root,and flouriſhes.

‡ 1 *Gra. montanum avenaceum.*
Mountaine Hauer-Graſſe.

‡ 2 *Gramen murorum ſpica longiſſima.*
Capon-taile Graſſe.

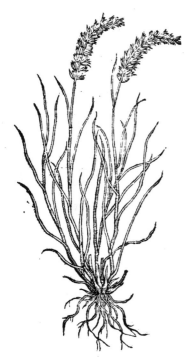

3   Next to this I thinke fit to place the *Gramen Criſtatum,*or Cocks-combe graſſe of *Bauhinus.*
This Graſſe hath for the root many white fibrous threds thicke packt together: the leaues are but
ſhort,about the bigneſſe of the ordinarie medow-graſſe; the ſtalks are ſome cubit and halfe high,
with ſome two or three knots a piece: the leaues of the ſtalke are ſome foure or fiue inches long:
the eare is ſmall,longiſh,of a pale greene colour,ſomewhat bending, ſo that in ſome ſort it reſembles the combe of a Cocke,or the ſeed-veſſell of that plant which is called *Caput Gallinaceum.* This
is ordinarily to be found in moſt medowes about Mid-ſummer.

4   There is alſo commonly about the ſame time in our medowes to be found a Graſſe growing to ſome cubit high,hauing a ſmall ſtalke,at the the top whereof there grows an eare ſome inch
and an halfe,or two inches long,conſiſting as it were of two rankes of corne: it very much reſembles Rie both in ſhape and colour, and in his ſhort bearded awnes,wherefore it may very fitly be
termed *Gramen ſecalinum,*or Rie-graſſe. Yet it is not *Gramen ſpica ſecalina* which *Bauhine* deſcribes
in the fifty ſeuenth place,in his *Prodromus,pag.*18. for that is much taller, and the eare much larger
than this of my deſcription.

5   In diuers places about hedges, in Iuly and Auguſt is to be found a fine large tall Graſſe,
which *Bauhine* (who alſo firſt deſcribed it) hath vnder the name of *Gramen ſpica Brizæ majus.* This
hath ſtalkes as tall as Rie, but not ſo thicke,neither are the leaues ſo broad: at the top of the ſtalke
grow diuers prettie little flattiſh eares conſiſting of two rankes of chaffie huskes or ſeed-veſſels,
which haue yellowiſh little floures like to thoſe of Wheat.

6   There is alſo commonly to be found about May or the beginning of Iune, in medowes and
ſuch

such places, that grasse which in the *Historia Lugdun.* is set forh vnder the name of *Gramen Lanatum Daleschampij* : the stalkes and leaues are much like the common medow grasse, but that they are more whitish and hairie ; the head or panicle is also soft and woolly, and it is commonly of a gray, or else a murrie colour.

7　There is to be found in some bogs in Summer time about the end of Iuly a prettie rushie grasse, some foote or better in height, the stalke is hard and rushie, hauing some three joints, at each whereof there comes forth a leafe as in other grasses, and out of the bosome of the two vppermost of these leaues comes forth a slender stalke being some 2 or 3 inches high, and at the top thereof growes as in a little vmble a pretty white chaffie floure ; and at, or nigh to the top of the maine stalke there grow three or foure such floures clustering together vpon little short and slender foot-stalkes: the leaues are but small, and some handfull or better long , the root I did not obserue. This seemes to haue some affinitie with the *Gramen iunceum aquaticum*, formerly described in the ninth chapter. I neuer found this but once, and that was in the company of Mr. *Thomas Smith*, and Mr. *Iames Clarke*, Apothecaries of London ; wee riding into Windsor Forrest vpon the search of rare plants, and wee found this vpon a bogge neere the high way side at the corner of the great parke. I thinke it may very fitly be called *Gramen iunceum leucanthemum* : White floured Rush-grasse.

8　The last yeare at Margate in the Isle of Tenet, neere to the sea side and by the chalky cliffe I obserued a pretty little grasse which from a small white fibrous roote sent vp a number of stalkes of an vnequall height ; for the longest, which were those that lay partly spred vpon the ground, were some handfull high, the other that grew streight vp were not so much ; and of this, one inch and halfe was taken vp in the spike or eare, which was no thicker than the rest of the stalke, and seemed nothing else but a plaine smooth stalke, vnlesse you looked vpon it earnestly, and then you might perceiue it to be like Darnell grasse: wherefore in the iournall that I wrote of this Simpling voyage, I called it *pag.* 3. *Gramen paruum marinum spica Loliacea.* I iudge it to be the same that *Bauhine* in his *Prodromus, pag.* 19. hath set forth vnder the name of *Gramen Loliaceum minus spica simplici.* It may be called in English, Dwarfe Darnell-grasse.

9　The Darnell-grasse that I compared the eare of this last described vnto, is not the *Gramen sorghinum* (which our Authour called Darnell-grasse) but another grasse growing in most places with stalkes about some span high, but they seldome stand vpright, the eare is made iust like that which hereafter *chap.* 58. is called *Lolium rubrum*, Red Darnell, of which I iudge this is a variety, differing little there from but in smalnesse of growth.

10　Vpon Hampstead heath I haue often obserued a small grasse whose longest leaues are seldome about two or three inches high, and these leaues are very greene, small, and perfectly round like the *Spartum Austriacum*, or Feather-grasse : I could neuer finde any stalke or eare vpon it : wherefore I haue brought it into the Garden to obserue it better. In the forementioned iournall, *pag.* 33. you may finde it vnder the name of *Gramen Spartium capillaceo folio minimum.* It may be this is that grasse which *Bauhine* set forth in his *Prodromus, pag.* 11. vnder the title of *Gramen sparteum Monspeliacum capillaceo folio minimum.* I haue thought good in this place to explaine my meaning by these two names, to such as are studious in plants, which may happen to light by chance (for they were not intended for publicke) vpon our Iournall, that they need not doubt of my meaning.

11　I must not passe ouer in silence two or three Grasses, which for any thing that I know are strangers with vs, the one I haue seene with Mr. *Parkinson*, and it is set forth by *Bauhine, pag.* 30. of his *Prodromus.* The other by *Lobell* in the second part of his *Aduersaria, pag.* 468. The first (which *Bauhine* fitly cals *Gramen alopecuroides spica aspera*, and thinks it to be *Gram. Echinatum Daleschampij,* described *Hist. Lugd. pag.* 432.) hath a fibrous and white root, from which ariseth a stiffe stalke diuided by many knots, or knees : the leaues are like to the other fox-taile-grasses, but greener: the eare is rough, of some inch in length, and growes as it were vpon one side of the stalke: the eare at first is greene, and shewes yellowish little floures in August.

12　This other Grasse which *Lobell* in the quoted place figures and describes by the name of *Gramen Scoparium Ischæmi panniculis Gallicum*, hath roots some cubit long, slender and very stiffe, (for of these are made the head brushes which are vulgarly vsed) the straw is slender, and some cubit high, being here and there ioynted like to other Grasses : the top hath foure or fiue eares standing after the manner of Cocks foot Grasse, whereof it is a kinde. It growes naturally about Orleance, and may be called in English, Brush-grasse. ‡

CHAP.

Chap. 23. Of Cotton-Graſſe.

¶ *The Deſcription.*

1   THis ſtrange Cotton-Graſſe, which *L'Obelius* hath comprehended vnder the kindes of Ruſhes; notwithſtanding that it may paſſe with the Ruſhes, yet I finde in mine owne experience, that it doth rather reſemble graſſe than ruſhes, and may indifferently be taken for either, for that it doth participate of both. The ſtalke is ſmall and ruſhy, garniſhed with many graſſie leaues alongſt the ſame, bearing at the top a buſh or tuft of moſt pleaſant downe or cotton like vnto the moſt fine and ſoft white ſilke. The root is very tough, ſmall and thready.

2   This Water Gladiole, or graſſy ruſh, of all others is the faireſt and moſt pleaſant to behold, and ſerueth very well for the decking and trimming vp of houſes, becauſe of the beautie and brauerie thereof: conſiſting of ſundry ſmall leaues, of a white colour mixed with carnation, growing at the top of a bare and naked ſtalke, fiue or ſix foot long, and ſometime more. The leaues are long and flaggie, not much vnlike the common reed. The root is thready, and not long.

1 *Gramen Tomentarium.*	2 *Gladiolus paluſtris Cordi.*
Cotton Graſſe.	Water Gladiole.

 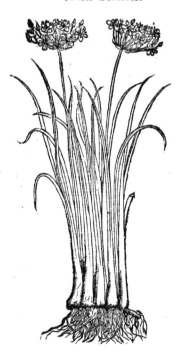

¶ *The Place and Time.*

1   Cotton graſſe groweth vpon bogy and ſuch like mooriſh places, and it is to be ſeene vpon the bogs in Hampſted heath. It groweth likewiſe in Highgate parke neere London.

2   Water Gladiole groweth in ſtanding pooles, motes, and water ditches. I found it in great plenty being in company with a worſhipfull Gentleman Mr. *Robert Wilbraham*, at a Village fifteene miles from London called Buſhey. It groweth likewiſe neere Redriffe by London, and many other places: the ſeaſon anſwereth all others.

¶ *The Names.*

1   *Gramen Tomentoſum* is called likewiſe *Iuncus bombicinus*: of *Cordus, Linum pratenſe*, and *Gnaphalium, Hicronymi Bockij.* In Engliſh, Cotton-Graſſe.

C

2 Water

2   Water Gladiole is called of *L'Obelius, Iuncus Cyperoides floridus paludoſus,* Flouring Cy-preſſe Ruſh : *Iuncus,* for that his ſtalke is like the ruſh : *Cyperoides,* becauſe his leaues reſemble *Cyperus : Floridus,* becauſe it hath on the top of euery ſtalke a fine vmble or tuft of ſmall floures in faſhion of the Lilly of Alexandria, the which it is very like, and therefore I had rather call it Lilly graſſe.

¶ *The Nature and Vertues.*

A   *Cordus* ſaith, That *Iuncus bombicinus* ſodden in wine, and ſo taken, helpeth the throwes and gri-pings of the belly, that women haue in their childing.

There be alſo ſundry kinds of Graſſes wholly vnknowne, or at leaſt not remembred of the old Writers, whereof ſome few are touched in name onely by the late and new Writers : now for as much as they haue onely named them, I will referre the better conſideration of them to the indu-ſtrie and diligence of painefull ſearchers of nature, and proſecute my purpoſed labour, to vnfold the diuers ſorts and manifold kindes of *Cyperus,* Flags, and Ruſhes : and becauſe that there is added vnto many of the Graſſes before mentioned, this difference, *Cyperoides,* that is to ſay, reſembling *Cyperus,* I thought it therefore expedient to joyne next vnto the hiſtory of Graſſes the diſcourſe of *Cyperus,* and his kinds, which are as follow.

CHAP. 24.   *Of Engliſh Galingale.*

1 *Cyperus longus.*   2 *Cyperus rotundus vulgaris.*
Engliſh Galingale.   Round Galingale.

¶ *The Deſcription.*

1   Engliſh Galingale hath leaues like vnto the common Reed, but leſſer and ſhorter. His ſtalke is three ſquare, two cubits high : vpon whoſe top ſtand ſundry branches, euery little branch bearing many ſmall chaffie ſpikes. The root is blacke and very long, cree-ping hither and thither, occupying much ground by reaſon of his ſpreading : it is of a moſt ſweet and pleaſant ſmell when it is broken.

2 The

2   The common round *Cyperus* is like the former in leaues and tops, but the roots are here and there knotty and round, and not altogether ſo well ſmelling as the former.

‡ 3   There is alſo another *Cyperus* which growes in Syria and Ægypt, whoſe roots are round, blackiſh, and large, many hanging vpon one ſtring, and hauing a quicke and aromaticke ſmell : the leaues and ſpoky tufts reſemble the former.

4   There is ſaid to be another kinde of this laſt deſcribed, which is leſſer, and the roots are blacker, and it growes in Creet, now called Candy.

5   There is alſo another round *Cyperus* which growes about ditches and the bankes of Riuers whereas the ſalt water ſometimes comes : the roots of this are hard and blacke, without ſmell, many hanging ſometimes vpon one ſtring : the ſtalke and leaues are much like the former, but the heads vnlike, for they are rough and blackiſh about the bigneſſe of a filbert, and hang ſome ſix or ſeuen at the top of the ſtalke. It floures in Iuly and Auguſt. ‡

¶ *The Place and Time.*

1 2   The firſt and ſecond of theſe grow naturally in fenny grounds, yet will they proſper exceedingly in gardens, as expreience hath taught vs.

3 4   The former of theſe growes naturally in Syria and Ægypt, the later in Candy.

5   This growes plentifully in the Mariſhes below Grauefend, in Shipey, Tenet, and other places.

¶ *The Names in generall.*

Cyperus is called in Greeke, Κύπειρος, or Κύπειρις: of the Latines as well *Cypirus* as *Cyperus*: of ſome, *Iuncus quadratus*: of *Pliny*, *Iuncus Angulosus*, and *Triangularis*: of others, *Aſpalathum* and *Eryſiſceptron*: in French, *Souchet*: in Dutch, **Galgan**: in Spaniſh, *Iunco odoroſa*: by vs, *Cyperus* and Engliſh Galingale.

† ¶ *The Names in particular.*

1   This is called, *Cyperus longus*, and *Cyperus longus Odoratior*: in Engliſh, common *Cyperus*, and Engliſh Galingale. 2   This is called, *Cyperus rotundus vulgaris*, Round Engliſh Galingale. 3   *Cyperus rotundus Syriacus*, or *Ægyptiacus*, Syrian, or Ægyptian round *Cyperus*. 4   *Cyperus minor Creticus*, Candy round *Cyperus*. 5   *Cyperus rotundus inodorus Littoreus*, Round ſalt marſh *Cyperus*, or Galingale. †

¶ *The Nature.*

Dioſcorides ſaith, That *Cyperus* hath an heating qualitie. *Galen* ſaith, the roots are moſt effectuall in medicine, and are of an heating and drying qualitie : and ſome doe reckon it to be hot and dry in the ſecond degree.

¶ *The Vertues.*

It maketh a moſt profitable drinke to breake and expell grauell, and helpe the dropſie.

If it be boiled in wine, and drunke, it prouoketh vrine, driueth forth the ſtone, and bringeth down the naturall ſickneſſe of women.

The ſame taken as aforeſaid, is a remedie againſt the ſtinging and poyſon of Serpents.

Fernelius ſaith, The root of *Cyperus* vſed in Baths helpeth the coldneſſe and ſtopping of the matrix, and prouoketh the tearmes.

He writeth alſo, that it increaſeth bloud by warming the body, and maketh good digeſtion; wonderfully refreſhing the ſpirits, and exhilerating the minde, comforting the ſenſes, and encreaſing their liuelineſſe, reſtoring the colour decayed, and making a ſweet breath.

The powder of *Cyperus* doth not onely dry vp all moiſt vlcers, either of the mouth, priuy members and fundament, but ſtaieth the humor and healeth them, though they be maligne and virulent, according to the iudgement of *Fernelius*.

5 *Cyperus rotundus littoreus.*
Round Salt-marſh *Cyperus.*

A

B

C

D

E

F

CHAP.

‡ Chap. 24.   *Of Italian Trasi, or Spanish Galingale.*

1 *Cyperus Esculentus sine Caule & flore.*
Italian Trasi, or Spanish Galingall,
without stalke and floure.

2 *Cyperus Esculentus, sine Trasi Italorum.*
Italian Trasi, or Spanish Galingall.

‡ 1   THe Italian Trasi, which is here termed Spanish Galingale, is a plant that hath many small roots, hanging at stringy fibres like as our ordinary Dropwort roots doe, but they are of the bignesse of a little Medlar, and haue one end flat and as it were crowned like as a Medlar, and it hath also sundry streakes of lines, seeming to diuide it into seuerall parts; it is of a brownish colour without, and white within; the tast thereof is sweet almost like a Chesnut. The leaues are very like those of the garden *Cyperus*, and neuer exceed a cubit in length. Stalkes, floures, or seed it hath none, as *Iohn Pona* an Apothecary of Verona, who diligently obserued it nigh to that city whereas it naturally growes, affirmes; but hee saith there grows with it much wild *Cyperus*, which as he judges hath giuen occasion of their error who gaue it the stalkes and floures of *Cyperus*, or English Galingale, as *Matthiolus* and others haue done. It is encreased by setting the roots, first steeped in water, at the beginning of Nouember. I haue here giuen you the figure of it without the stalke, according to *Pona*, and with the stalke, according to *Matthiolus* and others.

¶ *The Names.*

The Italian Trasi is called in Greeke by *Theophrastus* Μαλινάδια, *Hist. plant.*4.*cap.*10.as *Fabius Columna* hath proued at large: *Pliny* tearmes it, *Anthalium:* the later writers, *Cyperus Esculentus*, and *Dulcichinum:* The Italians, *Trasi*, and *Dolzolini*, by which names in Italy they are cryed vp and downe the streets, as Oranges and Lemmons are here.

¶ *The Temperature and Vertues.*

A   The milke or creame of these Bulbous roots being drunke, mundifies the brest and lungs, wherefore it is very good for such as are troubled with coughs. Now you must beat these roots, and macerate them in broth, and then presse out the creame through a linnen cloath, which by some late Writers is commended also to be vsed in venereous potions.

B   The same creame is also good to be drunke against the heate and sharpenesse of the vrine, especially if you in making it do adde thereto the seeds of Pompions, Gourds, and Cucumbers. The Citifens of Verona eate them for dainties, but they are somewhat windy. ‡

CHAP.

‡ CHAP. 26.   Of the true Galingale, the greater and the leſſer.

‡ 1 Galanga major.
The greater Galingale.

‡ 2 Galanga minor.
The leſſer Galingale.

THe affinitie of name and nature hath induced me in this place to inſert theſe two, the bigger and the leſſer Galingale ; firſt therefore of the greater.

¶ The Deſcription.

1   THe great Galingale, whoſe root onely is in vſe, and brought to vs from Iava in the Eaſt Indies, hath flaggie leaues ſome two cubits high, like thoſe of Cats-taile or Reed-mace: the root is thicke and knotty, reſembling thoſe of our ordinary Flagges, but that they are of a more whitiſh colour on the inſide, and not ſo large. Their taſte is very hot and biting, and they are ſomewhat reddiſh on the outſide.

2   The leſſer growing in China, and commonly in ſhops called Galingale, without any addition, is a ſmall root of a browniſh red colour both within and without; the taſte is hot and biting, the ſmell aromaticall, the leaues (if we may beleeue Garcias ab Horto) are like thoſe of Myrtles.

¶ The Names.

1   The firſt is called by Matthiolus, Lobel, and others, Galanga major. Some thinke it to be the Acorus of the Ancients : and Pena and Lobel in their Stirp. Aduerſ. queſtion whether it be not the Acorus Galaticus of Dioſcorides. But howſoeuer, it is the Acorus of the ſhops, and by many vſed in Mithridate in ſtead of the true. The Indians call it Lancuaz.

2   The leſſer is called Galanga, and Galanga minor, to diſtinguiſh it from the precedent. The Chinois call it, Lauandon : the Indians, Lancuaz : we in Engliſh tearme it, Galingale, without any addition.

¶ The Temperature and Vertues.

Theſe roots are hot and dry in the third degree, but the leſſer are ſomewhat the hotter.

They ſtrengthen the ſtomacke, and mitigate the paines thereof ariſing from cold and flatulencies.   A

The ſmell, eſpecially of the leſſer, comforts the too cold braine ; the ſubſtance thereof being chewed ſweetens the breath. It is good alſo againſt the beating of the heart.   B

They are vſefull againſt the Collicke proceeding of flatulencies, and the flatulent affeꞓts of the wombe ; they conduce to venery, and heate the too cold reines. To conclude, they are good againſt all cold diſeaſes. ‡   C

---

‡ CHAP. 27.   Of Turmericke.

THis alſo challengeth the next place, as belonging to this Tribe, according to Dioſcorides ; yet the root, which onely is brought vs, and in vſe, doth more on the outſide reſemble Ginger, but that it is yellower, and not ſo flat, but rounder. The inſide thereof is of a Saffron colour, the taſte hot and bitteriſh : it is ſaid to haue leaues larger than thoſe of Millet, and a leaſie ſtalke. There is ſome varietie of theſe roots, for ſome are longer and ſome rounder, and the later are the hotter, and they are brought ouer oft times together with Ginger.

¶ The Place.

It growes naturally in the Eaſt-Indies about Calecut, as alſo at Goa.

¶ The Names.

This without doubt is the Cyperus Indicus of Dioſcorides, Lib. 1. Cap. 4. It is now vulgarly by moſt Writers, and in ſhops called by the name of *Terra merita*, and *Curcuma:* yet ſome terme it *Crocus Indicus*, and we in Engliſh call it, Turmericke.

¶ *The Temperature and Vertues.*

A   This root is certainely hot in the third degree, and hath a qualitie to open obſtructions, and it is vſed with good ſucceſſe in medicines againſt the yellow Iaundiſe, and againſt the cold diſtempers of the liuer and ſpleene.

---

‡ Chap. 28. *Of Zedoarie.*

‡ *Zerumbeth, ſiue Zedoaria rotunda.*
Round Zedoarie.

Zedoarie is alſo a root growing naturally in the woods of Malaver about Calecut and Cananor in the Indies; the leaues thereof are larger than Ginger, and much like them ; the root is alſo as large, but conſiſting of parts of different figures, ſome long and ſmall, others round ; their colour is white, and oft times browniſh on the inſide, and they haue many fibres comming out of them, but they are taken away together with the outward rinde before they come to vs. Theſe roots haue a ſtrong medicine-like ſmell, and ſomewhat an vngratefull taſte.

¶ *The Names.*

Some call the long parts of theſe roots *Zedoaria*, and the round (whoſe figure we here giue you) *Zerumbeth*, and make them different, whenas indeed they are but parts of the ſame root, as *Lobel* and others haue well obſerued. Some make *Zedoaria* and *Zerumbeth* different, as *Auicen:* others confound them and make them one, as *Rhaſes* and *Serapio*. Some thinke it to be Ἀμώμου of *Ægineta:* but that is not ſo ; for he ſaith, Ἀμώμου τὴν ἄττ οῦτ, ὅτι τοῖς μύροις μάλιςα μίγνυται, It is an Aromaticke, and therefore chiefely mixed in ointments : which is as much as if he ſhould haue ſaid, That it was put into ointments for the ſmells ſake, which in this is no waies gratefull, but rather the contrary.

¶ *The Temperature and Vertues.*

A   It is hot and dry in the ſecond degree ; it diſcuſſes flatulencies, and fattens by a certaine hidden qualitie. It alſo diſſipates and amends the vngratefull ſmell which Garlicke, Onions, or too much wine infect the breath withall, if it be eaten after them. It cures the bites and ſtings of venomous creatures, ſtops laskes, reſolues the Abſceſſes of the wombe, ſtaies vomiting, helpes the Collicke, as alſo the paine of the ſtomacke.

B   It kils all ſorts of wormes, and is much vſed in Antidotes againſt the plague, and ſuch like contagious diſeaſes. ‡

Chap. 29. *Of Ruſhes.*

‡   I Do not here intend to trouble you with an accurate diſtinction and enumeration of Ruſhes; for if I ſhould, it would be tedious to you, laborious to me, and beneficiall to neither. Therefore I will onely deſcribe and reckon vp the chiefe and more note-worthy of them, beginning with the moſt vſuall and common. ‡

¶ *The Deſcription.*

1   The roots of our common Ruſhes are long and hairy, ſpreading largely in the ground, from which, as from one entire tuft, proceed a great company of ſmall ruſhes ; ſo exceedingly well knowne, that I ſhall not need to ſpend much time about the deſcription thereof.

2   There is ſundry ſorts of Ruſhes beſides the former, whoſe pictures are not here expreſt, and the rather for that the generall deſcription of Ruſhes, as alſo their common vſe and ſeruice are ſufficient to leade vs to the knowledge of them. This great Water-Ruſh or Bul-Ruſh, in ſtead of leaues bringeth forth many ſtrait twiggie ſhoots or ſprings, which be round, ſmooth, ſharpe pointed, and without knots. Their tuft or floure breaketh forth a little beneath the top, vpon the one ſide of the Ruſh, growing vpon little ſhort ſtems like Grape cluſters, wherein is contained the ſeed after the faſhion of a ſpeares point. The roots be ſlender and full of ſtrings. *Pliny* and *Theophraſtus* before him, affirme that the roots of the Ruſh doe die euery yeare, and that

                it groweth

it groweth againe of the ſeed. And they affirme likewiſe that the male is barren, and groweth againe of the young ſhoots ; yet I could neuer obſerue any ſuch thing.

‡ 3   There growes a Ruſh to the thickneſſe of a Reed, and to ſome two yards and an halfe, or three yards high, in diuers fenny grounds of this kingdome ; it is very porous and light, and they vſually make mats, and bottom chaires therewith. The ſeeds are contained in reddiſh tufts, breaking out at the top thereof. The roots are large and joynted, and it grows not, vnleſſe in waters. ‡

4   *Iuncus acutus*, or the ſharpe Ruſh, is likewiſe common and well knowne ; not much differing from *Iuncus lænis*, but harder, rougher, and ſharper pointed, fitter to ſtraw houſes and chambers than any of the reſt ; for the others are ſo ſoft and pithy, that they turne to duſt and filth with much treading ; where contrariwiſe this ruſh is ſo hard that it laſts ſound much longer.

‡ 5   There is alſo another pretty ſmall kinde of Ruſh growing to ſome foot in height, hauing ſmooth ſtalkes which end in a head like to that of the ordinary Horſe-taile. This ruſh hath alſo one little joynt toward the bottome thereof. It growes in watery places, but not ſo frequently as the former. ‡

1 *Iuncus lænis.*
Common Ruſhes

4 *Iuncus acutus.*
Sharpe Ruſh, or hard Ruſh.

3 *Iuncus aquaticus maximus.*
Great Water-Ruſh, or Bul-Ruſh.

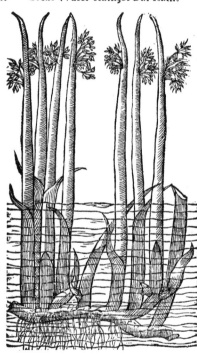

¶ *The Place.*

1   *Iuncus lænis* groweth in fertile fields, and medowes that are ſomewhat moiſt.
2   3   5   Grow in ſtanding pooles, and by riuers ſides in ſundry places.
4   *Iuncus acutus* groweth vpon dry and barren grounds, eſpecially neere the furrows of plowed land. I need not ſpeake of their time of growing, they being ſo common as they are.

¶ *The Time.*

The Ruſh is called in Greeke, ϰοῖνε : in Latine, *Iuncus* : in high Dutch, **Binken** : in low Dutch, **Bieſen** : in Italian, *Giunco* : in Spaniſh, *Iunco* : in French, *Ionc* : in Engliſh, Ruſhes.
2   3   The Grecians haue called the Bul-Ruſh, ὀξυϰοῖνος The greater are commonly in many places termed Bumbles.
1   *Iuncus lænis* is that Ruſh which *Dioſcorides* called ϰοῖνος ὀχία.
4   *Iuncus acutus* is called in Greeke ὀξυϰοῖνος : In Dutch, **Steren Bieſen.**
5   This is called by *Lobel, Iuncus aquaticus minor Capitulis Equiſeti* : By *Daleſchampius, Iuncus clauatus,* or Club ruſh.

¶ *The*

¶ *The Nature and Vertues.*

A   Theſe Ruſhes are of a dry nature.

B   The ſeed of Ruſhes dried at the firſt, and drunke with wine allaied with water, ſtayeth the laske and the ouermuch flowing of womens tearmes.

C   *Galen* yeeldeth this reaſon thereof, becauſe that their temperature conſiſteth of an earthly eſ-fence, moderately cold and waterie, and meanely hot, and therefore doth the more eaſily drie vp the lower parts, and by little and little ſend vp the cold humors to the head, whereby it prouoketh drowſineſſe and deſire to ſleepe, but cauſeth the head-ache; whereof *Galen* yeeldeth the reaſon as before.

D   The tender leaues that be next the root make a conuenient ointment againſt the biting of the Spider called *Phalangium.*

E   The ſeed of the Bull-Ruſh is moſt ſoporiferous, and therefore the greater care muſt be had in the adminiſtration thereof, leaſt in prouoking ſleepe you induce a drowſineſſe or dead ſleep.

Cʜᴀᴘ. 30.   *Of Reeds.*

¶ *The Kindes.*

OF Reeds the Ancients haue ſet downe many ſorts. *Theophraſtus* hath brought them all firſt into two principall kindes, and thoſe hath he diuided againe into moe ſorts. The two prin-cipall are theſe, *Auletica*, or *Tibiales Arundines*, and *Arundo vallatoria.* Of theſe and the reſt we will ſpeake in their proper places.

1 *Arundo vallatoria.*   2 *Arundo Cypria.*
Common Reed.   Cypreſſe Canes.

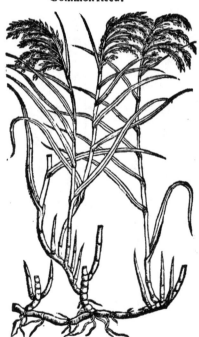 

¶ *The Deſcription.*

1   THe common Reed hath long ſtrawie ſtalkes, full of knotty joints or knees like vnto corne, whereupon doe grow very long rough flaggy leaues. The tuft or ſpokie eare doth grow at the top of the ſtalkes, browne of colour, barren and without ſeed, and doth reſemble a buſh of feathers, which turneth into fine downe or cotton which is caried away with the winde. The root is thicke, long, and full of ſtrings, diſperſing themſelues farre abroad, whereby

whereby it doth greatly iucreaſe. ‡ *Bauhine* reports, That he receiued from D. *Cargill* a Scottiſh man a Reed whoſe leaues were a cubit long, and two or three inches broad, with ſome nerues apparantly running alongſt the leaſe ; theſe leaues at the top were diuided into two, three, or foure points or parts ; as yet I haue not obſerued it. *Bauhine* termes it *Arundo Anglica folijs in ſummitate diſſectis.* ‡

1　The Cypreſſe Reed is a great Reed hauing ſtalkes exceeding long, ſometimes twenty or thirtie foot high, of a woody ſultance, ſet with very great leaues like that of Turky VVheate. It carrieth at the top the like downie tuft that the former doth.

3　*Arundo farcta.*
　Stuffed Canes.
4　*Calamus ſagittalis Lobelij.*
　Small ſtuffed Reed.
5　*Naſtos Cluſij,*
　Turky walking ſtaues.
6　*Arundo ſcriptoria.*
　Turkie writing Reed.

3　Theſe Reeds *Lobelius* hath ſeene in the Low coun-tries brought from Conſtantinople, where, as it is ſaid, the people of that countrey haue procured them from the parts of the Adriaticke ſea ſide where they doe grow. They are full ſtuft with a ſpongeous ſubſtance, ſo that there is no hollowneſſe in the ſame, as in Canes and other Reeds, except here and there certaine ſmall pores or paſ-ſages of the bigneſſe of a pinnes point ; in manner ſuch a pith as is to be found in the Bull-Ruſh, but more firme and ſolid.

4　The ſecond differeth in ſmalneſſe, and that it will winde open in fleakes, otherwiſe they are very like, and are vſed for darts, arrowes, and ſuch like.

5　This great ſort of Reeds or Canes hath no parti-cular deſcription to anſwer your expectation, for that as yet there is not any man which hath written thereof, eſpe-cially of the manner of growing of them, either of his owne knowledge or report from others , ſo that it ſhall ſuffice that ye know that that great cane is vſed eſpecial-ly in Conſtantinople and thereabout, of aged and weal-thy Citiſens, and alſo Noblemen and ſuch great perſo-nages, to make them walking ſtaues of, caruing them at the top with ſundry Scutchions , and pretty toyes of imagerie for the beautifying of them ; and ſo they of the better ſort doe garniſh them both with ſiluer and gold, as the figure doth moſt liuely ſet forth vnto you.

6　In like manner the ſmaller ſort hath not as yet beene ſeene growing of any that haue beene curious in herbariſme, whereby they might ſet downe any certainty thereof ; onely it hath beene vſed in Conſtantinople and thereabout, euen to this day to make writing pens with-all, for the which it doth very fitly ſerue, as alſo to make pipes, and ſuch like things of pleaſure.

¶ *The Place.*

The common Reed groweth in ſtanding waters and in the edges and borders of riuers almoſt euery where ; and the other being the angling Cane for fiſhers groweth in Spaine and thoſe hot Regions.

¶ *The Time.*

They flouriſh and floure from Aprill to the end of September, at what time they are cut downe for the vſe of man, as all do know.

¶ *The Names.*

The common Reed is called *Arundo,* and *Harundo vallatoria :* in French, *Roſeau :* in Dutch, 𝕽𝖎𝖊𝖙 : in Italian, *Canne a far ſiepo :* of *Dioſc. Phragmitis :* in Engliſh, Reed.

Arundo Cypria ; or after *Lobelius, Arundo Donax :* in French, *Canne :* in Spaniſh, *Cana,* in Italian, *Calami a far Connochia :* in Engliſh, Pole reed, and Cane, or Canes.

¶ *The Nature.*

Reeds are hot and dry in the ſecond degree, as *Galen* ſaith.

¶ *The Vertues.*

The roots of reed ſtamped ſmall draw forth thorns and ſplinters fixed in any part of mans body.　A
The ſame ſtamped with vinegre eaſe all luxations and members out of joynt.　　　　　　　　　B
And likewiſe ſtamped they heale hot and ſharpe inflammations. The aſhes of them mixed with　C
vinegre helpe the ſcales and ſcurfe of the head, and the falling of the haire.

The

D The great Reed or Cane is not vſed in phyſicke, but is eſteemed to make ſlears for Weauers, ſundry ſorts of pipes, as alſo to light candles that ſtand before Images, and to make hedges and pales, as we do of lats and ſuch like ; and alſo to make certaine diuiſions in ſhips to diuide the ſweet oranges from the ſowre, the Pomecitron and lemmons likewiſe in ſunder, and many other purpoſes.

Chap. 31. *Of Sugar-Cane.*

¶ *The Deſcription.*

1 SVgar Cane is a pleaſant and profitable Reed, hauing long ſtalkes ſeuen or eight foot high, joynted or kneed like vnto the great Cane ; the leaues come forth of euery joynt on euery ſide of the ſtalke one, like vnto wings, long, narrow, and ſharpe pointed. The Cane it ſelfe, or ſtalke is not hollow as the other Canes or Reeds are, but full, and ſtuffed with a ſpongeous ſubſtance in taſte exceeding ſweet. The root is great and long, creeping along within the vpper cruſt of the earth, which is likewiſe ſweet and pleaſant, but leſſe hard or woody than other Canes or Reeds ; from the which there doth ſhoot forth many young ſiens, which are cut away from the maine or mother plant, becauſe they ſhould not draw away the nouriſhment from the old ſtocke, and ſo get vnto themſelues a little moiſture, or elſe ſome ſubſtance not much worth, and cauſe the ſtocke to be barren, and themſelues little the better ; which ſhoots do ſerue for plants to ſet abroad for encreaſe.

Arundo Saccharina.
Sugar Cane.

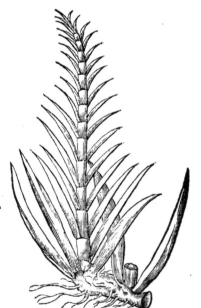

A

¶ *The Place.*

The Sugar Cane groweth in many parts of Europe at this day, as in Spaine, Portugal, Olbia, and in Prouence. It groweth alſo in Barbarie, generally almoſt euery where in the Canarie Iſlands, and in thoſe of Madera, in the Eaſt and Weſt Indies, and many other places. My ſelfe did plant ſome ſhoots thereof in my garden, and ſome in Flanders did the like : but the coldneſſe of our clymat made an end of mine, and I think the Flemmings will haue the like profit of their labour.

¶ *The Time.*

This Cane is planted at any time of the yeare in thoſe hot countries where it doth naturally grow, by reaſon they feare no froſts to hurt the young ſhoots at their firſt planting.

¶ *The Names.*

The Latines haue called this plant *Arundo Saccharina*, with this additament, *Indica*, becauſe it was firſt knowne or brought from India. Of ſome it is called, *Calamus Saccharatus* : in Engliſh, Sugar Cane : in Dutch, 𝕾𝖚𝖕𝖎𝖈𝖐𝖊𝖗𝖗𝖎𝖊𝖉𝖙.

¶ *The Nature and Vertue.*

The Sugar or juice of this Reed is of a temperate qualitie ; it drieth and cleanſeth the ſtomacke, maketh ſmooth the roughneſſe of the breſt and lungs, cleareth the voice, and putteth away hoarſeneſſe, the cough, and all ſoureneſſe and bitterneſſe, as *Iſaac* ſaith in *Dictis.*

¶ *The Vſe.*

Of the juyce of this Reed is made the moſt pleaſant and profitable ſweet, called Sugar, whereof is made infinite confections, confectures, Syrups and ſuch like, as alſo preſeruing and conſeruing of ſundry fruits, herbes, and floures, as Roſes, Violets, Roſemary floures, and ſuch like, which ſtill retaine with them the name of Sugar, as Sugar Roſet, Sugar Violet, &c. The which to write of would require a peculiar volume, and not pertinent vnto this hiſtorie, for that it is not my purpoſe to make of my booke a Confectionary, a Sugar Bakers furnace, a Gentlewomans preſeruing pan, nor yet an Apothecaries ſhop or Diſpenſatorie ; but onely to touch the chiefeſt matter that I purpoſed to handle in the beginning, that is, the nature, properties, and deſcriptions of plants. Notwithſtanding I thinke it not amiſſe to ſhew vnto yon the ordering of theſe reeds when

when they be new gathered, as I receiued it from the mouth of an Indian my ſeruant : he ſaith, They cut them in ſmall pieces, and put them into a trough made of one whole tree, wherein they put a great ſtone in manner of a mill-ſtone, whereunto they tie a gorſe, buffle, or ſome other beaſt which draweth it round : in which trough they put thoſe pieces of Canes, and ſo cruſh and grind them as we doe the barkes of trees for Tanners, or apples for Cyder. But in ſome places they vſe a great wheele wherein ſlaues doe tread and walke as dogs do in turning the ſpit : and ſome others doe feed as it were the bottome of the ſaid wheele, wherein are ſome ſharpe or hard things which doe cut and cruſh the Canes into powder. And ſome likewiſe haue found the inuention to turne the wheele with water works, as we doe our Iron mills. The Canes being thus brought into duſt or powder, they put them into great cauldrons with a little water, where they boile vntill there be no more ſweetneſſe left in the cruſhed reeds. Then doe they ſtraine them through mats or ſuch like things, and put the liquor to boile againe vnto the conſiſtence of hony, which being cold is like vnto ſand both in ſhew and handling, but ſomewhat ſofter ; and ſo afterwards it is carried into all parts of Europe, where it is by the Sugar Bakers artificially purged and refined to that whiteneſſe as we ſee.

CHAP. 31.   *Of Flouring Reed.*

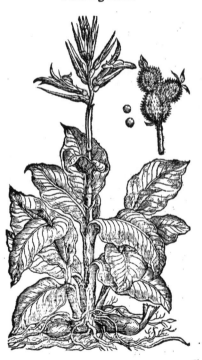

Arundo florida.
Flouring Reed.

¶ *The Deſcription.*

FLouriſhing Reed hath a thicke and fat ſtalke of foure or fiue foot high, great below neere the ground, and ſmaller toward the top, taper-wiſe ; whereupon do grow very faire broad leaues full of ribs or ſinewes like vnto Plantaine, in ſhape repreſenting the leaues of white Hellebor, or the great Gentian, but much broader and larger euery way ; at the top of which ſtalkes do grow phantaſticke floures of a red or vermilion colour ; which being faded, there follow round, rough, and prickly knobs, like thoſe of *Sparganium*, or water-Burre, of a browne colour, and from the middle of thoſe knobs three ſmall leaues. The ſeed contained in thoſe knobs is exceeding black, of a perfect roundneſſe, of the bigneſſe of the ſmalleſt peaſe. The root is thicke, knobby, and tuberous, with certain ſmall threds fixed thereto. ‡ There is a variety of this, hauing floures of a yellow or Saffron colour, with red joints.

¶ *The Place.*

It groweth in Italy in the garden of Padua, and many other places of thoſe hot regions. My ſelfe haue planted it in my garden diuers times, but it neuer came to flouring or ſeeding, for that it is very impatient to endure the injurie of our cold clymat. It is a natiue of the Weſt Indies.

¶ *The Time.*

It muſt be ſet or ſowen in the beginning of Aprill, in a pot with fine earth, or in a bed made with horſe-dung, and ſome earth ſtrawed thereon, in ſuch manner as Cucumbers and Muske-Melons are.

¶ *The Names.*

The name *Arundo Indica* is diuerſly attributed to ſundry of the Reeds, but principally vnto this, called of *Lobelius, Cannacorus* : of others, *Arundo florida*, and *Harundo florida* : in Engliſh, the Flouring Reed.

¶ *The Nature and Vertues.*

There is not any thing ſet downe as touching the temperature and vertues of this Flouriſhing Reed, either of the Ancients, or of the new or later Writers.

CHAP,

C H A P. 33.   *Of Paper Reed.*

PAper Reed hath many large flaggie leaues somewhat triangular and smooth, not much vnlike those of Cats-taile, rising immediately from a tuft of roots compact of many strings, amongst the which it shooteth vp two or three naked stalkes, square, and rising some six or seuen cubits high aboue the water: at the top wherof there stands a tuft or bundle of chaffie threds set in comely order, resembling a tuft of floures, but barren and void of seed.

Papyrus Niloitca.
Paper Reed.

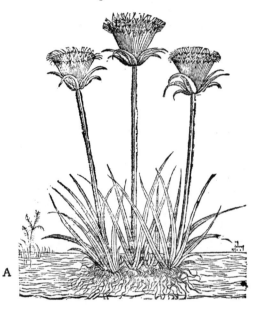

A

¶ *The Place.*

This kinde of Reed growes in the Riuers about Babylon, and neere the city Alcaire, in the riuer Nilus, and such other places of those countries.

¶ *The Time.*

The time of springing and flourishing answereth that of the common Reed.

¶ *The Names.*

This kinde of Reed, which I haue Englished Paper Reed, or Paper plant, is the same (as I doe reade) that Paper was made of in Ægypt, before the inuention of paper made of linnen clouts was found out. It is thought by men of great learning and vnderstanding in the Scriptures, and set downe by them for truth, that this plant is the same Reed mentioned in the second chapter of *Exodus*; whereof was made that basket or cradle, which was dawbed within and without with slime of that countrey, called *Bitumen Iudaicum*, wherein *Moses* was put being committed to the water, when *Pharaoh* gaue commandement that all the male children of the Hebrewes should be drowned.

¶ *The Nature, Vertues, and Vse.*

The roots of Paper Reed doe nourish, as may appeare by the people of Ægypt, which doe vse to chew them in their mouthes, and swallow downe the juice, finding therein great delight and comfort.

B   The ashes burned aswage and consume hard apostumes, tumors, and corasiue vlcers in any part of the body, but chiefely in the mouth.

C   The burnt Paper made hereof doth performe those effects more forcibly.

D   The stalkes hereof haue a singular vse and priuiledge in opening the channels and hollow passages of a Fistula, being put therein; for they doe swell as doth the pith of Elder, or a tent made of a sponge.

E   The people about Nilus do vse to burne the leaues and stalkes, but especially the roots.

F   The frailes wherein they put Raisins and figs are sometimes made hereof; but generally with the herbe *Spartum*, described in the next Chapter.

C H A P. 34.   *Of Mat-Weed.*

¶ *The Kindes.*

There be diuers kindes of Mat-weeds, as shall be declared in their seuerall descriptions.

¶ *The Description.*

THe herbe *Spartum*, as *Pliny* saith, groweth of it selfe, and sendeth forth from the root a multitude of slender rushie leaues of a cubit high, or higher, tough and pliable, of a whitish colour, which in time drawe narrow together, making the flat leafe to become round, as is the Rush. The stub or stalke thereof beareth at the top certaine feather-like tufts comming
forth

forth of a ſheath or huſke, among the which chaffie huskes is contained the ſeed, long and chaffie. The root conſiſteth of many ſtrings folding one within another, by meanes whereof it commeth to the forme of a tuft or haſſocke.

    1 *Spartum Plinij Cluſio.*   2 *Spartum alterum Plinij.*
    Plinies Mat-Weed.   Hooded Mat-Weed.

    2   The ſecond likewiſe *Pliny* deſcribeth to haue a long ſtalke not much vnlike to Reed, but leſſer, whereupon doe grow many graſſie leaues, rough and pliant, hard in handling as are the Ruſhes. A ſpokie chaffie tuft groweth at the top of the ſtalke, comming forth of a hood or ſinewie ſheath, ſuch as encloſeth the floures of Onions, Leekes, Narciſſus, and ſuch like, before they come to flouring, with ſeed and roots like the precedent.

    3   Engliſh Mat-weed hath a ruſhie root, deeply creeping and growing in heapes of ſand and grauell, from the which ariſe ſtiffe and ſharpe pointed leaues a foot and a halfe long, of a whitiſh colour, very much reſembling thoſe of Camels hay. The ſtalke groweth to the height of a cubit or more, whereupon doth grow a ſpike ‡ or eare of ſome fiue or ſix inches long, ſomewhat reſembling Rie; it is the thickeneſſe of a finger in the midſt, and ſmaller towards both the ends. The ſeed is browne as ſmall as Canarie ſeed, but round and ſomewhat ſharpe at the one end ‡. Of this plant neither Sheepe nor any other Cattel will taſte or eate.

    4   This other Engliſh Mat-Weed is like vnto the former, ſauing that the roots of this are long, not vnlike to Dogs Graſſe, but do not thruſt deepe into the ground, but creepe onely vnder the vpper cruſt of the earth. The tuft or eare is ſhorter, and more reſembling the head of Canary ſeed than that of Rie.

    ‡   5   *Lobel* giues a figure of another ſmaller Ruſh-leaued *Spartum* with ſmall heads, but hee hath not deſcribed it in his Latine Workes, ſo that I can ſay nothing certainely of it.

    6   To this kindred muſt be added the Feathered Graſſe, though not pertaking with the former in place of growth. It hath many ſmall leaues of a foots length, round, greene and ſharpe pointed, not much in forme vnlike the firſt deſcribed Mat-weed, but much leſſe: amongſt theſe leaues riſe vp many ſmall ſtalkes not exceeding the height of the leaues, which beare a ſpike vnlike the forementioned Mat-weeds, hauing three or foure ſeeds ending in, or ſending vp very fine white Feathers, reſembling the ſmaller ſort of feathers of the wings of the Bird of Paradiſe. The root conſiſts of many ſmall graſſie fibres.

                         D   ¶ *The*

3 *Spartum Anglicanum.*
Engliſh Mat-weed or Helme.

2 *Spartum Anglicanum alterum.*
Small Engliſh Mat-weed or Helme.

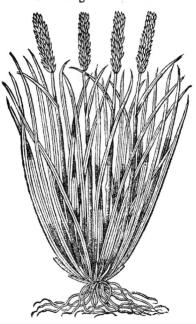

† 6 *Spartum Auſtriacum.*
Feather-graſſe.

¶ *The Place.*

1 2 Theſe grow in diuers places of Spain.

‡ 3 I being in company with M. *Thom. Hicks,
W. Broad*, and three other London Apothecaries
beſides, in Auguſt 1632, to find out rare plants in
the Iſle of Tenet, found this bigger Engliſh one
in great plenty, as ſoone as wee came to the ſea
ſide, going betweene Margate and Sandwich.

4, 5. Theſe it may be grow alſo vpon our
coaſts ; howeuer, they grow neere the ſea ſide in
diuers parts of the Low-countries.

6 This elegant plant *Cluſius* firſt obſerued to
grow naturally in the mountaines nigh to the
Baths of Baden in Germany, and in diuers places
of Auſtria and Hungary. It is nouriſhed for the
beauty in ſundry of our Engliſh gardens. ‡

¶ *The Time.*

Theſe beare their heads in the middle, & ſome
in the later end of Sommer.

¶ *The Names.*

‡ 1 This is called *Spartum primum Plinÿ*,
that is, the firſt Mat-weed deſcribed by *Pliny* : in
Spain they call it *Sparto* : the French in Prouince
terme it *Olpho.*

2 This is *Spartum alterum Plinÿ* , *Plinies* ſecond Mat-weed, or Hooded Matweed: it is called
Albardin in Spain.

3 This is *Spartum* 3. *Cluſij*, and *Gramen Sparteum ſecund. Schœnanthinum* of *Taber.* Our Author
gaue

gaue *Cluſius* his figure for the firſt, and *Tabernamontanus* figure for the ſecond *Spartum Anglicanum* : but I will thinke them both of one plant (though *Bauhine* diſtinguiſh them)vntill ſome ſhal make the contrary manifeſt.This the Dutch call **Halme** ; and our Engliſh in Tenet,Helme.*Turner* cals it Sea-Bent.

4   This is *Spartum herba* 4.*Batavicum* of *Cluſius* ; *Gramen Sparteum*, or *Iunci Spartium* of *Tabern.* and our Author gaue *Tabern.* figure,Chap. 2 3. of this Booke, vnder the title of *Iuncus marinus gramineus : Lobel* calls it *Spartum noſtras alterum.*   5   *Lobel* calls this, *Spartum noſtras parvum* : of which ſee the figure and deſcription at the end of the booke.

6   *Cluſius* calls this, *Spartum Auſtriacum : Daleſchampius, Gramen pinnatum* : we in England call it *Gramen plumoſum*, or Feathered-graſſe. ‡

¶ *The Temperature,Vertues,and Vſe.*

Theſe kinds of graſſie or rather ruſhy Reeds haue no vſe in phyſicke, but ſerue to make Mats **A** and hangings for chambers, frailes,baskets,and ſuch like. The people of the countries where they grow do make beds of them, ſtraw their houſes and chambers in ſtead of Ruſhes, for which they do excell, as my ſelfe haue ſeen. *Turner* affirmes,That they made hats of the Engliſh one in Northumberland in his time.

They do likewiſe in ſundry places of the Iſlands of Madera,Canaria,S.*Thomas*,& other of the **B** Iſlands in the tract vnto the Weſt Indies,make of them their boots,ſhooes,herdmens coats, fires, and lights. It is very hurtfull for Cattell,as Sheere-graſſe is.

The Feather-graſſe is worne by ſundry Ladies and Gentlewomen in ſtead of a Feather,which it **C** exquiſitly reſembles.

CHAP. 53.   *Of Camels-Hay.*

1 *Scœnanthum.*
Camels Hay.

2 *Scœnanthum adulterinum.*
Baſtard Camels Hay.

¶ *The Deſcription.*

1   CAmels Hay hath leaues very like vnto Mat-Weed or Helme. His roots are many, in quantitie meane, full of ſmall haires or threads proceeding from the bigger Root, deepely growing in the ground,hauing diuers long ſtalkes like Cyperus Graſſe, ſet

D 2   with

with some smaller leaues euen vnto the top,where do grow many small chaffie tufts or pannicles, like those of wilde Oats, of a reasonable good smell and sauour when they are broken,like vnto a Rose, with a certaine biting and nipping of the tongue.

† 2 *Francis Penny* (of famous memorie) a good Physition and skilfull Herbarist, gathered on the coast of the Mediterranean sea,betweene Aigues Mortes and Pescaire, this beautifull Plant, whose roots are creeping,and stalks and leaues resemble Squinanth.The floures are soft,pappous, and thicke compact,and some fiue or six inches in length,like to Fox-taile : they in colour resemble white silke or siluer. Thus much *Lobel*. Our Author described this in the first place,Chap.23. vnder *Iuncus marinus gramineus* ; for so *Lobel* calls it.

¶ *The Place.*

1   This growes in Africa, Nabathæa,and Arabia,and is a stranger in these Northern regions.
2   The place of the second is mentioned in the description.

¶ *The Time.*

Their time answereth the other Reeds and Flags.

¶ *The Names.*

Camels Hay is called in Greeke,ϰ〒ⓝ ἀϱϫῆνϣ : in Latine , *Iuncus odoratus*,and *Scœnanthum* : in shops,*Squinantbum*,that is, *Flos Iunci* : in French, *Pasteur de Chammeau* : in English, Camels Hay, and Squinanth.

2   This *Lobel* calls *Iuncus marinus gramineus*,and *Pseudoschœnanthum* : We call it bastard Squinanth,and Fox-taile Squinanth.

¶ *The Temperature.*

This plant is indifferently hot,and a little astrictiue.

¶ *The Vertues.*

A
B   Camels Hay prouoketh vrine,moueth the termes,and breaketh winde about the stomacke.

It causeth aking and heauinesse of the head : *Galen* yeeldeth this reason thereof, because it heateth moderatly, and bindeth with tenuitie of parts.

C
D   According to *Dioscorides*, it dissolues,digests,and opens the passages of the veins.

The floures or chaffie husks are profitable in drinke for them that pisse bloud any wayes. It is giuen in medicines that are ministred to cure the paines and griefes of the guts, stomacke, lungs, liuer,and reins,the fulnesse,loathsomnesse,and other defects of the stomacke, the dropsie, convulsion or shrinking of sinues,giuen in the quantitie of a drame, with a like quantitie of pepper for some few dayes.

E
The same boiled in wine helpeth the inflammation of the matrice,if the woman do sit ouer the fume thereof,and bathe her selfe often with it also.

Chap. 36.   *Of Burre-Reed.*

¶ *The Description.*

1   THe first of these plants hath long leaues, which are double edged , or sharpe on both sides,with a sharpe crest in the middle,in such manner raised vp that it seemeth to be triangle or three-square. The stalks grow among the leaues,and are two or three foot long, being diuided into many branches,garnished with many prickly husks or knops of the bignesse of a nut. The root is full of hairy strings.

2   The great water Burre differeth not in any thing from the first kinde in roots or leaues, saue that the first hath his leaues rising immediatly from the tuft or knop of the root ; but this kinde hath a long stalke comming from the root, whereupon a little aboue the root the leaues shoot out round about the stalke successiuely, some leaues still growing aboue others, euen to the top of the stalke,and from the top thereof downward by certaine distances. It is garnished with many round wharles or rough coronets, hauing here and there among the said wharles one single short leafe of a pale greene colour.

¶ *The Place.*

Both these are very common,and grow in moist medowes and neere vnto water courses. They plentifully grow in the fenny grounds of Lincolnshire and such like places ; in the ditches about S.*Georges* fields,and in the ditch right against the place of execution at the end of Southwark, called S.*Thomas* Waterings.

¶ *The Time.*

They bring forth their burry bullets or seedy knots in August.

1 *Sparganium ramoſum.*
Branched Burre Reed.

2 *Sparganium latifolium.*
Great water Burre.

¶ *The Names.*

Theſe plants of ſome are called *Sparganium* : *Theophraſtus* in his fourth booke and eighteenth chap. calleth them *Butomus* : of ſome, *Platanaria* : I call them Burre-Reed : in the Arabian tongue they are called *Safarhe Bamon* : in Italian, *Sparganio* : of *Dodonæus, Carex.* Some call the firſt *Sparganium ramoſum,* or branched Burre Reed. The ſecond, *Sparganium non ramoſum,* Not-branched Bur Reed.

¶ *The Temperature.*

They are cold and dry of complexion.

¶ *The Vertues.*

Some write, that the knops or rough burres of theſe plants boiled in wine, are good againſt the bitings of venomous beaſts, if they be either drunke, or the wound waſhed therewith.

Chap. 37. *Of Cats-taile.*

¶ *The Deſcription.*

CAts-taile hath long and flaggy leaues full of a ſpongeous matter or pith, amongſt which leaues groweth vp a long ſmooth naked ſtalke without knot, faſhioned like a ſpeare, of a firm or ſollid ſubſtance, hauing at the top a browne knop or eare, ſoft, thick, and ſmooth, ſeeming to be nothing elſe but a deale of flocks thicke ſet and thruſt together, which being ripe turns into a downe and is carried away with the winde. The roots be hard, thicke, and white, full of ſtrings, and good to burne, where there is plenty thereof to be had.

¶ *The Place.*

It groweth in pooles and ſuch like ſtanding waters, and ſometimes in running ſtreames.

I haue found a ſmaller kinde hereof growing in the ditches and mariſhie grounds in the Iſle of Shepey, going from Sherland houſe to Feuerſham.

¶ *The Time.*

They floure and beare their mace or torch in Iuly and Auguſt.

¶ *The*

Typha.
Cats-taile.

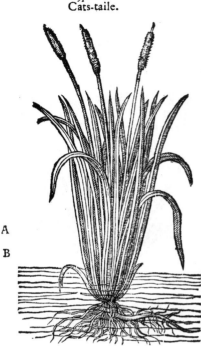

A
B

C

¶ *The Names.*

It is called in Greeke, *Typhe* : in Latine , *Typha* : of some, *Cestrum Morionis* : in French, *Marteau Masses* : in Dutch, **Lischdodem** and **Donsen** : in Italian, *Mazzaforda* : in Spanish, *Behordo,* and *Iunco amacorodato* : in English, Cats-taile or Reed-mace. Of this Cats-taile *Aristophanes* makes mention in his Comedy of Frogs, where he bringeth them forth one talking with another, being very glad that they had spent the whole day in skipping and leaping *inter Cyperum & Phleum* , among Galingale and Cats-taile. *Ovid* seemes to name this plant *Scirpus;* for hee termeth the mats made of the leaues, Cats-taile mats : as in his sixt booke *Fastorum,*

At Dominus,discedite,ait,planstroque morantes
Sustulit, in planstro scirpea matta fuit.

¶ *The Nature.*
It is cold and dry of complexion.

¶ *The Vertues.*

The soft downe stamped with Swines greafe wel washed, healeth burnes or scalds with fire or water.

Some Practitioners by their experience haue found, that the downe of the Cats taile beaten with the leaues of Betony, the roots of Gladiole, and the leaues of *Hyppoglosson* into pouder, and mixed with the yelks of egges hard sodden, & so eaten, is a most perfect remedy against the disease in children called Ἐπιφυσια, which is, when the gut called *Intestinum cæcum* is fallen into the cods. This medicine must be ministred euery day fasting for thirty dayes space : the quantitie thereof to be ministred at once is 1. 3. This being vsed as before is specified, doth not onely helpe children and striplings, but growne men also, if in time of their cure they vse conuenieut ligature or trussings, and fit confounding plaisters vpon the grieued place, according to art appointed for that purpose in Chirurgerie.

This downe in some places of the Isle of Ely and the Low-countries adioyning thereto, is gathered and well sold to make mattresses thereof for plow-men and poore people.

It hath been also often proued to heale kibed or humbled heeles, (as they are termed) being applied to them either before or after the skin is broken.

CHAP. 38. *Of Stitch-wort.*

¶ *The Description.*

1 STitch-wort, or as *Ruellius* termeth it, *Holosteum,* is of two kindes, and hath round tender stalks full of joints leaning toward the ground : at euery ioynt grow two leaues one against another. The floures be white, consisting of many small leaues set in the manner of a star. The roots are small, ioynted, and threddy. The seed is contained in small heads somewhat long, and sharp at the vpper end ; and when it is ripe, it is very small, and browne.

2 The second is like the former in shape of leaues and floures, which are set in form of a star ; but the leaues are orderly placed, and in good proportion, by couples two together, being of a whitish colour. When the floures be vaded, then follow the seeds, which are inclosed in bullets like the seed of flax, but not so round. The chiues or threds in the middle of the floure are sometimes of a reddish or blackish colour. ‡ There are more differences of this plant, or rather varieties, as differing little but in the largenesse of the leaues, floures, and stalks. ‡

¶ *The Place.*
They grow in the borders of fields vpon banke sides and hedges almost euery where.

¶ *The Time.*
They flourish all the Summer, especially in May and June.

¶ *The*

Gramen Leucanthemum.
Stitch-wort.

¶ *The Names.*

Some (as *Ruellius* for one) haue thought this to be the plant which the Grecians call ὀλόϛεον in Latine, *Tota ossea* : in English, All-bone. Wherof I see no reason, vnlesse it be by the figure *Antonomia* ; as when we say in English, he is an honest man, our meaning is, he is a knaue : for this is a tender herbe, hauing no such bony substance. ‡ *Dodonæus* questions whether this plant be not *Crataeogonon* ; and he calls it *Gramen Leucanthemum*, or White floured Grasse. The qualitie here noted with B, is by *Dioscorides* giuen to *Crataeogonon*, but it is with his ἱϛορεῖταί ὑπό τινων, that is, some say or report so much. Which phrase of speech hee often vseth when as hee writes faculties by heare-say, and doubts himselfe of the truth of them.

¶ *The Nature.*

The seed of Stitch-wort, as *Galen* writeth, is sharpe and biting to him that tastes it ; and to him that vseth it very like to Mill.

¶ *The Vertues.*

They are wont to drinke it in wine with the pouder A of Acornes, against the paine in the side, stitches, and such like.

Diuers report, saith *Dioscorides*, That the Seed of Stitchwort being drunke, causeth a woman to bring forth a man childe, if after the purgation of her sicknesse, before she conceiue, shee doe drinke it fasting thrice in a day, halfe a dram at a time, in three ounces of water many dayes together.

CHAP. 39. *Of Spider-wort.*

¶ *The Description.*

1   THe obscure description which *Dioscorides* and *Pliny* haue set down for *Phalangium*, hath bred much contention among late writers. This plant hath leaues much like Couch grasse, but they are somewhat thicker and fatter, and of a more whitish green colour. The stalks grow to the height of a cubit. The top of the stalke is beset with small branches, garnished with many little white floures, compact of six leaues. The threds or thrums in the middle are whitish, mixed with a faire yellow : which being fallen, there follow blacke seeds inclosed in smal round knobs which be three cornered. The roots be many, tough, and white of colour.

2   The second is like the first, but that his stalke is not branched as the first, and floureth a moneth before the other.

3   The third kinde of Spiderwort, which *Clusius* nameth *Asphodelus minor*, hath a root of many threddy strings, from the which immediatly rise vp grassie leaues narrow and sharp pointed : among which come forth diuers naked straight stalks diuided toward the top into sundry branches, garnished on euery side with faire starre-like floures of colour white, with a purple vein diuiding each leafe in the midst : they haue also certaine chiues or threds in them. The seed followeth inclosed in three square heads like vnto the kindes of Asphodils.

‡ 4   This Spiderwort hath a root consisting of many thick long and white fibres, not much vnlike the precedent, out of which it sends forth some fiue or six greene and firme leaues, somewhat hollow in the middle, and mutually involving each other at the root. Amongst these there riseth vp a round greene stalke, bearing at the top thereof some nine or ten floures, more or lesse ; These consist of six leaues a piece, of colour white, (the three innermost leaues are the broader, and more curled, and the three outmost are tipt with greene at the tops.) The whole floure much
resembles

resembles a white Lilly, but much smaller. Three square heads containing a dusky and vnequall seed, follow after the floure..

1 *Phalangium ramosum.*
Branched Spiderwort.

2 *Phalangium non ramosum.*
Vnbranched Spiderwort.

‡ 3 *Phalangium Cretæ.*
Candy Spider-wort.

‡ 4 *Phalangium Antiquorum.*
The true Spiderwort of the Antients.

5 *Phalangium*

‡ 5 *Phalangium Virginianum Tradesc.*
Tradescants Virginian Spider-wort.

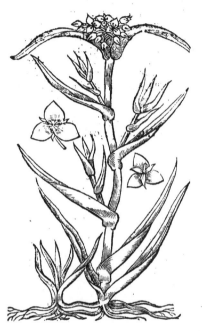

5   This plant in my iudgment cannot be fitlier ranked with any than these last described ; therefore I haue here giuen him the fitt place, as the last commer. It hath many creeping stringy roots, which here and there put vp green leaues in shape resembling those of the last described : amongst these there riseth vp a pretty stiffe stalke jointed, & hauing at each joint one leafe incompassing the stalke, and out of whose bosome ofttimes little branches arise : now the stalke at the top vsually diuides it selfe into two leaues, much after the manner of *Cyperus*; between which come forth many flours, consisting of three prety large leaues apiece, of colour deepe blew, with reddish chiues tipt with yellow standing in their middle. These fading, (as vsually they do the same day they shew themselues) there succeed little heads couered with the three little leaues that sustained the floure. In these heads there is contained a long blackish seed.

¶ *The Place.*

1. 2. 3.   These grow only in gardens with vs, and that very rarely. 4 This growes naturally in some places of Sauoy. 5 This Virginian is in many of our English gardens, as with Mr. *Parkinson*, Mr. *Tradescant*, and others.

¶ *The Time.*

1. 4. 5.   These floure in Iune; the second about the beginning of Iune, and the third about August.

¶ *The Names.*

The first is called *Phalangium ramosum*, branched Spider-wort.   2, *Phalangium non ramosum*, Vnbranched Spider-wort. *Cordus* calls it *Liliago.* 3, This *Clusius* calls *Asphodelus minor : Lobel, Phalangium Creta*, Candy Spider-wort.   4 This is thought to be the *Phalangium* of the Antients, and that of *Matthiolus.* It is *Phalangium Allobrogicum* of *Clusius*, Sauoy Spider-wort. This by Mr. *Parkinson* (who first hath in writing giuen the figure and description thereof) is aptly termed *Phalangium Ephemerum Virginianum*, Soon-fading Spiderwort of Virginia : or *Tradescants* Spiderwort, for that Mr. *Iohn Tradescant* first procured it from Virginia. *Bauhine* hath described it at the end of his *Pinax*, and very vnfitly termed it *Allium, siue Moly Virginianum.* ‡

¶ *The Nature.*

Galen saith, *Phalangium* is of a drying qualitie, by reason of the tenuitie of parts.

¶ *The Vertues.*

Dioscorides saith, That the leaues, seed, and floures, or any of them drunk in wine, preuaile against   **A** the bitings of Scorpions, and against the stinging and biting of the Spider called *Phalangium*, and all other venomous beasts.

The roots tunned vp in new ale and drunke for a moneth together, expell poyson, yea although   **B** it haue vniuersally spread it selfe through the body.

CHAP. 40.   *Of the Floure de-luce.*

¶ *The Kindes.*

THere be many kindes of Iris or Floure de-luce, whereof some are tall and great, some little, small, and low ; some smell exceeding sweet in the root, some haue no smell at all. Some floures are sweet in smell, and some without : some of one colour, some of many colours mixed : vertues attributed to some, others not remembred; some haue tuberous or knobby roots, others bulbous or Onion roots ; some haue leaues like flags, others like grasse or rushes.

¶ *The*

¶ *The Deſcription.*

1　THe common Floure de-luce hath long and large flaggy leaues like the blade of a ſword with two edges, amongſt which ſpring vp ſmooth and plaine ſtalks two foot long, bearing floures toward the top compact of ſix leaues ioyned together, wherof three that ſtand vpright are bent inward one toward another ; and in thoſe leaues that hang downeward there are certaine rough or hairy welts, growing or riſing from the nether part of the leafe vpward, almoſt of a yellow colour. The roots be thicke, long, and knobby, with many hairy threds hanging thereat.

2　The water Floure de-luce, or water Flag, or *Acorus*, is like vnto the garden Floure de-luce in roots, leaues, and ſtalkes, but the leaues are much longer, ſometimes of the height of foure cubits, and altogether narrower. The floure is of a perfect yellow colour, and the root knobby like the other ; but being cut, it ſeemes to be of the colour of raw fleſh.

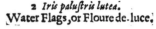

1 *Iris vulgaris.*　　　　　　2 *Iris paluſtris lutea.*
Floure de-luce.　　　Water Flags, or Floure de-luce.

¶ *The Place.*

The water Floure de-luce or yellow Flag proſpereth well in moiſt medowes, and in the borders and brinks of Riuers, ponds, and ſtanding lakes. Although it be a watery plant of nature, yet being planted in gardens it proſpereth well.

¶ *The Names.*

Floure de-luce is called in Greeke, Ἶρις: *Athenæus* and *Theophraſtus* reade Ἱερὴ: as though they ſhould ſay *Conſecratrix* : by which name it is called of the Latines, *Radix Marica*, or rather *Radix Naronica*, of the riuer Naron, by which the beſt and greateſt ſtore do grow. Whereupon *Nicander* in his Treacles commendeth it thus :

　　　　*Iridem quam aluit **Drilon**, & Naronis ripa.*
Which may thus be Engliſhed;

　　　　Iris, which Drilon water feeds,
　　　　And Narons banks, with other weeds.

The Italians, *Giglio azurro* : in Spaniſh, *Lilio Cardeno* : in French, *Flambe* : the Germans, **Gilgen, Schwertel** ; in Dutch, **Liſch**.

The ſecond is called in Latine, *Iris paluſtris lutea, Pſeudoacorus*, and *Acorus paluſtris* : in Engliſh, Water

Water flags, baſtard Floure de-luce, or Water Floure de-luce : and in the North they call them Seggs.

¶ *The Nature.*

1 The roots of the Floure de luce being as yet freſh and greene, and full of juyce, are hot almoſt in the fourth degree. The dried roots are hot and dry in the third degree, burning the throat and mouth of ſuch as taſte them.

2 The baſtard Floure de-luce his root is cold and dry in the third degree, and of an aſtringent or binding facultie.

¶ *The Vertues.*

The root of the common Floure de-luce cleane waſhed, and ſtamped with a few drops of Roſe-water, and laid plaiſterwiſe vpon the face of man or woman, doth in two daies at the moſt take away the blackneſſe or blewneſſe of any ſtroke or bruſe : ſo that if the skinne of the ſame woman or any other perſon be very tender and delicate, it ſhall be needfull that ye lay a piece of ſilke, ſindall, or a piece of fine laune betweene the plaiſter and the skinne ; for otherwiſe in ſuch tender bodies it often cauſeth heat and inflammation. — A

The juyce of the ſame doth not onely mightily and vehemently draw forth choler, but moſt eſpecially watery humors, and is a ſpeciall and ſingular purgation for them that haue the Dropſie, if it be druuke in whay or ſome other liquor that may ſomewhat temper and alay the heate. — B

The dry roots attenuate or make thinne thicke and tough humours, which are hardly and with difficultie purged away. — C

They are good in a loch or licking medicine for ſhortneſſe of breath, an old cough and all infirmities of the cheſt which riſe hereupon. — D

They remedie thoſe that haue euill ſpleenes, and thoſe that are troubled with convulſions or cramps, biting of ſerpents, and the running of the reines, being drunke with vinegre, as ſaith *Dioſcorides* ; and drunke with wine it bringeth downe the monethly courſes of women. — E

The decoction is good in womens baths, for it mollifieth and openeth the matrix. — F

Being boyled very ſoft, and laid to plaiſter-wiſe it mollifieth or ſoftneth the kings euill, and old hard ſwellings. — G

‡ The roots of our ordinarie flags are not (as before is deliuered) cold and dry in the third degree, nor yet in the ſecond, as *Dodonæus* affirmes ; but hot and dry, and that at the le he ſecond degree, as any that throughly taſts them will confeſſe. Neither are the facultu vſe (as ſome would perſuade vs) to be neglected ; for as *Pena* and *Lobel* affirme, though it hath no ſmell, nor great heat, yet by reaſon of other faculties it is much to be preferred before the *Galanga major*, or forreine *Acorus* of ſhops, in many diſeaſes ; for it imparts more heat and ſtrength to the ſtomacke and neighbouring parts than the other, which rather preyes vpon and diſſipates the innate heate and implanted ſtrength of thoſe parts. It bindes, ſtrengthens, and condenſes : it is good in bloudy flixes, and ſtaies the courſes. ‡ — H

CHAP. 41. *Of Floure de-luce of Florence.*

¶ *The Deſcription.*

1 THe Floure de-luce of Florence, whoſe root in ſhops and generally euery whete are called *Ireos*, or *Orice* (whereof ſweet waters, ſweet pouders, and ſuch like are made) is altogether like vnto the common Floure de-luce, ſauing that the floures of the *Ireos* is of a white colour, and the roots exceeding ſweet of ſmell, and the other of no ſmell at all.

2 The white Floure de-luce is like vnto the Florentine Floure de-luce in roots, flaggy leaues, and ſtalkes ; but they differ in that, that this *Iris* hath his floure of a bleake white colour declining to yellowneſſe ; and the roots haue not any ſmell at all ; but the other is very ſweet, as we haue ſaid.

3 The great Floure de-luce of Dalmatia hath leaues much broader, thicker, and more cloſely compact together than any of the other, and ſet in order like wings or the fins of a Whale fiſh, greene toward the top, and of a ſhining purple colour toward the bottome, euen to the ground : amongſt which riſeth vp a ſtalke of foure foot high, as my ſelfe did meaſure oft times in mv garden : whereupon doth grow faire large floures of a light blew, or as we terme it a watchet colour. The floures do ſmell exceeding ſweet, much like the Orenge floure. The ſeeds are contained in ſquare cods, wherein are packed together many flat ſeeds like the former. The root hath no ſmell at all.

1 *Iris Florentina.*
Floure de-luce of Florence.

2 *Iris alba.*
White floure de-luce.

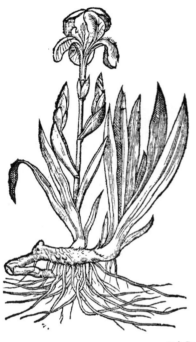

3 *Iris Dalmatica major.*
Great Floure de-luce of Dalmatia.

4 *Iris Dalmatica minor.*
Small Dalmatian *Iris.*

5 *Iris Biflora.*
Twice-flouring Floure de-luce.

6 *Iris Violacea.*
Violet Floure de-luce.

7 *Iris Pannonica.*
Austrian Floure de-luce.

E

† 8 *Iris Cameraríj.*
Germane Floure de-luce.

4   This ſmal Floure de-luce of Dalmatia is in ſhew. like to the precedent, but rather reſembling *Iris biflora*, being both of one ſtature, ſmall and dwarfe plants in reſpeƈt of the greater. The floures be of a more blew colour: they floure likewiſe in May, as the others doe; but beware that ye neuer caſt any cold water vpon them preſently taken out of the Well; for their tendernes is ſuch that they wither away and rot immediatly, as I my ſelfe haue proued: but thoſe which I left vnwatred at the ſame time liue and proſper to this day.

5   This kind of Floure de-lucé came firſt from Portugall to vs. It bringeth forth in the Spring time floures of a purple or violet colour, ſmelling like a violet, with a white hairy welt downe the middle. The root is thicke and ſhort, ſtubborne or hard to breake. In leaues and ſhew it is like to the leſſer Floure de-luce of Dalmatia, but the leaues are more ſpred abroad, and it commonly hath but one ſtalke, which in Autumne floureth againe, and bringeth forth the like floures; for which cauſe it is called *Iris biflora.*

6   *Iris violacea* is like vnto the former, but much ſmaller, and the floure is of a more deepe violet colour.

7   *Carolus Cluſius* that excellent and learned Father of Herbariſts, hath ſet forth in his Panonick Obſeruations, the piƈture of this beautifull Floure de-luce with great broad leaues thick and fat, of a purple colour neere vnto the ground, like the great Dalmatian Floure de-luce, which it very well reſembles.. The root is very ſweet being dry, ſtriuing with the Florentine *Iris* in ſweetneſſe. The floure is of all other moſt confuſedly mixed with ſundry colours, inſomuch that my pen cannot ſet downe euery line or ſtreake as it deſerueth. The three leaues that ſtand vpright do claſpe or embrace one another, and are of a yellow colour. The leaues that looke downward about the edges are of a pale colour, the middle part of white mixed with a line of purple, & it hath many ſmal lines ſtriped ouer the ſaid white floure, euen to the brim of the pale coloured edge. It ſmelleth like the Hauthorne floures, being lightly ſmelled vnto.

8   The Germane Floure de-luce, which *Camerarius* hath ſet forth in his booke named *Hortus Medicus*, hath great thicke and knobby roots. The ſtalke is thicke and full of juice: the leaues be very broad, in reſpeƈt of all the reſt of the Floure de-luces. The floure groweth at the top of the ſtalke, conſiſting of ſix great leaues blew of colour, welted downe the middle with white tending to yellow; at the bottome next the ſtalke it is white of colour, with ſome yellowneſſe fringed about the ſaid white, as alſo about the brims or edges, which greatly ſetteth forth his beauty; the which *Ioachimus Camerarius* the ſon of old *Camerarius* of Noremberg, had ſent him out of Hungary, and did communicate one of the plants thereof to *Cluſius*; whoſe figure he hath moſt liuely ſet forth with this deſcription, differing ſomewhat from that which *Ioachimus* himſelfe did giue vnto me at his being in London. The leaues, ſaith he, are very large, twice ſo broad as any of the others. The ſtalk is ſingle and ſmooth, the floure groweth at the top, of a moſt bright ſhining blew colour, the middle rib tending to whiteneſſe, the three vpper leaues ſomewhat yellowiſh. The root is likewiſe ſweet like *Ireos*.

¶ *The Place.*

Theſe kinds of Floure de-luces do grow wilde in Dalmatia, Goritia, and Piedmont; notwithſtanding our London gardens are very well ſtored with euery one of them.

¶ *The Time.*

Their time of flouring anſwereth the other Floure de-luces.

¶ *The Names.*

The Dalmatian Floure de-luce is called in Greeke of *Athenæus* and *Theophraſtus*, *Ieris*: it is named alſo *Ourania*, of the heauenly Bow or Rainbow: vpon the like occaſion, *Thaumaſtos*, or Admirable: for the Poets ſometime do call the Rainbow, *Thaumantias*: in Latine, *Iris*: in Engliſh, Floure de-luce. Their ſeuerall titles do ſufficiently diſtinguiſh them, whereby they may be knowne one from another.

¶ *The*

¶ *The Nature.*

The nature of theſe floure de-luces are anſwerable to thoſe of the common kinde, that is to ſay, the roots are hot and dry in the later end of the ſecond degree.

¶ *The Vertues.*

The juice of theſe Floure de-luces doth not only mightily and vehemently draw forth choler, **A** out eſpecially waterie humors, & is a ſingular good purgation for them that haue the dropſie, if it be drunke in ſweet wort or whay.

The ſame are good for them that haue euill ſpleens, or that are troubled with cramps or convul- **B** ſions, and for ſuch as are bit with ſerpents. It profiteth alſo much thoſe that haue the Gonorrhea, or running of the reins, being drunke with vineger, as *Dioſc.* ſaith ; and drunke with wine they bring downe the monethly termes.

CHAP. 42.   *Of variable Floure de-luces.*

1 *Iris lutea variegata.*
Variable Floure de-luce.

‡ 2 *Iris Chalcedonica.*
Turky Floure de-luce.

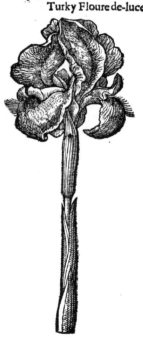

¶ *The Deſcription.*

1   THat which is called the Floure de-luce of many colours, loſeth his leaues in Winter, and in the Spring time recouereth them anew. I am not able to expreſſe the ſundrie colours and mixtures contained in this floure : it is mixed with purple, yellow, black, white, and a fringe or blacke thrum downe the middle of the lower leaues, of a whitiſh yellow, tipped or frized, and as it were a little raiſed vp, of a deep purple colour neere the ground.

2   The ſecond kinde hath long and narrow leaues of a blackiſh green like ſtinking Gladdon ; among which riſe vp ſtalks two foot long, bearing at the top of each ſtalke one floure compact of ſix great leaues : the three that ſtand vpright are confuſedly and very ſtrangely ſtriped, mixed with white and a duskiſh blacke colour. The three leaues that hang downward are like a gaping hood, and are mixed in like manner, (but the white is nothing ſo bright as of the other) and are as it were ſhadowed ouer with a darke purple colour ſomwhat ſhining ; ſo that according to my iudgment, the whole floure is of the colour of a Ginny hen, a rare and beautifull floure to behold.

‡ 3 *Iris maritima Narbonenſis.*
The ſea Floure de-luce.

4 *Iris ſylveſtris Biȝantina.*
Wild Bizantine Floure de-luce.

5 *Chamæiris anguſtifolia.*
Narrow leafed Floure de-luce.

6 *Chamæiris tenuifolia.*
Graſſe Floure de-luce.

‡ 8 *Chamæiris nivea aut Candida.*
White Dwarfe *Iris.*

‡ 7 *Iris flore cærulco obſoleto polyanthos.*
Narrow-leafed many-floured *Iris.*

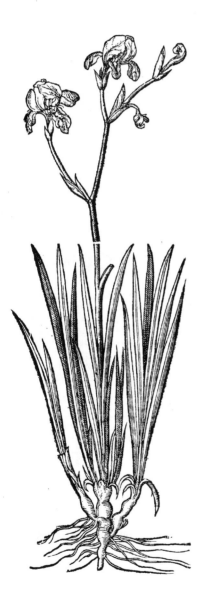

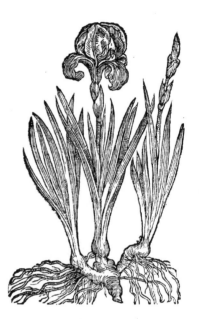

‡ 9 *Chamæiris latifolia flore rubello.*
Red floured Dwarfe *Iris.*

‡ 10 *Cha:*

‡ 10 *Chamæiris Lutea.*
Yellow Dwarfe *Iris.*

‡ 11 *Chamæiris variegata.*
Variegated Dwarfe *Iris.*

3   The French or rather Sea Floure de-luces (whereof there is alſo another of the ſame kinde altogether leſſer) haue their root without any ſauour. In ſhew they differ little from the garden Floure de-luce, but that the leaues of theſe are altogether ſlenderer, and vnpleaſant in ſmell, grow-ing plentifully in the rough crags of the rockes vnder the Alpes, and neere vnto the ſea ſide. The learned Dr. *Aſſatius* a long time ſuppoſed it to be *Medium Dioſc. Matthiolus* deceiued himſelfe and others, in that he ſaid, That the root of this plant hath the ſent of the Peach : for my ſelfe haue proued it to be without ſauour at all. It yeeldeth his floures in Iune, which are of all the reſt moſt like vnto the graſſe Floure de-luce. The taſte of his root is hot, bitter, and with much tenuitie of parts, as hath been found by Phyſicall proofe.

‡ 4   This *Iris Bizantina* hath long narrow leaues like thoſe of the laſt deſcribed ; very narrow, ſharpe pointed, hauiug no vngratefull ſmell ; the ſtalkes are ſome cubit and an halfe in length, and ſometime more ; at the top they are diuided into 2 or 3 branches that haue 2 or 3 floures a piece, like in ſhape to the floures of the broad leafed variegated bulbous *Iris* ; they haue alſo a good ſmell: the ends of the hanging-downe leaues are of a darke colour, the other parts of them are va-riegated with white, purple or violet colour. The three other leaues that ſtand vp are of a deepe violet or purple colour. The root is blackiſh, ſlender, hard, knotty. ‡

5   Narrow leafed Floure de-luce hath an infinite number of graſſie leaues much like vnto Reed, among which riſe vp many ſtalkes: on the ends of the ſame ſpring forth two, ſometimes three right ſweet and pleaſant floures, compact of nine leaues. Thoſe three that hang downeward are greater than the reſt, of a purple colour, ſtripped with white and yellow; but thoſe three ſmall leaues that appeare next, are of a purple colour without mixture : thoſe three that ſtand vpright are of an horſe-fleſh colour, tipped with purple, and vnder each of theſe leaues appeare three ſmall browne aglets like the tongue of a ſmall bird.

6   The ſmall graſſie Floure de-luce differeth from the other in ſmalneſſe and in thinneſſe of leaues, and in that the ſtalkes are lower than the leaues, and the floures in ſhape and colour are like thoſe of the ſtinking Gladdon, but much leſſe.

‡   There are many other varieties of the broad leafed Floure de-luces beſides theſe mentio-ned by our Authour; as alſo of the narrow leafed, which here wee doe not intend to inſiſt vpon, but referre ſuch as are deſirous to trouble themſelues with theſe nicities, to *Cluſius* and others.

Not-

Notwithstanding I judge it not amisse to giue the figures and briefe descriptions, of some more of the Dwarfe Floure de-luces, as also of one of the narrower leaued.

7   This therefore which we giue you in the seuenth place is *Iris flore cæruleo obsoleto, &c. Lobelij*. The leaues of this are small and long like those of the wild *Bizantine* Floure de-luce ; the root (which is not very big) hath many strong threds or fibres comming out of it : the stalke (which is somewhat tall) diuides it selfe into two or three branches, whereon grow floures in shape like those of the other Floure de-luces, but their colour is of an ouerworne blew, or ash colour.

8   Many are the differences of the *Chamæirides latifolia*, or broad leafed Dwarfe Floure de-luces, but their principall distinction is in their floures; for some haue floures of violet or purple colour, some of white, othersome are variegated with yellow and purple, &c. Therefore I will onely name the colour, and giue you their figures, because their shapes differ little. This eighth therfore is *Chamæiris niuea, aut candida*, White Dwarfe *Iris* : The ninth, *Chamæiris latifolia flore rubello*, Red floured Dwarfe *Iris* : The tenth, *Chamæiris lutea*, Yellow Dwarfe *Iris* : The eleuenth, *Chamæiris variegata*, Variegated *Iris*. The leaues and stalkes of these plants are vsually about a foot high ; the floures, for the bignesse of the plants, large, and they floure betimes, as in Aprill. And thus much I thinke may suffice for the names and descriptions of these Dwarfe varieties of Floure de-luces.

¶ *The Place.*

These plants doe grow in the gardens of London, amongst Herbarists and other Louers of Plants.

¶ *The Time.*

They floure from the end of March to the beginning of May.

¶ *The Names.*

The Turky Floure de-luce is called in the Turkish tongue, *Alaisa Susiani*, with this additament from the Italians, *Fiore Bellepintate* : in English, Floure de-luce. The rest of the names haue beene touched in their titles and histories.

¶ *Their Nature and Vertues.*

The faculties and temperature of these rare and beautifull floures are referred to the other sorts of Floure de-luces, whereunto they do very well accord.

There is an excellent oyle made of floures and roots of Floure de-luce, of each a like quantitie, A called *Oleum Irinum*, made after the same manner that oyle of Roses, Lillies and such like be made : which oyle profiteth much to strengthen the sinewes and joynts, helpeth the crampe proceeding of repletion, and the disease called in Greeke *Peripneumonia*.

The floures of French floure de-luce distilled with *Diatrion Santalon*, and Cinnamon, and the B water drunke, preuaileth greatly against the Dropsie, as *Hollerius* and *Gesner* testifie.

CHAP. 43.   *Of stinking Gladdon.*

¶ *The Description.*

STinking Gladdon hath long narrow leaues like *Iris*, but smaller, of a darke greene colour, and being rubbed, of a stinking smell very lothsome. The stalkes are many in number, and round toward the top, out of which doe grow floures like the Floure de-luce, of an ouerworne blew colour, or rather purple, with some yellow and red streakes in the midst. After the floures be vaded there come great huskes or cods, wherein is contained a red berry or seed as big as a pease. The root is long, and threddy vnderneath.

¶ *The Place.*

Gladdon groweth in many gardens : I haue seene it wild in many places, as in woods and sha-dowie places neere the sea.

¶ *The Time.*

The stinking Gladdon floureth in August, the seed thereof is ripe in September.

¶ *The Names.*

Stinking Gladdon is called in Greeke ἶρις, by *Dioscorides* ; and ἶρις ἀγρία by *Theophrastus*, according to *Pena* : in Latine, *Spatula fœtida* among the Apothecaries : it is called also *Xyris*: in English, stin-king Gladdon, and Spurgewort.

¶ *The Nature.*

Gladdon is hot and dry in the third degree.

¶ *The Vertues.*

Such is the facultie of the roots of all the Irides before named, that being pounding they pro- A uoke sneesing, and purge the head: generally all the kinds haue a heating and extenuating quality.

They

B

Xyris.
Stinking Gladdon.

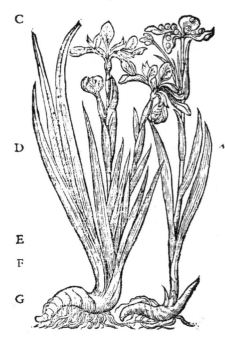

C

D

E

F

G

They are effectuall againſt the cough, they eaſily digeſt and conſume the groſſe humors which are hardly concocted : they purge colour and tough flegme : they procure ſleepe, and helpe the gripings within the belly.

It helpeth the Kings Euill, and Buboes in the groine, as *Pliny* ſaith. If it be drunke in Wine it prouoketh the termes, and being put in Baths for women to ſit ouer, it prouoketh the like effects moſt exquiſitely. The root put in manner of a peſſarie haſtneth the birth. They couer with fleſh, bones that be bare, being vſed in plaiſters. The roots boiled ſoft, and vſed plaiſterwiſe, ſoften all old hard tumours, and the ſwellings of the throat called *Strumæ*, that is, the Kings Euill, and emplaiſtered with honey it breaketh out broken bones.

The meale thereof healeth all the rifts of the fundament, and the infirmities thereof called *Condylomita*; and openeth Hemorrhoides. The juice ſnuffed or drawne vp into the noſe, prouoketh ſneeſing, and draweth down by the noſe great ſtore of filthy excrements, which would fall into other parts by ſecret and hidden waies, and conueiances of the channels.

It profiteth being vſed in a peſſarie, to prouoke the termes, and will cauſe abortion.

It preuaileth much againſt all euil affections of the breſt and lungs, being taken in a little ſweet wine, with ſome Spikenard; or in Whay with a little Maſticke.

The root of *Xyris* or Gladdon is of great force againſt wounds and fractures of the head : for it draweth out all thornes, ſtubs, prickles, and arrow-heads, without griefe ; which qualitie it effecteth (as *Galen* ſaith) by reaſon of his tenuitie of parts, and of his attracting, drying, and digeſting facultie, which chiefely conſiſteth in the ſeed or fruit, which mightily prouoketh vrine.

H   The root giuen in Wine, called in Phyſicke *Paſſum*, profiteth much againſt Convulſions, Ruptures, the paine of the huckle bones, the ſtrangurie, and flux of the bellie. Where note, That whereas it is ſaid that the potion aboue named ſtayeth the flux of the belly, hauing a purging qualitie ; it muſt be vnderſtood that it worketh in that manner as *Rhabarbarum* and *Aſarum* do, in that they concoct and take away the cauſe of the laske ; otherwiſe no doubt it moueth vnto the ſtoole, as *Rheubarb*, *Aſarum*, and the other Irides do. Hereof the country people of Somerſet-ſhire haue good experience, who vſe to drinke the decoction of this Root. Others doe take the infuſion thereof in ale or ſuch like, wherewith they purge themſelues, and that vnto very good purpoſe and effect.

I   The ſeed thereof mightily purgeth by vrine, as *Galen* ſaith, and the country people haue found it true.

CHAP. 44. *Of Ginger.*

¶ *The Deſcription.*

Ginger is moſt impatient of the coldneſſe of theſe our Northerne regions, as my ſelfe haue found by proofe, for that there haue beene brought vnto me at ſeuerall times ſundry plants thereof, freſh, greene, and full of juice, as well from the Weſt Indies, as from Barbary and other places ; which haue ſprouted and budded forth greene leaues in my garden in the heate of Summer, but as ſoone as it hath beene but touched with the firſt ſharpe blaſt of Winter, it hath preſently periſhed both blade and root. The true forme or picture hath not before this time been ſet forth by any that hath written ; but the world hath beene deceiued by a counterfeit figure, which the reuerend and learned Herbariſt *Matthias Lobel* did ſet forth in his Obſeruations. The forme whereof notwithſtanding I haue here expreſſed, with the true and vndoubted picture alſo, which

which I receiued from *Lobels* owne hands at the impreſſion hereof. The cauſe of whoſe former error, as alſo the meanes whereby he got the knowledge of the true Ginger, may appeare by his own
words ſent vnto me in Latine, which I haue here thus Engliſhed :

How hard and vncertaine it is to deſcribe in words the true proportion of Plants (hauing none
other guide than skilfull, but yet deceitfull formes of them, ſent from friends or other means) they
beſt do know who haue deeplieſt waded in this ſea of Simples. About thirty yeares paſt or more,
an honeſt and expert Apothecarie *William Dries*, to ſatisfie my deſire, ſent me from Antwerpe to
London the picture of Ginger, which he held to be truly and liuely drawne. I my ſelfe gaue him
credit eaſily, becauſe I was not ignorant, that there had beene often Ginger roots brought greene,
new, and full of juice, from the Indies to Antwerp : and further, that the ſame had budded & grown
in the ſaid *Dries* garden. But not many yeares after I perceiued, that the picture which was ſent
me by my friend was a counterfeit, and before that time had been drawne and ſet forth by an old
Dutch Herbariſt. Therefore not ſuffering this error any further to ſpread abroad (which I diſcouered not many yeares paſt at Fluſhing in Zeeland, in the garden of *William* of Naſſau Prince of
Orange, of famous memorie, through the means of a worthy perſon (if my memorie faile mee not)
Vander Mill; at what time he opened and looſed his firſt yong buds and ſhoots about the end of
Sommer, reſembling in leaues, and ſtalks of a foot high, the young and tender ſhoots of the common Reed called *Harundo vallatoria*) I thought it conuenient to impart thus much vnto Mr. *Iohn*
Gerrard an expert Herbariſt, and maſter of happy ſucceſſe in Surgerie, to the end he might let poſteritie know thus much, in the painfull and long laboured trauels which now he hath in hand, to
the great good and benefit of his countrey. The plant it ſelfe brought me to Middleborow and ſet
in my garden, periſhed through the hardneſſe of Winter.

Thus much haue I ſet downe, truly tranſlated out of his owne words in Latine ; though too fa-
uourably by him done to the commendation of my mean skill.

Zinziberis ficta icon.
The feigned figure of Ginger.

Zinziberis verior icon.
The true figure of Ginger.

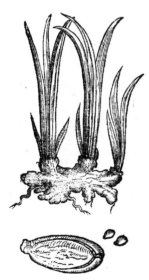

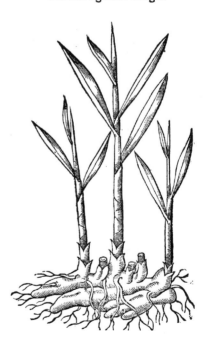

¶ *The Place.*

Ginger groweth in Spaine, Barbary, in the Canarie Iſlands, and the Azores. Our men who ſac-
ked Domingo in the Indies, digged it vp there in ſundry places wilde.

¶ *The*

¶ *The Time.*

Ginger flouriſheth in the hot time of Sommer, and loſeth his leaues in Winter.

¶ *The Names.*

Ginger is called in Latine *Zinziber*, and *Gingiber* : in Greeke, ζιγγίβερι and ριγγίβερι : in French, *Gingembre*.

¶ *The Nature.*

Ginger heateth and drieth in the third degree.

¶ *The Vertues.*

A   Ginger, as *Dioſcorides* reporteth, is right good with meat in ſauces, or otherwiſe in conditures ; for it is of an heating and digeſting qualitie, gently looſeth the belly, and is profitable for the ſtomacke, and effectually oppoſeth it ſelfe againſt all darkneſſe of the ſight ; anſwering the qualities and effects of Pepper. It is to be conſidered, That canded, greene, or condited Ginger is hot and moiſt in qualitie, prouoking Venerie : and being dried, it heateth and drieth in the third degree.

CHAP. 45.   *Of Aromaticall Reeds.*

1 *Acorus verus, officinis falſo Calamus, cum julo.*	*Acorus verus ſine julo.*
The true *Acorus* with his floure.	The true *Acorus* without his floure.

¶ *The Deſcription.*

1   THis ſweet ſmelling Reed is of a darke dun colour, full of joints and knees, eaſie to be broken into ſmall ſplinters, hollow, and full of a certaine pith cobweb-wiſe, ſomwhat gummy in eating, and hanging in the teeth, and of a ſharpe bitter taſte. It is of the thickeneſſe of a little finger, as *Lobel* affitmeth of ſome which he had ſeene in Venice.

2   Baſtard Calamus hath flaggy leaues like vnto the water Floure de-luce or Flagge, but narrower, three foot long ; of a freſh greene colour, and aromaticke ſmell, which they keepe a long time, although they be dried. Now the ſtalke which beares the floure or fruit is much like another

ther

ther leafe,but only from the fruit downwards,whereas it is fomwhat thicker,and not fo broad, but almoft triangular. The floure is a long thing refembling the Cats-tailes which grow on Hafels. It is about the thickneffe of an ordinary Reed,fome inch and halfe long,of a greenifh yellow colour, curioufly cheequered as if it were wrought with a needle with green and yellow filk intermixt. † I haue not as yet feene it beare his tuft in my garden, hauing read that it is barren, and by proofe haue feen it fo: yet for all that I beleeue *Clufius*, who faith he hath feene it beare his floure in that place where it doth grow naturally, although in England it is altogether barren. The root is fweet in fmell,bitter in tafte,and like vnto the common Flag,but fmaller,and not fo red.

3 *Calamus Aromaticus Antiquorum*.
The true Aromaticall Reed of the Antients.

‡ 3   I think it very fitting in this place to acquaint you with a Plant,which by the conjecture of the moft learned,and that not without good reafon, is iudged to be the true *Calamus* of the Antients. *Clufius* giues vs the hiftorie thereof in his notes vpon *Garcias ab Horto,lib.1.ca.32.* in thefe words : When as (faith he) this hiftorie was to be the third time printed,I very opportunely came to the knowledge of the true *Calamus Aromaticus* ; the which the learned *Bernard Paludanus* the Frifian, returning from Syria and Egypt, freely beftowed vpon me,together with the fruit Habhel, and many other rare feeds,about the beginning of the yeare 1579.Now we haue caufed a figure to be exactly drawne by the fragments thereof,for that it feemes fo exquifitly to accord with *Diofcorides* his defcription. In mine opinion it is rather to be iudged an vmbelliferous plant than a reedie ; for it hath a ftraight ftalke parted with many knots or ioynts,otherwife fmooth,hollow within,and inuefted on the in-fide with a flender filme like as a Reed, and it breaketh into fhiuers or fplinters,as *Diofcorides* hath written. It hath a fmell fufficiently ftrong, and the tafte not vngratefull,but bitter,and pertaking of fome aftriction : the leaues,as by remains of them might appeare, feeme by couples at euery ioynt to ingirt the ftalke : the root at the top is fomewhat tuberous,and then ends in fibres. Twenty fiue yeares after *Paludanus* gaue me this *Calamus*, the learned *Anthonie Coline* the Apothecarie (who lately tranflated into French thefe Commentaries, the fourth time fet forth, *Anno* 1593.) fent me from Lyons pieces of the like Reed ; certifying me withall,That he had made vfe thereof in his compofition of Treacle. Now thefe pieces, though they in forme refembled thofe I had from *Paludanus*,yet had they a more bitter tafte than his,neither did they pertake of any aftriction ; which peraduenture was to be attributed to the age of one of the two.Thus much *Clufius*.

¶ The Place.

The true *Calamus Aromaticus* groweth in Arabia,and likewife in Syria,efpecially in the moorifh grounds betwixt the foot of Libanus † and another little hill, not the mountaine Antilibanus, as fome haue thought, in a fmall vally neere to a lake whofe plafhes are dry in Summer. *Plin.12.22.* Baftard or falfe *Calamus* growes naturally at the foot of a hill neere Prufa a city of Bithynia,not far from a great lake. It profpers exceeding well in my garden,but as yet it beareth neither floure nor ftalke. It groweth alfo in Candy as *Pliny* reporteth ; in Galatia likewife, and in many other places.

¶ The Time.

They lofe their leaues in the beginning of Winter,and do recouer them again in the Spring of the yeare. ‡ In May this yeare 1632, I receiued from the worfhipfull gentleman Mr *Tho.Glynn* of Glynnlhiuon in Carnaruanfhire,my very good friend,the pretty *Iulus*,or floure of this plant;which I could neuer fee here about London,though it groweth with vs in many gardens,and that in great plenty. ‡

¶ The Names.

‡   The want of the true *Calamus* being fupplied by *Acorus* as a *fuccedaneum*, was the caufe (as *Pena* and *Lobel* probably conjecture) that of a fubftitute it tooke the prime place vpon it ; and being as it were made a Vice-Roy, would needs be a King. But the falfeneffe of the title was difcouered

uered by *Matthiolus* and others, and so it is sent backe to its due place again; though notwithstanding it yet in shops retaines the title of *Calamus*.

1   The figure that by our Author was giuen for this, is supposed (and that as I thinke truly) to be but a counterfeit of *Matthiolus* his inuention; who therein hath bin followed (according to the custome of the world) by diuers others. The description is of a small Reed called *Calamus odoratus Libani*, by *Lobel* in his Obseruations, and figured in his *Icones*, p. 54.

2   This is called Ἄκορον and Ἄκορον by the Greeks: by some, according to *Apuleius*, Ἀφροδισίας: and in Latine it is called *Acorus* and *Acorum*; and in shops, as I haue formerly said, *Calamus Aromaticus*: for they vsually take *Galanga maior* (described by me *Cap.* 26.) for *Acorus*. It may besides the former names be fitly called in English, the sweet Garden Flag.

3   This is iudged to be the Κάλαμος ἀρωματικός of *Dioscorides*, the Κάλαμος ἰνδικὸς of *Theophrastus*, that is, the true *Calamus Aromaticus* that should be vsed in Compositions. ‡

¶ *The Nature of the true Acorus or our sweet garden Flag.*

Dioscorides saith, the roots haue an heating facultie. *Galen* and *Pliny* doe affirme, that they haue thin and subtill parts both hot and dry.

¶ *The Vertues of the same.*

A   The decoction of the root of *Calamus* drunke prouoketh vrine, helpeth the paine in the side, liuer, spleen, and brest; convulsions, gripings, and burstings: it easeth and helpeth pissing by drops.

B   It is in great effect being put in broth, or taken in fumes through a close stoole, to prouoke womens naturall accidents.

C   The iuice strained with a little hony taketh away the dimnesse of the eyes, and helpeth much against poison, the hardnesse of the spleen, and all infirmities of the bloud.

D   The root boiled in wine, stamped and applied plaisterwise vnto the cods, wonderfully abateth the swelling of the same, and helpeth all hardnesse and collections of humors.

E   The quantitie of two scruples and a halfe of the root drunke in foure ounces of Muskadel, helpeth them that be bruised with grieuous beating or falls.

F   The root is with good successe mixed in counterpoysons. In our age it is put into Eclegma's, that is, medicines for the lungs, and especially when the lungs or chest are opprest with raw and cold humors.

G   ‡ The root of this preserued is very pleasant to the taste, and comfortable to the stomacke and heart, so that the Turks at Constantinople take it fasting in the morning against the contagion of the corrupt aire: and the Tartars haue it in such esteeme, that they will not drinke water (which is their vsuall drinke) vnlesse they haue first steeped some of this root therein. ‡

¶ *The Choice.*

The best *Acorus*, as *Dioscorides* saith, is that which is substantiall and well compact, white within, not rotten, full, and well smelling.

Pliny writeth, That those which grow in Candia are better than those of Pontus, and yet those of Candia worse than those of the Easterne countries, or those of England, although we haue no great quantitie thereof.

¶ *The Faculties of the true Calamus out of Dioscorides.*

H   ‡ It being taken in drink moueth vrine; wherefore boiled with the roots of grasse or Smallage seeds, it helpeth such as be hydropick, nephritick, troubled with the strangury, or bruised.

I   It moues the Courses either drunke or otherwise applied. Also the fume thereof taken by the mouth in a pipe, either alone or with dried turpentine, helps coughs.

K   It is boiled also in baths for women, and decoctions for Clisters, and it enters into plaisters and perfumes for the smells sake. ‡

CHAP. 46. *Of Corne.*

THus far haue I discoursed vpon Grasses, Rushes, Spartum, Flags, and Floure de-luces: my next labor is to set downe for your better instruction the historie of Corne, and the kinds therof, vnder the name of Graine, which the Latines call *Cerealea semina*, or Bread-corne: the Grecians, σῖτος, and ὄσπρια σπέρματα: of which we purpose to discourse. There belong to the history of grain all such things as be made of Corne, as *Far*, *Condrus*, *Alica*, *Tragus*, *Amylum*, *Ptisana*, *Polenta*, *Maza*, *Byne*, or Malt, Ζύθιον, and whatsoeuer are of that sort. There be also ioyned vnto them many seeds, which *Theophrastus* in his eighth booke placeth among the Graines, as Millet, Sorgum, Panicke, Indian wheat, and such like. *Galen* in his first booke of the Faculties of nourishments, reckoneth vp the diſeaſes of Graine, as well thoſe that come of the graine it ſelfe degenerating, or that are changed into ſome other kinde, and made worſe through the fault of the weather, or of the ſoile; as alſo ſuch as be cumberſome by growing among them, which doe likewiſe fitly ſucceed the graines. And beginning with corne, we will firſt ſpeake of wheat, and deſcribe it in the firſt place, becauſe it is preferred before all other corne.

1 *Triticum ſpica mutica.*
White Wheate.

¶ *The Deſcription.*

1 THis kinde of Wheat which *Lobelius*, diſtinguiſhing it by the eare, calleth *Spica Mutica*, is the moſt principall of all other, whoſe eares are altogether bare or naked, without awnes or chaffie beards. The ſtalke riſeth from a threddy root, compact of many ſtrings, joynted or kneed at ſundry diſtances; from whence ſhoot forth graſſie blades and leaues like vnto Rie, but broader. The plant is ſo well knowne to many, and ſo profitable to all, that the meaneſt and moſt ignorant need no larger deſcription to know the ſame by.

2 The ſecond kinde of Wheat, in roor, ſtalkes, joints and blades, is like the precedent, differing onely incare, and number of graines, whereof this kind doth abound, hauing an eare conſiſting of many ranks, which ſeemeth to make the eare double or ſquare. The root and graine is like the other, but not bare and naked, but briſtled or bearded, with many ſmall and ſharpe eiles or awnes not vnlike to thoſe of Barley.

3 Flat Wheate is like vnto the other kindes of Wheat in leaues, ſtalks, and roots, but is bearded and bordered with rough and ſharpe ailes, wherein conſiſts the difference. ‡ I know not what our author means by this flat Wheat, but I conjecture it to be the long rough eared Wheat, which hath blewiſh eares when as it is ripe, in other things reſembling the ordinary red Wheat. ‡

4 The fourth kinde is like the laſt deſcribed, and thus differeth from it, in that, this kind hath many ſhore ſmall ears comming forth of one great eare, & the beards hereof be ſhorter than of the former kind.

5 Bright Wheat is like the ſecond before deſcribed, and differeth from it in that, that this kind is foure ſquare, ſomewhat bright and ſhining; the other not.

‡ I thinke it a very fit thing to adde in this place a rare obſeruation, of the tranſmutation of one ſpecies into another, in plants; which though it haue beene obſerued in ancient times, as by *Theophraſtus, de cauſ. plant. lib.* 3. *cap.* 16. whereas among others hee mentioned the change of *ταιραφε ϛριμα*, Spelt into oates: and by *Virgil* in theſe words;

    Grandia ſæpe quibus mandauimus Hordea ſulcis,
    Infælix Lolium, & ſteriles dominantur avenæ.
       That is;
   In furrowes where great Barly we did ſow,
    Nothing but Darnell and poore Oats do grow.

yet none that I haue read haue obſerued, that two ſeuerall graines, perfect in each reſpect, did grow at any time in one eare: the which I ſaw this yeare 1632, in an eare of white Wheat which was found by my very good friend M.*Iohn Goodyer*, a man ſecond to none in his induſtrie and ſearching of plants, nor in his judgement or knowledge of them. This eare of wheat was as large and faire as moſt are, and about the middle thereof grew three or foure perfect Oats in all reſpects: which being hard to be found, I held very worthy of ſetting downe, for ſome reaſons not to be inſiſted vpon in this place. ‡

¶ *The Place.*

Wheat groweth almoſt in all the countries of the world that are inhabited and manured, and requireth a fruitfull and fat ſoile, and rather Sunny and dry, than watery grounds and ſhadowie: for in dry ground (as *Columella* reporteth) it groweth harder and better compact: in a moiſt and darke ſoile it degenerateth ſometime to be of another kinde.

                  F   ¶ *The*

2 Triticum aristis circumvallatum.
Bearded Wheat, or Red Wheat.

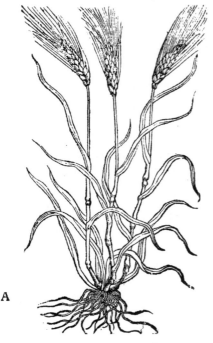

A

3 Triticum Typhinum.
Flat Wheat.

¶ *The Time.*

They are most commonly sowen in the fall of the leafe, or Autumne : sometime in the Spring.

¶ *The Names.*

Wheat is called of the Græcians, πυρὸς : of the Latines, *Triticum*, and the white Wheat *Siligo*. *Triticum* doth generally signifie any kinde of Corne which is threshed out of the eares and made clean by fanning or such ordinary means. The Germans call it **Weisen**: in low Dutch, **Terwe**: in Italian, *Grano* : the Spaniards, *Trigo* : the Frenchmen, *Bled, au Fourment* : in English we call the first, White-Wheat, and Flaxen Wheat. *Triticum lucidum* is called Bright Wheat : Red Wheat is called in Kent, Duck-bill Wheat, and Normandy Wheat.

¶ *The Nature.*

Wheat (saith *Galen*) is very much vsed of men, and with greatest profit. Those Wheats do nourish most that be hard, and haue their whole substance so closely compact as they can scarsely be bit asunder ; for such do nourish very much : and the contrary but little.

Wheat as it is a medicine outwardly applied, is hot in the first degree, yet can it not manifestly either dry or moisten. It hath also a certain clamminesse and stopping qualitie.

¶ *The Vertues.*

Raw Wheat, saith *Dioscorides*, being eaten, breedeth wormes in the belly : being chewed and applied, it doth cure the biting of mad dogs.

4 Triticum multiplici spica.
Double eared Wheat.

B   The floure of wheat being boiled with hony and water, or with oyle and water, taketh away all inflammations, or hot swellings.

C   The bran of Wheat boyled in strong Vinegre, clenseth away scurfe and dry scales, and dissolueth the beginning of all hot swellings, if it be laid vnto them. And boyled with the decoction of Rue, it slacketh the swellings in womens breasts.

D   The graines of white Wheat, as *Pliny* writeth in his two and twentieth booke, and seuenth chapter, being dried browne, but not burnt, and the pouder thereof mixed with white wine is good for watering eies, if it be laid thereto.

E   The dried pouder of red Wheat boyled with vinegre, helpeth the shrinking of sinewes.

F   The meale of Wheat mingled with the juice of Henbane, and plaisterwise applied, appeaseth

inflam-

5 *Triticum lucidum*.
Bright Wheat.

inflammations, as *Ignis ſacer*, or Saint *Anthonies* Fire, and ſuch like, ſtaying the flux of humors to the joynts, which the Græcians call *Rheumatiſmata*. Paſte made of fine meale, ſuch as Booke-binders vſe, helpeth ſuch as doe ſpit bloud, taken warme one ſpoonefull at once. The bran of wheat boiled in ſharpe vinegre, and rubbed vpon them that be ſcuruie and mangie, eaſeth the party very much.

The leauen made of Wheat hath vertue to   G heate and draw outward, it reſolueth, concoƈteth, and openeth all ſwellings, bunches, tumors, and felons, being mixed with ſalt.

The fine floure mixed with the yolke of an   H egge, honey, and a little ſaffron, doth draw and heale byles and ſuch like ſores, in children and in old people, very well and quickly. Take crums of wheaten bread one pound and an halfe, barley meale ℥ ij. Fennigreeke and Lineſeed of each an ounce, the leaues of Mallowes, Violets, Dwale, Sengreene, and Cotyledon, *ana* one handfull: boyle them in water and oyle vntill they be tender: then ſtampe them very ſmall in a ſtone morter, and adde thereto to the yolke of three egges, oyle of Roſes, and oyle of Violets, *ana* ℥ ij. Incorporate them altogether, but if the inflammation grow to an Eryſipelas, then adde thereto the juice of Nightſhade, Plantaine, and Henbane, *ana* ℥ ij. it eaſeth an Eryſipelas, or Saint *Anthonies* fire, and all inflammations very ſpeedily.

Slices of fine white bread laid to infuſe or   I ſteepe in Roſe water, and ſo applied vnto ſore eyes which haue many hot humours falling into them, doe eaſily defend the humour, and ceaſe the paine.

The oyle of wheat preſſed forth betweene two plates of hot iron, healeth the chaps and chinks   K of the hands, feet, and fundament, which come of cold, making ſmooth the hands, face or any other part of the body.

The ſame vſed as a Balſame doth excellently heale wounds, and being put among ſalues or vn-   L guents, it cauſeth them to worke more effeƈtually, eſpecially in old vlcers.

Chap. 47.   Of Rie.

¶ The Deſcription.

THe leafe of Rie when it firſt commeth vp, is ſomewhat reddiſh, afterward greene, as be the other graines. It groweth vp with many ſtalkes, ſlenderer than thoſe of wheat, and longer, with knees or joynts by certaine diſtances like vnto Wheat: the eares are orderly framed vp in rankes, and compaſſed about with ſhort beards, not ſharpe but blunt, which when it floureth ſtand vpright, and when it is filled vp with ſeed it leaneth and hangeth downeward. The ſeed is long, blackiſh, ſlender, and naked, which eaſily falleth out of the huskes of itſelfe. The roots be many, ſlender, and full of ſtrings.

¶ The Place.

Rie groweth very plentifully in the moſt places of Germany and Polonia, as appeareth by the great quantitie brought into England in time of dearth, and ſcarcitie of corne, as hapned in the yeare 1596. and at other times, when there was a generall want of corne, by reaſon of the aboundance of raine that fell the yeare before; whereby great penurie enſued, as well of cattell and all other viƈtuals, as of all manner of graine. It groweth likewiſe very well in moſt places of England, eſpecially towards the North.

The

Secale.
Rie.

A

B

¶ *The Time.*

It is for the moſt part ſowne in Autumne, and ſometimes in the Spring, which proueth to be a grain more ſubiect to putrifaction than that was ſowne in the fall of the leafe, by reaſon the Winter doth ouertake it before it can attain to his full maturitie and ripeneſſe.

¶ *The Names.*

Rie is called in high-Dutch, **Rocken**: in Low-Dutch, **Rogge**: in Spaniſh, *Centeno*: in Italian, *Segala*: in French, *Seigle*, which ſoundeth after the old Latine name which in *Pliny* is *Secale* and *Farrago, lib. 18. cap. 16.*

¶ *The Temperature.*

Rie as a medicine is hotter than wheat, & more forcible in heating, waſting, and conſuming away that whereto it is applied. It is of a more clammy and obſtructing nature than wheat, and harder to digeſt; yet to ruſtick bodies that can well digeſt it, it yeelds good nouriſhment.

¶ *The Vertues.*

Bread, or the leauen of Rie, as the Belgian phyſitians affirme vpon their practiſe, doth more forcibly digeſt, draw, ripen, and breake all apoſtumes botches, and biles than the leauen of wheat.

Rie meale bound vnto the head in a linnen cloath, doth aſſwage the long continuing paines thereof.

C H A P. 48.   *Of Spelt Corne.*

¶ *The Deſcription.*

S Pelt is like to Wheat in ſtalks and eare: it groweth vp with a multitude of ſtalkes, which are kneed and iointed higher than thoſe of Barly: it bringeth forth a diſordered eare for the moſt part without beards. The cornes be wrapped in certain dry huſks, from which they cannot eaſily be purged, and are ioyned together by couples in two chaffie huſks, out of which when they be taken they are like vnto wheat cornes: it hath alſo many roots as Wheat hath, whereof it is a kinde.

¶ *The Place*

It groweth in fat and fertile moiſt ground.

¶ *The Time.*

It is altered and changed into wheat it ſelfe, as degenerating from bad to better, contrary to all other that do alter or change; eſpecially (as *Theophraſtus* ſaith) if it be clenſed, and ſo ſowne, yet not forthwith, but in the third yeare.

¶ *The Names.*

The Grecians haue called it *Zeia* and *Zea*: the Latines, *Spelta*: in the German tongue, **Speltz**, and **Sinkel**: in low-Dutch, **Spelte**: in French, *Eſpeautre*: of moſt Italians, *Pirra, Farra*: of the Tuſcans, *Brada*: of the Millanois, *Alga*: in Engliſh, Spelt Corne. *Dioſcorides* maketh mention of two kinds of Spelt; one which he names *Aple*, or ſingle: another, *Dicoccos*, which brings forth two Cornes ioyned together in a couple of huskes, as before in the deſcription is mentioned. That Spelt which *Dioſcorides* calls *Dicoccos*, is the ſame which *Theoph.* and *Galen* do name *Zea*. The moſt antient Latines haue called *Zea* or *Spelta* by the name of *Far*, as *Dionyſius Halicarnaſſæus* doth ſufficiently teſtifie: The old Romans (ſaith he) did call ſacred mariages by the word *Farracia*, becauſe the

Zea, siue Spelta.
Spelt Corne.

the Bride and Bride-groome did eate of that *Far* which the Grecians call *Zea*. The same thing *Asclepiades* affirmeth in *Galen*, in his ninth book according to the places affected, writing thus ; *Farris, quod Zea appellant :* that is to say, *Far*, which is called *Zea*, &c. And this *Far* is also named of the Latines, *Ador, Adoreum* and *Semen adoreum.*

¶ *The Temper.*

Spelt, as *Dioscorides* reporteth, nourisheth more than Barley. *Galen* writeth in his books of the Faculties of simple Medicines, Spelt is in all his temperatute in a meane betweene wheat and barley, and may in vertue be referred to the kindes of Barly and Wheat, being indifferent to them both.

¶ *The Vertues.*

The floure or meale of Spelt corne boiled in wa- A ter with the pouder of Saunders, and a little Oile of Roses and Lillies, vnto the forme of a pultesse, and applied hot, takes away the swelling of the legs gotten by cold and long standing.

‡ Spelt (saith *Turner*) is common about Weisen- B burgh in high Almaine, eight Dutch miles on this side Strausburgh, and there all men vse it for wheat, for there groweth no wheat at all : yet I neuer saw fairer & pleasanter bread in any place in all my life, than I haue eaten there, made only of this Spelt. The corn is much lesse than Wheat, and somewhat shorter than Rie, but nothing so blacke. ‡

CHAP. 49.   *Of Starch-Corne.*

Triticum Amyleum.
Starch Corne.

¶ *The Description.*

THis other kinde of *Spelta* or *Zea* is called of the German Herbarists, *Amyleum frumentum*, or Starch-corne ; and it is a kinde of Graine sowne to that end, or a three-moneths graine, and is very like vnto wheat in stalke and seed ; but the eare thereof is set round about and made vp with two rankes, with certaine beards almost after the maner of Barly, and the seed is closed vp in chaffie husks, and is sowne in the Spring.

¶ *The Place.*

Amil corne or Starch corne is sowne in Germanie, Polonia, Denmarke, and other those Easterne regions, as well to feed their cattel and pullen with, as also to make starch; for the which purpose it very fitly serueth.

¶ *The Time.*

It is sowne in Autumne or the fall of the leafe, and oftentimes in the Spring ; and for that cause hath been called *Trimestre*, or three months grain : it bringeth his seed to ripenesse in the beginning of August, and is sown in the Low-Countries in the Spring of the yeare.

¶ *The Names.*

Because the Germanes haue great vse of it to make starch with, they do call it 𝕬𝖒𝖊𝖑𝖈𝖔𝖗𝖓𝖊: We thinke good to name it in Latine, *Amyleum frumentum* : in Engliſh it may be called Amelcorne, after the German word ; and may likewiſe be called Starch corne. *Tragus* and *Fuchſius* took it to be *Triticum trimeſtre*, or three moneths wheat : but it may rather be referred to the *Farra* ; for *Columella* ſpeaketh of a graine called *Far Halicaſtrum*, which is ſowne in the Spring, and for that cauſe it is named *Trimeſtre*, or three moneths Far. If any be deſirous to learne the making of ſtarch, let them read *Dodonæus* laſt edition, where they ſhall be fully taught ; my ſelfe not willing to ſpend time about ſo vain a thing, and not pertinent to the ſtory. It is vſed alſo to feed cattell and pullen, and is in nature ſomewhat like to Wheat or Barley.

Chap. 50.   *Of Barley*.

¶ *The Deſcription*.

BArley hath an helme or ſtraw which is ſhorter and more brittle than that of Wheat, and hath more joints : the leaues are broader and rougher ; the eare is armed with long rough & prickly beards or ailes, and ſet about with ſundry ranks, ſomtimes two, otherwhiles three, foure, or ſix at the moſt according to *Theophraſtus*, but eight according to *Tragus*. The grain is included in a long chaffie huske, the roots be ſlender, and grow thick together. Barley, as *Pliny* writeth, is of all grain the ſofteſt, and leaſt ſubiect to caſualtie, yeelding fruit very quickly and profitably.

| 1 *Hordeum diſtichon.* | 2 *Hordeum Polyſtichum vernum.* |
| Common Barley. | Beare Barley, or Barley Big. |

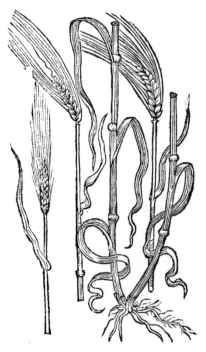

1   The moſt vſuall Barley is that which hath but two rowes of corne in the eare, each grain ſet iuſt oppoſit to other, and hauing his long awne at his end, is couered with an huske ſticking cloſe thereto.

2   This which commonly hath foure rowes of corne in the eare, and ſometimes more, as wee haue formerly deliuered, is not ſo vſually ſowne with vs : the eare is commonly ſhorter than the former, but the graine very like ; ſo that none who knowes the former but may eaſily know the later at the firſt ſight

¶ *The*

¶ *The Place.*

They are ſowne, as *Columella* teacheth, in looſe and dry ground, and are well knowne all Europe through.

2　　The ſecond is ſowne commonly in ſome parts of Yorke ſhire and the Biſhopricke of Durham.

¶ *The Names.*

1　　The firſt is called of the Grecians, κριϑή: in high Dutch, **Gerſten :** in Low-Dutch, **Gerſt :** in Italian, *Orzo :* in Spaniſh, *Ceuada :* in French, *Orge :* in Engliſh, Barley.

2　　The ſecond is called of the Grecians, πάλυριον, and alſo ἄραχος: *Columella* calleth it *Galaticum :* and *Hippocrates,* ἀχμαῖε κριϑή: of our Engliſh Northerne people, Big, and Barley Big. *Crimnon* (ſaith *Galen* in his Commentaries vpon the ſecond booke of *Hippocrates* his Prognoſtickes) is the groſſer part of Barley meale being groſſely ground. Malt is well knowne in England, inſomuch that the word needeth no interpretation : notwithſtanding becauſe theſe Workes may chance into ſtrangers hands that neuer heard of ſuch a word, or ſuch a thing, by reaſon it is not euery where made, I thought good to lay downe a word of the making thereof. Firſt, it is ſteeped in water vntil it ſwel, then is it taken from the water, and laid (as they terme it) in a Couch, that is, ſpred vpon an euen floore the thickneſſe of ſome foot and a halfe ; and thus it is kept vntill it Come, that is, til it ſend forth two or three little ſtrings or fangs at the end of each Corne. Then it is ſpred vſually twice a day, each day thinner than other, for ſome eight or ten dayes ſpace, vntill it be pretty dry, and then it is dried vp with the heate of the fire, and ſo vſed. It is called in high-Dutch, **Maltz :** in Low-Dutch, **Mout :** in Latine of later time, *Maltum ;* which name is borrowed of the Germans. *Aetius* a Greeke Phyſitian nameth Barley thus prepared, *Byne,* or *Bine :* and he alſo affirmeth, That a plaiſter of the meale of Malt is profitably laid vpon the ſwellings of the Dropſie. *Zythum,* as *Diodorus Siculus* affirmeth) is not only made in Egypt, but alſo in Galatia : The aire is ſo cold (ſaith he writing of Galatia) that the country bringeth forth neither wine nor oile, and therefore men are compelled to make a compound drinke of Barly, which they call *Zythum. Dioſcorides* nameth one kind of Barly drinke *Zythum :* another, *Curmi. Simeon Zethi* a later Grecian calls this kind of drink by an Arabicke name, φώνκκε : in Engliſh we call it Beere and Ale which is made of Barley Malt.

¶ *The Temperature.*

Barley, as *Galen* writeth in his booke of the Faculties of Nouriſhments, is not of the ſame temperature that wheat is, for wheat doth manifeſtly heate ; but contrariwiſe, what medicine or bread ſoeuer is made of Barly, is found to haue a certain force to coole and dry in the firſt degree, according to *Galen* in his booke of the Faculties of Simples. It hath alſo a little abſterſiue or clenſing qualitie, and drieth ſomewhat more than Bean meale.

¶ *The Vertues.*

Barly, ſaith *Dioſcorides,* doth clenſe, prouoke vrine, breedeth windineſſe, and is an enemie to the　A ſtomacke.

Barly meale boiled in an honied water with figs, taketh away inflammations : with pitch, roſin,　B and Pigeons dung, it ſoftneth and ripeneth hard ſwellings.

With Melilot and Poppy ſeeds it taketh away the pain in the ſides : it is a remedie againſt win-　C dineſſe in the guts, being applied with Lineſeed, Fœnugreek, and Rue : with tar, wax, oile, and the vrine of a yong boy, it doth digeſt, ſoften, and ripe hard ſwellings in the throat, called the Kings-Euill.

Boiled with wine, Myrtles, the barke of the Pomegranate, wilde peares, and the leaues of bram-　D bles, it ſtoppeth the laske.

Further, it ſerueth for *Ptiſana, Polenta, Maza,* Malt, Ale, and Beere : the making whereof if any be　E deſirous to learne, let them reade *Lobels Adverſaria,* in the chapter of Barly. But I think our London Beere Brewers are not to learne to make Beere of either French or Dutch, much leſſe of me that can ſay nothing therein of mine owne experience, more than by the writings of others. But I may deliuer vnto you a Confection made thereof, (as *Columella* did concerning ſweet Wine ſodden to the halfe) which is this ; Boile ſtrong Ale till it come to the thickneſſe of hony, or the forme of an vnguent or ſalue, which applied to the paines of the ſinues and joints (as hauing the propertie to abate Aches and pains) may for want of better remedies be vſed for old and new ſores, if made after this manner :

Take ſtrong Ale two pound, one Oxe gall, and boile them to one pound with a ſoft fire, conti-　F nually ſtirring it ; adding thereto of Vineger one pound, of *Olibanum* one ounce, floures of Camomil and melilot of each ℥ ſ. Rue in fine pouder ℥ſ. a little hony, and a ſmall quantity of the pouder of Comin ſeed ; boile them all together to the forme of an vnguent, and ſo apply it. There be ſundry ſorts of Confections made of Barley, as *Polenta, Ptiſana,* made of Water and husked or hulled Barley and ſuch like. *Polenta* is the meat made of parched Barley, which the Grecians doe properly

perly call *Alphiton*. Maza is made of parched Barley tempered with water, after *Hippocrates* and *Xe-nophon*. *Cyrus* hauing called his souldiers together, exhorted them to drinke water wherein parched Barley meale had been steeped, calling it by the same name *Maza*. *Hesychius* doth interpret *Maza* to be Barley meale mixed with water and oile.

Barley meale boiled in water, with garden Nightshade, the leaues of garden Poppy, the pouder of Fœnigreeke and Lineseed, and a little Hogs grease, is good against all hot & burning swellings, and preuaileth against the dropsie, being applied vpon the belly.

CHAP. 51.   *Of naked Barley.*

Hordeum nudum.
Naked Barley.

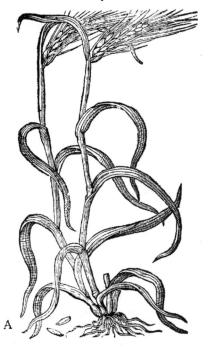

¶ *The Description.*

HOrdeum nudum is called *Zeopyrum*, and *Tritico Speltum*, because it is like to *Zea*, otherwise called *Spelta*, and is like to that which is called French Barley, whereof is made that noble drinke for sicke folks, called *Ptisana*. The plant is altogether like vnto Spelt, sauing that the eares are rounder, the eiles or beards rougher or longer, and the seed or graine naked without huskes, like to wheat, the which in its yellowish color it somwhat resembles.

¶ *The Place.*

‡ It is sown In sundry places of Germany, for the same vses as Barley is.

¶ *The Names.*

It is called *Hordeum nudum*, for that the corne is without huske, and resembleth Barly. In Greek it is called *Zeopyron*, because it participateth in similitude and nature with *Zea*, that is, *Spelt* and *Puros*, (that is) wheat. ‡

¶ *The Vertues.*

This Barley boiled in water cooleth vnnatural and hot burning choler. In vehement feuers you may adde thereto the seeds of white Poppy and Lettuce, not onely to coole, but also to prouoke sleepe.

B   Against the shortnesse of the breath, and paines of the brest, may be added to all the foresaid, figs, raisins of the Sun, liquorice, and Annise seed.

C   Being boiled in the whay of milke, with the leaues of Sorrel, Marigolds, and Scabious, it quencheth thirst, and cooleth the heate of the inflamed liuer, being drunke first in the morning, and last to bedward.

CHAP.

Hordeum Spurium.
Wall Barley.

Chap. 52.

Of Wall-Barley.

¶ *The Description.*

THis kinde of wilde Barley, is called of the Latines, *Hordeum Spurium* ; of *Pliny*, *Holcus* : in English, Wall Barley, Way Barley, or after old English writers, Way Bennet. It groweth vpon mud walls and ftony places by the wayes fide ; very wel refembling felf-fowed Barly, yet the blades are rather like graffe than Barly. ‡ This groweth fome foot or better in height, with graffie leaues; the eare is very like that of Rie, and the corne both in colour and fhape abfolutely refembles it ; fo that it cannot be fitlier named, than by calling it wilde Rie, or Rie-graffe. ‡

¶ *The Vertues.*

This baftard wilde Barley ftamped and ap- A plied vnto places wanting haire, caufeth it to grow and come forth ; whereupon in old time it was called *Riftida*.

Chap. 53.　Of S. Peters Corne.

1 *Briza Monococcos.* S. *Peters* Corne.　　　2 *Feftuca Italica.* Hauer-Graffe.

¶ *The*

¶ *The Description.*

1   BRifa is a Corne whofe leaues, ftalkes, and eares are leffe than Spelt ; the eare refembles our ordinary Barley, the Corne growing in two rowes, with awnes at the top, and huskes vpon it not eafily to be gotten off. In colour it much refembles Barley ; yet *Tragus* faith it is of a blackifh red colour.

2   This *Ægilops* in leaues and ftalkes refembles Wheat or Barley, and it growes fome two handfuls high, hauing a little eare or two at the top of the ftalke, wherein are inclofed two or three feeds a little fmaller than Barley, hauing each of them his awne at his end. Thefe feeds are wrapped in a crefted filme or skinne, out of which the awnes put themfelues forth.

Matthiolus faith, That he by his owne triall hath found this to be true, That as *Lolium*, which is our common Darnel, is certainely knowne to be a feed degenerate from wheat, being found for the moft part among wheat, or where wheat hath been: fo is *Feftuca* a feed or graine degenerating from Barley, and is found among Barley or where Barley hath been.

¶ *The Place.*

‡ 1   Briza is fowen in fome parts of Germany and France ; and my memorie deceiues me if I haue not oftentimes found many eares thereof amongft ordinary Barley, when as I liued in the further fide of Lincolnefhire, and they there call it Brant Barley.

2   This *Ægilops* growes commonly among their Barley in Italy and other hot countries. ‡

¶ *The Names.*

1   *Briza Monococcos*, after *Lobelius*, is called by *Tabernamontanus*, *Zea Monococcos* : in Englifh, Saint *Peters* Corne, or Brant Barley.

2   *Feftuca* of Narbone in France is called, ΑἰγἰΛω+ in Latine, *Ægilops Narbonenfis*, according to the Greeke : in Englifh, Hauer-graffe.

¶ *The Nature.*

They are of qualitie fomewhat fharpe, hauing facultie to digeft.

¶ *The Vertues.*

A   The juice of *Feftuca* mixed with Barley meale dried, and at time of need moiftned with Rofe water, applied plaifterwife, healeth the difeafe called *Ægilops*, or fiftula in the corner of the eye : it mollifieth and difperfeth hard lumps, and affwageth the fwellings of the joynts.

CHAP. 54. *Of Otes.*

¶ *The Description.*

1   AVena Vefca, common Otes, is called *Vefca, à Vefcendo*, becaufe it is vfed in many countries to make fundry forts of bread, as in Lancafhire, where it is their chiefeft bread corne for Iannocks, Hauer cakes, Tharffe cakes, and thofe which are called generally Oten cakes; and for the moft part they call the graine Hauer, whereof they do likewife make drinke for want of Barley.

2   *Auena Nuda* is like vnto the common Otes ; differing in that, that thefe naked Otes immediately as they be threfhed, without helpe of a Mill become Otemeale fit for our vfe. In confideration whereof in Northfolke and Southfolke they are called vnhulled or naked Otes. Some of thofe good houfe-wiues that delight not to haue any thing but from hand to mouth, according to our Englifh prouerbe, may (while their pot doth feeth) go to the barne, and rub forth with their hands fufficient for that prefent time, not willing to prouide for to morrow, according as the fcripture fpeaketh, but let the next day bring it forth.

¶ *The Nature.*

Otes are dry and fomewhat cold of temperature, as *Galen* faith.

¶ *The Vertues.*

A   Common Otes put into a linnen bag, with a little bay falt quilted handfomely for the fame purpofe, and made hot in a frying pan, and applied very hot, eafeth the paine in the fide called the ftitch, or collicke in the belly.

B   If Otes be boiled in water, and the hands and feet of fuch as haue the *Serpigo* or *Impetigo*, that is, certaine chaps, chinks, or rifts in the palmes of the hands or feet (a difeafe of great affinitie with the pocks) be holden ouer the fume or fmoke thereof in fome bowle or other veffell wherein the Otes are put, and the Patient couered with blankets to fweat, being firft annointed with that ointment or vnction vfually applied *contra Morbum Gallicum* : it doth perfectly cure the fame in fix times fo annointing fweating.

Otemeale

C Otemeale is good for to make a faire and wel coloured maid to looke like a cake of tallow, espe-cially if she take next her stomacke a good draught of strong vinegre after it.

D Otemeale vsed as a Cataplasme dries and moderately discusses, and that without biting; for it hath somewhat a coole temper, with some astriction, so that it is good against scourings.

1 *Auena Vesca.*
Common Otes.

2 *Auena Nuda.*
Naked Otes.

CHAP. 55. *Of Wilde Otes.*

¶ *The Description.*

1 BRomos sterilis, called likewise *Auena fatua*, which the Italians do call by a very apt name *Vena vana*, and *Auena Cassa*, (in English, Barren Otes or wilde Otes) hath like leaues and stalkes as our common Otes; but the heads are rougher, sharpe, many little sharpe huskes making each eare.

† 2 There is also another kinde of *Bromos* or wilde Otes, which *Dodonæus* calleth *Festuca alte-ra*, not differing from the former wilde Otes in stalkes and leaues, but the heads are thicker, and more compact, each particular eare (as I may tearme it) consisting of two rowes of seed handsome-ly compact and joyned together; being broader next the straw, and narrower as it comes to an end.

¶ *The Place and Time.*

‡ The first in Iuly and August may be found almost in euery hedge; the later is to be found in great plentie, in most Rie.

¶ *The Names.*

1 This is called in Greeke, βρόμος πῶ: in Latine, *Bromos sterilis* by *Lobel* : *Ægylops prima* by *Mat-thiolus* : in English, wilde-Otes, or Hedge-Otes.

2 *Lobel* calls this *Bromos sterilis altera*: *Dodonæus* termes it *Festuca altera*: in Brabant they call it, **Drauich** : in English, Drauke.

x *Bromos*

1 *Bromos ſterilis.*
Wilde Otes.

2 *Bromos altera.*
Drauke, or ſmall wilde Otes.

¶ *The Nature and Vertues.*

A   1   It hath a drying facultie (as *Dioſcorides* ſaith.) Boile it in water together with the roots vntill two parts of three be conſumed ; then ſtraine it out, and adde to the decoction a quantitie of
hony equall thereto : ſo boile it vntill it acquire the thickneſſe of thin hony. This medicine is good
againſt the *Ozæna* and filthy vlcers of the noſe, dipping a linnen cloath therein, and putting it vp
into the noſthrils ; ſome adde thereto Aloes finely poudred, and ſo vſe it.

B   Alſo boiled in Wine with dried Roſe leaues, it is good againſt a ſtinking breath. ‡

Cʜᴀᴘ. 56.   *Of Bearded Wilde Otes.*

¶ *The Deſcription.*

ÆGylops *Bromoides Belgarum* is a Plant indifferently partaking of the nature of *Ægilops* and
Bromos. It is in ſhew like to the naked Otes. The ſeed is ſharpe, hairy, and ſomewhat long,
and of a reddiſh colour, encloſed in yellowiſh chaffie huskes like as Otes, and may be
Engliſhed, Creſted or bearded Otes. I haue found it often among Barley and Rie in ſundry
grounds. This is likewiſe vnprofitable and hurtfull to Corne ; whereof is no mention made by the
Antients worthy the noting.

† Ægilops Bromoides.
Bearded wilde Otes.

CHAP. 57. Of Burnt Corne.

¶ The Deſcription.

1 HOrdeum vſtum, or Vſtilago Hordei, is that burnt or blaſted Barly which is altogether vnprofitable & good for nothing, an enemy vnto corne ; for that in ſtead of an eare with corne, there is nothing els but blacke duſt, which ſpoileth bread, or what-ſoeuer is made thereof.

2 Burnt Otes, or Vſtilago Avenæ or Avena-cea, is likewiſe an vnprofitable plant, degenera-ting from Otes, as the other from barly, rie and wheat. It were in vaine to make a long harueſt of ſuch euill corne, conſidering it is not poſſeſ-ſed with one good qualitie. And therefore thus much ſhall ſuffice for the deſcription.

3 Burnt Rie hath no one good property in phyſicke, appropriate either to Man, Birds, or Beaſt, and is an hurtfull maladie vnto all Corn where it groweth, hauing an eare in ſhape like to Corne, but in ſtead of graine it doth yeeld a blacke pouder or duſt, which cauſeth bread to looke black, and to haue an euill taſt : and that Corne where it is, is called ſmootie Corn, and the thing it ſelfe, Burnt Corn, or blaſted Corn.

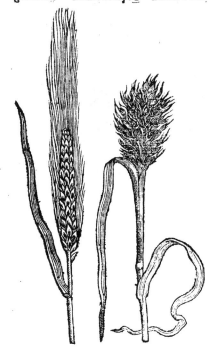

1 Hordenm vſtum, ſiue vſti-
ago hordei.　Burnt Barly.
2 Vſtilago Avenacea.
Burnt Otes.

3 Vſtilago Secalina.
Burnt Rie.

G　　　　　　　Chap.

Chap. 58.   *Of Darnell.*

1 *Lolium album.*
White Darnell.

2 *Lolium rubrum.*
Red Darnell.

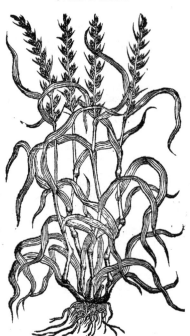 

¶ *The Deſcription.*

1   AMong the hurtfull weeds Darnell is the firſt. It bringeth forth leaues or ſtalkes like thoſe of wheat or barly, yet rougher, with a long eare made vp of many little ones, euerie particular one whereof containeth two or three grains leſſer than thoſe of Wheat, ſcarcely any chaffie huske to couer them with ; by reaſon whereof they are eaſily ſhaken out and ſcattered abroad.

2   Red Darnell is likewiſe an vnprofitable corne or graſſe, hauing leaues like barly. The joints of the ſtraw or ſtalke are ſometimes of a reddiſh colour, bearing at the top a ſmall and tender eare, flat, and much in forme reſembling the former.

¶ *The Place.*

They grow in fields among wheat and barley, of the corrupt and bad ſeed, as *Galen* ſaith, eſpecially in a moiſt and dankiſh ſoile.

¶ *The Time.*

They ſpring and flouriſh with the corne, and in Auguſt the ſeed is ripe.

¶ *The Names.*

1   Darnell is called in Greeke, ⲁⲓⲣⲁ : in the Arabian tongue, *Zizania*, and *Sceylen* : in French, *Iuray* : in Italian, *Loglio* : in Dutch, 𝕯𝖔𝖑𝖎𝖈𝖍 : in Engliſh, Darnell : of ſome, Iuray and Raye: and of ſome of the Latines, *Triticum temulentum*.

2   Red Darnell is called in Greeke, φοῖνιξ or *Phœnix*, becauſe of the crimſon colour: in Latine, *Lolium rubrum*, and *Lolium murinum* : of ſome, *Hordeum murinum*, and *Triticum murinum* : in Dutch 𝕸𝖚𝖞ſ𝖊 𝖈𝖔𝖟𝖊𝖓 : in Engliſh, Red Darnell, or great Darnell graſſe.

¶ *The Temperature.*

Darnell is hot in the third degree, and dry in the ſecond. Red Darnell drieth without ſharpnes, as *Galen* ſaith.

¶ *The*

¶ *The Vertues.*

The ſeed of Darnell, Pigeons dung, oile Oliue, and pouder of Line-ſeed, boiled to the forme of A
a plaiſter, conſume wens, hard lumps, and ſuch like excreſcenſes in any part of the body.

The new bread wherein Darnell is, eaten hot cauſeth drunkenneſſe; in like manner doth beere B
or ale wherein the ſeed is fallen, or put into the malt.

Darnel taken with red wine ſtayeth the flux of the belly, and the ouermuch flowing of womens C
termes.

Dioſcorides ſaith, That Darnel meale doth ſtay and keep backe eating ſores, gangrens, and putri- D
fied vlcers: and being boiled with Radiſh roots, ſalt, brimſtone, and vineger, it cureth ſpreading
ſcabs and dangerous tettars called in Greeke μίχητς, and leprous or naughty ſcurfe.

The ſeed of Darnell giuen in white or Rheniſh wine, prouoketh the fleures and menſes. E

A fume made thereof with parched barly meale, myrrh, ſaffron, and frankincenſe, made in forme F
of a pulteſſe and applied vpon the belly, helps conception, and cauſeth eaſie deliuerance of childe-
bearing.

Red Darnel (as *Dioſcorides* writeth) being drunke in ſowre or harſh Red wine, ſtoppeth the lask, G
and the ouermuch flowing of the fleures or menſes, and is a remedie for thoſe that piſſe in bed.

¶ *The Danger.*

Darnell hurteth the eyes and maketh them dim, if it happen in corne either for bread or drink:
which thing *Ouid. lib. 1. Faſtorum* hath mentioned in this verſe:

Et careant lolijs oculos vitiantibus agri.

And hereupon it ſeemeth that the old prouerb came, That ſuch as are dim ſighted ſhould be ſaid,
Lolio victitare.

CHAP. 59. *Of Rice.*

Oryza.
Rice.

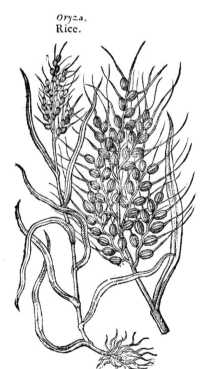

¶ *The Deſcription.*

RIce is like vnto Darnel in ſhew, as *Theophraſtus*
ſaith: it bringeth not forth an eare like corn,
but a certain mane or plume as Mill or Millet
or rather like Panick. The leaues, as *Pliny* writes, are
fat and full of ſubſtance like to the blades of leeks,
but broader; but (if neither the ſoile nor climat did
alter the ſame) the plants of Rice that did grow in
my garden had leaues ſoft and graſſie like barley.
the floure did not ſhew it ſelf with me, by reaſon of
the injurie of our vnſeaſonable yeare 1596. *Theophraſtus* concludeth, that it hath a floure of a purple
colour: but, ſaith my Author, Rice hath leaues like
vnto Dogs-graſſe or barly, a ſmall ſtraw or ſtem full
of ioynts like corn: at the top whereof groweth a
buſh or tuft far vnlike to barly or Darnel, garniſhed
with round knobs like ſmall gooſe-berries, wherein
the ſeed or graine is contained: euery ſuch round
knob hath one ſmall rough aile, taile, or beard like
vnto Barly hanging thereat. *Ariſtobulus*, as *Strabo* reporteth, ſheweth, that Rice growes in water in Ba-
ctria, and neere Babylon, and is two yards high, and
hath many eares, and bringeth forth plenty of ſeed.
It is reaped at the ſetting of the ſeuen ſtars, & purged as Spelt and Ote-meale, or hulled as French
Barley.

¶ *The Place.*

It groweth in the territorie of the Bactrians, in
Babylon, in Suſium, and in the lower part of Syria.
It groweth in theſe daies not only in thoſe countries before named, but alſo in the fortunat Iſlands
and in Spain, from whence it is brought to vs purged and prepared as wee ſee, after the manner of
French Barly. It proſpereth beſt in ſenny and wateriſh places.

G ¶ *The*

¶ *The Time.*

It is ſowne in the Spring in India,as *Eratoſthenes* witneſſeth, when it is moiſtned with Sommer ſhowers.

¶ *The Names.*

The Grecians call it ὄρυζα, or as *Theophraſtus* ſaith, ὄρυζη: the Latines keepe the Greeke word *Ory-za :* in French it is called *Riz :* in the German tongue, **Riſʒ,** and **Ryʒ :** in Engliſh,Rice.

¶ *The Temperature and Vertues.*

Galen ſaith,that all men vſe to ſtay the belly with this grain, being boiled after the ſame maner that *Chondrus* is. In England we vſe to make with milke and Rice a certain food or pottage,which doth both meanly binde the belly,and alſo nouriſh. Many other good kinds of food is made with this kind of grain,as thoſe that are skilfull in cookerie can tell.

Cʜᴀᴘ.60. *Of Millet.*

Milium.
Mill,or Millet.

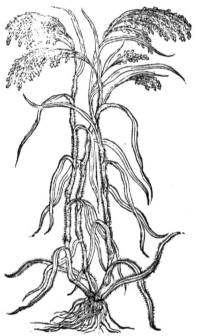

¶ *The Deſcription.*

Milium riſeth vp with many hairy ſtalks knot-ted or jointed like wheat. The leaues are long, and like the leaues of the common Reed. It bringeth forth on the top of the ſtalke a ſpoky buſh or mane called in Greek φόβη, like the plume or feather of the pole-reed,hanging down-ward,of colour for the moſt part yellow or white, in which groweth the feed,ſmall,hard, and gliſte-ring,couered with a few thin huskes,out of which it eaſily falleth. The roots be many, and grow deepe in the ground.

2   *Milium nigrum* is like vnto the former, ſa-uing that the eare or plume of this plant is more looſe and large,and the ſeed ſomwhat bigger,of a ſhining blacke colour.

¶ *The Place.*

It loueth a light and looſe mold, and proſpers beſt in a moiſt and rainy time. And after *Columella,* it groweth in great aboundance in Campania.

¶ *The Time.*

It is to be ſowne in Aprill and May, and not before,for it ioyeth in warme weather.

¶ *The Names.*

It is called of the Grecians,κέγχρος,of ſome,κέγχρις and of *Hippocrates, Paſpale,* as *Hermolaus* ſaith. In Spaniſh, *Mijo :* in Italian,*Miglio :* in high-Dutch, **Hirʒ :** in French, *Millet :* in low-Dutch, **Hirs :** in Engliſh,Mill,or Millet.

¶ *The Temperature.*

It is cold in the firſt degree, as *Galen* writeth, and dry in the third, or in the later end of the ſe-cond,and is of a thin ſubſtance.

¶ *The Vertues.*

A   The meale of Mill mixed with tar is layd to the bitings of Serpents and all venomous beaſts.

B   There is a drink made hereof bearing the name of *Syrupus Ambroſij,*or *Ambroſe* his Syrup,which procureth ſweat and quencheth thirſt, vſed in the city of Millan in tertian Agues. The receipt whereof *Henricus Rantzonius* in his booke of the Gouernment of health ſetteth down in this man-ner : Take (ſaith he) of vnhusked Mill a ſufficient quantitie,boile it vntill it be broken ; then take fiue ounces of the hot decoction,and adde thereto two ounces of the beſt White wine,and ſo giue it hot vnto the Patient being well couered with clothes,and then he will ſweat throughly. This is likewiſe commended by *Iohannes Heurneus,*in his booke of Practiſe.

C   Millet parched,and ſo put hot into a linnen bag,and applied,helps the griping pains of the bel-ly,or any other pain occaſioned by cold.

CHAP. 61. *Of Turkie Corne.*

1 *Frumentum Afiaticum.*
Corne of Afia.

2 *Frumentum Turcicum.*
Turky corne.

¶ *The Kindes.*

OF Turky Corns there be diuers forts, notwithftanding of one ftocke or kindred, confifting of fundry coloured Graines, wherein the difference is eafie to be difcerned; and for the better explanation of the fame, I haue fet forth to your view certain eares of different colours in their ful and perfect ripeneffe, and fuch as they fhew themfelues to be when their skin or filme doth open it felfe in the time of gathering.

The forme of the eares of Turky Wheat.

3 *Frumenti Indici fpica.*
Turky Wheat in the huske, as alfo naked or bare.

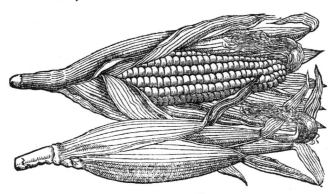

¶ *The*

¶ *The Description.*

1   COrne of Asia beareth a long great stem or stalke, couered with great leaues like the great Cane reed, but much broader, and of a darke brownish colour towards the bottome: at the top of the stalks grow idle or barren tufts like the common Reed, sometimes of one colour, and sometimes of another. Those eares which are fruitfull do grow vpon the sides of the stalks, among the leaues, which are thicke and great, so couered with skinnes or filmes, that a man cannot see them vntill ripenesse haue discouered them. The grain is of sundry colours, sometimes red, and sometimes white and yellow, as my selfe hath seen in myne own garden, where it hath come to ripenesse.

6 *Frumentum Indicum cæruleum.*
Blew Turky Wheat.

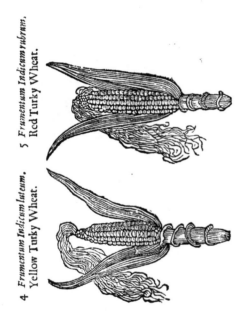

5 *Frumentum Indicum rubrum.*
Red Turky Wheat.

4 *Frumentum Indicum luteum.*
Yellow Turky Wheat.

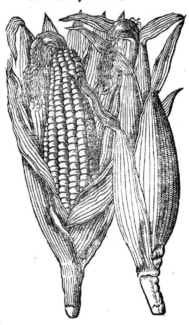

2   The stalk of Turky wheat is like that of the Reed, full of spongie pith, set with many joints fiue or six foot high, big beneath, and now and then of a purple colour, and by little and little small aboue: the leaues are broad, long, set with veins like those of the Reed. The eares on the top of the stalke be a spanne long, like vnto the feather-top of the common Reed, diuided into many plumes hanging downward, empty and barren without seed, yet blooming as Rie doth. The floure is either white, yellow, or purple, that is to say, euen as the fruit will be. The fruit is contained in veric big eares which grow out of the joints of the stalke, three or foure from one stalke, orderly placed one aboue another, couered with coats and filmes like husks & leaues, as if it were a certain sheath: out of which do stand long and slender beards, soft and tender, like those laces that grow vpon Sauorie, but greater and longer, euerie one fastned vpon his owne seed. The seeds are great, of the bignesse of common peason, cornered in that part whereby they are fastned to the eare, and in the outward part round: being of colour sometimes white, now and then yellow, purple, or red; of taste sweet and pleasant, very closely ioyned together in eight or ten orders or ranks. This graine hath many roots strong and full of strings.

¶ *The Place.*

These kinds of grain were first brought into Spaine, and then into other prouinces of Europe: not (as some suppose) out of Asia *minor*, which is the Turks dominions; but out of America and the Islands adioining, as out of Florida, and Virginia or Norembega, where they vse to sow or set it to make bread of it, where it growes much higher than in other countries. It is planted in the gardens of these Northern regions, where it commeth to ripenesse when the summer falleth out to be faire and hot; as my selfe haue seen by proof in myne owne garden.

¶ *The*

¶ *The Time.*

It is sowen in these countries in March and Aprill, and the fruit is ripe in September.

¶ *The Names.*

† Turky wheat is called of some *Frumentum Turcicum*, and *Milium Indicum*, as also *Ma zum*, and *Maiz*, or *Mays*. It in all probabilitie was vnknowne to the antient both Greeke and Latine Authors. In English it is called, Turky corne, and Turky wheat. The Inhabitants of America and the Islands adioyning, as also of the East and West Indies, do call it *Mais*: the Virginians, *Pagatowr*.

¶ *The Temperature and Vertues.*

Turky wheat doth nourish far lesse than either wheat, rie, barly, or otes. The bread which is made thereof is meanely white, without bran: it is hard and dry as Bisket is, and hath in it no clamminesse at all; for which cause it is of hard digestion, and yeeldeth to the body little or no nourishment; it slowly descendeth, and bindeth the belly, as that doth which is made of Mill or Panick. Wee haue as yet no certaine proofe or experience concerning the vertues of this kinde of Corne; although the barbarous Indians, which know no better, are constrained to make a vertue of necessitie, and thinke it a good food: whereas we may easily iudge, that it nourisheth but little, and is of hard and euill digestion, a more conuenient food for swine than for man.

Chap. 62.   *Of Turkie Millet*

Sorghum
Turky Millet.

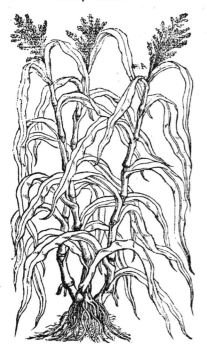

¶ *The Description.*

TVrky Millet is a stranger in England. It hath many high stalkes, thicke, and joynted commonly with some nine joynts, beset with many long and broad leaues like Turky wheat: at the top whereof groweth a great and large tuft or eare like the great Reed. The seed is round and sharpe pointed, of the bignesse of a Lentill, sometimes red, and now and then of a sullen blacke colour. It is fastned with a multitude of strong slender roots like vnto threds: the whole plant hath the forme of a Reed: the stalkes and eares when the seed is ripe are red.

¶ *The Place.*

It ioyeth in a fat and moist ground: it groweth in Italy, Spaine and other hot regions.

¶ *The Time.*

This is one of the Summer graines, and is ripe in Autumne.

¶ *The Names.*

The Millanois and other people of Lombardy call it *Melegua*, and *Meliga*: in Latine, *Melica*: in Hetruria, *Saggina*: in other places of Italy, *Sorgho*: in Portugal, *Milium Saburrum*: in English, Turky Mill, or Turky Hirsse.

‡ This seems to be the *Milium* which was brought into Italy out of India in the reigne of the Emperour *Nero*: the which is described by *Pliny, lib.*18.*cap.*7. ‡

¶ *The Temperature and Vertues.*

The seed of Turky Mill is like vnto Panicke in taste and temperature. The country People sometimes make bread hereof, but it is brittle, and of little nourishment, and for the most part it serueth to fatten hens and pigeons with.

CHAP. 63.   Of Panicke.

1 *Panicum Indicum.*
Indian Panick.

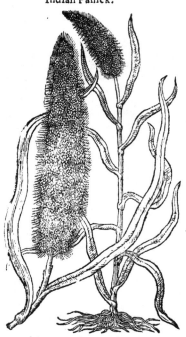

2 *Panicum cæruleum.* Blew Panick.

¶ *The Kindes.*

THere be ſundry ſorts of Panicke, although of the Antients there haue been ſet downe but two, that is to ſay, the wild or field Panick, and the garden or manured Panicke.

¶ *The Deſcription.*

1   THe Panick of India growes vp like Millet, whoſe ſtraw is knotty or ful of joints; the eares be round and hanging downeward; in which is contained a white or yellowiſh ſeed like Canarie ſeed, or *Alpiſti.*

2   Blew Panick hath a reddiſh ſtalk like to Sugar cane, as tall as a man, thicker than a finger, full of a fungous pith, of a pale colour: the ſtalks be vpright and knotty: thoſe that grow neere the root are of a purple colour. On the top or ſtalke commeth forth a ſpike or eare like the water Cats-taile, but of a blew or purple colour. The ſeed is like to naked Otes. The roots are very ſmal, in reſpeſt of the other parts of the plant.

‡ **3** *Panicum Americanum ſpica longiſſima.*
Weſt-Indian Panick with a very long eare.

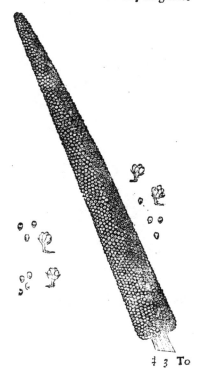

‡ 3   To

‡ ₃ To theſe may be added another Weſt-Indian Panick, ſent to *Cluſius* from Mr *Iames Garret* of London. The eare hereof was thicke, cloſe compact, and made taper-faſhion, ſmaller at the one end than at the other; the length thereof was more than a foot and a halfe. The ſhape of the ſeed is much like the laſt deſcribed, but that many of them together are contained in one hairy huske, which is faſtned to a very ſhort ſtalke, as you may ſee repreſented apart by the ſide of the figure. ‡

4 *Panicum vulgare*.
Common or Germane Panick.

5 *Panicum ſylveſtre*.
Wilde Panick.

4 Germane Panick hath many hairy roots growing thick together like vnto wheat, as is all the reſt of the plant, as well leaues or blades, as ſtraw or ſtalke. The eare groweth at the top ſingle, not vnlike to Indian Panick, but much leſſer. The graines are contained in chaffie skales, red declining to tawny.

5 The wild Panick groweth vp with long reeden ſtalks full of joints, ſet with long leaues like thoſe of *Sorghum* or Indian Panick: the tuft or feather-like top is like to the common reed, or eare of the graſſe called *Iſchæmon*, or Manna graſſe. The root is ſmall and thready.

The Place and Time.

The kindes of Panick are ſowen in the Spring, and are ripe in the beginning of Auguſt. They proſper beſt in hot and dry regions, and wither for the moſt part with much watering, as doth Mill and Turky wheat. They quickly come to ripeneſſe, and may be kept good a long time.

The Names.

Panick is called in Greeke ἔλυμος, and μελίνη: Diocles the Phyſitian nameth it *Mel Frugum*: the Spaniards, *Panizo*: the Latines, *Panicum*, of *Pannicula*: in Engliſh, Indian Panick, or Otemeale.

The Temperature.

Panicks nouriſh little, and are driers, as *Galen* ſaith.

The Vertues.

Panick ſtoppeth the lask, as Millet doth, being boiled (as *Pliny* reporteth) in Goats milke, and A. drunke twice in a day. Outwardly in pulteſſes or otherwiſe, it dries and cooles.

Bread made of Panicke nouriſheth little, and is cold and dry, very brittle, hauing in it neither B clammineſſe nor fatneſſe, and therefore it drieth a moiſt bell.

Chap.

CHAP: 64.   Of Canarie feed, or pety Panick.

1 *Phalaris.*
Canarie feed.

2 *Phalaris pratenfis.*
Quaking graffe.

¶ *The Defcription.*

1   CAnarie feed, or Canarie graffe after fome, hath many fmall hairy roots, from which arife fmall ftrawy ftalkes ioynted like corne,whereupon doe grow leaues like thofe of Barley,which the whole plant doth very well refemble. The fmall chaffie eare groweth at the top of the ftalk,wherein is contained fmall feeds like thofe of Panick,of a yellowifh colour and fhining.

2   Shakers or Quaking Graffe groweth to the height of halfe a foot, and fometimes higher, when it groweth in fertile medowes. The ftalke is very fmall and benty, fet with many graffie leaues like the common medow-graffe,bearing at the top a bufh or tuft of flat fcaly pouches, like thofe of Shepheards purfe,but thicker,of a browne colour,fet vpon the moft fmall and weak hairy foot-ftalks that may be found,whereupon thofe fmall pouches do hang; by means of which fmall hairy ftrings,the knaps which are the floures do continually tremble and fhake,in fuch fort that it is not poffible with the moft ftedfaft hand to hold it from fhaking.

‡ 3   There is alfo another graffie plant which may fitly be referred to thefe : the leaues and ftalks refemble the laft defcribed, but the heads are about the length and bredth of a fmall Hop, and handfomely compact of light fcaly filmes much like thereto; whence fome haue termed it, *Gramen Lupuli glumis.* The colour of this pretty head when it commeth to ripeneffe is white. ‡

¶ *The Place.*

1   Canarie feed groweth naturally in Spain, and alfo in the Fortunat or Canary Iflands, and alfo in England or any other of thefe cold regions,if it be fowne therein.

2 Quaking

3 *Phalaris pratensis altera.*
Pearle-grasse.

Alopecuros.
Fox-taile.

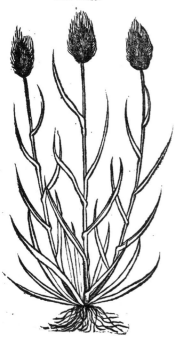

2 Quaking *Phalaris* groweth in fertile pastures, and in dry medowes.

3 This growes naturally in some parts of Spain, and is sowne yearely in many of our London gardens.

¶ *The Time.*

1 3 These Canarie seeds are sowne in May, and are ripe in August.

¶ *The Names.*

1 Canarie seed or Canarie Corne, is called of the Grecians, φαλαρις: the Latines retaining the same Name, *Phalaris :* in the Islands of Canary, *Alpisti :* in English, Canary seed, and Canarie Grasse.

2 *Phalaris pratensis* is called also *Gramen tremulum :* in Cheshire about Nantwich, Quakers, and Shakers : in some places, Cow-Quakes.

3 This by some is termed *Phalaris altera : Clusius* cals it *Gramen amourettes majus : Bauhine , Gramen tremulum maximum :* in English they call it Pearle grasse, and Garden Quakers.

¶ *The Nature and Vertues.*

I find nothing set downe as touching the temperature A of *Phalaris,* notwithstanding it is thought to be of the nature of Millet.

The juice and seed (as *Galen* saith) are thought to be B profitably drunke against the paines of the bladder. Apothecaries , for want of Millet, do vse the same with good successe in fomentations; for in dry fomentations it serueth in stead thereof, and is his *succedaneum,* or *quid pro quo.* We vse it in England also to feed Canary Birds.

CHAP. 65.

Of Fox-Taile.

¶ *The Description.*

1 FOx-taile hath many grassy leaues or blades rough and hairy like those of barly, but lesse and shorter. The stalk is likewise soft and hairy ; whereupon growes a small spike or eare, soft and very downy, bristled with very small haires in shape like vnto a Fox-taile, whereof it took his name, which dieth at the approch of winter, and recouereth it selfe the next yeare by falling of his seed.

‡ There is one or two varieties of this plant in the largenesse and smalnesse of the eare.

2 Besides these forementioned strangers, there is also another which grows naturally in many watry salt places of this kingdome, as in Kent by Dartford, in Essex, &c. The stalkes of this plant are grassy , and some two foot high, with leaues like Wheat or Dogs-grasse. The eare is very large, being commonly four or fiue inches long, downy, soft like silk, & of a brownish colour.

¶ *The*

¶ *The Place.*

1   This kinde of Fox-taile grasse groweth in England only in gardens.

¶ *The Time.*

1   This springeth vp in May, of the seed that was scattered the yere before, and beares his taile with his seed in Iune.

3   This beares his head in Iuly.

¶ *The Names.*

1   There hath not been more said of the antient or later writers, as touching the name, than is set downe, by which they call it in Greeke, *Alopecuros* : that is in Latine, *Cauda vulpis* : in English, Fox-taile.

2   This by *Lobel* is called *Alopecuros altera maxima Anglica paludosa* : That is, the large English Marish Fox-taile.

¶ *The Temperature and Vertues.*

I finde not any thing extant worthy the memorie, either of his nature or vertues.

Chap. 45.   Of Iobs Teares.

Lachrimæ Iobi.
Iobs Teares.

¶ *The Description.*

IObs Teares hath many knotty stalks, proceeding from a tuft of thready roots two foot high, set with broad leaues like to those of Reed ; amongst which leaues come forth many small branches like straw of corne : on the end wherof doth grow a gray shining seed or grain hard to break, and like in shape to the seeds of Gromel, but greater, and of the same colour, whereof I hold it a kinde : euery of which graines are bored through the middest like a bead, and out of the hole commeth a smal idle or barren chaffie eare like vnto that of Darnell.

¶ *The Place.*

It is brought from Italy and the countries adioyning, into these countries, where it doth grow very well, but seldom comes to ripenesse ; yet my self had ripe seed thereof in my garden, the Summer being very hot.

¶ *The Time.*

It is sowne early in the Spring, or else the Winter will ouertake it before it come to ripenesse.

¶ *The Names.*

Diuers haue thought it to be *Lithospermi species*, or a kinde of Gromell, which the seed doth very notably resemble, and doth not much differ from *Dioscorides* his Gromell. Some thinke it *Plinies Lithospermum*, and therefore it may very fitly be called in Latine, *Arundo Lithospermos*, that is in English, Gromell Reed, as *Gesner* saith. It is generally called *Lachrima Iob*, and *Lachrima Iobi* : of some it is called *Diospyros* : in English it is called Iobs Teares, or Iobs Drops, for that euery graine resembleth the drop or teare that falleth from the eye.

¶ *The Nature and Vertues.*

There is no mention made of this herbe for the vse of Physicke ; only in France and those places where it is plentifully growing, they make beads, bracelets, and chaines thereof, as we do with Pomander and such like.

CHAP.

Chap. 67. *Of Buck-wheat.*

Tragopyron.
Buckwheat, or Bucke.

¶ *The Deſcription.*

BVck-wheat may very well be placed among the kinds of graine or corne, for that oftentimes in time of neceſſitie bread is made thereof, mixed among other graine. It hath a round fat ſtalke ſomewhat creſted, ſmooth and reddiſh, which is diuided in many armes or branches, whereupon do grow ſmooth and ſoft leaues in ſhape like thoſe of Iuie or one of the Bindeweeds, not much vnlike Baſil, whereof *Tabernamontanus* called it *Ocymum Cereale:* The floures be ſmall, white, and cluſtred together in one or moe tufts or vmbels, ſlightly daſht ouer here & there with a flouriſh of light Carnation colour. The ſeeds are of a darke blackiſh colour, triangle, or three ſquare like the ſeed of blacke Bindeweed. The root is ſmall and threddy.

¶ *The Place.*

It proſpereth very wel in any ground, be it neuer ſo dry or barren, where it is commonly ſower to ſerue as it were in ſtead of a dunging. It quickly commeth vp, and is very ſoone ripe : it is very common in and about the Namptwich in Cheſhire, where they ſow it as well for food for their cattell, pullen, and ſuch like, as to the vſe aforeſaid. It groweth likewiſe in Lancaſhire, and in ſome parts of our South country, about London in Middleſex, as alſo in Kent and Eſſex.

¶ *The Time.*

This baſe kinde of graine is ſowen in Aprill and the beginning of May, and is ripe in the beginning of Auguſt.

¶ *The Names.*

Buck-wheat is called of the high Almaines, Heydencorn : of the baſe Almanes, Buckenweidt; that is to ſay, *Hirci triticum,* or Goats wheat : of ſome, *Fagi triticum,* Beech Wheat : In Greeke, ἐρύσιμον, by *Theophraſtus* ; and by late writers, πυρόεις : in Latine, *Fago triticum,* taken from the faſhion of the ſeed or fruit of the Beech tree. It is called alſo *Fegopyrum,* and *Tragopyrum* : In Engliſh, French wheat, Bullimong, and Buck-wheat : in French, *Dragee aux cheneaux.*

¶ *The Temperature.*

Buck-wheat nouriſheth leſſe than Wheat, Rie, Barley, or Otes; yet more than either Mill or Panicke.

¶ *The Vertues.*

Bread made of the meale of Buck-wheat is of eaſie digeſtion, and ſpeedily paſſeth through the belly, but yeeldeth little nouriſhment.

H

CHAP. 68. *Of Cow-Wheat.*

1 *Melampyrum album.*
White Cow-Wheat.

‡ 2 *Melampyrum purpureum.*
Purple Cow Wheat.

‡ 3 *Melampyrum cæruleum.*
Blew Cow-Wheat.

‡ 4 *Melampyrum luteam.*
Yellow Cow-Wheat.

¶ *The Deſcription.*

1 MElampyrum growes vpright with a ſtraight ſtalke, hauing other ſmall ſtalkes comming from the ſame, of a foot long. The leaues are long and narrow, and of a darke colour. On the tops of the branches grow buſhie or ſpikie ears full of floures and ſmall leaues mixed together, and much jagged, the whole eare reſembling a Foxe-taile. This
eare

eare beginning to floure below, and ſo vpward by little and little vnto the top : the ſmall leaues before the opening of the floures, and likewiſe the buds of the floures, are white of colour. Then come vp broad husks, wherein are encloſed two ſeeds ſomewhat like wheat, but ſmaller and browner. The root is of a woody ſubſtance.

‡ 2 3　Theſe two are like the former in ſtalkes and leaues, but different in the colour of their floures, the which in the one are purple, and in the other blew. *Cluſius* calls theſe, as alſo the *Crateogonon* treated of in the next chapter, by the name of *Parietariæ ſylueſtres*. ‡

4　Of this kinde there is another called *Melampyrum luteum*, which groweth neere vnto the ground, with leaues not much vnlike Harts horne, among which riſeth vp a ſmall ſtraw with an eare at the top like *Alopecuros*, the common Fox-taile, but of a yellow colour.

¶ *The Place.*

ι　The firſt groweth among corne, and in paſture grounds that be fruitfull : it groweth plentifully in the paſtures about London.

The reſt are ſtrangers in England.

¶ *The Time.*

They floure in Iune and Iuly.

¶ *The Names.*

Melampyrum is called of ſome *Triticum vaccinium* : in Engliſh, Cow-wheat, and Horſe-floure : in Greeke, μιλάμπυρον : The fourth is called *Melampyrum luteum* : in Engliſh, Yellow Cow-wheat.

¶ *The Danger.*

The ſeed of Cow-Wheat raiſeth vp fumes, and is hot and dry of nature, which being taken in meats and drinks in the manner of Darnell, troubleth the braine, cauſeth drunkenneſſe and head-ache.

CHAP. 69.　　Of Wilde Cow-Wheat.

1 *Crataogonon album.*
Wilde Cow-Wheat.

3 *Crataogonon Euphroſine.*
Eyebright Cow-Wheat.

¶ *The Deſcription.*

1 THe firſt kind of wilke Cow-Wheat *Cluſius* in his Pannonick hiſtory calls *Parietaria ſyl-veſtris*, or wilde Pellitory : which name, according to his owne words, if it do not fitly anſwer the Plant, he knoweth not what to call it, for that the Latines haue not giuen any name thereunto : yet becauſe ſome haue ſo called it, hee retaineth the ſame name. Notwith-ſtanding he referred it vnto the kindes of *Melampyrum*, or Cow-wheat, or vnto *Crataeogonon*, the wilde Cow-wheat, which it doth very well anſwer in diuers points. It hath an hairy foure ſquare ſtalke, very tender, weake and eaſie to breake, not able to ſtand vpright without the helpe of his neigh-bours that dwell about him, a foot high or more; whereupon do grow long thin leaues, ſharpe poin-ted, and oftentimes lightly ſnipt about the edges, of a darke purpliſh colour, ſometimes greeniſh, ſet by couples one oppoſite againſt the other ; among the which come forth two floures at one joynt, long and hollow ſomewhat gaping like the floures of a dead nettle, at the firſt of a pale yel-low, and after of a bright golden colour ; which do floure by degrees, firſt a few, and then more, by meanes whereof it is long in flouring. Which being paſt, there ſucceed ſmall cups or ſeed veſſels, wherein is contained browne ſeed not vnlike to wheat.

2 Red leafed wilde Cow-wheat is like vnto the former, ſauing that the leaues be narrower, and the tuft of leaues more jagged. The ſtalkes and leaues are of a reddiſh horſe-fleſh colour. The floures in forme are like the other, but in colour differing ; for that the hollow part of the floure with the heele or ſpurre is of a purple colour, the reſt of the floure yellow. The ſeed and veſſels are like the precedent.

3 This kinde of wilde Cow-wheat *Tabernamontanus* hath ſet forth vnder the title of *Odontites* : others haue taken it to be a kinde of *Euphraſia* or Eyebright, becauſe it doth in ſome ſort reſemble it, eſpecially in his floures. The ſtalkes of this plant are ſmall, woody, rough, and ſquare. The leaues are indented about the edges, ſharpe pointed, and in moſt points reſembling the former Cow-wheat ; ſo that of neceſſity it muſt be of the ſame kinde, and not a kinde of Eyebright as hath beene ſet downe by ſome.

¶ *The Place.*

Theſe wilde kindes of Cow-wheat doe grow commonly in fertile paſtures, and buſhie Copſes, or low woods, and among buſhes vpon barren heaths and ſuch like places.

The two firſt doe grow vpon Hampſted heath neere London, among the Iuniper buſhes and bil-berry buſhes in all parts of the ſaid heath, and in euery part of England where I haue trauelled.

¶ *The Time.*

They floure from the beginning of May, to the end of Auguſt.

¶ *The Names.*

1 The firſt is called of *Lobelius*, *Crataeogonon* : and of *Tabernamontanus*, *Milium Syluaticum*, or Wood Millet, and *Alſine ſyluatica*, or Wood-Chickweed.

2 The ſecond hath the ſame titles : in Engliſh, Wilde Cow-wheat.

3 The laſt is called by *Tabernamontanus*, *Odontites* : of *Dodonaeus*, *Euphraſia altera*, and *Euphro-ſine*. *Hippocrates* called the wilde Cow-wheat, *Polycarpum*, and *Polycritum*.

¶ *The Nature and Vertues.*

There is not much ſet downe either of the nature or vertues of theſe plants : onely it is reported that the ſeeds do cauſe giddineſſe and drunkenneſſe, as Darnell doth.

The ſeed of *Crataeogonon* made in fine pouder, and giuen in broth or otherwiſe, mightily prouo-keth Venerie.

Some write, that it will likewiſe cauſe women to bring forth male children.

† See the vertues attributed to *Crataeogonon* by *Dioſcorides* before, Chap. 38. B.

CHAP. 70. *Of White Aſphodill.*

¶ *The Kindes.*

HAuing finiſhed the kindes of corne, it followeth to ſhew vnto you the ſundry ſorts of Aſpho-dils, whereof ſome haue bulbous roots, other tuberous or knobby roots, ſome of yellow co-lour, and ſome of mixt colours : notwithſtanding *Dioſcorides* maketh mention but of one Aſpho-dill, but *Pliny* ſetteth downe two ; which *Dionyſius* confirmeth, ſaying, That there is the male and female Aſphodill. The later age hath obſerued many moe beſides the bulbed one, of which *Ga-len* maketh mention.

                                                    1 *Aſphodelus*

1 *Aſphodelus non ramoſus.*
White Aſphodill.

2 *Aſphodelus ramoſus.*
Branched Aſphodill.

¶ *The Deſcription.*

1   THe white Aſphodill hath many long and narrow leaues like thoſe of leeks, ſharpe
pointed. The ſtalke is round, ſmooth, naked, and without leaues, two cubits high, garniſhed from the middle vpward with a number of floures ſtarre-faſhion, made of fiue
leaues apiece ; the colour white, with ſome darke purple ſtreakes drawne downe the backe-ſide.
Within the floures be certaine ſmall chiues. The floures being paſt, there ſpring vp little round
heads, wherein are contained hard, blacke, and 3 ſquare ſeeds like thoſe of Buck-wheat or Staueſ-
acre. The root is compact of many knobby roots growing out of one head, like thoſe of the Peonie,
full of juice, with a ſmall bitterneſſe and binding taſte.

2   Branched Aſphodill agreeth well with the former deſcription, ſauing that this hath many
branches or armes growing out of the ſtalke, whereon the floures doe grow, and the other hath not
any branch at all, wherein conſiſteth the difference.

3   Aſphodill with the reddiſh floure groweth vp in roots, ſtalke, leaſe, and manner of growing
like the precedent, ſauing that the floures of this be of a dark red colour, and the other white, which
ſetteth forth the difference, if there be any ſuch difference, or any ſuch plant at all : for I haue conferred with many moſt excellent men in the knowledge of plants, but none of them can giue mee
certaine knowledge of any ſuch, but tell mee they haue heard it reported that ſuch a one there is,
and ſo haue I alſo ; but certainely I cannot ſet downe any thing of this plant vntill I heare more
certaintie ; for as yet I cannot credit my Authour, which for reuerence of his perſon I forbeare
to name.

4   The yellow Aſphodill hath many roots growing out of one head, made of ſundry tough,
fat, and oleous yellow ſprigs, or groſſe ſtrings, from the which riſe vp many graſſie leaues, thick and
groſſe, tending to ſquareneſſe ; among the which commeth vp a ſtrong thicke ſtalke ſet with the
like leaues euen to the floures, but leſſe : vpon the which do grow ſtarre-like yellow floures, otherwiſe like the white Aſphodill.

3 *Aſphodelus flore rubente.*
Red Aſphodill.

4 *Aſphodelus luteus.*
Yellow Aſphodill.

‡ 5 *Aſphodelus minimus.* Dwarfe Aſphodill.

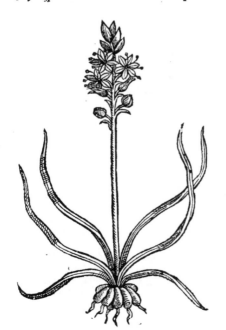

‡ 5 Beſides theſe, there is an Aſphodill which *Cluſius* for the ſmalneſſe cals *Aſphodelus minimus*. The roots thereof are knotty & tube-rous, reſembling thoſe of the laſt deſcribed, but leſſe: from theſe ariſe fiue or ſix very nar-row and long leaues ; in the middeſt of which grows vp a ſtalk of the height of a foot, round and without branches, bearing at the top ther-of a ſpoke of floures, conſiſting of ſix white leaues apiece, each of which hath a ſtreak run-ning alongſt it, both on the inſide and outſide like as the firſt deſcribed. It floures in the be-ginning of Iuly, when as the reſt are paſt their floures. It loſeth the leaues in winter, and gets new ones again in the beginning of Aprill. ‡

¶ *The Time and Place.*

They floure in May and Iune, beginning be-low and ſo flouring vpward : and they grow naturally in France, Italy, Spaine, and moſt of them in our London gardens.

¶ *The Names.*

Aſphodill is called in Latine, *Aſphodelus, Albucum, Albucus*, and *Haſtulus regia* : in Greek, ἀσφόδελος : in Engliſh, Aſphodil, not daffodil, for Daffodill is *Narciſſus*, another plant differing from Aſphodill. *Pliny* writeth, That the ſtalk with the floures is called *Anthericos* : and the roor, that is to ſay the bulbe, *Aſphodelus*.

Of

Of this Asphodill *Hesiod* maketh mention in his Works, where he saith, That fooles know not how much good there is in the Mallow and in the Asphodill ; because the roots of Asphodill are good to be eaten. Yet *Galen* doth not beleeue that he meant of this Asphodill, but of that bulbed one, whereof we will make mention hereafter. And he himselfe testifieth, That the bulbes thereof are not to be eaten without very long seething : and therefore it is not likely that *Hesiod* hath commended any such ; for he seemeth to vnderstand by the Mallow and the Asphodill, such kinde of food as is easily prepared, and soone made readie.

¶ *The Nature.*

These kindes of Asphodils be hot and dry almost in the third degree.

¶ *The Vertues.*

After the opinion of *Dioscorides* and *Aetius*, the roots of Asphodill eaten prouoke vrine and the **A** termes effectually, especially being stamped and strained with wine, and drunke.

One dram thereof taken in wine in manner before rehearsed, helpeth the pain in the sides, rup- **B** tures, convulsions, and the old cough.

The roots boiled in dregs of Wine cure foule eating vlcers, all inflammations of the dugges or **C** stones, and ease the fellon being put thereto as a pultesse.

The juice of the root boiled in old sweet Wine, together with a little myrrh and saffron, makes **D** an excellent Collyrie profitable for the eyes.

Galen saith, the roots burnt to ashes, and mixed with the grease of a Ducke, help the *Alopecia*, and **E** bring haire againe that was fallen by that disease.

The weight of a dram thereof taken with wine, helpeth the drawing together of sinues, cramps, **F** and burstings.

The like quantitie taken in broth prouoketh vomit, and helpeth those that are bitten with any **G** venomous beasts.

The iuice of the root cleanseth and taketh away the white morphew , if the face be annointed therewith ; but first the place must be chafed and well rubbed with a course linnen cloth.

CHAP. 71. *Of the Kings Speare.*

1 *Asphodelus luteus minor.* The Kings Speare.   2 *Asphodelus Lancastria.* Lancashire Asphodil. **H**

3 *Asphodelus*

1 3 *Aſphodelus Lancaſtriæ verus.*
The true Lancaſhire Aſphodill.

¶ *The Deſcription.*

1 THe leaues of the Kings Speare are long, narrow, and chamfered or furrowed, of a blewiſh greene colour. The ſtalke is round, of a cubit high. The floures which grow thereon from the middle to the top are very many, in ſhape like to the floures of the other ; which being paſt, there come in place therof little round heads or ſeed-veſſels, wherein the ſeed is contained. The roots in like manner are very many, long, and ſlender, ſmaller than thoſe of the other yellow ſort. Vpon the ſides whereof grow forth certaine ſtrings, by which the plant it ſelfe is eaſily encreaſed and multiplied.

2 There is found in theſe daies a certaine waterie or mariſh Aſphodill like vnto this laſt deſcribed, in ſtalkes and floures, without any difference at all. It bringeth forth leaues of a beautifull greene ſomwhat chamfered, like to thoſe of the floure de-luces, or corne-flag, but narrower, not full a ſpan long. The ſtalke is ſtraight, a foot high, whereupon grow the floures, conſiſting of ſix ſmall leaues : in the middle whereof come forth ſmall yellow chiues or threds. The ſeed is very ſmall, contained in long ſharpe pointed cods. The root is long, joynted, and creepeth as graſſe doth, with many ſmall ſtrings.

‡ 2 Beſides the laſt deſcribed (which our Author I feare miſtaking, termed *Aſphodelus Lancaſtriæ*) there is another water Aſphodill, which growes in many rotten mooriſh grounds in this kingdome, and in Lancaſhire is vſed by women to die their haire of a yellowiſh colour, and therefore by them it is tearmed Maiden-haire, (if we may beleeue *Lobel.*) This plant hath leaues of ſome two inches and an halfe, or three inches long, being ſomewhat broad at the bottome, and ſo ſharper towards their ends. The ſtalke ſeldome attaines to the height of a foot, and it is ſmooth without any leaues thereon ; the top thereof is adorned with pretty yellow ſtar-like floures, whereto ſucceed longiſh little cods, vſually three, yet ſometimes foure or fiue ſquare, and in theſe there is contained a ſmall red ſeed. The root conſiſts onely of a few ſmall ſtrings. ‡

¶ *The Place.*

1 The ſmall yellow Aſphodill groweth not of it ſelfe wilde in theſe parts, notwithſtanding we haue great plenty thereof in our London gardens.

2 The Lancaſhire Aſphodill groweth in moiſt and mariſh places neere vnto the Towne of Lancaſter, in the mooriſh grounds there, as alſo neere vnto Maudſley and Martom, two Villages not farre from thence ; where it was found by a Worſhipfull and learned Gentleman, a diligent ſearcher of ſimples, and feruent louer of plants, M*r.Thomas Hesket,* who brought the plants thereof vnto me for the encreaſe of my garden.

I receiued ſome plants thereof likewiſe from Maſter *Thomas Edwards,* Apothecary in Exceſter, learned and ſkilfull in his profeſſion, as alſo in the knowledge of plants. He found this Aſphodill at the foot of a hill in the Weſt part of England, called Bagſhot hill, neere vnto a Village of the ſame name.

‡ This Aſphodill figured and deſcribed out of *Dodonæus,* and called *Aſphodelus Lancaſtriæ* by our Author, growes in an heath ſome two miles from Bruges in Flanders, and diuers other places of the Low-countries ; but whether it grow in Lancaſhire or no, I can ſay nothing of certainetie : but I am certaine, that which I haue deſcribed in the third place growes in many places of the Weſt of England ; and this yeare 1632, my kinde friend M*r.George Bowles* ſent me ſome plant thereof, which I keepe yet growing. *Lobel* alſo affirmes this to be the Lancaſhire Aſphodill. ‡

¶ *The Time.*

They floure in May and Iune : moſt of the leaues thereof remaine greene in the Winter, if it be not extreme cold.

¶ *The Names.*

Some of the later Herbariſts thinke this yellow Aſphodill to be *Iphyon* of *Theophraſtus,* and others

others iudge it to be *Erizambac* of the Arabians. In Latine it is called *Afphodelus luteus*: of fome it is called *Haftula Regia*. We haue Englifhed it, the Speare for a King, or fmall yellow Afpho-dill.

2   The Lancafhire Afphodill is called in Latine, *Afphodelus Lancaftrie*: and may likewife be called *Afphodelus paluftris*, or *Pfeudoafphodelus luteus*, or Baftard yellow Afphodill.

‡ 3   This is *Afphodelus minimus luteus paluftris Scoticus & Lancaftrienfis*, of *Lobel*, and the *Pfeu-doafphodelus pumilio folijs Iridis*, of *Clufius*, as farre as I can iudge ; althouh *Bauhine* diftinguifheth them. ‡

¶ *The Temperature and Vertues.*

It is not yet found out what vfe there is of any of them in nourifhment or medicines.

Chap. 72.   *Of Onion Afphodill.*

Afphodelus Bulbofus.
Onion Afphodill.

¶ *The Defcription.*

THe bulbed Afphodil hath a round bulbous or Onion root, with fome fibres hanging there-at ; from the which come vp many graffie leaues, very well refembling the Leeke, among the which leaues there rifeth vp a naked or fmooth ftem, garnifhed toward the top with many ftar like floures, of a whitifh greene on the infide, and whol-ly greene without, confifting of fix little leaues fharpe pointed, with certaine chiues or threads in the middle. After the floure is paft there fuccee-deth a fmall knop or head three fquare, wherein li-eth the feed.

¶ *The Place.*

It groweth in the gardens of Herbarifts in Lon-don, and not elfewhere that I know of, for it is not very common.

¶ *The Time.*

It floureth in Iune and Iuly, and fomewhat after.

¶ *The Names.*

The ftalke and floures being like to thofe of the Afphodill before mentioned, do fhew it to be *A-fphodeli fpecies*, or a kinde of Afphodill ; for which caufe alfo it feemeth to be that Afphodil of which *Galen* hath made mention in the fecond book of the Faculties of nourifhments, in thefe words ; The root of Afphodill is in a manner like to the root of Squill, or Sea Onion, as well in fhape as bitternes. Notwithftanding, faith *Galen*, my felfe haue known certeine countrymen, who in time of famine could not with many boilings and fteepings make it fit to be eaten. It is called of *Dodonæus*, *Afphodelus fæmi-na*, and *Afphodelus Bulbofus*, *Hyacintho-Afphodelus*, and *Afphodelus Hyacinthinus* by *Lobel*, and that rightly ; for that the root is like the Hyacinth, and the floures like the Afphodill : and therefore as it doth participate of both kindes, fo likewife doth the name : in Englifh we may call it, Bulbed Afphodill. *Clufius* calls it *Ornithogalum majus*, and that fitly.

¶ *The Nature.*

The round rooted Afphodill, according to *Galen*, hath the fame temperature and vertue that *Aron*, *Arifarum*, and *Dracontium* haue, namely an abfterfiue and cleanfing quality.

¶ *The Vertues.*

The yong fprouts or fprings thereof are a fingular medicine againft the yellow Iaundife, for **A** that the root is of power to make thin and open.

Galen faith, that the afhes of this Bulbe mixed with oile or hens greafe cure the falling of the **B** haire in an *Alopecia* or fcalld bead.

CHAP.

Chap. 73. *Of Yellow Lillie.*

¶ *The Kindes.*

BEcauſe we ſhall haue occaſion hereafter to ſpeake of certaine Cloued or Bulbed Lillies, wee will in this chapter entreat onely of another kinde not bulbed, which likewiſe is of two ſorts, differing principally in their roots: for in floures they are Lillies, but in roots Aſphodils, partici-pating as it were of both, though neerer approching vnto Aſphodils than Lillies.

1 *Lilium non bulboſum.*   2 *Lilium non bulboſum Phœniceum,*
The yellow Lillie.   The Day-Lillie.

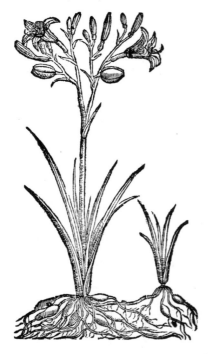 

¶ *The Deſcription.*

1   THe yellow Lillie hath very long flaggie leaues, chamfered or channelled, hollow in the middeſt like a gutter; among the which riſeth vp a naked or bare ſtalke, two cu-bits high, branched toward the top, with ſundry brittle armes or branches, whereon do grow many goodly floures like vnto thoſe of the common white Lillie in ſhape and proporti-on, of a ſhining yellow colour; which being paſt, there ſucceed three cornered huskes or cods, full of blacke ſhining ſeeds like thoſe of the Peonie. The root conſiſteth of many knobs or tuberous clogs, proceeding from one head, like thoſe of the white Aſphodill or Peonie.

2   The Day-Lillie hath ſtalkes and leaues like the former. The floures be like the white Lil-lie in ſhape, of an Orenge tawny colour: of which floures much might be ſaid which I omit. But in briefe, this plant bringeth forth in the morning his bud, which at noone is full blowne, or ſpred abroad, and the ſame day in the euening it ſhuts it ſelfe, and in a ſhort time after becomes as rot-ten and ſtinking as if it had beene trodden in a dunghill a moneth together, in foule and rainie weather: which is the cauſe that the ſeed ſeldome followes, as in the other of his kinde, not brin-ging forth any at all that I could euer obſerue; according to the old prouerbe, Soone ripe, ſoone rotten. His roots are like the former.

¶ *The*

¶ *The Place.*

Theſe Lillies do grow in my garden, as alſo in the gardens of Herbariſts, and louers of fine and rare plants ; but not wilde in England as in other countries.

¶ *The Time.*

Theſe Lillies do floure ſomewhat before the other Lillies, and the yellow Lillie the ſooneſt.

¶ *The Names.*

Diuers do call this kinde of Lillie, *Liliaſphodelus, Liliago*, and alſo *Liliaſtrum*, but moſt commonly *Lilium non bulboſum* : in Engliſh, Liriconfancie and yellow Lillie. The old Herbariſts name it, *Hemerocallis* : for they haue two kinds of *Hemerocallis* ; the one a ſhrub or wooddy plant, as witneſſeth *Theophraſtus*, in his ſixth booke of the hiſtory of Plants. *Pliny* ſetteth downe the ſame ſhrub among thoſe plants, the leaues whereof onely do ſerue for garlands.

The other *Hemerocallis* which they ſet downe, is a Floure which periſhes at night, and buddeth at the ſunne riſing, according to *Athenæus* ; and therefore it is fitly called ἡμεροκαλλὶς ; that is, Faire or beautifull for a day : and ſo we in Engliſh may rightly tearme it the Day-Lillie, or Lillie for a day.

¶ *The Nature.*

The nature is rather referred to the Aſphodills than to Lillies.

¶ *The Vertues.*

Dioſcorides ſaith, that the root ſtamped with honey, and a mother peſſarie made thereof with wooll, and put vp, bringeth forth water and bloud.   **A**

The leaues ſtamped and applied, allay hot ſwellings in the dugges, after womens trauell in childe-bearing, and likewiſe take away the inflammations of the eies.   **B**

The roots and the leaues be laid with good ſucceſſe vpon burnings and ſcaldings.   **C**

CHAP. 74.   *Of Bulbed Floure de-luce.*

‡ 1 *Iris Bulboſa latifolia.*
Broad leaued Bulbous Floure de-luce.

2 *Iris Bulboſa Anglica.*
Onion Floure de-luce.

¶ *The*

¶ *The Kindes.*

Like as we haue set downe sundry sorts of Floure de-luces, with flaggy leaues, and tuberous or knobby roots, varying very notably in sundry respects, which we haue distinguished in their proper Chapters : it resteth that in like manner we set forth vnto your view certaine bulbous or Onion-rooted Floure de-luces, which in this place doe offer themselues vnto consideration; whereof there be also sundry sorts, sorted into one chapter as followeth.

3 *Iris Bulbosa flore vario.*
Changeable Floure de-luce·

‡ 4 *Iris Bulbosa versicolor Polyclonos.*
Many branched changeable Floure de-luce.

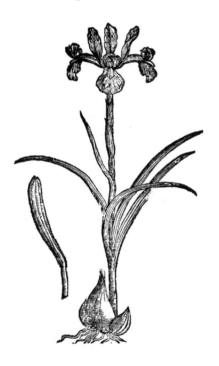

¶ *The Description.*

‡ 1   THe first of these, whose figure here we giue you vnder the name of *Iris bulbosa Latifolia,* hath leaues somewhat like those of the Day-Lillie, soft, and somewhat palish greene, with the vnder sides somewhat whiter ; amongst which there riseth vp a stalke bearing at the top thereof a floure a little in shape different from the formerly described Floure de-luces. The colour thereof is blew ; the number of the leaues whereof it consists, nine : three of these are little, and come out at the bottome of the Floure as soone as it is opened ; three more are large, and being narrow at their bottome, become broader by little and little, vntill they come to turne downeward, whereas then they are shapen somewhat roundish or obtuse. In the middest of these there runs vp a yellow variegated line to the place whereas they bend backe. The three other leaues are arched like as in other floures of this kinde, and diuided at their vpper end, and containe in them three threads of a whitish blew colour.

This is called *Iris bulbosa Latifolia,* by *Clusius* ; and *Hyacinthus Poetarum Latifolius,* by *Lobel.*

It floures in Ianuarie and Februarie, whereas it growes naturally, as it doth in diuers places of Portugall and Spaine. It is a tender plant and seldome thriues well in our gardens. ‡

2   Onion Floure de-luce hath long narrow blades or leaues, crested, chamfered, or streaked on the backe side as it were welted ; below somewhat round, opening it selfe toward the top, yet remaining as it were halfe round, whereby it resembleth an hollow trough or gutter. In the bottome of the hollownesse it tendeth to whitenesse ; and among these leaues doe rise vp a stalke of a cubit high ; at the top whereof groweth a faire blew Floure, not differing in shape from the com-

mon

mon Floure de-luce : the which being paſt,there come in the place thereof long thick cods or ſeed veſſels,wherein is contained yellowiſh ſeed of the bigneſſe of a tare or fitch. The root is round like an onion,couered ouer with certain browne skins or filmes. Of this kinde there are ſome fiue or ſix varieties, cauſed by the various colours of the floures.

5 Iris bulboſa Flore luteo cum flore & ſemine.
Yellow bulbed Floure de-luce in floure and ſeed.

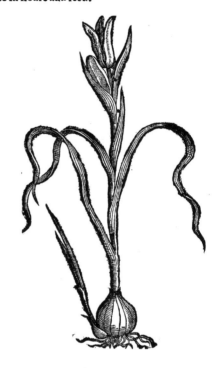

3   Changeable floure de-luce hath leaues,ſtalks,and roots like the former,but leſſer: the floure hath likewiſe the forme of the Floure de-luce,that is to ſay, it conſiſteth of ſix greater leaues, and three leſſer ; the greater leaues fold backward and hang downward ; the leſſer ſtand vpright, and in the middle of the leaues there riſeth vp a yellow welt white about the brimmes,and ſhadowed all ouer with a waſh of thin blew tending to a Watchet colour. Toward the ſtalke they are ſtripped ouer with a light purple colour,and likewiſe amongſt the hollow places of thoſe which ſtand vpright (which cannot be expreſſed in the figure) there is the ſame faire purple colour : the ſmel and ſauor is ſweet and pleaſant. The root is Onion-faſhion,or bulbous like the other.

‡ **4**   There is alſoanother variegated Floure de-luce, much like this laſt deſcribed in the colour of the Floure ; but each plant produceth more branches and floures, whence it is termed *Iris bulboſa verſicolor polyclonos*,Many-branched changeable Floure de-luce. ‡

5   Of which kinde or ſort there is another in my garden , which I receiued from my brother *Iames Garret* Apothecarie,far more beautifull than the laſt deſcribed ; the which is daſhed ouer, in ſtead of the blew or watchet colour,with a moſt pleaſant gold yellow colour, of ſmell exceeding ſweet,with bulbed roots like thoſe of the other ſort.

6   It is reported that there is in the garden of the Prince Electorthe Lantgraue of Heſſen,one of this kinde with white floures,the which as yet I haue not ſeene.

‡   Beſides theſe ſorts mentioned by our Author,there are of the narrow leafed bulbous Floure de-luces,ſome twenty foure or more varieties,which in ſhape of roots,leaues,and floures,differ verie little or almoſt nothing at all ; ſo that he which knowes one of theſe, may preſently know the reſt. Wherefore becauſe it is a thing no more pertinent to a generall hiſtorie of plants, to inſiſt vpon theſe accidentall nicities, than for him that writeth an hiſtorie of beaſts, to deſcribe all the colours,and their mixtures,in Horſes,Dogs,and the like ; I referre ſuch as are deſirous to informe

I   them-

themſelues of thoſe varieties, to ſuch as haue only and purpoſely treated of Floures and their di-
uerſities, as De-Bry, Swerts, and our countryman Mr Parkinſon, who in his Paradiſus terreſtris publi-
ſhed in Engliſh, Anno 1629. hath iudiciouſly and exactly comprehended all that hath bin deliuered by others in this nature.

‡ 6 Iris bulboſa flore cinereo.
Aſh-coloured Floure de-luce.

‡ 7 Iris bulboſa floro albido.
Whitiſh Floure de-luce.

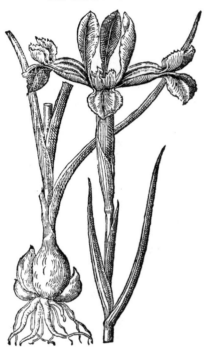

¶ The Place.
The ſecond of theſe bulbed Floure de-luces growes wilde of it ſelfe in the corne fields of the
Weſt part of England, as about Bathe and Wells & thoſe places adiacent, from whence they were
firſt brought into London, where they be naturaliſed, and encreaſe in great plenty in our London
gardens.
The other ſorts do grow naturally in Spaine and Italy wilde, from whence wee haue had plants
for our London gardens, whereof they do greatly abound.
¶ The Time.
They floure in Iune and Iuly, and ſeldome after.
¶ The Names.
The bulbed Floure de-luce is called of Lobel, Iris Bulboſa, and alſo Hyacinthus flore iridis: of ſome,
Hyacinthus Poetarum : and peraduenture it is the ſame that Apuleius mentioneth, Chap. 21. ſaying,
That Iris named among the old Writers Hieris, may alſo be called, and not vnproperly, Hierobulbus
or Hieribulbus ; as though you ſhould ſay, Iris bulboſa , or bulbed Ireos ; vnleſſe you would haue
ἱεροβολβὸς called a greater or larger Bulbe : for it is certain, that great and huge things were called of
the Antients, ἱερὰ, or Sacra · in Engliſh, Holy.
¶ The Nature.
The nature of theſe bulbed Floure de-luces are referred to the kindes of Aſphodils.
¶ The Vertues.
A   Take (ſaith Apuleius) of the herbe Hierobulbus ſix ℥. Goats ſuet as much , oile of Alcanna one
pound, mix them together, being firſt ſtamped in a ſtone mortar, it taketh away the paine of the
Gout.
B   Moreouer, if a woman do vſe to waſh her face with the decoction of the root, mixed with meale
of Lupines, it clenſeth away the freckles and morphew and ſuch like deformities.

Chap.

CHAP. 75. Of Spanish Nut.

1 *Sisynrichium majus.*
Spanish Nut.

‡ 2 *Sisynrichium minus.*
Small Spanish Nut.

3 *Iris tuberosa.*   Veluet Floure de-luce.

¶ *The Description.*

1 SPanish Nut hath smal graffy leaues like those of the Stars of Bethlem, or *Ornithogalum*; amongst which riseth vp a small stalke of halfe a foot high, garnished with the like leaues, but shorter. The floures grow at the top, of a sky colour, in shape resembling the Floure de-luce, or common *Iris*; but the leaues that turne downe are each of them marked with a yellowish spot: they faile quickely, and being past, there succeed small cods with seeds as small as those of Turneps. The root is round, composed of two bulbs, the one lying vpon the other as those of the Corn-flag vsually do; and they are couered with a skin or filme in shape like a Net. The Bulbe is sweet in tast, and may be eaten before any other bulbed root.

2 There is set forth another of this kinde somwhat lesser, with floures that smell sweeter than the former.

3 Velvet Floure de-luce hath many long square leaues spongeous or ful of pith, trailing vpon the ground, in shape like to the leaues of rushes: among which riseth vp a stalk of a foot high,

I 2   high,

high, bearing at the top a floure like a Floure de-luce. The lower leaues that turne downward are of a perfect blacke colour, soft and smooth as is blacke Veluet; the blackenesse is welted about with greenish yellow, or as we terme it, a Goose-turd green; of which colour the vppermost leaues do confist: which being past, there followeth a great knob or crested seed-vessell of the bignes of a mans thumbe, wherein is contained round white seed as big as the Vetch or tare. The root confists of many knobby bunches like fingers.

¶ The Place.

These bastard kindes of Floure de-luces are strangers in England, except it be among some few diligent Herbarists in London, who haue them in their gardens, where they encrease exceedingly, especially the last described, which is said to grow wild about Constantinople, Morea, & Greece: from whence it was transported into Italy, where it hath bin taken for *Hermodactylus*, and by some expressed and set forth in writing vnder the title of *Hermodactylus*, whereas in truth it hath no semblance at all with *Hermodactylus*.

¶ The Time.

The wilde or bastard Floure de-luces do floure from May to the end of Iuly.

¶ The Names.

1 2   These bulbed bastard Floure de-luces, which we haue Englished Spanish Nuts, are called in Spaine *Nozelhas*, that is, little Nuts; the lesser sort, *Parva Nozelha*, and *Macuca*: we take it to be that kinde of nourishing bulbe which is named in Greeke σισυριγχιον : of *Pliny, Sisynrichium*.

‡ 3   Some, as *Vlysses Aldroandus*, would haue this to be *Lonchitis prior* of *Dioscorides* : *Matthiolus* makes it *Hermodactylus verus*, or the true Hermodactill: *Dodonæus* and *Lobel* more fitly referre it to the Floure de-luces, and call it *Iris tuberosa*. ‡

¶ The Nature and Vertues.

Of these kinds of Floure de-luces there hath been little or nothing at all left in writing concerning their natures or vertues; only the Spanish nut is eaten at the tables of rich and delicious, nay vitious persons, in sallads or otherwise, to procure lust and lecherie.

C H A P. 76.   Of Corne-Flag.

1 *Gladiolus Narbonensis.* French Corn-Flag or Sword-Flag.	2 *Gladiolus Italicus.* Italian Corn-Flag or Sword-Flag.
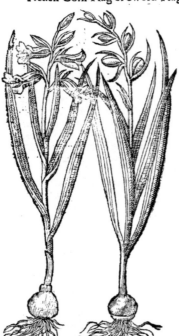	

‡ 4 Gladio-

‡ 4 *Gladiolus Lacuſtris.*
Water Sword-flag.

¶ *The Deſcription.*

1　FRench Corne-flagge hath ſmall ſtiffe
leaues ribbed or chamfered with long
nerues or ſinues running through the
ſame, in ſhape like thoſe of the ſmall Floure de-
luce, or the blade of a ſword, ſharpe pointed, of
an ouerworne green colour, among which riſeth
vp a ſtif brittle ſtalk two cubits high, wherupon
do grow in comly order many faire purple flours
gaping like thoſe of Snapdragon, or not much
differing from the Fox-gloue called in Latine
Digitalis. After them come round knobby ſeed-
veſſels full of chaffie ſeed, very light, of a brown
reddiſh colour. The root conſiſts of two bulbes
one ſet vpon the other; the vppermoſt whereof
in the beginning of the ſpring is leſſer, and more
full of juice; the lower greater, but more looſe
and lithie, which ſhortly after periſheth.

2　Italian Corn-flag hath long narrow leaues
with many ribs or nerues running through the
ſame: the ſtalk is ſtiffe and brittle, wherupon do
grow floures orderly placed vpon one ſide of the
ſtalk, whereas the precedent hath his floures placed on both ſides of the ſtalk, in ſhape & colour
like the former, as are alſo the roots, but ſeldom
ſeen one aboue another, as in the former.

3　There is a third ſort of Corne-flag, agreeing
with the laſt deſcribed in euery point, ſaue that
the floures of this are of a pale colour, as it were
betweene white and that which we call Maidens
bluſh.

‡ 4　This water Sword-flag, deſcribed by *Cluſius* in his *Cur. Poſt.* hath leaues about a ſpanne
long, thicke and hollow, with a partition in their middles, like as we ſee in the cods of ſtock Gillouers, and the like: their colour is green, and taſte ſweet, ſo that they are an acceptable food to the
wilde Ducks ducking to the bottom of the water; for they ſometimes lie ſome ells vnder water:
which notwithſtanding is ouer-topt by the ſtalke, which ſprings vp from among theſe leaues, and
beares floures of colour white, larger than thoſe of Stock-Gillouers: but in that hollow part which
is next the ſtalke they are of a blewiſh colour, almoſt in ſhape reſembling the floures of the Corn
flag, yet not abſolutely like them. They conſiſt of fiue leaues, whereof the two vppermoſt are refle-
cted toward the ſtalke: the three other being broader hang downward. After the floures there follow round pointed veſſels filled with red ſeed. It floures at the end of Iuly.

It was found in ſome places of Weſt-Friſeland, by *Iohn Dortman* a learned Apothecarie of Gro-
ning. It growes in waters which haue pure grauell at the bottom, and that bring forth no plant beſides.

Cluſius and *Dortman* who ſent it him, call it *Gladiolus Lacuſtris,* or *Stagnalis.*

¶ *The Place.*

Theſe kindes of Corne-flags grow in medowes and in earable grounds among corne, in many
places of Italy, as alſo in the parts of France bordering thereunto. Neither are the fields of Auſtria and Morauia without them, as *Cordus* writeth. We haue great plenty of them in our London
gardens, eſpecially for the garniſhing and decking them vp with their ſeemly floures.

¶ *The Time.*

They floure from May to the end of Iune.

¶ *The Names.*

Corne-Flag is called in Greeke, ξίφιον: in Latine, *Gladiolus,* and of ſome, *Enſis:* of others, φάσγανον:
and *Gladiolus ſegetalis. Theophraſtus* in his diſcourſe of *Phaſganum* maketh it the ſame with *Xiphion.*
Valerius Cordus calleth Corne-flag, *Victorialis fœmina:* others, *Victorialis rotunda:* in the Germane

I 3　　　　　　　　　　　　　　　　　Tongue,

Tongue,**Seigwurtz**; yet we muſt make a difference betweene *Gladiolus* and *Victorialis longa*, for that it is a kinde of Garlicke found vpon the higheſt Alpiſh mountaines, which is likewiſe called of the Germanes,**Seigwurtz**. The floures of the Corne-flag are called of the Italians, *Monacuccio* : in Engliſh, Corne-Flag, Corne-Sedge, Sword-Flag, Corne Gladin : in French, *Glais*.

¶ *The Temperature.*

The root of Corn-flag, as *Galen* ſaith, is of force to draw, waſte, conſume and dry, as alſo of a ſubtill and digeſting qualitie.

¶ *The Vertues.*

A   The root ſtamped with the pouder of Frankincenſe and wine, and applied, draweth forth ſplinters and thornes that ſticke faſt in the fleſh.

B   Being ſtamped with the meale of Darnell and honied water, it waſtes and makes ſubtill, hard lumps, nodes, and ſwellings, being emplaiſtred.

C   Some affirme, That the vpper root prouoketh bodily luſt, and the lower cauſeth barrenneſſe.

D   The vpper root drunke in water is profitable againſt that kinde of burſting in children called *Enterocele*.

E   The root of Corn-flag ſtamped with hogs greaſe and wheaten meale, hath been found by late Practitioners in Phyſicke and Surgerie, to be a certain and approued remedie againſt the *Scrophulæ*, and ſuch like ſwellings in the throat.

F   The cods with the ſeed dried and beaten into pouder, and drunk in Goats milke or Aſſes milke, preſently taketh away the paine of the Colique.

Chap. 77.   *Of Starry Hyacinths and their kindes.*

1 *Hyacinthus ſtellatus Fuchſij.*
Starry Iacinth.

‡ 2 *Hyacinthus ſtellaris albicans.*
The white floured ſtarry Iacinth.

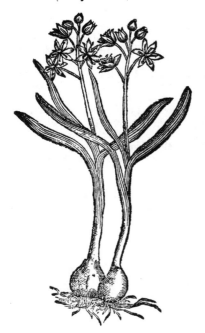
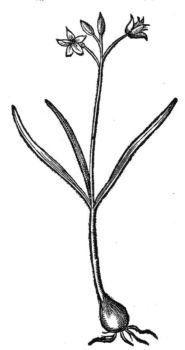

¶ *The Kindes.*

THere be likewiſe bulbous or Onion-rooted plants that do orderly ſucceed, whereof ſome are to be eaten, as Onions, Garlicke, Leekes, and Ciues ; notwithſtanding I am firſt to entreat

of

of those bulbed roots, whose faire and beautifull floures are receiued for their grace & ornament in gardens and garlands : the first is the Hyacinths, whereof there is found at this day diuers sorts, differing very notably in many points, as shall be declared in their seuerall descriptions.

† 3 *Hyacinthus stellatus bifolius.*
Two leaued starry Iacinth.

4 *Hyacinthus stellatus latifolius cum flore & semine.*
The Lilly leaued starry Iacinth in floure and seed.

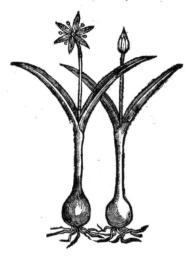

‡ 6 *Hyacinthus stellaris Byzantinus.*
The starry Iacinth of Constantinople.

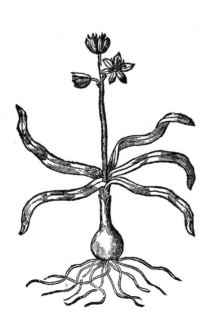

¶ *The Deſcription.*

1   THe firſt kinde of Iacinth hath three very fat thicke browne leaues, hollow like a little trough, very brittle, of the length of a finger : among which ſhoot vp fat thick browniſh ſtalks, ſoft and very tender, and ful of juice ; whereupon do grow many ſmal blew floures conſiſting of ſix little leaues ſpred abroad like a ſtar. The ſeed is contained in ſmal round bullets, which are ſo ponderous or heauy, that they lie trailing vpon the ground. The root is bulbous or onion-faſhion, couered with browniſh ſcales or filmes.

2   There is alſo a white floured one of this kinde.

3   There is found another of this kinde, which ſeldome or neuer hath more than two leaues. The roots are bulbed like the other : the floures be whitiſh, ſtar-faſhion, tending to blewneſſe ; the which I receiued of *Robinus* of Paris.

‡ 8 *Hyacinthus ſtellaris* Someri.   ‡ *9 Hyacinthus ſtellatus æſtiuus major.*
    Somers ſtarry Iacinth.   The greater ſtarry Summer Iacinth.

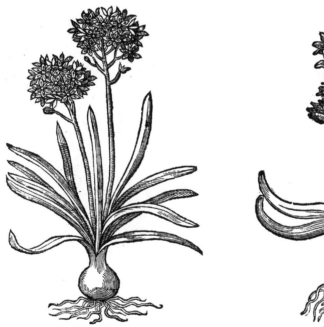 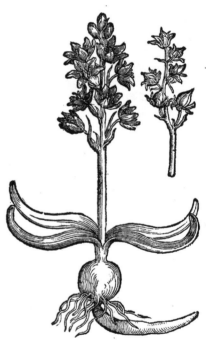

4   This kinde of Hyacinth hath many broad leaues ſpred vpon the ground like vnto thoſe of garden Lilly, but ſhorter. The ſtalks do riſe out of the midſt thereof, bare, naked, and very ſmooth, an handfull high ; at the top whereof do grow ſmall blew floures ſtarre-faſhion, very like vnto the precedent. The root is thicke and full of juice, compact of many ſcaly cloues of a yellow colour.

‡   There are ſome ten or eleuen varieties of ſtarry Iacinths beſides theſe two mentioned by our Author. They differ each from other either in the time of flouring, (ſome of them flouring in the Spring, other ſome in Summer) in their bigneſſe, or the colours of their floures. The leaues of moſt of them are much like to our ordinarie Iacinth or Hare-bells, and lie ſpred vpon the ground. Their floures in ſhape reſemble the laſt deſcribed, but are vſually more in number, and ſomewhat larger. The colour of moſt of them is blew or purple, one of them excepted, which is of an Aſh colour, and is knowne by the name of *Somers* his Iacinth. I thinke it not amiſſe to giue you their vſuall names, together with ſome of their figures ; for ſo you may eaſily impoſe them truly vpon the things themſelues whenſoeuer you ſhall ſee them.

5 *Hyacinthus ſtellaris Byzantinus nigra radice flore cæruleo.*

The blew ſtarry Iacinth of Conſtantinople with the blacke root.

6 *Hyacinthus ſtellatus Byzantinus major flore cærulo.*

The greater blew ſtarry Iacinth of Conſtantinople.

7 *Hyacinthus ſtellatus Byzantinus major flore boraginis.*

The other blew ſtarry Iacinth of Conſtantinople, with floures ſomwhat reſembling borage

8 *Hyacinthus ſtellaris æſtivus, ſive exoticus Someri flore cinereo.*

Aſh-coloured ſtarry Iacinth, or *Somers* Iacinth.

9 *Hyacinthus ſtellatus æſtivus maior.*

The greater ſtarry Summer Iacinth.

11 *Hyacinthus ſtellaris Poreti flore cæruleo ſtrijs purpureis.*

Porets ſtarry Iacinth with blew floures, hauing purple ſtreaks alongſt their middles.

12 *Hyacinthus Hiſpanicus ſtellaris flore ſaturè cæruleo.*

The Spaniſh ſtarry Iacinth with deepe blew floures.

13   There is another ſtarry Iacinth more large and beautifull than any of theſe before mentioned. The leaues are broad, and not verie long, ſpred vpon the ground, and in the midſt of them there riſeth vp a great ſpoke of faire ſtarry floures, which firſt begin to open themſelues below, and ſo ſhew themſelues by little and little to the top of the ſtalke. The vſuall ſort hereof hath blew or purple floures. There is alſo a ſort hereof which hath fleſh-coloured floures, and another with white floures : This is called *Hyacinthus ſtellatus Peruanus,* The ſtarry Iacinth of Peru.

10 *Hyacinthus ſtellatus æſtivus minor.*
The leſſer ſtarry Summer Iacinth.

13 *Hyacinthus Peruanus.*
Hyacinth of Peru.

Thoſe who are ſtudious in varieties of floures, and require larger deſcriptions of theſe, may haue recourſe to the Works of the learned *Carolus Cluſius* in Latine, or to Mr *Parkinſons* Worke in Engliſh, where they may haue full ſatisfaction. ‡

¶ *The Place.*

The three firſt mentioned plants grow in many places of Germany in Woods and mountaines, as *Fuchſius* and *Geſner* do teſtifie. In Bohemia alſo vpon diuers bankes that are full of Herbes. In
England

England we cheriſh moſt of theſe mentioned in this place, in our gardens, onely for the beautie of their floures.

¶ _The Time._

The three firſt begin to floure in the midſt of Ianuarie, and bring forth their ſeed in May. The other floures in the Spring.

¶ _The Names._

The firſt of theſe Hyacinths is called _Hyacinthus Stellatus_, or _ſtellaris Fuchſij_, of the ſtarre-like floures: _Narciſſus cæruleus, Bockij_ : of ſome, _Flos Martius ſtellatus_.

3   This by _Lobel_ is thought to be _Hyacinthus bifolius_ of _Theophraſtus_ : _Tragus_ calls it _Narciſſus cæruleus_ : and _Fuchſius, Hyacinthus cæruleus minor mas_. We may call it in Engliſh, the ſmall two lea- fed ſtarry Iacinth.

4   The Lilly Hyacinth is called _Hyacinthus Germanicus Liliflorus_, or German Hyacinth, taken from the countrey where it naturally groweth wilde.

¶ _The Vertues._

A   ‡ The faculties of the ſtary Iacinths are not written of by any : but the Lilly-leaued Hyacinth (which growes naturally in a hil in Aquitain called _Hos_, where the Herdmen call it _Sarahug_) is ſaid by them to cauſe the heads of ſuch cattell as feed thereon to ſwell exceedingly, and then killeth them ; which ſhewes it hath a maligne and poyſonous qualitie. _Cluſ._ ‡

Chap. 78.   _Of Autumne Hyacinths._

1 _Hyacinthus Autumnalis minor._
Small Autumne Iacinth.

2 _Hyacinthus Autumnalis maior._
Great Autumne Iacinth.

¶ _The Deſcription._

1   AVtumne Iacinth is the leaſt of all the Iacinths : it hath ſmal narrow graſſy leaues ſpred abroad vpon the ground : in the midſt whereof ſpringeth vp a ſmall naked ſtalke an handfull high, ſet from the middle to the top with many ſmall ſtar-like blew floures, hauing certain ſmall looſe chiues in the middle. The ſeed is black, contained in ſmall husks : the root is bulbous.

2   The great winter Iacinth is like vnto the precedent in leaues, ſtalkes, and floures, not differing in any point but in greatneſſe.

‡ 3   To theſe I thinke it not amiſſe to adde another ſmal Hyacinth, more differing from theſe laſt deſcribed in the time of the flouring, than in the ſhape. The root of it is little, ſmall, white, longiſh, with a few fibres at the bottom ; the leaues are ſmall and long like the laſt deſcribed : the ſtalke which is ſcarce an handfull high, is adorned at the top with thtee or foure ſtarry floures of a blewiſh Aſh colour, each floure conſiſting of ſix little leaues with ſix chiues, and their pointals of a darke blew, and a peſtill in the midſt. It floures in Aprill. ‡

¶ *The Place.*

† The greater Autumne Hyacinth groweth not wild in England, but it is to be found in ſome gardens.

The firſt or leſſer growes wilde in diuers places of England, as vpon a bank by the Thames ſide between Chelſey and London.

¶ *The Time.*

They floure in the end of Auguſt, and in September, and ſometimes later.

¶ *The Names.*

1   The firſt is called *Hyacinthus Autumnalis minor*, or the leſſer Autumne Iacinth, or Winter Iacinth.

2   The ſecond, *Hyacinthus Autumnalis major*, the great Autumne Iacinth, or Winter Iacinth.

3   This is called by *Lobel*, *Hyacinthus parvulus ſtellaris vernus*, The ſmall ſtarry Spring Iacinth.

CHAP. 79.   *Of the Engliſh Iacinth, or Hare-bells.*

1 *Hyacinthus Anglicus.*   2 *Hyacinthus albus Anglicus.*
Engliſh Hare-bells.   White Engliſh Hare-bells.

¶ *The Deſcription.*

1   THe blew Hare-bells or Engliſh Iacinth is very common throughout all England. It hath long narrow leaues leaning towards the ground, among the which ſpring vp naked

꓿

or bare ftalks loden with many hollow blew floures,of a ftrong fweet fmell fomewhat ftuffing the head : after which come the cods or round knobs,containing a great quantitie of fmall blacke fhining feed. The root is bulbous,full of a flimie glewifh juice,which will ferue to fet feathers vpon arrowes in ftead of glew,or to pafte bookes with : hereof is made the beft ftarch next vnto that of Wake-robin roots.

4 *Hyacinthus Orientalis cæruleus.*   5 *Hyacinthus Orientalis polyanthos.*
The blew Orientall Iacinth.   Many-floured Oriental Iacinth

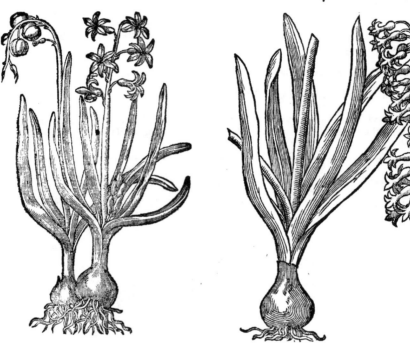

2   The white Englifh Iacinth is altogether like vnto the precedent, fauing that the leaues of this are fomewhat broader,the floures more open.and very white of colour.

3   There is found wilde in many places of England another fort,hauing floures of a faire Carnation colour,which maketh a difference from the other.

‡   There are alfo fundry other varieties of this fort , but I thinke it vnneceffarie to infift vpon them,their difference is fo little.confifting not in their fhape,but in the colour of their floures. ‡

The blew Hare-bels grow wilde in woods,Copfes,and in the borders of fields euery where thorow England.

The other two are not fo common,yet do they grow in the woods by Colchefter in Effex,in the fields & woods by South-fleet neere vnto Grauefend in Kent,as alfo in a piece of ground by Canturbury called the Clapper,in the fields by Bathe,about the woods by Warrington in Lancafhire and other places.

¶ *The Time.*

They floure from the beginning of May vnto the end of Iune.

¶ *The Names.*

1   The firft of our Englifh Hyacinths is called *Hyacinthus Anglicus,* for that it is thought to grow more plentifully in England than elfewhere : of *Dodonæus, Hyacinthus non fcriptus,* or the vnwritten Hyacinth.

2   The fecond,*Hyacinthus Belgicus candidus,* or Low-Country Hyacinth with white floures.

‡ 3   This third is called *Hyacinthus Anglicus aut Belgicus, flore incarnato,* Carnation Harebells.

4 ‡ *Hyacinthus*

‡ 6 *Hyacinthus Orientalis polyanthos alter.*
The other many-floured Orientall Iacinth.

‡ 7 *Hyacinthus Orientalis purpurorubens.*
Reddiſh purple Orientall Iacinth.

‡ 8 *Hyacinthus Orientalis albus.*
White Orientall Iacinth.

‡ 9 *Hyacinthus Brumalis.*
Winter Iacinth.

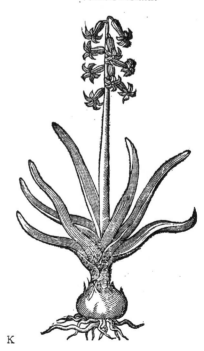

K

4   The Orientall Iacinth hath great leaues,thicke,fat,and full of iuice,deeply hollowed in the middle like a trough. From the middle of thoſe leaues riſeth vp a ſtalke two hands high,bare without leaues,very ſmooth, ſoft, and full of iuice, loden toward the top with many faire blew floures hollow like a bell,greater than the Engliſh Hyacinth,but otherwiſe like them. The root is great, bulbous or Onion-faſhion,couered with many ſcaly reddiſh filmes or pillings, ſuch as doe couer Onions.

5   The Iacinth with many floures (for ſo doth the word *Polyanthos* import) hath verie many large and broad leaues ſhort and very thicke,fat,or full of ſlimie iuice : from the middle whereof riſe vp ſtrong thicke groſſe ſtalks, bare and naked, ſet from the middle to the top with many blew or sky coloured floures growing for the moſt part vpon one ſide of the ſtalke. The root is great, thicke,and full of ſlimie iuice.

‡ 10 *Hyacinthus Orientalis caule folioſo.*
Orientall Iacinth with leaues on the ſtalke.

‡ 11 *Hyacinthus Orientalis flore pleno.*
The double floured Oriental Iacinth.

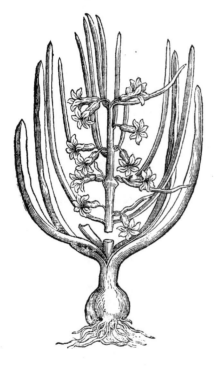 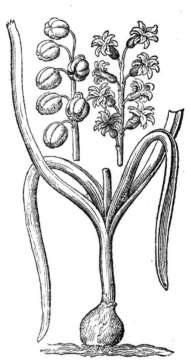

‡ 6   There is another like the former in each reſpect, ſauing that the flours are wholly white on the inſide,and white alſo on the outſide,but three of the out leaues are of a pale whitiſh yellow. Theſe floures ſmell ſweet as the former,and the heads wherin the ſeeds are contained,are of a lighter green colour. ‡

7   There is come vnto vs from beyond the ſeas diuers other ſorts, whoſe figures are not extant with vs ; of which there is one like vnto the firſt of theſe Oriental Iacinths, ſauing that the floures thereof are purple coloured,whence it is termed *Hyacinthus purpuro rubeus.*

8   Likewiſe there is another called *Orientalis albus,* differing alſo from the others in colour of the floures,for that theſe are very white,and the others blew.

9   There is another called *Hyacinthus Brumalis,*or winter Iacinth : it is like the others in ſhape, but differeth in the time of flouring.

‡ 10   There is another Hyacinth belonging rather to this place than any other , for that in root, leaues,floures,and ſeeds,it reſembles the firſt deſcribed Oriental Iacinth ; but in one reſpect it differs not onely from them , but alſo from all other Iacinths ; which is , it hath a leaſie ſtalke, hauing ſometimes one,and otherwhiles two narrow long leaues comming forth at the bottom of

the

‡ 14 *Hyacinthus obsoleto flore Hispanicus major.*
The greater dusky floured Spanish Iacinth.

‡ 15 *Hyacinthus minor Hispanicus.*
The lesser Spanish Iacinth.

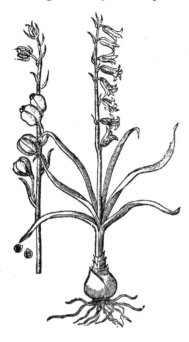

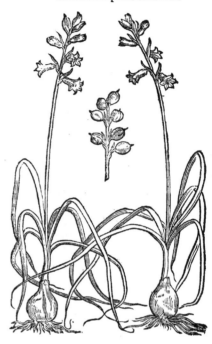

‡ 16 *Hyacinthus Indicus tuberosus.*
The tuberous rooted Indian Iacinth.

the setting of the floure. Whereupon *Clusius* calls it *Hyacinthus Orientalis caule folioso*: That is, the Oriental Hyacinth with leaues on the stalke.

¶ *Of double floured Oriental Hyacinths.*

Of this kindred there are two or three more varieties, whereof I will giue you the description of the most notable, and the names of the other two; which, with that I shall deliuer of this, may serue for sufficient description. The first of these (which *Clusius* calls *Hyacinthus Orientalis subvirescente flore*, or, the greenish floured double Orientall Iacinth) hath leaues, roots, and seeds like vnto the formerly described Orientall Iacinths; but the floures (wherin the difference consists) are at the first, before they be open, greene, and then on the out side next to the stalke of a whitish blew: and they consist of six leaues whose tips are whitish, yet retaining some manifest greenes: then out of the midst of the floures comes forth another floure consisting of three leaues, whitish on their inner side, yet keeping the great veine or streake vpon the outer side, each floure hauing in the middle a few chiues with blackish pendants. It floures in Aprill.

12　This varietie of the laſt deſcribed is called *Hyacinthus Orientalis flore cæruleo pleno*, The double blew Orientall Iacinth.

13　This, *Hyacinthus Orientalis candidiſſimus flore pleno*, The milke-white double Orientall Iacinth.

14　This, which *Cluſius* calls *Hyacinthus obſoletior Hiſpanicus*, hath leaues ſomewhat narrower, and more flexible than the *Muſcari*, with a white veine running alongſt the inſide of them : among theſe leaues there riſeth vp a ſtalke of ſome foot high, bearing ſome fifteene or ſixteene floures, more or leſſe, in ſhape much like the ordinary Engliſh, conſiſting of ſix leaues, three ſtanding much out, and the other three little or nothing. Theſe floures are of a very dusky colour, as it were mixt with purple, yellow, and greene : they haue no ſmell. The ſeed, which is contained in triangular heads, is ſmooth, blacke, ſcaly, and round. It floures in Iune.

15　The leſſer Spaniſh Hyacinth hath leaues like the Grape-floure, and ſmall floures ſhaped like the Orientall Iacinth, ſome are of colour blew, and other ſome white. The ſeeds are contained in three cornered ſeed-veſſels. I haue giuen the figure of the white and blew, together with their ſeed-veſſels.

16　This Indian Iacinth with the tuberous root (ſaith *Cluſius*) hath many long narrow ſharpe pointed leaues ſpread vpon the ground, being ſomewhat like to thoſe of Garlicke, and in the middeſt of theſe riſe vp many round firme ſtalkes of ſome two cubits high, and oft times higher, ſometimes exceeding the thickneſſe of ones little finger ; which is the reaſon that oftentimes, vnleſſe they be borne vp by ſomething, they lie along vpon the ground. Theſe ſtalkes are at certaine ſpaces ingirt with leaues which end in ſharpe points. The tops of theſe ſtalkes are adorned with many white floures, ſomewhat in ſhape reſembling thoſe of the Orientall Iacinth. The roots are knotty or tuberous, with diuers fibres comming out of them. ‡

¶ *The Place.*

Theſe kindes of Iacinths haue beene brought from beyond the Seas, ſome out of one countrey, and ſome out of others, eſpecially from the Eaſt countries, whereof they tooke the name *Orientalis*.

¶ *The Time.*

They floure from the end of Ianuarie vnto the end of Aprill.

¶ *The Nature.*

The Hyacinths mentioned in this Chapter do lighrly cleanſe and binde ; the ſeeds are dry in the third degree ; but the roots are dry in the firſt, and cold in the ſecond.

¶ *The Vertues.*

A　The Root of Hyacinth boyled in Wine and drunke, ſtoppeth the belly, prouoketh vrine, and helpeth againſt the venomous biting of the field Spider.

B　The ſeed is of the ſame vertue, and is of greater force in ſtopping the laske and bloudy flix. Being drunke in wine it preuaileth againſt the falling ſickneſſe.

C　The roots, after the opinion of *Dioſcorides*, being beaten and applied with white Wine, hinder or keepe backe the growth of haires.

D　† The ſeed giuen with Southerne-wood in Wine is good againſt the Iaundiſe. ‡

Cʜᴀᴘ. 80.　*Of Faire haired Iacinth.*

¶ *The Deſcription.*

1　THe Faire haired Iacinth hath long fat leaues, hollowed alongſt the inſide, trough faſhion, as are moſt of the Hyacinths, of a darke greene colour tending to redneſſe. The ſtalke riſeth out of the middeſt of the leaues, bare and naked, ſoft and full of ſlimie juyce, which are beſet round about with many ſmall floures of an ouerworne purple colour : The top of the ſpike conſiſteth of a number of faire ſhining purple floures, in manner of a tuft or buſh of haires, whereof it tooke his name *Comoſus*, or faire haired. The ſeed is contained in ſmall bullets, of a ſhining blacke colour, as are moſt of thoſe of the Hyacinths. The roots are bulbous or Onion faſhion, full of ſlimie juice, with ſome hairie threds faſtened vnto their bottome.

2　White haired Iacinth differeth not from the precedent in roots, ſtalkes, leaues, or ſeed. The floures hereof are of a darke white colour, with ſome blackneſſe in the hollow part of them, which ſetteth forth the difference.

3　Of this kinde I receiued another ſort from Conſtantinople, reſembling the firſt hairy Hyacinth very notably : but differeth in that, that this is altogether greater, as well in leaues, roots, and floures, as alſo it is of greater beautie without all compariſon.

1 *Hya-*

1 *Hyacinthus comoſus.*
Faire haired Iacinth.

2 *Hyacinthus comoſus albus.*
White haired Iacinth.

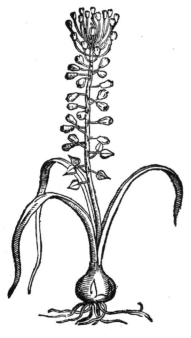

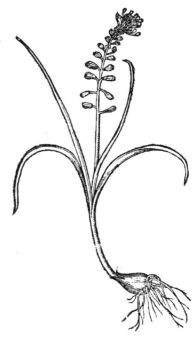

3 *Hyacinthus comoſus Bizantinus.*
Faire-haired Iacinth of Conſtantinople.

‡ 5 *Hyacinthus comoſus ramoſus elegantior.*
Faire curld-haired branched Iacinth.

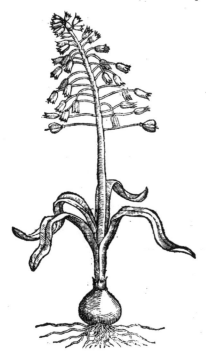

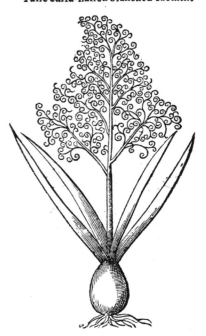

‡ 4 There

‡ 4   There are two other more beautifull haired Iacinths nouriſhed in the gardens of our prime Floriſts. The firſt of theſe hath roots and leaues reſembling the laſt deſcribed : the ſtalke commonly riſeth to the height of a foot, and it is diuided into many branches on euery ſide, which are ſmall and threddy ; and then at the end as it were of theſe threddy branches there come forth many ſmaller threds of a darke purple colour, and theſe ſpread and diuaricate themſelues diuers waies, much after the manner of the next deſcribed ; yet the threds are neither of ſo pleaſing a colour, neither ſo many in number, nor ſo finely curled. This is called *Hyacinthus comoſus ramoſus purpureus*, The faire hairy branched Iacinth.

5   This is a moſt beautifull and elegant plant, and in his leaues and roots he differs little from the laſt deſcribed ; but his ſtalke, which is as high as the former, is diuided into very many ſlender branches, which ſubdiuided into great plenty of curled threads variouſly ſpread abroad, make a very pleaſant ſhew. The colour alſo is a light blew, and the floures vſually grow ſo, that they are moſt dilated at the bottome, and ſo ſtraiten by little and little after the manner of a Pyramide. Theſe floures keepe their beautie long, but are ſucceeded by no ſeeds that yet could be obſerued. This by *Fabius Columna* (who firſt made mention hereof in writing) is called *Hyacinthus Sanneſius paniculoſa coma* : by others, *Hyacinthus comoſus ramoſus elegantior*, The faire curld-haire Iacinth.

Theſe floure in May and Iune. ‡

6   *Hyacinthus botryoides cæruleus.*
Blew Grape-floure.

7   *Hyacinthus botryoides cæruleus major.*
Great Grape-floure.

6   The ſmall Grape-floure hath many long fat and weake leaues trailing vpon the ground, hollow in the middle like a little trough, full of ſlimy juice like the other Iacinths ; amongſt which come forth thicke ſoft ſmooth and weake ſtalkes, leaning this way and that way as not able to ſtand vpright by reaſon it is ſurcharged with very heauie floures on the top, conſiſting of many little bottle-like blew floures, cloſely thruſt or packed together like a bunch of grapes, of a ſtrong ſmell, yet not vnpleaſant, ſomewhat reſembling the ſauour of the Orenge. The root is round and bulbous, ſet about with infinite young cloues or roots, whereby it greatly increaſeth.

7   The great Grape-floure is very like vnto the ſmaller of his kinde. The difference conſiſteth, in that this plant is altogether greater, but the leaues are not ſo long.

8   The sky-coloured Grape-floure hath a few leaues in reſpeⳤ of the other Grape-floures, the which are ſhorter, fuller of juice, ſtiffe and vpright, whereas the others traile vpon the ground.
The

The floures grow at the top, thruſt or packt together like a bunch of Grapes, of a pleaſant bright sky colour, euery little bottle-like floure ſet about the hollow entrance with ſmall white ſpots not eaſily to be perceiued. The roots are like the former.

8 *Hyacinthus Botryoides cæruleus major.*
Great Grape-floure.

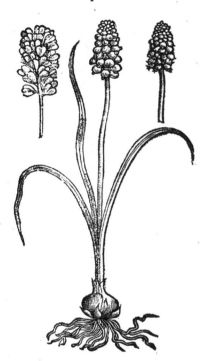

9   The white Grape-floure differeth not from the sky-coloured Iacinth, but in colour of the floure: for this Iacinth is of a pleaſant white colour tending to yellownes, tipped about the hollow part with White, whiter than White it ſelfe; otherwiſe there is no difference.

¶ *The Place.*

Theſe plants are kept in gardens for the beautie of their floures, wherewith our London gardens do abound.

¶ *The Time.*

They floure from Februarie to the end of May.

¶ *The Names.*

The Grape-floure is called *Hyacinthus Botryoides,* and *Hyacinthus Neotericorum, Dodonæi:* of ſome, *Bulbus Eſculentus, Hyacinthus ſylueſtris, Cordi: Hyacinthus exiguus, Tragi.* Some iudge them to be *Bulbinæ,* of *Pliny.*

†   The faire haired Iacinth deſcribed in the firſt place is, the *Hyacinthus* of *Dioſcorides* and the Antients. †

¶ *The Nature and vertues.*

†   The vertues ſet downe in the precedent Chapter properly belong to that kinde of Hyacinth which is deſcribed in the firſt place in this Chapter. †

CHAP. 81. *Of Muſcari, or Musked Grape-floure.*

¶ *The Deſcription.*

1   YEllow Muſcarie hath fiue or ſix long leaues ſpread vpon the ground, thicke fat, and full of ſlimie juyce, turning and winding themſelues crookedly this way and that way, hollowed alongſt the middle like a trough, as are thoſe of faire haired Iacinth, which at the firſt budding or ſpringing vp are of a purpliſh colour; but being growne to perfection, become of a darke greene colour: amongſt the which leaues riſe vp naked, thicke, and fat ſtalkes, infirme and weake in reſpect of the thickneſſe and greatneſſe thereof, lying alſo vpon the ground as do the leaues; ſet from the middle to the top on euery ſide with many yellow floures, euery one made like a ſmall pitcher or little box, with a narrow mouth, exceeding ſweet of ſmell like the ſauour of muſke, whereof it tooke the name *Muſcari.* The ſeed is cloſed in puffed or blowne vp cods, confuſedly made without order, of a fat and ſpongeous ſubſtance, wherein is contained round blacke ſeed. The root is bulbous or onion faſhion, whereunto are annexed certaine fat and thicke ſtrings like thoſe of Dogs-graſſe.

2   Aſh-coloured *Muſcari* or grape-floure, hath large and fat leaues like the precedent, not differing in any point, ſauing that the leaues at their firſt ſpringing vp are of a pale duſky colour like aſhes. The floures are likewiſe ſweet, but of a pale blacke colour, wherein conſiſteth the difference.

¶ *The*

1 *Muſcari flavum.*
Yellow musked Grape-floure.

2 *Muſcari Cluſij.*
Aſh-coloured Grape-floure.

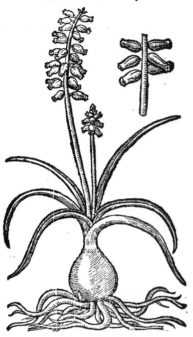

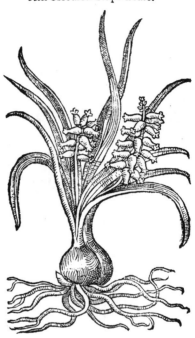

Muſcari caulis ſiliquis onuſtus.
The ſtalke of Muſcari hanged with the ſeed veſſels.

¶ *The Place.*

Theſe plants came from beyond the Thra-cian Boſphorus, out of Aſia, and from about Conſtantinople, and by the meanes of Friends haue been brought into theſe parts of Europe, whereof our London gardens are poſſeſſed.

¶ *The Time.*

They floure in March and Aprill, and ſome-times after.

¶ *The Names.*

They are called generally *Muſcari:* in the Turky Tongue, *Muſchoromi, Muſcurimi, Tipcadi,* and *Dipcadi,* of their pleaſant ſweet ſmell: of *Matthiolus, Bulbus Vomitorius.* Theſe plants may be referred vnto the Iacinths, whereof vndoub-tedly they be kindes.

¶ *The Nature and Vertues.*

There hath not as yet any thing beene tou-ched concerning the nature or vertues of theſe Plants, onely they are kept and maintained in gardens for the pleaſant ſmell of their floures, but not for their beauty, for that many ſtinking field floures do in beautie farre ſurpaſſe them. *Vomitorius,* in that he ſuppoſed they procure vo-miting; which of other Authors hath not bin remembred.

CHAP.

CHAP. 82.   Of Woolly Bulbus.

Bulbus Eriophorus.
Woolly Iacinth

¶ *The Deſcription.*

THere hath fallen out to be here inſerted a bulbous plant conſiſting of many Bulbes, which hath paſſed currant amongſt all our late Writers. The which I am to ſet forth to the view of our Nation, as others haue done in ſundry languages to theirs, as a kind of the Iacinths, which in roots and leaues it doth very well reſemble; called of the Grecians, Ἐριοφορος : in Latine, *Laniferus*, becauſe of his aboundance of Wooll-reſembling ſubſtance, wherewith the whole Plant is in euery part full fraught, as well roots, leaues, as ſtalkes. The leaues are broad, thicke, fat, full of juice, and of a ſpider-like web when they be broken. Among theſe leaues riſeth vp a ſtalke two cubits high, much like vnto the ſtalke of Squilla or Sea-Onion ; and from the middle to the top it is beſet round about with many ſmall ſtarre-like blew floures without ſmell, very like to the floures of Aſphodill ; beginning to floure at the bottome and ſo vpward by degrees, whereby it is long before it haue done flouring: which floures the learned Phyſitian of Vienna, *Iohannes Aicholzius*, deſired long to ſee ; who brought it firſt from Conſtantinople, and planted it in his Garden, where he nouriſhed it tenne yeares with great curioſitie ; which time being expired, thinking it to be a barren plant, he ſent it to *Carolus Cluſius*, with whom in ſome few yeres it did beare ſuch floures as are before deſcribed, but neuer ſince to this day. This painefull Herbariſt would gladly haue ſeene the ſeed that ſhould ſucceed theſe floures ; but they being of a nature quickly ſubject to periſh, decay, and fade, began preſently to pine away, leauing onely a few chaffie and idle ſeed-veſſels without fruit. My ſelfe haue beene poſſeſſed with this plant at the leaſt twelue yeares, whereof I haue yearely great encreaſe of new roots, but I did neuer ſee any token of budding or flouring to this day : notwithſtanding I ſhall be content to ſuffer it in ſome baſe place or other of my garden, to ſtand as the cipher o at the end of the figures, to attend his time and leiſure, as thoſe men of famous memorie haue done. Of whoſe temperature and vertues there hath not any thing been ſaid, but kept in gardens to the end aforeſaid.

CHAP. 83.   Of two feigned Plants.

¶ *The Deſcription.*

1   I Haue thought it conuenient to conclude the hiſtorie of the Hyacinths with theſe two bulbous Plants, receiued by tradition from others, though generally holden for feigned and adulterine. Their pictures I could willingly haue omitted in this hiſtorie, if the curious eye could elſewhere haue found them drawne and deſcribed in our Engliſh Tongue : but becauſe I finde them in none, I will lay them downe here, to the end that it may ſerue for excuſe to others who ſhall come after, which liſt not to deſcribe them, being as I ſaid condemned for feined and adulterine nakedly drawne onely. And the firſt of them is called *Bulbus* ιριφορος : by others, *Bulbus Bombicinus Commentitius*. The deſcription conſiſteth of theſe points, *viz.* The floures (ſaith the Author) are no leſſe ſtrange than wonderfull. The leaues and roots are like to thoſe of Hyacinths,

cinths, which hath cauſed it to occupie this place. The floures reſemble the Daffodils or Narciſſus. The whole plant conſiſteth of a woolly or flockie matter: which deſcription with the Pi-
&ture was ſent vnto *Dodonæus* by *Iohannes Aicholzius*. It may be that *Aicholzius* receiued inſtru&ti-
ons from the Indies, of a plant called in Greeke, αϛ϶λόϛ, which groweth in India, whereof *Theophra-*
ſtus and *Athenæus* do write in this manner, ſaying, The floure is like the *Narciſſus*, conſiſting of a
flockie or woolly ſubſtance, which by him ſeemeth to be the deſcription of our bombaſt Iacinth.

1 *Bulbus Bombicinus Commentitius.*　　　　　2 *Tigridis flos.*
　Falſe bumbaſt Iacinth.　　　　　　　　　　 The floure of Tygris.

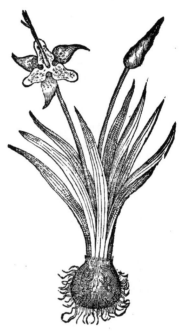

2　　The ſecond feigned picture hath beene taken of the Diſcouerer and others of late time, to
be a kinde of Dragons not ſeene by any that haue written thereof; which hath moued them to
thinke it a feigned picture likewiſe; notwithſtanding you ſhall receiue the deſcription thereof as
it hath come to my hands. The root (ſaith my Author) is bulbous or Onion faſhion, outwardly
blacke; from the which ſpring vp long leaues, ſharpe pointed, narrow, and of a freſh greene colour:
in the middeſt of which leaues riſe vp naked or bare ſtalkes, at the top whereof groweth a pleaſant
yellow floure, ſtained with many ſmall red ſpots here and there confuſedly caſt abroad: and in the
middeſt of the floure thruſteth forth a long red tongue or ſtile, which in time groweth to be the
cod or ſeed-veſſell, crooked or wreathed, wherein is the ſeed. The vertues and temperature are
not to be ſpoken of, conſidering that we aſſuredly perſuade our ſelues that there are no ſuch plants,
but meere fi&tions and deuices, as we terme them, to giue his friend a gudgeon.

‡　Though theſe two haue beene thought commenticious or feigned, yet *Bauhinus* ſeemeth to
vindicate the later, and *Iohn Theodore de Bry* in his *Florilegium* hath ſet it forth. He giues two Figures thereof, this which we here giue you being the one; but the other is farre more elegant, and
better reſembles a naturall plant. The leaues (as *Bauhine* ſaith) are like the ſword-flag, the root
like a leeke, the floures (according to *De Bryes* Figure) grow ſometimes two or three on a ſtalke:
the floure conſiſts of two leaues, and a long ſtile or peſtill: each of theſe leaues is diuided into
three parts, the vttermoſt being broad and large, and the innermoſt much narrower and ſharper:
the tongue or ſtile that comes forth of the midſt of the floure is long, and at the end diuided into
three crooked forked points. All that *De Bry* ſaith thereof is this; *Flos Tigridis rubet egregiè, circa*
medium tamen pallet, albuſque eſt & maculatus; ex Mexico à Caſparo Bauhino. That is; *Flos Tigridis* is
wondrous red, yet it is pale and whitiſh about the middle, and alſo ſpotted; it came from about
Mexico, I had it from *Caſpar Bauhine.* ‡

　　　　　　　　　　　　　　　　　　　　　　　　　　　　　　　CHAP.

CHAP. 84. *Of Daffodils.*

¶ *The Kindes.*

D Affodill, or *Narcissus* according to *Dioscorides*, is of two sorts; the floures of both are white, the one hauing in the middle a purple circle or coronet; the other with a yellow cup, circle or coronet. Since whose time there haue been sundry others described, as shall be set forth in their proper places.

1 *Narcissus medio purpureus.*
Purple circled Daffodill.

‡ 4 *Narcissus medio croceus serotinus polyanthos.*
The late many-floured Daffodill with
the Saffron coloured middle.

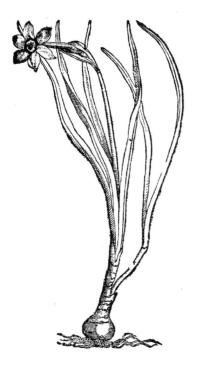

¶ *The Description.*

1   THe first of the Daffodils is that with the purple crowne or circle, hauing small narrow leaues, thicke, fat, and full of slimie juice; among the which riseth vp a naked stalke smooth and hollow, of a foot high, bearing at the top a faire milke white floure growing forth of a hood or thin filme such as the flours of onions are wrapped in : in the midst of which floure is a round circle or small coronet of a yellowish colour, purfled or bordered about the edge of the said ring or circle with a pleasant purple colour; which being past, there followeth a thicke knob or button, wherein is contained blacke round seed. The root is white, bulbous or Onion-fashion.

2   The second kinde of Daffodil agreeth with the precedent in euery respect, sauing that this Daffodill floureth in the beginning of Februarie, and the other not vntill Aprill, and is somewhat lesser. It is called *Narcissus med o purpureus præcox* ; That is, Timely purple ringed Daffodill. The next may haue the addition *præcocior*, More timely : and the last in place, but first in time, *præcocissimus*, Most timely, or very early flouring Daffodill.

3 The

‡ 5 *Narciſſus medio purpureus flore pleno.*
Double floured purple circled Daffodill.

6 *Narciſſus minor ſerotinus.*
The late flouring ſmall Daffodil.

7 *Narciſſus medioluteus.*
Primroſe Pearles, or the common white Daffodill.

8 *Narciſſus medioluteus polyanthos.*
French Daffodill.

9 *Narciſſus Piſanus.*
Italian Daffodill.

10 *Narciſſus albus multiplex.*
The double white Daffodill of Conſtantinople.

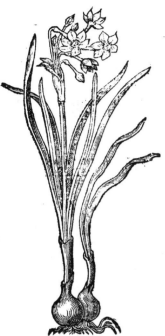

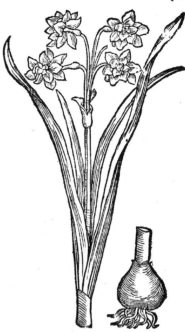

‡ 11 *Narciſſus flore pleno albo.*
The other double white Daffodill.

‡ 12 *Narciſſus flore pleno, medio luteo.*
Double white Daffodil with the middle
yellow.

3   The third kinde of Daffodill with the purple ring or circle in the middle, hath many ſmall narrow leaues, very flat, crookedly bending toward the top, among which riſeth vp a ſlender bare ſtalke, at whoſe top doth grow a faire and pleaſant floure like vnto thoſe before deſcribed, but leſ-ſer, wherein conſiſteth the difference.

‡   There is alſo another ſomewhat leſſe, and flouring ſomwhat earlier than the laſt deſcribed.

4   This in roots, leaues and ſtalks, differeth very little from the laſt mentioned kindes, but it beares many floures vpon one ſtalke, the out leaues being like the former, white, but the cup or ring in the middle of a ſaffron colour, with diuers yellow threds contained therein.

5   To theſe may be added another mentioned by *Cluſius*, which differeth from theſe only in the floures ; for this hath floures conſiſting of ſix large leaues fairely ſpred abroad, within which are other ſix leaues not ſo large as the former, and then many other little leaues mixed with threds comming forth of the middle. Now there are purple threds which run between the firſt & ſecond ranke of leaues, in the leaues, and ſo in the reſt. This floures in May ; and it is *Narciſſus pleno flore quintus* of *Cluſius*. ‡

‡ 13   *Narciſſus flore pleno, medio verſicolore.*
Double Daffodill with a diuers coloured middle.

14   *Narciſſus totus albus.*
Milke white Daffodill.

6   This late flouring Daffodill hath many fat thicke leaues, full of juice ; among which riſeth vp a naked ſtalke, on the top whereof groweth a faire white floure, hauing in the middle a ring or yellow circle. The ſeed groweth in knobby ſeed-veſſels. The root is bulbous or onion-faſhion. It floureth later than others before deſcribed, that is to ſay, in Aprill and May.

7   The ſecond kinde of Daffodill is that ſort of *Narciſſus* or Primroſe peereleſſe that is moſt common in our country gardens, generally knowne euerie where. It hath long fat and thick leaues, full of a ſlimie juice ; among which riſeth vp a bare thicke ſtalke, hollow within and full of juice. The floure groweth at the top, of a yellowiſh white colour, with a yellow crowne or circle in the middle, and floureth in the moneth of Aprill, and ſometimes ſooner. The root is bulbous faſhion.

8   The eighth Daffodill hath many broad and thicke leaues, fat and full of juice, hollow, and ſpongeous. The ſtalks, floures, and roots are like the former, and differeth in that, that this plant
bringeth

bringeth forth many floures vpon one ſtalk, and the other fewer, and not of ſo perfect a ſweet ſmel, but more offenſiue and ſtuffing the head. It hath this addition, *Polyanthos*, that is, of many floures, wherein eſpecially conſiſteth the difference.

9   The Italian Daffodill is very like the former, the which to diſtinguiſh in words, that they may be knowne one from another, is impoſſible. Their floures, leaues, and roots are like, ſauing that the floures of this are ſweeter, and more in number.

15 *Narciſſus juncifolius præcox.*
　Ruſh Daffodill, or *Iunquilia.*

26 *Narciſſus juncifolius ſerotinus.*
　Late flouring Ruſh Daffodill.

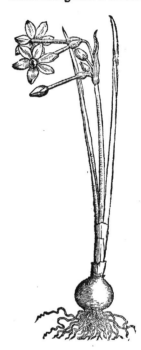

10   The double white Daffodill of Conſtantinople was ſent into England vnto the right ho-nourable the Lord Treaſurer, among other bulbed floures; whoſe roots when they were planted in our London gardens, did bring forth beautifull floures very white and double, with ſome yellow-neſſe mixed in the middle leaues, pleaſant and ſweet in ſmell; but ſince that time we neuer could by any induſtrie or manuring bring them vnto flouring againe. So that it ſhould appeare, when they were diſcharged of that birth or burthen which they had begotten in their own country, and not finding that matter, ſoile, or clymat to beget more floures, they remaine euer ſince barren and fruitleſſe. Beſides, we found by experience, that thoſe plants which in Autumne did ſhoot forth leaues, did bring forth no floures at all; and the others that appeared not vntill the Spring, did flou-riſh and beare their floures. The ſtalkes, leaues, and roots are like vnto the other kindes of Daf-fodils. It is called of the Turkes, *Giul Catamer lale*, that is, Narciſſus with double floures. Not-withſtanding we haue receiued from beyond the ſeas, as wel from the Low-Countries, as alſo from France, another ſort of greater beauty, which from yeare to yeare doth yeeld forth moſt pleaſant double floures, and great encreaſe of roots, very like as well in ſtalks as other parts of the plant, vnto the other ſorts of Daffodils. It differeth onely in the floures, which are very double and thicke thruſt together, as are the floures of our double Primroſe; hauing in the middle of the floure ſome few chiues or welts of a bright purple colour, and the other mixed with yellow, as a-foreſaid.

‡ 11   This alſo with double white floures, which *Cluſius* ſets forth in the ſixt place, is of the ſame kinde with the laſt deſcribed, but it beares but one or two floures vpon a ſtalke, whereas the other hath many.

12   This, which is *Cluſius* his *Narciſſus flore pleno* 2. is in roots, leaues, and ſtalkes very like the precedent;

precedent ; but the floures are composed of six large white out-leaues ; but the middle is filled with many faire yellow little leaues, much like to the double yellow wall-floure. They smel sweet like as the last mentioned.

13   This differs from the last mentioned only in that it is lesse, & that the middle of the floure within the yellow cup is filled with longish narrow little leaues as it were crossing each other. Their colour is white, but mixed with some greene on the out side, and yellow on the inside. ‡

14   The milke white Daffodill differeth not from the common white Daffodill, or Primrose Peerlesse, in leaues, stalks, roots, or floures, sauing that the floures of this plant haue no other colour but white, whereas all the others are mixed with one colour or other.

‡ 17 *Narcissus juncifolius Roscoluteus.*
Rose or round floured *Iunquilia.*

‡ 18 *Narcissus juncifolius amplo calice.*
White *Iunquilia* with the large cup.

‡ 19 *Narcissus juncifolius reflexus flore albo.*
The white reflex *Iunquilia.*

15   The Rush daffodill hath long narrow & thick leaues very smooth and flexible , almost round like Rushes, whereof it tooke his syrname *Iuncifolius* , or Rushy. It springeth vp in the beginning of Ianuarie, at which time also the floures shoot forth their buds at the top of small rushy stalks, sometimes two, and often more vpon one stalke, made of six small yellow leaues. The cup or crowne in the middle is likewise yellow, in shape resembling the other Daffodils, but smaller, and of a strong sweet smell. The root is bulbed, white within, and couered with a blacke skin or filme.

16   This Rush Daffodill is like vnto the precedent in each respect, sauing that it is altogether lesse, and longer before it come to flouring. There is also a white floured one of this kinde.

‡ 17   There

‡ 17   There is also another Rush Daffodill or *Iunquilia*,with floures not sharpe pointed,but round,with a little cup in the middle : The colour is yellow,or else white. This is *Lobels Narcissus juncifolius flore rotundæ circinitatis roseo.*

18   There is also another *Iunquilia*,whose leaues and stalks are like those of the first described rushy Daffodill,but the cup in the midst of the floure is much larger. The colour of the floure is commonly white. *Clusius* calls this, *Narcissus 1 juncifolius amplo calice.*

19   There are three or foure reflex *Iunquilia's*, whose cups hang downe, and the six incompassing leaues turne vp or backe, whence they take their names. The floures of the first are yellow ; those of the second all white,the cup of the third is yellow,and the reflex leaues white.The fourth hath a white cup,and yellow reflex floures. This seemes to be *Lobels Narcissus montanus minimus eoronatus.*

20   This is like to the ordinary lesser *Iunquilia*,but that the floures are very double,consisting of many long and large leaues mixed together ; the shorter leaues are obtuse, as if they were clipt off. They are wholly yellow.

‡ 19 *Narcissus juncifolius reflexus minor.*
    The lesser reflex *Iunquilia.*

‡ 20 *Narcissus juncifolius multiplex.*
    The double *Iunquilia.*

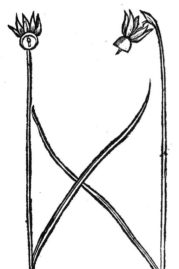
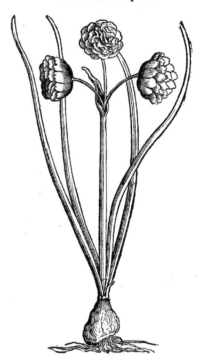

21   The Persian Daffodill hath no stalke at all but onely a small and tender foot-stalke of an inch high,such as the Saffron floure hath : vpon which short and tender stalk doth stand a yellow-ish floure consisting of six small leaues,of which the three innermost are somwhat narrower than those on the outside. In the middle of the floure doth grow forth a long stile or pointall, set about with many small chiues or threds. The whole floure is of an vnpleasant smell much like to Poppy:the leaues rise vp a little before the floure,long,smooth,& shining:the root is bulbed,thick and grosse,blackish on the out side,and pale within,with some threds hanging at the lower part.

22   The Autumne Daffodill bringeth forth long smooth glittering leaues of a deep green co-lour, among which riseth vp a short stalke bearing at the top one floure and no more, resembling the floure of Mead Saffron or common Saffron,consisting of six leaues of a bright shining yellow colour ; in the middle whereof stand six threds or chiues, and also a pestel or clapper yellow like-wise. The root is thicke and grosse like vnto the precedent.

‡   23   To this last may be adioyned another which in shape somewhat resembles it. The

leaues

leaues are ſmooth,green,growing ſtraight vp,and almoſt a fingers bredth ; among which riſeth vp a ſtalke a little more than halfe a foot in height, at the top of which groweth forth a yellow floure not much vnlike that of the laſt deſcribed Autumne Narciſſe : it conſiſts of ſix leaues ſome inch and halfe in length, and ſome halfe inch broad, ſharpe pointed,the three inner leaues.being ſomewhat longer than the outer. There grow forth out of the midſt of the floure three whitiſh chiues tipt with yellow,and a peſtell in the midſt of them longer than any of them. The root conſiſts of many coats, with fibres comming forth of the bottom thereof like others of this kinde. It floures in Februarie. ‡

21 *Narciſſus Perſicus.*
The Perſian Daffodill.

22 *Narciſſus Autumnalis major.*
The great Winter Daffodill.

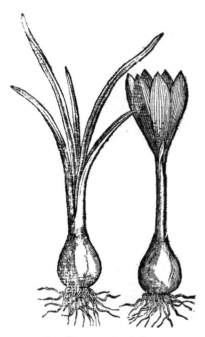
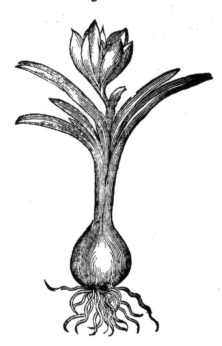

24   Small Winter Daffodill hath a bulbous root much like vnto the root of Ruſh Daffodill, but leſſer ; from the which riſeth vp a naked ſtalke without leaues, on the top whereof groweth a ſmall white floure with a yellow circle in the middle, ſweet in ſmell,ſomthing ſtuffing the head as do the other Daffodils.

¶ *The Place.*

The Daffodils with purple coronets grow wilde in ſundry places, chiefly in Burgondie, and in Suitzerland in medowes.

The Ruſh Daffodill groweth wilde in ſundry places of Spaine,among graſſe and other herbes, *Dioſcorides* ſaith that they be eſpecially found vpon mountaines. *Theocritus* affirmeth the Daffodils to grow in medowes,in his 19 Eidyl,or 20 according to ſome editions : where he writeth,That the faire Lady *Europa* entring with her Nymphs into the medowes, did gather the ſweet ſmelling daffodils ; in theſe verſes :

Αἰδ', ἰσπὶ ιω, &c.

Which we may Engliſh thus :

But when the Girles were come into
The medowes flouring all in ſight,
That Wench with theſe, this Wench with thoſe
Trim floures, themſelues did all delight :
She with the Narciſſe good in ſent,
And ſhe with Hyacinths content.

But

But it is not greatly to our purpoſe, particularly to ſeeke out their places of growing wilde, ſeeing we haue them all & euerie one of them in our London gardens, in great aboundance. The common wilde Daffodill groweth wilde in fields and ſides of woods in the Weſt parts of England.

¶ *The Time.*

They floure for the moſt part in the Spring, that is, from the beginning of February vnto the end of Aprill.

The Perſian and Winter Daffodils do floure in September and October.

‡ 23 *Narciſſus vernus præcocior flavo flore.*
The timely Spring yellow Daffodill.

24 *Narciſſus Autumnalis minor.*
Small Winter Daffodill.

¶ *The Names.*

Although their names be ſet forth in their ſeuerall titles, which may ſerue for their appellations and diſtinctions; notwithſtanding it ſhall not be impertinent to adde a ſupply of names, as alſo the cauſe why they are ſo called.

The Perſian Daffodill is called in the Sclauonian or Turkiſh tongue, *Zaremcada Perſiana*, and *Zaremcatta*, as for the moſt part all other ſorts of Daffodils are. Notwithſtanding the double floured Daffodill they name *Giul catamer lale*: which name they generally giue vnto all double floures.

The common white Daffodil with the yellow circle they call *Serin Cade*, that is to ſay, the kings Chalice; and *Deue bohini*, which is to ſay, Camels necke, or as we do ſay of a thing with long ſpindle ſhinnes, Long-ſhanks; vrging it from the long necke of the floure.

The Ruſh Daffodill is called of ſome *Iunquilias*, of the ſimilitude the leaues haue with Ruſhes: of *Dioſcoriaes, Bulbus Vomitorius*, or Vomiting Bulbe, according to *Dodonæus.*

Generally all the kindes are comprehended vnder this name *Narciſſus*; called of the Grecians ·····: in Dutch, **Barciſſen**: in Spaniſh, *Iennetten*: in Engliſh, Daffodilly, Daffodowndilly, and Primroſe peereleſſe.

Sophocles nameth them the Garland of the infernall gods, becauſe they that are departed & dulled with death, ſhould worthily be crowned with a dulling floure.

Of the firſt and ſecond Daffodill *Ovid* hath made mention in the third booke of his *Metamorphoſis,*

phoſis, where hee deſcribeth the transformation of the faire boy *Narciſſus* into a floure of his owne name ; ſaying,

Nuſquam corpus erat,croceum pro corpore florem
Inueninnt, folys medium cingentibus albis.

But as for body none remain'd ; in ſtead whereof they found
A yellow floure , with milke white leaues ingirting of it round.

Pliny and *Plutarch* affirme, as partly hath been touched before, that their narcoticke quality was the very cauſe of the name *Narciſſus*, that is, a qualitie cauſing ſleepineſſe ; which in Greeke is Νάρκαω · or of the fiſh Torpedo called Νάρκη, which benummes the hands of them that touch him, as being hurtfull to the ſinues,and bringeth dulneſſe to the head,which properly belongeth to the Narciſſes,whoſe ſmell cauſeth drowſineſſe.

¶ *The Nature.*
The roots of Narciſſus are hot and dry in the ſecond degree.

¶ *The Vertues.*

A   *Galen* ſaith,That the roots of Narciſſus haue ſuch wonderfull qualities in drying,that they con-ſound and glew together very great wounds,yea and ſuch gaſhes or cuts as happen about the veins, ſinues,and tendons. They haue alſo a certaine clenſing and attracting facultie.

B   The root of Narciſſus ſtamped with hony and applied plaiſter-wiſe,helpeth them that are bur-ned with fire,and joineth together ſinues that are cut in ſunder.

C   Being vſed in manner aforeſaid it helpeth the great wrenches of the ancles, the aches and pains of the joints.

D   The ſame applied with hony and nettle ſeed helpeth Sun burning and the morphew.

E   The ſame ſtamped with barrowes greaſe and leuen of rie bread,haſtneth to maturation hard im-poſtumes which are not eaſily brought to ripeneſſe.

F   Being ſtamped with the meale of Darnel and hony,it draweth forth thorns and ſtubs out of any part of the body.

G   The root,by the experiment of *Apuleius*,ſtamped and ſtrained,and giuen in drinke, helpeth the cough and cholique,and thoſe that be entred into a ptiſicke.

H   The roots whether eaten or drunken,do moue vomit,and being mingled with vineger and Net-tle ſeed,take away lentiles and ſpots in the face.

Chap. 85.   *Of the baſtard Daffodill.*

¶ *The Deſcription.*

1   The double yellow Daffodill hath ſmall ſmooth narrow leaues of a dark green colour ; among which riſeth vp a naked hollow ſtalke of two hands high , bearing at the top a faire and beautifull yellow floure of a pleaſant ſweet ſmell : it ſheddeth his floure, but there followeth no ſeed at all , as it hapneth in many other double floures. The root is ſmall, bulbous or onion-faſhion like vnto the other Daffodils,but much ſmaller.

2   The common yellow Daffodill or Daffodowndilly is ſo well knowne to all, that it needeth no deſcription.

3   We haue in our London gardens another ſort of this common kind,which naturally grow-eth in Spaine,very like vnto our beſt knowne Daffodil in ſhape and proportion,but altogether fai-rer,greater,and laſteth longer before the floure doth fall or fade.

‡ 4   This hath leaues and roots like the laſt deſcribed, but ſomewhat leſſe, the floure alſo is in ſhape not vnlike that of the precedent,but leſſe, growing vpon a weake ſlender greene ſtalke, of ſome fingers length : the ſeed is contained in three cornered,yet almoſt round heads. The root is ſmall,bulbous,and blacke on the out ſide.

5   This hath a longiſh bulbous root,ſomwhat blacke on the out ſide,from which riſe vp leaues not ſo long nor broad as thoſe of the laſt deſcribed : in the midſt of theſe leaues ſprings vp a ſtalk ſlender,and ſome halfe foot in height ; at the top of which, out of a whitiſh filme breakes forth a floure like in ſhape to the common Daffodill,but leſſe,and wholly white,with the brim of the cup welted about. It floures in Aprill,and ripens the ſeed in Iune.

¶ *The Place.*
The double yellow Daffodill I receiued from *Robinus* of Paris, which he procured by means of friends,from Orleance and other parts of France.

The

1 *Pſeudonarciſſus luteus multiplex.*
Double yellow Daffodill.

2 *Pſeudonarciſſus Anglicus.*
Common yellow Daffodill.

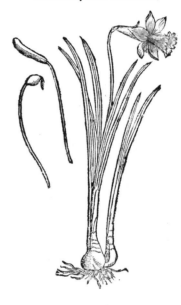

‡ 3 *Pſeudonarciſſus Hiſpanicus.*
The Spaniſh yellow Daffodill.

‡ 4 *Pſeudonarciſſus minor Hiſpanicus.*
The leſſer Spaniſh Daffodill.

‡ 5 *Pfeudonarciſſus albo flore.*
White baftard Daffodill.

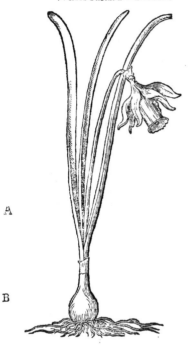

A

B

The yellow Englifh Daffodill groweth almoft euerie where through England. The yellow Spanifh Daffodill doth likewife decke vp our London Gardens, where they encreafe infinitely.

¶ *The Time.*

The double Daffodill fends forth his leaues in the beginning of Februarie, and his floures in Aprill.

¶ *The Names.*

The firft is called *Pfeudonarciſſus multiplex,* and *Narciſſus luteus polyanthos :* in Englifh, the double yellow Daffodill or Narciſſus.

The common fort are called in Dutch, **Geel Spozckel bloemen :** in Englifh, yellow Daffodil, Daffodilly, and Daffodowndilly.

¶ *The Temper.*

The temperature is referred vnto the kinds of Narciſſus.

¶ *The Vertues.*

Touching the vertues hereof, it is found out by experiment of fome of the later Phyfitians, that the decoction of the roots of this yellow Daffodil do purge by fiege tough and flegmatick humors, and alfo waterifh, and is good for them that are full of raw humors, efpecially if there be added thereto a little anife feed and Ginger, which will correct the churlifh hardneffe of the working.

The diftilled water of Daffodils doth cure the palfie, if the Patient be bathed and rubbed with the fayd liquor by the fire. It hath been proued by an efpeciall and trufty friend of mine, a man learned, and a diligent fearcher of nature, M<small>r</small> *Nicolas Belfon,* fomtime of Kings Colledge in Cambridge.

C<small>HAP</small>. 86.   *Of diuers other Daffodils or Narciſſes.*

‡ **T**Here are befides the forementioned forts of Daffodils, fundry others, fome of which may be referred to them, other fome not. I do not intend an exact enumeration of them, it being a thing not fo fit for a hiftorie of plants, as for a Florilegie, or booke of floures. Now thofe that require all their figures, and more exact defcriptions, may finnde full fatisfaction in the late Worke of my kinde friend M<small>r</small> *Iohn Parkinfon,* which is intituled *Paradifus terreftris :* for in other Florilegies, as in that of *De Bry, Swertz,* &c. you haue barely the names and figures, but in this are both figures and an exact hiftorie or declaration of them. Therefore I in this place will but onely briefely defcribe and name fome of the rareft that are preferued in our choice gardens, and a few others whereof yet they are not poffeffed.

¶ *The Defcriptions.*

1   The firft of thefe, which for the largeneffe is called *Nonpareille,* hath long broad leaues and roots like the other Daffodils. The floure confifts of fix very large leaues of a pale yellow colour, with a very large cup, but not very long : this cup is yellower than the incompaffing leaues, narrower alfo at the bottome than at the top, and vneuenly cut about the edges. This is called *Narciſſus omnium maximus,* or *Nonpareille ;* the figure well expreffeth the floure, but that it is fomwhat too little. There is a varietie of this, with the open leaues and cup both yellow, which makes the difference. There is alfo another *Nonpareille,* whofe floures are all white, and the fix leaues that ftand fpred abroad are vfually a little folded or turned in at their ends.

2   Befides thefe former, there are foure or fiue double yellow Daffodils which I cannot paffe ouer in filence ; the firft is that which is vulgarly amongft Florifts knowne by the name of *Robines*
Narciſſe

Narciſſus,and it may be was the ſame our Author in the precedent chapter mentions hee receiued from *Robine* ; but he giuing the figure of another, and a deſcription not well fitting this, I can af-firme nothing of certaintie. This double Narciſſe of *Robine* growes with a ſtalke ſome foot high, and the floure is very double,of a pale yellow colour, and it ſeems commonly to diuide it ſelf into ſome ſix partitions,the leaues of the floure lying one vpon another euen to the midſt of the floure. This may be called *Narciſſus pallidus multiplex Robini*, *Robines* double pale Narciſſe.

‡ 1 *Narciſſus omnium maximus.*
The *Nonpareille* Daffodill.

‡ 3 *Pſeudonarciſſus flore pleno.*
The double yellow Daffodill.

2　The next to this is that which from our Author,the firſt obſeruer thereof, is vulgarly called *Gerrards* Narciſſe. The leaues and root do not much differ from the ordinarie Daffodil ; the ſtalk is ſcarce a foot high,bearing at the top thereof a floure very double ; the ſix outmoſt leaues are of the ſame yellow colour as the ordinarie one is ; thoſe that are next are commonly as deepe as the tube or trunk of the ſingle one,and amongſt them are mixed alſo other paler coloured leaues,with ſome green ſtripes here & there among thoſe leaues. Theſe floures are ſomtimes all contained in a trunk like that of the ſingle one,the ſix out-leaues excepted : otherwhiles this incloſure is broke, and then the floure ſtands faire open like as that of the laſt deſcribed. *Lobel* in the ſecond part of his *Adverſaria* tels, That our Author Mr *Gerrard* found this in Wiltſhire, growing in the garden of a poore old woman,in which place formerly a cunning man (as they tetme him)had dwelt.

This may be called in Latine,according to the Engliſh, *Narciſſus multiplex Gerrardi*, *Gerrards* double Narciſſe.

The figure we here giue you is expreſſed ſomewhat too tall, and the floure is not altogether ſo double as it ought to be.

4　There are alſo two or three double yellow Daffodils yet remaining.The firſt of theſe is cal-led *Wilmots* Narciſſe,from Mr *Wilmot* late of Bow,and this hath a very faire double & large yellow floure compoſed of deeper and paler yellow leaues orderly mixed.

The ſecond (which is called *Tradeſcants* Narciſſe, from Maſter *Iohn Tradeſcant* of South Lam-beth)is the largeſt and ſtatelieſt of all the reſt : in the largeneſſe of the floures it exceeds *Wilmots*, which otherwiſe it much reſembles ; ſome of the leaues wherof the floure conſiſts are ſharp poin-ted,

ted, and theſe are of a paler colour; otherſome are much more obtuſe, and theſe are of a deeper and fairer yellow.

This may be called *Narciſſus Roſeus* Tradeſcanti, *Tradeſcants* Roſe Daffodill.

The third M^r.*Parkinſon* challengeth to himſelfe; which is a floure to be reſpected, not ſo much for the beautie, as for the various compoſure thereof, for ſome of the leaues are long and ſharpe pointed, others obtuſe and curled, a third ſort long and narrow, and vſually ſome few hollow, and in ſhape reſembling a horne; the vtmoſt leaues are commonly ſtreaked, end of a yellowiſh green, the next to them fold themſelues vp round, and are vſually yellow, yet ſometimes they are edged with greene. There is a deepe yellow peſtle diuided into three parts, in the midſt of this floure. It floures in the end of March. I vſually (before M^r.*Parkinſon* ſet forth his Florilegie, or garden of floures) called this floure *Narciſſus* πολψμορφος, by reaſon of its various ſhape and colour: but ſince I thinke it fitter to giue it to the Author, and terme it *Narciſſus multiplex varius* Parkinſoni, *Parkinſons* various double Narciſſe.

‡ 5   *Narciſſus Iacobæus Indicus.*   ‡ 6   *Narciſſus juncifolius montanus minimus.*
       The Indian or Iacobæan Narciſſe.   The leaſt Ruſh-leaued Mountaine Narciſſe.

 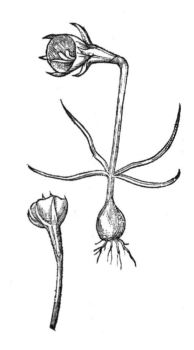

5   Now come I to treat of ſome more rarely to be found in our gardens, if at all. That which takes the firſt place is by *Cluſius* called *Narciſſus Iacobæus Indicus*, the Indian or Iacobæan Narciſſe. The root hereof is much like vnto an ordinary onion, the leaues are broad like the other Narciſſes: the ſtalke is ſmooth, round, hollow, and without knots, at the top whereof, out of a certaine skinny huske comes forth a faire red floure like that of the flouring Indian reed, but that the leaues of this are ſomewhat larger, and it hath ſix chiues or threds in the middle thereof, of the ſame colour as the floure, and they are adorned with browniſh pendants; in the midſt of theſe there ſtands a little farther out than the reſt, a three forked ſtile, vnder which ſucceeds a triangular head, after the falling of the floure.

This giues his floure in Iune or Iuly.

6   This *Lobel* calls *Narciſſus montanus juncifolijs minimus*, The leaſt Ruſh-leaued mountaine Narciſſe. The leaues of this are like the *Iunquilia*; the ſtalke is ſhort, the floure yellow, with the ſix winged leaues ſmall and paler coloured, the cup open and large to the bigneſſe of the floure.

7 This

7   This alſo is much like the former, but the ſix incompaſſing leaues are of a greeniſh ſcint yellow colour; the cup is indented or vnequally curled about the edges, but yellow like the precedent. *Lobel* calls this, *Narciſſus montanus juncifolius flore fimbriato*, The mountain Ruſh leaued Narciſſe with an indented or curled cup.

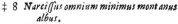

‡ 7 *Narciſſus montanus juncifolius fiore fimbriato.*
The mountaine Ruſh leaued Narciſſe with an indented or curled cup.

‡ 8 *Narciſſus omnium minimus montanus albus.*
The leaſt mountaine white Narciſſe.

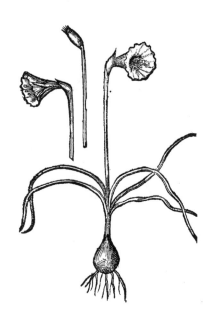

8   The leaues of this are ſmall as the Autumne Iacinth, the ſtalke ſome handfull high, and the floure like the laſt deſcribed, but it is of a whitiſh colour. *Lobel* calls this laſt deſcribed, *Narciſſus omnium minimus montanus albus*, The leaſt mountaine white Narciſſe. Theſe three laſt vſually floure in Februarie. ‡

CHAP. 87.   Of *Tulipa*, or the *Dalmatian Cap*.

¶ *The Kindes.*

TVlipa or the Dalmatian Cap is a ſtrange and forrein floure, one of the number of the bulbed floures, whereof there be ſundry ſorts, ſome greater, ſome leſſer, with which all ſtudious and painefull Herbariſts deſire to be better acquainted, becauſe of that excellent diuerſitie of moſt braue floures which it beareth. Of this there be two chiefe and generall kindes, *vtz. Præcox*, and *Serotina*; the one doth beare his floures timely, the other later. To theſe two we will adde another ſort called *Media*, flouring betweene both the others. And from theſe three ſorts, as from their heads, all other kindes doe proceed, which are almoſt infinite in number. Notwithſtanding, my louing friend Mr *Iames Garret*, a curious ſearcher of Simples, and learned Apothecarie of London, hath vndertaken to finde out, if it were poſſible, their infinite ſorts, by diligent ſowing of their ſeeds, and by planting thoſe of his owne propagation, and by others receiued from his friends

M   beyond

1 *Tulipa Bononienſis.*
Italian Tulipa.

2 *Tulipa Narbonenſis.*
French Tulipa.

3 *Tulipa præcox tota lutea.*
Timely flouring Tulipa.

4 *Tulipa coccinea ſerotina.*
Late flouring Tulipa.

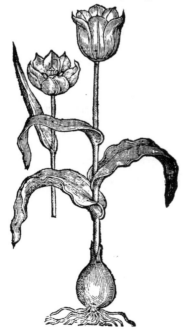

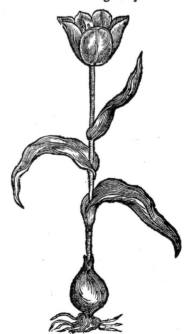

5 *Tulipa media ſanguinea albis oris.*
Apple bloome Tulipa.

6 *Tulipa candida ſuavis rubentibus oris.*
Bluſh coloured Tulipa.

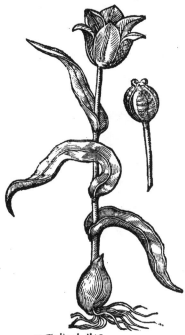
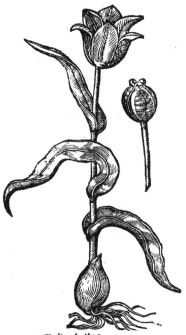
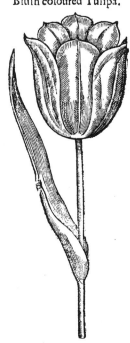

7 *Tulipa bulbifera.*
Bulbous ſtalked Tulipa.

‡ 8 *Tulipa ſanguinea luteo fundo.*
The blond-red Tulip with a yellow bottome.

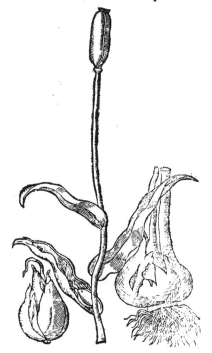

beyond the ſeas for the ſpace of twenty yeares,not being yet able to attaine to the end of his trauel,for that each new yeare bringeth forth new plants of ſundry colours not before ſeen; all which
to deſcribe particularly were to rolle *Siſiphus* ſtone,or number the ſands. So that it ſhall ſuffice to
ſpeake of and deſcribe a few,referring the reſt to ſome that meane to write of Tulipa a particular
volume.

‡ 9 *Tulipa purpurea.*   ‡ 10 *Tulipa rubra amethiſtina.*
The purple Tulip.   The bright red Tulip.

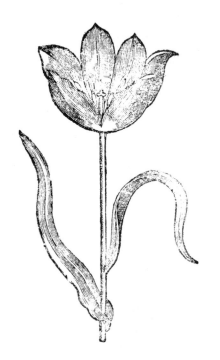 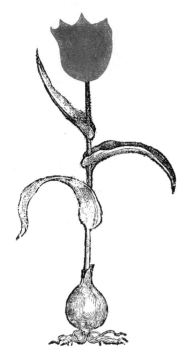

¶ *The Deſcription.*

1   THe Tulipa of Bolonia hath fat thicke and groſſe leaues,hollow,furrowed or chanelled,
bended a little backward,and as it were folded together:which at their firſt comming
vp ſeeme to be of a reddiſh colour,and being throughly growne turne into a whitiſh
greene. In the midſt of thoſe leaues riſeth vp a naked fat ſtalke a foot high,or ſomthing more ; on
the top whereof ſtandeth one or two yellow floures,ſomtimes three or more,conſiſting of ſix ſmal
leaues,after a ſort like to a deepe wide open cup,narrow aboue,and wide in the bottome. After it
hath been ſome few dayes floured,the points and brims of the floure turn backward,like a Dalmatian or Turkiſh Cap,called Tulipan,Tolepan,Turban,and Turfan,whereof it tooke his name. The
chiues or threds in the middle of the floure be ſomtimes yellow,otherwhiles blackiſh or purpliſh,
but commonly of one ouerworne colour or other, Nature ſeeming to play more with this floure
than with any other that I do know. This floure is of a reaſonable pleaſant ſmell,and the other of
his kinde haue little or no ſmel at all. The ſeed is flat,ſmooth,ſhining,and of a griſtly ſubſtance.
The root is bulbous,and very like to a common onion of S. *Omers.*

2   The French Tulipa agreeth with the former,except in the blacke bottome which this hath
in the middle of the floure,and is not ſo ſweet of ſmell,which ſetteth forth the difference.

3   The yellow Tulipa that floureth timely hath thicke and groſſe leaues full of juice,long,hol-
low,or gutter faſhion,ſet about a tender ſtalke, at the top whereof doth grow a faire and pleaſant
ſhining yellow floure,conſiſting of ſix ſmall leaues without ſmell.The root is bulbous or like an
onion.

4   The

‡ 11 *Tulipa flore albo strijs pur-*
pureis.
The white Tulip with pur-
ple streakes.

‡ 15 *Tulipa polyclados minor serotina flore rubro vel flauo*, Clusij.
The lesser many-branched late Tulip of Clusius, with red, or
else yellow floures.

‡ 12 *Tulipa flore albo oris dilute rubentibus.*
The white Tulip with light red edges.
‡ 13 *Tulipa flore pallido.* The straw-coloured Tulip.
‡ 14 *Tulipa flammea strijs flauescentibus.*
The flame coloured Tulip with yellowish streakes.

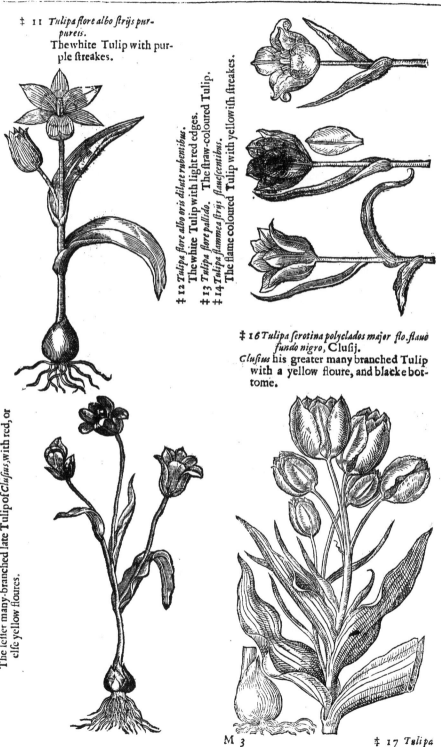

‡ 16 *Tulipa serotina polyclados major flo. flauò*
fundo nigro, Clusij.
Clusius his greater many branched Tulip
with a yellow floure, and blacke bottome.

‡ 17 *Tulipa pumilio obſcure rubens or is virentibus.*
The dwarfe Tulip with darke red floures edged with greene.
‡ 18 *Tulipa pumilio flore purpuraſcente intus candido.*
The Dwarfe Tulip with a purpliſh floure, white within.

‡ 19 *Tulipa pumilio lutea,*
The yellow Dwarfe Tulip.

‡ 20 *Tulipa Perſica flore rubro, oris albia ïs*
elegans.
The pretty Perſian Tulip hauing a red
floure with whitiſh edges.

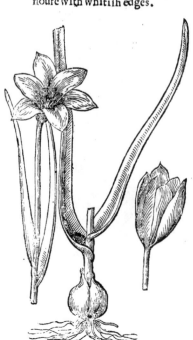

‡ 21 *Tulipa aurea oris rubentibus.*
The gold yellow with red edges.

4   The fourth kinde of Tulip, that floureth later, hath leaues, ſtalkes, and roots like vnto the precedent. The floures hereof be of a skarlet colour, welted or bordered about the edges with red. The middle part is like vnto a heart, tending to whiteneſſe, ſpotted in the ſame whiteneſſe with red ſpeckles or ſpots. The ſeed is contained in ſquare cods, flat, tough, and finewie.

5   The fift ſort of Tulipa, which is neither of the timely ones, nor of the later flouring ſort, but one that bringeth forth his moſt beautifull floures betweene both. It agreeth with the laſt deſcribed Tulipa, in leaues, ſtalkes, roots, and ſeed, but differeth in the floures. The floure conſiſteth of ſix ſmall leaues joyned together at the bottome : the middle of which leaues are of a pleaſant bloudy colour, the edges be bordered with white, and the bottome next vnto the ſtalke is likewiſe white ; the whole floure reſembling in colour the bloſſomes of an Apple tree.

‡ 22 *Tulipa miniata.*
The vermilion Tulip.

‡ 23 *Tulipa albo & rubro ſtriatus.*
The white and red ſtriped Tulip.

 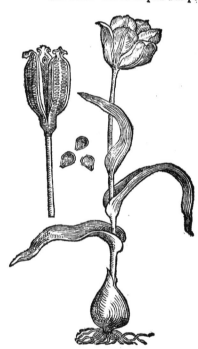

6   The ſixth hath leaues, roots, ſtalkes, and ſeed like vnto the former, but much greater in euery point. The floures hereof are white, daſht about the brimmes or edges with a red or bluſh colour. The middle part is ſtripped confuſedly with the ſame mixture, wherein is the difference.

7   *Carolus Cluſius* ſetteth forth in his Pannonicke hiſtory a kinde of Tulipa that beareth faire red floures, blacke in the bottome, with a peſtell in the middle of an ouer-worne greeniſh colour; of which ſort there happeneth ſome to haue yellow floures, agreeing with the other before touched: but this bringeth forth encreaſe of roots in the boſome of his loweſt leafe next to the ſtalke, contrary to the other kindes of Tulipa.

8   *Lobelius* in his learned Obſeruations hath ſet forth many other ſorts ; one he calleth *Tulipa Chalceaonica,* or the Turky Tulipa, ſaying it is the leaſt of the ſmall kindes or Dwarfe Tulipa's, whoſe floure is of a ſanguine red colour, vpon a yellow ground, agreeing with the other in root, leafe, and ſtalke.

9   He hath likewiſe ſet forth another ; his floure is like the Lilly in proportion, but in colour of a fine purple.

10   We may alſo behold another ſort altogether greater than any of the reſt, whoſe floure is in colour like the ſtone called *Amethyſt,* not vnlike to the floures of Peonies.

11   We haue likewiſe another of greater beautie, and very much deſired of all, with white floures daſht on the back ſide, with a light waſh of watched colour.

             12 There

‡ 24 *Tulipa luteo & rubro ſtriatus.*
The red and yellow Fooles coat.

‡ 25 *Tulipa flore coloris ſulphurei.*
The ſulpher coloured Tulip.

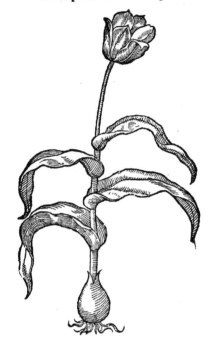

‡ 26 *Tulipa rubra oris pallidis.*
The red Tulip with pale edges.

12   There is another alſo in our London gardens, of a ſnow white colour ; the edges ſlightly waſht ouer with a little of that we call bluſh colour.

13   We haue another like the former, ſauing that his floure is of a ſtraw colour.

14   There is another to be ſeene with a floure mixed with ſtreaks of red and yellow, reſembling a flame of fire, wherupon we haue called it Flambant.

There be likewiſe ſo many more differing ſo notably in colour of their floures, although in leaues, ſtalke, and roots for the moſt part one like another, that (as I ſaid before) to ſpeake of them ſeuerally would require a peculiar volume.

‡   Therefore not to trouble you any further, I haue giuen you onely the figures and names of the notableſt differences which are in ſhape ; as the dwarfe Tulipa's, and the branched ones, together with the colour of their floures, contained in their titles, that you need not far to ſeeke it. ‡

There be a ſort greater than the reſt, which in forme are like ; the leaues whereof are thicke, long, broad, now and then ſomewhat folded in the edges; in the middeſt whereof doth riſe vp a ſtalk a foot high, or ſomthing higher, vpon which ſtandeth onely one floure bolt vpright, conſiſting of ſix leaues, after a ſort like to a deepe wide cup of this forme, *viz.* the bottome turned vpwards, with threads threds or chiues in the middle, of the colour of Saffron. The colour of the floure is fomtimes yellow, fometimes white, now and then as it were of a light purple, and many times red; and in this there is no fmall varieties of colours, for the edges of the leaues, and oftentimes the nails or lower part of the leaues are now and then otherwife colored than the leaues themfelues, and many times there doth run all along thefe ftreakes fome other colours. They haue no fmell at all that can be perceiued. The roots of thefe are likewife bulbed or onion fafhion, euerie of which to fet forth feuerally would trouble the writer, and wearie the reader; fo that what hath beene faid fhall fuffice touching the defcription of Tulipa's.

‡ True it is that our Author here affirmes, that the varietie of thefe floures are fo infinite, that it would both tire the writer and reader to recount them. Yet for that fome are more in loue with floures than with plants in generall, I haue thought good to direct them where they may finde fomewhat more at large of this plant. Let fuch therefore as defire further fatisfaction herein, haue recourfe to the Florilegies of *De Bry, Swertz, Robine*, or to M*r Parkinfon*, who hath not only treated of the floures in particular, but alfo of the ordering of them. ‡

‡ 27 *Tulipa lutea ferotina.*
The late flouring yellow Tulip.

‡ 28 *Tulipa ferotina lutea gutis fanguineis, fundo nigro.*
The late Yellow, with fanguine fpots and a blacke bottome.

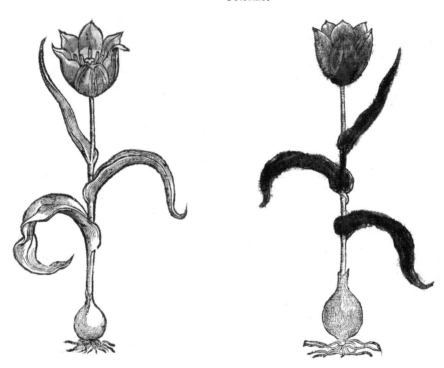

¶ *The Place.*

Tulipa groweth wilde in Thracia, Cappadocia, and Italy; in Bizantia about Conftantinople; at Tripolis and Aleppo in Syria. They are now common in all the Englifh gardens of fuch as affect floures.

¶ *The Time.*

They floure from the end of Februarie vnto the beginning of May, and fomwhat after: although *Augerius Busbequius* in his journey to Conftantinople, faw betweene Hadrianople and Conftantinople, great aboundance of them in floure euerie where, euen in the midft of Winter, in the moneth of Ianuarie, which that warme and temperat clymat may feeme to performe.

¶ *The*

¶ *The Names.*

The later Herbariſts by a Turkiſh or ſtrange name call it Tulipa, of the Dalmatian cap called Tulipa, the forme whereof the floure when it is open ſeemeth to repreſent.

It is called in Engliſh after the Turkiſh name Tulipa, or it may be called Dalmatian Cap, or the Turks Cap. What name the antient Writers gaue it is not certainely knowne. A man might ſuſpect it to be μικρὸν, if it were a Bulbe that might be eaten, and were of force to make milke cruddie; for *Theophraſtus* reckoneth it among thoſe Bulbes that may be eaten: and it is an herb, as *Heſychius* ſaith, wherewith milke is crudded. *Conradus Geſnerus* and diuers others haue taken Tulipa to be that *Satyrium* which is ſyrnamed *Erythronium*, becauſe one kind hath a red floure; or altogether a certaine kinde of *Satyrium*: with which it doth agree reaſonable well, if in *Dioſcorides* his deſcription we may in ſtead of Ἀπαριτημω, reade κ.γτοαπτημω, or Ἀπρατημω; for ſuch miſtakes are frequent in antient and moderne Authors, both in writing and printing. In the Turky tongue it is called *Calé lalé, Cavále lalé*, and likewiſe *Turban* and *Turfan*, of the Turkes Cap ſo called, as beforeſaid of *Lobelius*.

<table>
<tr><td>‡ 29 Tulipa Holias alba ſtrijs & punctis ſanguineis.
The white Holias with ſanguine ſpots and ſtreakes.</td><td>‡ 30 Tulipa media ſature purpurea fundo ſubcæruleo.
A middle Tulip of a deepe purple colour, with a blewiſh bottom.</td></tr>
</table>

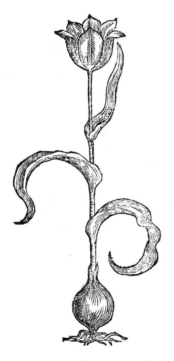 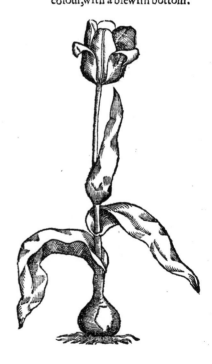

‡ I do verily thinke that theſe are the κρίνα τῦ ἀγρῦ, the Lillies of the field mentioned by our Sauiour, *Mat.6.28,29.* for he ſaith, That *Solomon* in all his royaltie was not arrayed like one of theſe. The reaſons that induce me to thinke thus, are theſe; Firſt, their ſhape: for their floures reſemble Lillies; and in theſe places whereas our Sauiour was conuerſant they grow wilde in the fields. Secondly, the infinite varietie of colour, which is to be found more in this than any other ſort of floure. And thirdly, the wondrous beautie and mixtures of theſe floures. This is my opinion, and theſe my reaſons, which any may either approue of or gainſay, as he ſhall thinke good. ‡

¶ *The Temperature and Vertues.*

There hath not been any thing ſet downe of the antient or later Writers, as touching the Nature or Vertues of the Tulipa, but they are eſteemed ſpecially for the beauty of their floures.

‡ The

‡ The roots preferued with fugar, or otherwife dreffed, may be eaten, and are no vnpleafant nor any way offenfiue meat,but rather good and nourifhing. ‡

Chap. 88.   Of bulbous Violets.

¶ The Kindes.

THeophraftus hath mentioned one kind of bulbous Leucoion, which Gaza tranflates, Viola alba, or the white Violet. Of this Viola Theophrafti,or Theophraftus his Violet, wee haue obferued three forts,whereof fome bring forth many floures and leaues,others fewer ; fome floure very early,and others later,as fhall be declared.

1 Leucoium bulbofum præcox minus.
   Timely flouring bulbous Violet.

‡ 2 Leucoium bulbofum præcox Byzantinum.
   The Byzantine early bulbous Violet.

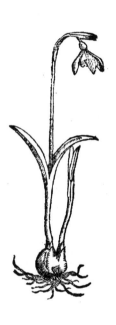 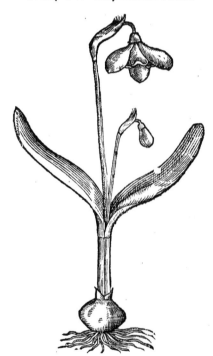

¶ The Defcription.

1   THe firft of thefe bulbous Violets rifeth out of the ground, with two fmall leaues flat and crefted,of an ouerworne greene colour, betweene the which rifeth vp a fmall and tender ftalke of two hands high ; at the top whereof commeth forth of a skinny hood a fmall white floure of the bigneffe of a Violet,compact of fix leaues, three bigger, and three leffer,tipped at the points with a light greene ; the fmaller are fafhioned into the vulgar forme of an heart,and prettily edged about with green : the other three leaues are longer, and fharpe pointed. The whole floure hangeth downe his head,by reafon of the weake foot-ftalke whereon it groweth. The root is fmall,white,and bulbous.

‡ 2   There are two varieties of this kind which differ little in fhape, but the firft hath a floure as bigge againe as the ordinarie one,and Clufius calls it Leucoium bulbofum præcox Byzantinum, The greater early Conftantinopolitan bulbous Violet. The other is mentioned by Lobel, and differs onely in colour of floures ; wherefore he calls it Leucoium triphyllum flore cæruleo, The blew floured bulbous Violet. ‡

3 The

3 *Leucoium bulboſum ſerotinum.*
Late flouring bulbous Violet.

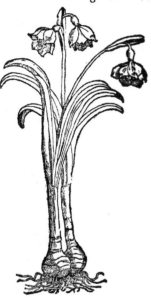

4 *Leucoium bulboſum majus polyanthemum.*
The many floured great bulbous Violet.

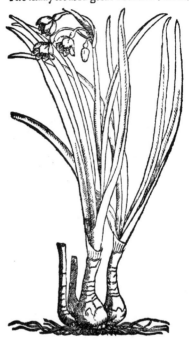

‡ 5 *Leucoium bulboſum Autumnale minimum.*
The leaſt Autumne bulbous Violet.

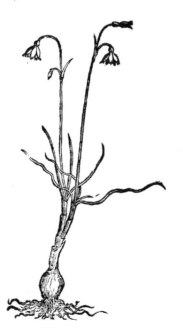

⚓3 The third ſort of bulbed Violets hath
narrow leaues like thoſe of the Leeke, but leſſer
and ſmoother, not vnlike to the leaues of the
baſtard Daffodill. The ſtalks be ſlender and naked, two hands high ; whereupon do grow faire
white floures tipped with a yellowiſh green colour, with many ſmall chiues or threds in the
middeſt of the floure. The ſeed is contained in
ſmall round buttons. The root is white and bulbous.

4   The great bulbed Violet is like vnto the
third in ſtalke and leaues, yet grearer & higher.
It bringeth forth on euerie ſtalke not one floure
onely, but fiue or ſix blowing or flouring one after another, altogether like the other floures in
forme and bigneſſe.

‡   5   This ſmall bulbous plant may be annexed to the former : the root is ſmall, compact
of many coats : the leaues are alſo ſmal, and the
ſtalke an handfull high, at the top whereof there
hang down one or two white floures conſiſting
of ſix leaues apiece much reſembling the laſt
deſcribed, but far leſſe. It floures in Autumne.

⚓   Beſides theſe, *Cluſius* makes mention of a
ſmall one much like this, which floureth in the
Spring, and the floures are ſomewhat reddiſh
nigh the ſtalk, and ſmell ſweet. *Cluſius* calls this,
Leucoium bulboſum vernum minimum, The ſmalleſt Spring bulbous Violet.

¶ *The*

¶ *The Place.*

These plants doe grow wilde in Italy and the places adiacent. Notwithstanding our London gardens haue taken possession of most of them many yeares past.

¶ *The Time.*

The first floureth in the beginning of Ianuary ; the second in September ; and the third in May; the rest at their seasons mentioned in their descriptions.

¶ *The Names.*

†   The first is called of *Theophrastus*, Λευκόιον; which *Gaza* renders *Viola alba*, and *Viola Bulbosa*, or bulbed Violet. *Lobelius* hath from the colour and shape called it *Leuconarcissolirion*, and that very properly, considering how it doth as it were participate of two sundry plants, that is to say, the root of the *Narcissus*, the leaues of the small Lillie, and the white colour ; taking the first part *Leu-co*, of his whitenesse ; *Narcisso*, of the likenesse the roots haue vnto *Narcissus* ; and *Lirium*, of the leaues of Lillies, as aforesaid. In English we may call it the bulbous Violet ; or after the Dutch name, 𝖲𝗈𝗆𝖾𝗋 𝖿𝗈𝗍𝗍𝖾𝗄𝖾𝗇𝗌 ; that is, Sommer fooles, and 𝕯𝔯𝔲𝔶𝔣𝔨𝔢𝔫𝔤. Some call them also Snow drops. This name *Leucoium*, without his Epithite *Bulbosum*, is taken for the Wall-floure, and stocke Gillofloure, by all moderne Writers.

¶ *The Nature and Vertues.*

Touching the faculties of these bulbous Violets we haue nothing to say, seeing that nothing is set downe hereof by the antient Writers, nor any thing obserued by the moderne ; onely they are maintained and cherished in gardens for the beautie and rarenesse of the floures, and sweetnesse of their smell.

CHAP. 89.   *Of Turkie or Ginny-hen Floure.*

1 *Frittillaria.*
Checquered Daffodill.

2 *Frittillaria variegata.*
Changeable Checquered Daffodill.

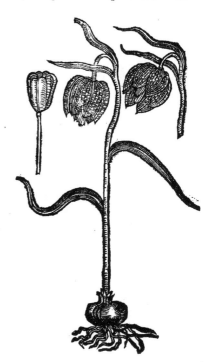

¶ *The*

¶ *The Deſcription.*

1   THe Checquered Daffodill, or Ginny-hen Floure, hath ſmall narrow graſſie leaues, among which there riſeth vp a ſtalke three hands high, hauing at the top one or two floures, and ſometimes three, which conſiſteth of ſix ſmall leaues checquered moſt ſtrangely : wherein Nature, or rather the Creator of all things, hath kept a very wonderfull order, ſurpaſſing (as in all other things) the curiouſeſt painting that Art can ſet downe. One ſquare is of a greeniſh yellow colour, the other purple, keeping the ſame order as well on the backſide of the floure as one the inſide, although they are blackiſh in one ſquare, and of a Violet colour in another ; inſomuch that euery leafe ſeemeth to be the feather of a Ginny hen, whereof it tooke his name. The root is ſmall, white, and of the bigneſſe of halfe a garden beane.

2   The ſecond kinde of Checquered Daffodill is like vnto the former in each reſpect, ſauing that this hath his floure daſht ouer with a light purple, and is ſomewhat greater than the other, wherein conſiſteth the difference.

‡ 3 *Fritillaria Aquitanica minor flore luteo obſoleto.*
The leſſer darke yellow Fritillarie.

‡ 9 *Fritillaria alba præcox.*
The early white Fritillarie.

‡ There are ſundry differences and varieties of the floure, taken from the colour, largenes, doubleneſſe, earlineſſe and lateneſſe of flouring, as alſo from the many or few branches bearing floures. We will onely ſpecifie their varieties by their names, ſeeing their forme differs little from thoſe you haue here deſcribed:

4   *Fritillaria maxima ramoſa purpurea.*   The greateſt branched purple checquered Daffodill.
5   *Fritillaria flore purpureo pleno.*   The double purple floured checquered Daffodill.
6   *Fritillaria polyanthos flauoviridis.*   The yellowiſh greene many floured checquered Daffodill.
7   *Fritillaria lutea Someri.*   Somers his yellow checquered Daffodill.
8   *Fritillaria alba purpureo teſſulata.*   The white Fritillarie checquered with purple.
9   *Fritillaria alba præcox.*   The early white Fritillarie or checquered Daffodill.
10   *Fritillaria minor flore luteo obſoleto.*   The leſſer darke yellow Fritillarie.
11   *Fritillaria anguſtifolia lutea variegata paruo flore, & Altera flore majore.*   Narrow leaued yellow variegated Fritillarie with ſmall floures ; and another with a larger floure.
12   *Fritillaria minima pluribus floribus.*   The leaſt Fritillarie with many floures.

13 *Fritilla-*

Fritillaria Hiſpanica vmbellifera, The Spaniſh Fritillaria with the floures ſtanding as it were in anvmbell. ‡

The Ginny hen floure is called of *Dodonæus, Flos Meleagris* : of *Lobelius, Lilio-narciſſus variegata*, for that it hath the floure of a Lilly, and the root of *Narciſſus* : it hath beene called *Frit llaria*, of the table or boord vpon which men play at Cheſſe, which ſquare checkers the floure doth very much reſemble ; ſome thinking that it was named *Fritillus* : whereof there is no certainty ; for *Martial* ſeemeth to call *Fritillus, Abacus*, or the Tables whereon men play at Dice, in the fifth booke of his Epigrams, writing to *Galla*.

> *Iam triſtis, Nucibus puer relictis,*
> *Clamoſo reuocatur à magiſtro :*
> *Et blando male proditus Fritillo*
> *Arcana modò raptus è popina*
> *Ædilem rogat vdus aleator,&c.*

The ſad Boy now his nuts caſt by,
Is call'd to Schoole by Maſters cry :
And the drunke Dicer now betray'd
By flattering Tables as he play'd,
Is from his ſecret tipling houſe drawne out,
Although the Officer he much beſought, &c.

In Engliſh we may call it Turky-hen or Ginny-hen Floure, and alſo Checquered Daffodill, and Fritillarie, according to the Latine.

Of the facultie of theſe pleaſant floures there is nothing ſet downe in the antient or later Writer, but are greatly eſteemed for the beautifying of our gardens, and the boſoms of the beautifull.

CHAP. 90.　*Of true Saffron, and the wilde or Spring Saffron.*

Crocus florens & ſine flore.　Saffron with and without floure.

 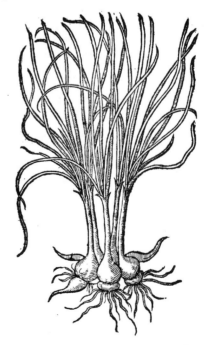

¶ *The Defcription.*

ALthough I haue expreffed two pictures of Saffron, as you fee, yet are you to vnderftand, that thefe two do but fet forth one kinde of plant, which could not fo eafily be perceiued by one figure as by two, becaufe his floure doth firft rife out of the ground nakedly in September, and his long fmal graffie leaues fhortly after the floure, neuer bearing floure and leafe at once. The which to expreffe, I thought it conuenient to fet downe two figures before you, with this defcription, *viz.* The root is fmall, round, and bulbous. The floure confifteth of fix fmall blew leaues tending to purple, hauing in the middle many fmall yellow ftrings or threds ; among which are two, three, or more thicke fat chiues of a fierie colour fomewhat reddifh, of a ftrong fmell when they be dried, which doth ftuffe and trouble the head. The firft picture fetteth forth the plant when it beareth floures, and the other expreffeth nothing but leaues.

¶ *The Place.*

Common or beft knowne Saffron groweth plentifully in Cambridge-fhire, Saffron-Waldon, and other places thereabout, as corne in the fields.

¶ *The Time.*

Saffron beginneth to floure in September, and prefently after fpring vp the leaues, and remaine greene all the Winter long.

¶ *The Names.*

Saffron is called in Greeke κρκος : in Latine, *Crocus* : in Mauritania, *Saffaran* : in Spanifh, *Açaffron* : in Englifh, Saffron : in the Arabicke tongue, *Zahafaran*.

¶ *The Temperature.*

Saffron is a little aftringent or binding ; but his hot qualitie doth fo ouer-rule in it, that in the whole effence it is in the number of thofe herbs which are hot in the fecond degree, and dry in the firft : therefore it alfo hath a certaine force to concoct, which is furthered by the fmall aftriction that is in it, as *Galen* faith.

¶ *The Vertues.*

A   *Avicen* affirmeth, That it caufeth head-ache, and is hurtfull to the braine, which it cannot do by taking it now and then, but by too much vfing of it ; for the too much vfing of it cutteth off fleep, through want whereof the head and fences are out of frame. But the moderat vfe thereof is good for the head, and maketh the fences more quicke and liuely, fhaketh off heauy and drowfie fleepe, and maketh a man merry.

B   Alfo Saffron ftrengthneth the heart, concocteth crude and raw humors of the cheft, opens the lungs, and remoueth obftructions.

C   It is alfo fuch a fpeciall remedie for thofe that haue confumption of the lungs, and are, as wee terme it, at deaths doore, and almoft paft breathing, that it bringeth breath again, and prolongeth life for certaine dayes, if ten or twenty graines at the moft be giuen with new or fweet Wine. For we haue found by often experience, that being taken in that fort, it prefently and in a moment remoueth away difficulty of breathing, which moft dangeroufly and fuddenly hapneth.

D   *Diofcorides* teacheth, That being giuen in the fame fort it is alfo good againft a furfet.

E   It is commended againft the ftoppings of the liuer and gall, and againft the yellow jaundife : and hereupon *Diofcorides* writeth, That it maketh a man well coloured. It is put into all drinkes that are made to helpe the difeafes of the intrals, as the fame Author affirmeth, and into thofe fpecially which bring downe the fleures, the birth, and the after-burthen. It prouoketh vrine, ftirreth flefhly luft, and is vfed in cataplafmes and pulteffes for the matrix and fundament, and alfo in plaifters and feare-cloathes which ferue for old fwellings and aches, and likewife for hot fwellings that haue alfo in them S. Anthonies fire.

F   It is with good fucceffe put into compofitions for infirmities of the eares.

G   The eyes being anointed with the fame diffolued in milke or fennel or rofe water, are preferued from being hurt by the fmall pox or meafels, and are defended thereby from humors that would fal into them.

H   The chiues fteeped in water ferue to illumine or (as we fay) limne pictures and imagerie, as alfo to colour fundry meats and confections. It is with good fucceffe giuen to procure bodily luft. The Confections called *Crocomagna, Oxycroceum,* and *Diacurcuma,* with diuers other emplaifters and electuaries, cannot be made without this Saffron.

I   The weight of ten grains of Saffron, the kernels of Walnuts two ounces, Figs two ounces, Mithridate one dram, and a few Sage leaues ftamped together with a fufficient quantitie of Pimpernel water, and made into a maffe or lumpe, and kept in a glaffe for your vfe, and thereof 12 graines giuen in the morning fafting, preferueth from the peftilence, and expelleth it from thofe that are infected.

¶ *The*

1 *Crocus vernus.*
Early flouring wilde Saffron.

2 *Crocus vernus minor.*
Small wilde Saffron.

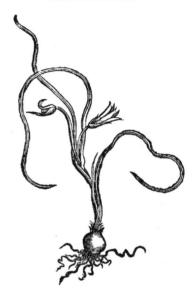

‡ 3 *Crocus vernus flore luteo.*
Yellow Spring Saffron.

‡ 4 *Crocus vernus flore albo.*
White Spring Saffron.

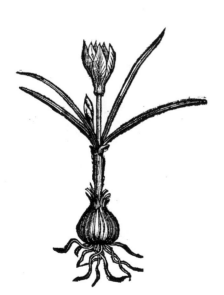

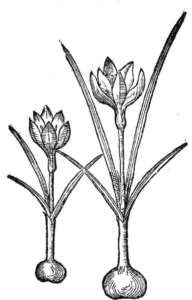

‡ 5 *Crocus*

‡ 5 *Crocus vernus flore purpureo.*
Purple Spring Saffron.

‡ 6 *Crocus montanus Autumnalis.*
Autumne mountaine Saffron.

‡ 7 *Crocus montanus Autumnalis flore maiore*
albido cæruleo.
Autumne mountaine Saffron with a
large whitiſh blew floure.

‡ 8 *Crocus Autumnalis flore albo.*
White Autumne Saffron.

¶ *The*

‡ 9 *Crocus vernus angustifolius flore vio-*
laceo.
Narrow leafed Spring Saffron
with a violet floure.

‡ 10 *Crocus vernus latifolius flore flavo*
strijs violaceis.
Broad leafed Spring Saffron with
yellow floure and purple stre·ks.

‡ 11 *Crocus vernus latifolius striatus flore duplici.*
Double floured streaked Spring Saffron.

¶ *The Kindes of Spring Saffron.*

OF wild Saffrons there be sundry sorts,
differing as well in the colour of the
floures, as also in the time of his flouring. Of which, most of the figures shall be
set forth vnto you.

¶ *The description of wilde Saffron.*

1 THe first kind of wilde Saffron
hath small short grassie leaues,
furrowed or channelled downe
the midst with a white line or streak: among
the leaues rise vp small floures in shape like
vnto the common Saffron, but differing in
color; for this hath floures of mixt colors;
that is to say, the ground of the floure is
white, stripped vpon the backe with purple,
and dasht ouer on the inside with a bright shining murrey color; the other not. In the middle of
the floures come forth many yellowish chiues, without any smell of Saffron at all. The root is
small, round, and couered with a browne skin or filme like vnto the roots of common Saffron.

2   The second wilde Saffron in leaues, roots, and floures is like vnto the precedent, but altogether lesser, and the floures of this are of a purple violet colour.

3   We haue likewise in our London gardens another sort, like vnto the other wilde Saffrons
in

‡ 12 *Crocus vernus latifolius flore purpureo.*
Broad leaued Spring Saffron with
the purple floure.

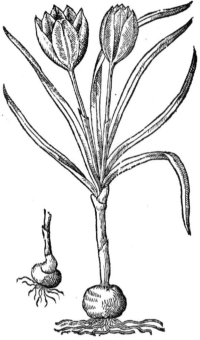

‡ 13 *Crocus vernus flore cinereo ſtriato.*
Spring Saffron with an Aſh-coloured ſtreaked floure.

‡ 14 *Crocus vernus latifolius flore flavo-vario
duplici.*
Broad leaued Spring Saffron with a double
floure yellow and ſtreaked.

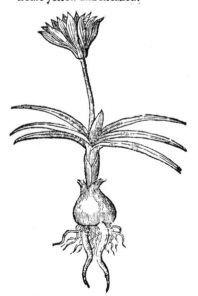

in euery point, ſauing that this hath floures of
amoſt perfect ſhining yellow colour, ſeeming
a far off to be a hot glowing cole of fire, which
maketh the difference.

4   There is found among Herbariſts another ſort, not differing from the others, ſauing
that this hath white floures, contrary to all the
reſt.

5   Louers of plants haue gotten into their
gardens one ſort hereof with purple or Violet
coloured flours, in other reſpects like vnto the
former.

6   Of theſe we haue another that floureth
in the fall of the leafe, with floures like to the
common Saffron, but deſtitute of thoſe chiues
which yeeld the colour, ſmell, or taſte that the
right manured Saffron hath.

‡ 7   And of this laſt kinde there is another with broader leaues, and the floure is alſo
larger, with the leaues thereof not ſo ſharpe
pointed, but more round ; the colour being at
the firſt whitiſh, but afterward intermixt with
ſome blewneſſe. ‡

8   There is alſo another of Autumne wild
Saffrons with white floures, which ſets forth
the diſtinction.

Many ſorts there are in our gardens beſides
thoſe before ſpecified, which I thought needleſſe to intreat of, becauſe their vſe is not great
‡ Therefore I will onely giue the figures and
names of ſome of the chiefe of them, and refer
ſuch as delight to ſee or pleaſe themſelues
with the varieties (for they are no ſpecifique
differences) of theſe plants, to the gardens and
the bookes of Floriſts, who are onely the pre-
ſeruers and admirers of theſe varieties , not
ſought after for any vſe but delight. ‡

¶ *The*

¶ *The Place.*

All thefe wilde Saffrons we haue growing in our London gardens. Thofe which doe floure in Autumne do grow vpon certaine craggy rocks in Portugall,not farre from the fea fide. The other haue been fent ouer vnto vs,fome out of Italy,and fome out of Spaine,by the labour and diligence of that notable learned Herbarift *Carolus Clufius*; out of whofe Obferuations, and partly by feeing them in our owne gardens,we haue fet downe their defcription.

That pleafant plant that bringeth forth yellow floures was fent vnto me from *Robinus* of Paris, that painfull and moft curious fearcher of Simples.

¶ *The Time.*

They floure for the moft part in Ianuarie and Februarie ; that of the mountain excepted,which floureth in September.

¶ *The Names.*

All thefe Saffrons are vnprofitable,and therefore they be truly faid to be *Croci fylveftres,* or wild Saffrons : in Englifh,Spring Saffrons,and vernall Saffrons.

¶ *The Nature and Vertues.*

Of the faculties of thefe we haue nothing to fet downe,for that as yet there is no knowne vfe of them in Phyficke.

Chap. 91. *Of Medow Saffron.*

¶ *The Kindes.*

THere be fundry forts of medow Saffrons, differing very notably as well in the colour of their floures,as alfo in nature and country from whence they had their being, as fhall be declared.

1 *Colchicum Anglicum purpureum.*
Purple Englifh Medow Saffron.

2 *Colchicum Anglicum album.*
White Englifh Medow Saffron.

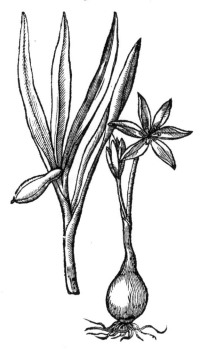

¶ *The*

¶ *The Deſcription.*

1   MEdow Saffron hath three or foure leaues riſing immediately forth of the ground, long, broad, ſmooth, fat, much like to the leaues of the white Lillie in forme and ſmoothneſſe: in the middle whereof ſpring vp three or foure thicke cods of the bigneſſe of a ſmall Wall-nut, ſtanding vpon ſhort tender foot-ſtalkes, three ſquare, and opening themſelues when they be ripe, full of ſeed ſomething round, and of a blackiſh red colour: and when this ſeed is ripe, the leaues together with the ſtalkes doe fade and fall away. In September the floures bud forth, before any leaues appeare, ſtanding vpon ſhort tender and whitiſh ſtemmes, like in forme and colour to the floures of Saffron, hauing in the middle ſmall chiues or threds of a pale yellow colour, altogether vnfit for meat or medicine. The root is round or bulbous, ſharper at the one end than at the other, flat on the one ſide, hauing a deepe clift or furrow in the ſame flat ſide when it floureth, and not at any time elſe: it is couered with blackiſh coats or filmes: it ſendeth downe vnto the loweſt part certaine ſtrings or threds. The root it ſelfe is full of a white ſubſtance, yeelding a juyce like milke, whileſt it is greene and newly digged out of the earth. It is in taſte ſweet, with a little bitterneſſe following, which draweth water out of the mouth.

3   *Colchicum Pannonicum florens & ſine flore.*
Hungary Mede Saffron with and without floure.

 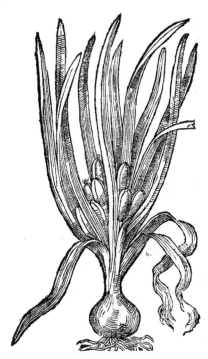

2   The ſecond kinde of Mede Saffron is like the precedent, differing onely in the colour of the floures, for that this plant doth bring forth white leaues, which of ſome hath beene taken for the true *Hermodactylus*; but in ſo doing they haue committed the greater error.

3   Theſe two figures expreſſe both but one and the ſelfe ſame plant, which is diſtinguiſhed becauſe it neuer beareth floures and leaues both at one time. So that the firſt figure ſets it forth when it is in leaues and ſeed, and the other when it floureth; and therefore one deſcription ſhall ſuffice for them both. In the Spring of the yeare it bringeth forth his leaues, thicke, fat, ſhining, and ſmooth, not vnlike the leaues of Lillies, which doe continue greene vnto the end of Iune; at which time the leaues do wither away, but in the beginning of September there ſhooteth forth of the ground naked milke white floures without any greene leafe at all: but ſo ſoone as the Plant hath done bearing of floures, the root remaines in the ground, not ſending forth any thing vntill Februarie in the yeare following.

It

‡ It beares plentifull ftore of reddifh feed in loofe triangular heads. The root hereof is bigger than that of the laft defcribed. ‡

† 4   The fmall Medow Saffron hath three or foure thicke fat leaues narrower than any of
the reft. The floure appeareth in the fall of the leafe, in fhape, colour and manner of growing like
the common mede Saffron, but of a more reddifh purple colour, and altogether leffer. The leaues
in this, contrary to the nature of thefe plants, prefently follow after the floure, and fo continue all
the Winter and Spring, euen vntill May or Iune. The root is bulbous and not great ; it is couered with many blackifh red coats, and is white within.

‡ 5   This medow Saffron hath roots and leaues like to thofe of the laft defcribed, but the
leaues of the floure are longer and narrower, and the colour of them is white on the infide, greene
on the middle of the backe part, and the reft thereof a certaine flefh colour.

4 *Colchicum montanum minus Hiſpanicum cum flore & ſemine.*
Small Spanifh Medow Saffron in floure and feed.

6   The medow Saffron of Illyria hath a great thicke and bulbous root, full of fubftance : from
which rifeth vp a fat, thicke, and groffe ftalke, fet about from the lower part to the top by equall
diftances, with long, thicke and groffe leaues, fharpe pointed, not vnlike to the leaues of leekes ;
among which leaues do grow yellowifh floures like vnto the Englifh medow Saffron, but fmaller.

7   The Affyrian medow Saffron hath a bulbous root, made as it were of two pieces ; from the
middle cleft whereof rifeth vp a foft and tender ftalke fet with faire broad leaues from the middle
to the top : among which commeth forth one fingle floure like vnto the common medow Saffron,
or the white Anemone of *Matthiolus* defcription.

8   The mountaine wilde Saffron, is a bafe and low plant, but in fhape altogether like the common medow Saffron, but much leffer. The floures are fmaller, and of a yellow colour, which fetteth forth the difference. ‡ The leaues and roots (as *Cluſius* affirmes) are more like to the Narcif-
fes ; and therefore he calls this *Narciſſus autumnalis minor*, the leffer Autumne Narciffe. ‡.

‡ 9   This, whofe figure we here giue you, is by *Cluſius* called *Colchicum Biȝantinum latifolium*,
The broad leaued *Colchicum* of Conftantinople. The leaues of this are not in forme and magnitude much vnlike to thofe of the white Hellebor, neither leffe neruous, yet more greene. It beares
many floures in Autumne, fo that there comes fometimes twenty from one root. Their forme and
colour are much like the ordinary fort, but that thefe are larger, and haue thicker ftalkes. They
are

are of a lighter purple without, and of a deeper on the infide, and they are marked with certaine veines running alongft thefe leaues. The roots and feeds of this plant are thrice as large as thofe of the common kinde.

10   This hath roots and leaues like to the firft defcribed, but the floure is fhorter, and growes vpon a fhorter ftalke, fo that it rifeth but little aboue the earth : the three inner leaues are of a reddifh purple ; the three out leaues are either wholly white, or purplifh on the middle in the infide, or ftreaked with faire purple veins, or fpotted with fuch coloured fpots : all the leaues of the floure are blunter and rounder than in the common kinde.

11   This in leaues, roots, manner and time of growing, as alfo in the colour of the floures, differs not from the firft defcribed, but the floures, as you may perceiue by the figure here expreffed, are very double, and confift of many leaues.

‡ 5 *Colchicum montanum minus verficolore flore.*
The leffer mountaine Saffron with a various coloured floure.

6 *Colchicum Illyricum.*
Greeke medow Saffron.

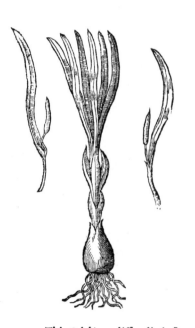
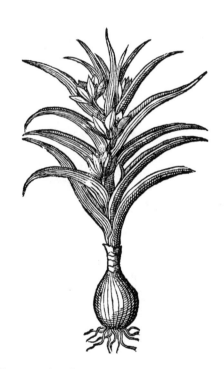

12   This *Colchicum* differs little from the firft ordinary one, but that the floures are fomewhat leffe, and the three out-leaues are fomewhat bigger than the three inner leaues ; the colour is a little deeper alfo than that of the common one ; but that wherein the principall difference confifts, is, That this floures twice in a yeare, to wit, in the Spring and Autumne : and hence *Clufius* hath called it *Colchicum biflorum*, Twice flouring Mede Saffron.

13   This alfo in the fhape of the root and leaues is not much differing from the ordinary, but the leaues of the floure are longer and narrower, the colour alfo when they begin to open and fhew themfelues, is white, but fhortly after they are changed into a light purple : each leafe of the floure hath a white thread tipt with yellow growing out of it, and in the middle ftands a white three forked one longer than the reft. The floure growes vp betweene three or foure leaues, narrower than thofe of the ordinary one, and broader than thofe of the fmall Spanifh kinde. *Clufius*, to whom we are beholden for this, as alfo for moft of the reft, calls it *Colchicum vernum*, or Spring Mede-Saffron, becaufe it then floures together with the Spring Saffrons and Dogs Tooth.

14   There are other Mede-Saffrons befides thefe I haue mentioned, but becaufe they may be
referred

7 *Colchicum Syriacum Alexandrinum.*
Aſſyrian Mede Saffron.

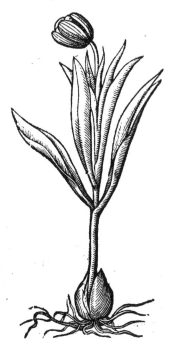

referred eaſily to ſome of theſe, for that their difference chiefely conſiſts either in the doubleneſſe or colour of the floures, whereof ſome are ſtriped, ſome fraided, others variegated, I will not inſiſt vpon them, but referre ſuch as deſire their further acquaintance to look into the gardens of our Floriſts, as Mr. *Parkinſons*, Mr. *Tuggies*, &c. or elſe into the booke of floures ſet forth not long ſince by Mr. *Parkinſon*, where they ſhall finde them largely treated of. Yet I cannot paſſe ouer in ſilence that curious *Colchicum* which is called by ſome, *Colchicum variegatum Chienſe*. The floure thereof is very beautiful, conſiſting of ſix pretty broad and ſharp pointed leaues, all curiouſly checkered ouer with deepe blew or purple, the reſt of the floure being of a light whitiſh color: the leaues, that riſe vp in the Spring, are not very long, but ſomewhat broad and ſharpe pointed; the root is like others of this kinde. I haue giuen you an exact and large figure of this, as I tooke it from the growing floure ſome three yeares agone, it being at that time amongſt her Maieſties floures kept at Edgcome in Surry, in the Garden of my much honoured friend Sir *Iohn Tunſtall*, Gentleman Vſher vnto her Maieſtie.

15 I giue you here in this place the true Hermodactill of the ſhops, which probably by all is adiudged to this Tribe, though none can certainly ſay what floures or leaues it beares: the Roots are onely brought to vs, and from what place I cannot tell; yet I coniecture from ſome part of Syria or the adiacent countries. Now how hard it is to iudge of Plants by one part or particle, I ſhall ſhew you more at large when I come to treat of *Piſtolochia*, wherefore I will ſay nothing thereof in this place. Theſe roots, which wanting the maligne qualitie of *Colchicum*, either of their owne nature, or by dryneſſe, are commonly about the bigneſſe of a Cheſnut, ſmooth, flattiſh, and ſharpe at the one end, but ſomewhat full at the other, where the ſtalke of the floure comes vp. Their colour is either white, browne, or blackiſh on the outſide, and very white within, but thoſe are the beſt that are white both without and within, and may eaſily be made into a fine white meale or pouder. ‡

8 *Colchicum parvum montanum luteum.*
Yellow mountaine Saffron.

‡ 9 *Colchicum latifolium.*   Broad leaued Mede Saffron.

‡ 10 *Colchicum verſicolore flore.*   Party-coloured Mede Saffron.

¶ The Place.

Medow Saffron,or *Colchicum*,groweth in Meſſinia,and in the Iſle of Colchis, whereof it tooke his name. The titles of the reſt do ſet forth their natiue countries ; notwithſtanding our London gardens are poſſeſſed with the moſt part of them.

The two firſt do grow in England in great aboundance in fat and fertile medowes, as about Vilford and Bathe,as alſo in the medowes neere to a ſmall village in the Weſt part of England,called Shepton Mallet,in the medowes about Briſtoll,in Kingſtrop medow neere vnto a water mil as you go from Northampton to Holmby houſe, vpon the right hand of the way, and likewiſe in great plenty in Nobottle wood two miles from Northampton,and in many other places.

‡   The reſt for the moſt part may be found in the gardens of the Floriſts among vs.   ‡

¶ The Time.

The leaues of all the kindes of Mede Saffron doe begin to ſhew themſelues in Februarie. The ſeed is ripe in Iune. The leaues,ſtalks,and ſeed do periſh in Iuly,and their pleaſant floures do come forth of the ground in September.

¶ The Names.

Dioſcorides calleth Medow Saffron κολχικόν : ſome, Ἐφήμερον : notwithſtanding there is another *Ephemeron* which is not deadly.Diuers name it in Latine *Bulbus agreſtis*, or wild Bulbe : in high Dutch it is called **Zeitlooſen :** in low Dutch, **Tilelooſen :** in French, *Mort au chien.* Some haue taken it to be the true Hermodactyl,yet falſely. Other ſome call it *Filius ante patrem*, although there is a kinde of *Lyſimachia* or Looſe-ſtrife ſo called,becauſe it firſt bringeth forth his long cods with ſeed, and then the floure after,or at the ſame time at the end of the ſaid cod. But in this Mede Saffron it is far otherwiſe, becauſe it bringeth forth leaues in Februarie, ſeed in May, and floures in September ; which is a thing cleane contrarie to all other plants whatſoeuer, for that they doe firſt floure,and after ſeed : but this Saffron ſeedeth firſt,and foure moneths after brings forth floures : and therefore ſome haue thought this a fit name for it, *Filius ante Patrem:* and we accordingly may call

‡ 11 *Colchicum flore pleno.*
Double floured Mede Saffron.

‡ 12 *Colchicum biflorum.*
Twice flouring Mede Saffron.

‡ 13 *Colchicum vernum.*
Spring Mede Saffron.

‡ 14 *Colchicum variegatum Chienſe.*
Checquered Mede Saffron of Chio.

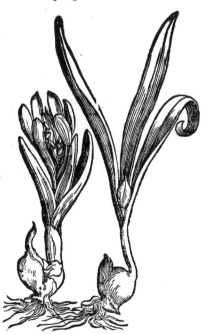

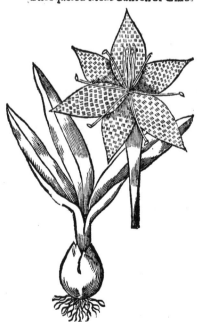

‡ 15 *Hermodactyli Officinarum.*
The true Hermodactyls of the ſhops.

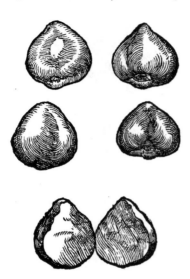

call it, The Sonne before the Father.

‡   Our Author in this chapter was of many mindes ; for firſt, in the deſcription of *Colchicum Anglicum*, being the ſecond, hee reproues ſuch as make that white floured *Colchicum* the true Hermodactyl. Then in the deſcription of the eighth hee hath theſe words, which being omitted in that place I here ſet downe. *Of all theſe kindes* (ſaith he) *of Medow Saffrons it hath not beene certainely knowne which hath been the true Hermodactyll ; notwithſtanding wee haue certaine knowledge that the Illyrian Colchicum is the Phyſicall Hermodactyll.* Yet when he comes to ſpeake of the names, after that out of *Dodonæus* he had ſet downe the truth in theſe words: *But notwithſtanding that Hermodactyl which we do vſe in compound medicines, differeth from this* (to wit, *Colchicum*) *in many notable points, for that the true Hermodactyll hath a bulbe or round root, which being dried continueth very white within, and without not wrinkled at all, but full and ſmooth, of a meane hardneſſe ;* and that he had out of the ſame Authour alledged the words of *Valerius Cordus* and *Auicen*, (which are here omitted) he concludes contrary to the truth, his firſt admonition, and ſecond aſſertion, That the white Medow Saffron which wee haue in the Weſt part of England, growing eſpecially about Shepton Mallet, is the Hermodactyll vſed in ſhops.

Thoſe we haue in ſhops ſeeme to be the Hermodactyls of *Paulus Ægineta* ; yet not thoſe of *Nicholaus* and *Actuarius*, which were cordial, and increaſers of ſperme ; the which the Authours of the *Aduerſaria. pag. 55.* thinke to be the *Behen album & rubrum* of the Arabians. And to theſe vnknowne ones are the vertues ſet downe by our Author in the third place vnder C, to be referred. ‡

¶ *The Temperature.*

Medow Saffron is hot and dry in the third degree.

¶ *The Vertues of Hermodactyls.*

A   †   The roots of Hermodactyls are of force to purge, and are properly giuen (ſaith *Paulus*) to thoſe that haue the Gout, euen then when the humors are in flowing. And they are alſo hurtfull to the ſtomacke.

B   The ſame ſtamped, and mixed with the whites of egges, barley meale, and crums of bread, and applied plaiſterwiſe, eaſe the paine of the Gout, ſwellings and aches about the joynts.

C   The ſame ſtrengthneth, nouriſheth, and maketh good juyce, encreaſeth ſperme or naturall ſeed, and is alſo good to cleanſe vlcers and rotten ſores.

¶ *The Correction.*

The pouder of Ginger, long Pepper, Anniſe ſeed or Cumine ſeed, and a little Maſticke, correct the churliſh working of that Hermodactyll which is vſed in Shops. But thoſe which haue eaten of the common Medow Saffon muſt drinke the milke of a cow, or elſe death preſently enſueth.

¶ *The Danger.*

The roots of all the ſorts of Mede Saffrons are very hurtfull to the ſtomacke, and being eaten they kill by choaking as Muſhromes do, according vnto *Dioſcorides* ; whereupon ſome haue called it *Colchicum ſtrangulatorium.*

†   That which was ſet forth by our Author in the fourth place, vnder the title of *Colchicum montanum minus*, was nothing but the former *Colchicum minus* expreſſed in ſeed: The nineth and tenth were the ſame with the firſt and ſecond. The ſixth and ſeuenth, which are *Colchicum Illyricum* and *Syriacum* I haue left with their figure and hiſtorie, though they be ſuſpected to be counterfeits : and *Cluſius* probably gueſſes, that the later is the Apennine Tulip, the Painter making the leaues of the floure too round, and thoſe of the plant too broad and ſhort †

CHAP. 92.   Of Star of Bethlem.

¶ The Kindes.

THere be ſundry kindes of wilde field Onions called Stars of Bethlem, differing in ſtature, taſt and ſmell, as ſhall be declared.

1 Ornithogalum.
Star of Bethlem.

2 Ornithogalum luteum, ſiue Cepe agraria.
Yellow or wilde Star of Bethlem.

¶ The Deſcription.

1   OVr common Starre of Bethlehem hath many narrow leaues, thicke, fat, full of juice, and of a very greene colour, with a white ſtreake downe the middle of each leafe: among the which riſe vp ſmall naked ſtalkes, at the top whereof grow floures com-paƈt of ſix little leaues, ſtripped on the backe ſide with lines of greene, the inſide being milke-white. Theſe floures open themſelues at the riſing of the Sunne, and ſhut againe at the Sun-ſet-ting; whereupon this plant hath been called by ſome, Bulbus Solſequius. The floures being paſt, the ſeed doth follow, incloſed in three cornered huskes. The root is bulbous, white both within and without.

† 2   The ſecond ſort hath two or three graſſy leaues proceeding from a clouen bulbous root. The ſtalke riſeth vp in the midſt naked, but toward the top there doe thruſt forth more-leaues like vnto the other, but ſmaller and ſhorter; among which leaues doe ſtep forth very ſmall weake and tender foot-ſtalks. The floures of this are on the backeſide of a pale yellow, ſtripped with greene, on the inſide of a bright ſhining yellow colour, with Saffron coloured threds in their middles. The ſeed is contained in triangular veſſels.

† 3   This Starre of Hungarie, contrarie to the cuſtome of other plants of this kinde, ſendeth forth before Winter fiue or ſix leaues ſpread vpon the ground, narrow, and of ſome fingers length, ſomewhat whitiſh greene, and much reſembling the leaues of Gillofloures, but ſomewhat rough-iſh. In Aprill the leaues beginning to decay, amongſt them riſes vp a ſtalke bearing at the top a

ſpoke

ſpoke of floures,which conſiſting of ſix leaues apiece ſhew themſelues open in May. They in colour are like the firſt deſcribed, as alſo in the greene ſtreake on the lower ſide of each leafe. The
ſeed is blacke,round,and contained in triangular heads. The root is bulbous, long, and white. †

† 4 This fourth,which is the Ornithogalum Hiſpanicum minus.of Cluſius,hath a little white root
which ſends forth leaues like the common one,but narrower, and deſtitute of the white line wherwith the other are marked. The ſtalke is ſome two handfulls high,bearing at the top thereof ſome
ſeuen or eight floures,growing each aboue other,yet ſo,as that they ſeem to make an vmbel : each
of theſe floures hath ſix leaues of a whitiſh blew colour,with ſo many white chiues or threds,and
a little blewiſh vmbone in the midſt. This floures in Aprill.

. 5   This fifth, firſt ſends vp one only leafe two or three inches long,narrow, and of a whitiſh colour,and of an acide taſte : nigh whereto riſeth vp a ſmall ſtalke ſome inch or two high, hauing one
or two leaues thereon,betweene which come forth ſmall ſtarre floures,yellow within,and of a greeniſh purple without. The ſeed,which is reddiſh and ſmall,is contained in triangular heads. The
root is white,round,and couered with an Aſh-coloured filme.

3 Ornithogalum Pannonicum.
   Star of Hungarie.

‡ 4 Ornithogalum Hiſpanicum minus.
    The leſſer Spaniſh Star-floure.

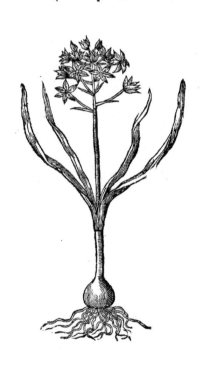

6 I thinke it not amiſſe hereto to adde another ſmall bulbous plant,which Cluſius calls Bulbus
uniunicus,The one leaued Bulbe. This from a ſmall root ſends forth one ruſh leafe of ſome foot in
length,which about two inches aboue the earth,being ſomewhat broader than in the other places,
and guttered,ſends forth a little ſtalke ſome three inches long, whoſe top is ſet with three little
floures,each ſtanding aboue other,about the bigneſſe here preſented vnto your view in the figure :
each of thoſe conſiſteth of ſix very white leaues,and are not much vnlike the floures of the Graſſe
of Parnaſſus,but yet without leaues to ſuſtain the floure,as it hath : ſix white threds tipt with yellow, and a three ſquare head with a white pointall poſſeſſe the midſt of the floure ; the ſmel thereof is ſomewhat like that of the floures of the Hawthorne. It floures in the midſt of Iune.

7   Hauing done with theſe two ſmall plants, I muſt acquaint you with three or foure larger,
belonging alſo to this Claſſis. The firſt of theſe is that which Dodonæus calls Ornithogalum majus,
and Cluſius,Ornithogalum Arabicum : This by Lobel and ſome others is called Lilium Alexandrinum,

‡ 5 *Ornithogalum luteum parvum.*
Dwarfe yellow Star of Bethlehem.

‡ 6 *Bulbus vnifolius.*
The one leaued Bulbe.

‡ 7 *Ornithogalum majus Arabicum.*
The great Arabicke Star-floure.

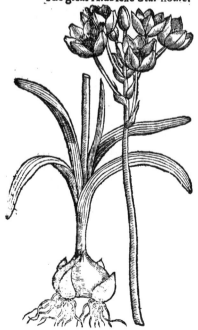

or the Lilly of Alexandria, as our Author calls it
in the Chapter of Cotton graſſe. This faire, but
tender plant, hath broad greene leaues comming
from a large white flat bottomed root ; among
which riſeth vp a ſtalke ſome cubit high , whoſe
top is garniſhed with ſundry pretty large floures
made of ſix pretty white leaues , with a ſhining
blackiſh head, ingirt with ſix white threds tipped
with yellow. This floures in May.

8   This, which is commonly called *Ornitho-*
galum ſpicatum, hath large leaues and roots, and the
ſtalke growes ſome cubit or more high, whereon
grow many ſtarre-floures in ſhape and colour like
thoſe of the ordinarie, but larger, and they begin
to floure below, and floure vpwards to the top.
There is a larger ſort of this *Spicatum,* whoſe floures
are not ſtreaked with greene on their backs. There
is alſo a leſſer, differing from the firſt of theſe only in bigneſſe.

9   This Neapolitan hath three or foure long
leaues not much vnlike thoſe of the Hyacinths,
but narrower. The ſtalke is pretty thicke, ſome
foot high, and hath vſually growing thereon ſome
fiue or ſix floures hanging one way , though their
ſtalkes grow alternately out of each ſide of the
maine ſtemme. Theſe floures are compoſed of
ſix leaues , being about an inch long , and ſome
quarter of an inch broad, white within, and of an
Aſh-coloured greene without, with white edges,
the

the middle of the floure is poſſeſſed by another little floure, conſiſting alſo of ſix little leaues, ha-uing in them ſix threds headed with yellow, and a white pointall. A blacke wrinkled ſeed is con-tained in three cornered heads, which by reaſon of their bigneſſe weigh downe the ſtalke. This floures in Aprill.

‡ 8 *Ornithogalum ſpicatum.*   ‡ 9 *Ornithogalum Neapolitanum.*
Spike faſhioned Star floure.   The Neapolitan Star-floure.

¶ *The Place.*

Stars of Bethlehem, or Star-floures, eſpecially the firſt and ſecond, grow in ſundry places that lie open to the aire, not onely in Germany and the Low-countries, but alſo in England, and in our gardens very common. The yellow kinde *Lobel* found in Somerſet-ſhire in the corne fields. The reſt are ſtrangers in England; yet we haue moſt of them, as the third, fourth, eighth, and ninth, in ſome of our choice gardens.

¶ *The Time.*

Theſe kindes of bulbed plants do floure from Aprill to the end of May.

¶ *The Names.*

Touching the names, *Dioſcorides* calls it Ὀρνιθόγαλον: *Pliny, Ornithogale:* in high Dutch it is called **Feldo; wibel, Acker; wibel:** as you ſhould ſay, *Cepa agraria:* in Engliſh, Stars of Bethlehem.

‡ The reſt are named in their titles & hiſtories: but *Cluſius* queſtions whether the *Bulbus vni-folius* be not *Bulbine* of *Theophraſtus, 7, hiſt. 13. Bauhinus* ſeemes to affirme the *Spicatum* to be the *Moly* of *Dioſcorides* and *Theophraſtus,* and *Epimedium* of *Pliny.* ‡

¶ *The Temperature.*

Theſe are temperate in heate and drineſſe.

¶ *The Vertues.*

A   The vertues of moſt of them are vnknown: yet *Hieronymus Tragus* writeth, That the root of the Star of Bethlehem roſted in hot embers, and applied with honey in manner of a cataplaſme or pul-teſſe, healeth old eating vlcers, and ſoftens and diſcuſſes hard tumors.

The roots, ſaith *Dioſcorides,* are eaten both raw and boiled.

† That which was the ſecond of our Author, vnder the title of *Cepa agraria,* and the third vnder *Ornithogalum luteum,* were figures of the ſame plant, but in the latter, as *Bauhine* obſerues, the bottome leaues are omitted, becauſe they fall away when as it is growne vp to a floure.

Chap.

CHAP. 93. *Of Onions.*

¶ *The Kindes.*

THere be, ſaith *Theophraſtus*, diuers ſorts of Onions, which haue their ſyr-names of the places where they grow : ſome alſo leſſer, others greater ; ſome be round, and diuers long, but none wilde, as *Pliny* writeth.

1 *Cepa alba.*
White Onions.

‡ 3 *Cepa Hiſpanica oblonga.*
Longiſh Spaniſh Onions.

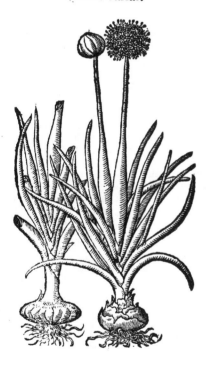
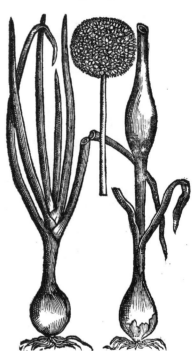

¶ *The Deſcription.*

1   THe Onion hath narrow leaues, and hollow within ; the ſtalke is ſingle, round, biggeſt in the middle, on the top whereof groweth a round head couered with a thin skin or film, which being broken, there appeare little white floures made up in form of a ball, and afterward blacke ſeed three cornered, wrapped in thin white skins. In ſtead of the root there is a bulbe or round head compact of many coats, which oftentimes becommeth great in manner of a Turnep, many times long like an egge. To be briefe, it is couered with very fine skins for the moſt part of a whitiſh colour.

2   The red Onion differeth not from the former but in ſharpneſſe and redneſſe of the roots, in other reſpects there is no difference at all.

‡ 3   There is alſo a Spaniſh kinde, whoſe root is longer than the other, but in other reſpects very little different.

4   There is alſo another ſmall kinde of Onion, called by *Lobel*, *Aſcalonitis Antiquorum*, or Scallions ; this hath but ſmall roots, growing many together. The leaues are like to Onions, but leſſe. It ſeldome beares either ſtalke, floure, or ſeed. It is vſed to be eaten in ſallads. ‡

¶ *The*

¶ *The Place.*

The Onion requireth a fat ground well digged and dunged , as *Palladius* faith. It is cheriſhed euerie where in kitchen gardens, now and then ſowne alone , and many times mixed with other herbs,as with Lettuce,Parſeneps, and Carrets. *Palladius* liketh well that it ſhould be ſowne with Sauorie,becauſe,faith *Pliny*,it proſpereth the better,and is more wholeſome.

‡ 4 *Aſcalonitides.*
Scallions.

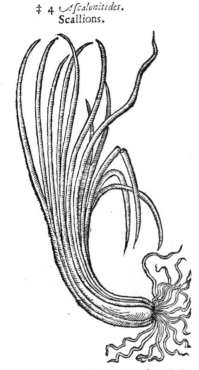

the Italians,*Cepolla*: the Spaniards,*Cebolla*,*Beba*,and *Cebola*.

¶ *The Time.*

It is ſowne in March or Aprill, and ſomtimes in September.

¶ *The Names.*

The Onion is called in Greeke, κεμμιο : in Latine *Cepa*,and many times *Cepe* in the neuter gender: the Shops keepe that name. The old Writers haue giuen vnto this many ſyr-names of the places where they grow, for ſome are named *Cypriæ, Sardiæ, Creticæ, Samothracia, Aſcaloniæ* of a towne in Iudæa otherwiſe called *Pompeiana*: in Engliſh,Onions. Moreouer, there is one named *Mariſco*,which the Countreymen call *Vnio*, faith *Columella*; and thereupon it commeth that the French men call it *Oignon*, as *Ruellius* thinketh: and peraduenture the Low-Dutch men name it 𝕬𝖚𝖊𝖚𝖎𝖒,of the French word corrupted.They are called *Setaniæ*,which are very little and ſweet;and theſe are thought to be thoſe which *Palladius* nameth *Cepullæ*, as though he called them *parua Cepæ*,or little Onions.

There is an Onion which is without an head or bulbe, and hath as it were a long necke , and ſpends it ſelf wholly in the leaues, and it is often cropped or cut for the pot, like the Leeks. This *Theophraſtus* names τησον : Of this *Pliny* alſo writeth,*Lib.19.Cap.6*. There is with vs two principall ſorts of Onions, the one ſeruing for a ſauce, or to ſeaſon meat with,which ſome call *Gethyon, Pállacana*:and the other is the headed or common Onion,which the Germanes call 𝕺𝖓𝖎𝖔𝖓 𝖟𝖜 𝖋𝖇𝖊𝖑:

¶ *The Temperature.*

All Onions are ſharpe,and moue teares by the ſmell. They be hot and dry,as *Galen* faith, in the fourth degree,but not ſo extreme hot as Garlicke. The iuice is of a thinne waterie and airie ſubſtance : the reſt is of thicke parts.

¶ *The Vertues.*

A   The Onions do bite,attenuate or make thin,and cauſe dryneſſe : being boiled they do loſe their ſharpneſſe,eſpecially if the water be twice or thrice changed,and yet for all that they doe not loſe their attenuating qualitie.

B   They alſo break wind, prouoke vrine,and be more ſoluble boiled than raw ; and rawt heynouriſh not at all,and but a little though they be boiled.

C   They be naught for thoſe that be cholericke , but good for ſuch as are replete with raw and phlegmaticke humors; and for women that haue their termes ſtayed vpon a cold cauſe, by reaſon they open the paſſages that are ſtopt.

D   *Galen* writeth,That they prouoke the hemorrhoids to bleed if they be laid vpon them either by themſelues,or ſtamped with vineger.

E   The iuice of Onions ſnuffed vp into the noſe,purgeth the head, and draweth forth raw flegmaticke humors.

F   Stamped with Salt, Rue, and Honey,and ſo applied, they are good againſt the biting of a mad Dog.

G   Roſted in the embers and applied,they ripen and breake cold Apoſtumes,Biles,and ſuch like.

The

The juice of Onions mixed with the decoction of Penniroyall, and anointed vpon the goutie H member with a feather,or a cloath wet therein,and applied,eaſeth the ſame very much.

The juice anointed vpon a pild or bald head in the Sun,bringeth the haire againe very ſpeedily. I

The juice taketh away the heat of ſcalding with water or oile, as alſo burning with fire & gun. K pouder,as is ſet forth by a very skilfull Surgeon Mr *William Clowes* one of the Queens Surgeon;and before him by *Ambroſe Parey*,in his treatiſe of wounds made by guu-ſhot.

Onions ſliced and dipped in the juice of Sorrell, and giuen vnto the Sicke of a tertian Ague, to L eat, takes away the fit in once or twice ſo taking them.

¶ *The Hurts.*

The Onion being eaten,yea though it be boiled, cauſeth head-ache, hurteth the eyes, and ma-keth a man dim ſighted,dulleth the ſences,ingendreth windineſſe,and prouoketh ouermuch ſleep, eſpecially being eaten raw.

CHAP. 94. *Of Squills or Sea Onions.*

‡ 1 *Scilla Hiſpanica vulgaris.*   The common Spaniſh Squill.

¶ *The Deſcription.*

‡ 1   THe ordinarie Squill or ſea Onion hath a pretty large root compoſed of ſundry white coats filled with a certain viſcous humiditie,and at the bottome thereof grow forth ſundry white and thicke fibres. The leaues are like thoſe of Lillies,broad, thicke, and very greene, lying ſpred vpon the ground,and turned vp on the ſides. The ſtalke groweth ſome cubit or more high,ſtraight,naked without leaues, beautified at the top with many ſtarre-faſhioned floures, very like thoſe of the bigger *Ornithogalum.* The ſeed is contained in chaffie three cornered ſeed-veſſels,being it ſelfe alſo blacke,ſmooth,and chaffie. It floures in Auguſt and September,and the ſeed is ripe in October. The leaues ſpring vp in Nouember and December, after that the ſeed is ripe,and ſtalke decayed. ‡

2   The great ſea Onion,which *Cluſius* hath ſet forth in his Spaniſh Hiſtorie, hath very great and broad leaues, as *Dioſcorides* ſaith, longer than thoſe of the Lilly, but narrower. The bulbe or headed root is very great,conſiſting of many coats or ſcaly filmes of a reddiſh colour. The floure is ſometimes yellow,ſomtimes purple,and ſomtimes of a light blew. ‡ *Cluſius* ſaith it is like that of the former,I thinke he means both in ſhape and colour. ‡

3   The ſea Onion of Valentia,or rather the ſea Daffodill,hath many long and fat leaues, and narrow like thoſe of Narciſſus,but ſmoother and weaker, lying vpon the ground; amongſt which riſeth vp a ſtalke a foot high,bare and naked,bearing at the top a tuft of white floures,in ſhape like vnto

vnto our common yellow Daffodill. The ſeed is incloſed in thicke knobby husks,blacke,fat,and thicke,very ſoft,in ſhape like vnto the ſeeds of *Ariſtolochia longa,* or long Birth-wort. The root is great,white,long,and bulbous.

4   Red floured Sea Daffodill or ſea Onion hath a great bulbe or root like to the precedent,the leaues long, fat, and ſharpe pointed, the ſtalke bare and naked, bearing at the top ſundry faire red floures,in ſhape like to the laſt deſcribed.

2 *Pancratium Cluſij.*   3 *Pancratium marinum.*
Great Squill or Sea Onion.   Sea Onion of Valentia.

5   The yellow floured Sea Daffodill or ſea Onion hath many thicke fat leaues like vnto the common Squill or ſea Onion,among the which riſeth vp a tender ſtraight ſtalke full of juice,bearing at the top many floures like the common yellow Daffodil. The ſeed and root is like the precedent.

‡ 6   To theſe may fitly be added that elegant plant which is knowne by the name of *Narciſſus tertius* of *Matthiolus,*and may be called white ſea Daffodill. This plant hath large roots, as big ſometimes as the ordinarie Squill. The leaues are like thoſe of other Daffodils, but broader, rounder pointed, and not very long. The ſtalke is pretty thicke, being ſometimes round , otherwhiles cornered,at the top whereof grow many large white floures : each floure is thus compoſed; it hath ſix long white leaues,in the midſt growes forth a white pointall, which is incompaſſed by a welt or cap diuided into ſix parts,which ſix are again by threes diuided into eighteen jags or diuiſions,a white thred tipt with greene,of an inch long,comming forth of the middle of each diuiſion. This floureth in the end of May. It is ſaid to grow naturally about the ſea coaſt of Illyria. ‡

¶ *The Place.*

The firſt is found in Spaine and Italy,not far from the ſea ſide.

The ſecond alſo neere vnto the ſea, in Italy, Spaine, and Valentia. I haue had plants of them brought me from ſundry parts of the Mediterranean ſea ſide, as alſo from Conſtantinople, where it is numbred among the kindes of Narciſſus.

The third groweth in the ſands of the ſea in moſt places of the coaſt of Narbone, and about Montpellier.

The fourth groweth plentifully about the coaſts of Tripolis and Aleppo, neere to the ſea, and alſo in ſalt marſhes that are ſandy and lie open to the aire.

                                                          The

¶ *The Time.*

They floure from May to the end of Iuly, and their ſeed is ripe in the end of Auguſt.

¶ *The Names.*

The firſt is called of the Grecians, ⲭⲓⲗⲗⲁ: and of the Latines alſo *Scilla* : the Apothecaries name it, *Squilla* : Diuers, *Cepamaris* : the Germanes, **Meer zwibel** : the Spaniards, *Cebolla albarrana* : the French-men, *Oignon de mer* : in Engliſh, Squill, and Sea Onion.

‡   The ſecond is called, ⲧⲁⲛⲅⲉⲛⲛ, and *Scilla rubra maior.*

3, 4, 5.   Theſe are all figures of the ſame plant, but the leaſt (which is the worſt) is the figure of the *Aduerſaria,* where it is called *Pancratium marinum* : *Doаonæus* calls it *Narciſſus marinus:* and *Cluſius, Hemerocallis Valentina* : and it is iudged to be the ⲏⲙⲉⲣⲟⲕⲁⲗ of *Theophraſtus, Lib.* 6. *Hiſt. cap.* 1. The Spaniards call this, *Amores mios* : the Turkes, *Con zambach* : the Italians, *Giglio marino.* Theſe three (as I ſaid) differ no otherwiſe than in the colour of their floures.

The ſixth is *Narciſſus tertius,* or *Conſtantinopolitanus,* of *Matthiolus* : *Cluſius* calls it, *Lilionarciſſus Hemerocallidis facie.* ‡

4 *Pancratium floribus rubris.*   ‡ 6 *Narciſſus tertius Matthioli.*
Red floured ſea Daffodill.   The white ſea Daffodill.

¶ *The Temperature.*

The ſea Onion is hot in the ſecond degree, and cutteth very much, as *Galen* ſaith. It is beſt when it is taken baked or roſted, for ſo the vehemency of it is taken away.

¶ *The Vertues of Squills.*

The root is to be couered with paſte or clay, (as *Dioſcorides* teacheth, and then put into an ouen   A to be baked, or elſe buried in hot embers till ſuch time as it be throughly roſted : for not being ſo baked or roſted it is very hurtfull to the inner parts.

It is likewiſe baked in an earthen pot cloſe couered and ſet in an ouen. That is to be taken eſpe- B cially which is in the midſt, which being cut in pieces muſt be boyled, but the water is ſtill to be changed, till ſuch time as it is neither bitter nor ſharpe : then muſt the pieces be hanged on a thread, and dried in the ſhadow, ſo that no one piece touch another. ‡ Thus vſed it loſeth moſt of the ſtrength , therefore it is better to vſe it lightly dried, without any other preparation ‡

P   Theſe

C   Theſe ſlices of the Squill are vſed to make oile, wine, or vineger of Squill. Of this vineger of Squil is made an Oxymel; the vſe whereof is to cut thicke tough and clammy humors, as alſo to be vſed in vomits.

D   This Onion roſted or baked is mixed with potions and other medicines which prouoke vrine, and open the ſtoppings of the liuer and ſpleene, and is alſo put into treacles. It is giuen to thoſe that haue the dropſie, the yellow jaundiſe, and to ſuch as are tormented with the gripings of the belly, and is vſed in a licking medicine againſt an old rotten cough, and ſhortneſſe of breath.

E   One part of this Onion being mixed with eight parts of ſalt, and taken in the morning faſting, to the quantitie of a ſpoonfull or two, looſeth the belly.

F   The inner part of Squilla boiled with oile and turpentine, is with great profit applied vnto the chaps or chilblanes of the feet or heeles.

G   It driueth forth long and round wormes, if it be giuen with hony and oile.

‡   The *Pancratium marinum*, or *Hemerocallis Valentina* (ſaith *Cluſius*) whenas I liued with *Rondeletius* at Montpellier, was called *Scilla*; and the Apothecaries made the trochiſces thereof for the compoſition of treacle. Afterwards it began to be called *Pancratium flore Lilij*. *Rondeletius* alſo was wont to tell this following ſtory concerning the poyſonous and maligne qualitie thereof. There were two fiſhermen, whereof the one lent vnto the other (whom he hated) his knife, poiſoned with the juice of this Hemerocallis, for to cut his meat withall: he ſuſpecting no treacherie, cut his victual therewith, and ſo eat them, the other abſtaining therefrom, and ſaying he had no ſtomacke. Some few dayes after, he that did eat the victuals died: which ſhewed the ſtrong and deadly qualitie of this plant, which therefore (as *Cluſius* ſaith) cannot be the *Scilla Epimenidia* of *Pliny*, which was eatable and without malignitie. ‡

C H A P. 95.   Of Leekes.

1 *Porrum capitatum*.
Headed or ſet Leeke.

‡ 2 *Porrum ſectiuum aut tonſile*.
Cut or vnſet Leeke.

 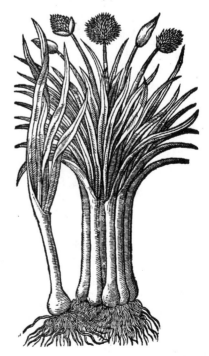

¶ *The Deſcription.*

1 THe leaues or the blades of the Leeke be long, ſomewhat broad, and very many, hauing a keele or creſt in the backſide, in ſmell and taſte like to the Onion. The ſtalkes, if the blades be not often cut, do in the ſecond or third yeare grow vp round, bringing forth on the top floures made vp in a round head or ball as doth the Onion. The ſeeds are like. The bulbe or root is long and ſlender, eſpecially of the vnſet Leeke. That of the other Leeke is thicker and greater.

‡ 2 Moſt Writers diſtinguiſh the common Leeke into *Porrum capitatum & ſectivum*; and *Lobel* giues theſe two figures wherewith we here preſent you. Now both theſe grow of the ſame ſeed, and they differ onely in culture; for that which is often cut for the vſe of the kitchen is called *Sectivum*: the other which is headed, is not cut, but ſpared, and remooued in Autumne. ‡

¶ *The Place.*

It requireth a meane earth, fat, well dunged and digged. It is very common euery where in other countries, as well as in England

¶ *The Time.*

It may be ſowne in March or Aprill, and is to be remooued in September, or October.

¶ *The Names.*

The Grecians call it πράσον: the Latines, *Porrum*. The Emperour *Nero* had great pleaſure in this root, and therefore he was called in ſcorne, *Porrophagus*. But *Palladius* in the maſculine gender called it *Porrus*: the Germanes, **Lauch**: the Brabanders, **Porreye**: the Spaniards, *Puerro*: the French, *Porreau*: the Engliſh-men, Leeke, or Leekes.

¶ *The Temperature.*

The Leeke is hot and dry, and doth attenuate or make thinne as doth the Onion.

¶ *The Vertues.*

Being boiled it is leſſe hurtfull, by reaſon that it loſeth a great part of his ſharpeneſſe: and yet being ſo vſed it yeeldeth no good juyce. But being taken with cold herbes his too hot quality is tempered. A

Being boyled and eaten with Ptiſana or barley creame, it concocteth and bringeth vp raw humors that lie in the cheſt. Some affirme it to be good in a loch or licking medicine, to clenſe the pipes of the lungs. B

The juyce drunke with hony is profitable againſt the biting of venomous beaſts, and likewiſe the leaues ſtamped and laid thereupon. C

The ſame juice, with vinegre, frankincenſe and milke, or oyle of roſes, dropped into the eares, mitigateth their paine, and is good for the noyſe in them. D

Two drams of the ſeed, with the like weight of myrtill berries drunke, ſtop the ſpitting of bloud which hath continued a long time. The ſame ingredients put into Wine keepe it from ſouring, and being already ſoure, amend the ſame, as diuers write. It cutteth and attenuateth groſſe and tough humors. E

‡ *Lobel* commends the following Loch as very effectuall againſt phlegmatick Squinancies, and other cold catarrhes which are like to cauſe ſuffocation. This is the deſcription thereof; Take blanched almonds three ounces, foure figs, ſoft *Bdellium* halfe an ounce, juyce of Liquorice, two ounces, of ſugar candy diſſolued in a ſufficient quantity of juyce of Leekes, and boyled in *Balneo* to the height of a Syrrup, as much as ſhall be requiſit to make the reſt into the forme of an *Eclegma*. ‡ F

¶ *The Hurts.*

It heateth the body, ingendreth naughty bloud, cauſeth troubleſome and terrible dreames, offendeth the eies, dulleth the ſight, hurteth thoſe that are by nature hot and cholericke, and is noyſome to the ſtomacke, and breedeth windineſſe.

CHAP. 96.

Of Ciues or Chiues, And wilde Leekes.

¶ *The Kindes.*

THere be diuers kindes of Leekes, ſome wilde, and ſome of the Garden, as ſhall be declared. Thoſe called Ciues haue beene taken of ſome for a kinde of wilde Onion: but all the Authors that I haue beene acquainted with, do accord that there is not any wilde Onion.

1 *Schœnoprason.*
Ciues or Chiues.

2 *Porrum vitigineum.*
French Leeks, or Vine Leekes.

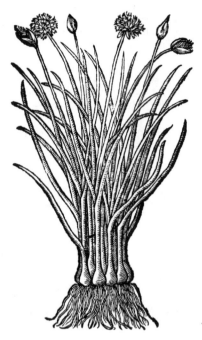

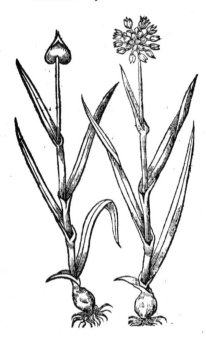

3 *Ampeloprason siue porrum siluestre.*
Wilde Leeke.

¶ *The Description.*

1 Clues bring forth many leaues about a hand-full high, long, slender, round like to little rushes ; amongst which grow vp small and tender stalkes, sending forth certaine knops with floures like those of the Onion, but much lesser. They haue many little bulbes or headed roots fastned together : out of which grow downe into the earth a great number of little strings, and it hath both the smell and taste of the Onion and Leeke, as it were participating of both.

2 The Vine Leeke or French Leeke groweth vp with blades like those of Leekes : the stalke is a cubit high, on the top whereof standeth a round head or button, couered at the first with a thinne skinne, which being broken, the floures and seeds come forth like those of the Onion. The bulbe or headed root is round, hard, and sound, which is quickly multiplied by sending forth many bulbes.

‡ 3 The wilde Leeke hath leaues much like vnto those of Crow-garlicke, but larger, and more acride. The floures and seeds also resemble those of the Crow-garlicke, the seeds being about the bignesse of cornes of wheat, with small strings comming forth at their ends. ‡

¶ *The*

¶ *The Place and Time.*

1 Ciues are ſet in gardens, they flouriſh long and continue many yeares, they ſuffer the cold of winter: they are cut and polled often, as is the vnſet Leeke.

2 The Vine-Leeke grows of it ſelfe in Vineyards, and neere vnto Vines in hot regions, wher-of it both tooke the name Vine-Leeke and French Leeke. It beareth his greene leaues in Winter, and withereth away in the Sommer. It groweth in moſt gardens of England.

‡ Thus far our Author deſcribes and intimates to you a garden Leek, much like the ordinary in all reſpects, but ſomewhat larger. But the following names belong to the wilde Leeke, which here we giue you in the third place. ‡

¶ *The Names.*

Ciues are called in Greeke σχοινοπρασον, *Shænopraſum* : in Dutch, **Bieſloack :** as though you ſhould ſay *Iunceum Porrum*, or Ruſh-leeke : in Engliſh, Ciues, Chiues, Ciuet, and Sweth : in French, *Brelles*.

† 2 The Vine-leeke, or rather wilde Leeke, is called in Greeke Ἀμπελοπρασον, of the place where it naturally groweth : it may be called in Latine *Porrum vitium*, or *Vitigineum Porrum* : in Engliſh, after the Greeke and Latine, Vine-leeke, or French Leeke.

¶ *The Temperature.*

Ciues are like in facultie vnto the Leeke, hot and dry. The Vine-Leek heateth more than doth the other Leeke.

¶ *The Vertues.*

Ciues attenuate or make thin, open, prouoke vrine, ingender hot and groſſe vapors, and are hurt- A full to the eyes and braine. They cauſe troubleſome dreames, and worke all the effects that the Leeke doth.

The Vine-leeke or Ampelopraſon prouoketh vrine mightily, and bringeth downe the floures. It B cureth the bitings of venomous beaſts, as *Dioſcorides* writeth.

† The figure of *Ampelopraſum* was in the firſt place, in the Chapter next but one, by the name of *Allium ſylueſtre*.

CHAP. 97. *Of Garlicke.*

¶ *The Deſcription.*

1 THe bulbe or head of Garlicke is couered with moſt thin skins or filmes of a very light white purple colour, conſiſting of many cloues ſeuered one from another, vnder which in the ground below groweth a taſſel of threddy fibres : it hath long greene leaues like thoſe of the Leeke, among which riſeth vp a ſtalke at the end of the ſecond or third yeare, where-upon doth grow a tuft of floures couered with a white skin, in which, being broken when it is ripe appeareth round blacke ſeeds.

‡ 2 There is alſo another Garlicke which groweth wilde in ſome places of Germanie and France, which in ſhape much reſembles the ordinarie, but the cloues of the roots are ſmaller and redder. The floure is alſo of a more dusky and darke colour than the ordinarie. ‡

¶ *The Place and Time.*

Garlick is ſeldome ſowne of ſeed, but planted in gardens of the ſmall cloues in Nouember and December, and ſometimes in Februarie and March.

¶ *The Names.*

It is called in Latine *Allium* : in Greeke Σκόροδον : The Apothecaries keepe the Latine name : the Germanes call it **Knoblauch :** the Low-Dutch **Look :** the Spaniards, *Aios, Alho* : the Italians *Aglio* : the French, *Ail* or *Aux* : the Bohemians *Czeſnec* : in Engliſh, Garlick, and poore mans treacle.

¶ *The Temperature.*

Garlicke is very ſharpe, hot and dry (as *Galen* ſaith) in the fourth degree, and exulcerates the skin by raiſing bliſters.

¶ *The Vertues.*

: Being eaten it heateth the body extremely, attenuateth and maketh thin thicke and groſſe hu- A mors, cutteth ſuch as are tough and clammy, digeſteth and conſumeth them, alſo openeth obſtru-ctions, is an enemie to all cold poyſons, and to the bitings of venomous beaſts : and therfore *Galen* nameth it *Theriacum Ruſtica*, or the Husbandmans Treacle.

It yeeldeth to the body no nouriſhment at all, it ingendreth naughty and ſharpe bloud. There- B fore

fore fuch as are of a hot complexion muſt eſpecially abſtaine from it. But if it be boiled in water vntill ſuch time as it hath loſt his ſharpeneſſe, it is the leſſe forcible, and retaineth no longer his euill juice, as *Galen* ſaith.

C   It taketh away the roughneſſe of the throat, it helpeth an old cough, it prouoketh vrine, it breaketh and conſumeth wind, and is alſo a remedie for the Dropſie which proceedeth of a cold cauſe.

D   It killeth wormes in the belly, and driueth them forth. The milke alſo wherein it hath bin ſodden is giuen to yong children with good ſucceſſe againſt the wormes.

     1 *Allium.*   ‡ 2 *Allium ſylveſtre rubentibus nucleis.*
     Garlicke.   Wilde Garlicke with red cloues.

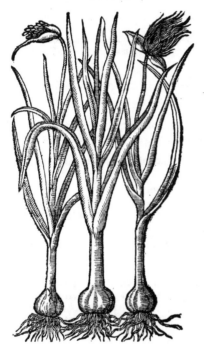 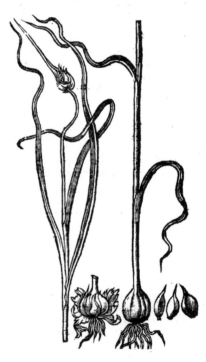

E   It helpeth a very cold ſtomacke, and is a preſeruatiue againſt the contagious and peſtilent aire.

F   The decoction of Garlicke vſed for a bath to ſit ouer, bringeth downe the floures & ſecondines or after-burthen, as *Dioſcorides* ſaith.

G   It taketh away the morphew, tettars or ring-wormes, ſcabbed heads in children, dandraffe, and ſcurfe, tempered with honey, and the parts anointed therewith.

H   With Figge leaues and Cumin it is laid on againſt the bitings of the Mouſe called in Greeke Μυγαλη : in Engliſh, a Shrew.

C H A P.

CHAP. 98. *Of Crow-Garlicke and Ramsons.*

¶ *The Description.*

1   THe wilde Garlicke or Crow garlick hath small tough leaues like vnto Rushes, smooth and hollow within, among which groweth vp a naked stalke, round, slipperie, hard and sound, on the top whereof, after the floures be gon, grow little seeds made vp in a round cluster like small kernels, hauing the smell and taste of garlicke. In stead of a root there is a bulbe or round head, without any cloues at all.

2   Ramsons do send forth two or three broad longish leaues sharpe pointed, smooth, and of a light greene colour. The stalke is a span high, smooth and slender, bearing at the top a cluster of white star-fashioned floures. In stead of a root it hath a long slender bulbe, which sendeth down a multitude of strings, and is couered with skins or thicke coats.

† 1 *Allium sylvestre.*   2 *Allium vrsinum.*
  Crow Garlicke.   Ramsons.

¶ *The Time.*

They spring vp in Aprill and May : their seed is ripe in August.

¶ *The Place.*

The Crow Garlicke groweth in fertile pastures in all parts of England. I found it in great plentie in the fields called the Mantels, on the backe side of Islington by London.

Ramsons grow in the woods and borders of fields vnder hedges, amongst the bushes. I found it in the next field vnto Boobies barn, vnder that hedge that bordereth vpon the lane; and also vpon the left hand, vnder an hedge adioining to a lane that leadeth to Hampsted, both places neere vnto London.

¶ *The*

¶ *The Names.*

Both of them be wilde Garlicke, and may be called in Latine *Allia ſylveſtria* : in Greeke,
Σκόροδα ἄγρια : the fitſt by *Dodonæus* and *Lobel* is called *Allium ſylveſtre tenuifolium.*

Ramſons are named of the later practitioners *Allium ſylveſtre,* or Beares Garlicke : *Allium latifolinm,* and *Moly Hippocraticum* : in Engliſh, Ramſons, Ramſies, and Buckrams.

¶ *The Nature.*

The temperatures of the wilde Garlicks are referred vnto thoſe of the gardens.

¶ *The Vertues.*

H   Wild Garlicke or Crow-Garlicke, as *Galen* ſaith, is ſtronger and of more force than the garden
Garlicke.

B   The leaues of Ramſons be ſtamped and eaten of diuers in the Low-Countries, with fiſh for a
ſauce, euen as we do eat green-ſauce made with ſorrell.

C   The ſame leaues may very well be eaten in Aprill and May with butter, of ſuch as are of a ſtrong
conſtitution, and laboring men.

D   The diſtilled water drunke breaketh the ſtone, and driueth it forth, and prouoketh vrine.

CHAP. 99. *Of Mountaine Garlicke.*

1 *Scorodopraſum.*
Great mountaine Garlicke.

‡ 2 *Scorodopraſum primum Cluſij.*
Cluſius his great mountain Garlick.

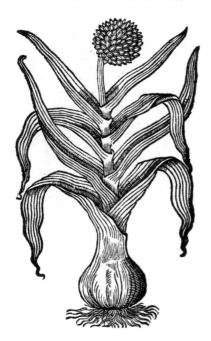 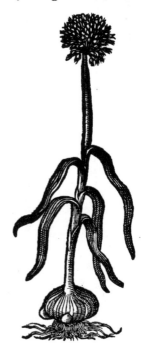

¶ *The Deſcription.*

1 2   THe great mountaine Garlicke hath long and broad leaues like thoſe of Leeks,
but much greater and longer, embracing or claſping about a great thicke ſtalke,
ſoft and full of iuyce, bigger than a mans finger, and bare towards the top ; vpon
which is ſet a great head bigger than a tenniſe ball, couered with a skinne after the manner of an
Onion. The skinne when it commeth to perfection breaketh, and diſcouereth a great multitude
of whitiſh floures : which being paſt, blacke ſeeds follow incloſed in a three cornered huske. The
root is bulbous, of the bigneſſe of a great Onion. The whole plant ſmelleth very ſtrong like vnto
Garlicke,

Garlicke and is in shew a Leeke, whereupon it was called *Scorodoprasum*, as if we should say, Gar-licke Leeke; participating of the Leeke and Garlicke, or rather a degenerate Garlicke growne monstrous.

‡ I cannot certainely determine what difference there may be betweene the plants expressed by the first figure, which is our Authors, and the second figure which is taken out of *Clusius*. Now the history which *Clusius* giues vs to the second, the same is (out of him) giuen by our Author to the first: so that by this reason they are of one and the same plant. To the which opinion I rather incline, than affirme the contrary with *Bauhine*, who distinguishing them, puts the first amongst the Leekes, vnder the name of *Porrum folio latissimo*: following *Tabernamontanus*, who first gaue this figure, vnder the name of *Porrum Syriacum*.

3　This plant is lesser in all the parts than the former; the root is set about with longer and slenderer bulbes wrapped in brownish skinnes; the floures and leaues are like, yet smaller than Garlicke.

‡ 3. *Scorodoprasum minus*.　　　　　　　‡ 4 *Ophioscoridon*.
The lesser Leeke leaued Garlicke.　　　　　Vipers Garlicke.

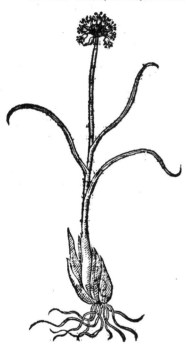 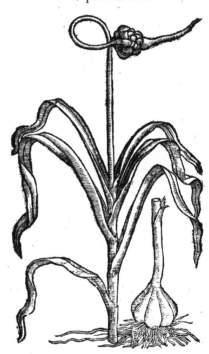

4　The third, which *Clusius* makes his second *Scorodoprasum*, hath stalkes some two cubits high, hauing many leaues like those of Leekes from the bottome of the stalke to the middle thereof; their smell is betweene that of Leekes and Garlicke; the rest of the stalke is naked, greene, smooth, sustaining at the top a head composed of many bulbes, couered with a whitish skinne ending in a long greene point; which skinne by the growth of the bulbe being broken, they shew themselues, being first of a purplish, and afterwards of a whitish colour, amongst which are some floures. The top of the stalke at first twines it selfe, so that it in some sort represents a serpent; then by little it vntwines againe, and beares the head streight vp. The root consists of many cloues much like that of Garlicke. ‡

5　The broad leaued mountaine Garlicke, or rather the Mountaine Ramsons, riseth vp with a stalke a cubit high, a finger thicke, yet very weake, full of a spongeous substance, neere to the bot-tome of a purplish colour, and greene aboue, bearing at the top a multitude of small whitish floures, somwhat gaping, star-fashion. The leaues are three or foure, broad ribbed like the leaues of great Gentian, resembling those of Ramsons, but greater. The root is great and long, couered with many scaly coats and hairy strings.

¶ The

5 *Allium Alpinum latifolium, ſeu Victorialis.*
Broad leaued Mountaine Garlicke.

A

B

¶ *The Place.*

The great mountaine Garlicke growes about Conſtantinople, as ſaith *Cluſius.* I receiued a plant of it from Mr *Thomas Edwards* Apothecary of Exceſter, who found it growing in the Weſt part of England.

Victorialis groweth in the mountaines of Germany, as ſaith *Carolus Cluſius*, and is yet a ſtranger in England for any thing that I do know.

¶ *The Time.*

‡ Moſt of theſe plants floure in the months of Iune and Iuly.

¶ *The Names.*

Of the firſt and ſecond I haue ſpoken already. The third is *Scorodopraſum minus* of *Lobel.* The fourth is *Allium ſativum ſecundum* of *Dodonæus*, and *Scorodopraſum ſecundum* of *Cluſius.* The fifth is *Allium anguinum* of *Matthiolus; Ophioſcoridon* of *Lobel*, and *Victorialis* of *Cluſius* and others, as alſo *Allium Alpinum.* The Germanes call it, **Seig=wurtz.**

¶ *The Temper.*

They are of a middle temper between Leekes and Garlicke.

¶ *Their Vertues.*

Scorodapraſum, as it partakes of the temper, ſo alſo of the vertues of Leekes and Garlicke ; that is, it attenuates groſſe and rough matter, helpes expectoration, &c.

Victorialis is like Garlicke in the operation thereof. Some (as *Camerarius* writeth) hang the root thereof about the necks of their cattell being falne blinde, by what occaſion ſoeuer it happen, and perſuade themſelues that by this meanes they will recouer their ſight. Thoſe that worke in the mines of Germany affirme, That they find this root very powerfull in defending them from the aſſaults of impure ſpirits or diuels, which often in ſuch places are troubleſome vnto them. *Cluſ.* ‡

CHAP. 100.   *Of Moly, or the Sorcerers Garlicke.*

¶ *The Deſcription.*

1 THe firſt kinde of Moly hath for his root a little whitiſh bulbe ſomewhat long, not vnlike to the root of the vnſet Leeke, which ſendeth forth leaues like the blades of corne or graſſe : among which doth riſe vp a ſlender weake ſtalke, fat, and full of juyce : at the top whereof commeth forth of a skinny filme a bundle of milke-white floures, not vnlike to thoſe of Ramſons. The whole plant hath the ſmell and taſte of Garlicke, whereof no doubt it is a kinde.

2 Serpents Moly hath likewiſe a ſmall bulbous root with ſome fibres faſtned to the bottom, from which riſe vp weake graſſie leaues of a ſhining greene colour, crookedly winding and turning themſelues towards the point like the taile of a Serpent, whereof it tooke his name : the ſtalke is tough, thicke, and full of juyce, at the top whereof ſtandeth a cluſter of ſmall red bulbes, like vnto the ſmalleſt cloue of Garlicke, before they be pilled from their skinne. And among thoſe bulbes there doe thruſt forth ſmall and weake foot-ſtalkes, euery one bearing at the end one ſmall white floure tending to a purple colour : which being paſt, the bulbes doe fall downe vpon the ground, where they without helpe do take hold and root, and thereby greatly encreaſe, as alſo by the infinite bulbes that the root doth caſt off : all the whole plant doth ſmell and taſte of Garlicke, whereof it is alſo a kinde.

3 *Homers* Moly hath very thicke leaues, broad toward the bottome, ſharpe at the point, and
hollowed

1 *Moly Dioſcoridcum.*
Dioſcorides his Moly.

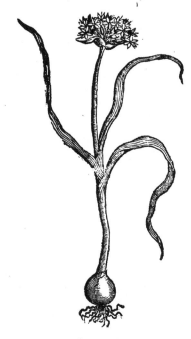

2 *Moly Serpentinum.*
Serpents Moly.

3 *Moly Homericum.*
Homers Moly.

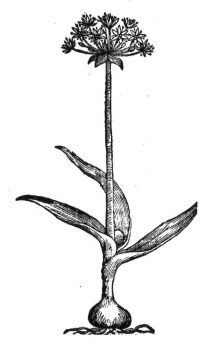

hollowed like a trough or gutter, in the boſome of which leaues neere vnto the bottom commeth forth a certain round bulbe or ball of a gooſe turd greene colour: which being ripe and ſet in the ground, groweth and becommeth a faire plant ſuch as is the mother. Among thoſe leaues riſes vp a naked ſmooth thicke ſtalke two cubits high, as ſtrong as a ſmall walking ſtaffe: at the top of the ſtalke ſtandeth a bundle of faire whitiſh floures, daſhed ouer with a waſh of purple colour ſmelling like the floures of Onions. When they be ripe there appeareth a black ſeed wrapped in a white skin or huske. The root is great and bulbous, couered with a blackiſh skin on the out ſide, and white wirhin, and of the bigneſſe of a great onion.

4 Indian Moly hath very thick fat ſhort leaues, ſharp pointed, in the boſome whereof commeth forth a thicke knobby bulbe like that of *Homers* Moly. The ſtalke is alſo like the precedent, bearing at the top a cluſter of ſcaly bulbs, included in a large thin skin or filme. The root is great, bulbous faſhion, and full of juice.

5 *Caucaſon,* or withering Moly, hath a very great bulbous root greater than that of *Homers* Moly, and fuller of a ſlimie juice; from which do ariſe three or foure great thick and broad leaues withered alwaies at the point;
wherein

wherein conſiſteth the difference betweene theſe leaues and thoſe of *Homers* Moly, which are not ſo. In the middle of the leaues riſeth vp a bunch of ſmooth greeniſh bulbes ſet vpon a tender foot-ſtalke, in ſhape and bigneſſe like vnto a great garden Worme, which being ripe and planted in the earth, do alſo grow vnto a faire plant like vnto their mother.

‡ Theſe two laſt mentioned (according to *Bauhine*, and I thinke the truth) are but figures of one and the ſame plant ; the later whereof is the better, and more agreeing to the growing of the plant.

6 To theſe may be fitly added two other Molyes: the firſt of theſe which is the yellow Mo-ly, hath roots whitiſh and round, commonly two of them growing together ; the leaues which it ſends forth are long and broad, and ſomwhat reſemble thoſe of the Tulip, and vſually are but two in number ; betweene which riſeth vp a ſtalke a foot high, bearing at the top an vmbell of faire yellow ſtar-like floures tipt on their lower ſides with a little greene. The whole plant ſmelleth of Garlicke.

‖ 4 *Moly Indicum.*
Indian Moly.

‖ 5 *Caucaſon.*
Withering Moly.

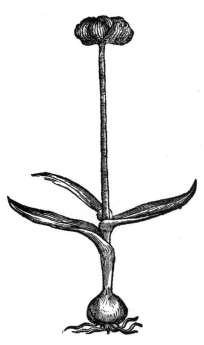

7 This little Moly hath a root about the bigneſſe of an Haſell nut, white, with ſome fibres hanging thereat ; the ſtalke is of an handfull or little more in height, the top thereof is adorned with an vmbel of ten or twelue white floures, each of which conſiſts of ſix leaues, not ſharpe poin-ted, but turned round, and pretty large, conſidering the bigneſſe of the plant. This plant hath alſo vſually but two leaues, but thoſe like thoſe of Leekes but far leſſe. ‡

¶ *The Place.*

† Theſe plants grow in the garden of M͛ *Iohn Parkinſon* Apothecarie, and with M͛ *Iohn Trade-ſcant* and ſome others, ſtudious in the knowledge of plants.

¶ *The Time.*

They ſpring forth of the ground in February, and bring forth their floures, fruit, and ſeed in the end of Auguſt.

¶ *The Names.*

† Some haue deriued the name *Moly* from the Greeke words, Μωλυήν παύσωσι : that is, to driue away diſeaſes. It may probably be argued to belong to a certaine bulbous plant, and that a kinde of

of Garlicke, by the words ᴍᴏʟʏ and ᴍᴏʟʏ. The former, *Galen* in his *Lexicon* of ſome of the more difficult words vſed by *Hippocrates*, thus expounds : Σκόροδι ἄνω ἦ κεχαλόν ἴχει, ἢ μιάδιαλνοφήσκι ἴτε δζλιδμε τσκε δ᾽ π᾽ μολυ
That is, *Moly Ζι* is a Garlicke hauing a ſimple or ſingle head, and not to be parted or diſtinguiſhed
into cloues. Some terme it *Moly : Erotianus* in his *Lexicon* expounds it thus : ᴍᴏʟʏ (ſaith hee)
Ͻʜͽͷʜͽͷʜͽͷ ϲϻϻͽϲ, &c. That is, *Moly* is a head of garlick round, and not to be parted into cloues. ‡

¶ *The Names in particular.*

‡ 1 This is called *Moly* by *Matthiolus* : *Moly Anguſtifolium* by *Dodonæus* ; *Moly Dioſcorideum*, by
Lobel and *Cluſius*.

2 This, *Moly Serpentinum vocatum*, by *Lobel* and the Author of the *Hiſt. Lugd.*

3 This ſame is thought to be the *Moly* of *Theophraſtus* and *Pliny* by *Dodonæus*, *Cluſius*, &c. and
ſome alſo would haue it to be that of *Homer* mentioned in the twentieth *Odyſſ. Lobel* calls it *Moly
Liliflorum*.

4. 5 The fourth and fifth being one, are called *Caucaſon*, and *Moly Indicum*, by *Lobel*, *Cluſius*,
and others.

6 This is *Moly montanum latifolium flavo flore* of *Cluſius*, and *Moly luteum* of *Lobel*, *Aduerſ.par.2.*

7 This ſame is *Moly minus* of *Cluſius*. ‡

‡ 6 *Moly latifolium flore flauo.*
Broad leaued Moly with the yellow floure.

‡ 7 *Moly minus flore albo.*
Dwarfe white floured Moly.

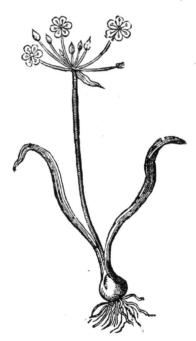

¶ *The Temperature and Vertues.*

Theſe Molyes are very hot, approching to the nature of Garlicke, and I doubt not but in time
ſome excellent man or other will find out as many good vertues in them, as their ſtately & comely proportion ſhould ſeeme to be poſſeſſed with. But for my part, I haue neither proued, nor heard
of others, nor found in the writings of the Antients, any thing touching their faculties. Only *Dioſcorides* reporteth, That they are of a maruellous efficacie to bring down the termes, if one of them
be ſtamped with oile of Floure de-luce according to art, and vſed in maner of a peſſarie or mother
ſuppoſitorie.

‡ C H A P. 101.   *Of diuers other Molyes.*

‡ BEſides the Garlickes and Molyes formerly mentioned by our Author, and thoſe I haue
iu this edition added,there are diuers others, which,mentioned by *Cluſius*,and belonging to this tribe, I thought good here to ſet forth. Now for that they are more than
conueniently could be added to the former chapters, (which are ſufficiently large) I thought it
not amiſſe to allot them a place by themſelues.

‡ 1 *Moly Narciſsinis folijs primum.*
The firſt Narciſſe-leaued Moly.

‡ 2 *Moly Narciſsinis folijs secundum.*
The ſecond Narciſſe-leaued Moly.

 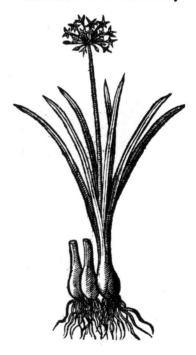

¶ *The Deſcription.*

‡ 1 THis,which in face nigheſt repreſenteth the Molyes deſcribed in the laſt **Chapter,**
hath a root made of many ſcales,like as an onion in the vpper part,but the lower
part is knotty,and runnes in the ground like as *Solomons* Seale; the Onion-like
part hath many fibres hanging thereat; the leaues are like thoſe of the white Narciſſe,very greenē
and ſhining,amongſt which riſeth vp a ſtalke of a cubit high, naked, firme, greene,and creſted. At
the top come forth many floures conſiſting of ſix purpliſh leaues,with as many chiues on their inſides : after which follow three-ſquare heads,opening when they are ripe, and containing a round
blacke ſeed.

2   This other being of the ſame kinde,and but a varietie of the former,hath ſofter & more aſh
coloured leaues,with the floures of a lighter colour. Both theſe floure at the end of Iune, or in
Iuly.

3   This hath fiue or ſix leaues equally as broad as thoſe of the laſt deſcribed, but not ſo long,
being ſomewhat twined,greene, and ſhining. The ſtalke is ſome foot in length, ſmaller than that
of the former,but not leſſe ſtiffe,creſted, and bearing in a round head many floures, in manner of
growing and ſhape like thoſe of the former,but of a more elegant purple colour. In ſeed and root
it

‡ 3 *Moly Narciſſinis folijs tertium.*
The third Narciſſe-leaued Moly.

‡ 4 *Moly montanum latifolium* 1 *Cluſij.*
The firſt broad leaued mountain Moly.

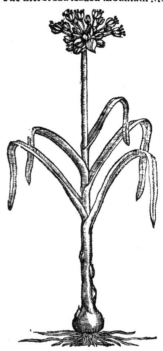

‡ 5 *Moly montanum ſecundum Cluſij.*
The ſecond mountain Moly.

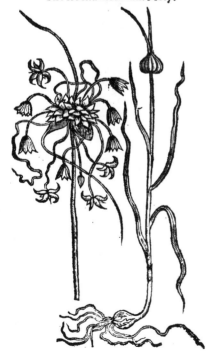

it reſembles the precedent. There is alſo a varietie of this kinde, with leaues longer and narrower, neither ſo much twined, the ſtalkes weaker, and floures much lighter coloured.

This floures later than the former, to wit, in
Iuly and Auguſt.

All theſe plants grow naturally in Leiten-
berg and other hils neere to Vienna in Auſtria,
where they were firſt found and obſerued by
Carolus Cluſius.

4   This hath a ſtalk ſome two cubits high,
which euen to the midſt is incompaſſed with
leaues much longer and broader than thoſe of
Garlick, and very like thoſe of the Leek: on the
top of the ſmooth & ruſh-like ſtalke groweth
a tuft conſiſting of many dark purple coloured
bulbes growing cloſe together, from amongſt
which come forth pretty long ſtalkes bearing
light purple ſtar faſhioned floures, which are
ſucceeded by three cornered ſeed-veſſels. The
root is bulbous, large, conſiſting of many cloues
and hauing many white fibres growing forth
therof. Moreouer, there grow out certain round
bulbes about the root, almoſt like thoſe which
grow in the head, and being planted apart, they
produce plants of the ſame kind. This is *Allium, ſiue Moly montanum latifolium* 1. *Cluſij.*

5   This hath a ſmooth round greene ſtalke
ſome cubit high, whereon doe grow moſt commonly

monly three leaues narrower than thoſe of the former, and as it were graſſy. The top of the ſtalke ſuſtaines a head wrapped in two lax filmes, each of them running out with a ſharpe point like two hornes, which opening themſelues, there appeare many ſmall bulbs heaped together, among which are floures compoſed of ſix purpliſh little leaues, and faſtned to long ſtalks. The root is round and white, with many long white fibres hanging thereat. *Cluſius* calls this *Allium, ſiue Moly montanum ſecundum.* And this is *Lobels Ampelopraſon proliferum.*

6   Like to the laſt deſcribed is this in height and ſhape of the ſtalke and leaues, as alſo in the forked or horned skin involving the head, which conſiſteth of many ſmall bulbs of a reddiſh green colour, and ending in a long green point ; amongſt which vpon long and ſlender ſtalks hang down floures like in forme and magnitude to the former , but of a whitiſh colour, with a darke purple ſtreake alongſt the middle, and vpon the edges of each leafe. The root is round and white like that of the laſt deſcribed. This *Cluſius* giueth vnder the title of *Allium, ſiue Moly montanum tertium.*

‡ 6 *Moly montanum 3. Cluſ.*   ‡ 7 *Moly montani quarti ſpec. 1.Cluſ.*
The third mountaine Moly.   The fourth mountaine Moly, the
                                         firſt ſort thereof.

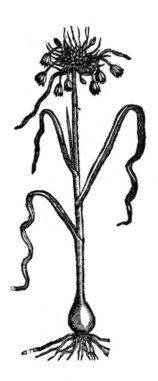 

7   This alſo hath three ruſhy leaues, with a round ſtalk ſome cubit high, whoſe top is likewiſe adorned with a forked membrane, containing many pale coloured floures hanging on long ſtalks, each floure conſiſting of ſix little leaues, with the like number of chiues, and a peſtill in the midſt. This tuft of floures cut off with the top of the ſtalke, and carried into a chamber, will yeeld a pleaſant ſmel (like that which is found in the floures of the earlier *Cyclamen*) but it quickly decays. After theſe floures are paſt, ſucceed three cornered heads containing a black ſmall ſeed not much vnlike Gillofloure ſeed. The root is round like the former, ſometimes yeelding off-ſets. This is *Alij montani 4. ſpecies* of *Cluſius.*

8   There is another kinde of this laſt deſcribed, which growes to almoſt the ſame height , and hath like leaues, and the head ingirt with the like skinny long pointed huskes ; but the floures of this are of a very darke colour. The roots are like the former, with off-ſets by their ſide. This is
                                                    Cluſius

Cluſius his *Moly montani quarti ſpecies ſecunda.* The roots of the three laſt deſcribed ſmel of garlicke, but the leaues haue rather an herby or graſſe-like ſmell.

The fiſt and ſixt of theſe grow naturally in the Styrian and Auſtrian Alps. The ſeuenth growes about Preſburg in Hungarie, about Niclaſpurg in Morauia, but moſt abundantly about the baths in Baden.

‡ 8 *Moly montani quarti ſpecies ſecunda Cluſij.*
The ſecond kinde of the fourth
mountaine Moly.

‡ 9 *Moly montanum quintum Cluſij.*
The fifth mountain Moly.

9   This growes to the like height as the former, with a green ſtalke, hauing few leaues thereupon, and naked at the top, where it carrieth a round head conſiſting of many ſtar-like ſmall floures, of a faire purple colour, faſtned to ſhort ſtalks, each floure being compoſed of ſix little leaues, with as many chiues, and a peſtill in the middle. The root is bulbous and white, hauing ſomtimes his off-ſets by his ſides. The ſmell of it is like Garlick. This groweth alſo about Preſburgh in Hungarie, and was there obſerued by *Cluſius* to beare his floure in May and Iune. He calleth this *Allium, ſeu Moly montanum quintum.*

Cʜᴀᴘ. 102. *Of White Lillies.*

¶ *The Kindes.*

THere be ſundry ſorts of Lillies, whereof ſome be wilde or of the field, others tame or of the garden; ſome white, others red; ſome of our own countries growing, others from beyond the ſeas: and becauſe of the variable ſorts, we wil diuide them into chapters, beginning with the two white Lillies, which differ little but in the natiue place of growing.

¶ *The Description.*

1   THe white Lilly hath long smooth and full bodied leaues of a graffie or light green co-
       lour. The ftalks be two cubits high, and fomtimes more, fet or garniſhed with the like
       leaues, but growing fmaller and fmaller toward the top ; and vpon them do grow faire
white floures ftrong of fmell, narrow toward the foot of the ftalke whereon they do grow, wide or
open in the mouth like a bell. In the middle part of them doe grow fmall tender pointals tipped
with a dufty yellow colour, ribbed or chamfered on the back fide, confifting of fix fmall leaues
thicke and fat. The root is a bulb made of fcaly cloues, full of tough and clammy juice, wherewith the whole plant doth generally abound.

2   The white Lilly of Conftantinople hath very large & fat leaues like the former, but narrower
and leffer. The ftalke rifeth vp to the height of three cubits, fet and garniſhed with leaues alfo like
the precedent, but much leffe. Which ftalke oftentimes doth alter and degenerat from his naturall roundneffe to a flat forme, as it were a lath of wood furrowed or chanelled alongft the fame, as
it were ribs or welts. The floures grow at the top like the former, fauing that the leaues doe turne
themfelues more backward like the Turks cap, and beareth many more floures than our Englifh
white Lilly doth.

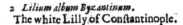

1   *Lilium album.*   2   *Lilium album Byzantinum.*
    The white Lilly.   The white Lilly of Conftantinople.

¶ *The Place.*

Our Englifh white Lilly groweth in moft gardens of England. The other groweth naturally in
Conftantinople and the parts adiacent, from whence we had plants for our Englifh gardens, where
they flouriſh as in their owne countrey.

¶ *The Time.*

The Lillies floure from May to the end of Iune.

¶ *The Names.*

The Lilly is called in Greeke κρίνον : in Latine, *Lilium*, and alfo *Rofa Iunonis*, or *Iuno's* Rofe, becaufe as it is reported it came vp of her milke that fell vpon the ground. But the Poets feign, That
Hercules, who *Iupiter* had by *Alcumena*, was put to *Iuno's* breafts whileft fhee was afleepe ; and after
the fucking there fell away aboundance of milk, and that one part was fpilt in the heauens, and the
other vpon the earth ; and that of this fprang the Lilly, and of the other the circle in the heauens
called

called *Lacteus Circulus*, or the Milky way, or otherwise in English Watling street. *S. Basil* in the explication of the 44 Psalm saith, That no floure so liuely sets forth the frailty of mans life as the Lilly. It is called in high-Dutch, **Weiſz Gilgen**: in low-Dutch, **witte Lilien**: in Italian, *Giglio*: in Spanish, *Lirio blanco*: in French, *Lys blanc*: in English, the white Lilly.

The other is called *Lilium album Byzantinum*, and also *Martagon album Byzantinum*: in English, the white Lilly of Constantinople: of the Turks themselues, *Sultan Zambach*; with this addition, (that they might be the better knowne which kinde of Lilly they meant when they sent Roots of them into these countries) *Fa fiora grandi Bianchi*: so that *Sultan Zambach fa fiora grandi Bianchi* is as much to say as, Sultans great Lilly with white floures.

¶ The Nature.

The white Lilly is hot, and partly of a subtill substance. But if you regard the root, it is dry in the first degree, and hot in the second.

¶ The Vertues.

The root of the garden Lilly stamped with hony gleweth together sinues that be cut in sunder. A It consumeth or scoureth away the vlcers of the head called Achores, and likewise all scuruinesse of the beard and face.

The roots stamped with vineger, the leaues of Henbane, or the meale of Barley, cureth the tu- B mors and impostumes of the priuy parts. It bringeth the haire again vpon places which haue bin burned or scalded, if it be mingled with oile or grease, and the place anointed therewith.

The same root rosted in the embers, and stamped with some leauen of Rie bread & hogs grease, C breaketh pestilential botches. It ripeneth apostumes in the flanks, comming of venerie and such like.

The floures steeped in oile Oliue, and shifted two or three times during Sommer, and set in the D Sun in a strong glasse, is good to harden the softnesse of sinues, and the hardnesse of the matrix.

Florentinus a writer of Husbandry saith, That if the root be curiously opened, and therein be put E some red, blew, or yellow colour that hath no causticke or burning qualitie, it will cause the floure to be of the same colour.

Iulius Alexandrinus the Emperours Physitian saith, That the water thereof distilled and drunke, F causeth easie and speedy deliuerance, and expelleth the secondine or after-burthen in most speedy manner.

He also saith, The leaues boiled in red wine, and applied to old wounds or vlcers, do much good G and forward the cure, according to the doctrine of *Galen, lib. 7. de simpl. med. facultat.*

The root of a white Lilly stamped and strained with wine, and giuen to drinke for two or three H dayes together, expelleth the poison of the pestilence, and causeth it to breake forth in blisters in the outward part of the skin, according to the experience of a learned Gentleman Mr *William Godorus*, Sergeant Surgeon to the Queens Majestie; who also hath cured many of the dropsie with the juice thereof tempered with Barly meale, and baked in cakes, and so eaten ordinarily for some moneth or six weeks together with meat, but no other bread during that time.

CHAP. 103.   Of Red Lillies.

¶ The Kindes.

THere be likewise sundry sorts of Lillies, which we do comprehend vnder one generall name in English, Red Lillies, whereof some are of our owne countries growing, and others of beyond the seas, the which shall be distinguished seuerally in this chapter that followeth.

¶ The Description.

1   THe gold-red Lilly groweth to the height of two, and somtimes three cubits, and often higher than those of the common white Lilly. The leaues be blacker and narrower, set very thicke about the stalke. The floures in the top be many, from ten to thirty, according to the age of the plant, and fertilitie of the soile, like in forme and greatnesse to those of the white Lilly, but of a white colour tending to a Saffron, sprinckled or poudred with many little blacke specks, like to rude vnperfect draughts of certaine letters. The roots be great bulbs, consisting of many cloues, as those of the white Lilly.

‡ 2   In

‡   2   In ſtead of the Plantain leaued Red Lilly, deſcribed and figured in this ſecond place by our Author out of *Tabernamontanus*, for that I iudge both the figure and deſcription counterfeit, I haue omitted them, and here giue you the many-floured red Lilly in his ſtead. This hath a root like that of the laſt deſcribed, as alſo leaues and ſtalks ; the floure alſo in ſhape is like that of the former, but of a more light colour, and in number of floures it exceedeth the precedent, for ſometimes it beares ſixty floures vpon one ſtalke. ‡

†   3   This red Lilly is like vnto the former, but not ſo tall ; the leaues be fewer in number, broader, and downy towards the top of the ſtalke, where it bears ſome bulbes. The floures in ſhape are like the former, ſauing that the colour hereof is more red, and thicke daſht with black ſpeckes. The root is ſcaly like the former.

4   There is another red Lilly which hath many leaues ſomewhat ribbed, broader than the laſt mentioned, but ſhorter, and not ſo many in number. The ſtalke groweth to the height of two cubits, and ſomtimes higher, whereupon do grow floures like the former : among the foot-ſtalkes of which floures come forth certaine bulbes or cloued roots, browne of colour, tending vnto rednes, which do fall in the end of Auguſt, vpon the ground, taking root and growing in the ſame place, whereby it greatly encreaſeth ; for ſeldome or neuer it bringeth forth ſeed for his propagation.

1 *Lilium aureum.*
Gold-red Lilly.

2 † *Lilium rubrum.*
The red Lilly.

5   There is another ſort of red Lilly hauing a faire ſcaly or cloued root, yellow aboue, and browne toward the bottom ; from which riſeth vp a faire ſtiffe ſtalk creſted or furrowed, of an ouer-worne browne colour, ſet from the lower part to the branches, whereon the floures do grow, with many leaues, confuſedly placed without order. Among the branches cloſe by the ſtem grow forth certaine cloues or roots of a reddiſh colour, like vnto the cloues of garlicke before they be pilled : which beeing fallen vpon the ground at their time of ripeneſſe, doe ſhoot forth certaine tender ſtrings or roots that do take hold of the ground, whereby it greatly encreaſeth. The floures are in ſhape like the other red Lillies, but of a darke Orange colour, reſembling a flame of fire ſpotted with blacke ſpots.

‡   6   This hath a much ſhorter ſtalke, being but a cubit or leſſe in height, with leaues blackiſh

kith, and narrower than those aforesaid. The floures, as in the rest, grow out of the top of the stalk, and are of a purplish Saffron colour, with some blackish spots. The root in shape is like the precedent. ‡

¶ *The Place.*

These Lillies do grow wilde in the plowed fields of Italy and Languedocke, in the mountaines and vallies of Hetruria and those places adiacent. They are common in our English gardens, as also in Germanie.

¶ *The Time.*

These red Lillies do floure commonly a little before the white lillies, and sometimes together with them.

3 *Lilium cruentum latifolium.*
The fiery red Lilly.

‡ 4 *Lilium cruentum bulbiferum.*
Red bulbe-bearing Lilly.

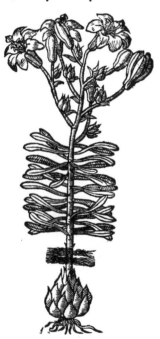

¶ *The Names.*

‡ 1   The first of these is thought by some to be the *Bulbus cruentus* of *Hippocrates* : as also the *Lilium purpureum* of *Dioscorides.* Yet *Matthiolus* and some others would haue it his *Hemerocallis. Dodonæus* and *Bapt. Porta* think it the *Hyacinthus* and *Cosmosandalos* of the Poets, of which you shal find more hereafter.   It is the *Martagon Chymistarum* of *Lobel,* and the *Lilium aureum majus* of *Tabernamontanus.*

2   This is *Martagon Chymistarum alterum* of *Lobel.*   3   This is *Clusius* his *Martagon bulbiferum secundum.*   4   *Martagon bulbiferum primum* of *Clusius.*   5   This *Dodonæus* cals *Lilium purpureum tertium,* and it is *Martagon bulbiferum tertium* of *Clusius.*   6   This last *Lobel* and *Dodonæus* call *Lilium purpureum minus.*

I haue thought good here also to giue you that discourse touching the Poets Hyacinth, which being translated out of *Dodonæus,* was formerly vnfitly put into the chapter of Hyacinths, which therefore I there omitted, and haue here restored to his due place, as you may see by *Dodon. Pempt. 2 lib. 2. cap. 2.* ‡

†   There is a Lilly which *Ouid, Metamorph. lib. 10.* calls *Hyacinthus,* of the boy *Hyacinthus,* of whose bloud he feigned that this floure sprang, when hee perished as he was playing with *Apollo* :

for

for whoſe ſake he ſaith that *Apollo* did print certain letters and notes of his mourning. Theſe are his words :

> *Ecce cruor, qui fuſus humo ſignauerat herbas,*
> *Deſinit eſſe cruor, Tyrioque nitentior oſtro*
> *Flos oritur, formamque capit, quam Lilia, ſi non*
> *Purpureus color his argenteus eſſet in illis.*
> *Non ſatis hoc Phæbo eſt, (is enim fuit auctor honoris)*
> *Ipſe ſuos gemitus folijs inſcribit, & ai ai,*
> *Flos habet inſcriptum, funeſtaque literæ ducta eſt.*

Which lately were elegantly thus rendred in Engliſh by Mr *Sands* :

> Behold ! the bloud which late the graſſe had dy'de
> Was now no bloud : from whence a floure full blowne,
> Far brighter than the Tyrian skarlet ſhone :
> Which ſeem'd the ſame, or did reſemble right
> A Lilly, changing but the red to white.
> Nor ſo contented, (for the youth receiv'd
> that grace from *Phæbus*) in the leaues he weav'd
> The ſad impreſſion of his ſighs, Ai, Ai,
> They now in funerall characters diſplay.

‡ 5 *Lilium cruentum ſecundum caulem*
        bulbulis donatum.
     Red Lilly with bulbs growing alongſt
     the ſtalke.

‡ 6 *Lilium purpureum minus.*
     The ſmall red Lilly.

Theocritus alſo hath made mention of this Hyacinth in *Bions* Epitaph, in the 19 Eidyl, which Eidyl by ſome is attributed to *Moſchus*, and made his third. The words are theſe :

Νῦν ὑάκινθε λάλει τὰ σὰ γράμματα ἡ πλέον αἶ αἶ.
Λαμβάνε τοῖς πετάλοισι.

     In Engliſh thus :
> Now Iacinth ſpeake thy letters, and once more
> Imprint thy leaues with Ai, Ai, as before.

                                     Likewiſe

Likewise *Virgil* hath written hereof in the third *Eclog* of his *Bucolicks.*

Et me Phœbus amat, Phœbo sua semper apud me
Munera sunt, Lauri & suaue rubens Hyacinthus.

Phœbus loues me, his gifts I alwayes haue,
Th e e're green Laurel and the Iacinth braue.

In like maner also *Nemetianus* in the 2 *Eclog* of his *Bucolicks :*

Te sine me misero mihi Lilia nigra videntur,
Pallentesque Rosæ, nec dulce rubens Hyacinthus :
At si tu venias, & candida Lilia fient
Purpureæque Rosæ & dulce rubens Hyacinthus.

Without thee, Loue, the Lillies blacke do seem,
The Roses pale, and Hyacinths I deeme
Not louely red. But if thou com'st to me,
Lillies are white, red Rose and Iacinths be.

The Hyacinths are said to be red which *Ovid* calleth purple, for the red colour is sometimes termed purple. Now it is thought that *Hyacinthus* is called *Ferrugineus*, for that it is red of a rustie iron colour : for as the putrifaction of brasse is named *Ærugo*, so the corruption of iron is called *Ferrugo*, which from the reddish colour is stiled also *Rubigo*. And certainly they are not a few that would haue *Color ferrugineus* to be so called from the rust which they thinke *Ferrugo*. Yet this opinion is not allowed of by all men ; for some iudge, that *Color ferrugineus* is inclining to a blew, for that when the best iron is heated and wrought, when as it is cold againe it is of a colour neere vnto blew, which from *Ferrum* (or iron) is called *Ferrugineus*. These later ground themselues vpon *Virgils* authoritie, who in the sixt of his *Æneids* describeth *Charons* ferrugineous barge or boat, and presently calleth the same blew. His words are these :

Ipse ratem conto subigit velisque ministrat,
Et ferruginea subuectat corpora Cymba,

He thrusting with a pole, and setting sailes at large,
Bodies transports in ferrugineous barge.

And then a little after he addes,

Cæruleam aduertit puppim, ripæque propinquat.

He then turnes in his blew barge, and the shore
Approches nigh to.

And *Claudius* also in his second booke of the carrying away of *Proserpina*, doth not a little confirme their opinions ; who writeth, That the Violets are painted *ferrugine dulci*, with a sweet iron colour.

Sanguineo splendore rosas, vaccinea nigro
Induit, & dulci violas ferrugine pingit.

He trimmes, the Rose with bloudy bright,
And Prime-tree berries blacke he makes,
And decks the Violet with a sweet
Darke iron colour which it takes.

But let vs returne to the proper names from which we haue digressed. Most of the later herbarists call this plant *Hyacinthus Poeticus*, or the Poets Iacinth. *Pausanias* in his second booke of Corinthiacks hath made mention of *Hyacinthus* called of the Hermonians *Comosandalos*, setting down the ceremonies done by them on their festiuall dayes in honor of the goddesse *Chthonia*. The Priests (saith he) and the magistrats for that yeare do leade the troupe of the pomp ; the women & men follow after; the boyes solemnly leade forth the goddesse with a stately shew : they go in white vestures, with garlands on their heads made of a floure which the inhabitants cal *Comosandalos*, which is the blew or sky-coloured Hyacinth, hauing the marks and letters of mourning as aforesaid.

¶ *The Temperature.*

The floure of the red Lilly, as *Galen* saith, is of a mixt temperature, partly of a thin and partly of an earthly essence. The root and leaues doe dry and clense, and moderatly digest and waste or consume away.

¶ *The Vertues.*

The leaues of the herbe applied are good against the stinging of Serpents.   A
The same boiled and tempered with vineger are good against burnings, and heale green wounds   B
and vlcers.
The root rosted in the embers, and pouned with oile of Roses, cureth burnings, and softneth the   C
... rdnesse of the matrice.

The

D   The ſame ſtamped with honey cureth the wounded ſinewes and members out of joint. It takes
     away the morphew,wrinkles,and deformities of the face.
E   Stamped with Vineger, the leaues of Henbane, and wheat meale, it remoueth hot ſwellings of
     the ſtones,the yard,and matrice.
F   The roots boiled in Wine,ſaith *Pliny*, cauſeth the cornes of the feet to fall away within few
     dayes,with remouing the medicine vntill it haue wrought his effect.
G   Being drunke in honied water,they driue out by ſiege vnprofitable bloud.

Chap. 104.   *Of Mountaine Lillies.*

¶ *The Deſcription.*

THe great mountain Lilly hath a cloued bulb or ſcaly root like thoſe of the red Lilly,
yellow of colour,very ſmall in reſpect of the greatneſſe of the plant ; from the which
riſeth vp a ſtalke,ſomtimes two or three,according to the age of the plant,whereof the
middle ſtalke commonly turneth from his roundneſſe into a flat forme, as thoſe of the white Lilly
of Conſtantinople. Vpon theſe ſtalks do grow faire leaues of a blackiſh greene colour, in roundles and ſpaces as the leaues of Woodroofe,not vnlike to the leaues of white Lilly,but ſmaller at
the top of the ſtalkes. The floures be in number infinite, or at the leaſt hard to be counted,very
thicke ſet or thruſt together,of an ouerworne purple, ſpotted on the inſide with many ſmal ſpecks
of the colour of ruſty iron. The whole floure doth turne it ſelfe backward at ſuch time as the ſun
hath caſt his beames vpon it,like vnto the Tulipa or Turks cap, as the Lilly or Martagon of Conſtantinople doth ; from the middle whereof do come forth tender pendants hanging thereat of the
colour the floure is ſpotted with.

1 *Lilium montanum majus.*   2 *Lilium montanum minus.*
The great mountaine Lilly.   Small mountaine Lilly.

2   The ſmall mountain Lilly is very like vnto the former in root, leafe, ſtalk, and floures: diffe-ring in theſe points, The whole plant is leſſe, the ſtalke neuer leaueth his round forme, and beareth fewer floures.

‡   There are two or three more varieties of theſe plants mentioned by *Cluſius*; the one of this leſſer kinde, with floures on the out ſide of a fleſh colour, and on the inſide white, with blackiſh ſpots; as alſo another wholly white without ſpots. The third varietie is like the firſt, but differs in that the floures blow later, and ſmell ſweet.

Theſe plants grow in the woody mountaines of Styria and Hungarie.   ſuch like pla-ces on the North of Francfort, vpon the Mœne. ‡

The ſmall ſort I haue had many yeares growing in my gar   gre   I haue   t had til of late, giuen me by my louing friend Mr *Iames Garret*   arie of London.

¶ *The Time.*

Theſe Lillies of the m   re at ſuch time as the common white Lilly doth, and   times ſooner.

¶ *The Names.*

The great mountain Lilly is called of *Tabernamontanus*, *Lilium Saracenicum*, receiued by Mr aforeſaid from Liſle in Flanders, by the name of *Martagon Imperiale* : of ſome, *Lilium Saraceni-cum maꝭ*. It is *Hemerocallis flore rubello* of *Lobel*.

The ſmall mountain Lilly is called in Latine *Lilium montanum*, & *Lilium ſylueſtre* : of *Dodonæu* *Hemerocallis* : of others, *Martagon* : but neither truly, for that there is of either, other plants proper-ly called by the ſame names. In high-Dutch it is called **Golꝭ wtꝣ**, from the yellowneſſe of the root; in low-Dutch, **Lilikens van Caluarien:** in Spaniſh, *Liꝭ to Amarillo* : in French, *Lys Sauuage:* in Engliſh, Mountain Lilly.

¶ *The Nature and Vertues.*

There hath not bin any thing left in writing either of the nature or vertues of theſe plants: not-withſtanding we may deem, that God which gaue them ſuch ſeemely and beautifull ſhape, hath not left them without their peculiar vertues, the finding out whereof we leaue to the learned and induſtrious ſearcher of Nature.

CHAP. 105.   *Of the Red Lilly of Conſtantinople.*

1 *Lilium Byzantinum.*
The red Lilly of Conſtantinople.

‡ 2 *Lilium Byzantinum flo. purpui o ſanguineo.*
The Byzantine purpliſh ſanguine-coloured Lilly,

¶ *The Deſcription.*

1 THe red Lilly of Conſtantinople hath a yellow ſcaly or cloued root like to the mountain Lilly, but greater : from the which ariſeth vp a faire fat ſtalke a finger thicke, of a
darke purpliſh colour toward the top ; which ſometimes doth turne from his naturall
roundneſſe into a flat forme, like as doth the great mountain Lilly : vpon which ſtalk grow ſundry
faire and moſt beautiful floures, in ſhape like thoſe of the mountaine Lilly : but of farre greater
beauty, ſeeming as it were framed of red wax, tending to a red lead colour. From the middle of the
floure commeth forth a tender pointal or peſtell, and likewiſe many ſmall chiues tipped with looſe
pendants. The floure is of a reaſonable pleaſant ſauor. The leaues are confuſedly ſet about the ſtalk
like thoſe of the white Lilly, but broader and ſhorter.

‡ 2 This hath a large Lilly-like root, from which ariſeth a ſtalke ſome cubit or more high,
ſet confuſedly with leaues like the precedent. The floures alſo reſemble thoſe of the laſt deſcribed,
but vſually more in number, and they are of a purpliſh ſanguine colour.

‡ 3 *Lilium Byzantinum flo. dilute rubente.*   ‡ 4 *Lilium Byzantinum miniatum polyanthos.*
The light red Byzantine Lilly.   The Vermilion Byzantine many-floured Lilly.

 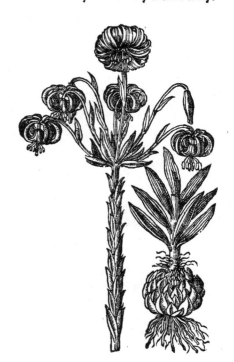

3 This differs little from the laſt, but in the colour of the floures, which are of a lighter red
color than thoſe of the firſt deſcribed. The leaues and ſtalks alſo, as *Cluſius* obſerueth, are of a lighter greene.

4 This may alſo more fitly be termed a variety from the former, than otherwiſe: for according
to *Cluſius*, the difference is only in this, that the floures grow equally from the top of the ſtalk, and
the middle floure riſes higher than any of the reſt, and ſomtimes conſiſts of twelue leaues as it
were a twin, as you may perceiue by the figure. ‡

¶ *The Time*
They floure and flouriſh with the other Lillies.

¶ *The Names.*

The Lilly of Conſtantinople is called likewiſe in England, Martagon of Conſtantinople : of *Lebel, Hemerocallis Chalcedonica,* and likewiſe *Lilium Byzantinum* : of the Turkes it is called *Zuſiniare* : of the Venetians, *Marocali.*

¶ *The Temperature and Vertues.*

Of the nature or vertues there is not any thing as yet ſet down, but it is eſteemed eſpecially for the beauty and rareneſſe of the floure : referring what may be gathered hereof to a farther conſideration.

---

‡ Chap. 106.   *Of the narrow leaued reflex Lillies.*

¶ *The Deſcription.*

‡ 1   THe root of this is not much vnlike that of other Lillies : the ſtalke is ſome cubit high or better ; the leaues are many and narrow, and of a darker greene than thoſe of the ordinarie Lilly ; the floures are refle Lily e thoſe treated of in the laſt Chapter, of a red or vermilion colour. This floures in the end of M.. wherefore *Cluſius* calls it *Lilium rubrum præcox,* The early red Lilly.

‡ 1   *Lilium rubrum anguſtifolium.*   ‡ 3   *Lilium mont. flore flavo punctato.*
The red narrow leaued Lilly.   The yellow mountain Lilly with
                                      the ſpotted floure.

2   This plant is much more beautifull than the laſt deſcribed ; the roots are like thoſe of Lillies, the ſtalke ſome cubit and an halfe in height, beeing thicke ſet with ſmall graſſie leaues. The floures grow out one aboue another, in ſhape and colour like thoſe of the laſt deſcribed, but often-

times are more in number, ſo that ſome one ſtalke hath borne 48 floures. The root is much like the former.

‡ 4 *Lilium mont. flore flavo non punctato.*
The yellow mountain Lilly with the vn-ſpotted floure.

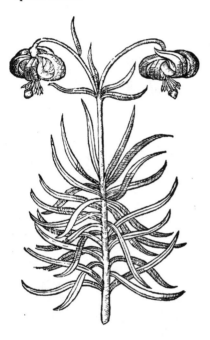

3　This in roots is like thoſe afore de-ſcribed; the ſtalke is ſome 2 cubits high, ſet confuſedly with long narrow leaues, with three conſpicuous nerues running a-long them. The floures are at firſt pale co-loured; afterwards yellow, conſiſting of ſix leaues bended back to their ſtalks, & mar-ked with blackiſh purple ſpots.

4　There is alſo another, differing from the laſt deſcribed ouly in that the floure is not ſpotted as that of the former.

¶ *The Place.*
Theſe Lillies are thought Natiues of the Pyrenæan mountains, and of late yeres are become denizons in ſome of our Eng-liſh gardens

¶ *The Time.*
The firſt (as I haue ſaid) floures in the end of May: the reſt in Iune.

¶ *The Names.*
1　This is called by *Cluſius*, *Lilium ru-brum præcox.*

2　*Cluſius* names this, *Lilium rubrum præcox 3 anguſtifol. Lobel* ſtiles it, *Hemero-callis Macedonica*, and *Martagon Pomponeum.*

3　This is *Lilium flavo flore maculis di-ſtinctum* of *Cluſius*: and *Lilium montanum flavo flo. of Lobel.*

4　This being a varietie of the laſt, is called by*Cluſius*, *Lilium flavo flore maculis non diſtinctum.*

¶ *The Temperature and Vertues.*
Theſe in all likelihood cannot much differ from the temper and vertues of other Lillies, which in all their parts they ſo much reſemble. ‡

Chap. 107. *Of the Perſian Lilly.*

¶ *The Deſcription.*

THe Perſian Lilly hath for his root a great white bulbe, differing in ſhape from the other Lillies, hauing one great bulbe firme or ſolid, full of juice, which commonly each yere ſet-teth off or encreaſeth one other bulbe, and ſometimes more, which the next yere after is ta-ken from the mother root, and ſo bringeth forth ſuch floures as the old plant did. From this root riſeth vp a fat thicke and ſtraight ſtem of two cubits high, whereupon is placed long narrow leaues of a greene colour, declining to blewneſſe as doe thoſe of the woad. The floures grow a-longſt the naked part of the ſtalk like little bels, of an ouerworn purple colour, hanging down their heads, euery one hauing his own footſtalke of two inches long, as alſo his peſtell or clapper from the middle part of the floure; which being paſt and withered, there is not found any ſeed at all, as in other plants, but is encreaſed only in his root.

¶ *The Place.*
This Perſian Lilly groweth naturally in Perſia and thoſe places adiacent, whereof it tooke his name, and is now (by the induſtry of Trauellers into thoſe countries, louers of plants) made a deni-zon in ſome few of our London gardens.

¶ *The*

¶ *The Time.*

This plant floureth from the beginning of May to the end of Iune.

¶ *The Names.*

This Perfian Lilly is called in Latine, *Lilium Perficum, Lilium Sufianum, Pennacio Perfiano,* and *Pennaco Perfiano,* either by the Turks themfelues, or by fuch as out of thofe parts brought them into England, but which of both is vncertain. *Alphonfus Pancius,* Phyfitian to the duke of Ferrara, when as he fent the figure of this plant vnto *Carolus Clufius,* added this title, *Pennacio Perfiano è Pianta bellifsima & è fpecie di Giglio ò Martagon, diuerfo della corona Imperiale:* That is in Englifh, This moft elegant plant *Pennacio* of Perfia, is a kinde of Lilly or Martagon, differing from the floure called the Crowne Imperiall.

Lilium Perficum.
The Perfian Lilly.

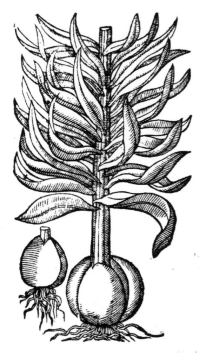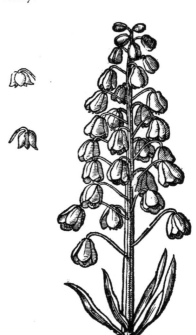

¶ *The Temperature and Vertues.*

There is not any thing known of the nature or vertues of this Perfian Lilly, efteemed as yet for his rareneffe and comely proportion; although (if I might bee fo bold with a ftranger that hath vouchfafed to trauell fo many hundreds of miles for our acquaintance) wee haue in our Englifh fields many fcores of floures in beauty far excelling it.

Chap. 108. *Of the Crowne Imperiall.*

¶ *The Defcription.*

THe Crowne Imperiall hath for his root a thicke firme and folid bulbe, couered with a yellowifh filme or skinne, from the which rifeth vp a great thicke fat ftalke two cubits high, in the bare and naked part of a darke ouerworne dusky purple colour. The leaues grow confufedly about the ftalke like thofe of the white Lilly, but narrower: the floures grow at the top of the ftalke, incompaffing it round, in forme of an Imperiall Crowne, (whereof it tooke his name) hanging their

heads

Corona Imperialis.
The Crowne Imperiall.

Corona Imperialis duplici corona.
The double Crowne Imperiall.

Corona Imperialis cum semine.
Crowne Imperiall with the seed.

heads downward as it were bels ; in colour it is yellowish ; or to giue you the true colour, which by words otherwise cannot be exprefsed, if you lay fap berries in steep in faire water for the space of two houres, and mix a little saffron in that infusion, and lay it vpon paper, it sheweth the perfect colour to limne or illumine the floure withall. The back side of the said floure is streaked with purplish lines, which doth greatly set forth the beauty therof. In the bottom of each of these bels there is placed sixe drops of most cleare shining sweet water, in taste like sugar, resembling in shew faire orient pearles ; the which drops if you take away, there do immediatly appeare the like : notwithstanding if they may be suffered to stand still in the floure according to his own nature, they will neuer fall away, no not if you strike the plant vntill it be broken. Among these drops there standeth out a certain pestel, as also sundry small chiues tipped with small pendants like those of the Lilly : aboue the whole floures there groweth a tuft of green leaues like those vpon the stalke, but smaller. After the floures be faded, there follow cods or seed-vessels six square, wherein is contained flat seeds tough & limmer, of the colour of Mace: the whole plant, as wel roots

as

as floures do fauor or smell very like a fox. As the plant groweth old, so doth it wax rich, bringing forth a Crowne of floures amongst the vppermost green leaues, which some make a second kinde, although in truth they are but one and the selfe same, which in time is thought to grow to a triple crowne, which hapneth by the age of the root, and fertilitie of the soile; whose figure or tipe I haue thought good to ioyne with that picture also which in the time of his infancie it had.

¶ *The Place.*

This plant likewise hath been brought from Constantinople amongst other bulbous roots, and made denizons in our London gardens, whereof I haue great plenty.

¶ *The Time.*

It floureth in Aprill, and sometimes in March, when as the weather is warme and pleasant. The seed is ripe in Iune.

¶ *The Names.*

This rare and strange Plant is called in Latine, *Corona Imperialis*, & *Lilium Byzantinum*: the Turks do call it *Cauale lale*, and *Tusai*. And as diuers haue sent into these parts of these roots at sundry times, so haue they likewise sent them by sundry names: some by the name *Tusai*: others, *Tousai*, and *Tuyschiachi*, and likewise *Turfani*, and *Turfanda*. ‡ *Clusius*, and that not without good reason, iudgeth this to be the *Hemerocallis* of *Dioscorides*, mentioned, *lib. 3. cap.* 120. ‡

¶ *The Temperature and Vertues.*

The vertue of this admirable plant is not yet knowne, neither his faculties or temperature in working.

† If this be the *Hemerocallis* of *Dioscorides*, you may finde the vertues thereof specified, pag. 99. of this Worke, wherein my iudgment they are not so fitly placed as they might haue beene here: yet we at this day haue no knowledge of the physicall operation, either of those plants mentioned in that place, or of this treated of in this Chapter.

Chap. 109.   *Of Dogs Tooth.*

¶ *The Description.*

1   THere hath not long since bin found out a goodly bulbous rooted plant, termed Satyrion, which was supposed to be the true Satyrion of *Dioscorides*, after that it was cherished, and the vertues thereof found out by the studious searchers of nature. Little difference hath bin found betwixt that plant of *Dioscorides*, and this *Dens Caninus*, except in the color, which (as you know) doth commonly vary according to the diuersitie of places where they grow, as it falls out in Squilla, Onions, and the other kinds of bulbed plants. It hath most commonly two leaues, very seldom three; which leafe in shape is very like to *Allium Vrsinum*, or Ramsons, though far lesse. The leaues turn down to the groundward; the stalk is tender and flexible like to *Cyclamen* or Sow-bread, about an handfull high, bare and without leaues to the root. The proportion of the floure is like that of Saffron or the Lilly floure, full of streams of a purplish white colour: the root is big, and like vnto a date, with some fibres growing from it: vnto the said root is a small flat halfe round bulb adioyning, like vnto *Gladiolus* or Corn-flag.

2   The second kind is far greater and larger than the first, in bulb, stalke, leaues, floure, and cod. It yeeldeth two leaues for the most part, which do close one within another, and at the first they do hide the floure (for so long as it brings not out his floure) it seemes to haue but one leafe like the Tulipa's, and like the Lillies, though shorter, and for the most part broader; wherefore I haue placed it and his kinds next vnto the Lillies, before the kinds of *Orchis* or stones. The leaues which it beareth are spotted with many great spots of a darke purple colour, and narrow below, but by little and little toward the top wax broad, and after that grow to be sharp pointed, in form somewhat neere Ramsons, but thicker and more oleous. When the leaues be wide opened the floure sheweth it selfe vpon his long weake naked stalke, bowing toward the earth-ward; which floure consists of six very long leaues of a fine delayed purple colour, which with the heate of the Sun opens it selfe, and bendeth his leaues backe againe after the manner of the Cyclamen floure, within which there are six purple chiues, and a white three forked stile or pestell. This floure is of no pleasant smell, but commendable for the beauty. When the floure is faded, there succeedeth a three square huske or head wherein are the seeds, which are very like them of *Leucoium bulbosum præcox*, but longer, slenderer, and of a yellow colour. The root is long, thicker below than aboue, set with many white fibres, waxing very tender in the vpper part, hauing one or more off-sets or yong shoots, from which the stalke riseth out of the ground (as hath bin said) bringing forth two leaues, and not three, or only one, saue when it will not floure.

3   The third kind is in all things like the former, ſaue in the leaues, which are narrower, and in the colour of the floure, which is altogether white, or conſiſting of a color mixt of purple & white. Wherefore ſith there is no other difference, it ſhall ſuffice to haue ſaid thus much for the Deſcription.

1 *Dens Caninus.*
Dogs tooth.

2 *Dens Caninus flore albo anguſtioribus folijs.*
White Dogs tooth.

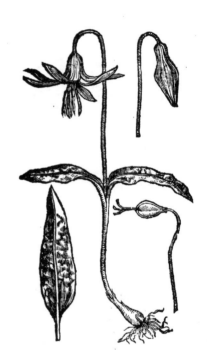

¶ *The Place.*

Theſe three plants grow plentifully at the foot of certain hils in the greene and moiſt grounds of Germany and Italy, in Stiria not far from Gratz, as alſo in Modena and Bononia in Italy, & likewiſe in ſome of the choice gardens of this Country.

¶ *The Time.*

They floure in Aprill, and ſomtimes ſooner, as in the middle of March.

¶ *The Names.*

This plant is called in Latine, *Dens Caninus*; and ſome haue iudged it *Satyrium Erythronium.* *Matthiolus* cals it *Pſeudohermodactylus :* the men of the Countrey where it groweth call it 𝕾𝖈𝖍𝖔𝖋𝖋= 𝖜𝖚𝖗𝖙𝖘; and the phyſitians about Styria call it *Dentali.* The ſecond may for diſtinctions ſake be termed *Dens caninus flore albo, anguſtioribus folijs*; that is, Dogs tooth with the white floure and narrow leaues.

¶ *The Nature.*

Theſe are of a very hot temperament, windy, and of an excrementitious nature, as may appear by the Vertues.

¶ *The Vertues.*

A   The women that dwel about the place where theſe growhaue with great profit put the meale or pouder of it into their childrens pottage, againſt the worms of the belly.

B   Being drunk with wine, it hath bin proued maruelouſly to aſſwage the Colique paſſion.

C   It ſtrengthneth and nouriſheth the body in great meaſure, and being drunke with water, it cures children of the falling ſickneſſe.

Chap.

CHAP. 110.   *Of Dogs Stones.*

¶ *The Kindes.*

STones or Teſticles,as *Dioſcorides* ſaith,are of two ſorts, one named *Cynoſorchis*, or Dogs Stones ; the other,*Orchis Serapias*,or Serapias ſtones. But becauſe there be many and ſundry other ſorts differing one from another,I ſee not how they may be contained vnder theſe two kinds only:therfore I haue thought good to diuide them as followeth ; the firſt kind I haue named *Cynoſorchis*, or Dogs ſtones : the ſecond,*Teſticulus Morionis*,or Fools ſtones : the third,*Tragorchis*,or goats ſtones : the fourth,*Orchis Serapias*,or Serapias ſtones : the fiſt,*Teſticulus odoratus*,or ſweet-ſmelling Stones, or after *Cordus,Teſticulus Pumilio*,or Dwarfe ſtones.

† 1 *Cynoſorchis major.*
  Great Dogs ſtones.

† 2 *Cynoſorchis major altera.*
  White Dogs ſtones.

¶ *The Deſcription.*

1   GReat Dogs ſtones hath foure and ſometimes fiue great broad thick leaues, ſomewhat like thoſe of the garden lilly,but ſmaller.The ſtalk riſeth vp a foot or more in height ; at the top whereof groweth a thick tuft of carnation or horſe-fleſh coloured floures, thick and cloſe thruſt together,made of many ſmall floures ſpotted with purple ſpots,in ſhape like to an open hood or helmet.And from the hollow place there hangeth forth a certain ragged chiue or taſſel,in ſhape like to the skin of a dog or ſome ſuch other fourfooted beaſt. The roots be round like vnto the ſtones of a dog,or two oliues,one hanging ſomewhat ſhorter than the other, whereof the higheſt or vppermoſt is the ſmaller,but fuller and harder. The loweſt is the greateſt,lighteſt, and moſt wrinkled or ſhriueled,not good for any thing.

2   Whitiſh Dogs ſtones hath likewiſe ſmooth long broad leaues,but leſſer and narrower than thoſe of the firſt kinde. The ſtalk is a ſpan long, ſet with fiſſe or ſix leaues claſping or embracing the ſame round about. His ſpiky floure is ſhort, thicke, buſhy, compact of many ſmall whitiſh
purple

purple colored floures, spotted on the inside with many small purple spots & little lines or streaks. The small floures are like an open hood or helmet, hauing hanging out of euerie one as it were the body of a little man without a head, with arms stretched forth, and thighs stradling abroad, after the same maner almost that the little boyes are wont to be pictured hanging out of *Saturns* mouth. The roots be like the former.

3   Spotted Dogs stones bring forth narrow leaues, ribbed in some sort like vnto the leaues of narrow Plaintain or Rib-wort, dasht with many black streaks and spots. The stalke is a cubit and more high, at the top whereof doth grow a tuft or eare of violet coloured floures, mixt with a dark purple, but in the hollownesse thereof whitish, not of the same forme or shape that the others are of, but lesser, and as it were resembling somwhat the floures of Larkes-spur. The roots be like the former.

4   Marish Dogs stones haue many thicke blunt leaues next the root, thicke streaked with lines or nerues like those of Plantain. The floure is of a whitish red or carnation : the stalk and roots be like the former.

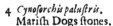

† 3 *Cynosorchis maculata.*
Spotted Dogs stones.

4 *Cynosorchis palustris.*
Marish Dogs stones.

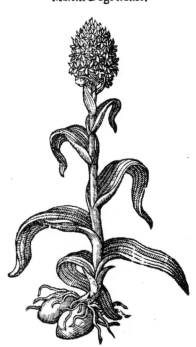

‡ 5   This hath fiue or six little leaues : the stalk is some handfull or better in height, set a-bout with somewhat lesse leaues : the tuft of floures at the top of the stalke are of a purple colour, small, with a white lip diuided into foure partitions hanging downe, which also is lightly spotted with purple : it hath a little spur hanging down on the hinder part of each floure. The seed is smal, and contained in such twined heads as in other plants of this kinde. The roots are like the former, but much lesse. ‡

¶ *The Place.*

These kinds of Dogs stones do grow in moist and fertil medows. The marish Dogs stones grow for the most part in moist and waterish woods, and also in marish grounds. ‡ The fift groweth in many hilly places of Austria, Germany, and England. ‡

¶ *The Time.*

They floure from the beginning of May to the end of August.

¶ *The*

¶ *The Names.*

The first and second are of that kinde which *Dioscorides* calleth *Cynosorchis*; that is in English, Dogs stones, after the common or vulgar speech; the one the greater, the other the lesser.

‡ 1   This is *Cynosorchis prior of Dodonæus; Cynosorchis nostra maior of Lobel.*

2   *Dodonæus* names this, *Cynosorchis altera : Lobel, Cynosorchis maioris secunda species.*

3   This *Lobel* calls *Cynosorchis Delphinia, &c. Tabern. Cynosorchis maculata.*

4   *Dodonæus* cals this, *Cynosorchis tertia : Lobel, Cynosorchis maior altera nostras : Tabern. Cynosorchis maior quarta.*

5   This is *Clusius* his *Orchis Pannonica quarta.*

‡ 5 *Cynosorchis minor Pannonica.*
The lesser Austrian Dogs stones.

¶ *The Temperature.*

These kInds of Dogs stones be of temperature hot and moist, but the greater or fuller stone seemes to haue much superfluous windinesse, and therefore being drunk it stirreth vp fleshly lust.

The second, which is lesser, is quite contrary in nature, tending to a hot and dry temperature; therefore his root is so farre from mouing venerie, that contrariwise it stayeth and keepeth it back, as *Galen* teacheth:

He also affirmeth, that Serapias stones are of a more dry faculty, & do not so much preuaile to stir vp the lust of the flesh.

¶ *The Vertues.*

Dioscorides writes, that it is reported, That A if men do eat of the great full or fat roots of these kinds of Dogs stones, they cause them to beget male children; and if women eat of the lesser dry or barren root which is withered or shriueled, they shall bring forth females. These are some Doctours opinions only.

It is further reported, That in Thessalia B the women giue the ful and tender root to be drunk in goats milk, to moue bodily lust; and the dry to restrain the same.

¶ *The Choice.*

Our age vseth all the kindes of stones to stir vp venery, and the Apothecaries mix any of them indifferently with compositions seruing for that purpose. But the best and most effectuall are these Dogs stones, as most haue deemed : yet both the bulbs or stones are not to be taken indifferently, but the harder and fuller, and that which containes most quantitie of juice; for that which is wrinkled is lesse profitable, or not fit at all to be vsed in medicine. And the fuller root is not alwayes the greater, but often the lesser, especially if the roots be gathered before the plant hath shed his floure, or when the stalke first commeth vp; for that which is fuller of iuyce is not the greatest before the seed be perfectly ripe. For seeing that euery other yeare by course, one stone or bulb waxeth full, the other empty and perisheth, it cannot be that the harder and fuller of juice should be alwaies the greater; for at such time as the leaues come forth, the fuller then begins to encrease, and whilest the same by little and little encreaseth, the other doth decrease and wither till the seed be ripe : then the whole plant, together with the leaues aud stalkes, doth forthwith fall away & perish, and that which in the meane time encreased, remaineth stil fresh and full vnto the next yeare.

† The figures of the first and second were transposed in the former edition : the third was of the *Cynosorchis Morio mas*, following in the next Chapter.

CHAP. III.   *Of Fooles Stones.*

¶ *The Deſcription.*

1   THe male Foole ſtones hath fiue, ſomtimes ſix long broad and ſmooth leaues, not vnlike
     to thoſe of the Lilly, ſauing that they are daſht & ſpotted in ſundry places with black
     ſpots and ſtreaks. The floures grow at the top, tuft or ſpike faſhion, ſomewhat like the
former, but thruſt more thicke together, in ſhape like to a fooles hood or cocks-combe, wide open,
or gaping before, and as it were creſted aboue, with certain eares ſtanding vp by euery ſide, and a
ſmall taile or ſpur hanging downe, the back ſide declining to a violet colour, of a pleaſant ſauour
or ſmell.

† 1 *Cynoſorchis Morio mas.*   2 *Cynoſorchis Morio fæmina.*
  The male Foole ſtones.   The female Foole ſtones.

2   The female Fooles ſtones haue alſo ſmooth narrow leaues, ribbed with nerues like thoſe of
Plantain. The floures be likewiſe gaping, and like the former, as it were open hoods, with a little
horne or heele hanging behind euerie one of them, and ſmall green leaues ſorted or mixed among
them, reſembling cocks-combes with little eares, not ſtanding ſtraight vp, but lying flat vpon the
hooded floure, in ſuch ſort that they cannot at the ſudden view be perceiued. The roots are a paire
of ſmall ſtones like the former. The floures of this ſort do vary infinitely in colour, according to
the ſoile or countrey where they doe grow : ſome bring forth their floures of a deep violet colour,
ſome as white as ſnow , ſome of a fleſh colour , and ſome garniſhed with ſpots of diuers colours,
which are not poſſible to be diſtinguiſhed.

‡   3   This hath narrow ſpotted leaues, with a ſtalke ſome foot or more high, at the top wherof growes a tuft of purple floures, in ſhape much like thoſe of the laſt deſcribed, each floure conſiſting of a little hood, two ſmall wings or ſide leaues, and a broad lip or leafe hanging downe. ‡

¶ *The*

‡ 3 *Cynosorchis Morio minor.*
The lesser spotted Fooles stones.

† The first was of *Cynosorchis maculata*, being the third in the former chapter.

¶ *The Place.*

These kinds of Fools stones grow naturally to their best liking in pastures and fields that seldom or neuer are dunged or manured.

¶ *The Time.*

They floure in May and Iune : their stones are to be gathred for medicine in September, as are those of the Dogs stones.

¶ *The Names.*

The first is called *Cynosorchis Morio*: of *Fuchsius*, *Orchis mas angustifolia* : of *Apuleius*, *Satyrion* : and also it is the *Orchis Delphinia* of *Cornelius Gemma.*

‡ The second is *Cynosorchis Morio foemina* of *Lobel* : *Orchis angustifolia foemin.* of *Fuchsius* : *Testiculus Morionis foemina* of *Dodonaus.*

3 This is *Cynosorchis minimis & secundum caulem,&c. maculosis folijs*, of *Lobel.* ‡

¶ *The Temperature.*
Fools stones both male and female are hot and moist of nature.

¶ *The Vertues.*
These Fooles stones are thought to haue the vertues of Dogs stones, whereunto they are referred.

CHAP. 112. *Of Goats stones.*

¶ *The Description.*

‡ THe greatest of the Goats stones bringeth forth broad leaues, ribbed in some sort like vnto the broad leaued Plantaine, but larger : the stalke groweth to the height of a cubit, set with such great leaues euen to the top of the stalke by equal distances. The tuft or bush of floures is small and flat open, with many tender strings or laces comming from the middle part of those small flours, crookedly tangling one with another like to the small tendrels of the Vine, or rather the laces or strings that grow vpon the herb Sauory. The whole floure is of a purple colour. The roots are like the rest of the Orchides, but greater.

2   The male Goats stones haue leaues like to those of the garden Lilly, with a stalke a foot long, wrapped about euen to the tuft of the floure with those his leaues. The floures which grow in this bush or tuft be very small, in form like to a Lizard, becaufe of the twisted or writhen tailes, and spotted heads. Euery of these small floures is at the first like a round close husk, of the bignes of a peafe, which when it openeth, there commeth out of it a little long and tender spurre or taile, white toward the setting of it to the floure ; the rest spotted with red dashes, hauing vpon each side a small thing adioyning vnto it like to a little leg or foot ; the rest of the said taile is twisted crookedly about, and hangeth downward. The whole plant hath a ranke or stinking smell or sauor like the smell of a goat, whereof it tooke his name.

3   The female Goats stones haue leaues like the male kinde, sauing that they be much smaller hauing many floures on the tuft resembling the flies that feed vpon flesh, or rather ticks. In stones or roots and in smell it is like the former.

1 *Tragorchis maximus.*
The greatest Goat stones.

2 *Tragorchis mas.*
The male Goat stones.

3 *Tragorchis fœmina.*
The female Goat stones.

‡ 4 *Tragorchis minor Batavica.*
The small Goat stones of Holland.

‡ 4   This alſo becauſe of the vnpleaſant ſmell may fitly be referred to this Claſſis. The roots hereof are ſmall, and from them ariſe a ſtalke ſome halfe a foot high, beſet with three or foure narrow leaues. The tuft of floures which groweth on the top of this ſtalke is ſmall, and the colour of them is red without, but ſomwhat paler within; each floure hanging down a lip parted in three. ‡

¶ *The Place.*

1. 2. 3.   Theſe kinds of Goats ſtones delight to grow in fat clay grounds, and ſeldom in any other ſoile to be found.

‡ 4   This grows vpon the ſea banks in Holland, and alſo in ſome places neere vnto the Hage.‡

¶ *The Time.*

They floure in May and Iune with the other kinds of Orchis.

¶ *The Names.*

† 1   Some haue named this kind of Goat ſtones in Greeke, τραγίται: in Latine, *Teſticulus Hirci-nus*, and alſo *Orchis Saurodes*, or *Scincophora*, by reaſon that the floures reſemble Lizards.

The ſecond may be called *Tragorchis mas*, male Goats ſtones; and *Orchis Saurodes*, or *Scincopho-ra*, as well as the former.

The third, *Tragorchis fæmina*, as alſo *Corioſmites*, and *Coriophora*, for that the floures in ſhape and their vngratefull ſmell reſemble Ticks, called in Greeke κόρις. †

¶ *The Temperature and Vertues.*

The temperature and vertues of theſe are referred to the Fooles ſtones, notwithſtanding they are ſeldome or neuer vſed in phyſick, in regard of the ſtinking and loathſome ſmell and ſauor they are poſſeſſed with.

CHAP. II3.   *Of Fox Stones.*

1 *Orchis hermaphroditica.*
Butterfly Satyrion.

† 2 *Teſticulus pſycodes.*
Gnat Satyrion.

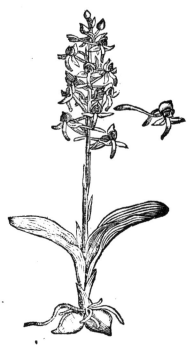

¶ *The*

¶ *The Kindes.*

THere be diuers kindes of Fox-ſtones, differing very much in ſhape of their leaues, as alſo in floures : ſome haue floures,wherein is to be ſeen the ſhape of ſundry ſorts of liuing creatures; ſome the ſhape and proportion of flies, in other gnats, ſome humble bees, others like vnto honey Bees ; ſome like Butter-flies,and others like Waſpes that be dead ; ſome yellow of colour, others white ; ſome purple mixed with red, others of a brown ouerworne colour : the which ſeuerally to diſtinguiſh,as well thoſe here ſet downe,as alſo thoſe that offer themſelues daily to our view and conſideration,would require a particular volume ; for there is not any plant which doth offer ſuch varietie vnto vs as theſe kinds of ſtones,except the Tulipa's,which go beyond all account:for that the moſt ſingular Simpleſt that euer was in theſe later ages,*Carolus Cluſius* (who for his ſingular induſtry and knowledge herein is worthy triple honor) hath ſpent at the leaſt 35 yeares, ſowing the ſeeds of Tulipa's from yeare to yeare,and to this day he could neuer attain to the end or certainty of their ſeuerall kinds of colours. The greateſt reaſon thereof that I can yeeld is this,That if you take the ſeeds of a Tulipa that bare white floures , and ſow them in a pan or tub with earth , you ſhal recciue from that ſeed plants of infinite colours. Contrariwiſe,if you ſow the ſeeds of a plant that beareth flours of variable colours,the moſt of thoſe plants will be nothing like the plant from whence the ſeed was taken. It ſhall be ſufficient therefore to ſet downe moſt of the varieties, and comprehend them in this chapter.

¶ *The Deſcription.*

1   BVtterfly Orchis or Satyrion beares next the root two very broad leaues like thoſe of the Lilly,ſeldome three : the floures be white of colour,reſembling the ſhape of a Butter-fly : the ſtalke is a foot high : the root is two ſtones like the other kindes of Stones or Cullions,but ſomwhat ſharper pointed.

† 3 *Teſticulus Vulpinus 2 ſphegodes.*   4 *Teſticulus Vulpinus major ſphegodes.*
Humble-bee Orchis.   Waſpe Orchis.

2   Gnat Satyrion commeth forth of the ground,bearing two, ſometimes three leaues like the former,but much ſmaller. The ſtalke groweth to the height of an hand, whereon are placed verie orderly ſmall floures like in ſhape to Gnats,and of the ſame colour. The root is like the former.

3　The Humble-Bee Orchis hath a few ſmall weake and ſhort leaues, which grow ſcatteringly about the ſtalk : the floures grow at the top among the ſmall leaues, reſembling in ſhape the humble Bee. The root conſiſteth of two ſtones or bulbes, with ſome few threds anexed thereunto.

4　The Waſpe Satyrion groweth out of the ground, hauing ſtalks ſmall and tender: the leaues are like the former, but ſomwhat greater, declining to a brown or dark colour. The flours be ſmall, of the colour of a dry oken leaſe, in ſhape reſembling the great Bee called in Engliſh an Hornet, or drone Bee. The root is like the other.

5　The leaues of Bee Satyrion are longer than the laſt before mentioned, narrower, turning themſelues againſt the Sun as it were round. The ſtalk is round, tender, and very fragile. At the top grow the floures, reſembling in ſhape the dead carkaſſe of a Bee. The ſtones or bulbes of the roots be ſmaller and rounder than the laſt deſcribed.

6　The Fly Satyrion is in his leaues like the other, ſauing that they be not of ſo dark a colour : the floures be ſmaller, and more plentifully growing about the ſtalke, in ſhape like vnto Flies, of a darke greeniſh colour, euen almoſt blacke.

† 5 *Orchis Melittias.*　　　　　　　† 6 *Orchis Myodes.*
Bee Orchis.　　　　　　　　　　　Fly Satyrion.

7　Yellow Orchis riſeth out of the ground with brown leaues ſmaller than the laſt before mentioned : the ſtalk is tender and crooked : the floures grow at the top, yellow of colour, in ſhape reſembling the yellow flies bred in the dung of kine after raine.

8　The ſmall yellow Satyrion hath leaues ſpred vpon the ground at the firſt comming vp ; the ſlender ſtalke riſeth vp in the midſt, halfe a hand high ; the floures grow ſcatteringly towards the top, reſembling the flies laſt before mentioned, dark or ruſty of colour : the ſtones or bulbs are very round.

9　Birds Orchis hath many large ribbed leaues, ſpred vpon the ground like vnto thoſe of Plantain : among the which riſe vp tender ſtalkes couered euen to the tuft of the floures with the like leaues, but leſſer, in ſuch ſort as the ſtalks cannot be ſeen for the leaues. The flours grow at the top, not ſo thick ſet or thruſt together as the others, purple of colour, like in ſhape vnto little birds, with their wings ſpred abroad ready to fly. The roots be like the former.

10　Spotted Birds Satyrion hath leaues like vnto the former, ſauing that they be daſhed or ſpotted here and there with darke ſpots or ſtreakes, hauing a ſtalke couered with the like leaues, ſo that the plants differ not in any point, except the black ſpots, which this kind is daſht with.

11   White Birds Satyrion hath leaues riſing immediately forth of the ground like vnto the blades or leaues of Leeks, but ſhorter ; among the which riſeth vp a ſlender naked ſtalk two hand-fuls high ; on the top whereof be white floures reſembling the ſhape or form of a ſmall bird ready to fly, or a white Butterfly with her wings ſpred abroad. The roots are round, and ſmaller than any of the former.

12   Soldiers Satyrion bringeth forth many broad large and ribbed leaues, ſpred on the ground like vnto thoſe of the great Plantain : amongſt the which riſeth vp a fat ſtalke full of ſap or juice, cloathed or wrapped in the like leaues euen to the tuft of floures, whereupon do grow little floures reſembling a little man hauing an helmet vpon his head, his hands and legs cut off, white vpon the inſide, ſpotted with many purple ſpots, and the back part of the floure of a deeper colour tending to redneſſe.

<table>
<tr><td>† 7 Orchis Myodes Lutea.
Yellow Satyrion.</td><td>† 8 Orchis Myodes minor.
Small yellow Satyrion.</td></tr>
</table>

13   Soldiers Cullions hath many leaues ſpread vpon the ground, but leſſer than the ſouldiers Satyrion, as is the whole plant. The backſide of the floures are ſomewhat mixed with whiteneſſe, and ſometimes are aſh-coloured. The inſide of the floure is ſpotted with white likewiſe.

14   Spider Satyrion hath many thin leaues like vnto thoſe of the Lilly, ſcatteringly ſet vpon a weake and feeble ſtalk ; whereupon doth grow ſmall floures, reſembling as well in ſhape as colour, the body of a dead humble Bee, ‡ or rather of a Spider ; and therefore I thinke *Lobel*, who was the Author of this name, would haue ſaid *Arachnitis*, of ἀράχνης, a Spider. ‡

‡   15   Thisby right ſhould haue bin put next the Gnat Satyrion, deſcribed in the ſecond place. It hath ſhort, yet pretty broad leaues, and thoſe commonly three in number, beſides thoſe ſmall ones ſet vpon the ſtem. The floures are ſmall, and much like thoſe of the ſecond formerly de-ſcribed.

‡   16   Our Author gaue you this figure in the fourteenth place, vnder the title of *Orchis An-drachnitis :* but it is of the *Orchis 16 minor* of *Tabern.* or *Orchis Anguſtifolia* of *Bauhinus.* This Or-chis is of the kinde of the *Myodes,* or Fly Satyrions, but his leaues are farre longer and narrower
than

† 9 *Orchis Ornithopora.*
Birds Satyrion.

† 10 *Orchis Ornithopora folio maculoſo.*
Spotted Birds Orchis.

† 11 *Orchis Ornithopora candida.*
White Birds Orchis.

† 12 *Orchis Strateumatica.*
Souldiers Satyrion.

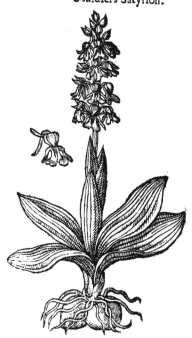

than any of the reſt of that kind, and therein conſiſts the only and chiefeſt difference.‡

¶ *The Place.*

Theſe kindes of Orchis grow for the moſt part in moiſt medowes and fertile paſtures, as alſo in moiſt woods.

The Bee, the Fly, and the Butter-fly Satyrions grow vpon barren chalky hils, & heathy grounds, vpon the hils adioyning to a village in Kent named Greenhithe, vpon Long-field downs by South-fleet, two miles from the ſame place, and in many other places of Kent : likewiſe in a field adioyning to a ſmall groue of trees, halfe a mile from S. Albons, at the South end thereof. They grow likewiſe at Hatfield neere S. Albons, by the relation of a learned Preacher there dwelling, Mr *Robert Abbot*, an excellent and diligent Herbariſt.

† 13 *Orchis Stratenmatica minor.*
Souldiers Cullions.

† 14 *Orchis Andrachnitis.*
Spider Satyrion.

That kind which reſembleth the white Butter-fly groweth vpon the declining of the hill at the end of Hampſted heath, neere to a ſmall cottage there in the way ſide, as yee goe from London to Henden a village there by. It groweth in the fields adioyning to the fold or pin-fold without the gate, at a village called High-gate, neere London : and likewiſe in the wood belonging to a Wor-ſhipfull gentleman of Kent named Mr *Sidley*, of Southfleet : where doe grow likewiſe many other rare and dainty ſimples, that are not to be found elſewhere in a great circuit.

¶ *The Time.*

They floure for the moſt part from May to the end of Auguſt, and ſome of them ſooner.

¶ *The Names.*

Theſe kindes of Orchis haue not bin much written of by the Antients, neither by the late wri-ters to any purpoſe; ſo that it may content you for this time to receiue the names ſet down in their
ſeueral

ſeuerall titles, reſeruing what elſe might be ſaid as touching the Greeke, French, or Dutch names, or any generall definition, vntill a further conſideration.

‡ 15 *Orchis trifolia minor.*
Small Gnat Satyrion.

‡ 16 *Orchis anguſtifolia.*
Narrow leaued Satyrion.

¶ *The Temperature and Vertues.*

The nature and vertues of theſe kinds of Orchis are referred vnto the others, namely to thoſe of the Fox ſtones; notwithſtanding there is no great vſe of theſe in phyſicke, but they are chiefly regarded for the pleaſant and beautifull floures wherewith Nature hath ſeemed to play and diſport her ſelfe.

† Theſe figures in this Chapter were formerly much diſplaced: as thus; The ſecond was of *Orchis Ornith. fol. macul.* being the tenth. The third was of *Triorchis mas minor* of *Tabernam.* being a varietie of *Cynoſorchis morio fœmina.* The fift was of *Orchis Batrachitis.* The ſixt, of *Orchis Melittias.* The ſeuenth and eighth were only tranſpoſed, or put the one for the other. The ninth was of the ſecond, calle i formerly *Teſticulus ſphegodes.* The tenth was of the third, called *Teſticulus Vulpinus.* The eleuenth was of *Strateumatics.* The twelfth was of *Strateumatics minor.* The thirteenth was a varietie of the fourth. The fourteenth was of *Orchis Anguſtifol.* which we here giue you in the ſixteenth place.

CHAP. 114. *Of ſweet Cullions.*

¶ *The Kindes.*

THere be ſundry ſorts of ſweet ſmelling Teſticles or Stones, whereof the firſt is moſt ſweet and pleaſant in ſmell, the others of leſſe ſmell or ſauour, differing in floure and roots. Some haue white floures, others yellow; ſome fleſh-colored, ſome daſht vpon white with a little reddiſh waſh: ſome haue two ſtones, others three, and ſome foure, wherein their difference conſiſteth.

¶ *The Deſcription.*

1 THe firſt kinde of ſweet Stones is a ſmall baſe and low plant in reſpect of all the reſt: The leaues be ſmall, narrow, and ſhort, growing flat vpon the ground; amongſt the which riſeth vp a ſmall weake and tender ſtalke of a finger long, whereupon doe grow ſame

ſmall white floures ſpike-faſhion, of a pleaſant ſweet ſmel. The roots are two ſmal ſtones in ſhape like the other.

  2   Triple Orchis hath commonly three, yet ſomtimes foure bulbs or tuberous roots, ſomwhat long, ſet with many ſmall fibres or ſhort threds ; from the which roots riſe immediatly many flat and plain leaues, ribbed with nerues alongſt them like thoſe of Plantain: among the which come forth naked ſtalks ſmall and tender, wherupon are placed certain ſmal white floures, trace-faſhion, not ſo ſweet as the former in ſmel and ſauor. ‡ The top of the ſtalke whereon the floures do grow, is commonly as if it were twiſted or writhen about. ‡

  3   Frieſeland Lady traces hath two ſmall round ſtones or bulbes, of the bigneſſe of the peaſe that we call Rounciſals ; from the which riſe vp a few hairy leaues leſſer than thoſe of the Triple Stones, ribbed as the ſmall leafed Plantain : among the which commeth forth a ſmall naked ſtalke ſet round about with little yellow floures, not trace-faſhion as the former.

  4   Liege Lady traces hath for his roots two greater ſtones, and two ſmaller ; from the which come vp two and ſomtimes more leaues, furrowed or made hollow in the middeſt like to a trough, from the which riſeth vp a ſlender naked ſtalke, ſet with ſuch floures as the laſt deſcribed, ſauing that they be of an ouerworne yellow colour.

    1 *Teſticulus odoratus.*   2 *Triorchis.*
      Lady-traces.   Triple Lady-traces.

¶ *The Place.*

    Theſe kinds of Stones or Cullions do grow in dry paſtures or heaths, and likewiſe vpon chalky hils, the which I haue found growing plentifully in ſundry places, as in the field by Iſlington neere London, where there is a bowling place vnder a few old ſhrubby Okes. They grow likewiſe vpon the heath at Barn-elmes, neere vnto the head of a conduit that ſendeth water to the houſe belonging to the late Sr *Francis Walſingham.* They grow in the field next vnto a village called Thiſtleworth, as you go from Branford to her Maieſties houſe at Richmond ; alſo vpon a common heath by a village neere London called Stepney, by the relation of a learned merchant of London named Mr *Iames Cole,* exceedingly well experienced in the knowledge of Simples.

    The yellow kindes grow in barren paſtures and borders of fields about Ouenden and Clare in Eſſex

Eſſex. Likewiſe neer vnto Muche Dunmow in Eſſex, where they were ſhewed me by a learned gentleman Mr *Iames Twaights*, excellently well ſeen in the knowledge of plants.

‡ I receiued ſome roots of the ſecond from my kind friend Mr *Thomas Wallis* of Weſtminſter, the which he gathered at Dartford in Kent, vpon a piece of ground commonly called the Brimth : but I could not long get them to grow in a garden, neither do any of the other Satyrions loue to be pent vp in ſuch ſtraight bounds. ‡

3 *Orchis Friſia lutea.*
Frieſeland Lady-traces.

4 *Orchis Leodienſis.*
Liege Lady-traces.

¶ *The Time.*

Theſe kinds of ſtones do floure from Auguſt to the end of September.

¶ *The Names.*

The firſt is called in Latine *Teſticulus odoratus* : in Engliſh, ſweet-ſmelling Teſticles or ſtones ; not of the ſweetneſſe of the roots, but of the floures. It is called alſo *Orchis ſpiralis*, or *Autumnalis*; for that this, as alſo that which is ſet forth in the next place, hath the top of the ſtalk as it were twiſted or twined ſpire faſhion, and for that it comes to flouring in Autumne : of our Engliſh women they be called Lady-traces : but euery countrey hath a ſeuerall name ; for ſome call them Sweet-Ballocks, ſweet Cods, ſweet Cullions, and Stander graſſe : in Dutch, **knabencraut**, and **Stondelcraut** : in French, *Satyrion*.

The ſecond ſort is called *Triorchis*, and alſo *Tetrorchis* : in Engliſh, Triple Lady-traces, or white Orchis.

The third is called *Orchis Friſia* : in Engliſh, Frieſeland Traces.

The laſt of theſe kindes of Teſticles or Stones is called of ſome in Latine, *Orchis Leodienſis*, and *Orchis lutea*, as alſo *Baſilica minor Serapias*, and *Triorchis Ægineta* : in Engliſh, yellow Lady-traces.

¶ *The Temperature.*

Theſe kinds of ſweet Cullions are of nature and temperature like the Dogs ſtones, although not vſed in phyſick in times paſt : notwithſtanding late writers haue attributed ſome vertues vnto them as followeth.

¶ *The Vertues.*

The full and ſappy roots of Lady-traces eaten or boiled in milke, and drunke, prouoke venerie, A nouriſh and ſtrengthen the body, and be good for ſuch as be fallen into a Conſumption or Feauer Hectique.

Chap.

CHAP. 115.   *Of Satyrion Royall.*

¶ *The Description.*

1   THe male Satyrion Royal hath large roots, knobbed, not bulbed as the others, but branched or cut into sundry sections like an hand, from the which come vp thicke and fat
stalks set with large leaues like those of Lillies, but lesse ; at the top whereof groweth
a tuft of floures spotted with a deep purple colour.

<table>
<tr><td>1 *Palma Christi mas.*
The male Satyrion royall.</td><td>2 *Palma Christi fœmina.*
The female Satyrion royall.</td></tr>
</table>

2   The female Satyrion hath clouen or forked roots, with some fibres ioyned thereto. The
leaues be like the former, but smaller and narrower, and confusedly dashed or spotted with blacke
spots : from which springeth vp a tender stalke, at the top whereof groweth a tuft of purple floures
in fashion like vnto a friers hood, changing and varying according to the soile and clymat, sometimes red, sometimes white, and sometimes light carnation or flesh colour.

3   This in roots & leaues is like the former, but that the leaues want the black spots, the stalk
is but low, and the top thereof hath floures of a whitish colour, not spotted : they on the foreside
resemble gaping hoods, with ears on each side, and a broad lip hanging down; the backe part ends
in a broad obtuse spur. These floures smell like Elder blossoms.

¶ *The Place.*

The Royal Satyrions grow for the most part in moist and fenny grounds, medowes, and Woods
that are very moist and shadowie. I haue found them in many places, especially in the midst of a
wood in Kent called Swainescombe wood neere to Grauesend, by the village Swainescombe, and
likewise in Hampsted wood foure miles from London.

¶ *The Time.*

They floure in May and Iune, but seldome later.

¶ *The*

‡ 3 *Orchis Palmata Pannonica S.Cluſ.*
The Auſtrian handed Satyrion.

¶ *The Names.*

1 Royal Satyrion or finger Orchis is cal-led in Latine *Palma Chriſti*; notwithſtanding there is another hearbe or plant called by the ſame name, which otherwiſe is called *Ricinus*. This plant is called likewiſe of ſome, *Satyrium Baſilicum*, or *Satyrium regium*, Some would haue it to be *Buzeiden*, or *Buzidan Arabum*. But *Avicen* ſaith, *Buzeiden* is a woody Indian medicin : and *Serapio* ſaith, *Buzeiden* be hard white roots like thoſe of *Behen album*, and that it is an Indi-an drug : but contrariwiſe the roots of *Palma Chriſti* are nothing leſſe than woody, ſo that it cannot be the ſame. *Matthiolus* would haue Sa-tyrion royal to be the *Digiti Citrini* of *Avicen*; finding fault with the Monkes which ſet forth Commentaries vpon *Meſues* Compoſitions, for doubting, and leauing it to the iudgement of the diſcreet Reader. Yet do we better allow of the Monks doubt, than of *Matthiolus* his aſ-ſertion : for *Avicens* words be theſe ; What is *Aſabaſuffra*, or *Digiti Citrini* ? and anſwering the doubt himſelfe, he ſaith, It is in floure or ſhape like the palm of a mans hand, of a mixt colour between yellow and white, and it is hard, in which there is a little ſweetneſſe, and there is a Citrine ſort duſty & without ſweetneſſe. *Rha-fis* alſo in the laſt booke of his Continent calls theſe, *Digiti Crocei*, or Saffron fingers ; and hee ſaith it is a gum or vein for Diers. Now theſe roots are nothing leſſe than of a Saffron colour, and wholly vnfit for dying. Wherefore doubtleſſe theſe words of *Avicen* and *Rhaſis* in the eares of men of iudgement do confirme, That Satyrion roi-all, or *Palma Chriſti*, are not thoſe *Digiti Citrini*. The Germans call it **Creutſblum** : in low-Dutch, **Handekens crupt :** the French, *Satyrion Royal.*

¶ *The Temperature and Vertues.*

The roots of Satyrion royall are like to *Cynoſorchis*, or Dogs ſtones, both in ſauor and taſte, and therefore are thought by ſome to be of like faculties. Yet *Nicolaus Nicolus*, in the Chapter of the cure of a quartan Ague, ſaith, That the roots of *Palma Chriſti* are of force to purge vpward & down-ward ; and that a piece of the root as long as ones thumbe ſtamped and giuen with wine before the fit commeth, is a good remedie againſt old Quartans after purgation : and reporteth, That one *Ba-liolus* after he had endured 44 fits, was cured therewith.

† This facultie of purging and vomiting, which our Author out of *Dodonæus*, and he out of *Nicolus*, giue to the root of *Palma Chriſti*, I doubt is miſtaken & put in the wrong place : for I iudge it to belong to the *Ricinus*, which alſo is called *Palma Chriſti* : for that *Nicolus* ſaith a piece or root muſt be taken as long as one thumbe : now the whole root of this plant is not ſo long. And beſides, *Ricinus* is knowne to haue a vomitorie purging facultie.

Chap. 116.   *Of Serapia's ſtons.*

¶ *The Kindes.*

THere be ſundry ſorts of Serapia's ſtones, whereof ſome be male, others female ; ſome great, and ſome of a ſmaller kind ; varying likewiſe in colour of the floures, wherof ſome be white, others purple, altering according to the ſoile, or clymat, as the greateſt part of bulbous roots do. Moreo-uer, ſome grow in marſhie and fenny grounds, and ſome in fertil paſtures lying open to the Sun ; va-rying likewiſe in the ſhape of their floures ; retaining the form of flies, butterflies, and gnats, like thoſe of the Fox-ſtones.

T

¶ *The*

1 *Serapias candido flore.*
White handed Orchis.

2 *Serapias minor nitente flore.*
Red handed Orchis.

3 *Serapias paluſtris latifolia.*
Mariſh Satyrion.

4 *Serapias paluſtris leptophylla.*
Fenny Satyrion.

† 5 *Serapias Montana.*
Mountaine Satyrion.

† 6 *Serapias Gariophyllata cum rad. & sem.*
Sweet-smelling Satyrion, with the root
and feed expreft at large.

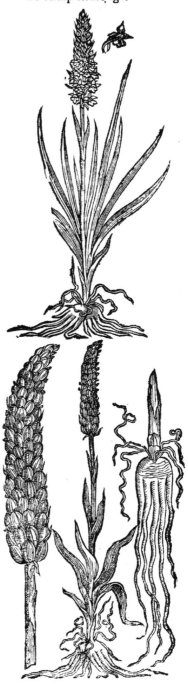

7 *Serapias castrata.*
Gelded Satyrion.

¶ *The Defcription.*

1   THe white handed Orchis or Satyrion hath long and large leaues , fpotted and dafhed
with blacke fpots,from the which doth rife vp a fmall fragile or brittle ftalke of two
hands high, hauing at the top a bufh or fpoky tuft of white floures,like in fhape vnto
thofe of *Palma Chrifti*,whereof this is a kinde. The root is thicke, fat, and full of iuyce, fafhioned
like the hand and fingers of a man , with fome tough and fat ftrings faftened vnto the vpper part
thereof.

2   Red handed Satyrion is a fmall low and bafe herb, hauing a fmall tender ftalk fet with two
or three fmall leaues like vnto thofe of the Leeke,but fhorter. The floure groweth at the top tuft-
fafhion,of a gliftering red colour,with a root fafhioned like an hand, but leffer than the former.

3   Serapia's ftones or marifh Satyrion hath a thick knobby root,diuided into fingers like thofe
of *Palma Chrifti*,whereof it is a kind : from which rife thick fat and fpongeous ftalks fet with broad
leaues like thofe of plantain,but much longer,euen to the top of the tuft of floures;but the higher
they rife toward the top , the fmaller they are. The floure confifts of many fmall hooded floures
fomewhat whitifh, fpotted within with deep purple fpots ; the back fide of thefe little floures are
Violet mixt with purple.

† 8 *Serapias Batrachites.*   ‡ 9 *Serapias Batrachites altera.*
 Frog Satyrion.   The other Frog Satyrion,

4   Fenny Satyrion(or Serapia's ftones) differeth little from the former,fauing that the leaues
are fmaller,and fomewhat fpotted,and the tuft of floures hath not fo many greene leaues, nor fo
long,mixed with the floures, neither are they altogether of fo dark or purplifh a colour as the former,The roots are like thofe of the laft defcribed.

5   Mountaine Orchis or Satyrion hath thicke fat and knobby roots , the one of them for the
moft part being handed,and the other long. It growes like the former in ftalks, leaues,and floures,
but is fomewhat bigger,with the leaues fmoother and more fhining.

6   Cloue Satyrion, or fweet-fmelling Orchis, hath flat and thicke roots diuided into fingers
                          like

like those of *Palma Christi*, sauing that the fingers are longer, smaller, and more in number: from the which rise vp long and narrow leaues like those of Narcissus or Daffodil: among which commeth forth a small tender stalke, at the top whereof groweth a purple tuft compact of many small floures resembling Flies, but in fauor and smell like the Cloue or Cloue Gillofloure; but far sweeter and pleasanter, as my selfe with many others can witnesse now liuing, that haue both seene and smelt them in my garden. ‡ After the floure is past, come many seed-vessels filled with a small seed, and growing after the manner as you see them heere at large expressed in a figure, together with the root also set forth at full. ‡

7   Gelded Satyrion hath leaues with nerues and sinues like those of Daffodill, set vpon a weak and tender stalk, with floures at the top white of colour, spotted within the floure, and in shape they are like gnats and little flies. The stalk is gelded as it were, or the stones and hands cut off, leauing for the root two long legs or fingers, with many strings fastned vnto the top.

8   Frog Satyrion hath smal flat leaues set vpon a slender weak stem: at the top wherof growes a tuft of floures compact of sundry small floures, which in shape do resemble little frogs, whereof it took his name. The root is likewise gelded, only reseruing two small mishapen lumps, with certaine fibres anexed thereto.

‡ 9   This also may fitly be added to the last described, the root shewing it to be of a kinde between the Serapia's and Orchis. It groweth to the height of the former, with short leaues ingirting the stalke at their setting on. The floures on the top resemble a Frog, with their long leaues; and if you looke vpon them in another posture, they will somwhat resemble little flies; wherefore *Lobel* calls it as well *Myodes*, as *Batrachites*. ‡

¶ *The Time.*

These plants flourish in the moneth of May and Iune, but seldome after, except some degenerat kinde, or that it hath had some impediment in the time when it should haue floured, as often happeneth.

¶ *The Names.*

We haue called these kindes Serapia's stones, or Serapiades, especially for that sundry of them do bring forth floures resembling flies and such like fruitful and lasciuious Insects, as taking their name from *Serapias* the god of the citisens of Alexandria in Egypt, who had a most famous temple at Canopus, where he was worshipped with all kind of lasciuious wantonnesse, songs, and dances; as we may reade in *Strabo*, in his seuenteenth booke. *Apuleius* confounds the Orchides and Serapiades, vnder the name of both the Satyrions; and withall saith it is called *Entaticos, Panion,* and of the Latines, *Testiculus Leporinus*: in English wee may call them Satyrions, and finger Orchis, and Hares stones.

¶ *The Temperature and Vertues.*

Serapia's stones are thought to be in nature, temperature, and vertues like vnto the Satyrion roiall; and although not so much vsed in physicke, yet doubtlesse they worke the effect of the other Stones.

† The fifth was the figure of *Satyrium trifolium* of *Tabern.* and is a kind of *Testiculus pseudes.*   6   In this place formerly was the figure of the last before, to wit, *Serapias montana.*   ‡ Here was the figure of *Myodes,* which should haue bin in the sixth place in the 101 Chapter of the former edition, being the 113 of this.

Chap. 117.

Of Fenny stones.

¶ *The Description.*

† 1   THis hath cleft or diuided roots like fingers, much like vnto the roots of other *Palma Christi's*; whereof this is a kinde: from the which riseth vp a stalke of a foot high, set here and there with very faire Lilly-like leaues, of colour red, the which do clip or embrace the stalkes almost round about, like the leaues of Thorow-wax. At the

top of the stalke groweth a faire bush of very red floures, among the which floures do grow many small sharpe pointed leaues. The seed I could neuer obserue, being a thing like dust that flieth in the winde.

2   The other marish handed Satyrion differeth little from the precedent, but in the leaues and floures, for that the leaues are smaller and narrower, and the floures are faire white, gaping wide open; in the hollownesse whereof appeare certain things obscurely hidden, resembling little helmets, which setteth forth the difference.

† 1 *Serapias Dracontias palustris.*   † 2 *Serapias palustris leptophylla altera.*
   Marish Dragon Satyrion.   The other marish handed Satyrion.

3   This third handed Satyrion hath roots fashioned like an hand, with some strings fastned to the vpper part of them; from which riseth vp a faire stiffe stalke armed with large leaues, very notably dasht with blackish spots, clipping or embracing the stalke round about: at the top of the stalke standeth a faire tuft of purple floures, with many greene leaues mingled amongst the same, which maketh the bush or tuft much greater. The seed is nothing else but as it were dust, like the other of his kinde: ‡ And it is contained in such twined vessels as you see exprest apart by the side of the figure; which vessels are not peculiar to this, but common to most part of the other Satyrions. ‡

4   The creeping rooted Orchis or Satyrion without testicles, hath many long roots dispersing themselues, or creeping far abroad in the ground, contrarie to all the rest of the Orchides : which roots are of the bignesse of strawes, in substance like those of Sope-wort ; from the which immediatly doth rise foure or fiue broad smooth leaues like vnto the small Plantaine : from the which shooteth vp a smal and tender stalke, at the top whereof groweth a pleasant spiky care of a whitish colour, spotted on the inside with little speckes of a bloudy colour. The seed also is verie small.

‡ 5   This from handed roots like others of this kinde, sendeth vp a large stalke, somtimes attaining to the height of two cubits ; the leaues are much like vnto those of the marish Satyrions; the floures are of an elegant purple, with little hoods like the top of an helmet (whence *Gemma* termed

3 *Palma Chriſti paluſtris*.
The third handed mariſh Satyrion.

4 *Palma Chriſti radice repente*.
Creeping Satyrion.

‡ 5 *Palma Chriſti maxima*.
The greateſt handed Satyrion.

termed the plant, *Cynoſorch. conopſæa* ; and from the height he called it *Macrocaulos*.) Theſe floures ſmell ſweet, & are ſucceeded by ſeeds like thoſe of the reſt of this kindred. ‡

It delights to grow in grounds of an indifferent temper, not too moiſt nor too dry. It flours from mid May to mid-Iune. ‡

¶ *The Place.*

They grow in mariſh and fenny grounds, & in ſhadowie woods that are very moiſt.

The fourth was found by a learned Preacher called Mr *Robert Abbot*, of Biſhops Hatfield, in a boggie groue where a conduit head doth ſtand, that ſendeth water to the queens houſe in the ſame towne. ‡ It grows alſo plentifully in Hampſhire, within a mile of a market towne called Peters-field, in a moiſt medow named Wood-mead, neere the path leading from Peters-field toward Beryton. ‡

¶ *The Time.*

They flour and flouriſh about the months of May and Iune.

‡ ¶ *The Names.*

‡ 1 This is *Cynoſorchis Dracuntias* of Lobel and *Gemma*.

2 This

2 This is *Cynosorchis palustris altera Leptaphylla* of *Lobel* : *Testiculus Galericulatus*, of *Tabernamontanus*.

3 *Lobell* and *Gemma* terme this, *Cynosorchis palustris altera Lophodes, vel nephelodes*.

4 This is *Orchis minor radice repente* of *Camerarius*.

5 This by *Lobel* and *Gemma* is called *Cynosorchis macrocaulos, sive Conopsea*.

¶ *The Nature and Vertues.*

There is little vse of these in physicke ; only they are referred vnto the handed Satyrions, wherof they are kinds: notwithstanding *Dalescampius* hath written in his great volume, that the marish Orchis is of greater force than any of the Dogs stones in procuring lust.

Camerarius of Noremberg, who was the first that described this kind of creeping Orchis, hath set it forth with a bare description only ; and I am likewise constrained to doe the like, because as yet I haue had no triall thereof.

CHAP. 118. *Of Birds nest.*

1 *Satyrium abortivum, sive Nidus avis.* Birds nest.

¶ *The Description.*

1 BIrds nest hath many tangling roots platted or crossed one ouer another very intricately, which resembleth a Crows nest made of sticks ; from which riseth vp a thicke soft grosse stalk of a browne colour, set with small short leaues of the colour of a dry oken leafe that hath lien vnder the tree all the winter long. On the top of the stalke groweth a spiky eare or tuft of floures, in shape like vnto maimed Satyrion, wherof doubtlesse it is a kinde. The whole plant, as well stickes, leaues, and floures, are of a parched brown colour.

‡ I receiued out of Hampshire from my often remembred friend Mr *Goodyer*, this following description of a *Nidus avis*, found by him the 29 of Iune, 1621.

¶ *Nidus avis flore & caule violaceo purpureo colore* ; an *Pseudoleimodoron Clus.hist.rar.plant.pag.270.*

This riseth vp with a stalke about nine inches high, with a few small narrow sharpe pointed short skinny leaues set without order, very little or nothing at all wrapping or inclosing the stalk, hauing a spike of floures like those of *Orobanche*, without tailes or leaues growing amongst them: which fallen there succeed small seed-vessels. The lower part of the stalke within the ground is not round like *Orobanche*, but slender or long, and of a yellowish white colour, with many smal brittle roots growing vnderneath confusedly, wrapped or folded together like those of the common *Nidus avis*. The whole plant as it appeareth aboue ground, both stalkes, leaues, and floures, is of a violet or deepe purple colour. This I found wilde in the border of a field called Marborne, neere Habridge in Haliborn, a mile from a towne called Alton in Hampshire, being the land of one *William Baldin*. In this place also groweth wilde the thistle called *Corona fratrum. Ioh. Goodyer.*

¶ *The Place.*

This bastard or vnkindely Satyrion is very seldome seene in these Southerly parts of England.

It

It is reported, that it groweth in the North parts of England, neere vnto a village called Knaesbo-rough. I found it growing in the middle of a wood in Kent two miles from Grauefend, neere vnto a worſhipfull gentlemans houſe called Mr *William Swan*, of Howcke green. The wood belongs to one Mr *Iohn Sidley*. Which plant I did neuer ſee elſewhere; and becauſe it is very rare, I am the more willing to giue you all the markes in the wood for the better finding it, becauſe it doth grow but in one piece of the wood: that is to ſay, The ground is couered all ouer in the ſame place neere about it with the herb Sanicle, as alſo with the kind of Orchis called *Hermaphroditica*, or Butterfly Satyrion.

¶ The Time.

It floureth and flouriſheth in Iune and Auguſt. The duſty or mealy ſeed (if it may bee called ſeed) falleth in the end of Auguſt: but in my iudgment it is an vnprofitable or barren duſt, and no ſeed at all.

¶ The Names.

It is called *Satyrium abortivum*: of ſome, *Nidus avis*: in French, *Nid d'oiſeau*: in Engliſh, Birds neſt, or Gooſe neſt: in Low-Dutch, **Uogels neſt**: in High-Dutch, **Margen dꝛehen.**

¶ The Nature and Vertues.

It is not vſed in phyſicke, that I can finde in any authoritie either of the antient or later writers, but is eſteemed as a degenerat kind of Orchis, and therefore not vſed.

THE SECOND BOOKE OF
THE HISTORIE OF PLANTS.

*Containing the Deſcription, Place, Time, Names, Nature, and
Vertues of all ſorts of Herbes, for meat, medicine, or
ſweet-ſmelling vſe, &c.*

E E haue in our firſt booke ſufficiently deſcribed the Graſſes, Ruſhes, Flags, Corne, and bulbous rooted Plants, which for the moſt part are ſuch as with their braue and gallant floures deck and beautiſie gardens, and feed rather the eies than the belly. Now there remain certain other bulbs, whereof the moſt, though not all, ſerue for food: of which we will alſo diſcourſe in the firſt place in this booke, diuiding them in ſuch ſort, that thoſe of one kinde ſhal be ſeparated from another. ‡ In handling theſe and ſuch as next ſucceed them, we ſhal treat of diuers, yea the moſt part of thoſe herbs that the Greeks call by a generall name Λάχανα: and the Latines, *Olera*: and we in Engliſh, Sallet-herbs. When we haue paſt ouer theſe, we ſhal ſpeake of other plants, as they ſhal haue reſemblance each to other in their externall form.

CHAP. I. *Of Turneps.*

¶ *The Kindes.*

THere be ſundry ſorts of Turneps, ſome wild, ſome of the garden; ſome with round roots globe faſhion, other ouall or peare-faſhion; and another ſort longiſh or ſomewhat like a Radiſh: and of all theſe there are ſundry varieties, ſome being great, and ſome of a ſmaller ſort.

¶ *The Deſcription.*

1   He Turnep hath long rough and greene leaues, cut or ſnipt about the edges with deepe gaſhes. The ſtalke diuideth it ſelfe into ſundry branches or armes, bearing at the top ſmall floures of a yellow colour, and ſometimes of a light purple: which being paſt, there do ſucceed long cods full of ſmall blackiſh ſeed like Rape ſeed: the root is round like a bowle, and ſometimes a little ſtretched out in length, growing very ſhallow in the ground, and often ſhewing it ſelfe aboue the face of the earth.

‡   2   This is like the precedent in each reſpect, but that the root is not made ſo globous or bowle-faſhioned as the former, but ſlenderer, and much longer, as you may perceiue by the figure we here giue you. ‡

3   The ſmall Turnep is like vnto the firſt deſcribed, ſauing that it is leſſer. The root is much ſweeter in taſte, as my ſelfe haue often proued.

4   There is another ſort of ſmall Turnep ſaid to haue red roots; ‡ and there are other-ſome whoſe roots are yellow both within and without; ſome alſo are greene on the out-ſide, and other-ſome blackiſh.

¶ *The Place.*

The Turnep proſpereth well in a light looſe and fat earth, and ſo looſe, as *Petrus Creſcentius* ſaith,
that

that it may be turned almoſt into duſt. It groweth in fields and diuers vineyards and hop-gardens in moſt places of England.

The ſmall Turnep growes by Hackney in a ſandy ground, and thoſe that are brought to Cheap-ſide market from that village are the beſt that euer I taſted.

¶ The Time.

Turneps are ſown in the ſpring, as alſo in the end of Auguſt. They floure and ſeed the ſecond yeare after they are ſowne : for thoſe that floure the ſame yeare that they are ſown, are a degenerat kind, called in Cheſhire about the Namptwich, Madneps, of their euill qualitie in cauſing frenſie and giddineſſe of the brain for a ſeaſon.

<table>
<tr><td>1   Rapum majus.
Great Turnep,</td><td>‡ 2   Rapum radice oblonga.
Longiſh rooted Turnep.</td></tr>
</table>

¶ The Names.

The Turnep is called in Latine, <i>Rapum</i> : in Greek, γογγύλη : the name commonly vſed in ſhops and euery where, is <i>Rapa.</i> The Lacedemonians call it γάσϊε the Boëtians, ζεκελτίς, as <i>Athenæus</i> reporteth : in high-Dutch, **Ruben** : in low-Dutch, **Rapen** : in French, <i>Naveau rond</i> : in Spaniſh, <i>Nabo</i> : in En-gliſh, Turnep, and Rape.

¶ The Temperature and Vertues.

The bulbous or knobbed root, which is properly called <i>Rapum</i> or Turnep, and hath giuen name to the plant, is many times eaten raw, eſpecially of the poore people in Wales, but moſt commonly boiled. The raw root is windy, and engendreth groſſe and cold bloud : the boiled doth coole leſſe, and ſo little, that it cannot be perceiued to coole at all, yet is it moiſt and windy.

B   It auaileth not a little after what maner it is prepared ; for being boiled in water, or in a certain broth, it is more moiſt, and ſooner deſcendeth, and maketh the body more ſoluble; but being roſted or baked, it drieth, and engendreth leſſe winde, and yet it is not altogether without wind : but how-ſoeuer they be dreſſed, they yeeld more plenty of nouriſhment than thoſe that are eaten raw : they do encreaſe milke in womens breſts, and naturall ſeed, and alſo prouoke vrin.

C   The decoction of Turneps is good againſt the cough and hoarſeneſſe of the voice, being drunke in the euening with a little ſugar, or a quantitie of clarified hony.

D   <i>Dioſcorides</i> writeth, That the Turnep it ſelfe being ſtamped, is with good ſucceſſe applied vpon
mouldy

mouldy or kibed heeles, and that alſo oile of Roſes boiled in a hollow turnep vnder the hot embers doth cure the ſame.

The yong and tender ſhoots or ſprings of Turneps at their firſt comming forth of the ground, boiled and eaten as a ſallad prouoke vrin.　E

The ſeed is mixt with Counterpoiſons and Treaeles, and beeing drunke it is a remedie againſt poiſons.　F

They of the Low-countries do giue the oile which is preſſed out of the ſeed, againſt the after-throwes of women newly brought to bed, and alſo miniſter it to yong children againſt the worms, which it both killeth and driueth forth.　G

The oile waſhed with water doth allay the feruent heat and ruggedneſſe of the skin.　H

Chap. 2.　Of wilde Turneps.

¶ The Kindes.

THere be three ſorts of wild Turneps; one our common Rape which beareth the ſeed whereof is made Rape oile, and feedeth ſinging birds: the other the common enemy to corn, which we call Charlock; whereof there be two kinds, one with a yellow or els purple floure, the other with a white floure: there is alſo another of the water and mariſh grounds.

1　*Rapum ſylveſtre.*
　Wilde Turneps.

2　*Rapiſtrum arvorum.*
　Charlock or Chadlock.

¶ The Deſcription.

i　WIld Turneps or Rapes haue long broad and rough leaues like thoſe of Turneps, but not ſo deeply gaſhed in the edges, The ſtalks are ſlender and brittle, ſomwhat hai-ty, of two cubits high, diuiding themſelues at the top into many arms or branches, whereon doe grow little yellowiſh floures: which being paſt, there doe ſucceed ſmall long cods which containe the ſeed like that of the Turnep, but ſmaller, ſomewhat reddiſh, and of a fiery hot

V　　and

and biting taſte as is the muſtard, but bitterer. The root is ſmall, and periſheth when the ſeed is ripe.

2   Charlocke or wild Rape hath leaues like vnto the former, but leſſer, the ſtalk and leaues being alſo rough. The ſtalks be of a cubit high, ſlender and branched : the floures are ſomtimes purpliſh, but more often yellow. The roots are ſlender, with certaine threds or ſtrings hanging vpon them.

‡ There is alſo another varietie hereof, with the leaues leſſe diuided, and much ſmoother than the two laſt deſcribed, hauing yellow floures and cods not ſo deeply ioynted as the laſt deſcribed: this is that which is ſet forth by *Matthiolus* vnder the name of *Lampſana*.

2   Water Chadlock groweth vp to the height of three foot and ſomwhat more, with branches ſlender and ſmooth in reſpect of any of the reſt of his kinde, ſet with rough ribbed leaues deepely indented about the lower part of the leafe. The floures grow at the top of the branches, vmble or tuft faſhion, ſometimes of one colour, and ſometimes of another. ‡ The root is long, tough, and full of ſtrings, creeping and putting forth many ſtalks : the ſeed-veſſels are ſhort and ſmal. *Bauhine* hath this vnder the title of *Raphanus aquaticus alter*. ‡

<table>
<tr><td>2 Rapiſtrum arvenſe alterum.
Another wilde Charlocke.</td><td>2 Rapiſtrum aquaticum.
Water Chadlocke.</td></tr>
</table>

¶ *The Place.*

Wilde Turneps or Rapes doe grow of themſelues in fallow fields, and likewiſe by high-wayes neere vnto old walls, vpon ditch banks, and neere vnto towns and villages, and in other vntoiled and rough places.

The Chadlocke groweth for the moſt part among corn in barren grounds, and often by the borders of fields and ſuch like places.

Water Chadlock groweth in moiſt medowes and mariſh grounds, as alſo in water ditches and ſuch like places.

¶ *The Time.*

Theſe do floure from March, till Summer be far ſpent, and in the mean ſeaſon the ſeed is ripe.

¶ *The*

¶ *The Names.*

Wilde Turnep is called in Latine, *Rapistrum, Rapum sylvestre* ; and of some, *Sinapi sylvestre* , or wilde mustard : in high-Dutch, 𝕳𝖊𝖉𝖊𝖗𝖎𝖈𝖍 : in Low-Dutch, 𝕳𝖊𝖗𝖎𝖈𝖐 : in French, *Velar :* in English, Rape, and Rape seed. *Rapistrum arvorum* is called Charlock, and Carlock.

¶ *The Temperature.*

The seed of these wild kinds of Turneps, as also the water Chadlock, are hot and dry as mustard seed is. Some haue thought that Charlock hath a drying and clensing qualitie, and somewhat digesting.

¶ *The Vertues.*

Diuers vse the seed of Rape in stead of Mustard seed, who either make herof a sauce bearing the A name of Mustard, or else mix it with mustard seed ; but this kind of sauce is not so pleasant to the taste, because it is not so bitter.

Galen writeth, That these being eaten ingender euill bloud ; yet *Dioscorides* saith, they warm the B stomacke, and nourish somewhat.

CHAP. 3. *Of Nauewes.*

¶ *The Kindes.*

THere be sundry kinds of Nape or Nauewes degenerating from the kinds of Turnep; of which, some are of the garden, and other wild or of the field.

¶ *The Description.*

1 NAuew gentle is like vnto Turneps in stalks, floures, and seed, as also in the shape of the leaues, but those of the Nauew are much smoother ; it also differs in the root: the Turnep is round like a globe, the Nauew root is somewhat stretched forth in length.

† 1 *Bunias.*
Nauew gentle.

† 2 *Bunias sylvestris Lobelij.*
Wilde Nauew.

2 The small or wilde Nauew is like vnto the former, sauing that it is altogether lesser. The root is small, somewhat long, with threds long and tough at the end thereof.

¶ *The Place.*

Nauew-gentle requireth a loose and yellow mould euen as doth the Turnep, and prospers in a fruitfull soile : it is sowne in France, Bauaria, and other places, in the fields, for the seeds sake, as is likewise that wilde Colewort called of the old writers *Crambe* ; for the plentifull encrease of the seeds bringeth no small gain to the husbandmen of that country, because that being pressed they yeeld an oile, which is vsed not only in lamps, but also in the making of sope ; for of this oile and a lie made of certain ashes, is boiled a sope which is vsed in the Low-Countries euerie where to scoure and wash linnen cloathes. I haue heard it reported, that it is at this day sowne in England for the same purpose.

This wilde Nauew groweth vpon ditch banks neere vnto villages and good townes, as also vpon fresh marshie banks in most places.

¶ *The Time.*

The Nauew is sown, floureth, and seedeth at the same time that the Turnep doth.

¶ *The Names.*

The Nauew is called in Latine *Napus*, and *Bunias* : in Greeke, Βούνιας: the Germanes call it **Steckruben** : the Brabanders, **Steckropen** : in Spanish, *Naps* : in Italian, *Nauo* : the Frenchmen, *Naveau* : in English, Nauew gentle, or French Naueau. The other is called *Napus sylvestris*, or Wild Nauew.

¶ *The Temperature and Vertues.*

A   The Nauew and the Turnep are all one in temperature and vertues, yet some suppose that the Nauew is a little drier, and not so soone concocted, nor passeth downe so easily and doth withall ingender lesse winde. In the rest it is answerable vnto the Turnep.

B   ‡ The seeds of these taken in drinke or broth are good against poyson, and are vsually put into antidotes for the same purpose.

† The figure that was in the first place is a kinde of the long Turnep, described by me in the second place of the first chapter of this second booke. And that in the second place was a lesser kinde of the same.

Chap. 4. *Of Lyons Turnep or Lions leafe.*

Leontopetalon.
Lions Leafe.

¶ *The Description.*

Lions turnep or Lions leafe hath broad leaues like vnto Coleworts, or rather like the Peonies, cut & diuided into sundry great gashes : the stalke is two foot long, thick, and full of juice, diuiding it selfe into diuers branches or wings, in the tops whereof there stand red floures: afterward there appeareth long cods, in which lie the seeds, like vnto tares or wild Chichs. The root is great, bumped like a Turnep, and black without.

¶ *The Place.*

It groweth among corn in diuers places of Italy, in Candy also, and in other prouinces towards the South and East. The right noble Lord *Zouch* brought a plant hereof from Italy at his returne into England , the which was planted in his garden. But as far as I know it perished.

¶ *The Time.*

It floureth in winter, as witnesseth *Petrus Bellonius.*

¶ *The Names.*

The Grecians call it Λεοντοπέταλον : that is, *Leonis folium*, or Lyons Leafe. *Pliny* doth also call it *Leontopetalon* : *Apuleius*, *Leontopodion* : yet there is another plant also called by the same name. There be many bastard names giuen vntoit, as
Rapeium

Rapeium, Papanerculum, Semen Leoninum, Pes Leoninus, and *Brumaria :* in Engliſh, Lions leafe, & Lions Turnep.

¶ *The Nature.*

Lions Turnep is of force to digeſt. It is hot and dry in the third degree, as *Galen* teacheth.

¶ *The Vertues.*

The root (ſaith *Dioſcorides*) taken in wine, helpeth them that are bitten of Serpents, and ſpeedily allaieth the paine. It is put into Cliſters which are made for them that be tormented with the Sciatica.

CHAP. 5.   *Of Radiſh.*

¶ *The Kindes.*

THere be ſundry ſorts of Radiſh, whereof ſome be long and white, others long and black; ſome round and white, others round or of the form of a peare, and blacke of colour; ſome wild or of the field, and ſome tame or of the garden; whereof we will treat in this preſent chapter.

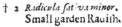

† 1 *Raphanus ſativus.*
Garden Radiſh.

† 2 *Radicula ſativa minor.*
Small garden Radiſh.

¶ *The Deſcription.*

1   THe garden Radiſh ſends forth great and large leaues, greene, rough, cut on both ſides with deepe gaſhes not vnlike to the garden Turnep, but greater. The ſtalks be round, and parted into many branches, out of which ſpring ſmall flours of a light purple colour, made of foure little leaues : and when they be paſt, there come in place ſharpe pointed cods puft or blown vp toward the ſtalke, full of ſpungeous ſubſtance, wherein is contained the ſeed, of a light brown colour, ſomwhat greater than the ſeeds of Turneps or Coleworts. The root is groſſe, long, and white both within and without, and of a ſharp taſte.

The

2   The ſmall garden Radiſh hath leaues like the former, but ſmaller, and more brittle in hand-ling. The ſtalke is two cubits high, whereon be the floures like the former. The ſeed is ſmaller, and not ſo ſharpe in taſte. The root is ſmall, long, white both within and without, except a little that ſhewes it ſelfe aboue the ground, of a reddiſh colour.

3   Radiſh with a round root hath leaues like the garden Turnep : among which leaues ſprings vp a round and ſmooth ſtalke, diuiding it ſelfe toward the top into two or three branches, whereon grow ſmall purpliſh floures made of foure leaues apiece : which beeing paſt, there come in place ſmall cods puft vp or bunched in two, and ſometimes three places, full of pith as the common Ra-diſh; wherein is contained the ſeed, ſomewhat ſmaller than the Colewort ſeed, but of an hotter taſte. The root is round and firme, nothing wateriſh like the common Radiſh, more pleaſant in taſt, wholſomer, not cauſing ſuch ſtinking belchings as the garden radiſh doth.

4   The Radiſh with a peare-faſhioned root groweth to the height of three or foure cubits, of a bright reddiſh colour. The leaues are deepely cut or jagged like thoſe of the Turnep, ſomewhat rough. The floures are made of foure leaues of a light carnation or fleſh colour. The ſeed is con-tained in ſmall bunched cods like the former. The root is faſhioned like a Peare or long Turnep, black without, and white within, of a firme and ſolid ſubſtance. The taſt is quick and ſharp, biting the tongue as the other kinds of Radiſh, but more ſtrongly.

3 *Raphanus orbiculatus.*
Round Radiſh.

              4 *Raphanus pyriformis, ſiue radice nigra.*
              The blacke or Peare-faſhion Radiſh.

¶ *The Place.*

All the kindes of Radiſh require a looſe ground which hath bin long manured, and is ſomewhat fat. They proſper wel in ſandy ground, where they are not ſo ſubiect to worms as in other grounds.

¶ *The Time.*

Theſe kinds of Radiſh are moſt fitly ſown after the Summer ſolſtice, in Iune or Iuly; for being ſown betimes in the ſpring, they yeeld not their roots ſo kindely nor profitably, for then they doe for the moſt part quickly run vp to the ſtalk and ſeed, where otherwiſe they do not floure and ſeed till the next ſpring following. They may be ſown ten moneths in the yeare, but as I ſaid, the beſt time is in Iune and Iuly.

¶ *The*

¶ *The Names.*

Radish is called in Greek, of *Theophrastus*, *Dioscorides*, *Galen*, & other old writers ᵢₙ ₚₐᵤₑ; in shops, *Raphanus*, and *Sativa radicula* : in high-Dutch, **Rettich** : in low-Dutch, **Radus** : in French, *Raifort* : in Italian, *Raphano* : in Spanish, *Rauano* : in English, Radish, and Rabone : in the Bohemian tongue, **Rzedkew**. *Cælius* affirmeth, that the seed of Radish is called of *Marcellus Empericus*, *Bacanon*: and so also of *Aëtius*, *lib.2.cap.2.* of his Tetrabible : yet *Cornarius* doth not reade *Bacanon*, but *Cacanon*. The name of *Bacanum* is also found in *N. Myrepsus*, in the 255 Composition of his first booke.

¶ *The Temperature.*

Radish doth manifestly heat and dry, open and make thin by reason of the biting qualitie that ruleth in it. *Galen* makes them hot in the third degree, and dry in the second, and sheweth that it is rather a sauce than a nourishment.

¶ *The Vertues.*

Radish are eaten raw with bread in stead of other food; but in that manner they yeeld very little **A** nourishment, and that faulty and il. But for the most part they are vsed as sauce with meats to procure appetite, and in that sort they ingender lesse bad bloud, than eaten alone or with bread only : but seeing they be of harder digestion than meats, they are oftentimes troublesom to the stomack; neuerthelesse they serue to distribute and disperse the nourishment, especially beeing eaten after meat : taken before meat, they cause belchings, and ouerthrow the stomacke.

Before meat they cause vomiting, especially the rind ; which as it is more biting than the inner **B** substance, so doth it with more force cause that effect, if it be giuen with Oxymel, a syrrup made with vineger and hony.

Moreouer, Radish prouoketh vrine, and dissolueth cluttered sand, driuing it forth, if a good **C** draught of the decoction thereof be drunke in the morning. *Pliny* and *Dioscorides* write, That it is good against an old Cough, and to make thinne, thicke and grosse flegme which sticketh in the chest.

In stead whereof, the Physitions of our age vse the distilled water, which likewise procures vrin **D** mightily, and driues forth stones in the kidnies.

The root sliced and laid ouer-night in white or Rhenish wine, and drunke in the morning, driues **E** out vrine and grauell mightily, but in taste and smell it is very loathsome.

The root stamped with honey and the pouder of a sheepes heart dried, causeth haire to grow in **F** short space.

The seed causeth vomit, prouoketh vrin, and being drunke with honied vineger, it killeth & dri- **G** ueth forth wormes.

The root stamped with the meale of Darnel and a little white wine vineger, takes away all black **H** and blew spots, and bruised blemishes of the face.

The root boiled in broth, and the decoction drunke, is good against an old cough; it moues wo- **I** mens sicknesse, and causeth much milke.

† Those figures that were in the first and second place were varieties of the long Turnep, described in the second place, *Cap.1.* of this second booke.

CHAP. 6. *Of wilde Radish.*

¶ *The Description.*

1 Wilde Radish hath a shorter narrower leafe than the common Radish, & more deeply cut or jagged, almost like the leaues of Rocket, but much greater. The stalke is slender and rough, of two cubits high, diuided toward the top into many branches. The floures are small and white, the cod is long, slender, and jointed, wherein is the seed. The root is a finger thick, white within and without, of a sharp and biting taste.

2 The water Radish hath long and broad leaues deepely indented or cut euen to the middle rib. The stalke is long, weake, and leaneth this way and that way, being not able to stand vpright without a prop, insomuch that ye shall neuer find it, no not when it is very yong, but leaning down vpon the mud or mire where it groweth. The floures grow at the top made of foure small yellow leaues. The root is long, set in sundry spaces with small fibres or threds like the rowell of a spurre, hot and burning in taste more than any of the garden Radishes.

¶ *The Place.*

The first growes vpon the borders of banks and ditches cast vp, and in the borders of fields.

The

The second growes by ditches, standing waters, and riuers ; as on the stone wall that bordereth vpon the riuer Thames by the Sauoy in London.

1 *Raphanus sylvestris.*
Wilde Radish.

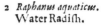

2 *Raphanus aquaticus.*
Water Radish.

¶ *The Time.*

They floure in Iune, and the seed is ripe in August.

¶ *The Names.*

† The first of these is *Rapistrum flore albo Erucæ folijs* of Lobel : *Armoratia*, or *Rapistrum album* of *Tabern.* and *Raphanus sylvestris* of our Author : in English, Wilde Radish.

The second is *Radicula sylvestris* of *Dodonæus: Raphanus aquaticus* or *palustris* of others: in English, Water Radish.

¶ *The Temperature.*

The wild Radishes are of like temperature with the garden Radish, but hotter and drier.

¶ *The Vertues.*

Dioscorides writeth, That the leaues are receiued among the pot-herbes, as also the boiled root, which, as he saith, doth heate, and prouoke vrine.

Cʜᴀᴘ. 7. *Of Horse-Radish.*

¶ *The Description.*

1   **H**Orse Radish brings forth great leaues, long, broad, sharpe pointed, and snipt about the edges, of a deepe greene colour like those of the great garden Dock, (called of some, Monks Rubarb, of others Patience) but longer and rougher. The stalke is slender and brittle, bearing at the top small white floures : which being past, there follow smal cods, wherein is the seed. The root is long and thick, white of colour, in tast sharp, and very much biting the tongue like Mustard.

2   Dittander or Pepperwort hath broad leaues long and sharpe pointed, of a blewish greene colour like Woad, somewhat snipt or cut about the edges like a Saw. The stalke is round and tough:

tough: vpon the branches whereof grow little white floures. The root is long and hard, creeping far abroad in the ground, in ſuch ſort that when it is once taken in a ground, it is not poſſible to root it out, for it wil vnder the ground creep and ſhoot vp and bud forth in many places far abroad. The root alſo is ſharpe and biteth the tongue like pepper, whereof it tooke the name Pepperwort.

‡ 3　This which we giue you in the third place hath a ſmall fibrous root, the ſtalke groweth vp to the height of two cubits, and it is diuided into many branches furniſhed with white floures; after which follow ſeeds like in ſhape and taſte to Thlaſpi or Treacle muſtard. The leaues are ſomewhat like thoſe of Woad. This is nouriſhed in ſome gardens of the Low-Countries, and Lobel was the firſt that gaue the figure hereof, and that vnder the ſame title as we here giue you it. ‡

1 Raphanus Ruſticanus.　　　　　2 Raphanus ſylueſtris Offic. Lepidium Ægineta Lob.
Horſe-Radiſh.　　　　　　　　　Dittander, or Pepperwort.

¶ The Place.

Horſe Radiſh for the moſt part groweth and is planted in gardens; yet haue I found it wilde in ſundry places, as at Namptwich in Cheſhire, in a place called the Milne eye; as alſo at a ſmall village neere London called Hogſdon, in the field next to a farme houſe leading to Kings-land, where my very good friend Mr Bredwel, practioner in phiſicke, a learned, and diligent ſearcher of ſimples, and Mr William Martin one of the fellowſhip of Barbers Surgeons, my deare and louing friend, in company with him found it, and gaue me knowledge of the place, where it flouriſheth to this day.

Dittander is planted in gardens, and is to be found wild alſo in ſundry places of England, as at Clare by Ouenden in Eſſex, at the Hall of Brinne in Lancaſhire, and neere to Exceſter in the Weſt parts of England. It delighteth to grow in ſandy and ſhadowie places ſomewhat moiſt.

¶ The Time.

Horſe-Radiſh floureth for the moſt part in Aprill or May, and the ſeed is ripe in Auguſt, & that ſo rare or ſeldome ſeen, as that Petrus Placentius hath written, that it brings forth no ſeed at all. Dittander floures in Iune and Iuly.

¶ The Names.

Horſe-Radiſh is commonly called Raphanus ruſticanus, or magnus, and of diuers ſimply Raphanus
ſylueſtris:

sylvestris: of the high-Dutch men, **Merrettich, Krain,** or **Kren:** in French, *Grand Raisort* in low-Dutch, **Merradus:** in English, Mountain Radish, great Raisort, and Horse Radish, It is called in the North part of England, Redcole.

Diuers thinke that this Horse-Radish is an enemy to Vines, and that the hatred between them is so great, that if the root hereof be planted neere to the Vine, it bendeth backward from it, as not willing to haue fellowship with it.

It is also reported, That the root hereof stamped, and cast into good and pleasant wine, turneth it forthwith to vineger. But the old writers doe ascribe this enmitie to the Vine and Brassica, our Coleworts, which the Antients haue named *japans.*

Pliny, lib. 19. cap. 9, describes Dittander by the name of *Lepidium*; and *Ægineta* also names it so: in shops, *Raphanus sylvestris,* and *Piperitis:* the Germans call it **Pfefferkraut:** the low-Dutch men, **Pepper cruyt:** in English, Dittander, Dittany, and Pepperwort.

3 *Lepidium Anuum.*
Annual Dittander.

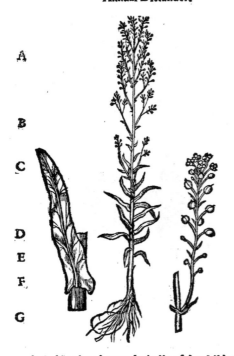

A
B
C
D
E
F
G

¶ *The Nature.*

These kinds of wilde Radishes are hot and dry in the third degree: they haue a drying and clensing qualitie, and somewhat digesting.

¶ *The Vertues.*

Horse Radish stamped, with a little vineger put thereto, is commonly vsed among the Germanes for sauce to eat fish with, and such like meats, as we do mustard: but this kind of sauce doth heat the stomack better, and causeth better digestion than mustard.

Oxymel, or syrrup made with vineger & hony, in which the rinds of horse-Radish haue been infused three daies, causeth vomit, and is commended against the quartan ague.

The leaues boiled in wine, and a little oile Oliue added thereto, & laid vpon the grieued parts in manner of a pultis, doe mollifie and take away the hard swellings of the liuer and milt, & being applied to the bottome of the belly is a remedie for the strangurie.

It profiteth much in the expulsion of the secondine or after-birth.

It mitigateth and aswageth the paine of the hip or haunch, commonly called Sciatica.

It profiteth much against the colique, strangurie, and difficultie of making water, vsed in stead of Mustard as aforesaid.

The root stamped and giuen to drinke killeth the worms in children. The juice giuen doth the same. An ointment made thereof doth the like, being anointed vpon the belly of the childe.

H   The leaues of Pepperwort, but especially the roots, be extreme hot, for they haue a burning and bitter taste. It is of the number of scorching and blistring simples, saith *Pliny, lib. 20. cap. 17.* and therefore by his hot qualitie it mendeth the skin in the face, and taketh away scabs, scars, & manginesse, if any thing remain after the healing of vlcers and such like.

CHAP. 8. *Of Winter-Cresses.*

¶ *The Description.*

THe Winter-Cresses haue many greene broad smooth and flat leaues like vnto the common Turneps, whose stalkes be round and full of branches, bringing forth at the top small yellow floures: after them follow small cods, wherein is contained small reddish seed.

¶ *The*

Barbarea.
Winter Creſſes.

¶ *The Place.*

It groweth in gardens among pot herbes, and very common in the fields neere vnto paths and highwaies almoſt euery where.

¶ *The Time.*

This herbe is greene all winter long, it floureth in May and ſeedeth in Iune.

¶ *The Names.*

Winter Creſſe is called of the Latines, *Cardamum*, or *Naſturtium Hibernum* : of ſome, *Barbarea*, and *Pſeudobunium* : the Germans call it **S. Barberen kraut** : in Low-Dutch, **Winter kerſſe.**

It ſeems to be *Dioſcorides* his ψευδοβούνιον : that is to ſay, falſe or baſtard *Bunium* : in Engliſh, winter Creſſes, or herb S. Barbara.

¶ *The Nature.*

This herbe is hot and dry in the ſecond degree.

¶ *The Vertues.*

The ſeed of Winter Creſſe cauſeth one to A
make water, and driueth forth grauell, and helps the ſtrangurie.

The iuyce thereof mundifieth corrupt and B
filthy vlcers, being made in form of an vnguent with wax, oile, and turpentine.

In winter when ſalad herbes be ſcarce, this C
herbe is thought to be equall with Creſſes of the garden, or Rocket.

This herbe helpeth the ſcuruy, being boiled D
among ſcuruy graſſe, called in Latine *Cochlearia*, cauſing it to worke the more effectually.

CHAP. 9. *Of Muſtard.*

¶ *The Deſcription.*

1 THe tame or garden Muſtard hath great rough leaues like to thoſe of the Turnep, but rougher and leſſer : the ſtalke is round, rough, and hairy, of three cubits high, diuided into many branches, whereon do grow ſmall yellow floures, & after them ſucceed cods ſlender and rough, wherin is contained round ſeed bigger than Rape ſeed, of colour yellow, of taſt ſharpe and biting the tongue, as doth our common field muſtard.

‡ 2 Our ordinarie Muſtard hath leaues like Turneps, but not ſo rough, the ſtalks are ſmooth and grow ſometimes to three, foure, or fiue cubits high, they haue many branches, and the leaues vpon theſe branches, eſpecially the vppermoſt, are long and narrow, and hang downeward on ſmall ſtalks ; the cods are ſhort, and lie flat and cloſe to the branches, and are ſomwhat ſquare : the ſeed is reddiſh or yellow. ‡

3 The other tame Muſtard is like to the former in leaues and branched ſtalks, but leſſer, and they are more whitiſh and rough. The floures are likewiſe yellow, and the ſeed browne like Rape ſeed, which is alſo not a little ſharp or biting.

‡ 4 This which I giue you here (being the *Sinapi ſatiuum alterum* of *Lobel*, and the *Sinapi album* of the ſhops) growes but low, and it hath rough crooked cods, and whitiſh ſeeds ; the ſtalkes, floures, and leaues are much like the firſt deſcribed. ‡

5 The wilde Muſtard hath leaues like thoſe of Shepheards purſe, but larger, and more deeply indented, with a ſtalke growing to the height of two foot, bearing at the top ſmall yellow floures made of two leaues : the cods be ſmall and ſlender, wherein is contained reddiſh ſeed, much ſmaller than any of the others, but not ſo ſharpe and biting.

† 1 *Sinapi ſativum.*
Garden Muſtard.

† 2 *Sinapi ſativum alterum,* Dod.
Field Muſtard.

‡ 4 *Sinapi album.*
White Muſtard.

† 5 *Sinapi ſylveſtre minus.*
Small wild Muſtard.

¶ *The Place.*

‡ Our ordinarie Muſtard (whoſe deſcription I haue added) as alſo the wilde and ſmall grow wilde in many places of this kingdome, and may all three be found on the banks about the backe of Old-ſtreet, and in the way to Iſlington. ‡

¶ *The Time.*

Muſtard may be ſowne in the beginning of the Spring : the ſeed is ripe in Iuly or Auguſt ; It commeth to perfection the ſame yeare that it is ſowne.

¶ *The Names.*

The Greekes call Muſtard, ᷎᷎᷎᷎ : the Athenians called it ᷎᷎᷎᷎ the Latines, *Sinapi* : the rude and barbarous, *Sinapium* : the Germans, **Senff** : the French, *Seneue*, and *Mouſtarde* : the low Dutchmen, **Moſtaert ſaet** : the Spaniards, *Moſtaza*, and *Moſtallas* the Bohemians, *Horcice* : *Pliny* calls it *Thlaſpi*, whereof doubtleſſe it is a kind : and ſome haue called it *Saurion.*

‡ Theſe kinds of Muſtard haue been ſo briefly treated of by all writers, that it is hard to giue the right diſtinctions of them, and a matter of more difficultie than is expected in a thing ſo vulgarly knowne and vſed. I will therefore endeauor in a few words to diſtinguiſh thoſe kinds of Muſtard which are vulgarly written of.

1   The firſt is *Sinapi primum* of *Matthiolus* and *Dodonæus* : and *Sinapi ſativum Erucæ aut Rapi folio* of *Lobel.*

2   The ſecond I cannot iuſtly refer to any of thoſe that are written of by Authors ; for it hath not a cod like Rape, as *Pena* and *Lobel* deſcribe it ; nor a ſeed bigger than it, as *Dodonæus* affirmeth ; yet I ſuſpect, & almoſt dare affirm, that it is the ſame with the former mentioned by them, though much differing from their figures and deſcription.

3   The third (which alſo I ſuſpect is the ſame with the fourth) is *Sinapi alterum* of *Matthiolus*, and *Sinapi agreſte Apÿ, aut potius Laueris folio*, of *Lobel* : and *Sinapi ſativum alterum* of *Dodonæus.*

4   The fourth is by *Lobel* called *Sinapi alterum ſativum* ; and this is *Sinapi album officinarum*, as *Pena* and *Lobel* affirme, *Adverſ.pag.68.*

5   The fift is *Sinapi ſylveſtre* of *Dodonæus* : and *Sinapi ſylveſtre minus Burſæ paſtoris folio*, of *Lobel.* It is much like Rocket, and therefore *Bauhine* fitly calls it *Sinapi Erucæ folio* : in Engliſh it may be called ſmall wilde Muſtard. ‡

¶ *The Temperature.*

The ſeed of Muſtard, eſpecially that which we chiefly vſe, doth heate and make thin, and alſo draweth forth. It is hot and dry in the fourth degree, according to *Galen*.

¶ *The Vertues.*

The ſeed of Muſtard pound with vineger is an excellent ſauce, good to be eaten with any groſſe   A meats either fiſh or fleſh, becauſe it doth help digeſtion, warmeth the ſtomacke, and prouoketh appetite.

It is giuen with good ſucceſſe in like manner to ſuch as be ſhort winded, and are ſtopped in the   B breſt with tough flegme from the head and brain.

It appeaſeth the tooth-ache, being chewed in the mouth.   C

They vſe to make a gargariſme with hony, vineger, and muſtard ſeed, againſt the ſwelling of the   D uvula, and the almonds about the throat and root of the tongue.

Muſtard drunke with water and hony prouoketh the terms and vrin.   E

The ſeed of muſtard beaten and put into the noſthrils cauſeth ſneeſing, and raiſeth women ſicke   F of the Mother out of their fits.

It is good againſt the falling ſickneſſe, and ſuch as haue the Lethargie, if it be laid plaiſterwiſe   G vpon the head (after ſhauing) being tempered with figs.

It helpeth the Sciatica, or ache in the hip or huckle bone : it alſo cureth all maner of pains pro-   H ceeding of a cold cauſe.

It is mixed with good ſucceſſe with drawing plaiſters, and with ſuch as waſt and conſume nodes   I and hard ſwellings.

It helpeth thoſe that haue their haire pulled off ; it taketh away the blew and black marks that   K come of bruiſings.

‡ The ſeed of the white muſtard is vſed in ſome Antidotes, as *Electuarium de ovo, &c.*   L

† The three figures in the former edition were all falſe : the firſt was of *Barbarea*, deſcribed in the precedent chapter : the ſecond, of *Eruca aquatica maior* of *Tabern.* The third, of *Eruca agreſt minor* of *Tabern*,

X   Chap.

CHAP. 10. *Of Rocket.*

¶ *The Kindes.*

THere be ſundry kinds of Rocket ; ſome tame or of the garden ; ſome wild or of the field ; ſome of the water, and of the ſea.

† 1 *Eruca ſativa.*
Garden Rocket.

2 *Eruca ſylveſtris.*
Wilde Rocket.

¶ *The Deſcription.*

1   GArden Rocket or Rocket gentle hath leaues like thoſe of turneps, but not neer ſo great nor rough. The ſtalks riſe vp of a cubit and ſomtimes two cubits high, weak and brittle ; at the top whereof grow the floures of a whitiſh colour, and ſomtimes yellowiſh : which being paſt, there ſucceed long cods which containe the ſeed, not vnlike to Rape ſeed, but ſmaller.

2   The common Rocket, which ſome keepe in gardens, and which is vſually called the Wild Rocket, is leſſer than the Roman Rocket, or Rocket gentle, the leaues and ſtalks narrower and more jagged. The floures be yellow, the cods alſo ſlenderer, the ſeed thereof is reddiſh, and biteth the tongue.

3   This kind of Rocket hath long narrow leaues almoſt ſuch as thoſe of Tarragon, but thicker and fatter, reſembling rather the leaues of Myagrum, altogether vnlike any of the reſt of the Rockets, ſauing that the branch, floure, and ſeed are like the garden Rocket.

4   There is another kinde of Rocket, thought by that reuerend and excellent herbariſt *Carolus Cluſius* to be a kind of Creſſes, if not Creſſes it ſelfe, yet couſin germane at leaſt. Vnto whoſe cenſure *Lobel* is indifferent, whether to call it Rocket with thin and narrow leaues, or to joine it with the kindes of Creſſes, hauing the taſte of the one, and the ſhape of the other. The leaues are much diuided, and the floures yellow.

5   There is a wilde kinde of ſea-Rocket which hath long weake and tender branches trailing

vpon the ground,with long leaues like vnto common Rocket, or rather Groundſwell, hauing ſmall
and whitiſh blew floures,in whoſe place commeth ſmall cods, wherein is contained ſeed like that
of Barley.

‡ 6   Beſides theſe there is another plant, whoſe figure which here I giue was by our Author
formerly ſet forth in the precedent Chapter,vnder the title of *Sinapi ſylueſtre*; together with a large
kinde thereof,vnder the name of *Sinapi ſatiuum alterum*. Now I will onely deſcribe the later,which
I haue ſometimes found in wet places : the root is wooddy : the ſtalke ſome foot long, creſted, and
hauing many branches lying on the ground : the leafe is much diuided, and that after the manner
of the wilde Rocket : the floures are of a bright yellow, and are ſucceeded by ſhort crooked cods,
wherein is contained a yellowiſh ſeed. ‡

† 3 *Eruca ſylueſtris Anguſtifolia.*   † 4 *Eruca naſturtio cognata tenuifolia.*
Narrow leaued wilde Rocket.   Creſſy-Rocket.

¶ *The Place.*

Romane Rocket is cheriſhed in gardens.

Common or wilde Rocket groweth in moſt gardens of it ſelfe : you may ſee moſt brick and ſtone
walls about London and elſewhere couered with it.

The narrow leaued Rocket groweth neere vnto water ſides, in the chinkes and creuiſes of ſtone
walls among the morter. I found it as you go from Lambeth bridge to the village of Lambeth, vnder a ſmall bridge that you muſt paſſe ouer hard by the Thames ſide.

I found Sea Rocket growing vpon the ſands neere vnto the ſea in the Iſle of Thanet, hard by a
houſe wherein Sir *Henry Criſpe* did ſometimes dwell,called Queakes houſe.

¶ *The Time.*

Theſe Kindes of Rocket floure in the moneths of Iune and Iuly, and the ſeed is ripe in September.

The Romane Rocket dieth euery yeare, and recouereth it ſelfe againe by the falling of his owne
ſeed.

X 2   ¶ *The*

¶ *The Names.*

Rocket is called in Greeke; ιυζομοι : in Latine, *Eruca* : in high Dutch, **kauckenkraut :** in French, *Roquette* : in Low-Dutch, **Rakette :** in Italian, *Ruchetta* : in Spanish, *Oruga*, in English, Rocket, and Racket. The Poets do oft times name it *Herbafalax* : *Eruca* doth signifie likewise a certaine canker worme, which is an enemy to pot-herbes, but especially to Coleworts.

‡   The first is called *Eruca fativa*, or *Hortenfis major* : Great Garden Rocket.
2   The second, *Eruca fyluestris* : Wilde Rocket.
3   This third is by *Lobel* called *Eruca fyluestris angustifolia* : narrow leaued wilde Rocket.
4   *Clufius* fitly calls this, *Nasturtium fyluestre* : and hee reprehendeth *Lobel* for altering the name into *Eruca Nasturtio cognata tenuifolia* : Cressy-Rocket.
5   The fifth is *Eruca marina*, (thought by *Lobel* and others to be *Cakile Serapionis*,) Sea Rocket.
6   *Eruca aquatica :* Water Rocket.

‡ 5 *Eruca marina.*
Sea Rocket.

† 6 *Eruca aquatica.*
Water Rocket.

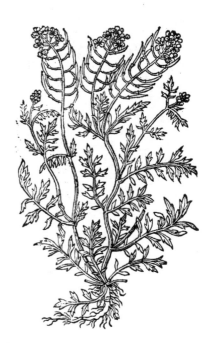

¶ *The Temperature.*

Rocket is hot and dry in the third degree, therefore faith *Galen* it is not fit nor accustomed to be eaten alone.

¶ *The Vertues.*

A   Rocket is a good fallet herbe, if it be eaten with Lettuce, Purflane, and such cold herbes ; for being fo eaten it is good and wholefome for the ftomacke, and caufeth that fuch cold herbes doe not ouer-coole the fame : otherwife, to be eaten alone, it caufeth head-ache, and heateth too much.

B   The vfe of Rocket ftirreth vp bodily luft, efpecially the feed.

C   It prouoketh vrine, and caufeth good digeftion.

D   *Pliny* reporteth, That whofoeuer taketh the feed of Rocket before he be whipt, fhall be fo hardened, that he fhall eafily indure the paines.

E   The root and feed ftamped, and mixed with vineger and the gall of an Oxe, taketh away freckles, lentiles, blacke and blew fpots, and all fuch deformities of the face.

† The figure that was in the third place, vnder the title of *Eruca fyl. angustifolia*, is of the fame plant that in the Chapter of *Turritis* is called *Camelina*, where you shall finde it treated of at large. And that in the first place is *Eryfimum fecundum* of *Tabern.* and I queftion whether it be not of *Sinapi fyluestre* minus.

CHAP.

CHAP. II. *Of Tarragon.*

Draco herba.
Tarragon.

¶ *The Deſcription.*

TArragon the ſallade herbe hath long and narrow leaues of a deep green colour, greater and longer than thoſe of common Hyſſope, with ſlender brittle round ſtalkes two cubites high : about the branches whereof hang little round floures, neuer perfectly opened, of a yellow colour mixed with blacke, like thoſe of common Wormewood. The root is long and fibrous, creeping far abroad vnder the earth, as doe the roots of Couch-graſſe, by which ſprouting forth it increaſeth, yeelding no ſeed at all, but as it were a certaine chaffie or duſtie matter that flieth away with the winde.

¶ *The Place.*
Tarragon is cheriſhed in gardens, and is encreaſed by the young ſhoots : *Ruellius* and ſuch others haue reported many ſtrange tales hereof ſcarce worth the noting, ſaying, that the ſeed of flax put into a Raddiſh root or ſea Onion, and ſo ſet, doth bring forth this herbe Tarragon.

¶ *The Time.*
It is greene all Summer long, and a great part of Autumne, and floureth in Iuly.

¶ *The Names.*
It is called in Latine, *Draco, Dracunculus hortenſis,* and *Tragum vulgare* by *Cluſius* ; of the Italians, *Dragoncellum,* in French, *Dragon,* in Engliſh, Tarragon.
It is thought to be that *Tarchon* which *Auicen* mentioneth in his 686. chapter: but he writeth ſo little thereof, as that nothing can certainly be affirmed of it. *Simeon Sethi* the Greeke alſo maketh mention of *Tarchon.*

¶ *The Temperature and Vertues.*
Tarragon is hot and dry in the third degree, and not to be eaten alone in ſallades, but ioyned with other herbs, as Lettuce, Purſlain, and ſuch like, that it may alſo temper the coldneſſe of them, like as Rocket doth, neither do we know what other vſe this herbe hath.

CHAP. 12. *Of Garden Creſſes.*

¶ *The Deſcription.*

1   GArden Creſſes or Towne Creſſes hath ſmall narrow jagged leaues, ſharpe and burning in taſte. The ſtalkes be round, a cubite high, which bring forth many ſmall white floures, and after little flat husks or ſeed veſſels, like to thoſe of ſhepheards purſe, wherin are contained ſeeds of a browne reddiſh colour. The root dieth when the ſeed is ripe.

2   There is another kinde in taſte like the former, but in leaues far different, which I recouered of ſeeds, ſent me from *Robinus* dwelling in Paris. The ſtalkes riſe vp to the height of a foot, garniſhed with many broad leaues deeply cut or indented about the edges: the middle of the leafe is decked and garniſhed with many little ſmall leaues or rather ſhreds of leaues, which make the ſame like a curlde fanne of feathers. The ſeed is like the former in ſhape.

3   Spaniſh Creſſes riſeth forth of the ground like vnto Baſill ; afterwards the leaues grow larger and broader, like thoſe of Marigolds ; amongſt the which riſeth vp a crooked ſymmer ſtalk,
where-

whereupon doe grow small tufts or spokie rundles of white floures. The seed followeth, browne of colour, and bitter in taste. The whole plant is of a loathsome smell and sauour.

4   Stone-Cresse groweth flat vpon the ground, with leaues iagged and cut about the edges like the Oke leafe, resembling well the leaues of Shepheards purse. I haue not seene the floures, and therefore they be not exprest in the figure; notwithstanding it is reported vnto mee, that they be smal and white of colour, as are those of the garden Cresses. The seed is contained in smal pouches or seed vessels, like those of Treacle mustard or Thlaspi.

¶ *The Place.*

Cresses are sowne in gardens, it skils not what soile it be; for that they like any ground, especially if it be well watered. ‡ Mr. *Bowles* found the fourth growing in Shropshire in the fields about Birch in the parish of Elesmere, in the grounds belonging to Mr. *Richard Herbert,* and that in great plenty. As also on the further side of Blacke heath, by the highway side leading from Greenewich to Lusam. ‡

¶ *The Time.*

It may be sowne at any time of the yeare, vnlesse it be in Winter; it groweth vp quickely, and bringeth forth betimes both stalke and seed: it dieth euery yeare, and recouereth it selfe of the fallen or shaken seed.

1 *Nasturtium hortense.*
Garden Cresses.

¶ *The Names.*

Cresses is called in Greeke κάρδαμον: in Latine *Nasturtium :* in English, Cresses: the Germans call it **Kerffe :** and in French, *Cresson :* the Italians *Nasturtio,* and *Agretto :* of some, Towne Cresses, and garden Karsse. It is called *Nasturtium,* as *Varro* and *Pliny* thinke, *à narribus torquendis,* that is to say, of writhing the nosthrils, which also by the loathsome smell and sharpenesse of the seed doth cause sneezing. ‡ The first is called *Nasturtium hortense,* Garden Cresses. 2 *Nasturtium hortense Crispum,* Garden Cresses with crispe, or curled leaues. 3 *Nasturtium Hispanicum,* or *Latifolium:* Spanish Cresses or broad leaued Cresses. 4 This is *Nasturtium petræum* of *Tabernamontanus* (and not of *Lobel,* as our Author termed it.) Stone Cresses. ‡

¶ *The Temperature.*

The herb of Garden Cresses is sharpe and biting the tongue; and therefore it is very hot and drie, but lesse hot whilest it is young and tender, by reason of the waterie moisture mixed therewith, by which the sharpnesse is somwhat allaied.

The seed is much more biting than the herbe, and is hot and dry almost in the fourth degree.

¶ *The Vertues.*

A   *Galen* saith that the Cresses may be eaten with bread *Veluti obsonium,* and so the antient Spartanes vsually did; and the low Country men many times do, who commonly vse to feed of Cresses with bread and Butter. It is eaten with other sallade herbs, as Tarragon and Rocket : and for this cause it is chiefely sowne.

B   It is good against the disease which the Germanes call **Scorbuch** and **Scorbuye :** in Latine, *Scorbutus :* which we in England call the Scuruie, and Scurby, and vpon the seas the Skyrby: it is as good and as effectuall as the Scuruie grasse, or water Cresses.

Dioscorides saith, if the seed be stamped and mixt with hony, it cureth the hardnesse of the milt: with vineger and Barley meale parched it is a remedy against the Sciatica, and taketh away hard swellings and inflammations. It scoureth away tettars mixed with brine : it ripeneth felons, called in Greeke, Ἄσιτις : it forcibly cutteth and raiseth vp thicke and tough humors of the chest, if it be mixed with things proper against the stuffing of the lungs.

D   *Dioscorides* saith it is hurtfull to the stomacke, and troubleth the belly.

Ip

3 *Naſturtium Hiſpanicum.*
Spaniſh Creſſes.

4 *Naſturtium petraum.*
Stone Creſſes.

It driueth forth wormes, bringeth downe the floures, killeth the childe in the mothers wombe, E and prouoketh bodily luſt.

Being inwardly taken, it is good for ſuch as haue fallen from high places: it diſſolueth clutte- F red bloud, and preuenteth the ſame that it do not congeale and thicken in any part of the body: it procureth ſweat, as the later Phyſitians haue found and tried by experience.

Chap. 13. *Of Indian Creſſes.*

¶ *The Deſcription.*

Creſſes of India haue many weake and feeble branches, riſing immediatly from the ground, di-
ſperſing themſelues far abroad; by meanes whereof one plant doth occupie a great circuit of ground, as doth the great Bindeweede. The tender ſtalks diuide themſelues into ſundry bran-
ches, trailing likewiſe vpon the ground, ſomewhat bunched or ſwollen vp at euery joint or knee, which are in colour of a light red, but the ſpaces betweene the joints are greene. The leaues are round like wall peniwort, called Cotyledon, the foot-ſtalke of the leafe commeth forth on the backeſide almoſt in the middeſt of the leafe, as thoſe of Frogbit, in taſte and ſmell like the garden Creſſes. The flours are diſperſed throughout the whole plant, of colour yellow, with a croſſed ſtar ouerthwart the inſide, of a deepe Orange colour: vnto the backe-part of the ſame doth hang a taile or ſpurre, ſuch as hath the Larkes heele, called in Latine *Conſolida Regalis*; but greater, and the ſpur or heele longer; which beeing paſt there ſucceed bunched and knobbed coddes or ſeed veſſells, wherein is contained the ſeed, rough, browne of colour, and like vnto the ſeeds of the beete, but ſmaller.

¶ *The Place.*

The ſeeds of this rare and faire plant came from the Indies into Spaine, and thence into France and Flanders, from whence I receiued ſeed that bore with mee both floures & ſeed, eſpecially thoſe I receiued from my louing friend *Iohn Robin* of Paris.

¶ *The Time.*

The ſeedes muſt be ſowen in the beginning of Aprill, vpon a bed of hot horſe dung, and ſome

fine

fine fifted earth caft thereon of an handful thicke. The bed muft be couered in fundry places with hoopes or poles, to fuftaine the mat or fuch like thing that it muft be couered with in the night, and laied open to the Sunne in the day time. The which being fprung vp, and hauing gotten three leaues, you muft replant them abroad in the hotteft place of the garden, and moft fine and fertile mold. Thus may you do with Muske-Melons, Cucumbers, and all cold fruits that require hafte; for that otherwife the froft will ouertake them before they come to fruit-bearing.

‡ They may alfo be fowne in good mold like as other feeds, and vfually are. ‡

Nafturtium Indicum cum flore & femine.
Indian Creffes with floure and feed.

¶ *The Names.*

This beautifull Plant is called in Latine, *Nafturtium Indicum* : in Englifh, Indian Creffes. Although fome haue deemed it a kind of *Convolvulus*, or Binde-weed; yet I am well contented that it retaine the former name, for that the fmell and tafte fhew it to be a kinde of Creffes.

¶ *The Nature and Vertues.*

We haue no certaine knowledge of his nature and vertues, but are content to referre it to the kindes of Creffes, or to a further confideration.

CHAP. 14. *Of Sciatica Creffes.*

¶ *The Defcription.*

1   SCiatica Creffes hath many flender branches growing from a ftalk of a cubit high, with fmall long and narrow leaues like thofe of Garden Creffes. The floures be very fmall, and yellow of colour: the feed veffels be little flat chaffie huskes, wherein is the feed of a reddifh gold colour, fharp and very bitter in taft. The root is fmal, tough, white within and without, and of a biting tafte.

‡ The plant whofe figure I here giue you in ftead of that with the narrower leaues of our Author, hath leaues fomewhat like Rocket, but not fo deep cut in, being only fnipt about the edges :

the

the vpper leaues are not ſnipt, nor diuided at all, and are narrower. The floures decking the tops of the branches are ſmall and white, the ſeed veſſels are leſſe than thoſe of Creſſes, and the ſeed it ſelfe exceeding ſmall, and of a blackiſh colour; the root is wooddy, ſometimes ſingle, otherwhiles diuided into two branches. ‡.

¶ _The Place._

It groweth vpon old wals and rough places by high waies ſides, and ſuch like : I haue found it in corne fields about Southfleet neare to Grauefend in Kent.

Iberis Cardamantica.
Sciatica Creſſes.

¶ _The Time._

It floureth according vnto the late or early ſowing of it in the fields, in Iune and Iuly.

¶ _The Names._

Sciatica Creſſes is called in Greeke ἰηρις and ὑψιλογανικι : in Latine _Iberis_ : of _Pliny, Heberis,_ and _Naſturtium ſylueſtre,_ and in like manner alſo _Lepidium_ : There is another _Lepidium_ of _Pliny_ : in Engliſh, Sciatica Creſſe. ‡ The firſt deſcribed may bee called _Iberis Cardimantica tenuifolia,_ Small leaued Sciatica Creſſes. The ſecond, _Iberis latiore folio,_ broad leaued Sciatica Creſſes. ‡

¶ _The Nature._

Sciatica Creſſe is hot in the fourth degree, and like to Garden Creſſes both in ſmell and in taſte.

¶ _The Vertues._

The roots gathered in Autumne, ſaith _Dioſcorides,_ doe heat and burne, and are with good ſucceſſe with ſwines greaſe made vp in manner of a plaiſter, and put vpon ſuch as are tormented with the Sciatica : it is to lie on the grieued place but foure houres at the moſt, and then taken away, and the patient bathed with warme water, and the place afterwards annointed with oile and wooll laied on it; which things _Galen_ in his ninth booke of medicines, according to the place grieued, citeth out of _Democrates,_ in certaine verſes tending to that effect.

CHAP. 15. _Of Banke Creſſes._

¶ _The Deſcription._

1   BAnke Creſſes hath long leaues, deeply cut or jagged vpon both ſides, not vnlike to thoſe of Rocket, or wilde muſtard. The ſtalkes be ſmall, limber, or pliant, yet very tough, and will twiſt and writhe as doth the Ozier or water Willow, whereupon do grow ſmall yellow floures, which being paſt there do ſucceed little ſlender cods, full of ſmall ſeeds, in taſte ſharpe biting the tongue as thoſe of Creſſes.

2   The ſecond kinde of banke Creſſes hath leaues like vnto thoſe of Dandelion, ſomewhat reſembling Spinach. The branches be long, tough, and pliant like to the other. The floures be yellowiſh, which are ſucceeded by ſmall long cods, hauing leaues growing amongſt them : in theſe cods is contained ſmall biting ſeed like the other of this kinde. The ſmell of this plant is very vngratefull.

¶ _The Place._

Banke Creſſes is found in ſtonie places among rubbiſh, by path waies, vpon earth or mud walls, and in other vntoiled places.

The

The second kinde of banke Cresses groweth in such places as the former doth: I found it growing at a place by Chelmes ford in Essex called little Baddowe, and in sundry other places.

‡ If our Author meant this which I haue described and giuen you the figure of, (as it is probable he did) I doubt he scarce found it wilde: I haue seen it in the Garden of Master *Parkinson*, and
it groweth wilde in many places of Italy. ‡

¶ *The Time.*

They floure in Iune and Iuly, and the seed is ripe in August and September.

¶ *The Names.*

Banke Cresses is called in Latine *Irio* and *Erysimum*: in Greeke ἰρύμμα, and of some, χαμσπλιον, according to *Dioscorides*: *Theophrastus* hath another *Erysimum*. ‡ The first is called *Irio*, or *Erysimum* by
Matthiolus Dodonæus, and others. *Turner*, *Fuchsius* and *Tragus* call it *Verbena fæmina* or *recta*. The second is *Irio alter* of *Matthiolus*, and *Saxifraga Romanorum*, *Lugd.* It may bee called Italian Banke
Cresses: or Romane Saxifrage. ‡

1 *Erysimum Dioscoridis, Lobelij.*
Bancke Cresses.

† 2 *Erysimum alterum Italicum.*
Italian Bancke Cresses.

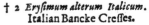

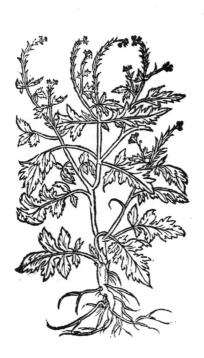

¶ *The Nature.*

The seed of bancke Cresses is like in taste to garden Cresses, and is as *Galen* saith of a fierie temperature, and doth extreamly attenuate or make thinne.

¶ *The Vertues.*

A   The seed of banke Cresses is good against the rheume that falleth into the chest, by rotting the
same.

B   It remedieth the cough, the yellow jaundise, and the Sciatica or ache of the huckle-bones, if it
be taken with hony in manner of a lohoch, and often licked.

C   It is also drunke against deadly poisons, as *Dioscorides* addeth: and beeing made vp in a plaster
with water and hony and applied, it is a remedy against hidden cankrous apostumes behind the ears,
hard swellings and inflammations of the paps and stones.

D   ‡ The seeds of the Italian Banke Cresses, or Roman Saxifrage taken in the weight of a dram,

in

in a decoction of graſſe roots, effectually clenſe the reins, and expel the ſtone, as the Author of the *Hiſt. Lugd.* affirmeth. ‡

† The figure that was here in the ſecond place was of the *Sonchus ſylvaticus, or Libanotis Theophraſti ſterilis* of *Tabern.* You ſhall find mention of it among the S. or Sow-Thiſtles.

CHAP. 16.　*Of Dock-Creſſes.*

† *Lampſana.*
Dock-Creſſes.

¶ *The Deſcription.*

† DOck-Creſſes is a wilde Wort or pot-herbe, hauing roughiſh hairy leaues of an ouerworn green colour, deepely cut or indented vpon both ſides like the leaues of ſmall Turneps.   The ſtalkes grow to the height of two or three cubits, and ſomtimes higher, diuiding themſelues toward the top into ſundry little bran-ches, whereon grow many ſmall floures like thoſe of *Hieracium* or Hawk-weed ; which decaying, are ſucceeded by little creſted heads containing a lon-giſh ſmall ſeed ſomewhat like Lettice ſeed, but of a yellowiſh colour : the plant is alſo milky, the ſtalk woody, and the root ſmall, fibrous, and white.

¶ *The Place.*
Dock-Creſſes grow euery where by highwayes, vpon walls made of mud or earth, and in ſtony pla-ces.

¶ *The Time.*
It floureth from May to the end of Auguſt : the ſeed is ripe in September.

¶ *The Names.*

Dock-Creſſes are called in Greeke, Λαμψάνη : in Latine, *Lampſana,* and *Napium,* by *Dodonæus : Taber-namontanus* calls this, *Sonchus ſylvaticus : Camerarius* affirmeth, That in Pruſſia they call it *Papillaris.*

¶ *The Temperature.*
Dock-Creſſes are of nature hot, and ſomwhat abſterſiue or clenſing.

¶ *The Vertues.*
Taken in meat, as *Galen* and *Dioſcorides* affirme, it ingendreth euill juice and bad nouriſhment.　**A**
‡ *Camerarius* affirmeth, That it is vſed with good ſucceſſe in Pruſſia againſt vlcerated or ſore **B** breaſts. ‡

† The figure that was here, was of the *Rapiſtrum arvorum* deſcribed in the ſecond chapter of this booke ; and the true figure of this plant here deſcribed, was p. 231. vnder the name of *Sonchus ſylvaticus.*

CHAP. 17.
Of water-Parſenep and water-Creſſes.

1　GReat water-Parſenep groweth vpright, and is deſcribed to haue leaues of a pleaſant ſa-uor, fat and full of juice, as thoſe of Alexanders, but ſomewhat leſſe, reſembling the garden Parſenep : the ſtalke is round, ſmooth, and hollow, like to Kex or Caſhes: the root conſiſteth of many ſmall ſtrings or threds faſtened vnto the ſtalke within the water or myrie ground :

1 *Sium majus latifolium.*
Great water-Parſenep.

† 2 *Sium majus anguſtifolium.*
The leſſer water Parſnep.

‡ 4 *Sium alterum Olufatri facie.*
Long leaued water-Creſſes.

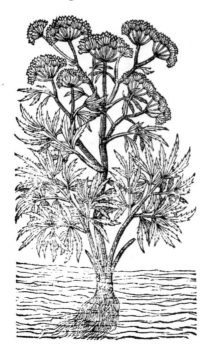

ground: at the top doe grow many white
flours, in ſpoky rundles like fennel; which
being bruiſed, doe yeeld a very ſtrong fauour, ſmelling like *Petroleum,* as doth the
reſt of the plant.

‡ 2   This plant much reſembles the
laſt deſcribed, and groweth vp ſome cubit
and a halfe, with many leaues finely ſnipt
about the edges, growing vpon one ribbe,
and commonly they ſtand bolt vpright.
The vmbel conſiſts of little white floures:
the root is ſmall, and conſiſteth of many
ſtrings.

‡ 3   There is another very like this,
but they thus differ; the ſtalks and leaues
of this later are leſſe than thoſe of the precedent, and not ſo many vpon one rib: the
other grows vpright, to ſome yard or more
high: this neuer growes vp, but alwayes
creepes, & almoſt at euery joint puts forth
an vmbel of floures.

4   To theſe may bee added another,
whoſe root conſiſts of aboundance of writhen and ſmall blacke fibres. The ſtalkes
are like Hemlock, ſome three cubits high;
the leaues are long, narrow, and ſnipped
about

about the edges, growing commonly two or three together : the vmbel of floures is commonly of a yellowiſh green : the ſeed is like parſly ſeed, but in taſte ſomwhat reſembles Cumin, *Daucus Creticus*, and the rind of a pomegranat.

5   Water-Creſſe hath many fat and weake hollow branches trailing vpon the grauell & earth where it groweth, taking hold in ſundry places as it creepeth ; by means whereof the plant ſpreads ouer a great compaſſe of ground.   The leaues are likewiſe compact and winged with many ſmall leaues ſet vpon a middle rib one againſt another, except the point leafe, which ſtands by it ſelfe, as doth that of the aſh, if it grow in his naturall place, which is in a grauelly ſpring.   The vpper face of the whole plant is of a browne colour, and greene vnder the leaues, which is a perfect marke to know the phyſicall kinde from the others.   The white floures grow alongſt the ſtalkes, and are ſucceeded by cods wherein the ſeed is contained.   The root is nothing elſe but as it were a thrum or bundle of threds.

† 5 *Naſturtium aquaticum, ſiue Cratena Sium.*
Common water-Creſſes.

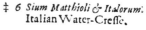

‡ 6 *Sium Matthioli & Italorum.*
Italian Water-Creſſe.

6   There is alſo another kinde hereof hauing leaues growing many on one ſtalke, ſnipt about the edges, being in ſhape betwixt the garden Creſſes and Cuckow floures : the ſtalk is creſted and diuided into many branches ; the floures white, and are ſucceeded by cods like thoſe of our ordinarie water-Creſſe laſt deſcribed.

¶ *The Place.*

‡   1   The firſt of theſe I haue not found growing, nor as yet heard of within this kingdome.
    2   The ſecond I firſt found in the company of M. *Robert Lorkin*, going betweene Redriffe and Deptford, in a rotten boggy place on the right hand of the way.
    3   The third growes almoſt in euery watery place about London.
    4   This is more rare, and was found by M. *Goodyer* in the ponds about Moore Parke ; and by M. *George Bowles* in the ditches about Elleſmere, and in diuers ponds in Flint-ſhire.
    5   The fifth is as frequent as the third, and commonly they grow neere together.
    6   This *Lobel* ſaith he found in Piemont in riuelets amongſt the hills : I haue not yet heard that it growes with vs. ‡

¶ _The Time._

They ſpring and wax green in Aprill, and floure in Iuly.

The water-Creſſe to be eaten in ſallads ſheweth it ſelf in March when it is beſt, and floureth in Summer with the reſt.

¶ _The Names._

‡ 1   The firſt of theſe is _Sium majus latifolium_ of _Tabernamontanus._

2   This is _Sion odoratum Tragi_; _Sium_ of _Matthiolus, Dodonæus,_ and others: it is taken to be _Sium_ or _Lauer_ of _Dioſcorides. Lobel_ calls it alſo _Paſtinaca aquatica,_ or water Parſenep.

3   This may be called _Sium vmbellatum repens,_ Creeping water Parſenep. Of this there is a reaſonable good figure in the _Hiſt.Lugdunenſis,pag.1092._ vnder the title of _Sium verum Matthioli_; but the deſcription is of that we here giue you in the ſixt place.

4   This is _Sium alterum_ of _Dodonæus_ : and _Sium alterum Oluſatri facie_ of _Lobel._

5   Many iudge this to be the _Siſymbrium alterum,_ or _Cardamine_ of _Dioſcorides_ : as alſo the _Sion_ of _Crateuas_ : and therefore _Lobel_ termes it _Sion Cratiuæ erucæ folium._ It is called by _Dodonæus,_ and vulgarly in ſhops known by the name of _Naſturtium aquaticum,_ or water Creſſes.

6   This is called _Sium vulgare_ by _Matthiolus : Lobel_ alſo termes it _Sium Matthioli & Italorum._ This was thought by our countryman Dr _Turner_ to be no other than the ſecond here deſcribed : of which opinion I muſt confeſſe I alſo was; but vpon better conſideration of that which _Lobel_ and _Bauhine_ haue written, I haue changed my mind.

¶ _The Temperature._

Water-Creſſe is euidently hot and dry.

¶ _The Vertues._

Water-Creſſe being boiled in wine or milke, and drunke for certain daies together, is very good againſt the Scuruy or Scorbute.

Being chopped or boiled in the broth of fleſh, and eaten for thirty daies together at morning, noone, and night, it prouoketh vrine, waſts the ſtone, and driueth it forth. Taken in the ſame manner it cureth yong maidens of the green ſickneſſe, bringeth down the termes, and ſendeth into the face their accuſtomed liuely colour loſt by the ſtopping of their _Menſtrua._

Cʜᴀᴘ. 18.   _Of wilde water-Creſſes or Cuckow-floures._

¶ _The Deſcription._

1   THe firſt of the Cuckow flours hath leaues at his ſpringing vp ſomwhat round, & thoſe that ſpring afterward grow jagged like the leaues of Greek Valerian; among which riſeth vp a ſtalk a foot long, ſet with the like leaues, but ſmaller and more jagged, reſembling thoſe of Rocket. The floures grow at the top in ſmall bundles, white of colour, hollow in the middle, reſembling the white ſweet-Iohn : after which come ſmall chaffie huskes or ſeed-veſſels, wherein the ſeed is contained. The root is ſmall and threddy.

2   The ſecond ſort of Cuckow floures hath ſmall jagged leaues like thoſe of the ſmall Valerian, agreeing with the former in ſtalks and roots: the floures be white, ouerdaſht or declining toward a light carnation.

‡ 3   The leaues and ſtalks of this are like thoſe of the laſt deſcribed, nor are the floures which firſt ſhew themſelues much vnlike them; but when as they begin to faile, in their middle riſe vp heads of pretty double floures made of many leaues, like in colour to theſe of the ſingle. ‡

4   The fourth ſort of Cuckow floures groweth creeping vpon the ground, with ſmall threddy ſtalks, whereupon grow leaues like thoſe of the field Clauer or three-leaued Graſſe : among which do come vp ſmall and tender ſtalks two handfuls high, hauing floures at the top in greater quantitie than any of the reſt, of colour white; and after them follow cods containing a ſmall ſeed. The root is nothing elſe but as it were a bundle of thrums or threds.

5   Milk-white Lady-ſmock hath ſtalks riſing immediately from the root, diuiding themſelues into ſundry ſmall twiggy and hard branches, ſet with leaues like thoſe of _Serpillum._ The floures grow at the top, made of foure leaues of a yellowiſh colour : the root is tough and wooddy, with ſome fibres anexed thereto.   ‡ This is no other than the firſt deſcribed, differing only therefrom in that the floures are milke white, as our Author truly in the title of his figure made them. Yet forgetting himſelfe in his deſcription, he maketh them yellowiſh, contrary to himſelfe and the truth. ‡

1 *Cardamine.*   Cuckow floures.

2 *Cardamine altera.*   Ladies smocks.

‡ 3 *Cardamine altera flore pleno.*
Double floured Lady-smocke.

4 *Cardamine trifolia.*
Three leaued Lady-smocke.

6 *Cardamine Alpina.*
Mountain Lady-ſmock.

‡ 7 *Sium minus impatiens.*
The impatient Lady-ſmocke.

8 *Cardamine pumila Bellidis folio Alpina.*
**The dwarfe Daiſie-leaued Lady-ſmock
of the Alps.**

6 Mountain Lady-ſmock hath many roots,
nothing elſe but as it were a bundle of threddy
ſtrings, from the which doe come forth three or
foure ſmall weake and tender leaues , made of
ſundry ſmal leaues, in ſhew like to thoſe of ſmal
water Valerian. The ſtalks be ſmal and brittle;
whereupon doe grow ſmall floures like the firſt
kind.

‡ 7 I ſhould be blame-worthy if in this place
I omitted that prety conditioned *Sium* which is
kept in diuers of our London gardens , and was
firſt brought hither by that great Treaſurer of
Natures rarities, Mr *Iohn Tradeſcant*. This Plant
hath leaues ſet many on a rib like as the other
Sium deſcribed in the ſecond place hath, but are
cut in with two or three pretty deep gaſhes: The
ſtalke is ſome cubit high, & diuided into many
branches, which haue ſmall white floures growing vpon them: after theſe floures are paſt, there
follow ſmall long cods containing a ſmal white
ſeed. Now the nature of this plant is ſuch, that
if you touch but the cods when the ſeed is ripe,
though you doe it neuer ſo gently, yet will the
ſeed fly al abroad with violence, as diſdaining to
be touched : whence they vſually call it *Noli me
tangere*; as they for the like quality name the *Perſicaria ſiliquoſa*. The nature of this plant is ſomewhat admirable, for if the ſeeds, as I ſaid, be fully
ripe,

ripe,though you put but your hand neere them,as profering to touch them , though you doe not, yet will they fly out vpon you,and if you expect no such thing, perhaps make you afraid,by reason of the suddennesse thereof. This herbe is written of only by *Prosper Alpinus*, vnder the title of *Sium minimum* ; and it may be called in English,Impatient Lady-smock,or Cuckow floure. It is an annual,and yearely sowes it selfe by the falling seeds. ‡

‡ 8   The leaues of this somewhat resemble those of Daisies, but lesse, and lie spred vpon the ground,amongst which rises vp a weake and slender stalke set with three or foure leaues at certain distances,it being some handful high:the top is adorned with smal white flours consisting of foure leaues apiece,after which follow large and long cods,considering the smalnesse of the plant:within these in a double order is contained a smal reddish seed of somwhat a biting tast.The root creepeth vpon the top of the ground, putting vp new buds in diuers places. *Clusius* found this growing vpon the rocks on the Etscherian mountain in Austria,and hath giuen vs the historie and figure of it vnder the name of *Plantula Cardamines æmula*, and *Sinapi pumulum Alpinum*.

¶ *The Place and Time*.

That of the Alpish mountains is a stranger in these cold countries : the rest are to be found eue-..e where as aforesaid,especially in the castle ditch at Clare in Essex. ‡ 7 This grows naturally in ...ome places of Italy:also I found it and the eighth about Bath & other parts of this kingdome. ‡

These floure for the most part in Aprill and May, when the Cuckow begins to sing her pleasant notes without stammering.

¶ *The Names*.

They are commonly called in Latine *Flos Cuculi*, by *Brunfelius* and *Dodonæus*, for the reason beforesaid ; and also some call them *Nasturtium aquaticum minus*,or lesser water-Cresse : of some,*Cardamine*,and *Sisymbrium alterum* of *Dioscorides : it is* called in the Germane tongue **Wildercress** : in French,*Passerage sauvage :* in English,Cuckow-flours:in Norfolk,Canturbury bels:at the Namptwich in Cheshire my natiue country,Lady-smockes ; which hath caused me to name it after their fashion.

¶ *The Nature and Vertues*.

These herbs be hot and dry in the second degree : wee haue no certaine proofe or authoritie of their vertues,but surely from the kinds of water-Cresse they cannot much differ, and therefore to them they may be referred in their vertues.

† The figure that was in the fourth place,being of the same plant that is described in the first place ; the counterfeit stalkes and heads being taken away, as *Bauhine* rightly hath obserued,its also the description thereof,which (as many other) our Author frames by looking vpon the figure,and the strength of his owne fance,I haue omitted as impertinent.

CHAP. 19. *Of Treacle Mustard*.

¶ *The Description*.

1   TReacle mustard hath long broad leaues, especially those next the ground , the others lesser,sleightly indented about the edges like those of Dandelion.The stalks be long and brittle,diuided into many branches euen from the ground to the top, where grow many small idle flours tuft-fashion:after which succeed large flat thin chaffie husks or seed vessels heart-fashion,wherein are contained brown flat seeds,sharp in tast,burning the tongue as doth mustard seed,leauing a tast or sauor of garlick behind for a farewell.

2   Mithridate mustard hath long narrow leaues like those of Woad or rather Cow Basil. The stalks be inclosed with small snipt leaues euen to the branches, pyramidis fashion, that is, smaller and smaller toward the top,where it is diuided into sundry branches,whereon do grow smal flours: which being past,the cods,or rather thin chaffie husks do appeare, full of sharpe seed like the former. The root is long and slender.

3   The third kind of Treacle mustard,named Knaues mustard (for that it is too bad for honest men)hath long fat and broad leaues like those of Dwale or deadly Nightshade, in taste like those of *Vulvaria* or stinking Orach,set vpon a round stalke two cubits high,diuided at the top into smal arms or branches,whereon grow small foolish white spoky floures. The seed is contained in flat pouches like those of Shepheards purse,brown,sharp in taste,and of an ill sauor.

4   Bowyers mustard hath the lower leaues resembling the ordinarie Thlaspi, but the vpper are very small like Tode-flax,but smaller. The leaues be small,slender,and many ; the floures be small and white,each consisting of foure leaues : the seeds be placed vpon the branches from the lowest part of them vnto the top,exceeding sharp and hot in taste, and of a yellowish colour. The root is small and wooddy.

5   Grecian mustard hath many leaues spred vpon the ground, like those of the common Daisie, of a darke greenish colour : from the middest whereof spring vp stalkes two foot long,diuided

1 *Thlaſpi Dioſcoridis.*
Treacle Muſtard.

2 *Thlaſpi vulgatiſſimum.*
Mithridate Muſtard.

3 *Thlaſpi majus.*
Knaues Muſtard.

4 *Thlaſpi minus.*
Bowyers Muſtard.

5 *Thlaſpi Græcum.*
Grecian Muſtard.

5 *Thlaſpi amarum.*
Clownes Muſtard.

7 *Thlaſpi Clypeatum Lobelij.*
Buckler Muſtard.

6 *Thlaſpi minus Clypeatum.*
Small Buckler muſtard.

into many ſmall branches,whereupon grow ſmall white floures compoſed of 4 leaues, after which ſucceed round flat husks or ſeed veſſels,ſet vpon the ſtalke by couples, as it were ſundry paires of ſpectacles,wherein the ſeed is contained, ſharpe and biting as the other. This is ſometimes ſeene with yellow floures.

† 6　Clownes muſtard hath a ſhort white fibrous root,from whence ariſeth vp a ſtalke of the height of a foot,which a little aboue the root diuides it ſelfe into ſome four or fiue branches,and theſe again are ſubdiuided into other ſmaller , ſo that it reſembles a little ſhrub : longiſh narrow leaues notched after the manner of Sciatica Creſſes by turnes garniſh theſe branches , and theſe leaues are as bitter as the ſmaller Centaurie. The floures ſtand thick together at the tops of theſe branches in manner of little vmbels,and are commonly of a light blew and white mixed together, being ſeldome only white or yellow. After the flours ſucceed ſeed-veſſels after the maner of other plants of this kind,and in them is contained a ſmall hot ſeed. †

7　Buckler muſtard hath many large leaues ſpred vpon the ground like *Hieracium* or Hawke-weed,ſomewhat more toothed or ſnipt about the edges : among which comes vp ſtalks ſmall and brittle,a cubit high,garniſhed with many ſmall pale yellowiſh flours : in whoſe place ſucceed many round flat cods or pouches,buckler-faſhion,containing a ſeed like vnto the others.

8　Small Buckler muſtard is a very ſmall baſe or low plant,hauing whitiſh leaues like thoſe of wild Time, ſet vpon ſmall weake and tender branches. The floures grow at the top like the other Buckler muſtard. The ſeed-veſſels are like,not ſo round, ſomwhat ſharpe pointed,ſharp in taſt,and burning the tongue. The whole plant lieth flat vpon the ground like wild Tyme.

¶ *The Place.*

Treacle or rather Mithridate muſtard growes wild in ſundry places in corne fields,ditch banks, and in ſandy dry and barren ground. I haue found it in corn fields betwixt Croidon and Godſtone in Surrey,at Southfleet in Kent,by the path that leads from Harnſey (a ſmall village by London) vnto Waltham croſſe,and in many other places.

The other grow vnder hedges oftentimes in fields,and in ſtony and vntoiled places. They grow plentifully in Bohemia and Germany:they are ſeen likewiſe on the ſtony banks of the riuer Rhine. They are likewiſe to be found in England in ſundry places wilde, the which I haue gathered into my garden. ‡ I haue found none but the firſt and ſecond growing wilde in any part of England as yet ; yet I deny not,but that ſome of the other may be found,but not all. ‡

¶ *The Time.*

Theſe Treacle muſtards are found with their flours from May to Iuly,and the ſeed is ripe in the end of Auguſt.

¶ *The Names.*

The Grecians call theſe kinds of herbs, θλάσπι, θλασπίδιον, Σίναπι ἄγριον, of the husk or ſeed-veſſell, which is like a little ſhield. They haue alſo other names which be found amongſt the baſtard words : As *Scandulaceum,Capſella,Pes gallinaceus*.Neither be the later writers without their names,as *Naſturtium tectorum*,and *Sinapi ruſticum* : it is called in Dutch,𝖜𝖎𝖑𝖉𝖊 𝖐𝖊𝖗𝖘𝖊 : in French,*Seneue ſauvage*:in Engliſh, Treacle muſtard, diſh Muſtard, Bowyers muſtard : of ſome, *Thlaſpi*, after the Greeke name, Churles muſtard,and wild Creſſes.

‡ 1　This is *Thlaſpi Dioſcoridis Draba, aut Chamelinæ folio* of Lobel : *Thlaſpi Latius* of *Dodonæus :* and the ſecond,*Thlaſpi* of *Matthiolus*.

2　This,*Thlaſpi vulgatiſſimum vaccariæ folio* of *Lobel :* the firſt *Thlaſpi* of *Matthiolus*, and the ſecond of *Dodonæus* ; and this is that *Thlaſpi* whoſe ſeed is vſed in ſhops.

3　This is *Thlaſpi majus* of *Tabernamontanus*.

4　This is *Thlaſpi minus* of *Dodonæus*,*Thlaſpi anguſtifolium* of *Fuchſius : Thlaſpi minus hortenſe Oſyridis folio,&c.* of *Lobel :* and *Naſturtium ſylveſtre* of *Thalius*.

5　This is *Alyſſon* of *Matthiolus : Thlaſpi Græcum polygonati folio* of *Lobel* and *Tabern*.

6　This the Author of the *Hiſt. Lugd.* calls *Naſturtium ſylveſtre : Tabern.* calls it *Thlaſpi amarum.*

7　*Lobel* termes this *Thlaſpi parvum Hieracifolium*, and *Lunaria lutea Monſpelienſium*.

8　This is *Thlaſpi minus clypeatum Serpillifolio* of *Lobel*. ‡

The figures of theſe two laſt mentioned were tranſpoſed in the former edition.

¶ *The Temperature.*

The ſeeds of theſe kinds of Treacle muſtards be hot and dry in the end of the third degree.

¶ *The Vertues.*

A　The ſeed of Thlaſpi or Treacle muſtard eaten,purgeth choler both vpward and downward, prouoketh floures,and breaketh inward apoſtumes.

B　The ſame vſed in clyſters helpeth the Sciatica, and is good vnto thoſe purpoſes for which Muſtard ſeed ſerueth.

¶ *The Danger.*

The ſeeds of theſe herbes be ſo extreme hot and vehement in working, that beeing taken in too
great

great quantitie, purgeth and ſcoureth euen vnto blood, and is hurtfull to women with childe, and therefore great care is to be had in giuing them inwardly in any great quantitie.

Chap. 20.   Of Candy Muſtard.

¶ The Deſcription.

Candy Muſtard excelleth all the reſt, as well for the comely floures that it brings forth for the decking vp of gardens and houſes, as alſo for that it goeth beyond the reſt in his phyſical vertues: it riſeth vp with a very brittle ſtalke of a cubit high, which diuideth it ſelfe into ſundry boughs or branches ſet with leaues like thoſe of ſtock gilliſloures, of a gray or ouerworn green colour. The ſloures grow at the top of the ſtalk, round, thick cluſtering together, like thoſe of Scabious or Diuels bit, ſometimes blew, often purple, carnation, or horſe-fleſh, but ſeldome white for any thing that I haue ſeen, varying according to the ſoile or clymat. The ſeed is reddiſh, ſharp, aud biting the tongue, wrapped in little husks faſhioned like an heart. ‡ There is a leſſe varietie of this, with white well ſmelling floures, in other reſpects little differing from the ordinarie. ‡

Thlaſpi Candiæ.
Candy Muſtard.

‡ *Thlaſpi Candiæ paruum flore albo.*
Small Candy Muſtard with a white floure.

¶ The Place.

This growes naturally in ſome places of Auſtria, as alſo in Candy, Spain, and Italy, whence I receiued ſeeds of the right honorable the lord *Ed. Zouch*, at his return into England from thoſe parts. ‡ *Cluſius* found the later as he trauelled through Switzerland into Germany. ‡

¶ The Time.

It floureth from the beginning of May to the end of September, at which time you ſhall haue floures and ſeeds vpon one branch, ſome ripe, and ſome that will not ripen at all.

¶ The Names.

† This plant is called by *Dodonæus* (but not rightly) *Arabis* and *Draba*; as alſo *Thlaſpi Candiæ:* which laſt name is retained by moſt writers: in Engliſh, Candy Thlaſpi, or Candy Muſtard. †

¶ The Temperature.

The ſeed of Candy Muſtard is hot and dry in the end of the ſecond degree, as is that called *Scorodothlaſpi*, or Treacle muſtard.

CHAP. 21. *Of Treacle Muſtard.*

¶ *The Deſcription.*

1 ROund leaued Muſtard hath many large leaues laid flat on the ground like the leaues of the wild Cabbage, and of the ſame colour ; amongſt which riſe vp many ſlender ſtalks of ſome two handfulls high or thereabouts, which are ſet with leaues far vnlike thoſe next the ground, encloſing or embracing the ſtalks as doe the leaues of *Perfoliatum*, or Thorow-wax. The floures grow at the top of the branches, white of colour : which being paſt, there do ſucceed flat husks or pouches like to thoſe of Shepheards purſe, with hot ſeed biting the tongue.

| 1 *Thlaſpi rotundifolium.* | 2 *Thlaſpi Pannonicum Cluſij.* |
| Round leaued Muſtard. | Hungary Muſtard. |

2 Hungary Muſtard bringeth forth ſlender ſtalkes of one cubit high : the leaues which firſt appeare are flat, ſomwhat round like thoſe of the wild Beet ; but thoſe leaues which after doe garniſh the ſtalks are long and broad like thoſe of the garden Colewort, but leſſer and ſofter, green on the vpper ſide, and vnder declining to whiteneſſe, ſmelling like Garlicke. The floures be ſmall and white, conſiſting of foure ſmall leaues, which in a great tuft or vmbell do grow thicke thruſt together : which being paſt, there followes in euery ſmall huske one duskiſh ſeed and no more, bitter, and ſharpe in taſte. The root is white and ſmal, creeping vnder the ground far abroad like the roots of Couch-graſſe ; preparing new ſhoots and branches for the yeare following, contrarie to all the reſt of his kind, which are encreaſed by ſeed, and not otherwiſe.

3 Churles muſtard hath many ſmall twiggy ſtalks, ſlender, tough, and pliant, ſet with ſmall leaues like thoſe of Cudweed or Lauander, with ſmall white floures, the husks and ſeeds are ſmall, few, ſharpe, bitter, and vnſauorie : the whole plant is of a whitiſh colour.

4 Peaſants muſtard hath many pretty large branches, with thin and jagged leaues like thoſe of Creſſes, but ſmaller, in ſauor and taſte like to the ordinarie *Thlaſpi* : the floures be whitiſh, & grow in a ſmall ſpoky tuft. The ſeed in taſte and ſauor is equall with the other of his kind and countrey, or rather exceeds them in ſharpneſſe.

5 Yellow

3 *Thlaſpi Narbonenſe Lobely.*
Churles Muſtard.

4 *Thlaſpi vmbellatum Narbonenſe.*
Peaſants Muſtard of Narbone.

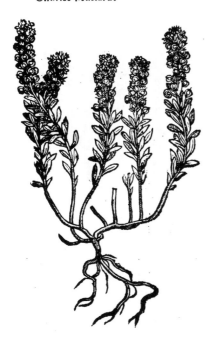

‡ 5 *Thlaſpi ſupinum luteum.*
Yellow Muſtard.

5 Yellow Muſtard hath an exceeding number of whitiſh leaues ſpred vpon the ground in manner of a turſe or haſſocke; from the midſt whereof riſeth vp an vpright ſtalke of three foot high, putting forth many ſmall branches or armes: at the top whereof grow many ſmal yellow floures like thoſe of the wall floure, but much leſſer: which being paſt, the husks appeare flat, pouch-faſhion, wherein is the ſeed like Treacle-Muſtard, ſharp alſo and biting.

6 White Treacle Muſtard hath leaues ſpred vpon the ground like the other, but ſmaller: the ſtalks riſe vp from the middeſt thereof, branched, ſet with leaues ſmaller than thoſe that lie vpon the ground euen to the top, where groweth a tuft of white floures in faſhion like to thoſe of the other Thlaſpies: the ſeed is like the other: ‡ The cods of this are ſometimes flat, and otherwhiles round: the floures alſo grow ſometimes ſpikefaſhion, otherwhiles an vmbel. I haue giuen you two figures expreſſing both theſe varieties. ‡

7 The

6 *Thlaspi album supinum, & eius varietas*
White Treacle Mustard.

7 *Thlaspi minus Clusij.*
Clusius his small Mustard.

‡ 8 *Thlaspi petræum minus.*
Small Rock Mustard.

- This ſmall kinde of muſtard hath a few ſmall leaues ſpred vpon the ground like thoſe of the leſſer Daiſie, but of a blewiſher green colour: from which riſe vp ſmall ſlender ſtalkes ſet with three, and ſomtimes foure ſmall ſharpe pointed leaues; the floures grow at the top, ſmal & white: the cods are flat, pouch-faſhion like thoſe of Shepheards purſe, and in each of them there is contained two or three yellowiſh ſeeds.

‡ 8 To theſe we may fitly adde another ſmall mountain Thlaſpi, firſt deſcribed by that diligent and learned Apothecary *Iohn Pona* of Verona, in his deſcription of mount Baldus. This from a threddy root bringeth forth many ſmall whitiſh leaues lying ſpred vpon the ground, and a little nicked about their edges. Among theſe riſeth vp a ſtalke ſome two or three handfulls high, diuaricated toward the top into diuers ſmall branches, vpon which grow white little floures conſiſting of foure leaues apiece: which fading, there follow round ſeed-veſſels like vnto thoſe of *Myagrum*; whence *Pona* the firſt deſcriber thereof calls it *Thlaſpi petræum Myagrodes*. The ſeed is as ſharpe and biting as any of the other Thlaſpies. This groweth naturally in the chinks of the rocks, in that part of Baldus which is termed *Vallis frigida*, or the cold Valley.

¶ *The Place.*

Theſe kinds of Treacle muſtards grow vpon hils and mountains in corne fields, in ſtony barren and grauelly grounds.

¶ *The Time.*

Theſe floure in May, Iune, and Iuly. The ſeed is ripe in September.

¶ *The Names.*

‡ 1 This is *Thlaſpi oleraceum* of *Tabernamontanus*: *Thlaſpi primum* of *Daleſchampius* : *Thlaſpi mitius rotundifolium* of *Columna*. Our Author confounded it with that whoſe figure is the firſt in the enſuing chapter, and called it *Thlaſpi incanum*.

2　*Thlaſpi montanum peltatum* of *Cluſius* : and *Thlaſpi Pannonicum* of *Lobel* and *Tabern.*

3　*Thlaſpi Narbonenſe centunculi anguſtifolio* of *Lobel* : and *Thlaſpi maritimum* of *Daleſchamp.*

4　*Thlaſpi vmbellatum Naſturtij hortenſis folio Narbonenſe* of *Lobel*. The figures of this and the precedent were tranſpoſed in the former edition.

5　*Thlaſpi ſupinum luteum* of *Lobel*. Our Authors figure was a varietie of the next following.

6　*Thlaſpi album ſupinum* of *Lobel* : *Thlaſpi montanum ſecundum* of *Cluſius*.

7　*Thlaſpi pumilum* of *Cluſius* : *Thlaſpi minimum* of *Tabern.*

8　*Thlaſpi petræum Myagrodes* of *Pona* : *Thlaſpi tertium ſaxatile* of *Camerarius*, in his *Epit.* of *Matthiolus.* ‡

¶ *The Temperature and Vertues.*

The ſeeds of theſe churliſh kinds of Treacle muſtard haue a ſharp or biting qualitie, breake inward apoſtumes, bring down the floures, kil the birth, and help the Sciatica or pain in the hip. They purge choler vpward and downward, if you take two ounces and a halfe of them, as *Dioſcorides* writeth. They are mixed in counterpoiſons, as Treacle, Mithridate, and ſuch like compoſitions.

Chap. 22.

Of wooddy Muſtard.

¶ *The Deſcription.*

2　WOoddy muſtard hath long narrow leaues declining to whiteneſſe, like thoſe of the ſtock Gillofloure, very like the leaues of Roſemary, but ſomewhat broader, with rough ſtalks very tough & pliant, being of the ſubſtance of wood: the flours grow at the top, white of colour : the ſeeds follow, in taſte ſharp and biting. The husks or ſeed-veſſels are round and ſomewhat longiſh.

Small wooddy muſtard groweth to the height of two cubits, with many ſtalks ſet with ſmal narrow leaues like thoſe of Hyſſop, but rougher ; and at the top grow floures like thoſe of Treacle muſtard, or Thlaſpi. The whole plant groweth as a ſhrub or hedge-buſh.

3　Thorny muſtard groweth vp to the height of foure cubits, of a wooddy ſubſtance like vnto a hedge-buſh or wild ſhrub, with ſtalks beſet with leaues, floures, and ſeeds like the laſt before mentioned ; agreeing in all points, ſaue in the cruell pricking ſharpe thornes wherewith this plant is armed ; the other not. The root is tough, woody, and ſome ſtrings or fibres anexed thereto.

Z　　　　　　　　　　4 There

1 *Thlaspi fruticosum incanum.*
Hoary wooddy mustard.

2 *Thlaspi fruticosum minus.*
Small wooddy Mustard.

3 *Thlaspi spinosum.*
Thorny Mustard.

‡ 4 *Thlaspi fruticosum folio Leucoij marini.*
Bushy Mustard.

‡ 5 *Thlaspi hederacinm.*
Ivy muſtard.

4   There is another ſort of wooddy muſtard growing in ſhadowy and obſcure mountaines and rough ſtony places, reſembling the laſt deſcribed ; ſauing that this plant hath no prickes at all , but many ſmal branches ſet thicke with leaues, reſembling thoſe of the leſſer ſea *Leucoion* : the floures are many and white : the ſeed like the other Thlaſpies : the root is wooddy and fibrous.

‡   5   There is (ſaith *Lobel*) in Portland and about Plimouth, and vpon other rockes on the ſea coaſt of England , a creeping little herbe hauing ſmall red creſted ſtalkes about a ſpanne high : The leaues are thicke, and faſhioned like Ivy : the white floures and ſmall ſeeds do in taſt and ſhape reſemble the Thlaſpies. ‡

¶ *The Place.*

‡   1   The firſt of theſe groweth about Mechline.

2 . 3 . 4.   Theſe plants grow vpon the Alpiſh and Pyrene mountaines : in Piemont and in Italy, in ſtony and rocky grounds.

¶ *The Time.*

They floure when the other kinds of Thlaſpies do ; that is, from May to the end of Auguſt.

¶ *The Names.*

‡   1   This *Cluſius* and *Lobel* call *T Maſpi incanum Mechlinienſe : Bauhine* thinks it to be the *Iberis prima* of *Tabernamontanus*, whoſe figure retained this place in the former edition.

2   This is *Thlaſpi fruticoſum alterum* of *Lobel* : *Thlaſpi 5 Hiſpanicum* of *Cluſius*.

3   *Lobel* calls this, *Thlaſpi fruticoſum ſpinoſum.*

4   *Camerarius* calls this, *Thlaſpi ſemperuirens biflorum folio Leucoÿ*, &c. *Lobel*, *Thlaſpi fruticoſum folio Leucoÿ* &c.

5   This *Lobel* calls *Thlaſpi hederaceum.*   †

¶ *The Nature and Vertues.*

I finde nothing extant of their nature or vertues ; but they may be referred to the kinds of Thlaſpies, whereof no doubt they are of kindred and affinitie, as well in facultie as forme.

Chap. 23.   *Of Toures Muſtard.*

¶ *The Deſcription.*

1   TOwers Muſtard hath bin taken by ſome for a kind of Muſtard, and referred by them to it : of ſome, for one of the Muſtards , and ſo placed amongſt the Thlaſpies as a kinde thereof ; and therefore my ſelfe muſt needs beſtow it ſomewhere with others.   Therefore I haue with *Cluſius* and *Lobel* placed it among the Thlaſpies as a kind thereof. It comes out of the ground with many long and large rough leaues like thoſe of Hounds tongue, eſpecially thoſe next the ground : among which riſeth vp a long ſtalke a cubit or more high, ſet about with ſharpe pointed leaues like thoſe of Woad. The floures grow at the top, if I may terme them floures, but they are as it were a little duſty chaffe driuen vpon the leaues and branches with the winde : after which come very ſmall cods, wherein is ſmal reddiſh ſeed like vnto that of Chameline or Engliſh Worm-ſeed, with a root made of a tuſt full of innumerable threds or ſtrings.

Z 2

‡ 2 This

‡ 2   This second kinde hath a thicker and harder root than the precedent, hauing also fewer fibres ; the leaues are bigger than those of the last described, somwhat curled or sinuated, yet lesse, rough, and of a lightish green : in the midst of these there rise vp one or two stalkes or more, vsually some two cubits high, diuided into some branches, which are adorned with leaues almost ingirting them round at their setting on. The floures are like those of the former, but somewhat larger; & the colour is either white, or a pale yellow : after these succeed many long cods filled with a seed somwhat larger than the last described. ‡

3   Gold of pleasure is an herbe with many branches set vpon a streight stalke, round, and diuided into sundry wings, in height two cubits. The leaues be long, broad, and sharp pointed, somwhat snipt or indented about the edges like those of Sow-Thistles. The floures along the stalkes are white : the seed contained in round little vessels is fat and oily.

1 *Turritis.*
Towers Mustard.

‡ 2 *Turritis maior.*
Great Tower Mustard.

4 Treacle Wormseed riseth vp with tough and pliant branches, wherupon grow many smal yellow floures ; after which come long slender cods like Flix-weed or *Sophia*, wherein is contained small yellowish seed bitter as wormseed or Colliquintida. The leaues are small, & dark of color, shaped like those of wild stock Gillouers, but not so thick nor fat. The root is small and single.

¶ *The Place.*

Towers Treacle groweth in the West part of England, vpon dunghils and such like places. I haue likewise seene it in sundry other places, as at Pyms by a village called Edmonton neere London, by the city walls of West-Chester in corn fields, and where flax did grow about Cambridge.
‡ The second is a stranger with vs, yet I am deceiued if I haue not seen it growing in Mr *Parkinsons* garden. ‡

The other grow in the territorie of Leiden in Zeeland, and many places of the Low-countries; and likewise wild in sundry places of England.

¶ *The Time.*

These herbs do floure in May and Iune, and their seed is ripe in September.

¶ *The Names.*

‡ 1   This is *Turritis* of *Lobel* : *Turrita vulgatior* of *Clusius.*
2   This is *Turrita maior* of *Clusius* ; who thinks it to be *Brassica virgata* of *Cordus.*

3 *Mat-*

3   *Matthiolus* calls this,*Pseudomyagrum* : *Tragus* calls it,*Sesamum* : *Dodonæus*, *Lobel*, and others call it *Myagrum*.

4   This *Lobel* calls *Myagrum thlaspi effigie*. *Tabernamontanus* hath it twice; first vnder the name of *Erysimum tertium* : secondly,of *Myagrum secundum*. And so also our Author (as I formerly noted) had it before vnder the name of *Eruca syluestris angustifolia*; and here vnder the name of *Camelina*. ‡

3 *Myagrum*.
Gold of pleasure.

4 *Camelina*.
Treacle Worme-feed.

¶ *The Temperature.*
These plants be hot and dry in the third degree.

¶ *The Vertues.*
It is thought,saith *Dioscorides*,That the roughnesse of the skinne is polished and made smooth A with the oilie fatnesse of the seed of *Myagrum*.

Ruellius teacheth,That the juice of the herbe healeth vlcers of the mouth; and that the poore B peasant doth vse the oile in banquets,and the rich in their lamps.

The seed of *Camelina* stamped,and giuen children to drinke, killeth the Wormes, and driueth C them forth both by siege and vomit.

† The two Drabaes here omitted are treated of at large in the following Chapter.

---

‡ Chap. 24. *Of Turky Cresses.*

‡ OVr Author did briefely in the precedent Chapter make mention of the two plants wee first mention in this Chapter; but that so briefely, that I thought it conuenient to discourse more largely of them,as also to adde to them other two, being by most Writers adjuged to be of the same Tribe or kindred. The vertues of the first were by our Author out of *Dodonæus* formerly put to the *Thlaspi Candia*, Chapter 20. from whence I haue brought them to their proper place,in the end of this present Chapter.

¶ *The*

¶ *The Description.*

† 1   The first hath crested slender, yet firme stalkes of some foot long; which are set with leaues of some inch in length, broad at the setting on, sinuated about the edges, and sharpe pointed; their colour is a whitish greene, and **taste acride**; the leaues that are at the bottome of the stalk are many, and larger. The tops of the stalkes are diuided into many branches of an vnequall length, and sustaine many floures: each whereof consists of foure little white leaues, so that together they much resemble the vmbell of the Elder when it is in floure. Little swolne seed vessels diuided into two cells follow the fading floures: the seed is whitish, about the bignesse of Millet; the root also is white, slender and creeping.

† 2   This hath creeping roots, from which arise many branches lying vpon the ground heere and there, taking root also: the leaues, which vpon the lower branches are many, are in forme and colour much like those of the last described, but lesse, and somewhat snipt about the edges. The stalkes are about a handfull high, or somewhat more, round, greene, and hairy, hauing some leaues growing vpon them. The floures grow spoke fashion at the top of the stalks, white, and consisting of foure leaues: which fallen, there follow cods containing a small red seed.

1 *Draba Dioscoridis.*   ‡ 2 *Draba prima repens.*
Turkie Cresses.   The first creeping Cresse.

3   From a small and creeping root rise vp many shoots, which while they are young haue many thicke juicy and darke green leaues rose fashion adorning their tops, out of the midst of which spring out many slender stalkes of some foot high, which at certaine spaces are incompassed (as it were) with leaues somewhat lesser than the former, yet broader at the bottome: the floures, cods, and seed are like the last mentioned.

4   There is a plant also by some referred to this Classis: and I for some reasons thinke good to make mention thereof in this place. It hath a strong and very long root of colour whitish, and of as sharpe a taste as Cresses: the stalkes are many, and oft-times exceed the height of a man, yet slender, and towards their tops diuided into some branches, which make no vmbell, but carry their floures dispersed, which consist of foure small yellow leaues: after the floure is past there follow long slender cods containing a small, yellowish, acride seed. The leaues which adorne this plant are long, sharpe pointed, and snipt about the edges, somewhat like those of Saracens Confound, but that these towards the top are more vnequally cut in.

¶ *The*

‡ 3 *Draba altera repens.*
The other creeping Creſſe.

¶ *The Place.*

None of theſe (that I know of) are found naturally growing in this kingdom, the laſt excepted, which I thinke may be found in ſome places.

¶ *The Time.*

The firſt of theſe floure in May and the beginning of Iune : the 2 and 3 in Aprill. The fourth in Iune and Iuly.

¶ *The Names.*

1   This by a generall conſent of *Matthiolus, Anguillara, Lobel, &c.* is judged to be the *Arabis,* or *Draba* of the Antients.

2   *Draba altera* of *Cluſius.*

3   *Draba tertia ſucculento folio,* of *Cluſius. Eruca Muralis,* of *Daleſchampius.*

4   This by *Camerarius* is ſet forth vnder the name of *Arabis quorundam,* and hee affirmes in his *Hor. Med.* that he had it out of England vnder the name of *Solidago;* The which is verie likely, for without doubt this is the very plant that our Author miſtooke for *Solidago Sarracenica,* for hee bewraies himſelfe in the Chapter of *Epimedium,* whereas hee ſaith it hath cods like *Sarracens Conſound;* when as both he, and all other giue no cods at all to *Sarracens Conſound.* My very good friend Mr. *Iohn Goodyer* was the firſt, I thinke, that obſerued this miſtake in our Author; for which his obſeruation, together with ſome others formerly and hereafter to be remembred, I acknowledge my ſelfe beholden to him.

¶ *The Vertues attributed to the firſt.*

1   *Dioſcorides* ſaith, that they vſe to eat the dried ſeed of this herbe with meat, as we do pepper, especially in Cappadocia.   A

They vſe likewiſe to boyle the herbe with the decoction of Barley, called Ptiſana; which beeing ſo boiled, concocteth and bringeth forth of the cheſt tough and raw fleagme which ſticketh therein.   B

The reſt are hot, and come neere to the vertues of the precedent. ‡   C

CHAP. 25. *Of Shepheards purſe.*

¶ *The Deſcription.*

THe leaues of Shepheards purſe grow vp at the firſt long, gaſhed in the edges like thoſe of Rocket, ſpred vpon the ground : from theſe ſpring vp very many little weake ſtalkes diuided into ſundry branches, with like leaues growing on them, but leſſer; at the top whereof are orderly placed ſmall white floures : after theſe come vp little ſeed veſſels, flat, and cornered, narrow at the ſtemme like to a certaine little pouch or purſe, in which lieth the ſeed. The root is white not without ſtrings. ‡ There is another of this kinde with leaues not ſinuated or cut in. ‡

2   The ſmall Shepheards purſe commeth forth of the ground like the Cuckow floure, which I haue Engliſhed Ladie-ſmockes, hauing ſmall leaues deepely indented about the edges; amongſt which riſe vp many ſmall tender ſtalkes with floures at the top, as it were chaffe. The huskes and ſeed is like the other before mentioned.

¶ *The Place.*

Theſe herbes doe grow of themſelues for the moſt part, neere common high waies, in deſart and vntilled places, among rubbiſh and old walls.

¶ *The*

¶ *The Time.*

They floure,flouriſhand feed all the Summer long.

¶ *The Names.*

Shepheards purſe is called in Latine, *Paſtoris burſa,* or *Pera paſtoris :* in high **Dutch, Seckel : in**
low-Dutch,**Borſekens crupt :** in French,*Bourſe de paſteur ou Curè :* in Engliſh,Shepheards purſe or
ſcrip : of ſome,Shepheards pouch,and poore mans Parmacetie : and in the North part of England,Toy-wort,Pick-purſe,and Caſe-weed.

¶ *The Temperature.*

They are of temperature cold and dry,and very much binding,after the opinion of *Ruellius,Mat-*
*thiolus,*and *Dodonæus;*but *Lobel* and *Pæna* hold them to be hot and dry, iudging the ſame by their
ſharpe taſt : which hath cauſed me to inſert them here among the kindes of Thlaſpi, conſidering
the faſhion of the leaues,cods,ſeed,and taſte thereof : which do ſo wel agree together,that I might
very well haue placed them as kindes thereof.But rather willing to content others that haue written before,than to pleaſe my ſelfe,I haue followed their order in marſhalling them in this place,
where they may ſtand for couſine germanes.

¶ *The Vertues.*

A   Shepheards purſe ſtaieth bleeding in any part of the body, whether the iuice or the decoction
thereof be drunke,or whether it be vſed pulteſſe wiſe,or in bath,or any other way elſe.

B   In a Clyſter it cureth the bloudy flix : it healeth greene and bleeding wounds : it is maruellous
good for inflammations new begun,and for all diſeaſes which muſt be checked backe and cooled.

C   The decoction doth ſtop the laske,the ſpitting and piſſing of bloud,and all other fluxes of bloud.

C H A P.

CHAP. 26. *Of Italian Rocket.*

¶ *The Deſcription.*

1   ITalian Rocket hath long leaues cut into many parts or diuiſions like thoſe of the aſh tree, reſembling *Ruellius* his Bucks horne : among which riſe vp ſtalkes weake and tender, but thicke and groſſe, two foot high, garniſhed with many ſmall yellowiſh floures like the middle part of Tanſie floures, of a naughty ſauor or ſmell : the ſeed is ſmall like ſand or duſt, in taſt like Rocket ſeed, whereof in truth we ſuſpect it to be a kinde. The root is long and wooddy.

1 *Rheſeda Plinij.*
Italian Rocket.

2 *Rheſeda maxima.*
Crambling Rocket.

2   Crambling Rocket hath many large leaues cut into ſundry ſections, deeply diuided to the middle rib, branched like the horns of a ſtag or hart : among which riſe vp long fat & fleſhy ſtalks two cubits high, lying flat vpon the ground by reaſon of his weake and feeble branches. The floures grow at the top, cluſtering thicke together, white of colour, with browniſh threds in them. The ſeed is like the former. ‡ *Lobel* affirmes it growes in the Low-countrie gardens with writhen ſtalks, ſometimes ten or twelue cubits high, with leaues much diuided. ‡

¶ *The Place .*

Theſe plants grow in ſandy, ſtony, grauelly, and chalky barren grounds. I haue found them in ſundry places of Kent, as at Southfleet vpon Long-field downes, which is a chalky & hilly ground very barren. They grow at Greenhith vpon the hils, and in other places of Kent. ‡ The firſt grows alſo vpon the Wolds in Yorkſhire. The ſecond I haue not ſeen growing but in gardens, and much doubt whether it grow wild with vs or no. ‡

¶ *The Time.*

Theſe Plants do flouriſh in Iune, Iuly, and Auguſt.

¶ *The*

¶ The Names.

The firſt is called of *Pliny*, *Reſeda*, *Eruca peregrina*, & *Eruca Cantabrica:* in Engliſh, Italian Rocket.
The ſecond is called *Reſeda maxima:* of *Anguillara*, *Pignocomon* : whereof I finde nothing extant
worthy of memorie, either of temperature or vertues.

CHAP. 27.   Of Groundſell.

¶ The Deſcription.

1   THe ſtalke of Groundſell is round, chamfered and diuided into many branches. The
leaues be green, long, and cut in the edges almoſt like thoſe of Succorie, but leſſe, like
in a manner to the leaues of Rocket. The floures be yellow, and turn to down, which is
caried away with the wind. The root is full of ſtrings and threds.

<table>
<tr><td>1 Erigerum.
Groundſell.</td><td>2 Erigerum tomentoſum.
Cotton Groundſell.</td></tr>
</table>

2   Cotton Groundſel hath a ſtraight ſtalke of a brown purple colour, couered with a fine cotton or downy haire of the height of two cubits. The leaues are like thoſe of S.*Iames* wort or Rag-
wort; and at the top of the ſtalke grow ſmall knops, from which come floures of a pale yellow colour; which are no ſooner opened and ſpred abroad, but they change into down like that of the thiſtle, euen the ſame houre of his flouring, and is carried away with the winde : the root is ſmall and
tender.
   ‡ 3   There is another with leaues more jagged, and finelier cut than the laſt deſcribed, ſoft
alſo and downy : the floures are fewer, leſſe and paler than the ordinarie, but turne ſpeedily into
down like as the former. ‡

¶ The Place.
Theſe herbs are very common throughout England, and do grow almoſt euery where.

¶ The Time.
They flouriſh almoſt euery moneth of the yeare.

¶ The

‡ 3 *Erigeron tomentofum alterum.*
The other Cotton Groundfell.

Grounfel is called in Greeke, ἠριγέρων : in Latine, *Senecio*, becaufe it waxeth old quickly: by a baftard name *Herbntum* : in Germany , **Creußwurtz** : in low-Dutch, **Cruys cruyt**, and **Crupsken cruyt** : in Spanifh, *Yerva cana* : in Italian, *Cardoncello, Speliciofa* : in Englifh, Groundfel.

Cotton Groundfel feems to be all one with *Theophraftus* his *Aphace* : he maketh mention of *Aphace Lib.7.* which is not onely a kinde of pulfe, but an herbe alfo, vnto which this kinde of Groundfell is very like. For as *Theophraftus* faith, The herbe *Aphace* is one of the pot-herbs & a kind of Succorie: adding further , That it floureth in hafte , but yet foon is old, and turneth into down ; and fuch a one is this kind of Groundfell. But *Theophraftus* faith further, That it floureth all the winter, and fo long as the fpring lafteth ; as my felfe haue often feene this Groundfell do.

¶ *The Nature.*

Groundfel hath mixt faculties : it cooleth, and withall digefteth, as *Paulus Ægineta* writeth.

¶ *The Vertues.*

The leaues of Groundfel boiled in wine or wa- **A** ter, and drunke, heale the paine and ach of the ftomacke that proceeds of Choler.

The leaues and floures ftamped with a little **B** hogs greafe ceafe the burning heat of the ftones and fundament. By adding to a little Saffron or falt it helpeth the *Struma* or Kings euill.

The leaues ftamped and ftrained into milke and drunke , helpe the red gums and frets in Chil- **C** dren.

Diofcorides faith, That with the fine pouder of Frankincenfe it healeth wounds in the finues. The **D** like operation hath the downe of the floures mixed with vineger.

Boiled in ale with a little hony and vineger, it prouoketh vomit, efpecially if you adde thereto a **E** few roots of *Afarabacca.*

Chap. 28.   *Of Saint Iames his Wort.*

¶ *The Kindes.*

THe herb called Saint *Iames* his Wort is not without caufe thought to be a kind of Groundfel ; of which there be fundry forts, fome of the pafture, & one of the fea ; fome fweet fmelling, and fome of a loathfome fauor. All which kinds I will fet downe.

¶ *The Defcription.*

1   SAint Iames his wort or Rag-wort is very well known euery where, and bringeth forth at the firft broad leaues gafhed round about like to the leaues of common Wormewood, but broader, thicker, not whitifh or foft, of a deep green colour, with a ftalke which rifeth vp aboue a cubit high, chamfered, blackifh, and fomwhat red withall. The armes or wings are fet with leffer leaues like thofe of Groundfell or wilde Rocket. The floures at the top be of a yellow colour like Marigolds, as well the middle button, as the fmall floures that ftand in a pale round about, which turne into downe as doth Groundfel. The root is threddy.

‡ 2   This hath ftalks fome cubit high, crefted, and fet with long whitifh leaues ; the lower leaues are the fhorter, but the vpper leaues the longer , yet the narrower : at the top of the ftalke grow fome foure or fiue floures as in an vmbell, which are of a darke red colour before they open

themfelues,

1 *Iacobæa.* Rag-wort.   ‡ 2 *Iacobæa angustifolia.* Narrow leaued Ragweed.

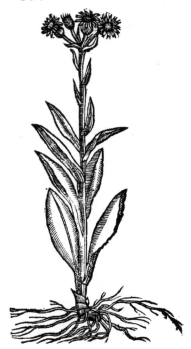

‡ 3 *Iacobæa latifolia.*
Broad leaued Ragweed.

4 *Iacobæa marina.*
Sea Rag-weed.

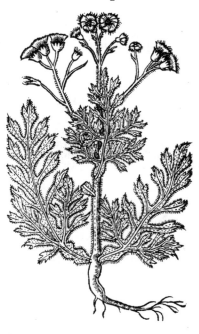

themſelues; but opened, of a bright golden colour, and thoſe are ingirt by fifteene or more little leaues, which are of a flame colour aboue, and red vnderneath. The floures flie away in downe, and the ſeed is blackiſh, and like that of the former. The roots are made of many ſtrings like thoſe of the precedent.

3   This broad leaued Rag-weed hath ſtiffe creſted ſtalkes, which are ſet with broad wrinckled ſharpe pointed leaues, of a greene colour: the bottome leaues are the larger and rounder, the top leaues the leſſe, and more diuided. The floures grow at the top of the ſtalkes, in ſhape and colour like thoſe of the common Rag-weed, but much bigger: They alſo turne into downe as the former. ‡

4   Sea Rag-wort groweth to the height of two cubits: the ſtalkes be not reddiſh as the other, but contrariwiſe Aſh-coloured, gray, and hoary: the leaues be greater and broader than the other: the floures grow at the top, of a pale yellow colour, couered on the cup or huske of the floure, as alſo the leaues, with a certaine ſoft white downe or freeſe: the floures vaniſh into downe, and flie away with the winde.

¶ The Place.

Land Rag-wort groweth euery where in vntilled paſtures and fields, which are ſomewhat moiſt eſpecially, and neare vnto the borders of fields.

‡ 2 3   Theſe grow vpon the Auſtrian and Heluetian Alpes. ‡

The fourth kinde of Ragwort groweth neere the ſea ſide in ſundry places: I haue ſeene it in the field by Margate, by Queakes houſe, and by Byrchenton in the Iſle of Tenet: likewiſe it groweth neere the Kings ferry in the Iſle of Shepey, in the way leading to Shirland houſe, where Sr. Edward Hobby dwelleth: and likewiſe at Queenborough caſtle in the ſame Iſle; and in other places. ‡ I haue been at the former and later of theſe places to finde out plants, yet could I not ſee this plant. It growes in the garden of Mr. Ralph. Tuggy, but I feare hardly wilde in this kingdome. ‡

¶ The Time.

They floure in Iuly and Auguſt, at which time they are carried away with the Down.

¶ The Names.

The firſt is called in Latine, Herba S. Iacobi, or S. Iacobi flos, and Iacobæa: in high Dutch, Sant Jacobs bloumen: in low-Dutch, Sant Jacobs crupt: in French, Fluer de S. Iacques: in Engliſh, S. Iames his Wort: the countrey people doe call it Stagger-wort, and Staner-wort, and alſo Rag-wort, ‡ and Rag-weed. In Holderneſſe in York-ſhire they call it Seggrum.

The ſecond is Iacobæa Pannonica 2. of Cluſius.

The third is his Iacobæa latifolia. Geſner calls it Coniza montana. ‡

The fourth is named Cineraria, or Aſh-coloured S. Iames Wort: ſome call it Erigeron Marinum, or Sea Groundſell: of ſome, Artemiſia marina. ‡ And by Proſper Alpinus, Artemiſia alba. ‡

¶ The Temperature.

S. Iames wort is hot and dry in the ſecond degree, and alſo clenſing, by reaſon of the bitterneſſe which it hath.

¶ The Vertues.

It is commended by the later Phyſitians to bee good for greene wounds, and old filthy vlcers   A which are not ſcoured, mundified, and made cleane; it alſo healeth them, with the juyce heere-of tempered with honey and May Butter, and boiled together vnto the forme of an Vnguent or ſalue.

It is much commended, and not without cauſe, to helpe old aches and paines in the armes, hips,   B and legs, boiled in hogs greaſe to the forme of an ointment.

Moreouer, the decoction hereof gargariſed is much ſet by as a remedy againſt ſwellings and im-   C poſtumations of the throat, which it waſteth away and throughly healeth.

The leaues ſtamped very ſmall, and boiled with ſome hogs greaſe vnto the conſumption of the   D juyce, adding thereto in the end of the boyling a little Maſticke and Olibanum, and then ſtrained, taketh away the old ache in the huckle bones called Sciatica.

‡   The Ægyptians (ſaith Proſper Alpinus) vſe the Sea Rag-wort, for many things: for they com-   E mend the decoction made with the leaues thereof againſt the ſtone in the kidnies and bladder, as alſo to helpe the old obſtructions of the inward parts, but principally thoſe of the wombe; as alſo the coldneſſe, ſtrangulation, barrenneſſe, inflation thereof, and it alſo brings downe the intercepted courſes: wherefore women troubled with the mother are much eaſed by baths made of the leaues and floures hereof. ‡

Aa

CHAP.

Cʜᴀᴘ.29.   *Of Garden Succorie.*

¶ *The Kindes.*

THere be ſundry ſorts of plants comprehended vnder the title of *Cichoracea*,that is to ſay, Cicho-
rie, Endiue, Dandelion,&c.differing not ſo much in operation and working, as in ſhape and
forme,which hath cauſed many to deeme them diuers,who haue diſtinguiſhed them vnder the titles aforeſaid: of euery which kinde there be diuers ſorts,the which ſhall be diuided in their ſeuerall chapters,wherein the differences ſhall be expreſt.

¶ *The Deſcription.*

1   GArden Succory is of two ſorts,one with broad leaues,and the other with narrow,deep-
ly cut and gaſhed on both ſides.The firſt hath broad leaues ſomwhat hairie,not much
vnlike to Endiue,but narrower,amongſt which do riſe vp ſtalkes,whereon are placed,
the like leaues,but ſmaller.The ſtalke diuideth it ſelfe toward the top into many branches,where-
on do grow little blew floures conſiſting of many ſmall leaues,after which followeth white ſeed.
The root is tough, long,and white of colour,continuing many yeares ; from the which as from e-
uery part of the plant doth iſſue forth bitter and milkie juice. The whole plant is of a bitter taſte
likewiſe.

1 *Cichorium ſativum.*
    Garden Succorie.

3 *Intybus ſativa.*
    Garden Endiue.

2   The ſecond kind of Succorie is like vnto the former,but greater in euery point. That which
cauſeth the difference is,that this beareth floures white of colour tending to blewneſſe,the others
blew,as I haue ſaid.

3   Garden Endiue bringeth forth long leaues,broad,ſmooth, more greene than white, like almoſt to thoſe of lettuce,ſomething nicked in the edges.The ſtalke groweth vp among the leaues,
being round and hollow,diuided into branches ; out of which being broken or cut there iſſueth a
juice like milke,ſomewhat bitter: the floures vpon the branches conſiſt of many leaues, in colour
                                                        com-

‡ 5 *Cichorium spinosum.*
Thorny Succorie.

commonly blew, seldome white. The root is long, white, with strings growing thereat, which withers after the seed is ripe.

4　Curled Endiue hath leaues not vnlike those of the curled or Cabbage Lettuce, but much greater; among which rise vp strong and thicke stalkes, set with the like leaues, but lesse, and not so notably curled or crisped. The floures grow at the top, blew of colour. The root perisheth, as doth the whole plant, when it hath brought forth his ripe seed.

‡　5　To to these may fitly be added the thorny or prickly Succorie of Candy, being of this kindred, and there vsed in defect of the true Succory, in stead therof. The root is pretty long, white, with few fibres hanging thereat: the stalk is hard, woody, & diuaricated into many branches, which commonly end in two or three pricks like hornes. The leaues are bitter, long, narrow, and sharpe pointed, and lie spred vpon the ground, and are a little sinuated or cut about the edges. The floures, which vsually grow vpon little footstalkes, at the diuisions of the branches, are much like those of the ordinarie Succory, yet much lesse, consisting of 5 blew leaues with yellow chiues in the middle. The is like those of the common Succory. It floureth in Iuly and August. ‡

¶ *The Place and Time.*

This Succorie and these Endiues are only sowne in gardens.

Endiue being sowne in the spring quickly commeth vp to floure, which seedeth in haruest, and afterward dieth. But being sowne in Iuly, it remaineth till winter, at which time it is taken vp by the roots, and laid in the sun or aire for the space of two houres; then will the leaues be tough, and easily endure to be wrapped vpon an heap, and buried in the earth with the roots vpward, where no earth can get within, (which if it did would cause rottennesse) the which so couered, may be taken vp at times conuenient, and vsed in sallads all the winter, as in London and other places is to be seen; and then it is called white Endiue: Whereof *Pliny* seemeth not to be ignorant, speaking to the same purpose, *lib.* 20. *cap.* 8.

¶ *The Names.*

These herbs be called by one name in Greeke, xipiλn: notwithstanding for distinctions sake they called the garden Succorie, xipis huaese, and the wilde Succorie, xipis ἀχεία. *Pliny* nameth the Succorie, *Hedypnois*; and the bitterer *Dioscorides* calls mxpis: in Latine, *Intybum sylvestre, Intybum agreste, Intybum erraticum*, and *Cichorium*: in shops it is called *Cichorea*; which name is not onely allowed of the later Physitions, but also of the Poet *Horace, lib.* 1. *Ode* 31.

———— *Me pascunt Oliuæ,*
Me Cichorea, leuesque maluæ.

With vs, saith *Pliny, lib.* 20. *cap.* 8. they haue called it *Intybum erraticum*, or wilde Endiue, *Ambugia*, (others reade *Ambubeia*) and some there be that name it *Rostrum porcinum*: others, as *Gulielmus Placentinus* and *Petrus Crescentius* terme it *Sponsa solis*: the Germanes call it **wegwarten**, which is as much to say, as the keeper of the waies: the Italians, *Cichorea*: the Spaniards, *Almerones*: the English, Chicory, and Succory: the Bohemians, *Czakanka*.

Endiue is named in Greeke, xipis imese: in Latine, *Intybum satiuum*: of some, *Endiuia*: of *Auicen* and *Serapio, Taraxacon*: of the Italians, *Scariola*, which name remaineth in most shops; also *Seriola*, as though they should fitly call it *Seris*: but not so wel *Serriola*, with a double *r*; for *Serriola* is *Lactuca sylvestris*, or wilde Lettuce: it is called in Spanish, *Serraya Enuide*: in English, Endiue, & Scariole: and when it hath bin buried in the earth as aforesaid, it is called white Endiue.

‡　5　This was first set forth by *Clusius*, vnder this name, *Chondrylla genus elegans caruleo flore*: since, by *Lobel* and *Bauhine*, by the title we giue you, to wit, *Cichorium spinosum. Honorius Bellus* writes, that in Candy whereas it naturally groweth, they vulgarly terme it squaxixn, that is, *Hydria spina*, the

　Pitcher-

Pitcher-Thorne,becauſe the people fetch all their water in ſtone pots or pitchers,which they ſtop with this plant,to keepe mice and other ſuch things from creeping into them : and it groweth ſo round,that it ſeems by nature to be prouided for that purpoſe. ‡

¶ *The Temperature.*

Endiue and Succorie are cold and dry in the ſecond degree,and withal ſomewhat binding ; and becauſe they be ſomething bitter they do alſo clenſe and open.

Garden Endiue is colder, and not ſo dry or cleanſing, and by reaſon of theſe qualties they are thought to be excellent medicines for a hot liuer,as *Galen* hath written in his 8 booke of the compoſition of medicines according to the places affected.

¶ *The Vertues.*

A   Theſe herbs being green haue vertue to coole the hot burning of the liuer,to help the ſtopping of the gall,yellow jaundiſe,lacke of ſleep,ſtopping of vrine,and hot burning feuers.

B   A ſyrrup made thereof and ſugar is very good for the diſeaſes aforeſaid.

C   The diſtilled water is good in potions,cooling and purging drinks.

D   The diſtilled water of Endiue,Plantain,and Roſes,profiteth againſt the excoriations in the conduit of the yard, to be injected with a ſyringe,whether the hurt came by vncleanneſſe, or by ſmall ſtones and grauell iſſuing forth with the vrine ; as often hath been ſeene.

E   Theſe herbs eaten in ſallads or otherwiſe,eſpecially the white Endiue, doth comfort the weake and feeble ſtomacke,and cooleth and refreſheth the ſtomacke ouermuch heated.

F   The leaues of Succorie bruiſed are good againſt inflammation of the eies,being outwardly applied to the grieued place.

Cʜᴀᴘ. 30.

Of wilde Succorie.

† 1 *Cichorium ſylveſtre.*
Wilde Succorie.

† 2 *Cichorium luteum.*
Yellow Succorie.

¶ *The*

¶ *The Kindes.*

IN like manner as there be sundry sorts of Succories and Endiues, so is there wilde kindes of either of them.

¶ *The Description.*

1   WIlde Succorie hath long leaues somewhat snipt about the edges like the leaues of Sow-thistle, with a stalk growing to the height of two cubits, which is diuided toward the top into many branches. The floures grow at the top, blew of colour: The root is tough and wooddy, with many strings fastned thereto.

3 *Intybum sylvestre.*
Wilde Endiue.

2   Yellow Succorie hath long and large leaues deeply cut about the edges like those of Hawk-weed: the stalke is branched into sundry arms, wheron grow yellow floures very double, resembling the floures of Dandelion or Pisse-a-bed; the which beeing withered, it flieth away in down with euery blast of winde.

3   Wilde Endiue hath long smooth leaues sleightly snipt about the edges. The stalke is brittle and full of milky iuice, as is all the rest of the plant: the floures grow at the top, of a blew or sky colour: the root is tough & threddie.

4   Medow Endiue, or Endiue with broad leaues, hath a thick tough and woody root with many strings fastened thereto; from which rise vp many broad leaues spred vpon the ground like those of garden Endiue, but lesser, & somewhat rougher; among which rise vp many stalks immediatly from the root; euery of them are diuided into sundry branches, whereupon grow many floures like those of the former, but smaller.

¶ *The Place.*

These plants grow wilde in sundry places of England, vpon wild & vntilled barren grounds, especially in chalky and stony places.

¶ *The Time.*

They floure from the midst to the end of August.

¶ *The Names.*

‡   The first of these is *Seris Picris* of *Lobel*: or *Cichorium sylvestre*: or *Intybus erratica* of *Tabernæmontanus*. ‡

Yellow Succorie is not without cause thought to be *Hyosiris*, or (as some copies haue it) *Hyosciris*; of which, *Pliny, lib. 20. cap. 8.* writeth: *Hyosiris* (saith he) is like to Endiue, but lesser & rougher. It is called of *Lobel, Hedypnois*: the rest of the names set forth in their seuerall titles shall be sufficient for this time.

¶ *The Temperature.*

They agree in temperature with the garden Succorie or Endiue.

¶ *The Vertues.*

The leaues of these herbs are boiled in pottage or broths, for sicke and feeble persons that haue   A
hot weake and feeble stomacks, to strengthen the same.

They are iudged to haue the same vertues with those of the garden, if not of more force in working.   B

‡   The first figure was of *Cichoreum album satiuum* of *Tabern.* The second is *Cichoreum luteum.* But the true figures of those our Author meant were vnder these titles. The first of *Hieracium lenifolium.* The second, *Dens leonis Cichoriæta*; for that is *Lobels Hedypnois.*

Chap. 31. *Of Gum Succorie.*

¶ *The Description.*

1   GVm Succorie with blew floures hath a thick and tough root, with some strings anexed thereto, full of a milky juice, as is all the rest of the plant, the flours excepted: the leaues are great and long, in shape like those of garden Succorie, but deeplier cut or jagged, somewhat after the maner of wild Rocket: among which rise tender stalks very easie to be broken, branched toward the top in two or sometimes three branches, bearing very pleasant floures of an azure colour or deep blew: which being past, the seed flieth away in down with the wind.

z

1 *Chondrilla cærulea.*
Blew Gum Succorie.

2 *Chondrilla cærulea latifolia.*
Robinus **Gum Succorie.**

2   Gum Succorie with broad leaues, which I haue named *Robinus* Gum Succorie, (for that he was the first that made any mention of a second kind, which he sent me as a great dainty, as indeed it was) in root is like the former: the leaues be greater, not vnlike to those of Endiue, but cut more deeply euen to the middle rib: the stalks grow to the height of two foot: the floures likewise are of an azure colour, but sprinkled ouer as it were with siluer sand; which addeth to the floure great grace and beauty.

3   Yellow gum Succorie hath long leaues, like in form and diuision of the cut leaues to those of wilde Succorie, but lesser, couered all ouer with a hoary down. The stalk is two foot high, white and downy also, diuided into sundry branches, whereon grow torne floures like those of Succorie, but in colour yellow, which is turned into down that is carried away with the winde. The root is long, and of a mean thicknesse; from which, as from all the rest of the plant, issueth forth a milkie juice, which being dried is of a yellowish red, sharp or biting the tongue. There is found vpon the branches hereof a gum, as *Dioscorides* faith, which is vsed at this day in physick in the Isle Lemnos, as *Bellonius* witnesseth.

4   Spanish Gumme Succorie hath many leaues spred vpon the ground, in shape like those of
Groundsell,

Groundſel,but much more diuided,and not ſo thick nor fat : among which riſe vp branched ſtalks
ſet with leaues like thoſe of *Stæbe Salamantica minor,* or Siluer-weed, whereof this is a kinde. The
floures grow at the top,of an ouerworne purple colour,which do ſeldome ſhew themſelues abroad
blown. ‡ The ſeed is like that of *Carthamus* in ſhape,but black and ſhining. ‡

† 3 *Chondrilla lutea.*
Yellow Gum Succory.

† 4 *Chondrilla Hiſpanica.*
Spaniſh Gum Succorie.

5 Ruſhy gumme Succory hath a tough and hard root,with a few ſhort threds faſtned thereto,
from which riſe vp a few jagged leaues like thoſe of Succorie, but much more diuided : the ſtalke
groweth vp to the height of two foot, tough and limmer like vnto Ruſhes , whereon are ſet many
narrow leaues : the flours be yellow,ſingle,and ſmall,which being faded do fly away with the wind:
the whole plant hauing milky juice like vnto the other of his kinde.

‡ There is another ſort of this plant to be found in ſome places of this kingdome,mentioned
by *Bauhinus* vnder the name of *Chondrilla viſcoſa humilis.*

† 6 Sea gum Succorie hath many knobby or tuberous roots full of juice,of a whitiſh purple
colour, with long ſtrings faſtned to them ; from which immediatly riſe vp a few ſmall thin leaues
faſhioned like thoſe of Succorie,narrower below,and ſomewhat larger toward their ends ; among
which ſpring vp ſmall tender ſtalks, naked, ſmooth,hollow,round,of ſome foot high or thereabout:
each of theſe ſtalks haue one floure,in ſhape like that of the Dandelion,but leſſer. The whole plant
is whitiſh or hoary,as are many of the ſea plants. †

7 Swines Succorie hath white ſmall and tender roots,from which riſe many indented leaues
like thoſe of Dandelion,but much leſſe,ſpred or laid flat vpon the ground : from the midſt wherof
riſe vp ſmall ſoft and tender ſtalks,bearing at the top double yellow floures like thoſe of Dandelion or Piſſe-abed,but ſmaller : the ſeed with the downy tuft flieth away with the wind.

8 The male Swines Succorie hath a long and ſlender root , with ſome few threds or ſtrings
faſtned thereto ; from which ſpring vp ſmall tender leaues about the bigneſſe of thoſe of Daiſies,
ſpred vpon the ground, cut or ſnipt about the edges confuſedly, of an ouerworne colour, full of a
milky juice : amongſt which riſe vp diuers ſmall tender naked ſtalkes, bearing at the top of euerie
ſtalke one floure and no more, of a feint yellow colour, and ſomething double : which beeing ripe,

do

5 *Chondrilla juncea.*
Ruſhy Gum Succorie.

6 *Chondrilla marina Lobelij.*
Sea Gum Succorie.

7 *Hypochæris Porcellia.*
Swines Succorie.

8 *Hyoſeris maſcula.*
Male Swines Succorie.

do turn into down that is caried away with the wind : the ſeed likewiſe cleaueth to the ſaid down, and is alſo caried away with the wind. The whole plant periſheth when it hath perfected his ſeed, and recoucreth it ſelſe again by the falling thereof.

‡ 9 Cichorium verrucarium.
Wart-Succorie

‡ 9 I thinke it expedient in this place to deliuer vnto you the hiſtorie of the Cichorium verrucarium, or Zacintha of Matthiolus ; of which our Author maketh mention in his Names and Vertues, although hee neither gaue figure, nor the leaſt deſcription therof. This wart Succorie (for ſo I will cal it) hath leaues almoſt like Endiue, greene, with pretty deep gaſhes on their ſides: the ſtalks are much creſted, and at the top diuided into many branches; between which and at their ſides grow many ſhort ſtalkes with yellow floures like thoſe of Succorie, but that theſe turne not into down, but into cornered and hard heads, moſt commonly diuided into 8 cells or parts, wherein the ſeed is contained. ‡

¶ The Place.

† Theſe plants are found only in gardens in this country ; the ſeuenth and eighth excepted, which peraduenture may be found to grow in vntilled places, vpon ditches banks and the borders of fields, or the like.

¶ The Time.

They floure from May to the end of Auguſt.

¶ The Names.

Gum Succorie hath beene called of the Grecians, χονδρίλλη : of the Latines, Condrilla, and Chondrilla : Dioſcorides and Pliny call it Cichorion, and Seris, by reaſon of ſome likeneſſe they haue with Succorie, eſpecially the two firſt, which haue blew floures as thoſe of the Succories. Lobel maketh Cichorea verrucaria to be Zacintha of Matthiolus.

¶ The Names in particular.

‡ 1   This is called Condrilla cærulea Belgarum of Lobel : Apate of Daleſchampius.
2   Condrilla 2 of Matthiolus : Chondrilla latifolia cærulea of Tabernamontanus.
3   Chondrilla prior Dioſcoridis, of Cluſius and Lobel.
4   Chondrilla rara purpurea, &c. of Lobel : Chondrilla Hiſpanica Narbonenſis of Tabernamontanus : Seneccio carduus Apulus, of Columna.
5   Chondrilla prima Dioſcoridis, of Columna and Bauhine : Viminea viſcoſa of Lobel and Cluſius.
6   Chondrilla altera Dioſcoridis, of Columna : ſome thinke it to be ἀπάπη of Theophraſtus : Lobel calls it Chondrilla puſilla marina lutea bulboſa.
7   Hypocharis porcellia of Tabernamontanus.
8   Hieracium minimum 9 of Cluſius ; Hyoſeris latifolia of Tabern. The two laſt ſhould haue beene put among the Hieracia.
9   Cichorium verrucarium, and Zacinthus of Matthiolus and Cluſius. ‡

¶ The Temperature and Vertues.

Theſe kinds of gum Succorie are like in temperature to the common Succorie, but drier.

The root and leaues tempered with hony and made into trochisks or little flat cakes, with nitre **A** or ſalt-petre added to them, clenſe away the morphew, ſun-burnings, and all ſpots of the face.

The gum which is gathered from the branches, whereof it tooke his name, laieth down the ſtai- **B** ring haires of the eie-brows and ſuch like places : and in ſome places it is vſed for maſtick, as Bellonius obſerues.

The gum poudred with myrrh, and put into a linnen cloath, and a peſſary made thereof like a fin- **C** ger, and put vp, brings downe the terms in yong wenches and ſuch like.

The

D   The leaues of *Zacintha* beat to pouder, and giuen in the decreaſing of the Moon to the quantitie of a ſpoonfull, takes away werts and ſuch like excreſcence, in what part of the body ſoeuer they be; the which medicine a certaine Surgeon of Padua did much vſe, whereby he gained great ſums of mony, as reporteth that antient phyſition *Ioachim Camerarius* of Noremberg, a famous city in Germany. And *Matthiolus* affirmes, that he hath known ſome helped of werts, by once eating the leaues hereof in a ſallad.

†   The figure of the third was of the ſame plant as the firſt, and was *Chondrylla alba* of *Tabern.* The fourth was of *Hieracium montanum maius latifolium* of *Tabern.* which you ſhall finde in the tenth place, Chap. 34.

Cʜᴀᴘ. 32.   *Of Dandelion.*

¶ *The Deſcription.*

1   THe hearbe which is commonly called Dandelion doth ſend forth from the root long leaues deeply cut and gaſhed in the edges like thoſe of wild Succorie, but ſmoother; vpon euery ſtalke ſtandeth a floure greater than that of Succorie, but double, & thicke ſet together, of colour yellow, and ſweet in ſmell, which is turned into a round downy blowbal that is carried away with the wind. The root is long, ſlender, and full of milky juice, when any part of it is broken, as is the Endiue or Succorie, but bitterer in taſt than Succorie.

‡   There are diuers varieties of this plant, conſiſting in the largeneſſe, ſmallneſſe, deepeneſſe, or ſhallowneſſe of the diuiſions of the leafe, as alſo in the ſmoothneſſe, and roughneſſe thereof. ‡

1 *Dens Leonis.*
Dandelion.

‡ 3 *Dens Leonis bulboſus.*
Knotty rooted Dandelion.

2   There is alſo another kinde of Succorie which may be referred hereunto, whoſe leaues are long, cut like thoſe of broad leafed Succory: the ſtalks are not vnlike, being diuided into branches as thoſe of Dandelion, but leſſer, which alſo vaniſheth into down when the ſeed is ripe, hauing a long and white root.

‡ 3   There is another *Dens Leonis* or Dandelion, which hath many knotty and tuberous roots
like

like thoſe of the Aſphodil, the leaues are not ſo deepely cut in as thoſe of the common Dandelion, but larger, and ſomewhat more hairy : the floures are alſo larger, and of a paler yellow, which flie away in ſuch downe as the ordinary. ‡

¶ *The Place.*

They are found often in medowes neere vnto water ditches, as alſo in gardens and high wayes much troden.

¶ *The Time.*

They floure moſt times in the yeare, eſpecially if the winter be not extreme cold.

¶ *The Names.*

Theſe plants belong to the Succory which *Theophraſtus* and *Pliny* call *Aphaca*, or *Aphace Leonardus* : *Fuchſius* thinketh that Dandelion is *Hydipnois Pliny*, of which he writeth in his 20 Booke, and 8.chapter, affirming it to be a wilde kinde of broad leafed Succorie, and that Dandelion is *Taraxacon* : as *Auicen* teacheth in his 692.chapter, is garden Endiue, as *Serapio* mentions in his 143.chapter, who citing *Paulus* for a witneſſe concerning the faculties, ſetteth downe theſe words which *Paulus* writeth of Endiue and Succorie. Diuers of the later Phyſitians do alſo call it *Dens Leonis*, or Dandelion : it is called in high Dutch, 𝕶𝖔𝖑𝖇𝖗𝖆𝖚𝖙 : in low-Dutch, 𝕻𝖆𝖕𝖊𝖓𝖈𝖗𝖚𝖎𝖙 : in French, *Piſſenlit ou couronne de preſtre*, or *Dent de lyon* : in Engliſh, Dandelion : and of diuers, Piſſabed. The firſt is alſo called of ſome, and in ſhops *Taraxacon*, *Caput monachi*, *Roſtrum porcinum*, and *Vrinaria*. The other is *Dens Leonis Monſpelienſium* of *Lobel*, and *Cichoreum Conſtantinopolitanum*, of *Matthiolus*.

¶ *The Temperature and Vertues.*

Dandelion is like in temperature to Succorie, that is to ſay, to wilde Endiue. It is cold, but it drieth more, and doth withall cleanſe, and open by reaſon of the bitterneſſe which it hath joyned with it : and therefore it is good for thoſe things for which Succory is. ‡ Boiled it ſtrengthens the weake ſtomacke, and eaten raw it ſtops the belly, and helpes the Dyſentery, eſpecially beeing boiled with Lentiles; The juice drunke is good againſt the vnuoluntarie effuſion of ſeed : boyled in vineger, it is good againſt the paine that troubles ſome in making of water : a decoction made of the whole plant helps the yellow jaundiſe.

† The figure which was in the 2 place was of the *Cich. Luteum*, where you may find it, but to what plant the deſcription may be referred, I cannot yet determine.

Chap. 33.　Of Sow-Thiſtle.

1 † *Sonchus aſper*. Prickly Sow-thiſtle.　　‡ 2 *Sonchus aſperior*. The more prickly Sow-thiſtle.

¶ *The*

¶ *The Kindes.*

THere be two chiefe kindes of Sow-thiſtles ; one tenderer and ſofter ; the other more pricking and wilder: but of theſe there be ſundry ſorts more found by the diligence of the later Writers; all which ſhall be comprehended in this chapter, and euery one be diſtinguiſhed with a ſeuerall deſcription.

¶ *The Deſcription.*

1   THe prickly Sow-thiſtle hath long broad leaues cut very little in, but full of ſmal prickles round about the edges ſomething hard and ſharp, with a rough and hollow ſtalke the floures ſtand on the tops of the branches, conſiſting of many ſmall leaues, ſingle, and yellow of colour; and when the ſeed is ripe it turneth into downe, and is carried away with the winde. The whole plant is full of a white milky juice.

‡ 2   There is another kinde of this, whoſe leaues are ſometimes prettily deepe cut in like as thoſe of the ordinary Sow-thiſtle; but the ſtalkes are commonly higher than thoſe of the laſt deſcribed, and the leaues more rough and prickly; but in other reſpects not differing from the reſt of this kinde. It is alſo ſometimes to be found with the leaues leſſe diuided. ‡

‡ 3   *Sonchus Lauis.*   4 *Sonchus Lauis latifolius.*
Hares Lettuce.   Broad leaued Sow-thiſtle.

3   The ſtalke of Hares Lettuce, or ſmooth-Thiſtle is oftentimes a cubit high, edged and hollow, of a pale colour, and ſometimes reddiſh : the leaues be greene, broad, ſet round about with deepe cuts or gaſhes, ſmooth, and without prickles. The floures ſtand at the top of the branches, yellow of colour, which are carried away with the winde when the ſeed is ripe. ‡ This is ſometimes found with whitiſh, and with ſnow-white floures, but yet ſeldome: whence our Author made two kindes more, which were the fourth and fifth; calling the one, The white floured Sow-thiſtle; and the other, The ſnow-white Sow-thiſtle. But theſe I haue omitted as impertinent, and giue you others in their ſtead. ‡

4   Broad leaued Sow-thiſtle hath a long thicke and milky root, as is all the reſt of the plant, with many ſtrings or fibres; from the which commeth forth a hollow ſtalke branched or diuided into ſundry Sections. The leaues be great, ſmooth, ſharp pointed, and green of colour : the floures
be

be white, in ſhape like the former. ‡ The floures of this are for the moſt part yellow like as the former. ‡

‡ 5　Wall Sow-thiſtle hath a fibrous wooddy root, from which riſes vp a round ſtalke not creſted: the leaues are much like to thoſe of the other Sow-thiſtles, broad at the ſetting on, then narrower, and after much broader, and ſharpe pointed, ſo that the end of the leafe much reſembles the ſhape of an yvie leafe ; theſe leaues are very tender, and of ſomewhat a whitiſh colour on the vnder ſide; the top of the ſtalke is diuided into many ſmall branches, which beare little yellow floures that flie away in downe.

6　This hath longiſh narrow leaues ſoft and whitiſh, vnequally diuided about the edges. The ſtalkes grow ſome foot high, hauing few branches, and thoſe ſet with few leaues, broad at their ſetting on, and ending in a ſharpe point: the floures are pretty large like to the great Hawk-weed, and fly away in downe: the root is long, white, and laſting. It floures moſt part of Summer; and in Tuſcany, where it plentifully growes, it is much eaten in ſallets, with oile and vineger, hauing a ſweetiſh and ſomewhat aſtringent taſte. ‡

‡ 5 _Sonchus læuis muralis._
Wall (or yvie-leaued) Sow thiſtle.

‡ 6 _Sonchus læuis anguſtifolius._
Narrow leaued Sow-thiſtle.

† 7　This blew floured Sow-thiſtle is the greateſt of all the reſt of the kindes, ſomewhat reſembling the laſt deſcribed in leaues; but thoſe of this are ſomewhat rough and hairy on the vnder ſide: the floures are in ſhape like thoſe of the ordinary Sow-thiſtle, but of a faire blew colour; which fading, flie away in downe that carries with it a ſmall aſh-coloured ſeed. The whole plant yeeldeth milke as all the reſt doe. †

8　Tree Sow-thiſtle hath a very great thicke and hard root ſet with a few hairy threds ; from which ariſeth a ſtrong and great ſtalke of a wooddy ſubſtance, ſet with long leaues not vnlike to Languedebeefe, but more deeply cut in about the edges, and not ſo rough : vpon which do grow faire double yellow floures, which turne into downe, and are carried away with the winde. The whole plant is poſſeſt with ſuch a milkie juice as are the tender and hearby Sow-thiſtles, which certainly ſheweth it to be a kinde thereof: otherwiſe it might be referred to the Hawke-weeds, whereunto in face and ſhew it is like. ‡ This hath a running root, and the heads and tops of the ſtalkes are very rough and hairy. ‡

7 *Sonchus flore cæruleo.*
Blew-floured Sow-thistle.

8 *Sonchus Arborescens.*
Tree Sow-thistle.

‡ 9 *Sonchus arborescens alter.*
The other Tree Sow-thistle.

† 10 *Sonchus syluaticus.*
Wood Sow-thistle.

‡　9　This other Tree Sow-thiſtle growes to a mans height or more, hauing a firme creſted ſtalke, ſmooth, without any prickles, and ſet with many leaues incompaſſing the ſtalke at their ſetting on, and afterwards cut in with foure, or ſometimes with two gaſhes onely : the vpper leaues are not diuided at all: the colour of theſe leaues is green on the vpper ſide, and grayiſh vnderneath: the top of the ſtalke is hairy, and diuided into many branches, which beare the floures in an equall height, as it werd in an vmbel : the floures are not great, conſidering the largeneſſe of the plant, but viſually as big as thoſe of the common Sow-thiſtle, and yellow, hauing a hairy head or cap: the ſeed is creſted, longiſh, and aſhe coloured, and flies away with the downe: the root is thicke, whitiſh, hauing many fibres, putting out new ſhoots, and ſpreading euery yeare. *Bauhine* maketh this all one with the other, according to *Cluſius* his deſcription : but in my opinion there is ſome difference betweene them, which chiefely conſiſts, in that the former hath larger and fewer floures; the plant alſo not growing to ſo great a height. ‡

‡　10　This plant (whoſe figure our Authour formerly gaue, *pag.*148. vnder the title of *Eryſimum ſylueſtre*) hath long knotty creeping roots, from whence ariſeth a round ſlender ſtalk ſome two foot high, ſet at firſt with little leaues, which grow bigger and bigger as they come neerer the middle of the ſtalke, being pretty broad at their ſetting on, then ſomewhat narrower, and ſo broader againe, and ſharp pointed, being of the colour of the Wall (or Iuie leaued) Sow-thiſtle. The tops is diuided into many ſmall branches, which end in ſmall ſcaly heads like thoſe of the wilde Lettuce, containing floures conſiſting of foure blewiſh purple leaues, turned backe and ſnipped at their ends; there are alſo ſome threds in the middle of the floure, which turning into downe, carry away with them the ſeed, which is ſmall, and of an Aſh colour. *Bauhine* makes a bigger and a leſſer of theſe, diſtinguiſhing betweene that of *Cluſius* (whoſe figure I here giue you) and that of *Columna* ; yet *Fabius Columna* himſelfe could finde no difference, but that *Cluſius* his plant had fiue leaues in the floure, and his but foure: which indeed *Cluſius* in his deſcription affirmes, yet his figure (as you may ſee) expreſſes but foure: adding, That the root is not well expreſſed; which notwithſtanding *Cluſius* deſcribes according to *Columna's* expreſſion. ‡

¶ *The Place.*

The firſt foure grow wilde in paſtures, medowes, woods, and mariſhes neere the ſea, and among pot-herbes.

The fifth growes vpon walls, and in wooddy mountainous places.

The Tree Sow-thiſtle growes amongſt corne in watery places.

The ſixth, ſeuenth and tenth are ſtrangers in England.

¶ *The Time.*

They floure in Iune, Iuly, Auguſt, and ſomewhat later.

¶ *The Names.*

Sow-thiſtle is called in Greeke, ϲιγχοϲ: in Latine, *Sonchus* : of diuers, *Cicerbita, lactucella*, and *Lacterones* : *Apuleius* calleth it *Lactuca Leporina*, ' or Hares-thiſtle : of ſome, *Braſſica Leporina*, or Hares Colewort. The Engliſh names are ſufficiently touched in their ſeuerall titles : In Dutch it is called **Haſen Latouwe**: the French, *Palays de lieure.*

‡ ¶ *Names in particular.*

1　This is *Sonchus aſper major* of *Cordus* : *Sonchus tenerior aculeis aſperior* of *Lobel* : *Sonchus* 3. *aſperior* of *Dodonæus.*

2　This is *Sonchus aſper* of *Matthiolus, Fuchſius*, and others.

3　This, *Matthiolus, Dodonæus, Lobel*, and others call *Sonchus læuis : Tragus* calls it *Intybus erratica tertia.*

4　This *Tabernamontanus* only giues, vnder the title as you haue it here.

5　*Matthiolus* ſtiles this, *Sonchus læuis alter : Caſalpinus* calls it *Lactuca murorum :* and *Tabern. Sonchus ſylua:icus quartus : Lobel, Sonchus alter folio ſinuato hederaceo.*

6　*Lobel* calls this, *Sonchus læuis Matthioli :* it is *Terracrepulus* of *Caſalpinus :* and *Crepis* of *Daleſchampius.*

7　*Cluſius* and *Camerarins* giue vs this vnder the title of *Sonchus cæruleus.*

8　Onely *Tabern.* hath this figure, vnder the title our Author giues it : *Bauhine* puts it amongſt the *Hieracia*, calling it *Hieracium arboreſcens paluſtre.*

9　This *Bauhine* alſo makes an *Hieracium*, and would perſuade vs that *Cluſius* his deſcription belongs to the laſt mentioned, and the figure to this : to which opinion I cannot conſent. *Cluſius* giueth it vnder the name of *Sonchus* 3. *læuis altiſſimus.*

10　This *Cluſius* giues vnder the name of *Sonchius læuior Pannonicus* 4. *flore purp. Tabern.* calls it *Libanotis Theophraſti ſterilis: Columna* hath it by the name of *Sonchus montanus purpureus* πτερωδης: *Cordus, Geſner, Thalius*, and *Bauhine* refer it to the *Lactucæ ſylueſtres:* the laſt of them terming it *Lactuca montana purpuro cærulea.* ‡

　　　　　　　　　　　　　　　　　¶ *The*

¶ *The Temperature.*

The Sow-thiſtles, as *Galen* writeth, are of a mixt temperature ; for they conſiſt of a watery and earthie ſubſtance, cold, and likewiſe binding.

¶ *The Vertues.*

A   Whileſt they are yet young and tender they are eaten as other pot-herbes are: but whether they be eaten, or outwardly applied in manner of a pulteſſe, they do euidently coole : therefore they are good for all inflammations or hot ſwellings, if they be laied thereon.

B   Sow-thiſtle giuen in broth taketh away the gnawings of the ſtomacke proceeding of an hot cauſe : and increaſe milke in the breſts of Nurſes, cauſing the children whom they nurſe to haue a good colour : and of the ſame vertue is the broth if it be drunken.

C   The juyce of theſe herbes doth coole and temper the heat of the fundament aud priuy parts.

CHAP. 34:   *Of Hawke-weed.*

¶ *The Kindes.*

HAwke-weed is alſo a kinde of Succorie : of which *Dioſcorides* maketh two ſorts, and the later Writers more : the which ſhall be deſcribed in this chapter following, where they ſhall be diſtinguiſhed as well with ſeuerall titles as ſundry deſcriptions.

† 1 *Hieracium majus Dioſcoridis.*   2 *Hieracium minus, ſiue Leporinum.*
     Great Hawke-weed.   Small Hares Hawke-weed, or Yellow Diuels-bit

¶ *The Deſcription.*

1   THe great Hawke-weed hath large and long leaues ſpred vpon the ground, in ſhape like thoſe of Sow-thiſtle : the ſtalk groweth to the height of two cubits, branched into ſundry armes or diuiſions, hollow within as the yong Kexe, reddiſh of colour : whereupon do grow yellow floures thicke and double, which turne into Downe that flieth away with the winde when the ſeed is ripe. The root is thicke, tough and threddy.

2 The

2   The small Hawke-weed, which of most writers hath beene taken for Diuels-bit, hath long leaues deeply cut about the edges, with some sharp roughnesse thereon like vnto Sow-thistle. The stalks and floures are like the former : the root is compact of many smal strings, with a smal knob, or as it were the stump of an old root in the middle of those strings, cut or bitten off ; whereupon it took his name Diuels bit.

3   Blacke Hawk-weed hath very many long jagged leaues not much vnlike to those of Bucks-horne, spred flat and far abroad vpon the ground, which the picture cannot expresse as is requisit, in so little room : among which rise vp many stalks, slender and weake, the floures growing at the top yellow and very double : it hath also a threddy root.

‡   Our Author formerly gaue three figures and so many descriptions of this small *Hieracium*, which I haue contracted into two ; for the only difference that I can finde is, that the one hath the root as it were bitten off, with the leaues lesse cut in : the other hath a root somwhat longer, and fi-brous as the former : the leaues also in this are much more finely and deep cut in: in other respects there is no difference. ‡

3 *Hieracium nigrum.*
Black Hawk-weed.

4 *Hieracium Aphacoides.*
Succorie Hawkweed.

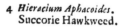

4   Succorie Hawke-weed hath many long and large leaues spred vpon the ground, deeply cut on both sides almost to the middle rib ; from which rise vp small stalkes and floures like those of the lesse Dandelion, but lesser. The root consisteth of many small threddy strings.

5   Endiue Hawk-weed hath many broad leaues indented about the edges very like garden En-diue, but narrower : among which rise vp stalks a foot or more high, slender, hairy, and brittle : the floures are yellow, and grow at the top double, and thicke set in a scaly huske like the Knapweed or *Iacea*, hauing great thicke and threddy roots. ‡ This hath a stalke sometimes more and other-whiles lesse rough, with the leaues sometimes more cut in, more long and narrow, and again other-whiles more short and broad. ‡

6   Long rooted Hawk-weed hath many broad leaues spred vpon the ground, sleightly and con-fusedly indented about the edges, with somwhat a bluntish point : among which leaues spring vp strong and tough stalks a foot and halfe high, set on the top with faire double yellow floures much like vnto a Pisse-abed. The root is very long, white, and tough.

7   Sharpe Hawke-weed hath leaues like those of Languebeefe or Ox-tongue, but much narro-wer, sharpe about the edges, and rough in the middle : the stalkes be long and slender, set with the like leaues, but lesser: the floures grow at the top, double and yellow: the root is tough & threddie.

5 *Hieracium intybaceum.*
Endiue Hawk-weed.

6 *Hieracium longius radicatum.*
Long rooted Hawk-weed.

7 *Hieracium aſperum.*
Sharpe Hawk-weed.

8 *Hieracium falcatum Lobelij.*
Crooked Hawk-weed.

† 8   Crooked or falked Hawke-weed hath leaues like vnto the garden Succorie, yet much ſmaller, and leſſe diuided, ſleightly indented on both ſides, with tender weake and crooked ſtalks, whereupon grow floures like thoſe of *Lampſana*, of a black or pale yellow colour, and the root ſmal and thready. The ſeeds are long, and falcated or crooked, ſo that they ſomewhat reſemble the foot or clawes of a bird, and from theſe ſeeds the plant hath this Epithit' *Falcatum*, or crooked in maner of a Sicle or Sithe.

‡ 9   This in leaues is not much vnlike the laſt deſcribed, but that they are ſomwhat broader, and leſſe cut in, hauing little or no bitterneſſe, nor milkineſſe : the ſtalkes are ſome foot high, commonly bending or falling vpon the ground : the floures are ſmal and yellow, and ſeem to grow out of the midſt of the ſeed, when as indeed they grow at the top of them, the reſt being but an empty huſke, which is falcated like that of the laſt deſcribed. This figure wee giue you was taken before the floures were blown, ſo that by that means the falcated or crooked ſeed-veſſels are not expreſt in this, but you may ſee their manner of growing by the former. ‡

‡ 9 *Hieracium falcatum alterum.*
The other crooked Hawkweed.

† 10 *Hieracium latifolium montanum.*
Broad leaued mountain Hawkweed.

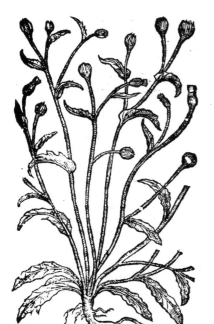

10   The broad leaued mountain Hawkeweed hath broad long ſmooth leaues deeply indented toward the ſtalke, reſembling the leaues of the greateſt Sow-thiſtle. The ſtalke is hollow & ſpungeous, full of a milky iuice, as is the reſt of the plant, as alſo all the other of his kinde : the floures grow at the top of the ſtalks, double and yellow.

11   The narrow leaued mountaine Hawkeweed hath leaues like thoſe of the laſt deſcribed, but narrower. The ſtalks be fat, hollow, and full of milke : the floures grow at the top double, and yellow of colour. The root is ſmall and thready.

There is a ſmall mountain Hawkweed hauing leaues like vnto the former, but more deeply cut about the edges, and ſharper pointed : the ſtalkes are tender and weake ; the floures be double and yellow like thoſe of Piloſella or great Mouſe-eare : the root is ſmall and thready.

¶ *The Place.*

Theſe kinds of herbs grow in vntoiled places neere vnto the borders of corn fields, in medowes, highwaies, woods, mountains, and hilly places, and neere to the brinks of ditches. † The two falcated Hawkweeds grow only in ſome few gardens. ‡

‡ *The*

11 *Hieracium montanum latifolium minus.*
The lesser broad leaued mountain Hawkweed.

‡ *The Time.*
They floure for the most part all Summer long, some sooner, some later.

¶ *The Names in generall.*
These plants are all contained vnder the name of *Hieracium*, which is called in Greek also ἱεράκιον : diuers name it in Latine, *Accipitrina* ; which is termed in French, *Cichoreé iaulne* : in English, Hawkeweed. These herbes took their name from a Hawke, which is called in Latine *Accipiter*, and in Greeke, ἱέραξ ; for they are reported to cleare their sight by conueying the juice hereof into their eyes. *Gaza* calleth it *Porcellia* ; for it is numbred amongst the Succories : they are called also *Lampuca.*

Yellow Hawkweed is called of some, *Morsus diaboli*, or yellow Diuels bit, for that the root doth very well resemble the bitten or cropt root of the common Diuels bit, being like Scabious.

‡ ¶ *The Names in particular.*
1   *Matthiolus*, *Fuchsius*, *Dodonaus* & others call this *Hieracium majus.*

2, 3.   These are varieties of the same plant, the first of them being called by *Fuchsius*, *Dodonaus*, and *Matthiolus*, *Hieracium minus.* *Lobel* calls it, *Hieracium minus præmorsa radice.* That sort of this with more cut leaues is by *Tabern.* called *Hieracium nigrum.*

4   *Lobel* calls this, *Hieracium folijs & facie Chondrilla* : *Bauhinus* maketh this to differ from that which our Author gaue in this fourth place out of *Tabern.* for he termes this, *Hieracium Chondrillæ folio hirsutum* ; and the other, *Hieracium Chondrilla folio Glabrum* ; the one smooth leaued, the other rough ; yet that which growes frequently with vs, and is very well represented by this figure, hath smooth leaues, as he also obserued it to haue in Italy, and about Montpelier in France.

5   This is *Hieracium alterum grandius*, and *Hieracium montanum angustifolium primum* of *Tabernamontanus.*

6   *Lobel* cals this, from the length of the root (though somtimes it be not so long) *Hieracium longius radicatum* : as also *Tabern.* *Hieracium macrorhizon* : it is thought to be the *Apargia* of *Theophrastus*, by *Daleschampius*, in the *hist. Lugd. pag.* 562. but the figure there that beares the title is of *Hieracium minus.*

7   *Tabernamontanus* first gaue this figure, vnder the name of *Hieracium intybaceum asperum* : *Bauhine* refers it to the wilde yellow Succories, and calls it *Cichoreum Montanum angustifolium hirsutie asperum.*

8   *Lobel* calls this, *Hieracium Narbonense falcata siliqua.*

9   He calls this, *Hieracium facie Hedypnois* : and *Cæsalpinus* termes this, *Rhagadiolus* : and the last mentioned, *Rhagadiolus alter.*

10   This by *Tabern.* is called *Hieracium montanum majus latifolinm.* The figure of this was giuen by our Author, Chap. 30. vnder the title of *Chondrilla Hispanica.*

11   *Tabernamontanus* also stiles this, *Hieracium montanum latifolium minus.*

¶ *The Temperature.*
The kindes of Hawkweed are cold and dry, and somewhat binding.

¶ *The Vertues.*

A   They are in vertue and operation like to *Sonchus* or Sow-Thistle, and being vsed after the same manner, be as good to all purposes that it doth serue vnto.

B   They be good for the eye-sight, if the iuyce of them be dropped into the eies, especially of that called Diuels bit, which is thought to be the best, and of greatest force.

Therefore

Therefore as *Diofcorides* writeth, it is good for an hot ftomacke, and for inflammations if it bee C laid vpon them.

The herbe and root being ftamped and applied, is a remedie for thofe that be ftung of the fcor- D pion; which effe&t not onely the greater Hawkeweeds, but the leffer ones alfo do performe.

Chap. 35. *Of Clufius his Hawkeweeds.*

¶ *The Kindes.*

THere be likewife other forts of Hawkeweeds, which *Carolus Clufius* hath fet forth in his Pannonicke obferuations, the which likewife require a particular chapter, for that they doe differ in forme very notably.

1 *Hieracium primum latifolium Clufij.*
   The firft Hawkeweed of *Clufius*.

2 *Hieracium 5.Clufij.*
   Clufius his 5. kinde of Hawkeweed.

¶ *The Defcription.*

1 THe firft of *Clufius* his Hawkeweeds hath great broad leaues fpread vpon the ground, fomewhat hairie about the edges, oftentimes a little jagged, alfo foft as is the leafe of Mullen or Higtaper, and fometimes dafht here and there with fome blacke fpots, in fhape like the garden Endiue, full of a milkie juice: among which rifeth vp a thicke hollow ftalke of a cubit high, diuiding it felfe at the top into two or three branches, whereupon do grow fweet fmelling floures not vnlike to thofe of yellow Succorie, fet or placed in a blacke hoarie and woolly cup or huske, of a pale bleake yellow colour, which turneth into a downie blowball that is carried away with the winde: the root entereth deeply into the ground, of the bigneffe of a finger, full of milke, and couered with a thicke blacke barke.

2 The fecond fort of great Hawkeweed according to my computation, and the 5. of *Clufius*, hath leaues like the former, that is to fay, foft and hoarie, and as it were couered with a kinde of white

white woollineſſe or hairineſſe, bitter in taſte, of an inch broad. The ſtalke is a foot high, at the top whereof doth grow one yellow floure like that of the great Hawkweed, which is carried away with the winde when the ſeed is ripe. The root is blacke and full of milkie juice, and hath certain white ſtrings annexed thereto.

3   This kinde of Hawkeweed hath blacke roots a finger thicke, full of milkie juice, deeply thruſt into the ground, with ſome ſmall fibres belonging thereto: from which come vp many long leaues halfe an inch or more broad, couered with a ſoft downe or hairineſſe, of an ouerworne ruſſet colour: and amongſt the leaues come vp naked and hard ſtalks, whereupon do grow yellow floures ſet in a woollie cup or chalice, which is turned into downe, and carried away with his ſeed by the winde.

4   The fourth Hawkeweed hath a thicke root aboue a finger long, blackiſh, creeping vpon the top of the ground, and putting out ſome fibres, and it is diuided into ſome heads, each whereof at the top of the earth putteth out ſome ſix or ſeuen longiſh leaues ſome halfe an inch broad, and ſomewhat hoarie, hairie, and ſoft as are the others precedent, and theſe leaues are ſnipt about the edges, but the deepeſt gaſhes are neereſt the ſtalkes, where they are cut in euen to the middle rib, which is ſtrong and large. The ſtalke is ſmooth, naked, and ſomewhat high: the floures be yellow and double as the other.

3 *Hieracium 6. Cluſij.*
　Cluſius his 6 Hawkeweed.

　　　　　　　　　　　　4 *Hieracium 7. Cluſij.*
　　　　　　　　　　　　　Cluſius his 7. Hawkeweed.

‡ 5 The ſame Author hath alſo ſet forth another *Hieracium*, vnder the name of *Hieracium paruum Creticum*, which he thus deſcribes; this is an elegant little plant ſpreading ſome ſix, or more leaues vpon the top of the ground, being narrower at that part whereas they adhere to the root, and broader at the other end, and cut about the edges, hauing the middle rib of a purple colour; amongſt theſe riſe vp two or three little ſtalkes about a foot high, without knot vntill you come almoſt to the top, whereas they are diuided into two little branches, at which place growes forth leaues much diuided; the floures grow at the top of a ſufficient bigneſſe, conſidering the magnitude of the plant, and they conſiſt of many little leaues lying one vpon another, on the vpper ſide wholly white, and on the vnder ſide of a fleſh colour, the root is ſingle, longiſh, growing ſmall
　　　　　　　　　　　　　　　　　　　　　　　　　　　　　　　towards

towards the end, and putting forth ftringy fibres on the fides. Thus much *Clufius*, who receiued this figure and defcription from his friend *Iaques Plateau* of Turnay. I conjecture this to be the fame plant that *Bauhine* hath fomewhat more accurately figured and defcribed in his *Prod.pag.68.*vnder the title of *Chondrilla purpurafcens fœtida*: which plant being an annuall, I haue feene growing fome yeares fince with M*r.Tuggy* at Weftminfter; and the laft Summer with an honeft and skilful Apothecarie one M*r.Nicholas Swayton* of Feuerfham in Kent: but I muft confeffe I did not compare it with *Clufius*, yet now I am of opinion, that both thefe figures and defcriptions are of one and the fame plant. It floures in Iuly and Auguft, at the later end of which moneth the feeds alfo come to ripeneffe.

‿ 6　This other (not defcribed by *Clufius*, but by *Lobel*) hath long rough leaues cut in and toothed like to Dandelion, with naked hairy ftalkes, bearing at their tops faire large and very double yellow floures, which fading fly away in downe. It growes in fome medowes. ‡

‡ 5 *Hieracium parvum Creticum.*　　　　‡ 6 *Hieracium Dentis leonis folio hirfutum.*
Small Candy Hawk-weed.　　　　　　　　Dandelion Hawk-weed.

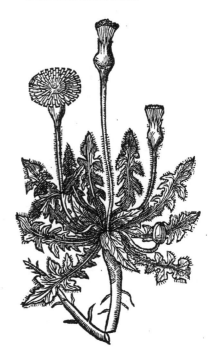

¶ *The Place.*

Thefe kindes of Hawke-weeds, according to the report of *Clufius*, do grow in Hungary and Auftria, and in the graffy dry hills, and herby and barren Alpifh mountaines, and fuch like places: notwithftanding if my memory faile me not I haue feene them growing in fundry places of England, which I meane, God willing, better to obferue hereafter, as opportunitie fhall ferue me.

¶ *The Time.*

He faith they floure from May to Auguft, at what time the feed is ripe.

¶ *The Names.*

The Author himfelfe hath not faid more than here is fet downe as touching the names, fo that it fhall fuffice what hath now been faid, referring the handling thereof to a further confideration.

¶ *The Nature and Vertues.*

I finde not any thing at all fet downe either of their Nature or Vertues, and therefore I forbeare to fay any thing elfe of them, as a thing not neceffary to write of their faculties vpon my own conceit and imagination.

Chap.

C H A P. 36.

‡ Of French, or Golden Lung-wort.

‡ 1 *Pulmonaria Gallica, ſiue aurea latifolia.*
Broad-leaued French or golden
Lung-wort.

‡ 2 *Pulmonaria Gallica, ſiue aurea anguſtifolia.*
Narrow leaued French or golden Lung-
wort.

¶ *The Deſcription.*

‡ 1   THis which I here giue you in the firſt place, as alſo the other two, are of the kinds of Hawke-weed, or *Hieracium* ; wherefore I thought it moſt fit to treat of them in this place, and not to handle them with the *Pulmonaria maculoſa*, or Sage of Ieruſalem : whereas our Author gaue the name *Pulmonaria Gallorum,* and pointed at the deſcription; but this figure being falſe, and the deſcription imperfect, I judged it the beſt to place it here next to thoſe plants which both in ſhape and qualities it much reſembles. This firſt hath a pretty large yet fibrous and ſtringy root; from the which ariſe many longiſh leaues, hairy, ſoft, and vnequally diuided, and commonly cut in the deepeſt neereſt the ſtalke ; they are of a darke greene colour, ſometimes broader and ſhorter, and otherwhiles narrower and longer (whence *Tabernamontanus* makes three forts of this, yet are they nothing but varieties of this ſame plant.) Amongſt theſe leaues grow vp one or two naked ſtalkes, commonly hauing no more than one leafe apiece, and that about the middle of the ſtalke ; theſe ſtalkes are alſo hairie, and about a cubit high, diuided at their tops into ſundry branches, which beare double yellow floures of an indifferent bigneſſe, which fading and turning into downe, are together with the ſeed carried away with the winde. This whole plant is milkie like as the other Hawk-weeds.

2   This Plant (though confounded by ſome with the former) is much different from the laſt deſcribed ; for the root is ſmall and fibrous; the leaues are ſmall, of the bigneſſe, and ſomewhat of the ſhape (though otherwiſe indented) of Daſie leaues, whitiſh and hoarie ; the ſtalke is not aboue an handfull high, creſted, hoary, and ſet with many longiſh narrow leaues; and at the top on ſhort foot-ſtalkes it beares foure or fiue floures of a bright yellow colour, and pretty large, conſidering

‡ 3 *Hieracium hortense latifolium, siue Pilosella maior.*
Golden Mouse-eare, or Grim the Colliar.

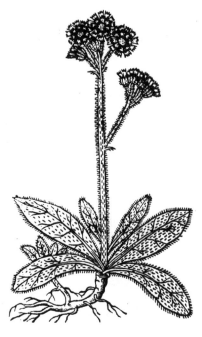

considering the smalnesse of the plant: the floures (like as others of this kind) fly away in down, and carry the seed with them.

3. This plant (which some also haue confounded with the first described) hath a root at the top, of a reddish or brownish colour, but whitish within the earth, and on the lower side sending forth whitish fibres: it bringeth forth in good and fruitfull grounds leaues about a foot long, & two or three inches broad, of a darke green colour and hairy, little or nothing at all cut in about the edges. Amongst these leaues riseth vp a stalke some cubit high, round, hollow, and naked, but that it somtimes hath a leafe or two toward the bottom, and towards the top it puts forth a branch or two. The floures grow at the top as it were in an vmbel, and are of the bignesse of the ordinary Mouseare, and of an orenge colour. The seeds are round and blackish, and are caried away with the downe by the wind. The stalks and cups of the flours are all set thicke with a blackish downe or hairinesse as it were the dust of coles ; whence the women who keepe it in gardens for nouelties sake, haue named it Grim the Colliar.

¶ *The Place.*

‡ 1 I receiued this from Mr. *Iohn Goodyer*, who found it May 27, 1631. in floure ; and the third of the following May, not yet flouring, in a Copse in Godlemen in Surrey, adioyning to the orchard of the Inne whose signe is the Antilope.

2 This I had from my kinde friend Mr *William Coote*, who wrot to mee, That hee found them growing on a hill in the Lady *Briget Kingsmills* ground at Sidmonton not far from Newberry, in an old Roman camp close by the Decuman port, on the quarter that regards the West South-West, vpon the skirts of the hill.

3 This is a stranger, and only to be found in some few gardens.

¶ *The Time.*

All these floure in Iune, Iuly, and August, about the later part of which month they ripen their seed.

¶ *The Names.*

1 This was first set forth by *Tragus*, vnder the name of *Auricula muris major* : and by *Tabern.* (who gaue three figures expressing the seuerall varieties thereof) by the name of *Pulmonaria Gallica siue aurea* : *Daleschampius* hath it vnder the name of *Corchorus.*

2 This was by *Lobel* (who first set it forth) confounded with the former, as you may see by the title of the figure in his Obseruations, *pag.* 317. yet his figure doth much differ from that of *Tragus*, who neither in his figure nor description allowes so much as one leafe vpon the stalke; and *Tabernamontanus* allowes but one, which it seldome wants. Now this by *Lobels* figure hath many narrow leaues, and by the Description, *Aduers. pag.* 253. it is no more than an handfull or handfull and halfe high : which very well agrees with the plant we here giue you, and by no means with the former, whose naked stalks are at least a cubit high. So it is manifest that this plant I haue heere described is different from the former, and is that which *Pena* and *Lobel* gaue vs vnder the title of *Pulmonaria Gallorum flore Hieracij. Bauhine* also confounds this with the former.

3 *Basil Besler*, in his *Hortus Eystettensis* hath wel exprest this plant vnder the title of *Hieracium latifolium peregrinum Phlomoides* : *Bauhinus* calls it *Hieracium hortense floribus atropurpurascentibus* ; and saith that some call it *Pilosella major* : and I iudge it to be the *Hieracium Germanicum* of *Fabius Columna.* This also seemeth rather to be the herbe *Costa* of *Camerarius*, than the first described ; and I dare almost be bold to affirme it the same : for he saith that it hath fat leaues lying flat vpon the ground, and as much as he could discern by the figure, agreed with the *Hieracium latifolium* of *Clusius* : to which indeed in the leaues it is very like, as you may see by the figure which is in the first place in the foregoing chapter, which very well resembles this plant, if it had more and smaller floures.

Cc　　　　　　　　　　　　　　¶ *The*

¶ *The Temperature and Vertues.*

A   I iudge theſe to be temperat in qualitie, and endued with a light aſtriction.

B   1   The decoction or the diſtilled water of this herb taken inwardly, or outwardly applied, conduce much to the mundifying and healing of green wounds ; for ſome boile the herb in wine, and ſo giue it to the wounded Patient, and alſo apply it outwardly.

C   It is alſo good againſt the internal inflammations and hot diſtempers of the heart, ſtomack, and liuer.

D   The iuyce of this herb is with good ſucceſſe dropped into the ears, when they are troubled with any pricking or ſhooting paine or noiſe.

E   Laſtly, the water hath the ſame qualitie as that of Succorie. *Tragus.*

     2   *Pena* and *Lobel* affirme this to be commended againſt Whitlowes, and in the diſeaſes of the lungs.

F   3   This (if it be the *Coſta* of *Camerarius*) is of ſingular vſe in the Pthiſis, that is, the vlceration or conſumption of the lungs : whereupon in Myſnia they giue the conſerue, ſyrrup, and pouder of it for the ſame purpoſe : and they alſo vſe it in broths and otherwiſe. *Cam.* ‡

CHAP. 37. *Of Lettuce.*

1 *Lactuca ſativa.*   2 *Lactuca criſpa.*
Garden Lettuce.   Curled Lettuce.

¶ *The Kindes.*

THere be according to the opinion of the Antients, of Lettuce two ſorts, the one wilde or of the field, the other tame or of the garden : but time, with the induſtrie of later writers, haue found out others both wilde and tame, as alſo artificiall, which I purpoſe to lay downe.

¶ *The Deſcription.*

1 GArden Lettuce hath a long broad leaſe, ſmooth, and of a light greene colour: the ſtalke is round, thicke ſet with leaues full of milky juice, buſhed or branched at the top: wherupon do grow yellowiſh floures, which turne into downe that is carried away with the winde. The ſeed ſticketh faſt vnto the cottony downe, and flieth away likewiſe, white of colour, and ſomewhat long : the root hath hanging on it many long tough ſtrings, which being cut or broken, do yeeld forth in like manner as doth the ſtalke and leaues, a juice like to milke. And this is the true deſcription of the naturall Lettuce, and not of the artificiall ; for by manuring, tranſplanting, and hauing a regard to the Moone and other circumſtances, the leaues of the artificiall Lettuce are oftentimes transformed into another ſhape : for either they are curled, or elſe ſo drawne together, as they ſeeme to be like a Cabbage or headed Colewort, and the leaues which be within and in the middeſt are ſomething white, tending to a very light yellow.

5 *Lactuca capitata.*
　Cabbage Lettuce.

6 *Lactuca intybacea.*
　Lumbard Lettuce.

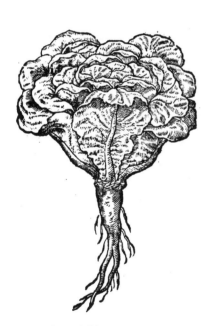

2 The curled Lettuce hath great and large leaues deeply cut or gaſhed on both the ſides, not plaine or ſmooth as the former, but intricately curled and cut into many ſections. The floures are ſmall, of a bleake colour, the which do turne into downe, and is carried away with the winde. The ſeed is like the former, ſauing that it changeth ſometime into blackneſſe, with a root like vnto the former.

3 This ſmall ſort of curled Lettuce hath many leaues hacked and torne in pieces very confuſedly, and withall curled in ſuch an admirable ſort, that euery great leafe ſeemeth to be made of many ſmall leaues ſet vpon one middle rib, reſembling a fan of curled feathers vſed among Gentlewomen : the floures, roots, and ſeeds agree with the former.

4 The Sauoy Lettuce hath very large leaues ſpred vpon the ground, at the firſt comming vp broad, cut or gaſht about the edges, criſping or curling lightly this or that way, not vnlike to the leaues of Garden Endiue, with ſtalkes, floures, and ſeeds like the former, as well in ſhape, as yeelding that milky iuice wherewith they do all abound.

5 Cabbage Lettuce hath many plain and ſmooth leaues at his firſt growing vp, which for the moſt part lie flat ſtill vpon the ground : the next that doe appeare are thoſe leaues in the middeſt, which turn themſelues together, embracing each other ſo cloſely, that it is formed into that globe

or

or round head, whereof the simplest is not ignorant. The seed hereof is blacke, contrary to all the rest, which may be as it were a rule whereby ye may know the seed of Cabbage Lettuce from the other sorts.

6　The Lumbard Lettuce hath many great leaues spred vpon the ground like vnto those of the garden Endiue, but lesser. The stalks rise vp to the height of three foot: the floures be yellowish, which turne into downe and flie away with the winde: the seed is white as snow.

¶ *The Place.*

Lettuce delighteth to grow, as *Palladius* saith, in a mannured, fat, moist, and dunged ground: it must be sowen in faire weather in places where there is plenty of water, as *Columella* saith, and prospereth best if it be sowen very thin.

¶ *The Time.*

It is certaine, saith *Palladius*, that Lettuce may well be sowen at any time of the yeare, but especially at euery first Spring, and so soone as Winter is done, till Summer be well nigh spent.

¶ *The Names.*

Garden Lettuce is called in Latine, *Lactuca sativa: Galen* names it θριδαξ: the Pythagorians ωηχεη some judge it to be called *Lactuca, à Lacteo succo*, of the milky iuice which issueth forth of the wounded stalks and roots: the Germanes name it **Lattich**: the low Dutch, **Latouwe**: the Spaniards, *Lechuga*, and *Alface*: the English, Lettuce: and the French, *Laictue*: when the leaues of this kinde are curled or crumpled, it is named of *Pliny, Lactuca crispa*: and of *Columella, Lactuca Ceciliana*: in English, curled or crompled Lettuce.

The Cabbage Lettuce is commonly called *Lactuca capitata*, and *Lactuca sessilis*: *Pliny* nameth it *Lactuca Laconica*: *Columella, Lactuca Batica*: *Petrus Crescentius, Lactuca Romana*: in English, red Lettuce, and Loued Lettuce.

There is another sort with reddish leaues, called of *Columella, Lactuca Cypria*: in English, red Lettuce.

¶ *The Temperature.*

Lettuce is a cold and moist pot-herbe, yet not in the extream degree of cold or moisture, but altogether moderatly; for otherwise it were not to be eaten.

¶ *The Vertues.*

A　Lettuce cooleth the heat of the stomacke, called the heart-burning; and helpeth it when it is troubled with choler: it quencheth thirst, causeth sleepe, maketh plenty of milke in nurses, who through heat and drinesse grow barren and drie of milke: for it breedeth milke by tempering the drinesse and heat. But in bodies that be naturally cold, it doth not ingender milke at all, but is rather an hinderance thereunto.

B　Lettuce maketh a pleasant sallad, being eaten raw with vineger, oile, and a little salt: but if it be boiled it is sooner digested, and nourisheth more.

C　It is serued in these daies, and in these countries in the beginning of supper, and eaten first before any other meat: which also *Martiall* testifieth to be done in his time, maruelling why some did vse it for a seruice at the end of supper, in these verses.

Claudere quæ cænas Lactuca solebat auorum,
Dic mihi, cur nostras incohat illa dapes?

Tell me why Lettuce, which our Grandsires last did eate,
Is now of late become, to be the first of meat?

D　Notwithstanding it may now and then be eaten at both those times to the health of the body: for being taken before meat it doth many times stir vp appetite: and eaten after supper it keepeth away drunkennesse which commeth by the wine; and that is by reason that it staieth the vapours from rising vp into the head.

E　The iuice which is made in the veins by Lettuce is moist and cold, yet not il, nor much in quantity: *Galen* affirmeth that it doth neither binde the belly no loose it, for it hath in it no harshnesse nor stipticke quality by which the belly is staied, neither is there in it any sharp or biting faculty, which scoureth and prouoketh to the stoole.

F　But howsoeuer *Galen* writeth this, and howsoeuer the same wants these qualities, yet it is found by experience, that it maketh the body soluble, especially if it be boiled; for by moistening of the belly it maketh it the more slippery: which *Martiall* very well knew, writing in his 11. Booke of Epigrams in this manner: *Prima tibi dabitur, ventri Lactuca mouendo*
————————— *Vtilis.*

G　Lettuce beeing outwardly applied mitigateth all inflammations; it is good for burnings and scaldings, if it be laid thereon with salt before the blisters do appeare, as *Pliny* writeth.

H　The iuice of Lettuce cooleth and quencheth the naturall seed if it be too much vsed, but procureth sleepe.

‡ Chap.

‡ CHAP. 38. *Of wilde Lettuce.*

¶ *The Description.*

‡ THere are three sorts of of wild Lettuce growing wild here with vs in England, yet I know not any one that hath mentioned more than two; yet I thinke all three of them haue bin written of, though two of them be confounded together and made but one (a thing often happening in the history of Plants) and vnlesse I had seene three distinct ones, I should my selfe haue been of the same opinion.

1   The first and rarest of these hath long and broad leaues, not cut in, but onely snipt about the edges, and those leaues are they that are on the lower part of the stalke almost to the middle ther-of: then come leaues from thence to the top, which are deeply diuided with large gashes : the stalk, (if it grow in good grounds, exceeds the height of a man (for I haue seene it grow in a garden to the height of eight or nine foot) it is large, round, and smooth, and towards the top diuided into many branches which beare yellow floures somewhat like to the garden Lettuce, after which also succeed blackish seeds like to other plants of this kinde. The whole plant is full of a clammy mil-ky juice, which hath a very strong and grieuous smell of *Opium*.

‡ 1 *Lactuca syl.major odore Opij.*
The greater wilde Lettuce smelling of *Opium*.

‡ 3 *Lactuca sylvestris folijs dissectis.*
The wilde Lettuce with the diuided leafe.

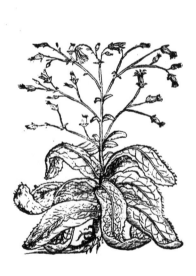   

2   This hath broad leaues onely cut about the edges, but not altogether so large as those of the last described : the stalk, which commonly is two cubits or better high, is also smooth, and diuided into many branches, bearing such floures and seeds as the last described; and this also hath a milkie juice of the same smell as the last described, from which it differs onely in the magnitude, and that this hath all the leaues whole, and not some whole and some diuided, as the former.

3   This in stalkes, floures and seeds is like to the last described, but the leaues are much diffe-rent, for they are all deeply gashed or cut in like as the leaues of Succory, or Dandelion. This also is full of a milky juice, but hath not altogether so strong a sent of *Opium* as the two former, though it partake much thereof. The stalke of this is sometimes a little prickly, and so also is the middle rib vpon the backside of the leafe, both in this and the former. All these three haue wooddy roots which die euery yeare, and so they come vp againe of the scattered seed.

¶ *The Place.*

The first of these was found in Hampshire by M^r. *Goodyer* & the seeds hereof sent to M^r, *Parkinson*

in

in whoſe garden I ſaw it growing ſome two yeares ago. The other grow plentifully betweene London and Pancridge Church, about the ditches and highway ſide.

¶ The Time.

They come vp in the Spring, and ſometimes ſooner, and ripen their ſeed in Iuly and Auguſt.

¶ The Names.

   1   I take the firſt of theſe to be the <i>Lactuca ſylueſtris</i> of <i>Dioſcorides</i> and the Antients, and that
which the Authors of the <i>Aduerſaria</i> gaue vs vnder the title of <i>L actuca agreſtis ſcari olæ hortenſis folio,
Lactucæ flore, Opij odore vehementi, ſoporifero & viroſo.

   2   This is the <i>Endiuia</i> of <i>Tragus, pag.</i> 268. and the <i>Theſion</i> of <i>Daleſchampius, pag.</i> 564. <i>Bauhine</i>
confounds this with the former.

   3   This is the <i>Lactuca ſylueſtris prior,</i> of <i>Tragus :</i> the <i>Lactuca ſylueſtris</i> of <i>Matthiolus, Fuchſius, Dodonæus,</i> and others : it is the <i>Scris Domeſtica</i> of <i>Lobel.</i>

¶ The Temperature.

Theſe certainly, eſpecially the two firſt, are cold, and that in the later end of the third or beginning of the fourth degree (if <i>Opium</i> be cold in the fourth.)

¶ The Vertues.

A   Some (ſaith <i>Dioſcorides</i>) mix the milkie juice hereof with <i>Opium ;</i> (for his <i>Meconium</i> is our <i>Opi-
um</i>) in the making thereof.

B   He alſo ſaith, that the juice hereof drunke in Oxycrate in the quantitie of 2 <i>obuli,</i> (which make
ſome one ſcruple) purgeth wateriſh humors by ſtoole; it alſo clenſeth the little vlcer in the eie called <i>Argemon</i> in Greeke, as alſo the myſtineſſe or darkeneſſe of ſight.

C   Alſo beaten and applied with womans milke it is good againſt burnes and ſcaldes.

D   Laſtly, it procures ſleepe, aſſwages paine, moues the courſes in women, and is drunke againſt the
ſtingings of Scorpions, and bitings of ſpiders.

E   The ſeed taken in drinke, like as the garden Lettuce, hindreth generation of ſeed and venereous
imaginations. ‡

C H A P. 39. Of Lambs Lettuce, or Corne ſallad.

  1 <i>Lactuca Agnina.</i>   2 <i>Lactuca Agnina latifolia.</i>
    Lambs Lettuce.   Corne ſallade.

¶ *The Description.*

1   THe Plant which is commonly called *Olus album*, or the white pot-herb (which of some hath bin set out for a kind of Valerian, but vnproperly, for that it doth very notably resemble the Lettuce, as well in for m, as in meat to be eaten; which property is not to be found in Valerian, and therefore by reason and authoritie I place it as a kind of Lettuce) hath many slender weak stalks trailing vpon the ground, with certain edges a foot high when it grows in most fertile ground; otherwise a hand or two high, with sundry ioynts or knees: out of euery one whereof grow a couple of leaues narrow and long, not vnlike to Lettuce at the first comming vp, as well in tendernesse as taste in eating; and on the top of the stalkes stand vpon a broad tuft as it were certaine white flourers that be maruellous little, which can scarsely bee knowne to bee floures, sauing that they grow many together like a tuft or vmbel: it hath in stead of roots a few slender threads like vnto haires.

2   The other kinde of Lettuce, which *Dodonæus* in his last edition setteth forth vnder the name of *Album olus*: the Low-country men call it **Veitmoes**, and vse it for their meat called Wetmose; with vs, Loblollie. This plant hath small long leaues a finger broad, of a pale green colour; among which shooteth vp a small cornered and slender stem halfe a foot high, ioynted with two or three ioynts or knees, out of which proceed two leaues longer than the first, bearing at the top of the branches tufts of very small white floures closely compact together, with a root like the former.

‡   Both these are of one plant, differing in the bignesse and broadnesse of the leafe and the whole plant besides. ‡

¶ *The Place.*

These herbes grow wilde in the corne fields; and since it hath growne in vse among the French and Dutch strangers in England, it hath bin sowne in gardens as a sallad herbe.

¶ *The Time.*

They are found greene almost all Winter and Summer.

¶ *The Names.*

The Dutch-men do cal it **Veitmoes**: that is to say, *Album olus*: of some it is called **Velterop**: the French terme it *Sallade de Chanoine*: it may be called in Greeke, Λιουλλχατω: in English, The White Pot-herbe; but commonly, Corne sallad.

¶ *The Temperature and Vertues.*

This herbe is cold and something moist, and not vnlike in facultie and temperature to the garden Lettuce; in stead whereof, in Winter and in the first moneths of the Spring it serues for a sallad herbe, and is with pleasure eaten with vineger, salt and oile, as other sallads be; among which it is none of the worst.

Chap. 40. *Of Coleworts.*

¶ *The Kindes.*

Dioscorides maketh two kindes of Coleworts; the tame and the wilde: but *Theophrastus* makes more kindes hereof: the ruffed or curled Cole, the smooth Cole, and the wilde Cole. *Cato* imitating *Theophrastus*, setteth downe also three Coleworts: the first hee describeth to bee smooth, great, broad leaued, with a big stalke: the second ruffed: the third with little stalks, tender and very much biting. The same distinction also *Pliny* maketh, in his 20. Booke, and the ninth chapter: where he saith, That the most antient Romanes haue diuided it into three kindes: the first roughed, the second smooth, and the third which is properly called ιιιμᾶτ, or Colewort. And in his nineteenth booke he hath also added to these, other moe kindes: that is to say, *Tritianum*, *Cumanum*, *Pompeianum*, *Brutianum*, *Sabellium*, and *Lacuturrium*.

The Herbarists of our time haue likewise obserued many sorts, differing either in colour, or els in forme: other headed with the leaues drawne together, most of them white, some of a deepe greene, some smooth leaued, and others curled or ruffed: differing likewise in their stalkes, as shall be expressed in their seuerall descriptions.

¶ *The*

1 *Braſſica vulgaris ſatiua.*
Garden Colewort.

2 *Braſſica ſatiua criſpa.*
Curled Garden Cole.

3 *Braſſica rubra.*
Red Colewort.

4 *Braſſica capitata alba.*
White Cabbage Cole.

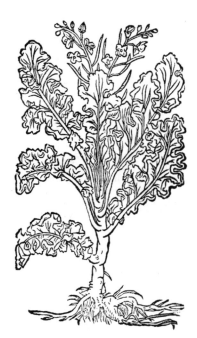

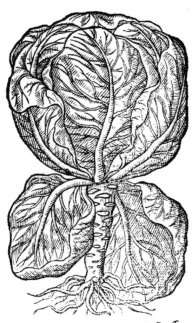

¶ *The Description.*

1   THe Garden Colewort hath many great broad leaues of a deepe blacke greene colour, mixed with ribs and lines of reddish and white colours : the stalke groweth out of the middest from among the leaues, branched with sundry armes bearing at the top little yellow floures : and after they be past, there do succeed long cods full of round seed like those of the Turnep, but smaller, with a wooddy root hauing many strings or threds fastned thereto.

2   There is another lesser sort than the former, with many deepe cuts on both sides euen to the middest of the rib, and very much curled and roughed in the edges ; in other things it differeth not.

3   The red kinde of Colewort is likewise a Colewort of the garden, and differeth from the common in the colour of his leaues, which tend vnto rednesse; otherwise very like.

4   There is also found a certaine kinde hereof with the leaues wrapped together into a round head or globe, whose head is white of colour, especially toward Winter when it is ripe. The root is hard, and the stalks of a wooddy substance. ‡ This is the great ordinary Cabbage knowne euery where, and as commonly eaten all ouer this kingdome. ‡

5 *Brassica capitata rubra.*
Red Cabbage Cole.

6 *Brassica patula.*
Open Cabbage Cole.

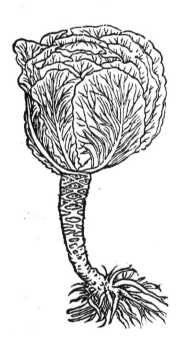

5   There is another sort of Cabbage or loued Colewort which hath his leaues wrapped together into a round head or globe, yet lesser than that of the white Cabbage, and the colour of the leaues of a lighter red than those of the former.

6   The open loued Colewort hath a very great hard or wooddy stalke, whereupon do grow very large leaues of a white greene colour, and set with thick white ribs, and gathereth the rest of the leaues closely together, which be lesser than those next the ground; yet when it commeth to the shutting vp or closing together, it rather dilateth it selfe abroad, than closeth all together.

7   Double Colewort hath many great and large leaues, whereupon doe grow here and there other small jagged leaues, as it were made of ragged shreds and jagges set vpon the small leafe, which giueth shew of a plume or fan of feathers. In stalk, root, and euery other part besides it doth agree with the Garden Colewort.

8 The

8   The double criſpe or curled Colewort agreeth with the laſt before deſcribed in euery re-ſpeſt,onely it differeth in the leaues,which are ſo intricately curled, and ſo thick ſet ouer with o-ther ſmall cut leaues,that it is hard to ſee any part of the leafe it ſelfe, except ye take and put aſide ſome of thoſe jagges and ragged leaues with your hand.

9 *Braſſica florida.*   10 *Braſſica Tophoſa.*
Cole-Florie.   Swollen Colewort.

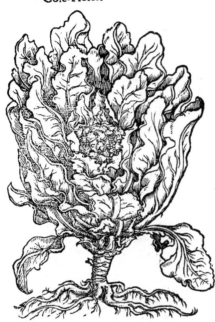 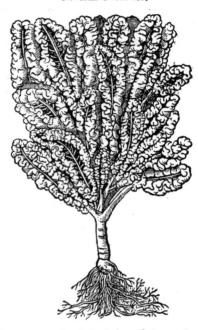

9   Cole flore,or after ſome Colieflore,hath many large leaues ſleightly indented about the edges,of a whitiſh greene colour, narrower and ſharper pointed than Cabbage: in the middeſt of which leaues riſeth vp a great white head of hard floures cloſely thruſt together,with a root full of ſtrings;in other parts like to the Coleworts

10   The ſwollen Colewort of all other is the ſtrangeſt,which I receiued of a worſhipful mer-chant of London maſter *Nicholas Lete,*who brought the ſeed thereof out of France ; who is greatly in loue with rare and faire floures and plants,for which he doth carefully ſend into Syria,hauing a ſeruant there at Aleppo, and in many other countries, for the which my ſelfe and likewiſe the whole land are much bound vnto him. This goodly Colewort hath many ſ.aues of a blewiſh green or of the colour of Woad,bunched or ſwollen vp about the edges as it were a piece of leather wet and broiled on a gridiron,in ſuch ſtrange ſort that I cannot with words deſcribe it to the full. The floures grow at the top of the ſtalkes,of a bleake yellow colour.The root is thicke and ſtrong like to the other kindes of Coleworts.

11   Sauoy Cole is alſo numbred among the headed Coleworts or Cabbages. The leaues are great and large,very like to thoſe of the great Cabbage,which turne themſelues vpwards as though they would imbrace one another to make a loued Cabbage,but when they come to the ſhutting vp they ſtand at a ſtay, and rather ſhew themſelues wider open,than ſhut any neerer together;in other reſpeſts it is like vnto the Cabbage.

12   The curled Sauoy Cole in euery reſpeſt is like the precedent, ſauing that the leaues here-of doe ſomewhat curle or criſpe about the middle of the plant : which plant if it be opened in the Spring time,as ſometimes it is,it ſendeth forth branched ſtalkes,with many ſmall white floures at the top,which beeing paſt,there follow long cods and ſeeds like the common or firſt kinde deſcri-bed.

13   This kinde of Colewoort hath very large leaues deeply jagged euen to the middle rib,in face reſembling great and ranke parſley.It hath a great and thicke ſtalke of three cubits high, whereupon doe grow floures,cods,and ſeed like the other Cole-woorts.

14 The

11 *Braſſica Sabauda.*
Sauoy Cole.

12 *Braſſica Sabauda criſpa.*
Curled Sauoy Cole.

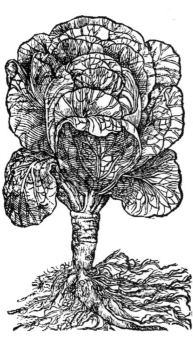

13 *Braſſica Selinoides.*
Parſley Colewort.

† 15 *Braſſica marina Anglica.*
Engliſh ſea Colewort.

14   The ſmall cut Colewort hath very large leaues,wonderfully cut,hackt, and hewen euen to the middle rib,reſembling a kind of curled Parſly that ſhal be deſcribed in his place(which is not common,nor hath bin knowne or deſcribed vntill this time) very wel agreeing with the laſt before mentioned,but differeth in the curious cutting and iagging of the leaues : in ſtalke,floure & ſeed not vnlike.

† 16 *Braſſica ſylveſtris.*
Wilde Coleworts.

15   Sea Colewort hath large and broad leaues very thick and curled, and ſo brittle that they cannot be handled without breaking, of an ouerworne green colour tending to grayneſſe : among which riſe vp ſtalks two cubits high,bearing ſmall pale floures at the top ; which being paſt,there follow round knobs wherein is contained one round ſeed and no more , blacke of colour, of the bigneſſe of a tare or vetch. ‡ And therefore *Pena* and *Lobel* call it *Braſſica marina monoſpermos.* ‡

16   The wilde Colewort hath long broad leaues not vnlike to the tame Colewort, but leſſer, as is all the reſt of the plant, and is of his owne nature wilde, and therefore not ſought after as a meat , but is ſowne and husbanded vpon ditch banks and ſuch like places, for the ſeeds ſake, by which oftentimes great gaine is gotten.

¶ *The Place.*

The greateſt ſort of Colewoorts grow in gardens, and do loue a ſoile which is fat, being throughly dunged and manured : they doe beſt proſper being remoued, and euery of them grow in our Engliſh gardens, except the wilde, which growes in fields and new digged ditch banks.

The ſea Colewort grows naturally vpon the baych and brims of the ſea,where there is no earth to be ſeen, but ſand and rolling pibble ſtones , which thoſe that dwell neere the ſea do call Bayche. I found it growing between Whitſtable and the Iſle of Tenet neere the brinke of the ſea,and in many places neere to Colcheſter,and elſewhere by the ſea ſide.

¶ *The Time.*

Petrus Creſcentius ſaith,That the Colewort may be ſowne and remoued at any time of the yeare, whoſe opinion I altogether miſlike. It is ſowne in the Spring,as in March,Aprill,and oftentimes in May, and ſometimes in Auguſt ; but the ſpeciall time is about the beginning of September.

The Colewort,ſaith *Columella*, muſt be remoued when it attaineth to ſix leaues after it is come vp from ſeed ; the which muſt be done in Aprill or May,eſpecially thoſe which were ſowne in Autumne,which afterwards flouriſh in the Winter monerhs,at what time they are fitteſt for meat.

But the Sauoy Cole and the Cole florey muſt be ſowne in Aprill, in a bed of hot horſe-dung, and couered with ſtraw or ſuch like , to keepe it from the cold and froſty mornings ; and hauing gotten ſix leaues in ſuch ſort,you ſhall remoue it as aforeſaid ; otherwiſe if you tarry for temperat weather before you ſow,the yeare will be ſpent before it come to ripeneſſe.

¶ *The Names.*

Euery of the Colewoorts is called in Greeke by *Dioſcorides* and *Galen*, κράμβη : it is alſo called ἀμέθυϲτον : ſo named not onely becauſe it driueth away drunkenneſſe, but alſo for that it is like in colour to the pretious ſtone called the Amethyſt ; which is meant by the firſt and garden Colewort. The Apothecaries and the common Herbariſts do call it *Caulis*,of the goodneſſe of the ſtalke. In the Germane Tongue it is called **koole kraut :** in French, *Des Choux :* in Engliſh, Coleworts.

Cole-florey is called in Latine *Craſſica Cypria* and *Cauliflora :* in Italian, *Cauliflore :* it ſeemes to agree with *Braſſica Pompeiana* of *Pliny*,whereof he writeth,*lib.* 19.*cap.* 8.

¶ *The Temperature.*

All the Coleworts haue a drying and binding facultie, with a certaine nitrous or ſalt quality, wherby they mightily clenſe,either in juice or in broth. The whole ſubſtance or body of the Cole-wort is of a binding and drying faculty,becauſe it leaueth in the decoction his ſalt quality ; which lieth in the juice and watry part thereof: the water wherein it is firſt boiled, draweth to it ſelfe all the quality ; for which cauſe the decoction thereof looſeth the belly,as doth alſo the iuice of it, if it be drunke: but if the firſt broth in which it was boiled be caſt away, then doth the Colewort dry and bind the belly. But it yeeldeth to the body ſmal nouriſhment,and doth not ingender good, but a groſſe and melancholicke bloud. The white Cabbage is beſt next vnto the Cole-florey ; yet *Cato* doth chiefely commend the ruſſet Cole : but he knew neither the white ones, nor the Cole-florey; for if he had,his cenſure had bin otherwiſe.

¶ *The Vertues.*

Dioſcorides teacheth,that the Colewoort beeing eaten is good for them that haue dim eies, and A that are troubled with the ſhaking palſie.

The ſame Author affirmeth,that if it be boiled and eaten with vineger, it is a remedy for thoſe B that be troubled with the ſpleene.

It is reported, that the raw Colewort beeing eaten before meate, doth preſerue a man from C drunkenneſſe: the reaſon is yeelded, for that there is a naturall enmitie betweene it and the vine, which is ſuch,as if it grow neere vnto it,forthwith the vine periſheth and withereth away : yea, if wine be poured vnto it while it is in boiling, it will not be any more boiled,and the colour thereof quite altered,as *Caſſius* and *Dionyſius Vticenſis* do write in their bookes of tillage: yet doth not *Athenæus* aſcribe that vertue of driuing away drunkenneſſe to the leaues, but to the ſeeds of Cole-wort.

Moreouer,the leaues of Coleworts are good againſt all inflammations,and hot ſwellings ; bee- D ing ſtamped with barley meale, and laied vpon them with ſalt: and alſo to breake carbuncles.

The iuyce of Coleworts,as *Dioſcorides* writeth,beeing taken with Floure-de-lys and nitre, doth E make the body ſoluble : and being drunke with wine, it is a remedy againſt the bitings of veno-mous beaſts.

The ſame being applied with the pouder of Fennugreeke,taketh away the paine of the gout,and F alſo cureth old and foule vlcers.

Being conueied into the noſthrils,it purgeth the head : being put vp with barley meale it brin- G geth downe the floures.

Pliny writeth,that the iuice mixed with wine, and dropped into the eares, is a remedy againſt H deafeneſſe.

The ſeed,as *Galen* ſaith,driueth forth wormes,taketh away freckles of the face, ſun-burning,and I what thing ſoeuer that need to be gently ſcoured or clenſed away.

They ſay that the broth wherein the herbe hath bin ſodden is maruellous good for the ſinewes K and ioints,and likewiſe for cankers in the eies,called in Greeke *Carcinomata*, which cannot be hea-led by any other meanes, if they be waſhed therewith.

† The fifteenth and ſixteenth figures were formerly tranſpoſed.

CHAP.41. *Of Rape-Cole.*

¶ *The Deſcription.*

1 THe firſt kinde of Rape Cole hath one ſingle long root, garniſhed with many threddy ſtrings : from which riſeth vp a great thick ſtalk, bigger than a great Cucumber or great Turnep : at the top whereof ſhooteth forth great broad leaues, like vnto thoſe of Cabbage Cole. The floures grow at the top on ſlender ſtalkes, compact of foure ſmall yellow floures : which being paſt,the ſeed followeth incloſed in little long cods,like the ſeed of Muſtard.

2 The ſecond hath a long fibrous root like vnto the precedent;the tuberous ſtalk is very great and long,thruſting forth in ſome few places here and there, ſome foot-ſtalks,whereupon doe grow ſmooth leaues, ſleightly indented about the edges : on the top of the long Turnep ſtalke grow leane ſtalks and floures like the former. ‡ This ſecond differs from the former onely in the length of the ſwolne ſtalke,whence they call it *Caulorapum longum*,or long Rape Cole. ‡

¶ *The Place.*

They grow in Italy,Spaine,and ſome places of Germanie, from whence I haue receiued ſeeds for my garden,as alſo from an honeſt and curious friend of mine called maſter *Goodman*,at the Mi-nories neere London.

　　　　　　　　　　　　　　　　　　　　　　　　¶ *The*

1 *Caulorapum rotundum*.
Round rape Cole.

¶ *The Time.*

They floure and flourish when the other Coleworts do, whereof no doubt they are kinds, and muſt be carefully ſet and ſowen, as Muske Melons and Cucumbers are.

¶ *The Names.*

They are called in Latine, *Caulorapum*, and *Rapocaulis*, bearing for their ſtalkes, as it were Rapes and Turneps, participating of two plants, the Colewort and Turnep, whereof they tooke their names.

¶ *The Temperature and Vertues.*

There is nothing ſet downe of the faculties of theſe plants, but are accounted for daintie meate, contending with the Cabbage Cole in goodneſſe and pleaſant taſte.

Chap. 42.   Of Beets.

¶ *The Deſcription.*

1 THE common white Beet hath great broad leaues, ſmooth, and plain: from which riſe thicke creſted or chamfered ſtalks: the floures grow along the ſtalks cluſtering together, in ſhape like little ſtars, which being paſt, there ſucceed round & vneuen prickly ſeed. The root is thicke, hard, and great.

1 *Beta alba.*   White Beets.

2 *Beta rubra.*   Red Beets.

† 3 *Beta rubra Romana.*
Red Roman Beet.

2   There is another fort like in fhape and proportion to the former fauing that the leaues of this
be ftreaked with red here and there confufedly,
which fetteth forth the difference.

3   There is likewife another fort hereof, that
was brought vnto me from beyond the feas, by that
courteous Merchant mafter *Lete,* before remembred, the which hath leaues very great, and red of
colour, as is all the reft of the plant, as well root, as
ftalke, and floures full of a perfect purple juyce
tending to redneffe : the middle rib of which leaues
are for the moft part very broad and thicke, like the
middle part of the Cabbage leafe, which is equall
in goodneffe with the leaues of Cabbage being
boyled. It grew with me 1596. to the height of
eight cubits, and did bring forth his rough and vneuen feed very plentifully : with which plant nature doth feeme to play and fport herfelfe : for the
the feeds taken from that plant, which was altogether of one colour and fowen, doth bring forth
plants of many and variable colours, as the worfhipfull Gentleman mafter *Iohn Norden* can very
well teftifie : vnto whom I gaue fome of the feeds
aforefaid, which in his garden brought forth many
other of beautifull colours.

¶ *The Place.*

The Beete is fowne in gardens: it loueth to grow
in a moift and fertile ground.   ‡   The ordinary
white Beet growes wilde vpon the fea-coaft of Te-
net and diuers others places by the Sea, for this is not a different kind as fome would haue it.   ‡

¶ *The Time.*

The fitteft time to fow it is in the fpring : it flourifheth and is green all Summer long, and likewife in Winter, and bringeth forth his feed the next yeare following.

¶ *The Names.*

The Grecians haue named it ʒɛυ̂λαι νῦ̂κλον : the Latines, *Beta* : the Germanes, **Mangolt** : the Spaniards, *Afelgas* : the French, *de la Porée, des Iotes,* and *Beets* : *Theophraftus* faith, that the white Beete
is furnamed ⲙⲁⲗⲁⲝⲏ, that is to fay, *Sicula,* or of Sicilia : hereof commeth the name *Sicla,* by which
the Barbarians, and fome Apothecaries did call the Beet ; the which word we in England doe vfe,
taken for the fame.

¶ *The Nature.*

The white Beets are in moifture and heate temperate, but the other kindes are dry, and all of
them abfterfiue : fo that the white Beet is a cold and moift pot-herbe, which hath joyned with it a
certaine falt and nitrous quality, by reafon whereof it cleanfeth and draweth flegme out of the
nofthrils.

¶ *The Vertues.*

Being eaten when it is boyled, it quickly defcendeth, loofeth the belly, and prouoketh to the   A
ftoole, efpecially being taken with the broth wherein it is fodden : it nourifheth little or nothing,
and is not fo wholefome as Lettuce.

The juyce conueighed vp into the nofthrils doth gently draw forth flegme, and purgeth the head.   B

The great and beautifull Beet laft defcribed may be vfed in Winter for a fallad herb, with vine-   C
gre, oyle, and falt, and is not only pleafant to the tafte, but alfo delightfull to the eie.

The greater red Beet or Roman Beet, boyled and eaten with oyle, vinegre and pepper, is a moft   D
excellent and delicat fallad : but what might be made of the red and beautifull root (which is to
be preferred before the leaues, as well in beautie as in goodneffe) I refer vnto the curious and cunning cooke, who no doubt when hee had the view thereof, and is affured that it is both good and
wholefome, will make thereof many and diuers difhes, both faire and good.

C H A P.

CHAP. 31. *Of Blites.*

¶ *The Deſcription.*

1 THe great white Blite groweth three or foure foot high, with grayiſh or white round ſtalkes : the leaues are plaine and ſmooth, almoſt like to thoſe of the white Orach, but not ſo ſoft nor mealy : the floures grow thruſt together like thoſe of Orach : after that commeth the ſeed incloſed in little round flat husky skinnes.

2 There is likewiſe another ſort of Blites very ſmooth and flexible like the former, ſauing that the leaues are reddiſh, mixed with a darke greene colour, as is the ſtalke and alſo the reſt of the plant.

3 There is likewiſe found a third ſort very like vnto the other, ſauing that the ſtalkes, branches, leaues, and the plant is altogether of a greene colour. But this growes vpright, and creepes not at all.

d There is likewiſe another in our gardens very like the former, ſauing that the whole Plant traileth vpon the ground: the ſtalkes, branches, and leaues are reddiſh : the ſeed is ſmall, and cluſtering together, greene of colour, and like vnto thoſe of *Ruellius* his coronopus, or Bucks-horne.

‡ 2 *Blitum majus album.*   2 *Blitum majus rubrum.*
The great white Blite.   The great red Blite.

¶ *The Place.*
The Blites grow in Gardens for the moſt part, although there be found of them wilde many times.

¶ *The Time.*
They flouriſh all the Summer long, and grow very greene in Winter likewiſe.

¶ *The Names.*
It is called in Greeke, ▨▧▨ : in Latine, *Blitum* : in Engliſh, Blite, and Blites : in French, *Blites,* or *Blitres.*

¶ *The*

‡ 3 *Blitum minus album.*
The small white Blite.

‡ 4 *Blitum minus rubrum.*
The small red Blite.

¶ *The Temperature.*

The Blite (saith *Galen* in his sixth booke of the faculties of simple medicines) is a pot-herbe seruing for meat, being of a cold moist temperature, and that chiefely in the second degree. It yeelds to the body small nourishment, as in his second booke of the faculties of nourishments he plainly shewes: for it is one of the pot-herbes that be vnsauorie or without taste, whose substance is waterish.

¶ *The Vertues.*

The Blite doth nourish little, and yet is fit to make the belly soluble, though not vehemently, seeing it hath no nitrous or sharpe qualitie whereby the belly should be prouoked. I haue heard many old wiues say to their seruants, Gather no Blites to put into my pottage, for they are not good for the eie-sight. Whence they had those words I know not, it may be of some Doctor that neuer went to schoole; for that I can find no such thing vpon record, either among the old or later writers.

CHAP. 44.   *Of Floure-Gentle.*

¶ *The Kindes.*

THere be diuers sorts of Floure-gentle, differing in many points very notably, as in greatnesse and smalnesse; some purple, and others of a skarlet colour; and one aboue the rest wherewith Nature hath seemed to delight her selfe, especially in the leaues, which in variable colours striues with the Parrats feathers for beauty.

1 *Amaranthus purpureus.*
Purple Floure-Gentle.

2 *Amaranthus Coccineus.*
Scarlet Floure-gentle.

3 *Amaranthus tricolor.*
Floramor,and Paſſeuelours.

4 *Amaranthus Pannicula ſparſa.*
Branched Floure-gentle.

¶ *The Deſcription.*

1   PVrple Floure-gentle riſeth vp with a ſtalke a cubit high, and ſomtimes higher, ſtreaked or chamfered alongſt the ſame, often reddiſh toward the root, and very ſmooth; which diuides it ſelf toward the top into ſmal branches, about which ſtand long leaues, broad, ſharpe pointed, ſoft, ſlipperie, of a greene colour, and ſometimes tending to a reddiſh : in ſtead of floures come vp eares or ſpoky tufts, very braue to look vpon, but without ſmel, of a ſhining light purple, with a gloſſe like Veluet, but far paſſing it : which when they are bruiſed doe yeeld a juice almoſt of the ſame colour, and being gathered, doe keep their beauty a long time after; inſomuch that being ſet in water, it will reuiue again as at the time of his gathering, and remaineth ſo, many yeares; whereupon likewiſe it hath taken it's name. The ſeed ſtandeth in the ripe eares, of colour blacke, and much glittering : the root is ſhort and full of ſtrings.

‡ 5 *Amaranthus pannicula incurua holoſerica.*
Veluet Floure-gentle.

2   The ſecond ſort of Floure-gentle hath leaues like vnto the former : the ſtalke is vpright, with a few ſmall ſlender leaues ſet vpon it : among which do grow ſmall cluſters of ſcaly floures, of an ouer-worne ſcarlet colour : the ſeed is like the former.

3   It farre exceedeth my skill to deſcribe the beauty and excellencie of this rare plant called *Floramor*; and I thinke the penſil of the moſt curious painter will be at a ſtay, when he ſhall come to ſet it downe in his liuely colours. But to colour it after my beſt manner, this I ſay, *Floramor* hath a thicke knobby root, whereon do grow many threddie ſtrings; from which riſeth a thicke ſtalke, but tender and ſoft, which beginneth to diuide it ſelfe into ſundry branches at the ground and ſo vpward, whereupon doth grow many leaues, wherein doth conſiſt his beauty : for in few words, euerie leaſe reſembleth in colour the moſt faire and beautifull feather of a Parat eſpecially thoſe feathers that are mixed with moſt ſundry colours, as a ſtripe of red, and a line of yellow, a daſh of white, and a rib of green colour, which I cannot with words ſet forth, ſuch are the ſundry mixtures of colours that Nature hath beſtowed in her greateſt jolitie, vpon this floure. The floure doth grow betweene the foot-ſtalks of thoſe leaues, and the body of the ſtalke or trunke, baſe, and of no moment in reſpect of the leaues, being as it were little chaffie husks of an o-uerworne tawny colour : the ſeed is black, and ſhining like burniſhed horne.

‡   I haue not ſeene this thus variegated as our Author mentions : but the leaues are commonly of three colours; the lower part or that next the ſtalke is greene; the middle red, and the end yellow; or elſe the end red, the middle yellow, & the bottom green. ‡

4   This plant hath a great many threds or ſtrings, of which his roots do conſiſt. From which riſe vp very thicke fat ſtalks creſted and ſtreaked, exceeding ſmooth, and of a ſhining red colour; which begin at the ground to diuide themſelues into branches, whereupon grow many great large leaues of a darke green colour tending to redneſſe, in ſhew like thoſe of the red Beet, ſtreaked and daſht here and there with red mixed with green. The flours grow alongſt the ſtalks, from the midſt thereof euen to the top in ſhape like *Panicum*, that is, a great number of chaffie confuſed ears thruſt hard together, of a deep purple colour. I can compare the ſhape thereof to nothing ſo fitly as to the veluet head of a ſtag, compact of ſuch ſoft matter as is the ſame : wherein is the ſeed, in colour white, round, and bored through the middle.

‡ 5   This in ſtalks and leaues is much like the purple Floure-gentle, but the heads are larger, bended round, and laced, or as it were wouen one with another, looking very beautifully like to Crimſon veluet : this is ſeldome to be found with vs, but for the beauties ſake is kept in the Gardens of Italy, whereas the women eſteemed it not only for the comelineſſe and beautious aſpect, but

but also for the efficacie thereof against the bloudy issues, and sanious vlcers of the wombe and kidneyes, as the Authors of the *Aduersaria* affirme. ‡

¶ *The Place and Time.*

These pleasant floures are sowne in gardens, especially for their great beautie.

They floure in August, and continue flourishing till the frost ouertake them, at what time they perish. But the Floramor would be sowne in a bed of hot horse-dung, with some earth strewed thereon in the end of March, and ordered as we doe muske Melons, and the like.

¶ *The Names.*

This plant is called in Greeke Ἀμάραντος, because it doth not wither and wax old : in Latine *Amaranthus purpureus* : in Dutch, **Samatbluomen** : in Italian, *Fior velluto* : in French, *Passe velours* : in English, floure Gentle, purple Veluet floure, Floramor ; and of some, floure Velure.

¶ *The Temperature and Vertues.*

Most attribute to floure Gentle a binding faculty, with a cold and dry temperature.

A   It is reported they stop all kindes of bleeding ; which is not manifest by any apparant quality in them, except peraduenture by the colour onely that the red eares haue : for some are of opinion, that all red things stanch bleeding in any part of the body : because some things, as *Bole armoniacke*, *sanguis Draconis, terra Sigillata*, and such like of red colour doe stop bloud : But *Galen, lib.* 2. & 4. *de simp. facult.* plainly sheweth, that there can be no certainty gathered from the colours, touching the vertues of simple and compound medicines : wherefore they are ill persuaded, that thinke the floure Gentle to stanch bleeding, to stop the laske or bloody flix, because of the colour onely, if they had no other reason to induce them thereto.

CHAP. 45.   *Of Orach.*

¶ *The Description.*

1   THe Garden white Orach hath an high and vpright stalke, with broad sharpe pointed leaues like those of Blite, yet smoother and softer. The floures are small and yellow, growing in clusters: the seed round, and like a leafe couered with a thin skin, or filme, and groweth in clusters. The root is wooddy and fibrous : the leaues and stalkes at the first are of a glittering gray colour, and sprinkled as it were with a meale or floure.

2   This differs from the former, only in that it is of an ouerworne purple colour.

‡ 3 This might more fitly haue beene placed amongst the Blites ; yet finding the figure here (though a contrary description) I haue let it injoy the place. It hath a white and slender root, and it is somewhat like, yet lesse than the Blite, with narrow leaues somewhat resembling Basill : it hath abundance of small floures, which are succeeded by a numerous sort of seeds, which are blacke and shining. ‡

4   There is a wilde kinde growing neere the sea, which hath pretty broad leaues, cut deeply about the edges, sharpe pointed, and couered ouer with a certaine mealinesse, so that the whole plant as well leaues, as stalkes and floures, looke of an hoary or gray colour. The stalkes lye spred on the shore or Beach, whereas it vsually growes.

‡ 5 The common wilde Orach hath leaues vnequally sinuated, or cut in somewhat after the manner of an oaken leafe, and commonly of an ouerworne grayish colour : the floures and seeds are much like those of the garden but much lesse.

6   This is like the last described, but the leaues are lesser and not so much diuided, the seeds grow also in the same manner as those of the precedent.

7   This also in the face and manner of growing is like those already described, but the leaues are long and narrow, sometimes a little notched : and from the shape of the leafe *Lobel* called it, *Atriplex Syluestris polygoni, aut Helxines folia.*

8   This elegant Orach hath a single and small root, putting forth a few fibres, the stalkes are some foot high, diuided into many branches, and lying along vpon the ground ; and vpon these grow leaues at certaine spaces whitish and vnequally diuided, somewhat after the manner of the wilde Orach ; about the stalke or setting on of the leaues grow as it were little berries, somewhat like a little mulberry, and when these come to ripenesse, they are of an elegant red colour, and make a fine shew. The seed is small, round, and ashcoloured. ‡

¶ *The Place.*

The Garden Oraches grow in most gardens. The wilde Oraches grow neere path-wayes and ditch sides ; but most commonly about dung-hils and such fat places. Sea Orach I haue found at Queeneborough, as also at Margate in the Isle of Thanet : and in most places about the sea side. ‡The eighth groweth only in some choise gardens: I haue seen it diuers times with Mr *Parkinson.* ‡

¶ *The*

1 *Atriplex satiua alba.*
White Orach.

3 *Atriplex sylueſtris, ſiue Polyſpermon.*
Wilde Orach, or All-ſeed.

† 4 *Atriplex marina.*
Sea Orach.

‡ 5 *Atriplex fylueftris vulgaris.*
Common wild Orach.

‡ 6 *Atriplex fylueftris altera.*
The other wilde Orach.

‡ 7 *Atriplex fylueftris anguftifolia.*
Narrow leaued wilde Orach.

‡ 8 *Atriplex baccifera.*
Berry-bearing Orach.

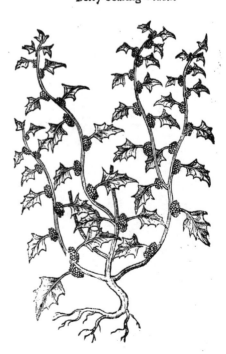

¶ *The Time.*

They floure and feed from Iune to the end of Auguſt.

¶ *The Names.*

Garden Orach is called in Greeke, πεῥꞷξꞷ: in Latine, *Atriplex*, and *Aureum Olus* : in Dutch, **meld**: in French, *Arrouches ou bonnes dames* : in Engliſh, Orach, and Orage : in the Bohemian tongue, *Leboda*: *Pliny* hath made ſome difference betweene *Atriplex* and *Chryſolachanum*, as though they differed one from another, for of *Atriplex* he writeth in his twentieth booke ; and of *Chryſolachanum* in his twenty eighth booke, and 8. chapter : where he writeth thus, *Chryſolachanum*, ſaith he, groweth in Pinetum like Lettuce : it healeth cut ſinewes if it be forthwith applied.

3   This wilde Orach hath been called of *Lobel, Polyſpermon Caſſani Baſſi*, or All ſeed.

¶ *The Temperature.*

Orach, ſaith *Galen*, is of temperature moiſt in the ſecond degree, and cold in the firſt.

¶ *The Vertues.*

Dioſcorides writeth, That the garden Orach is both moiſt and cold, and that it is eaten boiled as   A other ſallad herbes are, and that it ſofteneth and looſeth the belly.

It conſumeth away the ſwellings of the throat, whether it be laid on raw or ſodden.   B

The ſeed being drunken with mead or honied water, is a remedy againſt the yellow jaundiſe.   C

Galen thinketh, that for that cauſe it hath a clenſing quality, and may open the ſtoppings of the   D liuer.

† The figure which was in the ſecond place was of *Pes Anſerinus*. 2. of *Tabet*. The figure in the fourth place was of the wilde Orach, that I haue deſcribed in the fifth place.

Chap. 46. *Of ſtinking Orach.*

Atriplex olida.
Stinking Orach.

¶ *The Deſcription.*

STinking Orach growes flat vpon the ground and is a baſe and low plant with many weake and feeble branches, whereupon doe grow ſmall leaues of a grayiſh colour, ſprinckled ouer with a certaine kind of duſty mealineſſe, in ſhape like the leaues of Baſill : amongſt which leaues here and there confuſedly be the ſeeds diſperſed, as it were nothing but duſt and aſhes. The whole plant is of a moſt loathſome ſauour or ſmell ; vpon which plant if any ſhould chance to reſt and ſleepe, hee might very well report to his friends, that he had repoſed himſelfe among the chiefe of *Scoggins* heires.

¶ *The Place.*

It groweth vpon dunghils, and in the moſt filthy places that may bee found, as alſo about the common piſſing places of great princes and Noblemens houſes. Sometime it is found in places neere Bricke kilns and old walls, which doth ſomwhat alter his ſmell, which is like toſted cheeſe : but that which groweth in his natural place ſmels like ſtinking ſalt-fiſh, whereof it tooke his name *Garoſmus.*

¶ *The Time.*

It is an hearbe for a yeare, which ſpringeth vp, and when the ſeed is ripe it periſheth, and recouereth it ſelfe againe of his owne ſeed ; ſo that if it be gotten into a ground, it cannot bee deſtroyed.

¶ *The Names.*

Stinking Orach is called of *Cordus, Garoſmus*, becauſe it ſmelleth like ſtinking fiſh : it is likewiſe called

called *Tragium Germanicum*, and *Atriplex fœtida garum olens*, by *Pena* and *Lobel* : for it ſmelleth more ſtinking than the rammiſh male Goat : whereupon ſome by a figure haue called it *Vulvaria* : and it may be called in Engliſh, ſtinking Mother-wort.

¶ *The Nature and Vertues.*

There hath bin nothing ſet downe by the Antients, either of his nature or vertues, notwithſtanding it is thought profitable, by reaſon of his ſtinking ſmell, for ſuch as are troubled with the mother : for as *Hippocrates* ſaith, when the mother doth ſtifle or ſtrangle, ſuch things are to be applied vnto the noſe as haue a ranke and ſtinking ſmell.

C H A P . 47. *Of Gooſe-foot.*

¶ *The Deſcription.*

1 GOoſe-foot is a common herbe, and thought to be a kinde of Orach : it riſeth vp with a ſtalke a cubit high or higher, ſomewhat chamfered and branched: the leaues be broad, ſmooth, ſharpe pointed, ſhining, hauing certaine deepe cuts about the edges , reſembling the foot of a gooſe : the floures be ſmall, ſomething red : the ſeed ſtandeth in cluſters vpon the top of the branches, being very like the ſeed of wild Orach, and the root is diuided into ſundry ſtrings.

‡ 2 This differs from the laſt deſcribed, in that the leaues are ſharper cut, and more diuided, the ſeed ſomewhat ſmaller, and the colour of the whole plant is a deeper or darker greene. ‡

‡ 1 *Atriplex ſylueſtris latifolia, ſiue Pes Anſerinus.*
Gooſe-foot.

‡ 2 *Atriplex ſylueſtris latifolia altera.*
The other Gooſe-foot.

¶ *The Place.*

It growes plentifully in fat moiſt places, as vpon dung-hils and by high waies.

¶ *The Time.*

It flouriſheth when the Orach doth, whereof this is a wilde kinde.

¶ *The Names.*

The later Herbariſts haue called it *Pes anſerinus*, and *Chenopodium*, of the likeneſſe that the leaues haue with the foot of a Gooſe : in Engliſh, Gooſe-foot, and wilde Orach.

¶ *The*

¶ *The Temperature.*

This herb is cold and moiſt, and that no leſſe than Orach, but as it appeareth more cold.

¶ *The Vertues.*

It is reported that it killeth ſwine, if they do eat thereof: it is not vſed in phyſick, & much leſſe as a ſallad herbe.

CHAP. 48.
Of Engliſh Mercury.

Bonus Henricus.
Engliſh Mercury, or good Henry.

¶ *The Deſcription.*

GOod Henry, called *Tota bona*, ſo named of the later Herbariſts, is accounted of them to be one of the Docks, but not properly. This bringeth forth very many thick ſtalks ſet with leaues two foot high ; on the branches whereof toward the top ſtand greene floures in cluſters thicke thruſt together. The ſeed is flat like that of the Orach, whereof this is a kinde. The leaues be faſtned to long footſtalks, broad behind and ſharpe pointed, faſhioned like the leaues of Aron or Wake-robin, white or gray-iſh of colour, and as it were couered ouer with a fine meale : in handling it is ſoft and oleous, with a very thick root, and parted into many di-uiſions, of a yellow colour within, like the ſharp pointed Docke.

¶ *The Place.*

It is commonly found in vntilled places, and among rubbiſh neere common waies, old wals, and by hedges in fields.

¶ *The Time.*

It floureth in Iune and Iuly eſpecially.

¶ *The Names.*

It is called of ſome, *Pes Anſerinus*, and *Tota Bona*: in Engliſh, All-good, and Good Henrie : in Cambridge ſhire it is called Good King Harry : the Germans call it **Guter Heinrich**, of a cer-taine good qualitie it hath. As they alſo name another pernicious herb, *Malus Henricus*, or bad Henry. It is taken for a kinde of Mercurie, but vnproperly, for that it hath no participation with Mercurie either in forme or qualitie, except yee will call euery herbe Mercury that hath power to looſe the belly.

¶ *The Temperature.*

This plant is moderatly hot and dry, clenſing and ſcouring withall.

¶ *The Vertues.*

The leaues boiled with other pot-herbs and eaten, make the body ſoluble.

The ſame bruiſed and layd vpon green wounds or foule and old vlcers, doe ſcoure, mundifie, and heale them.

A
B

Ee

C H A P. 49. *Of Spinach.*

Spinachia.
Spinach.

¶ *The Description.*

1   S Pinach is a kind of Blite, after some, not-withstanding I rather take it for a kinde of Orach. It bringeth forth soft and tender leaues of a dark green colour, ful of juice, sharp pointed, & in the largest part or nether end square; parted oftentimes with a deepe gash on either side next to the stem or footstalke : the stalke is round, a foot high, hollow within, on the tops of the branches stand little floures in clusters, in whose places doth grow a prickely seed. The root consisteth of many small threds.

2   There is another sort found in our gardens, like vnto the former in goodnesse, as also in shape, sauing that the leaues are not so great, nor so deeply gasht or indented ; and the seed hath no prickes at all, wherefore it is called wilde Spinach.

¶ *The Place.*

It is sowne in gardens without any great labour or industrie, and forsaketh not any ground, beeing but indifferent fertill.

¶ *The Time.*

It may be sowne almost at any time of the yere, but being sowne in the spring, it quickely groweth vp and comes to perfection within two moneths ; but that which is sown in the fall of the leaf groweth not so soone to perfection, yet continueth all the winter, and seedeth presently vpon the first spring.

¶ *The Names.*

It is called in these daies *Spinachia*: of some, *Spinacheum olus* : of others, *Hispanicum olus*: *Fuchsius* nameth it *Spinachia*: the Arabians and *Serapio* call it *Hispane* : the Germanes, **Spinet** : in English, Spinage and Spinach : in French, *Espinas*.

¶ *The Temperature.*

Spinach is euidently cold and moist almost in the second degree, but rather moist. It is one of the pot-herbs whose substance is waterie and almost without taste, and therefore quickely descendeth and looseth the belly.

¶ *The Vertues.*

A   It is eaten boiled, but it yeeldeth little or no nourishment at all : it is something windie, & easily causeth a desire to vomit. It is vsed in sallads when it is yong and tender.

B   This herb of all other pot-herbs and sallad-herbes maketh the greatest diuersitie of meats and sallads.

C H A P. 50. *Of Pellitorie of the wall.*

¶ *The Description.*

P Ellitorie of the wall hath round tender stalks somwhat brown and reddish of colour, and somwhat shining : the leaues be rough like to the leaues of Mercury nothing snipt about the edges : the floures be small, growing close to the stems : the seed is black and very small, couered with a rough huske which hangeth fast vpon garments : the root is somewhat reddish.

¶ *The.*

Parietaria.
Pellitorie of the wall.

¶ *The Place.*

It groweth neere to old wals in the moist corners of Churches and stone buildings, amongst rubbish and such like places.

¶ *The Time.*

It commeth vp in May: it seedeth in Iuly and August: the root onely continueth and is to bee found in Winter.

¶ *The Names.*

It is commonly called *Parietaria,* or by a corrupt word *Paritaria,* because it groweth neere to walls: and for the same cause it is named of diuers *Muralis:* also *Muralium* of *Pliny* and *Celsus:* of the Grecians ἑλξίνη: There is also another *Helxine* syrnamed *Cissampelos:* some call it *Perdicium,* of Partridges which sometimes feed hereon: some *Vrceolaris,* and *Vitraria,* because it serueth to scoure Glasses, Pipkins, and such like: it is called in high-Dutch, **Tagvnonacht:** in Spanish, *Yerua del muro:* in English, Pellitorie of the Wall: in French, *Parietaire.*

¶ *The Temperature.*

Pellitorie of the Wall (as *Galen* saith) hath force to scoure, and is something cold and moist.

¶ *The Vertues.*

Pellitory of the wall boiled, and the decoction A of it drunken, helpeth such as are vexed with an old cough, the grauell and stone, and is good against the difficultie of making water, and stopping of the same, not onely inwardly, but also outwardly applied vpon the region of the bladder, in manner of a fomentation or warme bathing, with spunges or double clouts, or such like.

Dioscorides saith, That the juice tempered with Ceruse or white leade maketh a good ointment B against S. *Anthonies* fire and the shingles: and mixed with the Cerot of Alcanna, or with the male Goats tallow, it helpeth the gout in the feet: which *Pliny* also affirmeth, *Lib.22.cap.17.*

It is applied (saith he) to paines of the feet with Goats suet and wax of Cyprus; where in stead C of wax of Cyprus there must be put the cerot of Alcanna.

Dioscorides addeth, That the juice hereof is a remedy for old coughs, and taketh away hot swel- D lings of the almonds in the throat, if it be vsed in a gargarisme, or otherwise applied: it mitigateth also the paines of the eares, being poured in with oile of Roses mixed therewith.

It is affirmed, That if three ounces of the juice be drunke it procureth vrine out of hand. E

The leaues tempered with oile of sweet almonds in manner of a pultesse, and laid to the pained F parts, are a remedy for them that be troubled with the stone, and that can hardly make water.

Chap. 51.   *Of French Mercurie.*

¶ *The Kindes.*

THere be two kinds of Mercury reckoned for good, and yet both sometimes wilde; besides two wilde neuer found in gardens, vnlesse they be brought thither.

¶ *The Description.*

1 THe male garden Mercury hath tender stalks full of joints and branches, whereupon doe grow greene leaues like Pellitorie of the wall, but snipt about the edges: among which come forth two hairie bullets round, and ioined together like those of Goose-grasse or Cleuers, each containing in it selfe one small round seed: the root is tender, and full of white hairy strings.

2 The female is like vnto the former in leaues, stalkes, and manner of growing, differing but in
the

the floures and ſeed : for this kinde hath a greater quantitie of floures and ſeed growing together like little cluſters of grapes, of a yellowiſh colour. The ſeed for the moſt part is loſt before it can be gathered.

1 *Mercurialis mas.*
Male Mercurie.

2 *Mercurialis fœmina.*
Female Mercury.

¶ *The Place.*

French Mercurie is ſowen in Kitchen gardens among pot-herbes, in Vineyards, and in moiſt ſhadowie places : I found it vnder the dropping of the Biſhops houſe at Rocheſter; from whence I brought a plant or two into my garden, ſince which time I cannot rid my garden from it.

¶ *The Time.*

They floure and flouriſh all the Summer long.

¶ *The Names.*

It is called in Greeke, ληνόφυτον, and φυλλὸν ἀρρενικόν, or Mercurie his herbe; whereupon the Latines call it *Mercurialis* : it is called in Italian, *Mercorella* : in Engliſh, French Mercurie : in French, *Mercuriale,* *Vignoble,* and *Foirelle, quia Fluidam laxamue aluum reddit, Gallobelga enim foize & foizeus, ventris Fluorem vocant.*

¶ *The Temperature.*

Mercury is hot and drie, yet not aboue the ſecond degree : it hath a cleanſing faculty, and (as *Galen* writeth) a digeſting quality alſo.

¶ *The Vertues.*

A   It is vſed in our age in cliſters, and thought very good to clenſe and ſcour away the excrements and other filth contained in the guts. It ſerueth to purge the belly, being eaten or otherwiſe taken, voiding out of the belly not only the excrements, but alſo flegme and choler. *Dioſcorides* reporteth, that the decoction hereof purgeth wateriſh humors.

B   The leaues ſtamped with butter, and applied to the fundament; prouoke to the ſtoole; and the herbe bruiſed and made vp in manner of a peſſary, cleanſeth the mother, and helpeth conception.

C   *Coſtæus* in his booke of the nature of plants ſaith, that the iuice of Mercury, Hollihocks, & purſlane mixed together, and the hands bathed therein, defendeth them from burning, if they be thruſt into boyling lead.

                                                                    Chap,

CHAP. 52. *Of Wilde Mercurie.*

‡ 1 *Cynocrambe.*
Dogs Mercury.

† 2 *Phyllon arrhenogonon, ſiue mariſicum.*
Male childrens Mercury.

3 *Phyllon Theligonon, ſiue Fœminiſicum.*
Childrens Mercury the female.

¶ *The Deſcription.*

1   DOgs Mercurie is ſomewhat like vnto the garden Mercury, ſauing the leaues hereof are greater, and the ſtalke not ſo tender, and yet very brittle, growing to the height of a cubit, without any branches at all, with ſmall yellow ſloures. The ſeed is like the female Mercurie. ‡ it is alſo found like the male Mercurie, as you ſee them both expreſt in the figure; and ſo there is both male and female of this Mercury alſo. ‡

2   Male childrens Mercury hath three or foure ſtalkes, or moe: the leaues be ſomewhat long, not much vnlike the leaues of the Oliue tree, couered ouer with a ſoft downe or wooll gray of colour; and the ſeed alſo like thoſe of Spurge, growing two together, being firſt of an aſh-colour, but after turne to a blew.

‡ 3   This is much in ſhape like to the laſt deſcribed, but the ſtalkes are weaker, and haue more leaues vpon them; the ſloures alſo are ſmall and moſſy, and they grow vpon long ſtalkes, whereas the ſeeds of the other are faſt- ned to very ſhort ones: the ſeed is contained in round little heads, being ſometimes two, otherwhiles three or more in a cluſter. ‡

¶ *The Place.*

They grow in woods and copses, in the borders of fields, and among bushes and hedges. ‡ But the two last described are not in England, for any thing that I know. ‡

The Dogs Mercury I haue found in many places about Greene-hithe, Swaines.combe village, Grauesend, and Southfleet in Kent; in Hampsted wood, and all the villages thereabout, foure miles from London.

¶ *The Time.*

These flourish all the Summer long, vntill the extreame frost do pull them downe.

¶ *The Names.*

Dogs Mercury is called in Greeke, ꞷꞷꞷꞷꞷꞷꞷ : in Latine, *Canina*, and *Brassica Canina*, and *Mercuri-alis syluestris :* in English, Dogs Cole, and Dogs Mercury.

Childrens Mercury is called *Phyllon thelygonon*, and *Phyllon Arrhenogonon.*

¶ *The Temperature and Vertues.*

These wilde kindes of Mercury are not vsed in Physicke ; notwithstanding it is thought they agree as well in nature as quality with the other kindes of Mercury.

A   ‡ It is reported by the Antients, that the male *Phyllon* conduces to the generation of boies, and the female to girles.

B   At Salamantica they giue and much commend the decoction of either of these against the bitings of a mad dog.

C   The Moores at Granado vse them frequently in womens diseases. ‡

† The figure of the *Cynocrambe* was omitted, and in stead thereof was put the figure of *Phyllon marisicum.*

CHAP. 53.   Of Torne-sole.

1 *Heliotropium majus.*
Great Torne-sole.

† 2 *Heliotropium minus.*
Small Torne-sole.

¶ *The Kindes.*

THere bee foure sorts of Torne-sole, differing one from another in many notable points, as in greatnesse and smallnesse, in colour of floures, in forme and shape.

¶ *The*

¶ *The Deſcription.*

1   THE great Torneſole hath great ſtraight ſtalks couered with a white hairy cotton, eſpecially about the top; the leaues are ſoft and hairy in handling, in ſhape like the leaues
of Baſill : the floures grow at the top of the branches, in colour white, thicke together
in rowes vpon one ſide of the ſtalke, which ſtalke doth bend to turne backeward like the taile of a
ſcorpion : the root is ſmall and threddy.

2   The ſmall Torneſole hath many little and weake branches trailing vpon the ground, whereupon do grow ſmall leaues, like thoſe of the leſſer Baſill. The floures do grow without any certaine
order, amongſt the leaues and tender branches, gray of colour, with a little ſpot of yellow in the
middeſt, the which turne into crooked tailes like thoſe of the precedent, but not altogether ſo
much.

† 3 *Heliotropium ſupinum Cluſij & Lobelij.*   Hairie Torneſole.

4 *Heliotropium Tricoccum.*
Widow-waile Tornſole.

3   Hairy Torneſole hath many feeble and
weake branches trailing vpon the ground, ſet
with ſmall leaues, leſſer than the great Tornſole,
of which it is a kinde, hauing the ſeed in ſmall
chaffie husks, which do turne backe like the taile
of a ſcorpion, iuſt after the manner of the firſt deſcribed.

4   This kinde of Torneſole hath leaues very
like to thoſe of the great Torneſole, but of a
blacker greene colour : the floures be yellow, and
vnprofitable ; for they are not ſucceeded by the
fruit, but after them commeth out the fruit hanging vpon ſmall foot-ſtalkes three ſquare, and in
euery corner there is a ſmall ſeed like to thoſe of
the Tythimales; the root is ſmall and threddy.

¶ *The Place.*

Torneſole, as *Dioſcorides* ſaith, doth grow in
fennie grounds and neere vnto pooles and lakes.
They are ſtrangers in England as yet : It doth
grow about Montpelier in Languedock, where
it is had in great vſe to ſtaine and die clouts
withall, wherewith through Europe meat is coloured.

¶ *The Time.*

They flouriſh eſpecially in the Summer ſolſtice, or about the time when the ſun entreth into
Cancer.

¶ *The Names.*

The Græcians call it *Heliotropium :* the Latines keepe theſe names, *Heliotropium magnum,*
and

and *Scorpiurum* : of *Ruellius*, *Herba Cancri* : it is named *Heliotropium*, not becaufe it is turned about at the daily motion of the Sun, but by reafon it floureth in the former folftice, at which time the Sun being fartheft gone from the Equinoctiall circle, returneth to the fame : and *Scorpiurum* of the twiggie tops, that bow backeward like a fcorpions taile : of the Italians, *Tornefole bobo* ; in French, *Tournfol* : fome thinke it to be *Herba Clytia*, into which the Poëts feigne *Clytia* to be metamorpho-fed ; whence one hath thefe verfes :

Herba velut Clytiæ femper petit obuia folem,
Sic pia mens Chriftum, quo prece fpectet, habet.

¶ *The Temperature.*

Tornfole, as *Paulus Ægineta* writeth, is hot and dry, and of a binding faculty.

¶ *The Vertues.*

A   A good handfull of great Tornfole boyled in water, and drunke, doth gently purge the body of hot cholericke humours and tough clammie and flimie flegme.

B   The fame boyled in wine and drunke is good againft the ftinging of Scorpions, or other veno-mous beafts, and is very good to be applied outwardly vpon the griefe or wound.

C   The feed ftamped and laid vpon warts and fuch like excrefcences, or fuperfluous out-growings, caufeth them to fall away.

D   The fmall Tornefole and his feed boyled with Hyffope, Creffes, and falt-peter and drunke, dri-ueth forth flat and round wormes.

E   With the fmall Tornefole they in France doe die linnen rags or clouts into a perfect purple co-lour, wherewith cookes and confectioners do colour jellies, wines, meats, and fundry confectures: which clouts in fhops be called Tornefole, after the name of the herbe.

† The fecond and third figures were formerly tranfpofed : the fourth was the figure of the hairy Scorpion-graffe defcribed in the fourth place, in the follow-ing Chapter.

Chap. 54. *Of Scorpion Graffe.*

¶ *The Defcription.*

1   SCorpion graffe hath many fmooth, plaine, euen leaues, of a darke greene colour ; ftalkes fmall, feeble and weake, trailing vpon the ground, and occupying a great circuit in re-fpect of the plant. The floures grow vpon long and flender foot-ftalks, of colour yellow, in fhape like to the floures of broome; after which fucceed long, crooked, rough cods, in fhape and colour like vnto a Caterpiller ; wherein is contained yellowifh feed like vnto a kidney in fhape. The root is fmall and tender : the whole plant perifheth when the feed is ripe.

2   There is another Scorpion graffe, found among (or rather refembling) peafe, and thereupon called *Scorpioides Leguminofa*, which hath fmall and tender roots like fmall threds : branches many, weake and tender, trailing vpon the ground, if there be nothing to take hold vpon with his clafping and crooked feed veffels ; otherwife it rampeth vpon whatfoeuer is neere vnto it. The leaues be few and fmall : the floures very little and yellow of colour : the feed followeth, little and blackifh, con-teined in little cods, like vnto the taile Scorpion.

3   There is another fort almoft in euery fhallow grauelly running ftreame, hauing leaues like to *Becabunga* or Brooklime. The floures grow at the top of tender fat greene ftalkes, blew of colour, and fometimes with a fpot of yellow among the blew; the whole branch of floures do turne them-felues likewife round like the fcorpions taile.

There is alfo another growing in watery places, with leaues like vnto *Anagallis aquatica* or wa-ter Checkweed, hauing like flender ftalkes and branches as the former, and the floures not vnlike, fauing that the floures of this are of a light blew or watchet colour, fomewhat bigger, and layd more open, whereby the yellow fpot is feene.

4   There is likewife another fort growing vpon moft dry grauelly and barren ditch bankes, with leaues like thofe of Moufe-eare: this is called *Myofotes fcorpioides* ; it hath rough and hairy leaues, of an ouerworne ruffet colour : the floures doe grow vpon weake, feeble, and rough branches, as is all the reft of the plant. They likewife grow for the moft part at one fide of the ftalke, blew of colour, with a like little fpot of yellow as the others, turning themfelues backe againe like the taile of a Scorpion.

There

1 *Scorpioides Bupleuri folio, Pena & L'Obelij.*
Scorpion graſſe, or Caterpillers.

‡ 2 *Scorpioides Matthioli.*
Matthiolus his Scorpion graſſe.

‡ 3 *Myoſotis ſcorpioides paluſtris.*
Water Scorpion graſſe.

‡ 4 *Myoſotis ſcorpioides aruenſis hirſuta.*
Mouſe-eare Scorpion graſſe.

There is another of the land called *Myoſotis Scorpioides repens*, like the former : but the floures are thicker thruſt together, and do not grow all vpon one ſide as the other, and part of the floures are blew, and part purple, confuſedly mixt together.

¶ *The Place.*

1, 2 Theſe Scorpion graſſes grow not wilde in England, notwithſtanding I haue receiued ſeed of the firſt from beyond the ſeas, and haue diſperſed them through England, which are eſteemed of gentlewomen for the beautie and ſtrangneſſe of the crooked cods reſembling Caterpillers. The others do grow in waters and ſtreames, as alſo on dry and barren bankes.

¶ *The Time.*

The firſt floureth from May to the end of Auguſt: the others I haue found all the Summer long.

¶ *The Names.*

‡ 1 *Fabius Columna* iudges this to be the *Clymenon* of *Dioſcorides* : others call it *Scorpioides*, ınd *Scorpioides Bupleuri folio.*

2 This is the *Scorpioides* of *Matthiolus, Dod. Lobel*, and others ; and I iudge it was this plant our Author in this place intended, and not the *Scorpioides Leguminoſa* of the *Aduerſaria*, for that hath not a few leaues, but many vpon one rib ; and beſides, *Dodonæus*, whom in deſcriptions and hiſtory our Author chiefely followes, deſcribes this immediately after the other : *Guillandinus, Caſalpinus*, and *Bauhine* iudge it to be the *Telephium* of *Dioſcorides.*

3 This and the next want no names, for almoſt euery writer hath giuen them ſeuerall ones : *Brunfelſius* called it *Cynogloſſa minor* : *Tragus, Tabernamontanus*, and our Author (page 537. of the former edition) haue it vnder the name of *Euphraſia Carulea* : *Dodonæus* cals it *Scorpioides fæmina* : *Loniccrus, Leontopodium ; Caſalpinus, Heliotropium minus in paluſtribus* : *Cordus* and *Thalius, Echium paluſtre.*

4 This is *Auricula muris minor tertia, Euphraſia quarta*, and *Piloſella ſylueſtris* of *Tragus* : *Scorpioides mas* of *Dodonæus ; Alſine Myoſotis* : and *Myoſotis hirſuta repens* of *Lobel ; Heliotropium minus alterum* of *Caſalpinus ; Echium minimum* of *Columna ; and Echium paluſtre alterum* of *Thalius* : our Author had it thrice : firſt in the precedent chapter, by the name of *Heliotropium rectum*, with a figure : ſecondly in this preſent chapter, without a figure : and thirdly, *pag.* 514. alſo with a figure vnder the name of *Piloſella flore cæruleo.* ‡

¶ *The Nature and Vertues.*

A   There is not any thing remembred of the temperature : yet *Dioſcorides* ſaith, that the leaues of Scorpion graſſe applyed to the place, are a preſent remedy againſt the ſtinging of Scorpions : and likewiſe boyled in wine and drunke, preuaile againſt the ſaid bitings, as alſo of addars, ſnakes, and ſuch venomous beaſts : being made in aa vnguent with oile, wax, and a little gum *Elemni*, they are profitable againſt ſuch hurts as require an healing medicine.

CHAP. 55.   *Of* Nightſhade.

¶ *The Kindes.*

THere be diuers Nightſhades, whereof ſome are of the garden ; and ſome that loue the fields, and yet euery of them found wilde ; whereof ſome cauſe ſleepineſſe euen vnto death : others cauſe ſleepineſſe, and yet Phyſicall : and others very profitable vnto the health of man, as ſhall be declared in their ſeuerall vertues.

¶ *The Deſcription.*

1   GArden Nightſhade hath round ſtalkes a foot high, and full of branches, whereon are ſet leaues of a blackiſh colour, ſoft and full of juice, in ſhape like to the leaues of Baſill, but much greater : among which do grow ſmall white floures with yellow pointals in the middle; which being paſt, there ſucceed round berries, greene at the firſt, and black when they be ripe, like thoſe of Iuy : the root is white and full of hairy ſtrings.

‡ 2   The root of this is long, pretty thicke and hard, being couered with a browniſh skin; from this root grow vp many ſmall ſtalkes of the height of a cubit and better, ſomewhat thicke withall: the leaues that grow alongſt the ſtalkes are like thoſe of the Quince-tree, thicke, white, ſoft and downie. The floures grow about the ſtalke at the ſetting on of the leafe, ſomewhat long, and of a pale colour, diuided into foure parts, which are ſucceeded by ſeeds contained in hairy or woolly receptacles : which when they come to ripeneſſe are red, or of a reddiſh ſaffron colour. ‡

¶ *The Place,*

This Nightſhade commeth vp in many places, and not onely in gardens, of which notwithſtanding

ding it hath taken his ſyrname, and in which it is often found growing among other herbs; but alſo neere common highwaies, the borders of fields, by old walls and ruinous places.

‡ 2 This growes not with vs, but in hotter countries. *Cluſius* found it growing among rubbiſh at Malago in Spaine. ‡

1 *Solanum hortenſe.*
Garden Nightſhade.

‡ 2 *Solanum ſomniferum.*
Sleepie Nightſhade.

¶ *The Time.*

It floureth in Summer, and oftentimes till Autumne be wel ſpent; and then the fruit commeth to ripeneſſe.

‡ 2 This *Cluſius* found in floure and with the ſeed ripe in Februarie: for it liues many years in hot countries, but in cold it is but an annual. ‡

¶ *The Names.*

It is called of the Grecians στρύχνη: of the Latines, *Solanum*, and *Solanum hortenſe*: in ſhops, *Solarum* of ſome, *Morella*, *Vua Lupina*, and *Vua Vulpis*: in Spaniſh likewiſe *Morella*, and *Yerua Mora*: *Marcellus* an old phyſick writer, and diuers others of his time called it *Strumum*: *Pliny, lib. 27. ca. 8.* ſheweth that it is called *Cucubalus*: both theſe words are likewiſe extant in *Apuleius*, amongſt the confuſed names of Nightſhade; who comprehending all the kinds of Nightſhade together in one chapter, being ſo many, hath ſtrangely & abſurdly confounded their names. In Engliſh it is called garden Nightſhade, Morel, and petty Morel: in French, *Morelle, Gallobelgis: feu ardent, quia medetur igni ſacro.*

¶ *The Nature and Vertues.*

Nightſhade (as *Galen* ſaith, *Lib. de Facult. Simp.*) is vſed for thoſe infirmities that haue need of cooling and binding; for theſe two qualities it hath in the ſecond degree: which thing alſo he affirmeth in his booke of the faculties of nouriſhments, where he ſaith, that there is no pot-herb wee vſe to eat, that hath ſo great aſtriction or binding as Nightſhade hath; and therefore Phyſitians do worthily vſe it, and that ſeldome as a nouriſhment, but alwaies as a medicine.

1 *Dioſcorides* writeth, that Nightſhade is good againſt S. Anthonies fire, the ſhingles, paine of the head, the heart-burning or heate of the ſtomacke, and other like accidents proceeding of ſharp and biting humors. But although it hath theſe vertues, yet it is not alwaies good that it ſhould be applied vnto thoſe infirmities, for that many times there hapneth more dangers by applying theſe remedies, than by the diſeaſe it ſelf: for as *Hippocrates* writes, *Lib. 6. de Aphoriſm.* the 25 Particular,

It

It is not good that S. Anthonies fire fhould be driuen from the outward parts to the inward. And likewife in his Prognofticks he faith, It is neceffarie that S. Anthonies fire fhould break forth, and that it is death to haue it driuen in : which is to be vnderftood not only of S. *Anthonies* fire, but alfo of other like burftings out procured by nature. For by vfing thefe kindes of cooling and repelling medicines, the bad, corrupt, and fharpe humors are driuen backe inwardly to the chiefe and princi-pall parts, which cannot be done without great danger and hazard of life. And therefore wee muft not vnaduifedly, lightly, or rafhly adminifter fuch kinde of medicines, vpon the comming out of S. *Anthonies* fire, the fhingles, or fuch hot inflammations.

B   The iuice of the green leaues of garden Nightfhade mixed with Barly meale, is very profitably applied vnto S. *Anthonies* fire, and to all hot inflammations.

C   The iuice mixed with oile of rofes, Cerufe, and Litharge of gold, & applied, is more proper and effectuall to the purpofes afore fet downe.

D   † Neither the iuice hereof nor any other part is vfually giuen inwardly, yet it may without any danger.

E   The leaues ftamped are profitably put into the ointments of Poplar buds called *Vnguentum Po-puleon*, and it is good in all other ointments made for the fame purpofe.

F   ‡ 2   The barke of the root of fleepy Nightfhade taken in the weight of ʒ 1. hath a fomnife-rous qualitie, yet is it milder than *Opium*, and the fruit thereof vehemently prouokes vrine. But (as *Pliny* faith) the remedies hereof are not of fuch efteem that we fhould long infift vpon them, efpe-cially feeing we are furnifhed with fuch ftore of medicines leffe harmfull, yet feruing for the fame purpofe. ‡

† The figure in the fecond place was of the *Solanum Pomiferum*, or *Mala Æthiopica*, treated of at large in the 61 Chap. of this booke, and therefore it is omit-ted here, and in ftead thereof another put in the place.

Chap. 56.   Of fleepy Nightfhade.

Solanum Lœthale.
Dwale, or deadly Nightfhade.

¶ *The Defcription.*

DWale or fleeping Nightfhade hath round blackifh ftalkes fix foot high, whereupon do grow great broad leaues of a dark green colour : among which grow fmal hollow floures bel-fafhion, of an ouerworn purple colour; in the place whereof come forth great round berries of the bigneffe of the black chery, green at the firft, but when they be ripe of the colour of black jet or burnifhed horne, foft, and ful of purple iuice ; among which iuice lie the feeds, like the berries of Ivy : the root is very great, thick, and long la-fting.

¶ *The Place.*

It growes in vntoiled places neere highwaies and the fea marifhes, and in fuch like places.

It groweth very plentifully in Holland in Lincolnefhire, and in the Ifle of Ely at a place called Walfoken, neere vnto Wisbitch.

I found it growing without Highgate, neere vnto a pound or pinfold on the left hand.

¶ *The Time.*

This flourifheth all the Spring and Sum-mer, bearing his feed and floure in Iuly and Auguft.

¶ *The Names.*

It is called of *Diofcorides*, ʃτρύχνος ὑπνωτικὸς : of *Theophraftus*, ʃτρύχνος ὑπνώδης : of the Latines, *Solanum fomniferum*, *somniferum*, or sleeping Nightshade ; and *Solanum lethale*, or deadly Nightshade ; and *Solanum manicum*, raging Nightshade : of some, *Apollinaris minor vlticana*, and *Herba Opsago :* in English, Dwale, or sleeping Nightshade : the Venetians and Italians call it *Bella dona :* the Germanes, **Dollwurtz :** the low Dutch, **Dulle besien :** in French, *Morelle mortelle :* it commeth very neere vnto *Theophrastus* his *Mandragoras*, (which differeth from *Dioscorides* his *Mandragoras*.)

¶ The Nature.

It is cold euen in the fourth degree.

¶ The Vertues.

This kinde of Nightshade causeth sleep, troubleth the mind, bringeth madnesse if a few of the berries be inwardly taken, but if moe be giuen they also kill and bring present death. *Theophrastus* in his sixth booke doth likewise write of Mandrake in this manner ; Mandrake causeth sleepe, and if also much of it be taken it bringeth death.

The greene leaues of deadly Nightshade may with great aduice be vsed in such cases as Pettimorell : but if you will follow my counsell, deale not with the same in any case, and banish it from your gardens and the vse of it also, being a plant so furious and deadly : for it bringeth such as haue eaten thereof into a dead sleepe wherein many haue died, as hath beene often seene and proued by experience both in England and elsewhere. But to giue you an example hereof it shall not be amisse : It came to passe that three boies of Wisbich in the Isle of Ely did eate of the pleasant and beautifull fruit hereof, two whereof died in lesse than eight houres after that they had eaten of them. The third child had a quantitie of honey and water mixed together giuen him to drinke, causing him to vomit often : God blessed this meanes and the child recouered. Banish therefore these pernitious plants out of your gardens, and all places neere to your houses, where children or women with child do resort, which do oftentimes long and lust after things most vile and filthie ; and much more after a berry of a bright shining blacke colour, and of such great beautie, as it were able to allure any such to eate thereof.

The leaues hereof laid vnto the temples cause sleepe, especially if they be imbibed or moistened in wine vinegre. It easeth the intollerable paines of the head-ache proceeding of heat in furious agues, causing rest being applied as aforesaid.

Chap. 57.　*Of Winter Cherries.*

¶ The Description.

1　THe red Winter Cherrie bringeth forth stalkes a cubit long, round, slender, smooth and somewhat reddish, reeling this way and that way by reason of his weakenesse, not able to stand vpright without a supporter : whereupon do grow leaues not vnlike to those of common Nightshade, but greater ; among which leaues come forth white floures, consisting of fiue small leaues : in the middle of which leaues standeth out a berry, green at the first, and red when it is ripe, in colour of our common Cherry and of the same bignesse, inclosed in a thin huske or little bladder, it is of a pale reddish colour, in which berry is conteined many small flat seeds of a pale colour. The roots be long, not vnlike to the roots of Couch-grasse, ramping and creeping within the vpper crust of the earth farre abroad, whereby it encreaseth greatly.

2　The blacke Winter Cherry hath weake and slender stalkes somewhat crested, and like vnto the tendrels of the vine, casting it selfe all about, and taking hold of such things as are next vnto it : whereupon are set jagged leaues deepely indented or cut about the edges almost to the middle ribbe. The floures be very small and white standing vpon long footstalkes or stemmes. The skinnie bladders succeed the floures, parted into three sells or chambers, euery of the which conteineth one seed and no more, of the bignesse of a small pease, and blacke of colour, hauing a mark of white colour vpon each berry, in proportion of an heart. The root is very small and thready.

¶ The Place.

1　The red Winter Cherry groweth vpon old broken walls, about the borders of fields, and in moist shadowie places, and in most gardens, where some cherrish it for the beautie of the berries, and others for the great and worthy vertues thereof.

2　The blacke Winter Cherrie is brought out of Spaine and Italy, or other hot regions, from whence I haue had of those blacke seeds marked with the shape of a mans heart, white, as aforesaid : and haue planted them in my garden where they haue borne floures, but haue perished before the fruit could grow to maturitie, by reason of those vnseasonable yeares, 1594. 95. 96.

　　　　　　　　　　　　　　　　　　　　　　　¶ The

¶ *The Time.*

The red winter Cherrie beareth his floures and fruit in August.
The blacke beareth them at the same time, where it doth naturally grow.

¶ *The Names.*

The red winter Cherrie is called in Greeke, Στρύχνε : in Latine, *Vesicaria*, and *Solanum Vesicarium:* in shops, *Alkekengi : Plinie* in his 21.booke nameth it *Halicacabus,* and *Vesicaria,* of the little bladders : or as the same Authour writeth, because it is good for the bladder and the stone : it is called in Spanish, *Vexiga de Porro :* in French, *Alquequenges, Bagenauldes,* and *Cerises d'outre mer :* in English, red Nightshade, Winter Cherries, and Alkakengie.

   1 *Solanum Halicacabum.*   2 *Halicacabum Peregrinum.*
   Red Winter Cherries.   Blacke Winter Cherries.

The Blacke Winter Cherrie is called *Halicacabum Peregrinum,Vesicaria Peregrina,*or strange winter Cherrie:of *Pena* and *Lobel* it is called,*Cor Indum,Cor Indicum :* of others,*Pisum Cordatum :* in nglish,the Indian heart,or heart pease : some haue taken it to be *Dorycnion,*but they are greatly deceiued, being in truth not any of the Nightshades ; it rather seemeth to agree with the graine named of *Serapio, Abrong,* or *Abrugi,* of which he writeth in his 153. chapter in these words : It is a little graine spotted with blacke and white,round, and like the graine Maiz, with which notes this doth agree.

¶ *The Temperature.*

The red Winter Cherrie is thought to be cold and dry,and of subtile parts.
The leaues differ not from the temperature of the garden Night shade, as *Galen* saith.

¶ *The Vertues.*

A   The fruit brused and put to infuse or steepe in white wine two or three houres, and after boiled two or three bublings, straining it, and putting to the decoction a little sugar and cinnamon, and drunke,prauaileth very mightily against the stopping of vrine, the stone and grauell, the difficultie and sharpenesse of making water,and such like diseases:if the griefe be old,the greater quantity must be taken ; if new and not great, the lesse : it scoureth away the yellow jaundise also,as some write.

CHAP. 58.   Of the Maruell of the World.

Mirabilia Peruviana flore luteo.
The Maruell of Peru with yellowish floures.

‡ *Mirabilia Peruviana flore albo.*
The Maruell of Peru with white floures.

¶ *The Description.*

THis admirable Plant, called the Maruell of Peru, or the Maruell of the World, springs forth of the ground like vnto Basil in leaues ; among which it sendeth out a stalke two cubits and a halfe high, of the thicknesse of a finger, full of juice, very firme, and of a yellowish green colour, knotted or kneed with joints somewhat bunching forth, of purplish colour, as in the female Balsamina : which stalke diuideth it selfe into sundry branches or boughes, and those also knottie like the stalke. His branches are decked with leaues growing by couples at the joints like the leaues of wilde Peascods, greene, fleshy, and full of joints ; which being rubbed doe yeeld the like vnpleasant smell as wilde Peascods do, and are in taste also very vnsauory, yet in the later end they leaue a tast and sharp smack of Tabaco. The stalks toward the top are garnished with long hollow single floures, folded as it were into fiue parts before they be opened ; but being fully blown, do resemble the floures of Tabaco, not ending in sharp corners, but blunt & round as the flours of Bindweed, and larger than the floures of Tabaco, glittering oft times with a fine purple or crimson colour, many times of an horse-flesh, sometimes yellow, sometimes pale, and somtime resembling an old red or yellow colour ; sometime whitish, and most commonly two colours occupying halfe the floure, or intercoursing the whole floure with streaks or orderly streames, now yellow, now purple, diuided through the whole, hauing sometime great, somtime little spots of a purple colour, sprinkled and scattered in a most variable order and braue mixture. The ground or field of the whole floure is either pale, red, yellow, or white, containing in the middle of the hollownesse a pricke or pointal set round about with six small strings or chiues. The floures are very sweet and pleasant, resembling the Narcisse or white Daffodill, and are very suddenly fading ; for at night they are floured wide open, and so continue vntill eight of the clocke the next morning, at which time they begin to close (after the maner of Bindweed, especially if the weather be very hot : but the aire being temperat, they remain open the whole day, and are closed only at night, and so perish, one floure lasting

sting

ſting but onely one day, like the true Ephemerum or Hemerocallis. This maruellous variety doth not without caufe bring admiration to all that obferue it. For if the floures be gathered and referued in feuerall papers, and compared with thofe floures that will ſpring and flouriſh the next day, you ſhall eaſily perceiue that one is not like another in colour, though you ſhall compare one hundred which floure one day, and another hundred which you gather the next day, and fo from day to day during the time of their flouring. The cups and huskes which containe and embrace the floures are diuided into fiue pointed fections, which are green, and as it were, confifting of skinnes, wherein is conteined one feed and no more, couered with a blackifh skinne, hauing a blunt point whereon the floure groweth ; but on the end next the cup or huske it is adorned with a little fiue cornered crowne. The feed is as big as a pepper corne, which of it felfe fadeth with any light motion. Within this feed is contained a white kernell, which being bruifed, refolueth into a very white pulpe like ſtarch. The root is thicke and like vnto a great raddifh, outwardly black, and within white, fharpe in tafte, wherewith is mingled a fuperficiall fweetneffe. It bringeth new floures from Iuly vnto October in infinite number, yea euen vntill the frofts doe caufe the whole plant to perifh: notwithſtanding it may be referued in pots, and fet in chambers and cellars that are warme, and fo defended from the injurie of our cold climate ; prouided alwaies that there be not any water caft vpon the pot, or fet forth to take any moifture in the aire vntill March following ; at which time it muft be taken forth of the pot and replanted in the garden. By this meanes I haue preferued many (though to fmall purpofe) becaufe I haue fowne feeds that haue borne floures in as ample manner and in as good time as thofe referued plants.

Of this wonderfull herbe there be other forts, but not fo amiable or fo full of varietie, and for the moft part their floures are all of one color. But I haue fince by practife found out another way to keepe the roots for the yere following with very little difficultie, which neuer faileth. At the firft froft I dig vp the roots and put vp or rather hide the roots in a butter ferkin, or fuch like veffell, filled with the fand of a riuer, the which I fuffer ſtill to ſtand in fome corner of an houfe where it neuer receiueth moifture vntill Aprill or the midft of March, at which time I take it from the fand and plant it in the garden, where it doth flourifh exceeding well and increafeth by roots ; which that doth not which was either fowne of feed the fame yeere, nor thofe plants that were preferued after the other manner.

¶ *The Place,*

The feed of this ftrange plant was brought firft into Spaine, from Peru, whereof it tooke his name *Mirabilia Peruana*, or *Peruviana* : and fince difperfed into all the parts of Europe: the which my felfe haue planted many yeeres, and haue in fome temperate yeeres receiued both floures and ripe feed.

¶ *The Time.*

It is fowne in the midft of Aprill, and bringeth forth his variable floures in September, and perifheth with the firft froft, except it be kept as aforefaid.

¶ *The Names.*

It is called in Peru of thofe Indians there, *Hachal.* Of others after their name *Hachal Indi* : of the high and low Dutch, *Solanum Odoriferum* : of fome. *Iafminum Mexicanum* : and of *Carolus Clufius*, *Admirabilia Peruviana* : in Englifh rather the Maruell of the World, than of Peru alone.

¶ *The Nature and Vertues.*

We haue not as yet any inftructions from the people of India, concerning the nature or vertues of this plant : the which is efteemed as yet rather for his rareneffe, beautie, and fweetneffe of his floures, than for any vertues knowne ; but it is a pleafant plant to decke the gardens of the curious. Howbeit *Iacobus Antonius Cortufus* of Padua hath by experience found out, that two drams of the root thereof taken inwardly doth very notably purge waterifh humours.

CHAP. 59.   *Of Madde Apples.*

¶ *The Defcription.*

RAging Apples hath a round ftalke of two foot high, diuided into fundry branches, fet with broad leaues fomewhat indented about the edges, not vnlike the leaues of white Henbane, of a darke browne greene colour, fomewhat rough. Among the which come the floures
of

of a white colour,and somtimes changing into purple,made of six parts wide open like a star,with certain yellow chiues or thrums in the middle : which being past, the fruit comes in place, set in a cornered cup or huske after the manner of great Nightshade, great and somewhat long,of the bignesse of a Swans egge,and sometimes much greater,of a white colour,somtimes yellow,and often brown,wherein is contained small flat seed of a yellow colour.The root is thick,with many threds fastned thereto.

Mala Insana.
Mad or raging Apples.

¶ *The Place.*

This Plant growes in Egypt almost euery where in sandy fields euen of it selfe, bringing forth fruit of the bignesse of a great Cucumber,as *Petrus Bellonius* writeth, *lib.* 2. of his singular obseruations. We had the same in our London gardens,where it hath borne floures ; but Winter approching before the time of ripening,it perished:neuerthelesse it came to beare fruit of the bignes of a goose egg one extraordinarie temperate yeare, as I did see in the garden of a worshipfull merchant M.r *Haruy* in Limestreet ; but neuer to the full ripenesse.

¶ *The Time.*
This herb must be sowne in Aprill in a bed of horse-doung, as Muske-melons are,and floureth in August.

¶ *The Names.*
Petrus Bellonius hath iudged it to be *Malinathalla Theophrasti.* In the dukedome of Millain it is called *Melongena* ; and of some, *Melanzana :* in Latine, *Mala insana :* and in English, Mad Apples. In the Germane tongue,**Dollopffell** : in Spanish, *Verangenes.*

¶ *The Nature.*
The herb is cold almost in the fourth degree.

¶ *The Vse,and Danger.*

The people of Toledo eat them with great deuotion, being boiled with fat flesh, putting to it some scraped cheese,which they do keep in vineger,hony,or salt pickle all winter,to procure lust.
Petrus Bellonius and *Hermolaus Barbarus* report, That in Egypt & Barbary they vse to eat the fruit of *Mala insana* boiled or rosted vnder ashes,with oile,vineger,and pepper,as people vse to eat Mushroms. But I rather wish English men to content themselues with the meat and sauce of our owne country,than with fruit and sauce eaten with such perill ; for doubtlesse these Apples haue a mischieuous qualitie,the vse whereof is vtterly **to** bee forsaken. As wee see and know many haue eaten and do eat Mushroms more for wantonnesse than for need;for there are two kinds therof deadly,which being dressed by an vnskilfull cooke may procure vntimely death : it is therefore better to esteem this plant and haue it in the garden for your pleasure and the rarenesse thereof, than for any vertue or good qualities yet knowne.

Chap.60. *Of Apples of Loue.*

¶ *The Description.*

THe Apple of Loue bringeth forth very long round stalkes or branches, fat and full of juice, trailing vpon the ground,not able to sustain himselfe vpright by reason of the tendernesse of the stalkes,and also the great weight of the leaues and fruit wherewith it is surcharged. The leaues are great,and deeply cut or iagged about the edges, not vnlike to the leaues of Agrimonie, but greater,and of a whiter greene colour : Amongst which come forth yellow floures growing vpon ſhort ſtems or footſtalkes, cluſtering together in bunches : which being fallen there doe come in place faire and goodly apples, chamfered, vneuen, and bunched out in many places ; of a bright ſhining red colour, and the bigneſſe of a gooſe egge or a large pippin. The pulpe or meat is very full of moiſture, ſoft, reddiſh, and of the ſubſtance of a wheat plumme. The ſeed is ſmall, flat and rough : the root ſmall and thready : the whole plant is of a ranke and ſtinking ſauour.

There hath happened vnto my hands another ſort, agreeing very notably with the former, as well in leaues and ſtalkes as alſo in floures and roots, onely the fruit hereof was yellow of colour, wherein conſiſteth the difference.

Poma amoris.
Apples of loue.

¶ *The Place.*

Apples of Loue grow in Spaine, Italie, and ſuch hot Countries, from whence my ſelfe haue receiued ſeeds for my garden, where they doe increaſe and proſper.

¶ *The Time.*

It is ſowne in the beginning of Aprill in a bed of hot horſe-dung, after the maner of muske Melons and ſuch like cold fruits.

¶ *The Names.*

The Apple of Loue is called in Latine *Pomum Aureum, Poma Amoris,* and *Lycoperſicum :* of ſome, *Glaucium :* in Engliſh, Apples of Loue, and Golden Apples: in French, *Pommes d'amours.* Howbeit there be other golden Apples whereof the Poëts doe fable, growing in the Gardens of the daughters of *Heſperus,* which a Dragon was appointed to keepe, who, as they fable, was killed by *Hercules.*

¶ *The Temperature.*

The Golden Apple, with the whole herbe it ſelfe is cold, yet not fully ſo cold as Mandrake, after the opinion of *Dodonæus.* But in my iudgement it is very cold, yea perhaps in the higheſt degree of coldneſſe : my reaſon is, becauſe I haue in the hotteſt time of Summer cut away the ſuperfluous branches from the mother root, and caſt them away careleſly in the allies of my Garden, the which (notwithſtanding the extreme heate of the Sun, the hardneſſe of the trodden allies, and at that time when no rain at all did fal) haue growne

as freſh where I caſt them, as before I did cut them off ; which argueth the great coldneſſe contained therein. True it is, that it doth argue alſo a great moiſture wherewith the plant is poſſeſſed, but as I haue ſaid, not without great cold, which I leaue to euery mans cenſure.

¶ *The Vertues.*

A   In Spaine and thoſe hot Regions they vſe to eate the Apples prepared and boiled with pepper, ſalt, and oyle : but they yeeld very little nouriſhment to the body, and the ſame naught and corrupt.

B   Likewiſe they doe eate the Apples with oile, vinegre and pepper mixed together for ſauce to their meat, euen as we in theſe cold countries doe Muſtard.

Cʜᴀᴘ. 61. *Of the Æthiopian Apple.*

¶ *The Deſcription.*

THe Apple of Æthiopia hath large leaues of a whitiſh greene colour, deepely indented about the edges, almoſt to the middle rib; the which middle rib is armed with a few ſharpe prickles. The floures be white, conſiſting of ſix ſmall leaues, with a certain yellow pointel in the midſt.

The

Mala Æthiopica.
Apples of Æthiopia.

The fruit is round, and bunched with vneuen lobes or bankes lesser than the golden Apple, of colour red, and of a firme and sollid substance; wherein are contained small flat seeds. The root is small and threddy.

¶ *The Place.*

The seeds of this plant haue beene brought vnto vs out of Spaine, and also sent into France and Flaunders: but to what perfection it hath come vnto in those parts I am ignorant; but mine perished at the first approach of Winter. His first originall was from Æthiopia, wherof it tooke his name.

¶ *The Time.*

This plant must be sowne as Muske-melons, and at the same time. They floure in Iuly, and the fruit is ripe in September.

¶ *The Names.*

In English we haue thought good to call it the Æthiopian Apple, for the reason before alledged: in Latine, *Mala Æthiopica:* of some it hath been thought to be *Malinathalla.* ‡ This is the *Solanum Pomiferum* of *Lobel* and others; by which name our Author also formerly had it; in the fiftieth chapter of the former edition. ‡

¶ *The Nature.*

The temperature agreeth with the Apple of Loue.

¶ *The Vertues.*

These Apples are not vsed in Physicke that I can reade of; onely they are vsed for a sauce and seruice vnto rich mens tables to be eaten, being first boiled in the broth of fat flesh with pepper and salt, and haue a lesse hurtfull juyce than either mad Apples or golden Apples.

CHAP. 47. *Of Thornie Apples.*

¶ *The Description.*

1   THe stalkes of Thorny-apples are oftentimes aboue a cubit and a halfe high, seldome higher, an inch thicke, vpright and straight, hauing very few branches, sometimes none at all, but one vpright stemme; whereupon doe grow leaues smooth and euen, little or nothing indented about the edges, longer and broader than the leaues of Nightshade, or of the mad Apples. The floures come forth of long toothed cups, great, white, of the forme of a bell, or like the floures of the great Withwinde that rampeth in hedges; but altogether greater and wider in the mouth, sharpe cornered at the brimmes, with certaine white chiues or threds in the middest, of a strong ponticke sauour, offending the head when it is smelled vnto: in the place of the floure commeth vp round fruit full of short and blunt prickles of the bignesse of a green Wall-nut when it is at the biggest, in which are the seeds of the bignesse of tares or of the seed of Mandrakes, and of the same forme. The herbe it selfe is of a strong sauor, and doth stuffe the head, and causeth drowsinesse. The root is small and threddy.

2   There is another kinde hereof altogether greater than the former, whose seeds I receiued of the right honourable the Lord *Edward Zouch*; which he brought from Constantinople, and of his liberalitie did bestow them vpon me, as also many other rare and strange seeds; and it is that Thorn-apple that I haue dispersed through this land, whereof at this present I haue great vse in Surgery; as well in burnings and scaldings, as also in virulent and maligne vlcers, apostumes, and such like. The which plant hath a very great stalke in fertile ground, bigger then a mans arme, smooth and greene of colour, which a little aboue the ground diuideth it selfe into sundry branches or armes in manner of an hedge tree; whereupon are placed many great leaues cut and indented deepely about

about the edges, with many vneuen ſharpe corners : among theſe leaues come white round floures made of one piece in manner of a bell, ſhutting it ſelfe vp cloſe toward night, as doe the floures of the great Binde-weed, whereunto it is very like, of a ſweet ſmell, but ſo ſtrong, that it offends the ſences. The fruit followeth round, ſometimes of the faſhion of an egge, ſet about on euery part with moſt ſharpe prickles ; wherein is contained very much ſeed of the bigneſſe of tares, and of the ſame faſhion. The root is thicke, made of great and ſmall ſtrings : this plant is ſowen, beareth his fruit, and periſheth the ſame yeare. ‡ There are ſome varieties of this plant, in the colour and dou- bleneſſe of the floures. ‡

1 *Stramoninm Peregrinum.*
The Apple of Peru.

2 *Stramonium ſpinoſum.*
Thorny apples of Peru.

¶ *The Plate.*

1   This plant is rare and ſtrange as yet in England : I receiued ſeeds thereof from *Iohn Robin* of Paris, an excellent Herbariſt; which do grow and bare floures, but periſhed before the fruit came to ripeneſſe.

2   The Thorne-apple was brought in ſeed from Conſtantinople by the right honourable the Lord *Edward Zouch*, and giuen vnto me, and beareth fruit and ripe ſeed.

¶ *The Time.*

The firſt is to be ſowne in a bed of dorſe-dung, as we do Cucumbers and Muske-melons.
The other may be ſowne in March and Aprill, as other ſeeds are.

¶ *The Names.*

The firſt of theſe Thorne-apples may be called in Latine, *Stramonia*, and *Pomum*, or *Malum ſpi- noſum* : of ſome, *Corona regia*, and *Meloſpinum*: The Grecians of our time name it μαρχιαχαν, or rather cηνοιχαλον; as though they ſhould ſay, a nut ſtuffing, and cauſing drowſineſſe and diſquiet ſleepe: the Italians, *Paracoculi* : it ſeemeth to *Valerius Cordus* to be *Hyoſcyamus Peruvianus*, or Henbane of Peru: *Cardanus* doubteth whether it ſhould be inſerted among the Night-ſhades as a kinde there- of : of *Matthiolus* and others it is thought to be *Nux methel* : *Serapio, cap.* 375. ſaith, That *Nux methel* is like vnto *Nux vomica*; the ſeed whereof is like that of Mandrake : the huske is rough or full of prickles ; the taſte pleaſing and ſtrong : the qualitie thereof is cold in the fourth degree. Which deſcription agreeth therewith, except in the forme or ſhape it ſhould haue with *Nux vomica* : *An- guillara* ſuſpected it to be *Hippomanes* which *Theocritus* mentioned, wherewith in his ſecond *Eclog*
he

he sheweth that horses are made mad : for *Cratevas*, whom *Theocritus* his Scholiast doth cite, writeth, That the plant of *Hippomanes* hath a fruit full of prickles, as hath the fruit of wild Cucumber. In English it may be called Thorn-Apple, or the Apple of Peru.

‡   The words of *Theocritus*, *Eidyl* 2. are these :

Ιππομανὲς φυτὸν ἔστι παρ' Ἀρκάσιν, &c.

Which is thus in English :

Hippomanes 'mongst th' Arcadians springs, by which ev'n all
The Colts and agile Mares in mountaines mad do fall.

Now in the Greeke *Scholia* amongst the Expositions there is this, Κρατεύας φησὶ, &c.   That is, *Cratevas* saith, that the plant hath a fruit like the wild Cucumber, but blacker ; the leaues are like a poppy, but thorny or prickly. Thus I expound these words of the Greeke Scholiast, being *pag.* 51. of the edition set forth by *Dan. Heinsius*, *An. Dom.* 1603. *Iulius Scaliger* blames *Theocritus*, because he cals *Hippomanes*, φυτὸν, a Plant : but *Heinsius*, as you may see in his notes vpon *Theocritus*, *pag.* 120. probably iudges, that φυτὸν in this place signifies nothing but χρῆμα, a thing [*growing.*] Such as are curious may haue recourse to the places quoted, where they may finde it more largely handled than is fit for me in this place to insist vpon. There is no plant at this day known, in mine opinion, whereunto *Cratevas* his description may more fitly be referred, than to the *Papauer spinosum*, or *Ficus infernalis*, which we shall hereafter describe.   ‡

¶   *The Temperature.*
The whole plant is cold in the fourth degree, and of a drowsie and numming qualitie, not inferior to Mandrake.

¶   *The Vertues.*

The juice of Thorn-apples boiled with hogs grease to the form of an vnguent or salue, cures all   A
inflammations whatsoeuer, all manner of burnings or scaldings, as well of fire, water, boiling lead, gun-pouder, as that which comes by lightning, and that in very short time, as my selfe haue found by my daily practise, to my great credit and profit. The first experience came from Colchester, where Mistresse *Lobel* a merchants wife there being most grieuously burned by lightning, and not finding ease or cure in any other thing, by this found helpe and was perfectly cured when all hope was past, by the report of Mr *William Ram* publique Notarie of the said towne.

The leaues stamped small and boiled with oile Oliue vntill the herbs be as it were burnt, then   B
strained and set to the fire again, with some wax, rosin, and a little turpentine, and made into a salue, doth most speedily cure old vlcers, new and fresh wounds, vlcers vpon the glandulous part of the yard, and other sores of hard curation.

Chap. 63.
Of Bitter-sweet, or wooddy Nightshade.

¶   *The Description.*

Bitter-sweet bringeth forth wooddy stalks as doth the Vine, parted into many slender creeping branches, by which it climeth and taketh hold of hedges and shrubs next vnto it. The barke of the oldest stalks are rough and whitish, of the colour of ashes, with the outward rind of a bright green colour, but the yonger branches are green as are the leaues : the wood brittle, hauing in it a spongie pith : it is clad with long leaues, smooth, sharp pointed, lesser than those of the Bindweed. At the lower part of the same leaues doth grow on either side one smal or lesser leafe like vnto two eares. The floures be small, and somewhat clustered together, consisting of fiue little leaues apiece of a perfect blew colour, with a certain pricke or yellow pointal in the middle : which being past, there do come in place faire berries more long than round, at the first green, but very red when they be ripe ; of a sweet taste at the first, but after very vnpleasant, of a strong sauor, growing together in clusters like burnished coral. The root is of a mean bignesse, and full of strings.

I haue found another sort which bringeth forth most pleasant white floures, with yellow pointals in the middle : in other respects agreeing with the former.

¶   *The Place.*
Bitter-sweet growes in moist places about ditches, riuers, and hedges, almost euerie where.
The

Amara-dulcis.
Bitter-sweet.

The other sort with the white floures I found in a ditch side, against the right honourable the Earle of Suffex his garden wall, at his house in Bermondsey street by London, as you go from the court which is full of trees, vnto a ferm house neere thereunto.

¶ *The Time.*

The leaues come forth in the spring, the flours in Iuly, the berries are ripe in August.

¶ *The Names.*

The later Herbarists haue named this plant *Dul-camara, Amarodulcis,* & *Amaradulcis*; that is in Greek, λυκαδμησν: they call it also *Solanum lignosum* and *Siliquastrum* : *Pliny* calleth it *Melortum* : *Theophrastus, Vitis sylvestris* : in English we call it Bitter-sweet, and wooddy Nightshade. But euery Author must for his credit say something, although but to smal purpose; for *Vitis sylvestris* is that which wee call our Ladies Seale, which is no kinde of Nightshade: for *Tamus* and *Vitis sylvestris* are both one; as likewise *Solanum lignosum* or *fruticosum*, and also *Solanum rubrum* : whereas indeed it is no such plant, nor any of the Nightshades, although I haue followed others in placing it here. Therefore those that vse to mix the berries of it in compositions of diuers cooling ointments, in sted of the berries of Nightshade, haue committed the greater errour; for the fruit of this is not cold at all, but hot, as forthwith shal be shewed. *Dioscorides* saith it is *Cyclaminus altera*, describing it by the description of those with white floures aforesaid, whereunto it doth very well agree.

‡ *Dioscorides* describes his *Muscoso flore* with a mossy floure, that is, such an one as consists of small chiues or threads, which can by no meanes be agreeable to the floure of this plant. ‡

¶ *The Temperature.*

The leaues and fruit of Bitter-sweet are in temperature hot and dry, clensing and wasting away.

¶ *The Vertues.*

A   The decoction of the leaues is reported to remoue the stoppings of the liuer and gall, and to be drunke with good successe against the yellow jaundise.

B   The juice is good for those that haue fallen from high places, and haue been thereby bruised, or dry-beaten: for it is thought to dissolue bloud congealed or cluttered any where in the intrals, and to heale the hurt places.

C   *Tragus* teacheth to make a decoction of wine, with the wood finely sliced and cut into smal pieces : which he reporteth to purge gently both by vrine and siege, those that haue the Dropsie or jaundice.

D   *Dioscorides* ascribeth vnto *Cyclaminus altera*, or Bitter-sweet with white floures, as I suppose, the like faculties.

E   The fruit (saith he) being drunke in the weight of one dram, with three ounces of white wine for forty daies together, helpeth the spleen.

F   It is drunk against difficultie of breathing : it throughly clenseth women newly brought a bed.

Chap. 64.   *Of Bindeweed Nightshade.*

¶ *The Description.*

INchanters Night-shade hath leaues like to petty Morell, sharp at the point like vnto Spinage : the stalke is streight and vpright, very brittle, two foot high: The floures are white tending to Carnation, with certaine small browne chiues in the midst : the seed is contained in smal round
bullets

Circæa Lutetiana.
Inchanters Night-shade.

bullets, rough and very hairy. The roots are tough, and many in number, thrusting themselues deep into the ground, and dispersing far abroad; whereby it doth greatly encrease, insomuch that when it hath once taken fast rooting, it can hardly with great labour be rooted out or destroied.

¶ *The Place.*

It groweth in obscure and darke places, about dung-hills, and in vntoiled grounds, by path-waies and such like.

¶ *The Time.*

It flourisheth from Iune to the end of September.

¶ *The Names.*

It is called of *Lobel*, *Circæa Lutetiana*: in English, Inchanters Night-shade, or Binde-weed Nightshade.

¶ *The Nature and Vertues.*

There is no vse of this herbe either in Physicke or Surgerie, that I can reade of; which hath happened by the corruption of time, and the errour of some who haue taken *Mandragoras* for *Circæa*; in which errour they haue still persisted vnto this day, attributing vnto *Circæa* the vertues of *Mandragoras*; by which meanes there hath not any thing beene said of the true *Circæa*, by reason, as I haue said, that *Mandragoras* hath beene called *Circæa*: but doubtlesse it hath the vertue of Garden Night-shade, and may serue in stead thereof without error.

Chap. 65.   Of Mandrake.

¶ *The Description.*

THe male Mandrake hath great broad long smooth leaues of a darke greene colour, flat spred vpon the ground: among which come vp the floures of a pale whitish colour, standing euery one vpon a single small and weake foot-stalke of a whitish greene colour: in their places grow round Apples of a yellowish colour, smooth, soft, and glittering, of a strong smell: in which are contained flat and smooth seeds in fashion of a little kidney, like those of the Thorne-apple. The root is long, thicke, whitish, diuided many times into two or three parts resembling the legs of a man, with other parts of the body adjoyning thereto, as the priuie part, as it hath been reported; whereas in truth it is no otherwise than in the roots of carrots, parseneps, and such like, forked or diuided into two or more parts, which Nature taketh no account of. There hath beene many ridiculous tales brought vp of this plant, whether of old wiues, or some runnagate Surgeons or Physicke-mongers I know not, (a title bad enough for them) but sure some one or moe that sought to make themselues famous and skilfull aboue others, were the first brochers of that errour I speake of. They adde further, That it is neuer or very seldome to be found growing naturally but vnder a gallowes, where the matter that hath fallen from the dead body hath giuen it the shape of a man; and the matter of a woman, the substace of a female plant; with many other such doltish dreames. They fable further and affirme, That he who would take vp a plant thereof must tie a dog therunto to pull it vp, which will giue a great shreeke at the digging vp; otherwise if a man should do it, he should surely die in short space after. Besides many fables of louing matters, too full of scurrilitie to set forth in print, which I forbeare to speake of. All which dreames and old wiues tales you shall from henceforth cast out of your bookes and memory; knowing this, that they are all and euerie part of them false and most vntrue: for I my selfe and my seruants also haue digged vp, planted, and replanted very many, and yet neuer could either perceiue shape of man or woman, but sometimes one streight root, sometimes two, and often six or seuen branches comming from the maine

great

great root, euen as Nature lift to beftow vpon it, as to other plants. But the idle drones that haue little or nothing to do but eate and drinke, haue beftowed fome of their time in caruing the roots of Brionie, forming them to the fhape of men and women: which falfifying practife hath confirmed the errour amongft the fimple and vnlearned people, who haue taken them vpon their report to be the true Mandrakes.

The female Mandrake is like vnto the male, fauing that the leaues hereof be of a more fwart or darke greene colour: and the fruit is long like a peare, and the other like an apple.

Mandragoras mas & fœmina.
The male and female Mandrake.

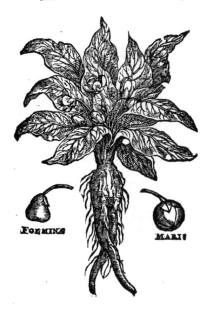

FOEMINA   MARIS

¶ *The Place.*

Mandrake groweth in hot Regions, in woods and mountaines, as in mount Garganus in Apulia, and fuch like places; we haue them onely planted in gardens, and are not elfewhere to be found in England.

¶ *The Time.*

They fpring vp with their leaues in March, and floure in the end of Aprill: the fruit is ripe in Auguft.

¶ *The Names.*

Mandrake is called of the Grecians Μανδραγόρας: of diuers, Κιρκαία, and *Circæa,* of *Circe* the witch, who by art could procure loue: for it hath beene thought that the Root hereof ferueth to win loue: of fome, ἀνθρωπόμορφος, *Anthropomorphos,* and *Morion:* fome of the Latines haue called it *Terræ malum,* and *Terreftre malum,* and *Canina malus:* Shops, and alfo other Nations doe receiue the Greeke name. *Diofcorides* faith, That the male is called of diuers *Morion:* and defcribeth alfo another Mandrake by the name of *Morion,* which, as much as can be gathered by the defcription, is like the male, but leffe in all parts: in Englifh we call it Mandrake, Mandrage, and Mandragon.

¶ *The Temperature.*

Mandrake hath a predominate cold facultie, as *Galen* faith, that is to fay, cold in the third degree, but the root is cold in the fourth degree.

¶ *The Vertues.*

A   *Diofcorides* doth particularly fet downe many faculties hereof; of which notwithftanding there be none proper vnto it, fauing thofe that depend vpon the drowfie and fleepie power thereof, which qualitie confifteth more in the root than in any other part.

B   The Apples are milder, and are reported that they may be eaten, being boyled with pepper and other hot fpices.

C   *Galen* faith that the Apples are fomething cold and moift, and that the barke of the root is of greateft ftrength, and doth not onely coole, but alfo dry.

D   The juice of the leaues is very profitably put into the ointment called *Populeon,* and all cooling ointments.

E   The juyce drawne forth of the roots dried, and taken in fmall quantitie, purgeth the belly exceedingly from flegme and melancholike humors.

F   It is good to be put into medicines and collyries that doe mitigate the paine of the eies; and put vp as a peffarie it draweth forth the dead childe and fecondine.

G   The greene leaues ftamped with barrowes greafe and barley meale, coole all hot fwellings and inflammations; and they haue vertue to confume apoftumes and hot vlcers, being bruifed and applied thereon.

H   A fuppofitorie made with the fame juyce, and put into the fundament caufeth fleepe.

I   The wine wherein the root hath been boyled or infufed prouoketh fleepe and affwageth paine.

K   The fmell of the Apples moueth to fleepe likewife; but the juice worketh more effectually if you take it in fmall quantitie.

Great

Great and ftrange effects are fuppofed to bee in Mandrakes, to caufe women to be fruitfull and I;
beare children, if they fhall but carry the fame neere to their bodies. Some do from hence ground
it, for that *Rahel* defired to haue her fifters Mandrakes (as the text is tranflated) but if we look well
into the circumftances which there we fhall finde, we may rather deem it otherwife. Yong *Ruben*
brought home amiable and fweet-fmelling floures, (for fo fignifieth the Hebrew word, vfed *Can-*
tic. 7. 13. in the fame fence) rather for their beauty and fmell, than for their vertue. Now in the
floures of Mandrake there is no fuch deleϭable or amiable fmell as was in thefe amiable floures
which *Ruben* brought home. Befides, we reade not that *Rahel* conceiued hereupon, for *Leah Iacobs*
wife had foure children before God granted that bleffing of fruitfulneffe vnto *Rahel*. And laft of
all, (which is my chiefeft reafon) *Iacob* was angry with *Rahel* when fhee faid, Giue mee children or
els I die; and demanded of her, whether he were in the ftead of God or no, who had withheld from
her the fruit of her body. And we know the Prophet *Dauid* faith, Children & the fruit of the womb
are the inheritance that commeth of the Lord, *Pfal. 127.*

Serapio, *Auicen*, and *Paulus Ægineta* write, That the feed and fruit of *Mandragoras* taken in drinke, M
do clenfe the matrix or mother: and *Diofcorides* wrot the fame long before them.

He that would know more hereof, may reade that chapter of D'*Turners* booke concerning this N
matter, where he hath written largely and learnedly of this Simple.

Chap. 66.   Of Henbane.

¶ *The Defcription.*

1   THe common blacke Henbane hath great and foft ftalkes, leaues very broad, foft, and
woolly, fomewhat jagged, efpecially thofe that grow neere to the ground, and thofe
that grow vpon the ftalke, narrower, fmaller, and fharper, the floures are bell-fafhion,
of a feint yellowifh white, and browne within towards the bottome: when the floures are gone:

there come hard knobby huſks like ſmall cups or boxes,wherein are ſmall brown ſeeds.

2   The white Henbane is not much vnlike the blacke, ſauing that his leaues are ſmaller, whiter,and more woolly,and the floures alſo whiter : the cods are like the other,but without pricks. It dieth in winter,and muſt likewiſe be ſowne again the next yeare.

‡ 3 *Hyoſcyamus albus minor.*
  The leſſer white Henbane.

‡ 4 *Hyoſcyamus albus Creticus.*
  White Henbane of Candy.

 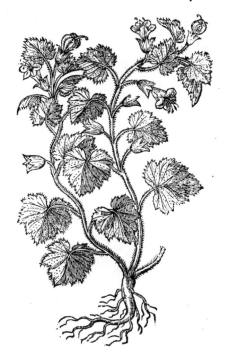

‡ 3   This other white Henbane is much like the laſt deſcribed,but that it is leſſe,the leaues ſmaller and rounder,hanging vpon pretty long ſtalks : the floures and ſeed-veſſels are like thoſe of the laſt mentioned.

4   This is ſofter and tenderer than the laſt deſcribed ; the leaues alſo hang vpon long footſtalks and are couered ouer with a ſoft downineſſe ; and they are ſomewhat broader, yet thinner & more ſinuated than thoſe of the white,and ſomewhat reſemble the forme of a Vine leafe beeing ſnipped. about the edges ; the ſtalks are alſo couered with a white colour : the floures are of a gold yellow, with a veluet-coloured circle in their middles ; the root is ſufficiently thicke and large. *Cluſius* had the figure and deſcription of this from his friend *Iaqnes Plateau,*who had the Plant growing of ſeed receiued from Candy.

5   The ſtalke of this growes ſome cubit high,being pretty ſtiffe,about the thickneſſe of ones little finger,and couered ouer with a ſoft & white downe : the leaues grow diſperſt vpon the ſtalk, not much vnlike thoſe of the common kind,but leſſer,and more diuided,and white(while they are yong)couered with a ſlender and long downineſſe : the top of the ſtalke is diuided into certaine branches that bend or hang down their heads,which alternatly among narrower, leſſer,and vndiuided leaues carry cups like as the common one,ending in fiue pretty ſtiffe points,in which are contained floures at firſt ſomewhat like the common kinde,but afterwards as they grow bigger, they change into an elegant red purpliſh colour,with deep colored veins : neither is the ring or middle part purple as in the common kind, but whitiſh, hauing a purpliſh pointall, and fiue thrcds in the middle : the ſeeds and ſeed-veſſels are like thoſe of the common kinde. *Cluſius* receiued the ſeed hereof from *Paludanus,*returning from his trauels into Syria and Egypt ; wherefore he calls it *Hyoſcyamus Ægyptius,*Egyptian Henbane. ‡

‡ 5 *Hyoſcyamus flore rubello.*
Henbane with a reddiſh floure.

¶ *The Place.*

Blacke Henbane grows almoſt euery where by highways, in the borders of fields about dunghils and vntoiled places : the white Henbane is not found but in the gardens of thoſe that loue phyſicall plants : the which groweth in my garden, and doth ſow it ſelfe from yeare to yeare.

¶ *The Time.*

They ſpring out of the ground in May, bring forth their floures in Auguſt, and the ſeed is ripe in October.

¶ *The Names.*

Henbane is called of the Grecians, ὑοσκύαμος : of the Latines, *Apollinaris*, and *Faba ſuilla* : of the Arabians, as *Pliny* ſaith, *Altercum* : of ſome, *Faba Iovis*, or *Iupiters* bean : of *Pythagoras, Zoroaſtes*, and *Apuleius, Inſana, Alterculum, Symphoniaca*, and *Caliculariſc* of the Tuſcanes, *Fabulonia*, and *Faba lupina* : of *Matthæus ſylvaticus, Dens Caballinus, Milimandrum, Caſſilago* : of *Iacobus à Manlijs, Herba pinnula* : in ſhops it is called *Iuſquiamus*, and *Hyoſcyamus* : in Engliſh, Henbane : in Italian, *Hyoſquiamo* : in Spaniſh, *Velenno* : in high-Dutch, 𝔅𝔦𝔩𝔰𝔢𝔫 𝔨𝔯𝔞𝔲𝔱 : in French, *Hannebane, Endormie* : the other is called *Hyoſcyamus albus*, or white Henbane.

¶ *The Nature.*

Theſe kinds of Henbane are cold in the fourth degree.

¶ *The Vertues.*

A　Henbane cauſeth drowſineſſe, and mitigateth all kinde of paine : it is good againſt hot & ſharp diſtillations of the eyes and other parts : it ſtaieth bleeding and the diſeaſe in women : it is applied to inflammations of the ſtones and other ſecret parts.

B　The leaues ſtamped with the ointment *Populeon*, made of Poplar buds, aſſwageth the pain of the gout, the ſwelling of the ſtones, and the tumors of womens breſts, and are good to be put into the ſame ointment, but in ſmall quantity.

C　To waſh the feet in the decoction of Henbane cauſeth ſleepe ; or giuen in a clyſter it doth the ſame, and alſo the often ſmelling to the floures.

D　The leaues, ſeed, and iuice taken inwardly cauſe an vnquiet ſleep like vnto the ſleepe of drunkenneſſe, which continueth long, and is deadly to the party.

E　The ſeed of white Henbane is good againſt the cough, the falling of watrie humors into the eys or breſt, againſt the inordinat flux of womens iſſues, & all other iſſues of bloud, taken in the weight of ten grains, with water wherein bony hath bin ſodden.

F　The root boiled with vinegre, & the ſame holden hot in the mouth, eaſeth the pain of the teeth. The ſeed is vſed by Mountibank tooth-drawers which run about the country, to cauſe worms come forth of the teeth, by burning it in a chafing diſh of coles, the party holding his mouth ouer the tume thereof : but ſome crafty companions to gain mony conuey ſmall lute-ſtrings into the water, perſuading the patient, that thoſe ſmall creepers came out of his mouth or other parts which he intended to eaſe.

CHAP. 67.   *Of yellow Henbane, or Engliſh Tabaco.*

Hyoſcyamus luteus.
Yellow Henbane.

¶ *The Deſcription.*

YEllow Henbane groweth to the height of two
cubits : the ſtalke is thicke, fat, and green of colour, ful of a ſpongeous pith, and is diuided into
ſundry branches, ſet with ſmooth and euen leaues,
thicke and full of juice. The floures grow at the tops
of the branches, orderly placed, of a pale yellow color, ſomthing leſſer than thoſe of the black Henbane.
The cups wherein the floures do ſtand, are like, but
leſſer, tenderer, and without ſharpe points, wherein is
ſet the huske or cod ſomwhat round, full of very ſmal
ſeed like the ſeed of marjerom. The root is ſmall and
threddy.

¶ *The Place.*

Yellow Henbane is ſowne in gardens, where it doth
proſper exceedingly, inſomuch that it cannot be deſtroied where it hath once ſown it ſelf, & it is diſperſed into moſt parts of London.

¶ *The Time.*

It floureth in the ſummer moneths, and oftentimes
till Autumne be farre ſpent, in which time the ſeed
commeth to perfection.

¶ *The Names.*

Yellow Henbane is called *Hyoſcyamus luteus :* of
ſome, *Petum*, and *Petun :* of others, *Nicotiana*, of *Nicot* a
Frenchman that brought the ſeeds from the Indies, as
alſo the ſeeds of the true Tabaco, whereof this hath
bin taken for a kind ; inſomuch that *Lobel* hath called
it *Dubius Hyoſcyamus*, or doubtfull Henbane, as a plant
participating of Henbane and Tabaco : and it is vſed
of diuers in ſtead of Tabaco, and called by the ſame name, for that it hath bin brought from Trinidada, a place ſo called in the Indies, as alſo from Virginia and other places, for Tabaco; and doubtleſſe, taken in ſmoke it worketh the ſame kind of drunkenneſſe that the right Tabaco doth.

‡ Some vſe to call this Nicotian in Engliſh, being a name taken from the Latine. ‡

¶ *The Temperature.*

This kinde of Henbane is thought of ſome to be cold and moiſt ; but after *Lobel* it rather heats
than cooles at all, becauſe of the biting taſt, as alſo that roſenninneſſe or gummineſſe it is poſſeſſed
of ; which is euidently perceiued both in handling and chewing it in the mouth.

¶ *The Vertues.*

A   This herb auaileth againſt all apoſtumes, tumors, inueterat vlcers, botches, and ſuch like, beeing
made into an vnguent or ſalue as followeth : Take of the greene leaues three pouñds and an halfe,
ſtampe them very ſmal in a ſtone mortar ; of oile Oliue one quart : ſet them to boile in a braſſe pan
or ſuch like, vpon a gentle fire, continually ſtirring it vntill the herbs ſeem blacke, and wil not boile
or bubble any more : then ſhall you haue an excellent green oile ; which beeing ſtrained from the
feces or droſſe, put the cleare and ſtrained oile to the fire again, adding therto of wax half a pound,
of roſen foure ounces, and of good turpentine two ounces : melt them all together, and keepe it in
pots for your vſe, to cure inueterat vlcers, apoſtumes, burnings, green wounds, and all cuts or hurts
in the head ; wherewith I haue gotten both crownes and credit.

B   It is vſed of ſome in ſtead of Tabaco, but to ſmall purpoſe or profit, although it doth ſtupifie or
dull the ſences, and cauſe that kind of giddines that Tabaco doth, and likewiſe ſpitting, which any
other herb of hot temperature will do, as Roſemary, Time, Winter-Sauorie, ſweet Marjerome, and
ſuch like : any of the which I like better to be taken in ſmoke, than this kind of doubtful Henbane.

CHAP. 68.

Of Tabaco, or Henbane of Peru.

¶ The Kindes.

THere be two forts or kinds of Tabaco, one greater, the other leffer, the greater was brought into Europe out of the prouinces of America, which we call the Weft Indies; the other from Trinidada, an Iland neere vnto the continent of the fame Indies. Some haue added a third fort, and others make the yellow Henbanq a kind thereof.

1 *Hyofcyamus Peruvianus.*
  Tabaco, or Henbane of Peru.

† 2 *Sana fan&a Indorum.*
  Tabaco of Trinidada.

¶ The Defcription.

1   TAbaco, or Henbane of Peru hath very great ftalkes of the bigneffe of a childes arme, growing in fertile and well dunged ground of feuen or eight foot high, diuiding it felfe into fundry branches of great length; whereon are placed in moft comly order very faire loug leaues, broad, fmooth, and fharp pointed, foft, and of a light green colour, fo faftned about the ftalke, that they feeme to embrace and compaffe it about. The floures grow at the top of the ftalks, in fhape like a bell-floure, fomewhat long and cornered, hollow within, of a light carnation colour, tending to whiteneffe toward the brims. The feed is contained in long fharpe pointed cods or feed-veffels like vnto the feed of yellow Henbane, but fomewhat fmaller, and browner of colour. The root is great, thicke, and of a wooddy fubftance, with fome threddy ftrings annexed thereunto.

2   Trinidada Tabaco hath a thicke tough and fibrous root, from which immediately rife vp long broad leaues and fmooth, of a greenifh colour, leffe than thofe of Peru: among which rifes vp a ftalk diuiding it felf at the ground into diuers branches, wheron are fet confufedly the like leaues but leffer. At the top of the ftalks ftand vp long necked hollow floures of a pale purple tending to a blufh colour: after which fucceed the cods or feed-veffels, including many fmall feeds like vnto the feed of Marjerom. The whole plant perifheth at the firft approch of winter.

‡ 3 *Tabacum minimum.*
Dwarfe Tabaco.

‡ 3   This third is an herb ſom ſpan or better long,not in face vnlike the precedent, neither defectiue in the hot and burning taſt.The
floures are much leſſe than thoſe of the yellow
Henbane,and of a greeniſh yellow. The leaues
are ſmall , and narrower than thoſe of Sage of
Ieruſalem.The root is ſmall and fibrous. ‡

¶ *The Place.*

Theſe were firſt brought into Europe out of
America,which is called the Weſt Indies , in
which is the prouince or countrey of Peru : but
being now planted in the gardens of Europe it
proſpers very well,and comes from ſeed in one
yeare to beare both floures and ſeed. The which
I take to be better for the conſtitution of our
bodies, than that which is brought from India,
& that growing in India better for the people
of the ſame country : notwithſtanding it is not
ſo thought of our Tabaconiſts,for acording to
the Engliſh prouerb, Far fetcht & dear bought
is beſt for Ladies.

¶ *The Time.*

Tabaco muſt be ſowne in the moſt fruitfull
ground that may be found,careleſly caſt abroad
in ſowing,without raking it into the ground,or
any ſuch pain or induſtry taken as is requiſit in
the ſowing of other ſeeds,as my ſelf haue found
by proof,who haue experimented euery way to
cauſe it quickly to grow:for I haue committed
ſome to the earth in the end of March, ſome in
Aprill,and ſome in the beginning of May, be-
cauſe I durſt not haſard all my ſeed at one time,leſt ſome vnkindely blaſt ſhould happen after the
ſowing,which might be a great enemie thereunto.

¶ *The Names.*

The people of America call it *Petun.*Some,as *Lobel* and *Pena,*haue giuen it theſe Latine names,
*Sacra herba,Sancta herba,*and *Sana ſancta Indorum.* Others,as *Dodonæus,* call it *Hyoſcyamus Peruvianus,*
or Henbane of Peru.*Nicolaus Monardus* names it *Tabacum.* That it is *Hyoſcyami ſpecies,*or a kinde of
Henbane,not only the forme being like to yellow Henbane,but the qualitie alſo doth declare ; for
it bringeth drowſineſſe, troubleth the ſences, and maketh a man as it were drunke by taking the
fume only ; as *Andrew Theuet* teſtifieth,and common experience ſheweth : of ſome it is called *Nicotiana,*the which I refer to the yellow Henbane for diſtinctions ſake.

¶ *The Temperature.*

It is hot and dry,and that in the ſecond degree,as *Monardis* thinketh, and is withall of power to
diſcuſſe or reſolue,and to clenſe away filthy humors,hauing alſo a ſmall aſtriction, and a ſtupifying or benumming qualitie, and it purgeth by the ſtoole : and *Monardis* writeth, that it hath a certain power to reſiſt poiſon. And to proue it to be of an hot temperature, the biting quality of the
leaues do ſhew,which is eaſily perceiued by taſte : alſo the green leaues laid vpon vlcers in ſinewie
parts may ſerue for a proofe of heate in this plant ; becauſe they doe draw out filth and corrupted
mattar,which a cold Simple would neuer do. The leaues likewiſe being chewed draw forth flegm
and water,as doth alſo the fume taken when the leaues are dried : which things declare that this is
not a little hot ; for what things ſoeuer,that being chewed or held in the mouth bring forth flegm
and water,the ſame be all accounted hot ; as the root of Pellitorie of Spaine,of Saxifrage,& other
things of like power. Moreouer,the benumming qualitie hereof is not hard to be perceiued , for
vpon the taking of the fume at the mouth there followeth an infirmitie like vnto drunkenneſſe,and
many times ſleepe,as after the taking of *Opium :* which alſo ſheweth in the taſte a biting qualitie,
and therefore is not without heate ; which when it is chewed and inwardly taken,it doth forthwith
ſhew,cauſing a certain heate in the cheſt,and yet withall troubling the wits, as *Petrus Bellonius* in
his

his third booke of Singularities doth declare : where alſo he ſheweth, That the Turks oftentimes vſe *opium*, and take one dram and a halfe thereof at one time, without any other hurt following, ſaying that they are thereupon as it were taken with a certain light drunkenneſſe. So alſo this Tabaco, being in taſt biting, and in temperature hot, hath notwithſtanding a benumming quality. Hereupon it ſeemeth to follow, that not only this Henbane of Peru, but alſo the juice of Poppie otherwiſe called *Opium*, conſiſteth of diuers parts, ſome biting and hot, others extreame cold, that is to ſay, ſtupifying, or benumniing. If ſo be that this benumming qualitie proceed of extreame cold, as *Galen* and all the old Phyſitions do hold opinion ; then ſhould this be cold : but if the benumming facultie doth not depend of an extreme cold qualitie, but proceedeth of the eſſence of the ſubſtance, then Tabaco is not cold and benumming, but hot and benumming, and the later not ſo much by reaſon of his temperature, as through the propertie of his ſubſtance ; no otherwiſe than a purging medicine, which hath his force not from the temperature, but from the eſſence of the whole ſubſtance.

¶ The Vertues.

Nicolaus Monardis ſaith, that the leaues hereof are a remedie for the paine of head called the Megram or Migram, that hath bin of long continuance; and alſo for a cold ſtomack, eſpecially in children; and that it is good againſt the pains in the kidnies.　A

It is a preſent remedie for the fits of the mother, it mitigateth the paine of the gout, if it be roſted in hot embers, and applied to the grieued part.　B

It is likewiſe a remedy for the tooth-ache, if the teeth and gumbs be rubbed with a linnen cloth dipped in the juice, and afterward a round ball of the leaues laid vnto the place.　C

The juice boiled in ſugar in form of a ſyrrup, and inwardly taken, driueth forth wormes of the belly, if withall a leaſe be layd to the nauell.　D

The ſame doth likewiſe ſcoure and clenſe old & rotten vlcers, and bringeth them to perfect digeſtion, as the ſame Author affirmeth.　E

In the Low-countries it is vſed againſt ſcabs and filthineſſe of the skin, and to cure wounds : but ſome hold opinion it is to be vſed only to hot and ſtrong bodies ; for they ſay, the vſe is not ſafe in weak and old folks ; and for this cauſe (as it ſeemeth) the women in America (ſaith *Theuet*) abſtain from the herb *Petun* or Tabaco, and do in no wiſe vſe it.　F

The weight of foure ounces of the juice hereof drunke doth purge both vpwards and downewards, and procureth afterward a long and ſound ſleepe, as wee haue learned of a friend by obſeruation, who affirmed, That a ſtrong countreyman of a middle age hauing a dropſie, took it, and being wakened out of his ſleepe called for meat and drinke, and after that became perfectly cured.　G

Moreouer, the ſame man reported, That he had cured many countreymen of agues, with the diſtilled water of the leaues drunke a little while before the fit.　H

Likewiſe there is an oile to be taken out of the leaues that healeth merri-galls, kibed heeles, and ſuch like.　I

It is good againſt poyſon, and taketh away the malignitie therof, if the juice be giuen to drink, or the wounds made by venomous beaſts be waſhed therewith.　K

The dry leaues are vſed to be taken in a pipe ſet on fire and ſuckt into the ſtomacke, and thruſt forth againe at the noſthrils, againſt the paines in the head, rheumes, aches in any part of the bodie, whereof ſoeuer the originall proceed, whether from France, Italy, Spaine, Indies, or from our familiar and beſt knowne diſeaſes. Thoſe leaues do palliate or eaſe for a time, but neuer perform any cure abſolutely : for although they empty the body of humors, yet the cauſe of the griefe cannot be ſo taken away. But ſome haue learned this principle, That repletion doth require euacuation ; that is to ſay, That fulneſſe craueth emptineſſe ; and by euacuation doe aſſure themſelues of health. But this doth not take away ſo much with it this day, but the next bringeth with it more. As for example, a Well doth neuer yeeld ſuch ſtore of water as when it is moſt drawn and emptied. My ſelfe ſpeake by proofe ; who haue cured of that infectious diſeaſe a great many, diuers of which had couered or kept vnder the ſickeneſſe by the helpe of Tabaco as they thought, yet in the end haue bin conſtrained to haue vnto ſuch an hard knot, a crabbed wedge, or elſe had vtterly periſhed.　L

Some vſe to drink it (as it is termed) for wantonneſſe, or rather cuſtome, and cannot forbeare it, no not in the midſt of their dinner ; which kinde of taking is vnwholſome and very dangerous : although to take it ſeldom, and that phyſically, is to be tolerated, and may do ſome good : but I commend the ſyrrup aboue this fume or ſmoky medicine.　M

It

N   It is taken of ſome phyſically in a pipe for that purpoſe once in a day at the moſt, and that in the morning faſting, againſt paines in the head, ſtomack, and griefe in the breſt and lungs: againſt catarrhs and rheums, and ſuch as haue gotten cold and hoarſeneſſe.

O   Some haue reported, That it doth little preuaile againſt an hot diſeaſe, and that it profiteth an hot complexion nothing at all. But experience hath not ſhewed as yet that it is injurious vnto either.

P   They that haue ſeene the proofe hereof, haue credibly reported, That when the Moores and Indians haue fainted either for want of food or reſt, this hath bin a preſent remedie vnto them, to ſupply the one, and to help them to the other.

Q   The prieſts and Inchanters of the hot countries do take the fume thereof vntil they be drunke, that after they haue lien for dead three or ſoure houres, they may tell the people what wonders, viſions, or illuſions they haue ſeen, and ſo giue them a prophetical direction or foretelling (if we may truſt the Diuell) of the ſucceſſe of their buſineſſe.

R   The iuyce or diſtilled water of the firſt kind is very good againſt catarrhs, the dizzineſſe of the head, and rheums that fall downe the eies, againſt the pain called the megram, if either you apply it vnto the temples, or take one or two green leaues, or a dry leafe moiſtned in wine, and dried cunningly vpon the embers, and laid thereto.

S   It cleeres the ſight, and taketh away the webs and ſpots thereof, being annointed with the juyce bloud-warme.

T   The oile or iuyce dropped into the eares is good againſt deafneſſe; a cloth dipped in the ſame and layd vpon the face, taketh away the lentils, redneſſe, and ſpots thereof.

V   Many notable medicines are made hereof againſt the old and inveterat cough, againſt aſthmaticall or pectorall griefes, all which if I ſhould ſet downe at large, would require a peculiar volume.

X   It is alſo giuen vnto ſuch as are accuſtomed to ſwoune, and that are troubled with the Colicke and windineſſe: and likewiſe againſt the Dropſie, the Wormes in children, the Piles, and the Sciatica.

Y   It is vſed in outward medicines, either the herbe boiled with oile, wax, roſin, and turpentine, as before is ſet downe in yellow Henbane, or the extraction thereof with ſalt, oile, balſam, the diſtilled water, and ſuch like, againſt tumours, apoſtumes, old vlcers of hard curation, botches, ſcabbes, ſtinging with nettles, carbuncles, poiſoned arrowes, and wounds made with gunnes or any other weapons.

Z   It is excellent good in burnings and ſcaldings with fire, water, oile, lightning, or ſuch like, boiled with hogges greaſe into the forme of an ointment, as I haue often prooued, and found moſt true; adding a little of the juice of Thorne-Apple leaues, ſpreading it vpon a cloth and ſo applying it.

A   I doe make hereof an excellent Balme to cure deep wounds and punctures made by ſome narrow ſharpe pointed weapon. Which Balſame doth bring vp the fleſh from the bottome verie ſpeedily, and alſo heale ſimple cuts in the fleſh according to the firſt intention, that is, to glew or ſoder the lips of the wound together, not procuring matter or corruption to it, as is commonly ſeene in the healing of wounds. The Receit is this: Take Oile of Roſes, Oile of S. Iohns Wort, of either one pinte, the leaues of Tabaco ſtamped ſmall in a ſtone mortar two pounds; boile them together to the conſumption of the juice, ſtraine it and put it to the fire againe, adding thereunto of Venice Turpentine two ounces, of Olibanum and Maſticke of either halfe an ounce, in moſt fine and ſubtil pouder: the which you may at all times make an vnguent or ſalue, by putting thereto wax and roſin to giue vnto it a ſtiffe body, which worketh exceeding well in malignant and virulent vlcers, as in wounds and punctures. I ſend this jewell vnto you women of all ſorts, eſpecially ſuch as cure and helpe the poore and impotent of your countrey without reward. But vnto the beggarly rabble of witches, charmers, and ſuch like couſeners, that regard more to get money, than to helpe for charitie, I wiſh theſe few medicines far from their vnderſtanding, and from thoſe deceiuers, whom I wiſh to be ignorant herein. But courteous gentlewomen, I may not for the malice that I doe beare vnto ſuch, hide any thing from you of ſuch importance: and therefore take one more that followeth, wherewith I haue done many and good cures, although of ſmall coſt; but regard it not the leſſe for that cauſe. Take the leaues of Tabaco two pounds, Hogs greaſe one pound, ſtampe the herbe ſmall in a ſtone morter, putting thereto a ſmall cup full of red or claret wine, ſtirre them well together, couer the morter from filth, and ſo let it reſt vntill morning; then put it to the fire and let it boile gently, continually ſtirring it vntill the conſumption of the wine: ſtraine it and ſet it to the fire againe, putting thereto the iuyce of the herbe one pound, of Venice turpentine foure onnces; boile them together to the conſumption of the iuice, then adde therto of

the

the roots of round *Ariſtolochia* or Birthworth in moſt fine pouder two ounces, ſufficient wax to giue it a body; the which keep for thy wounded poore neighbor, as alſo the old and filthy vlcers of the legs and other parts of ſuch as haue need of help.

† The figures were formerly tranſpoſed.

CHAP. 69. *Of Tree Nightſhade.*

Amomum Plinij.
Tree Nightſhade.

¶ *The Deſcription.*

THis rare and pleaſant plant, called tree Night-ſhade, is taken of ſome to bee a kinde of Ginny pepper, but not rightly: of others, for a kind of Night-ſhade, whoſe iudgement and cenſure I gladly admit; for that it doth more fitly anſwer it both in the form and nature. It groweth vp like vnto a ſmall ſhrub or wooddy hedge-buſh, two or three cubits high, coue-red with a greeniſh barke ſet with many ſmal twiggy branches, and garniſhed with many long leaues verie green, like vnto thoſe of the Peach tree. The floures are white, with a certaine yellow pricke or pointall in the middle, like to the floures of garden Nightſhade. After which ſucceed ſmall round berries very red of colour, and of the ſame ſubſtance with winter Cher-ries; wherein are contained little flat yellow ſeeds. The root is compact of many ſmall hairy yellow ſtrings.

¶ *The Place.*

It groweth not wild in theſe cold regions, but wee haue them in our gardens, rather for pleaſure than profit, or any good qualitie as yet knowne.

¶ *The Time.*

It is kept in pots and tubs with earth & ſuch like, in houſes during the extremitie of Winter, and is ſet abroad in the garden in March or Aprill, becauſe it cannot endure the coldneſſe of our climat: it floures in May, and the fruit is ripe in September.

¶ *The Names.*

Tree Nightſhade is called in Latine *Solanum arboreſcens*: of ſome, *Strychnodendron*; and ſome do iudge it to be *Amomum* of *Pliny*: it is *Pſeudocapſicum* of *Dodonæus*.

¶ *The Nature and Vertues.*

We haue not as yet any thing ſet downe as touching the temperature or vertues of this plant, but it is referred of ſome to the kinds of Ginny pepper, but without any reaſon at all; for Ginny pep-per though it bring forth fruit very like in ſhape vnto this plant, yet in taſte moſt vnlike, for that *Capſicum* or Ginny pepper is more ſharpe in taſte than our common pepper, and the other hath no taſte of biting at all, but is like vnto the berries of garden Nightſhade in taſt, although they differ in colour: which hath moued ſome to call this plant red Nightſhade, of the colour of the berries; and tree Nightſhade, of the wooddy ſubſtance which doth continue and grow from yeare to yeare: and Ginny Pepper dieth at the firſt approch of winter

CHAP. 70 *Of Balme-Apple, or Apple of Ierusalem.*

1 *Balsamina mas.*
The male Balsam Apple.

2 *Balsamina fœmina.*
The female Balsam Apple.

¶ *The Description.*

1   THe male Balme Apple hath long small and tender branches, set with leaues like those of the Vine, and the like small clasping tendrels wherewith it catcheth hold of such things as grow neere it, not able by reason of his weakenesse to stand vpright without some pole or other thing to support it. The floures consist of fiue small leaues of mean bignes, and are of a feint yellow colour : which being past, there come in place long Apples, something sharpe toward the point, almost like an egge, rough all ouer, as it were with small harmelesse prickles, red both within and without when they be ripe, and cleaueth in sunder of themselues : in the Apple lieth great broad flat seed, like those of Pompion or Citrull, but somthing blacke when they be withered. The root is threddy, and disperseth it selfe far abroad in the ground.

2   The female Balm-Apple doth not a little differ from the former : it brings forth stalks not running or climing like the other, but a most thick and fat trunk or stock full of juice, in substance like the stalks of Purslane, of a reddish colour, and somwhat shining. The leaues be long & narrow in shape like those of Willow or the Peach tree, somewhat toothed or notched about the edges : among which grow the floures, of an incarnat colour tending to blewnesse, hauing a small spurre or taile anexed thereto as hath the Larks heele, of a faire light crimson colour : in their places come vp the fruit or apples rough and hairy, but lesser than those of the former, yellow when they be ripe ; which likewise cleaue asunder of themselues, and cast abroad their seeds much like vnto Lentiles, saith mine Author. But those which I haue from yeare to yere in my garden bring forth seed like the Cole-flory or Mustard seed ; whether they be of two kinds, or the climat doth alter the shape, it resteth disputable.

¶ *The*

¶ *The Place.*

These plants prosper best in hot regions: they are strangers in England, and do with great labor and industrie grow in these cold countries.

¶ *The Time.*

They must be sowne in the beginning of Aprill in a bed of hot horse-dung, euen as Muske-melons, Cucumbers, and such like cold fruits are; and replanted abroad from the said bed, into the most hot and fertile place of the garden, at such time as they haue gotten three leaues apiece.

¶ *The Names.*

Diuersly hath this plant been named; some calling it by one name, and some by another, euery one as it seemed good to his fancie. *Baptista Sardus* calls it *Balsamina Cucumerina*: others, *Viticella*, & *Charantia*, as also *Pomum Hierosolymitanum*, or Apples of Ierusalem: in English, Balm Apple: in Italian, *Caranza*: in the German tongue, *Balsam opffel*: in French, *Merueille*: some of the Latins haue called it *Pomum mirabile*, or Maruellous Apples. It is thought to be *Balsamina*, because the oile wherein the ripe Apples be steeped or infused, is thought profitable for many things, as is *Opobalsamum*, or the liquor of the plant *Balsamum*.

The female Balsam Apple is likewise called *Balsamina*, and oftentimes in the Neuter gender *Balsaminum*: *Gesner* chuseth rather to name it *Balsamina amygdaloides*: *Valerius Cordus*, *Balsamella*: others, *Balsamina femina*: in English, the female Balme Apples.

¶ *The Temperature.*

The fruit or Apples hereof, as also the leaues, do notably dry, hauing withall a certaine moderate coldnesse very neere to a mean temperature, that is after some, hot in the first, and dry in the second degree.

¶ *The Vertues.*

The leaues are reported to heale greene wounds, if they be bruised and laid thereon; and taken **A** with wine they are said to be a remedy for the colique, and an effectuall medicine for burstings and convulsions or crampes.

The leaues of the male *Balsamina* dried in the shadow, and beaten into pouder and giuen in **B** wine vnto those that are mortally wounded in the body, cureth them inwardly, and helpeth also the Collicke.

The oile which is drawne forth of the fruit cureth all greene and fresh wounds as the true natu- **C** rall Balsam: it helpeth the cramps and convulsions, and the shrinking of sinewes, being annointed therewith.

It profiteth women that are in great extremitie of childe-birth, in taking away the paine of the **D** matrix, causing easie deliuerance beeing applied to the place, and annointed vpon their bellies, or cast into the matrix with a syring, and easeth the dolour of the inward parts.

It cureth the Hemorrhoides and all other paines of the fundament, being thereto applied with **E** lint of old clouts.

The leaues drunken in wine, heale ruptures. **F**

I finde little or nothing written of the property or vertues of the female kinde, but that it is **G** thought to draw neere vnto the first in temperament and vertue.

Oile oliue in which the fruit (the seed taken forth) is either set in the Sun, as we do when wee **H** make oile of roses, or boiled in a double glasse set in hot water, or else buried in hot horse dung, taketh away inflammations that are in wounds. It doth also easily and in short time consolidate or glew them together, and perfectly cure them.

It cureth the vlcers of the dugs or paps, the head of the yard or matrix, as also the inflammati- **I** on thereof being iniected or conueied into the place, with a syringe or mother pessarie.

This apple is with good successe applied vnto wounds, prickes and hurts of the sinewes. It hath **K** great force to cure scaldings and burnings: it taketh away scarres and blemishes, if in the meane time the pouder of the leaues be taken for certaine daies together.

It is reported that such as be barren are made fruitfull herewith, if the woman first be bathed in **L** a fit and conuenient bath for the purpose, and the parts about the share and matrix annointed herewith, and the woman presently haue the company of her husband.

Chap.

Chap. 71. *Of Ginny or Indian Pepper.*

1 *Capſicum longioribus ſiliquis.*
Long codded Ginny Pepper.

‡ 2 *Capſicum rotundioribus ſiliquis.*
Round codded Ginny Pepper.

3 *Capſicum minimis ſiliquis.*
Small codded Ginny Pepper.

‡ *Capſici ſiliqua varia.*
Varieties of the cods of Ginny pepper.

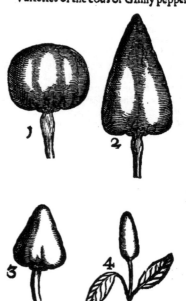

¶ *The Deſcription.*

1 THe firſt of theſe plants hath ſquare ſtalkes a foot high or ſomewhat more, ſet with many thicke and fat leaues, not vnlike to thoſe of garden Nightſhade, but narrower and ſharper pointed, of a darke greene colour. The floures grow along ſt the ſtalkes, out of the wings of the leaues, of a white colour, hauing for the moſt part fiue ſmall leaues blaſing out like a ſtar, with a greene button in the middle. After them grow the cods, greene at the firſt, and when they be ripe of a braue colour glittering like red corall, in which is contained little flat ſeeds, of a light yellow colour, of a hot biting taſte like common pepper, as is alſo the cod it ſelfe : which is long, and as big as a finger, and ſharpe pointed.

‡ 2 The difference that is betweene this and the laſt deſcribed is ſmall, for it conſiſts in nothing but that the cods are pretty large and round, after the faſhion of cherries, and not ſo long as thoſe of the former. ‡

3 The third kinde of Ginnie pepper is like vnto the precedent in leaues, floures, and ſtalkes. The cods hereof are ſmall, round, and red, very like to the berries of *Dulcamara* or wooddy Nightſhade, both in bigneſſe, colour, and ſubſtance, wherein conſiſteth the difference : notwithſtanding the ſeed and cods are very ſharpe and biting, as thoſe of the firſt kinde.

‡ *Capſici ſiliquæ variæ.*
Varieties of the cods of Ginnie Pepper.

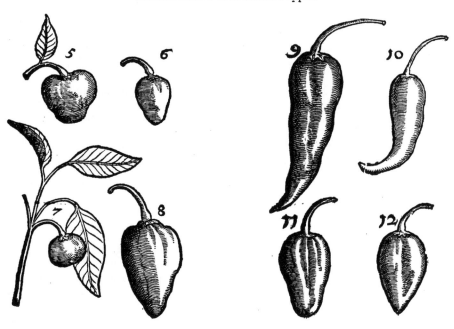

‡ There are many other varieties of Ginnie Pepper, which chiefly conſiſt in the ſhape and colour of the cods: wherefore I thought good (and that chiefly becauſe it is a plant that will hardly brooke our climate) only to preſent you with the figures of their ſeuerall ſhapes, whereof the cods of ſome ſtand or grow vpright, and otherſome hang downe : ſuch as deſire further information of this plant, may be abundantly ſatisfied in *Cluſius* his *Cure poſter.* from *pag.95.to pag.108.* where they ſhall find theſe treated of at large in a treatiſe written in Italian by *Gregory de Regio*, a Capuchine Fryer, and ſent to *Cluſius*, who tranſlating it into Engliſh, left it to be ſet forth with other his obſeruations, which was performed two yeares after his death, to wit *Anno Domini 1611.* The figures we here giue are the ſame which are in that tractat. ‡

¶ *The Place.*
Theſe plants are brought from forrein countries, as Ginnie, India, and thoſe parts, into Spaine

and Italy : from whence we haue receiued feed for our Englifh gardens, where they come to fruit-bearing : but the cod doth not come to that bright red colour which naturally it is poffeffed with, which hath happened by reafon of thefe vnindely yeeres that are paft : but we expect better when God fhall fend vs a hot and temperate yeere.

¶ *The Time.*

The feeds hereof muft be fowne in a bed of hot horfe-dung, as muske-Melons are, and remooued into a pot when they haue gotten three or foure leaues, that it may the more conueniently be cari-ed from place to place to receiue the heate of the Sunne : and are toward Autumne to be caried in-to fome houfe, to auoid the iniurie of the cold nights of that time of the yere, when it is to beare his fruit.

¶ *The Names.*

Actuarius calleth it in Greeke ϰαπσικον : in Latine, *Capficum:* and it is thought to be that which *Aui-cen* nameth *Zinziber caninum,* or dogs Ginger : and *Pliny, Siliquaftrum,* which is more like in tafte to pepper than is *Panax,* and it is therefore called *Piperitis,* as he hath written in his 19.booke, 12.chap-ter. *Panax* (faith he) hath the tafte of Pepper and *Siliquaftrum,* for which caufe it is called *Piperi-tis.* The later Herbarifts doe oftentimes call it *Piper Indianum,* or *Indicum,* fometimes *Piper Calicu-thium,* or *Piper Hifpanicum :* In Englifh it is called, Ginnie Pepper, and Indian Pepper : in the Ger-mane Tongue, Indianiſchar Pfeffer: in low Dutch, Bꝛeſilie Peper : in French, *Poivre d'Inde,* very well knowne in the fhops at Billingfgate by the name of Ginnie Pepper, where it is vfually to be bought.

¶ *The Temperature.*

Ginnie Pepper is extreame hot and dry euen in the fourth degree : that is to fay, far hotter and drier than *Auicen* fheweth dogs Ginger to be.

¶ *The Vertues.*

A   Ginnie Pepper hath the tafte of Pepper, but not the power or vertue, notwithftanding in Spaine and fundry parts of the Indies they do vfe to dreffe their meate therewith, as we do with Calecute Pepper : but (faith my Author) it hath in it a malitious quality, whereby it is an enemy to the liuer and other of the entrails. *Auicen* writeth that it killeth dogs.

B   It is faid to die or colour like Saffron; and being receiued in fuch fort as Saffron is vfually ta-ken, it warmeth the ftomacke, and helpeth greatly the digeftion of meats.

C   It diffolueth the fwellings about the throat, called the Kings euill, as kernels and cold fwellings, and taketh away fpots and lentiles from the face, being applied thereto with hony.

CHAP. 72. *Of horned Poppie.*

¶ *The Defcription.*

1   THe yellow horned Poppie hath whitifh leaues very much cut or jagged, fomewhat like the leaues of Garden Poppie, but rougher and more hairy. The ftalkes be long, round, and brittle. The floures be large and yellow, confifting of foure leaues ; which being paft, there come long huskes or cods, crooked like an horne or cornet, wherein is conteined fmall blacke feed. The root is great, thicke, fcaly, and rough, continuing long.

2   The fecond kinde of horned Poppie is much flenderer and leffer than the precedent, and hath leaues with like deepe cuts as Rocket hath, and fomething hairy. The ftalkes be very flender, brittle, and branched into diuers armes or wings ; the floures fmall, made of foure little leaues, of a red colour, with a fmall ftrake of blacke toward the bottome ; after which commeth the feed, inclofed in flender, long, crooked cods full of blackifh feed. The root is fmall and fingle, and dieth euery yeere.

‡ 3   This is much like the laft defcribed, and according to *Clufius,* rather a variety than diffe-rent. It is diftinguifhed from the laft mentioned by the fmoothneffe of the leaues, and the colour of the floures, which are of a pale yellowifh red, both which accidents *Clufius* affirmes happen to the former, toward the later end of Summer.

4   There is another fort of horned Poppie altogether leffer than the laft defcribed, hauing tenderer leaues, cut into fine little parcels : the floure is likewife leffer, of a blew purple colour like the double Violets.

¶ *The*

1 *Papauer cornutum flore luteo.*
Yellow horned Poppie.

2 *Papauer cornutum flore rubro.*
Red horned Poppie.

‡ 3 *Papauer corniculatum phæniceum glabrum.*
Red horned Poppie with ſmooth leaues.

4 *Papauer cornutum flore violaceo.*
Violet coloured horned Poppie.

¶ *The Place.*

The yellow horned Poppy groweth vpon the sands and banks of the sea : I haue found it growing neer vnto Rie in Kent, in the Isles of Shepey and Thanet, at Lee in Essex, at Harwich, at Whitestable, and many other places alongst the English coast.

The second groweth not wilde in England. *Angelus Palea*, and *Bartholomaus ab Vrbe-veterum*, who haue commented vpon *Mesue*, write that they found this red horned Poppy in the kingdomes of Arragon and Castile in Spaine, and the fields neere vnto common paths. They do grow in my garden very plentifully.

¶ *The Time.*

They floure from May to the end of August.

¶ *The Names.*

Most Writers haue taken horned Poppy, especially that with red floures to be *Glaucium* : neither is this their opinion altogether vnprobable ; for as *Dioscorides* saith, *Glaucium* hath leaues like those of horned Poppy, but Νπαχύτερα that is to say fatter, χαμαίζηλα, low, or lying on the ground, of a strong smell and of a bitter taste, the iuice also is much like in colour to Saffron. Now *Lobel* and *Pena* witnesse, that this horned Poppie hath the same kinde of iuice, as my selfe likewise can testifie. *Dioscorides* saith that *Glaucium* groweth about Hierapolis, a citie in Syria ; but what hindereth that it should not be found also somewhere else ? These things shew it hath a great affinity with *Glaucium*, if it be not the true and legitimate *Glaucium* of *Dioscorides*. Howbeit the first is the *Mecon Ceratites*, or *Papauer corniculatum* of the Antients, by the common consent of all late writers: in English, Sea Poppy, and Horned Poppy : in Dutch, **Geelheul** and **Hoorne Heule** : in the Germane Tongue, **Gelbomag** : ic French, *Pauot Cornu* : in Spanish, *Dormidera marina*.

¶ *The Temperature.*

Horned Poppies are hot and dry in the third degree.

¶ *The Vertues.*

A　　The root of horned Poppy boiled in water vnto the consumption of the one halfe, and drunke, prouoketh vrine, and openeth the stopping of the liuer.

B　　The seed taken in the quantitie of a spoonefull looseth the belly gently.

C　　The iuice mixed with meale and hony, mundifieth old rotten and filthy vlcers.

D　　The leaues and floures put into vnguents or salues appropriate for greene wounds, digest them, that is, bring them to white matter, with perfect quitture or sanies.

† The figure that formerly was in the fourth place of this chap. vnder the title of *Papauer cornutum luteum minus*, was of a Bindeweed called by *Clusius, Conuoluulus, i.e. Althea*. You shall finde it hereafter in the due place. The description as far as I can iudge was of the *Cuminum Corniculatum* which was *pag. 909.*

C H A P. 371　*Of Garden Poppies.*

¶ *The Description.*

1　　THe leaues of white Poppie are long, broad, smooth, longer then the leaues of Lettuce, whiter, and cut in the edges: the stem or stalke is straight and brittle, oftentimes a yard and a halfe high: on the top whereof grow white floures, in which at the very beginning appeareth a small head, accompanied with a number of threds or chiues, which being full growne is round, and yet something long withall, and hath a couer or crownet vpon the top; it is with many filmes or thin skin diuided into coffers or seuerall partitions, in which is contained abundance of small round and whitish seed. The root groweth deepe, and is of no estimation nor continuance.

2　　Like vnto this is the blacke garden Poppy, sauing that the floures are not so white and shining, but vsually red, or at least spotted or straked with some lines of purple. The leaues are greater, more iagged, and sharper pointed. The seed is likewise blacker, which maketh the difference.

‡ 3　　There also another garden Poppie whose leaues are much more sinuated, or crested, and the floure also is all iagged or finely cut about the edges, and of this sort there is also both blacke and white. The floures of the black are red, and the seed blacke; and the other hath both the floures and seed white:

4　　There are diuers varieties of double Poppies of both these kinds, and their colours are commonly either white, red, darke purple, scarlet, or mixt of some of these. They differ from the former onely in the doublenesse of their floures.

1 *Papauer ſativum album.*
White garden Poppy.

2 *Papauer ſativum nigrum.*
Blacke garden Poppy.

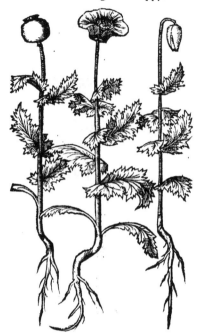

‡ 3 *Papauer fimbriatum album.*
White jagged Poppy.

4 *Papauer flo. multipl. albo & nigro.*
The double white and black Poppy.

‡ 5 *Papauer ſylveſtre.*
Wild Poppy.

5   There is alſo another kind of Poppy which oft times is found wild : the ſtalks, leaues, ſlours & heads are like, but leſſe than thoſe of the precedent : the floures are of an ouerworne blewiſh purple colour; after which follow heads ſhort and round, which vnder their couer or crowne haue little holes by which the ſeed may fall out; contrarie to the heads of the former, which are cloſe, and open not of themſelues: there is alſo a double one of this kinde.

¶ *The Place.*

Theſe kinds of Poppies are ſowne in gardens, and do afterward come of the fallings of their ſeed.

¶ *The Time.*

They floure moſt commonly in Iune. The ſeed is perfected in Iuly and Auguſt.

¶ *The Names.*

Poppy is called of the Grecians μηκων : of the La-tines, *Papauer :* the ſhops keep the Latine name : it is called in high-Dutch, **Magſamen ;** in low-Dutch, **Huel** and **Mancop:** in Engliſh, Poppy, and Cheeſe-bowles: in French, *Pauot,* and *Oliette,* by the Wallons.

The garden Poppy which hath blacke ſeeds, is ſur-named of *Dioſcorides,* ἡμερον, or wilde ; and is as he ſaith called ſatiue, becauſe *Opium* flowes from it : of *Pliny* and the Latines, *Papauer nigrum,* wherof there be many va-riable colours, and of great beauty, although of euill ſmell, whereupon our gentlewomen call it Ioan Sil-uer pin.

¶ *The Temperature.*

All the Poppies are cold, as *Galen* teſtifieth in his booke of the faculties of ſimple medicines.

¶ *The Vertues.*

A   This ſeed, as *Galen* ſaith in his booke of the faculties of Nouriſhments, is good to ſeaſon bread with, but the white is better than the black. He alſo addeth, that the ſame is cold and cauſeth ſleep, and yeeldeth no commendable nouriſhment to the body : it is often vſed in comfits, ſerued at the table with other iunketting diſhes.

B   The oile which is preſſed out of it is pleaſant and delightfull to be eaten, & is taken with bread or any other waies with meat, without any ſence of cooling.

C   A greater force is in the knobs or heads, which do ſpecially preuaile to moue ſleepe, and to ſtay and repreſſe diſtillations or rheums, and come neere in force to *Opium,* but more gentle. *Opium,* or the condenſed iuice of Poppy heads, is ſtrongeſt of all ; *Meconium* (which is the iuice of the heads and leaues) is weaker. Both of them any waies taken either inwardly, or outwardly applied to the head, prouoke ſleepe. *Opium* ſomewhat too plentifully taken doth alſo bring death, as *Pliny* truly writeth.

D   It mitigateth all kind of paines : but it leaueth behinde it oftentimes a miſchiefe worſe than the diſeaſe it ſelfe, and that hard to be cured, as a dead palſie and ſuch like.

E   The vſe of it, as *Galen lib.* 11. of medicines according to the places affected, ſaith, is ſo offen-ſiue to the firme and ſolide parts of the body, as that they had need afterwards to be reſtored.

F   So alſo collyries or eye medicines made with *Opium* haue bin hurtfull to many; inſomuch that they haue weakened the eies and dulled the ſight of thoſe that haue vſed it : what ſoeuer is com-pounded of *Opium* to mitigate the extreame paines of the eares, bringeth hardneſſe of hearing. Wherefore all thoſe medicines and compounds are to bee ſhunned that are to be made of *Opium,* and are not to be vſed but in extreame neceſſitie ; and that is, when no other mitigater or aſſwager of paine doth any thing preuaile, as *Galen* in his third booke of medicines, according to the places affected, doth euidently declare.

G   The leaues of Poppie boiled in water with a little ſugar and drunke, cauſe ſleep : or if it be boi-led without ſugar, and the head, feet, and temples bathed therewith, it doth effect the ſame.

H   The heads of Poppie boiled in water with ſugar to a ſyrrup cauſe ſleepe, and are good againſt rheumes and catarrhes that diſtil and fall down from the brain into the lungs, and eaſe the cough.

I   The green knops of Poppy ſtamped with barly meale and a little barrows greaſe, help S. *Antho-nies* fire, called *Ignis ſacer.*

The

The leaues,knops,and ſeed ſtamped with vineger,womans milke, and ſaffron, cure an *Eryſipelas,* K
(another kind of S. *Anthonies* fire)and eaſe the gout mightily, and put in the fundament as a Cli-
ſter,cauſe ſleepe.

The ſeed of blacke Poppy drunk in wine,ſtoppeth the flux of the belly,and the ouermuch flow- L
ing of womens ſickneſſe.

A Caudle made of the ſeeds of white Poppy,or made into Almond milk,and ſo giuen,cauſeth M
ſleepe.

†   It is manifeſt that this wilde Poppy (which I haue deſcribed in the fift place) is that wher- N
of the compoſition *Diacodium* is to be made ; as *Galen* hath at large treated in his ſeuenth book of
Medicines according to the places affected. *Crito* alſo,and after him *Themiſon* and *Democritus* doe
appoint *ἀγον* or the wilde Poppy to be in the ſame compoſition;and euen the ſame *Democritus* addeth,that it ſhould be that which is not ſowne : and ſuch an one is this,which grows without ſowing.   *Dod.*

Chap. 74.   *Of Corne-Roſe or wilde Poppy.*

1 *Papauer Rhœas.*
Red Poppy,or Corn-Roſe.

2 *Papauer ſpinoſum.*
Prickly Poppy.

¶   *The Deſcription.*

1   THe ſtalks of red Poppy be blacke, tender, and brittle, ſomewhat hairy : the leaues are
cut round about with deepe gaſhes like thoſe of Succorie or wild Rocket. The flours
grow forth at the tops of the ſtalks,being of a beautifull and gallant red colour, with
blackiſh threds compaſſing about the middle part of the head,which being fully growne, is leſſer
than that of the garden Poppy : the ſeed is ſmall and blacke.

† 2   There is alſo a kind hereof in all points agreeing with the former,ſauing that the flours
of this are very double and beautifull,and therein only conſiſts the difference. †

‡ 3   There

‡ 3   There is a small kinde of red Poppy growing commonly wilde together with the first deferibed, which is leffe in all parts, and the floures are of a fainter or ouerworn red, inclining fomewhat to orange.

4   Befides thefe there is another rare plant, which all men, and that very fitly, haue referred to the kinds of Poppy. This hath a flender long and fibrous root, from which arifes a ftalk fome cubit high, diuided into fundry branches, round, crefted, prickly, and full of a white pith. The leaues are diuided after the manner of horned Poppy, fmooth, with white veines and prickly edges. The floure is yellow, and confifts of foure or fiue leaues ; after which fucceeds a longifh head, being either foure, fiue, or fix cornered, hauing many yellow threds incompaffing it : the head whileft it is tender, is reddifh at the top, but being ripe it is blacke, and is fet with many and ftiffe pricks. The feed is round, blacke, and pointed, being fix times as big as that of the ordinarie Poppy. ‡

¶ *The Place.*

They grow in earable lands, among wheat, fpelt, rie, barley, oats, and other graine, and in the borders of fields. ‡ The double red and prickly Poppy are not to be found in this kingdome, vnleffe in the gardens of fome prime Herbarifts.

¶ *The Time.*

The fields are garnifhed and ouerfpred with thefe wilde Poppies in Iune and Auguft.

¶ *The Names.*

† Wilde Poppy is called in Greeke of *Diofcorides*, Μήκων ῥοιάς : in Latine, *Papauer erraticum : Gaza,* according to the Greeke, nameth it *Papauer fluidum* : as alfo *Lobel,* who calls it *Pap. Rhœas,* becaufe the floure thereof foon falls away : which name *Rhœas* may for the fame caufe be common not only to thefe, but alfo to the others, if it be fo called of the fpeedy falling of the floures : but if it be fyrnamed *Rhœas,* of the falling away of the feed (as it appeareth) then fhall it be proper to that which is defcribed in the fift place in the foregoing chapter, out of whofe heads the feed eafily & quickly falls ; as it doth alfo out of this, yet leffe manifeftly. They name it in French *Cocquelicot, Confanons, Panot fauvage :* in Dutch, **Collen bloemen, Cozen rofen :** in high-Dutch, **Klapper Roffen :** in Englifh, red Poppy, and Corn-rofe.

‡ 4   Some haue called this *Ficus infernalis,* from the Italian name *Figo del inferno.* But *Clufius* and *Bauhine* haue termed it *Papauer fpinofum :* and the later of them would haue it, & that not without good reafon, to be *Glaucium* of *Diofcorides, lib. 1. cap. 100.* And I alfo probably coniecture it to be the *Hippomanes* of *Cratevas,* mentioned by the Greeke Scholiaft of *Theocritus,* as I haue formerly briefly declared, Chap. 62.

¶ *The Temperature.*

The facultie of the wild Poppies is like that of the other Poppies, that is to fay, cold, and caufing fleepe.

¶ *The Vertues.*

Moft men being led rather by falfe experiments than reafon, commend the floures againft the Pleurifie, giuing to drinke as foon as the pain comes, either the diftilled water, or fyrrup made by often infufing the leaues. And yet many times it happens, that the paine ceafeth by that meanes, though hardly fometimes, by reafon that the fpittle commeth vp hardly and with more difficulty, efpecially in thofe that are weake and haue not a ftrong conftitution of body. *Baptifta Sardus* might be counted the author of this error, who hath written, That moft men haue giuen the floures of this Poppy againft the pain of the fides ; and that it is good alfo againft the fpitting of bloud.

C H A P. 75. *Of baftard wilde Poppie.*

¶ *The Defcription.*

THe firft of thefe baftard wild Poppies hath flender weake ftems a foot high, rough and hairie, fet with leaues not vnlike to thofe of Rocket, made of many fmal leaues deeply cut or jagged about the edges. The floures grow at the top of the ftalkes, of a red colour, with fome fmall blackneffe toward the bottom. The feed is fmall, contained in little round knobs. The root is fmal and threddy.

2   The fecond is like the firft, fauing that the cods hereof be long, and the other more round, wherein the difference doth confift.

¶ *The Place.*

Thefe plants grow in the Corne fields in Sommerfet fhire, and by the hedges and high-wayes, as ye trauell from London to Bathe. *Lobel* found it growing in the next field vnto a village in Kent called

called Southfleet, my selfe being in his company, of purpose to discouer some strange plants not hitherto written of.

‡ Mr *Robert Lorkin* and I found both these growing in Chelsey fields, as also in those belonging to Hamersmith : but the shorter headed one is a floure of a more elegant colour, and not so plentifull as the other.

1 *Argemone capitulo torulo.*
Bastard wilde Poppy.

2 *Argemone capitulo longiore.*
Long codded wilde Poppy.

¶ *The Time.*

They floure in the beginning of August, and their seed is ripe at the end thereof.

¶ *The Names.*

The Bastard wilde Poppy is called in Greeke, Αργεμωνη: in Latine, *Argemone, Argemonia, Concordia, Concordalis,* and *Herba liburnica :* of some, *Pergalium, Arsela,* and *Sacrocolla Herba :* in English, Windrose, and bastard wilde Poppy.

¶ *The Temperature.*

They are hot and dry in the third degree.

¶ *The Vertues.*

The leaues stamped, and the juyce dropped into the eies ease the inflammation thereof, and cure A the disease of the eye called *Argema*, whereof it tooke his name : which disease when it happeneth on the blacke of the eye it appeares white ; and contrariwise when it is in the white then it appeareth blacke of colour.

The leaues stamped and bound vnto the eies or face that are blacke or blew by meanes of some B blow or stripe, do perfectly take it away. The dry herbe steeped in warme water worketh the like effect.

The leaues and roots stamped, and the juice giuen in drinke, helpe the wringings or gripings of C the belly. The dry herbe infused in warme water doth the same effectually.

The herbe stamped cureth any wound, vlcer, canker, or fistula, being made vp into an vnguent or D salue, with oyle, wax, and a little turpentine.

The juyce taken in the weight of two drammes, with wine, mightily expelleth poyson or ve- E nome.

The

F   The juyce taketh away warts if they be rubbed therewith ; and being taken in meate it helpes the milt or ſpleene if it be waſted.

CHAP. 76.
Of Winde-floures.

¶ The Kindes.

THe ſtocke or kindred of the <i>Anemones</i> or Winde-floures, eſpecially in their varieties of colours, are without number, or at the leaſt not ſufficiently knowne vnto any one that hath written of plants. For <i>Dodonæus</i> hath ſet forth fiue ſorts ; <i>Lobel</i> eight ; <i>Tabernamontanus</i> ten : My ſelfe haue in my garden twelue different ſorts: and yet I do heare of diuers more differing very notably from any of theſe : which I haue briefely touched, though not figured, euery new yeare bringing with it new and ſtrange kindes ; and euery country his peculiar plants of this ſort, which are ſent vnto vs from far countries, in hope to receiue from vs ſuch as our country yeeldeth.

1   <i>Anemone tuberoſa radice.</i>   2   <i>Anemone coccinea multiplex.</i>
     Purple Winde-floure.   Double Skarlet Winde-floure.

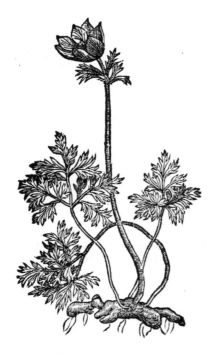

¶ The Deſcription.

1   THe firſt kinde of <i>Anemone</i> or Winde-floure hath ſmall leaues very much ſnipt or jag-ged almoſt like vnto Camomile, or Adonis floure : among which riſeth vp a ſtalke bare or naked almoſt vnto the top ; at which place is ſet two or three leaues like the other : and at the top of the ſtalke commeth forth a faire and beautifull floure compact of ſeuen leaues, and ſometimes eight, of a violet colour tending to purple. It is impoſſible to deſcribe the colour in his full perfection, conſidering the variable mixtures. The root is tuberous or knobby, and very brittle.

2 The

3 *Anemone maxima Chalcedonia Polyanthos.*
The great double VVinde-floure of Bithynia.

4 *Anemone Chalcedonica ſimplici flore.*
The ſingle Winde-floure of Bithynia.

5 *Anemone Bulbocaſtani radice.*
Cheſnut Winde-floure.

6 *Anemone*

2   The ſecond kind of *Anemone* hath leaues like to the precedent, inſomuch that it is hard to diſtinguiſh the one from the other but by the floures onely : for thoſe of this plant are of a moſt bright and faire skarlet colour, and as double as the Marigold ; and the other not ſo. The root is knobby and very brittle, as is the former.

3   The great *Anemone* hath double floures, vſually called the *Anemone* of Chalcedon (which is a city in Bithynia) and great broad leaues deeply cut in the edges, not vnlike to thoſe of the field Crow-foot, of an ouerworne greene colour : amongſt which riſeth vp a naked bare ſtalke almoſt vnto the top, where there ſtand two or three leaues in ſhape like the others, but leſſer ; ſometimes changed into reddiſh ſtripes, confuſedly mixed here and there in the ſaid leaues. On the top of the ſtalke ſtandeth a moſt gallant floure very double, of a perfect red colour, the which is ſometimes ſtriped amongſt the red with a little line or two of yellow in the middle ; from which middle commeth forth many blackiſh thrums. The ſeed is not to be found that I could euer obſerue, but is carried away with the wind. The root is thicke and knobby.

4   The fourth agreeth with the firſt kind of *Anemone*, in roots, leaues, ſtalkes, and ſhape of floures; differing in that, that this plant bringeth forth faire ſingle red floures, and the other of a violet colour, as is aforeſaid.

5   The fifth ſort of *Anemone* hath many ſmall jagged leaues like thoſe of Coriander, proceeding from a knobby root reſembling the root of *Bulbocaſtanum* or earth Cheſnut. The ſtalke riſes vp among the leaues, of two hands high, bearing at the top a ſingle floure, conſiſting of a pale or border of little purple leaues, ſometimes red, and often of a white colour ſet about a blackiſh pointall, thrummed ouer with many ſmall blackiſh haires.

6 *Anemone latifolia Cluſij.*	‡ 7 *Anemone latifolia duplo flauo flore.*
Broad leaued Winde-floure.	The double yellow Winde-floure.

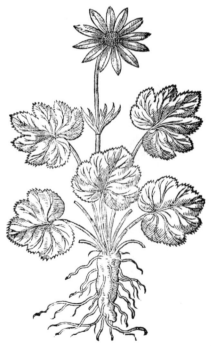 

6   The ſixth hath very broad leaues in reſpect of all the reſt of the *Anemones*, not vnlike to thoſe of the common Mallow, but greene on the vpper part, and tending to redneſſe vnderneath, like the leaues of Sow-bread. The ſtalke is like that of the laſt deſcribed, on the top whereof growes a faire yellow ſtar-floure, with a head ingirt with yellow thrums. The root (ſaith my Author) is a finger long, thicke and knobby.

‡ 7   There is alſo another whoſe lower leaues reſemble thoſe of the laſt deſcribed, yet thoſe which grow next aboue them are more diuided or cut in : amongſt theſe leaues riſeth vp a ſtalke

some

8 *Anemone Geranifolia.*
Storks bill Wind-floure.

9 *Anemone Matthioli.*
Matthiolus white Winde-floure.

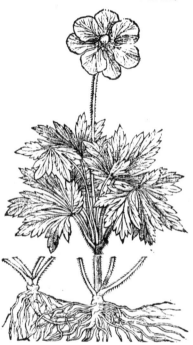

10 *Anemone trifolia.*
Three leaued Wind-floure.

11 *Anemone Paueracea.*
Poppy Winde-floure.

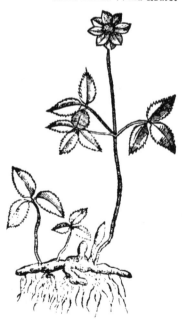

ſome foot high, the top whereof is adorned with a floure conſiſting of two ranks of leaues, whereof thoſe on the out ſide are larger, rounder pointed, and ſometimes ſnipt in a little; the reſt are narrower and ſharper pointed. The colour of theſe leaues is yellow, deeper on the inſide, and on the outſide there are ſome ſmall purple veins running alongſt theſe leaues of the floure. The root is ſome two inches long, the thickneſſe of ones little finger, with ſome tuberous knobs hanging thereat. ‡

8   The eighth hath many large leaues deeply cut and jagged, in ſhape like thoſe of the Storks bill or Pink-needle: among which riſeth vp a naked ſtalke ſet about toward the top with the like leaues, but ſmaller and more finely cut, bearing at the top of the ſtalke a ſingle floure conſiſting of many ſmall blew leaues, which do change ſometimes into purple, and oftentimes into white, ſet about a blackiſh pointall with ſome ſmall threds like vnto a pale or border. The root is thick and knobby.

9   The ninth ſort of Anemone hath leaues like vnto the garden Crow-foot: the ſtalk riſeth vp from amongſt the leaues, of a foot high, bearing at the top faire white floures made of fiue ſmall leaues; in the middle whereof are many little yellow chiues or threds. The root is made of many ſlender threds or ſtrings, contrarie to all the reſt of the Winde-floures.

10   The tenth ſort of Anemone hath many leaues like the common medow Trefoile, ſleightly ſnipt about the edges like a ſaw: on the top of the ſlender ſtalks ſtands a ſingle white flour tending to purple, conſiſting of eight ſmall leaues, reſembling in ſhape the floures of common field Crow-foot. The root is knobby, with certain ſtrings faſtned thereto.

11   The eleuenth kind of Anemone hath many jagged leaues cut euen to the middle rib, reſembling the leaues of *Geranium Columbinum* or Doues foot. The leaues that doe embrace the tender weake ſtalks are flat and ſleightly cut: the floures grow at the top of the ſtalks, of a bright ſhining purple colour, ſet about a blackiſh pointall with ſmall thrums or chiues like a pale. The root is knobby, thicke, and very brittle, as are moſt of thoſe of the Anemones.

¶ *The Place.*

All the ſorts of Anemones are ſtrangers, and not found growing wilde in England; notwithſtanding all and euery ſort of them do grow in my garden very plentifully.

¶ *The Time.*

They floure from the beginning of Ianuarie to the end of April, at what time the flours do fade, and the ſeed flieth away with the wind, if there be any ſeed at all; the which I could neuer as yet obſerue.

¶ *The Names.*

Anemone, or Wind-floure, is ſo called, ϰαὶ τῆ ἀνέμου; that is to ſay, of the wind: for the floure doth neuer open it ſelfe but when the wind doth blow, as *Pliny* writeth: whereupon it is named of diuers, *Herba venti*: in Engliſh, Wind-floure.

Thoſe with double floures are called in the Turky tongue, *Giul*, and *Gul Catamer*: and thoſe with ſmall jagged leaues and double floures are called *Lalé benᷤede*, and *Galipoli lalé*. They cal thoſe with ſmall jagged leaues and ſingle floures, *Biniᷤate, Biniᷤade*, and *Biniᷤante*.

¶ *The Temperature.*

All the kinds of Anemones are ſharp, biting the tongue, and of a binding facultie.

¶ *The Vertues.*

A   The leaues ſtamped, and the juice ſniffed vp into the noſe, purgeth the head mightily.

B   The root champed or chewed procureth ſpitting, and cauſeth water and flegme to run forth of the mouth, as Pellitorie of Spain doth.

C   It profiteth in collyries for the eies, to ceaſe the inflammation thereof.

D   The juice mundifieth and clenſeth malignant, virulent, and corroſiue vlcers.

E   The leaues and ſtalks boiled and eaten of nurſes, cauſe them to haue much milke: it prouoketh the terms, and eaſeth the leproſie, being bathed therewith.

---

‡ **C**ʜᴀᴘ. 77.   *Of diuers other Anemones or Winde-floures.*

¶ *The Kindes.*

‡   **T**Heſe floures which are in ſuch eſteeme for their beautie may well be diuided into two ſorts, that is, the *Latifolia* or broad leaued, and the *Tenuifolia*, or narrow leaues. Now of each of theſe ſorts there are infinite varieties, which conſiſt in the ſingleneſſe and doubleneſſe of the floures, and in their diuerſitie of colours, which would aske a large diſcourſe to handle exactly. Wherefore I onely intend (beſides thoſe ſet downe by our Authour) to giue you the

figures of some few others, with their description, briefly taken out of the Workes of the learned and diligent Herbarist *Carolus Clusius*; where such as desire further discourse vpon this subiect may be abundantly satisfied: and such as do not vnderstand Latine may finde as large satisfaction in the late worke of M.r *Iohn Parkinson*; whereas they shall not onely haue their history at large, but also learne the way to raise them of seed, which hath beene a thing not long knowne (except to some few;) and thence hath risen this great varietie of these floures, wherewith some Gardens so much abound.

¶ The Description.

1   THe root of this is like that of the great double red *Anemone* described in the third place of the precedent chapter; and the leaues also are like, but lesser and deeper coloured. The stalke growes some foot high, slender and greene, at the top whereof groweth a single floure, consisting of eight leaues of a bright shining skarlet colour on the inside, with a paler coloured ring incompassing a hairy head set about with purple thrums: the outside of the floure is hairy or downie. This is *Anem. latifol. simpl. flo.* 16. of *Clusius*.

‡ 1 *Anemone latifolia flore coccineo.*
The broad leaued skarlet Anemone.

‡ 2 *Anemone latifolia flore magno coccineo.*
The skarlet Anemone with the large floure.

2   This in shape of roots & leaues is like the former, but the leaues are blacker, and more shining on their vpper sides: the stalke also is like to others of this kinde, and at the top carrieth a large floure consisting of eight broad leaues, being at the inside of a bright skarlet colour, without any circle; and the thrums that ingirt the hairy head are of a sanguine colour. This head (as in others of this kindred) growes larger after the falling of the floure, and at length turns into a downie substance, wherein a smooth blacke seed is inclosed like as in other Anemones; which sowne as soone as it is ripe vsually comes vp before Winter. This is *Anem. latifol. simpl. flore* 17. of *Clusius*.

3   This differs not from the former but in floures, which are of an orange-tawny colour, like that of Corne-rose, or red Poppy; and the bottomes of the leaues of the floures are of a paler colour, which make a ring or circle about the hairy head. This is the eighteenth of *Clusius*.

Besides these varieties here mentioned, there are many others, which in the colour of the leaues of the floure, or the nailes which make a circle at the bottome thereof, doe differ each from other. Now let vs come to the narrow leaued ones, which also differ little but in colour of their floures.

‡ 3 *Anemone latifolia ByZantina.*
The broad leaued Anemone of Conſtantinople.

‡ 4 *Anemone tenuifolia flore amplo ſanguineo.*
Small leaued Anemone with the ſanguine floure.

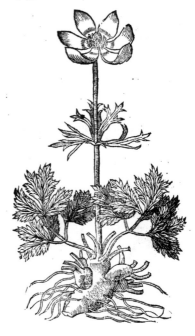

‡ 5 *Anemone tenuifolia flore coccineo.*
The ſmall leaued skarlet Anemone.

‡ 5 *Anemone tenuifol. flo. dilute purpureo.*
The light purple ſmall leaued Anemone.

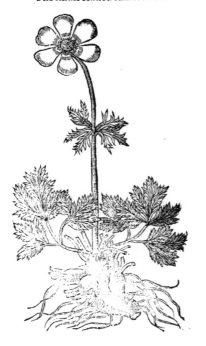

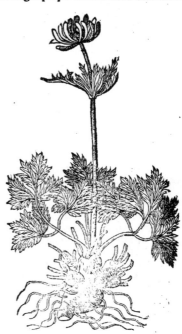

‡ 7 *Anemone tenuifol. flo. ex albido.*
The whitiſh ſmall leaued Anemone.

‡ 8 *Anemone tenuifolia flo. carneo ſtriato.*
The ſtriped fleſh-coloured Anemone.

‡ 9 *Anemone tenuifol. flo. pleno coccin.*
The ſmall leaued double crimſon Anemone.

‡ 10 *Anemone tenuifol. flo. pleno atropurpuracefente.*
The double darke purple Anemone.

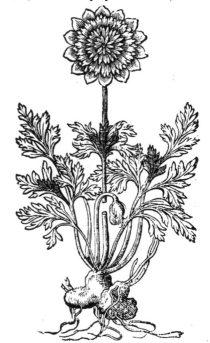

4   The root of this is knotty and tuberous like those of other Anemones, and the leaues are much diuided and cut in like to those of the first described in the former Chapter: the stalke (which hath three or foure leaues ingirting it, as in all other Anemones) at the top sustaineth a faire sanguine floure consisting of six large leaues with great white nailes. The seeds are contained in downie heads like as those of the former. This is *Anem. Tenuifol. simpl. flo.6. of Clusius.*

5   This differs from the former in the floure, which consists of six leaues made somewhat rounder than those of the precedent: their colour is betweene a skarlet and sanguine. And there is a varietie hereof also of a bricke colour. This is the eighth of *Clusius.*

6   This differs from the rest, in that the floure is composed of some fourteene or more leaues, and those of a light purple, or flesh-colour. This is the ninth of *Clusius.*

7   The floure of this is large, consisting of six leaues, being at the first of a whitish greene, and then tending to a flesh colour, with their nailes greene on the outside, and white within, and the threds in the middle of a flesh colour. There is a lesser of this kinde, with the floure of a flesh colour, and white on the outside, and wholly white within, with the nailes greenish. These are the tenth and eleuenth of *Clusius.*

8   This floure also consists of six leaues of a flesh colour, with whitish edges on the outside; the inside is whitish, with flesh-coloured veines running to the middest thereof.

Besides these single kindes, there are diuers double both of the broad and narrow leaued Anemones, whereof I will onely describe and figure two, and refer you to the forementioned Authors for the rest, which differ from these onely in colour.

9   This broad leaued double Anemone hath roots, stalkes, and leaues like those of the single ones of this kinde, and at the top of the stalke there stands a faire large floure composed of two or three rankes of leaues, small and long, being of a kinde of skarlet or orange-tawny colour, the bottomes of these leaues make a whitish circle, which giues a great beauty to the floure; and the downie head is ingirt with sanguine threds tipt with blew. This is the *Pauo major 1. of Clusius.*

10   This in shape of roots, leaues, and stalkes resemble the formerly described narrow leaued Anemones, but the floure is much different from them; for it consists first of diuers broad leaues, which incompasse a great number of smaller narrow leaues, which together make a very faire and beautifull floure: the outer leaues hereof are red, and the inner leaues of a purple Veluet colour.

Of this kinde there are diuers varieties, as the double white, crimson, blush, purple, blew, carnation, rose-coloured, &c.

⁋ *The Place and Time.*

These are onely to be found in gardens, and bring forth their floures in the Spring.

⁋ *Their Names.*

I iudge it nowaies pertinent to set downe more of the names than is already deliuered in their seuerall title and descriptions.

⁋ *The Nature and Vertues.*

A   These are of a hot and biting facultie, and not (that I know of) at this day vsed in medicines, vnlesse in some one or two ointments: yet they were of more vse amongst the Greeke Physitions, who much commend the juyce of them for taking away the scares and scales which grow on the eyes; and by them are called ελαι, and Διννώματα.

B   *Trallianus* also saith, That the floures beaten in oyle, and so anointed, cause haire to grow where it is deficient.

C   The vertues set downe in the former Chapter doe also belong to these here treated of, as these here deliuered are also proper to them. ‡

CHAP. 78.   *Of wilde Anemones, or Winde-floures.*

⁋ *The Kindes.*

Like as there by many and diuers sorts of the garden Anemones, so are there of the wilde kindes also, which do vary especially in their floures.

⁋ *The*

1 *Anemone nemorum lutea.*
Yellow wilde Winde-floure.

2 *Anemone nemorum alba.*
White Winde-floure.

‡ 3 *Anemone nemorum flo. pleno albo.*
The double white wood Anemone.

‡ 4 *Anemone nemorum flo. pleno purpuraſcente.*
The double purpliſh wood Anemone.

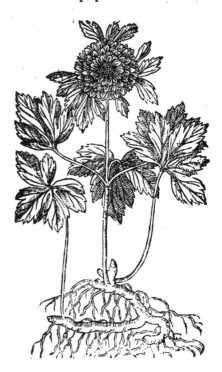

¶ *The Description.*

1   THe first of these wilde *Anemones* hath jagged leaues deepely cut or indented, which do grow vpon the middle part of a weake and tender stalke : at the top whereof doth stand a pretty yellow floure made of six small leaues, and in the middle of the floure there is a little blackish pointell, and certaine slender chiues or threds. The root is small, somewhat knottie and very brittle.

2   The second hath jagged leaues, not vnlike to water Crowfoot or mountaine Crowfoot. The floure groweth at the top of the stalke not vnlike to the precedent in shape, sauing that this is of a milke white colour, the root is like the other.

‡   There is also of this single kinde two other varieties, the one with a purple floure, which we may therefore call *Anemone nemorum purpurea*, the wilde purple Winde-floure. And the other with a Scarlet (or rather a blush) coloured floure, which wee may terme *Anemone nemorum coccinia*, The wilde Scarlet winde-floure. These two differ not in other respects from the white Winde-floure. ‡

3   There is in some choise gardens one of this kinde with white floures very double, as is that of the Scarlet *Anemone*, and I had one of them giuen me by a worshipfull Merchant of London, called Mr *Iohn Franqueuille*, my very good friend.

‡   4   This in roots and stalkes is like the last described wood *Anemones*, or winde-floures. But this and the last mentioned double one haue leaues on two places of their stalkes ; whereas the single ones haue them but in one, and that is about the middle of the stalkes. The floure of this double one consists of some forty or more little leaues, whereof the outermost are the biggest ; the bottomes or nailes of these leaues are of a deepe purple, but the other parts of a lighter blush colour. ‡

¶ *The Place.*

A   All these wilde single *Anemones* grow in most woods and copses through England, except that with the yellow floure, which as yet I haue not seene : notwithstanding I haue one of the greater kindes which beareth yellow floures, whose figure is not expressed nor yet described, for that it doth very notably resemble those with single floures, but is of small moment, either in beautie of the floure, or otherwise. ‡ The double ones grow onely in some few gardens. ‡

¶ *The Time.*

B   They floure from the middest of Februarie vnto the end of Aprill, or the midst of May.

¶ *The Names.*

C   ‡   The first of these by most Writers is referred to the *Ranunculi*, or Crowfeet ; and *Lobel* calls it fitly *Ranunculus nemorosus luteus* : only *Dodonæus, Cæsalpinus*, and our Authour haue made it an *Anemone.*

2   This with the varieties also, by *Tragus, Fuchsius, Cordus, Gesner, Lobel*, and others, is made a *Ranunculus* : yet *Dodonæus, Cæsalpinus*, and our Authour haue referred it to the *Anemones. Clusius* thinkes this to be *Anemone*, Ἀνεμώνη of *Theophrastus.*

3   *Clusius* cals this *Anemone Limonia*, or *Ranunculus syluarum flo. pleno albo.*

4   And he stiles this *Anem. Limonia*, or *Ranunc. syl. flore pleno purpurascente.* ‡

¶ *The Temperature and Vertues.*

The faculties and temperature of these plants are referred to the garden sorts of *Anemones.*

CHAP. 79.   *Of Bastard Anemones, or Pasque floures.*

¶ *The Description.*

1   THe first of these Pasque floures hath many small leaues finely cut or jagged, like those of Carrots : among which rise vp naked stalkes, rough and hairie ; whereupon doe grow beautifull floures bell fashion, of a bright delaied purple colour : in the bottome whereof groweth a tuft of yellow thrums, and in the middle of the thrums it thrusteth forth a small purple pointell : when the whole floure is past there succedeth an head or knob compact of many gray hairy lockes, and in the sollid parts of the knobs lieth the seed flat and hoary, euery seed hauing his owne small haire hanging at it. The root is thicke and knobby, of a finger long, running right downe, and therefore not vnlike to those of the *Anemone*, which it doth in all other parts very notably resemble, and whereof no doubt this is a kinde.

2   There is no difference at all in the leaues, roots, or seeds, betweene this red Pasque floure and the precedent, nor in any other point, but in the colour of the floures ; for whereas the other are

are of a purple colour, these are of a bright red, which setteth forth the difference.

3   The white Passe-floure hath many fine jagged leaues, closely couched or thrust together, which resemble an Holy-water sprinckle, agreeing with the others in roots, seeds, and shape of floures, sauing that these are of a white colour, wherein chiefely consisteth the difference.

‡ 4   This also in shape of roots and leaues little differs from the precedent, but the floures are lesser, of a darker purple colour, and seldome open or shew themselues so much abroad as the other of the first described, to which in all other respects it is very like.

5   There is also another kinde with leaues lesse diuided, but in other parts like those alreadie described, sauing that the floure is of a yellow colour something inclining to red. ‡

1 *Pulsatilla vulgaris.*
Purple Passe-floure.

2 *Pulsatilla rubra.*
Red Passe-floure.

¶ *The Place.*

Ruellius writeth, that the Passe-floure groweth in France in vntoiled places: in Germany they grow in rough and stony places, and oftentimes on rockes.

Those with purple floures do grow very plentifully in the pasture or close belonging to the parsonage house of a small village six miles from Cambrige, called Hildersham: the Parsons name that liued at the impression hereof was Mr *Fuller,* a very kind and louing man, and willing to shew vnto any man the said close, who desired the same.

¶ *The Time.*

They floure for the most part about Easter, which hath mooued mee to name it *Pasque-Floure,* or Easter floure: and often they doe floure againe in September. ‡ The yellow kinde floures in May. ‡

¶ *The Names.*

† Passe-floure is called commonly in Latine, *Pulsatilla:* and of some, *Apium risus,& herba ven-ti. Daleschampius* would haue it to be *Anemone Limonia* and *Samolus* of *Pliny:* in French, *Coquelourdes:* in Dutch, **Keckenschell:** in English, Pasque floure, or Passe-floure, and after the Latine name *Pulsatilla,* or Flaw floure : in Cambridge-shire where they grow, they are named Couentrie bels.

¶ *The*

3 *Pulsatilla flore albo.*
White Passe-floure.

‡ 4 *Pulsatilla flore minore.*
The lesser purple Passe-floure.

¶ *The Temperature.*

Passe-floure doth extremely bite, and exulcerateth and eateth into the skinne if it be stamped and applied to any part of the body, whereupon it hath beene taken of some to be a kinde of Crowfoot, and not without reason, for that it is not inferiour to the Crowfoots : and therefore it is hot and drie.

¶ *The Vertues.*

There is nothing extant in writing among Authors of any peculiar vertue, but they serue onely for the adorning of gardens and garlands, being floures of great beautie.

Chap. 80: *Of Adonis floure.*

¶ *The Description.*

1   THe first hath very many slender weake stalkes, trailing or leaning to the ground, set on euery part with fine jagged leaues very deepely cut like those of Camomill, or rather those of May-weed : vpon which stalkes doe grow small red floures, in shape like the field Crow-foot, with a blackish greene pointell in the middle, which being growne to maturity turneth into a small greenish bunch of seeds, in shape like a little bunch of grapes. The root is smal and threddy.

2   The second differeth not from the precedent in any one point, but in the color of the floures, which are of a perfect yellow colour, wherein consisteth the difference.

¶ *The Place.*

The red floure of Adonis groweth wilde in the West parts of England among their corne, euen as May-weed doth in other parts, and is likewise an enemy to corne as May-weed is : from thence I brought the seed, and haue sowne it in my garden for the beautie of the floures sake. That with the yellow floure is a stranger in England.

¶ *The*

1 *Flos Adonis flore rubro.*
Adonis with red floures.

¶ *The Time.*

They floure in the Summer moneths, May, Iune, and Iuly, and sometimes later.

¶ *The Names.*

Adonis floure is called in Latine, *Flos Adonis* and *Adonidis*: of the Dutchmen, **Feldroszlin**: in English we may cal it red Mayths, by which name it is called of them that dwell where it groweth naturally, and generally red Camomil: in Greeke, ἰσύδμοι: and *Eranthemum*: our London women do call it Rose-a-rubie.

¶ *The Nature.*

There hath not bin any that hath written of the temperature hereof; notwithstanding, so far as the taste thereof sheweth, it is somthing hot, but not much.

¶ *The Vertues.*

The seed of Adonis floure is thought to bee good against the stone: amongst the Antients it was not knowne to haue any other faculty; albeit experience hath of late taught vs, That the seed stamped, and the pouder giuen in wine, ale, or beere to drinke, doth wonderfully and with great effect helpe the colique.

Chap. 81. *Of Docks.*

¶ *The Kindes.*

Dioscorides setteth forth foure kinds of Docks; wilde or sharpe pointed Docke, garden Docke, round leafed Docke, and the soure Dock called Sorrel: besides these, the later Herbarists haue added certain other Docks also, which I purpose to make mention of.

¶ *The Description.*

1 THat which among the Latines signifies to soften, ease, or purge the belly, the same signification hath λαπάσσειν, among the Grecians; whereof *Lapathum* and *Λάπαθα* (as some do reade) tooke their names for herbs which are vsed in pottage and medicine, very wel knowne to haue the power of clensing: of these there be many kindes and differences, great store euerie where now growing; among whom is that which is now called sharp pointed Dock or sharp leafed Dock. It growes in most medowes and by running streams, hauing long narrow leaues sharp and hard pointed: among which come vp round hollow stalks of a brown colour, hauing joints like knees, garnished with such like leaues, but smaller: at the end whereof grow many floures of a pale colour, one aboue another; and after them comes a brownish three square seed, lapped in browne chaffie husks like Patience. The root is great, long, and yellow within.

‡ There is a varietie of this with crisped or curled leaues, whose figure was by our Author giuen in the second place in the following chapter, vnder the title of *Hydrolapathum minus*. ‡

2 The second kind of sharp pointed Dock is like the first, but much smaller, and doth bear his seed in rundles about his branches, in chaffie husks like Sorrell; not so much in vse as the former, called also sharp pointed Dock.

‡ 3 This in roots, stalks, and seeds is like to the precedent, but the leaues are shorter & rounder than those of the first described, wherein consists the chiefe difference betwixt this and it ‡

¶ *The Place.*

These kinds of Docks grow, as is beforesaid, in medowes and by riuers sides.

¶ *The*

† 1 *Lapathum acutum.*
Sharpe pointed Docke.

2 *Lapathum acutum minimum.*
Small ſharpe Docke.

‡ 3 *Lapathum ſylueſtre fol. minus acutum.*
The roundiſh leaued wilde Docke.

A
B
C

¶ *The Time.*

They floure in Iune and Iuly.

¶ *The Names.*

They are called in Latine, *Lapathum acutum, Rumex, Lapatium,* and *Lapathium* : of ſome, *Oxylapathum:* in Engliſh, Dock, and ſharp pointed dock, the greater and the leſſer : of the Greeks, ἱξανδικ in high Dutch, **Mengelwurtz, Streiſſwurtz:** in Italian, *Rombice* : in Spaniſh, *Romaza, Paradella:* in low-Dutch, **Patich** (which word is deriued of *Lapathum*) and alſo **Peerdick:** in French, *Pareille.*

‡ The third is *Lapathum folio retuſo,* or *minus acuto* of *Lobel ;* and *Hippolapathum ſylueſt,* of *Taber.*‡

¶ *The Nature and Vertues.*

Theſe herbs are of a mixture between cold and heate, and almoſt dry in the third degree, eſpecially the ſeed, which is very aſtringent.

The pouder of any of the kinds of docks drunk in wine, ſtoppeth the lask and bloudy flix, and eaſeth the pain of the ſtomacke.

The roots boiled til they be very ſoft, & ſtamped with barrowes greaſe, and made into an ointment, helps the itch and all ſcuruy ſcabs & mangineſſe : and for the ſame purpoſe it ſhall be neceſſarie to boile them in water as aforeſaid , and the party to be rubbed and bathed therewith.

† The firſt figure in the former edition was of *Hydrolapathum nigrum,* being the firſt in the next chapter ; and the figure of that wee giue you in the third place of this chapter, was that in the firſt place of the former Chap. vnder the forementioned title.

CHAP.

CHAP. 82. *Of water-Docke.*

† 1 *Hydrolapathum magnum.*
Great water-Docke.

† 2 *Hydrolapathum minus.*
Small water-Docke.

† 3 *Hippolapathum sativum.*
Patience, or Monks Rubarb.

4 *Hippolapathum rotundifolium.*
Baſtard Rubarb.

‡ 5 *Lapathum satiuum sanguineum.*
Bloudwort.

¶ *The Description.*

1   THe great water-dock hath very long
and great leaues, stiffe and hard, not
vnlike to the garden Patience, but
much longer. The stalke riseth vp to a great
height, oftentimes to the height of fiue foot or
more. The floure groweth at the top of the stalk
in spoky tufts, brown of colour. The seed is contained in chaffie husks three square, of a shining
pale colour. The root is very great, thick, brown
without and yellowish within.

2   The smal water-Dock hath short narrow
leaues set vpon a stiffe stalke. The floures grow
from the middle of the stalke vpward in spoky
rundles, set in spaces by certain distances round
about the stalk, as are the floures of Horehound:
which Docke is of all the kinds most common,
and of least vse, and takes no pleasure or delight
in any one soile or dwelling place, but is found
almost euery where, as well vpon the land as in
waterie places, but especially in gardens among
good and wholsome pot-herbs, being there better knowne, than welcome or desired: wherefore
I intend not to spend farther time about his description.

3   The garden Patience hath very strong
stalks furrowed or chamfered, of eight or nine
foot high when it groweth in fertile ground, set
about with great large leaues like to those of the
water-Docke, hauing alongst the stalkes toward the top floures of a light purple colour declining
to brownnesse. The seed is three square, contained in thin chaffie husks like those of the common
Docke. The root is very great, browne without and yellow within, in colour and taste like the true
Rubarb.

4   Bastard Rubarb hath great broad round leaues in shape like those of the garden Bur-docke.
The stalke and seeds are so like vnto the precedent that the one cannot be knowne from the other,
sauing that the seeds of this are somewhat lesse. The root is exceeding great and thicke, very like
vnto the Rha of Barbary, as well in proportion as colour and taste, and purgeth after the same manner, but must be taken in greater quantity, as witnesseth that famous learned physition now liuing,
Mr Dr *Bright*, and others who haue experimented the same.

5   This fift kind of Dock is best knowne vnto all of the stocke or kindred of Dockes: it hath
long thin leaues sometimes red in euery part thereof, and often striped here & there with lines and
strakes of a darke red colour: among which rise vp stiffe brittle stalkes of the same colour: on the
top whereof come forth such floures and seed as the common wild Docke hath. The root is likewise red, or of a bloudy colour.

¶ *The Place.*

They grow for the most part in ditches and water-courses, very common thorow England. The
two last saue one do grow in gardens: my selfe and others in London and elsewhere haue them
growing for our vse in physicke and Surgerie. The last is sown for a pot-herb in most gardens.

¶ *The Time.*

Most of the Docks doe rise vp in the Spring of the yeare, and their seed is ripe in Iune and August.

¶ *The Names.*

The Docke is called in Greeke Λάπαθον: in Latine, *Rumex*, and *Lapathum*: yet *Pliny, lib. 19. ca. 12.*
seems to attribute the name of *Rumex* only to the garden Docke.

The

The Monks Rubarb is called in Latine, *Rumex ſativus*, and *Patientia*, or Patience, which word is borrowed of the French, who call this herb *Patience* : after whom the Dutchmen alſo name this pot herb **Patientie** : of ſome, *Rhabarbarum Monachorum*, or Monks Rubarb, becauſe as it ſeemes ſome Monke or other hath vſed the root hereof in ſtead of Rubarb.

Bloudwort or bloudy Patience is called in Latine *Lapathum ſanguineum* : of ſome, *Sanguis Draconis*, of the bloudy colour wherewith the whole plant is poſſeſt : it is of pot-herbs the chiefe or principall, hauing the propertie of the baſtard Rubarb, but of leſſe force in his purging qualitie.

¶ The Temperature.

Generally all the Docks are cold, ſome little and moderatly, and ſome more : they do all of them dry, but not all after one manner ; yet ſome are of opinion, that they are dry almoſt in the third degree.

¶ The Vertues.

The leaues of the garden Dock or Patience may be eaten, and are ſomwhat cold but more moiſt,　A and haue withall a certain clamminneſſe, by reaſon whereof they eaſily and quickely paſſe through the belly when they be eaten : and *Dioſcorides* writeth, That all the Docks being boiled, do mollifie the belly : which thing alſo *Horace* hath noted in his ſecond booke of Sermons, *Satyr* 4. writing thus :

> ——————— *Si dura morabitur alvus*
> *Mugilus, & viles pellent obſtantia conchæ,*
> *Et Lapathi brevis herba.* ———

He calleth it a ſhort herb, being gathered before the ſtalke be growne vp, at which time it is fitteſt　B to be eaten.

And being ſodden, it is not ſo pleaſant to be eaten as either Beets or Spinach : it ingenders moiſt　C bloud of a mean thickneſſe, and nouriſheth little.

The leaues of the ſharp pointed Docks are cold and dry, but the ſeed of Patience and the water　D Docke do coole, with a certain thinneſſe of ſubſtance.

The decoction of the roots of Monks Rubarb is drunk againſt the bloudy flix, the lask, the wam-　E bling of the ſtomack comming of choler, and alſo againſt the ſtinging of ſerpents, as *Dioſcorides* writeth.

It is alſo good againſt the ſpitting of bloud, being taken with *Acacia* (or his *ſuccedaneum* the dri-　F ed juice of ſloes) as *Pliny* ſaith.

Monks Rubarb or Patience is an excellent wholſome pot-herb, for being put into the pottage in　G ſome reaſonable quantitie, it looſens the belly, helps the jaundice, the timpanie, and ſuch like diſeaſes proceeding of cold cauſes.

If you take the roots of Monks Rubarb and red Madder of each halfe a pound, Sena foure oun-　H ces, Aniſe ſeed and Licorice of each two ounces, Scabious and Agrimonie of each one handfull ; ſlice the roots of the Rubarb, bruiſe the Aniſe ſeed and Licorice, breake the herbs with your hands and put them into a ſtone pot called a ſtean, with foure gallons of ſtrong ale, to ſteep or infuſe the ſpace of three daies, and then drinke this liquor as your ordinary drink for three weeks together at the leaſt, though the longer you take it, ſo much the better ; prouiding in a readineſſe another ſtean ſo prepared, that you may haue one vnder another, being alwaies carefull to keep a good diet : it cureth the dropſie, the yellow jaundice, all manner of itch, ſcabs, breaking out and mangineſſe of the whole body : it purifieth the bloud from all corruption, preuaileth againſt the greene ſicknes very greatly, and all oppilations or ſtoppings, makes yong wenches look faire and cherry-like, bringing downe their termes, the ſtopping whereof hath cauſed the ſame.

The ſeed of Baſtard Rubarb is manifeſtly aſtringent, inſomuch that it cures the bloudy flix, mix-　I ed with the ſeed of Sorrell, and giuen to drink in red wine.

There haue not bin any other faculties attributed to this plant, either of the antient or later wri-　K ters, but generally of all it hath bin referred to the other Docks or Monks Rubard : of which number I aſſure my ſelfe this is the beſt, and doth approch neereſt vnto the true Rubarb. Many reaſons induce me ſo to thinke and ſay ; firſt, this hath the ſhape and proportion of Rubarb, the ſame color both within and without, without any difference, they agree as well in taſte as ſmell ; it coloureth the ſpittle of a yellow colour when it is chewed, as Rubarb doth ; and laſtly, it purgeth the belly after the ſame gentle manner as the right Rubarb doth ; only herein it differeth, that this muſt be giuen in three times the quantitie of the other. Other diſtinctions and differences, with the temperature and other circumſtances, I leaue to the learned Phyſitions of our London colledge, who are very well able to ſearch this matter, as a thing far aboue my reach, being no Graduat, but a Country Scholler, as the whole frame of this hiſtorie doth well declare : but I hope my good meaning will be well taken, conſidering I do my beſt : and I doubt not but ſome of greater learning wil perfect that which I haue begun according to my ſmall skill, eſpecially the ice being broken to him, and the wood rough-hewn to his hand. Notwithſtanding I thinke it good to ſay thus much more in my own defence, That although there be many wants and defects in mee, that were requiſite to performe ſuch a worke ; yet may my long experience by chance happen vpon ſome one thing or o-ther that may do the Learned good : conſidering what a notable experiment I learned of one *Iohn Benet* a Surgeon of Maidſtone in Kent, (a man as ſlenderly learned as my ſelfe) which he practiſed vpon a butchers boy of the ſaid towne, as himſelfe reported vnto me. His practiſe was this : Being deſired to cure the foreſaid Lad of an ague which did grieuouſly vex him, hee promiſed him a me-dicine ; and for want of one for the preſent (for a ſhift, as himſelfe confeſſed vnto me) he tooke out of his garden three or four leaues of this plant of Rubarb, which my ſelfe had among other ſimples giuen him, which he ſtamped and ſtrained with a draught of ale, and gaue it the lad in the morning to drinke : it wrought extreamely downeward and vpward within one houre after, and neuer ceaſed vntil night : in the end, the ſtrength of the boy ouercame the force of the phyſick, it gaue ouer wor-king, and the Lad loſt his Ague ; ſince which time (as he ſaith) he hath cured with the ſame medi-cine many of the like maladie, hauing euer great regard to the quantitie, which was the cauſe of the violent working in the firſt cure. By reaſon of which accident, that thing hath bin reuealed to poſteritie, which heretofore was not ſo much as dreamed of. Whoſe blunt attempt may incourage ſome ſharper Wit and greater Iudgement in the faculties of plants, to ſeeke farther into their na-ture than any of the Antients haue done ; and none fitter than the learned Phyſitions of the Col-ledge of London, where are many ſingularly well learned and experienced in naturall things.

L   The roots ſliced and boiled in the water of *Carduus benedictus*, to the conſumption of the third part, adding thereto a little hony, and eight or nine ſpoonfulls of the decoction thereof drunke be-fore the fit, cure the ague in two or three times ſo taking it at the moſt : vnto robuſtious or ſtrong bodies twelue ſpoonſuls may be giuen. This experiment was practiſed by a worſhipfull Gentle-woman Miſtreſſe *Anne Wylbraham* vpon diuers of her poore neighbours with good ſucceſſe.

†   The figure that was in the firſt place was of *Lapathum fol. minus acuto*, deſcribed by me in the third place of the preceding chapter. The ſecond was of *La-pathum acutum criſpum* of Tabern. The third was of *Hydrolapathum minus*.

CHAP. 83. Of Rubarb.

‡   I T hath hapned in this as in many other forrein medicines or Simples, which though they be of great and frequent vſe, as Hermodactyls, Muske, Turbith, &c. yet haue we no certain knowledge of the very place which produces them, nor of their exact manner of growing : which hath giuen occaſion to diuers to think diuerſly, and ſome haue bin ſo bold as to counterfeit figures out of their own fancies, as *Matthiolus*. So that this ſaying of *Ruellius* is found very true, *Nul-la medicinæ pars magis incerta, quam quæ ab alio quam noſtro orbe petitur.* But we will endeauor to ſhew you more certaintie of this here treated, than was known vntill of very late years.

¶ *The Deſcription.*

1   T His kind of Rubarb hath very great leaues ſomewhat ſnipt or indented about the ed-ges like the teeth of a Saw, not vnlike the leaues of *Enula campana*, called by the vulgar ſort Elecampane, but greater : amongſt which riſeth vp a ſtraight ſtalke of two cubits high, bearing at the top a ſcaly head like thoſe of Knap-weed or *Iacea major* : in the midſt of which knap or head thruſteth forth a faire floure conſiſting of many purple threds like thoſe of the Arti-choke : which being paſt, there followeth a great quantitie of down, wherein is wrapped long ſeed like vnto the great Centorie, which the whole plant doth very well reſemble : the root is long and thicke, blackiſh without, and of a pale colour within ; which being chewed makes the ſpittle verie yellow, as doth the Rubarb of Barbarie.

‡ 2   This other baſtard Rha, which is alſo of *Lobels* deſcription, hath a root like that of the laſt deſcribed ; but the leaues are narrower, almoſt like thoſe of the common Dock, but hoarie on the other ſide : the ſtalk growes vp ſtraight, and beares ſuch heads and floures as the precedent.

‡ 3   I haue thought good here to omit the counterfeit figure of *Matthiolus*, giuen vs in this place by our Author, as alſo the hiſtorie, which was not much pertinent, and in lieu of them to pre-ſent you with a perfect figure and deſcription of the true *Rha Ponticum* of the Antients, which was firſt of late diſcouered by the learned *Proſper Alpinus*, who writ a peculiar tract thereof, and it is alſo again figured and deſcribed in his Worke *de Plantis exoticis*. Our countreyman Mr *Iohn Parkinſon* hath alſo ſet forth very well both the figure and deſcription hereof in his *Paradiſus terreſtris*. This
Plant

1 *Rha Capitatum Lobelij.*
Turky Rubarb.

2 *Rha Capitatum angustifolium.*
The other bastard Rubarb.

‡ 3 *Rha verum antiquorum.*
The true Rubarb of the Antients.

Rhabarbarum siccatum.
The dry roots of Rubarb.

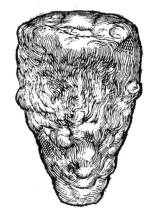

plant hath many large roots diuerſly ſpreading in the ground,of a yellow colour,from which grow
vp many very great leaues like thoſe of the Butter-burre,but of a freſh green colour,with great and
manifeſt red veins diſperſed ouer them.   The ſtalke alſo is large and creſted,ſending forth ſundrie
branches bearing ſmal white floures,which are ſucceeded by ſeeds three ſquare and browniſh, like
as thoſe of other Docks. Dr Liſter one of his Majeſties Phyſitions, was the firſt that enriched this
kingdome with this elegant and vſefull plant,by ſending the ſeeds thereof to Mr Parkinſon. Proſper
Alpinus proues this to be the true Rha of the Antients,deſcribed by Dioſcorides,lib.3.ca.2. yet neither he nor any other(that I know of) haue obſerued a fault, which I more than probably ſuſpect
to be in the text of Dioſcorides in that place , namely in the word μϵλαίν,which I iudge ſhould bee
μλϵ, that is, yellow , and not blacke, as Ruellius and others haue tranſlated it : now μλϵ is a word
often vſed by Dioſcorides, as may appeare by the chapters of Hieracium magnum & parvum, Conyza,
Peucedanum,Ranunculus,and diuers others, and I ſuſpect the like fault may be found in ſome other
places of the ſame Author.   But I will no further inſiſt vpon this , ſeeing the thing it ſelfe in all
other reſpects,as alſo in yellowneſſe,ſhews it ſelſe to be that deſcribed by Dioſcorides,and that my
conjecture muſt therefore be true. Beſides,the root whereto he compares it is vπϵρυθϵϵ , that is , Rubeſcens,or rather ex flavo rubeſcens, as any verſed in reading Dioſcorides may eaſily gather by diuers
places in him. Now I here omit his words, becauſe they are in the next deſcription alledged by
our Author, as alſo the deſcription of our ordinarily vſed Rubarb, for that it is ſufficiently deſcribed vnder the following title of the choice thereof. Mr Parkinſon is of opinion, that this is the true
Rubarb vſed in ſhops,only leſſe heauy,bitter,and ſtrong in working, by reaſon of the diuerſitie of
our Clymat from that whereas the dried Rubarb brought vs vſually growes. This his opinion is
very probable,and if you compare the roots together you may eaſily be enduced to be of the ſame
beleeſe. ‡
  † 4   The Pontick Rubarb is leſſe and ſlenderer than that of Barbary. Touching Pontick Rubarb Dioſcorides writeth thus : Rha,that diuers call Rheon,which growes in thoſe places that are beyond Boſphorus,from whence it is brought,hath yellow roots like to the great Centorie,but leſſe
and redder, ἄϵμϵς, that is to ſay,without ſmell (Dodonæus thinks it ſhould be ἄϵμϵς, that is,well ſmelling) ſpongie,and ſomething light. That is the beſt which is not worm-eaten, and taſted is ſomewhat viſcide with a light aſtriction,and ſhewed becomes of a yellow or Saffron colour.

¶ The Place.

It is brought out of the country of Sina (commonly called China)which is toward the Eaſt in
the vpper part of India,and that India which is without the riuer Ganges,and not at all ex Scenita-
rum prouincia,as many do vnaduiſedly think,which is in Arabia Fælix,& far from China:it growes
on the ſides of the riuer Rha, now called Volga, as Ammianus Marcellus ſaith,which riuer ſprings
out of the Hyperborean mountains,and running through Muſcouia,falls into the Caſpian or Hir-
can ſea.
  ‡ The Rha of the Antients growes naturally,as Alpinus ſaith,vpon the hil Rhodope in Thrace,
now called Romania. It growes alſo,as I haue bin informed,vpon ſome mountains in Hungary.It
is likewiſe to be found growing in ſome of our choice gardens.   ‡

¶ The choice of Rubarb.

The beſt Rubarb is that which is brought from China freſh & new,of a light purpliſh red,with
certain veins and branches of an vncertain varietie of colour, commonly whitiſh ; but when it is
old the colour becomes illfauored by turning yellowiſh or pale,but more if it be worme-eaten : being chewed in the mouth it is ſomthing gluie or clammy,and of a Saffron colour,which being rubbed vpon paper or ſome white thing ſheweth the colour more plainly : the ſubſtance thereof is neither hard or cloſely compact,nor yet heauy,but ſomthing light,and as it were in a middle betwixt
hard and looſe,and ſomthing ſpongie : it hath alſo a pleaſing ſmell. The ſecond in goodnes is that
which comes from Barbary. The laſt and worſt from Boſphorus and Pontus,

¶ The Names.

It is commonly called in Latine Rha Barbarum,or Rha Barbaricum : of diuers,Rheu Barbarum:the
Moors and Arabians do more truly name it Raued Seni,à Senenſi prouincia,from whence it is brought
into Perſia and Arabia,and afterward into Europe : and likewiſe from Tanguth through the land
of Cataia into the Sophy of Perſia his country,and from thence into Egypt,and ſo into Europ.It
is called of the Arabians and the people of China & the parts adiacent,Rauend Cini,Raved Seni,and
Raued Sceni : in ſhops,Rhabarbarum : in Engliſh,Rubarb,and Rewbarb.

¶ The

4　Rha Ponticum Siccatum.
Rubarb of Pontus dried.

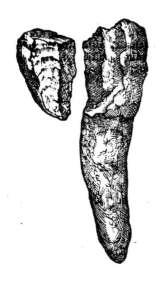

¶ *The Nature.*

Rubarb is of a mixt ſubſtance, temperature, and faculties : ſome of the parts thereof are earthy, binding and drying : others thin, airous, hot, and purging.

¶ *The Vertues.*

Rubarb is commended by *Dioſcorides*, a- **A**
gainſt windineſſe, weakneſſe of the ſtomack, and all griefes thereof, Convulſions, diſeaſes of the ſpleen, liuer, and kidnies, gripings, and inward gnawings of the guts, infirmities of the bladder and cheſt, ſwellings about the heart, diſeaſes in the matrix, pain in the huckle bones, ſpitting of bloud, ſhortneſſe of breath, yexing, or the hicket, the bloudy flix, the laske proceeding of raw humors, fits in agues, and againſt the bitings of venomous beaſts.

Moreouer hee ſaith, that it taketh away **B**
blacke and blew ſpots, and Tettars or Ringwormes, if it be mixed with vineger, and the place anointed therewith.

Galen affirmes it to be good for burſtings, **C**
cramps, and convulſions, & for thoſe that are ſhort winded and ſpit bloud.

But touching the purging faculty neither **D**
Dioſcorides **not** *Galen* haue written any thing, becauſe it was not vſed in thoſe days to purge with. *Galen* held opinion, That the thin airous parts do make the binding qualitie of more force, not becauſe it doth reſiſt the cold and earthy ſubſtance, but becauſe it carrieth the ſame, and maketh it deeply to pierce, and therby to work the greater effect ; the dry and thin eſſence containing in it ſelfe a purging force & quality to open obſtructions, but helped and made more facile by the ſubtill and airous parts. *Paulus Ægineta* ſeemeth to be the firſt that made triall of the purging facultie of Rubarb ; for in his firſt book, *cap.* 43 he makes mention thereof, where he reckons vp Turpentine among thoſe medicines which make the bodies of ſuch as are in health ſoluble : But when we purpoſe (ſaith he) to make the turpentine more ſtrong, we adjvnto it a little Rubarb. The Arabians that followed him brought it to a farther vſe in phyſick, as chiefly purging downward choler, and oftentimes flegme.

The purgation which is made with Rubarb is profitable and fit for all ſuch as be troubled with **E**
choler, and for thoſe that are ſicke of ſharpe and tertian feuers, or haue the yellow jaundice, or bad liuers.

It is a good medicine againſt the pleuriſie, inflammation of the lungs, the ſqinancy or Squincy, **F**
madneſſe, frenſie, inflammation of the kidnies, bladder, and all the inward parts, and eſpecially againſt S. *Anthonies* fire, as well outwardly as inwardly taken.

Rubarb is vndoubtedly an eſpeciall good medicine for the liuer and infirmities of the gall ; for **G**
beſides that it purgeth forth cholericke and naughty humors, it remooueth ſtoppings out of the conduits.

It alſo mightily ſtrengthneth the intrals themſelues, inſomuch that Rubarb is iuſtly termed of **H**
diuers, the life of the liuer : for *Galen, lib.* 11. of the method or manner of curing, affirmeth, that ſuch kinds of medicines are moſt fit and profitable for the liuer, as haue ioyned with a purging and opening qualitie an aſtringent or binding power. The quantitie that is to be giuen is from one dram to two ; and the infuſion from one and a halfe to three.

It is giuen or ſteeped, and that in hot diſeaſes, with the infuſion or diſtilled water of Succory, **I**
Endiue, or ſome other of the like nature, and likewiſe in Whay ; and if there be no heat it may be giuen in wine.

It is alſo oftentimes giuen beeing dried at the fire, but ſo, that the leaſt or no part thereof at all **K**
be burned ; and beeing ſo vſed, it is a remedie for the bloudy Flix, and for all kinds of Laskes : for

it both purgeth away naughty and corrupt humors, and likewise withall stoppeth the belly.

K　　The same being dried after the same manner, doth also stay the ouermuch flowing of the Monethly Sickenesse, and stoppeth bloud in any part of the body, especially that which commeth thorow the bladder; but it should bee giuen in a little quantitie, and mixed with some other binding thing.

L　　*Alesues* saith, That Rubarb is an harmlesse medicine, and good at all times, and for all ages, and likewise for children and women with childe.

M　　‡　My friend Mr *Sampson Iohnson* Fellow of *Magdalen* Colledge in Oxford assures me, That the Physitions of Vienna in Austria vse scarce any other at this day than the Rubarb of the Antients, which growes in Hungary not far from thence : and they prefer it before the dried Rubarb brought out of Persia and the East Indies, because it hath not so strong a binding facultie as it, neither doth it heate so much, only it must be vsed in somewhat a larger quantitie. ‡

<hr />

<h1 style="text-align:center">CHAP. 84.</h1>
<h2 style="text-align:center">Of Sorrell.</h2>

¶ *The Kindes.*

THere be diuers kinds of Sorrell, differing in many points, some of the garden, others wild; some great, and some lesser.

1 *Oxalis, siue Acetosa.*
Sorrell.

2 *Oxalis tuberosa.*
Knobbed Sorrell.

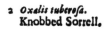

¶ *The Description.*

THough *Dioscorides* hath not expressed the *Oxalides* by that name, yet none ought to doubt but that they were taken and accounted as the fourth kinde of *Lapathum* : for although some
like

like it not wel that the seed should be said to be *Drimus* ; yet that is to bee vnderstood according to the common phrase, when acride things are confounded with those that be sharp and soure: else we might accuse him of such ignorance as is not amongst the simplest women. Moreouer, the word *Oxys* doth not only signifie the leafe, but the sauor and tartnesse, which by a figure drawn from the sharpnesse of kniues edges, is therefore called sharpe : for τὸ χύνως signifieth a sharpe or soure juyce which pierceth the tongue like a sharp knife : whereupon also *Lapathum* may be called *Oxalis*, as it is indeed. The leaues of this are thinner, tenderer, and more vnctuous than those of *Lapathum acutum*, broader next to the stem, horned and erected like Spinach & *Atriplex*. The stalk is much streaked, reddish, and full of juice : the root is yellow and fibrous, the seed sharp, cornered and shining, growing in chaffie husks like the other Docks.

2　　The second kind of *Oxalis* or Sorrel hath large leaues like Patience, confusedly growing together vpon a great tall stalke, at the top whereof grow tufts of a chaffie substance. The root is tuberous, much like the Peonie or rather *Filipendula*, fastned to the lower part of the stem with smal long strings or laces.

3　　The third kind of Sorrell groweth very small, branching hither and thither, taking hold (by new shoots) of the ground where it groweth, whereby it disperseth it selfe farre abroad : the leaues are little and thin, hauing two small leaues like eares fastned thereto, in shew like the herb *Sagittaria* : the seed in taste is like the other of his kinde.

4　　The fourth kinde of Sorrell hath leaues somwhat round and cornered, of a whiter color than the ordinarie, and hauing two short eares anexed vnto the same. The seed and root in taste is like the other Sorrels.

3　*Oxalis tenuifolia.*
Sheeps Sorrel.

4　*Oxalis Franca seu Romana.*
Round leaued or French Sorrel.

5　　This kind of curled Sorrell is a stranger in England, and hath very long leaues in shape like garden Sorrell, but curled and crumpled about the edges as is the curled Colewort. The stalke riseth vp among the leaues, set here and there with the like leaues, but lesser. The floures, seeds, and roots are like the common Sorrell or soure Docke.

6　　The small Sorrell that groweth vpon dry barren sandy ditch banks, hath small grassy leaues somewhat forked or crossed ouer like the crosse hilt of a Rapier. The stalks rise vp amongst the
leaues,

6 *Oxalis minor.*
Small Sorrell.

leaues, ſmall, weake, and tender, of the ſame ſoure taſte that the leaues are of. The floure, ſeed, and root is like the other Sorrels, but altogether leſſer.

7   The ſmalleſt ſort of Sorrel is like vnto the precedent, ſauing that the loweſt leaues that lie vpon the ground be ſomewhat round, and without the little ears that the other hath, which ſetteth forth the difference.

‡ 8   There is alſo kept in ſome gardens a verie large Sorrel, hauing leaues thicke, whitiſh, and as large as an ordinarie Docke, yet ſhaped like Sorrell, and of the ſame acide taſte. The ſtalks and ſeed are like thoſe of the ordinarie, yet whiter coloured. ‡

¶ *The Place.*

† The common Sorrel groweth for the moſt part in moiſt medowes and gardens. The ſecond by waters ſides, but not in this kingdome that I know of. The fourth and fift alſo are garden plants with vs ; but the third and ſixth grow vpon grauelly and ſandy barren ground and ditch banks.

¶ *The Time.*

They flouriſh at that time when as the other kindes of Docks do floure.

¶ *The Names.*

Garden Sorrel is called in Greeke, ιξανς, and αιαξυος : of *Galen,* ιξυλαπαθυ : that is to ſay, *Acidum Lapathum,* or *Acidus rumex,* ſoure Docke ; and in ſhops commonly *Acetoſa :* in the Germane Tongue, **Sawz ampffer :** in low-Dutch, **Surckele,** and **Surinck :** the Spaniards, *Azederas, Agrelles,* and *Azedas :* in French, *Ozeille,* and *Surelle, Aigrette :* in Engliſh, garden Sorrell.

The ſecond is called of the later Herbariſts, *Tuberoſa acetoſa,* and *Tuberoſum lapathum :* in Engliſh, Bunched or knobbed Sorrel.

The third is called in Engliſh, Sheeps Sorrell : in Dutch, **Schap Surkel.**

The fourth, Roman Sorrell, or round leaued Sorrell.

The fift, Curled Sorrell.

The ſixt and ſeuenth, Barren Sorrell, and dwarfe Sheeps Sorrell.

‡ The eighth is called *Oxalis,* or *Acetoſa maxima latifolia,* great broad leaued Sorrell. ‡

¶ *The Temperature.*

The Sorrels are moderatly cold and dry.

¶ *The Vertues.*

A   Sorrell doth vndoubtedly coole and mightily dry ; but becauſe it is ſoure, it likewiſe cutteth tough humors.

B   The iuice hereof in Sommer time is a profitable ſauce in many meats, and pleaſant to the taſte : it cooleth an hot ſtomacke, moueth appetite to meat, tempereth the heat of the liuer, and openeth the ſtoppings thereof.

C   The leaues are with good ſucceſſe added to decoctions which are vſed in agues.

D   The leaues of Sorrel taken in good quantitie, ſtamped and ſtrained into ſome ale, and a poſſet made thereof, coole the ſicke body, quench thirſt, and allay the heate of ſuch as are troubled with a peſtilent feuer, hot ague, or any great inflammation within.

E   The leaues ſodden, and eaten in manner of a Spinach tart, or as meat, ſoften and looſen the belly and attemper and coole the bloud exceedingly.

F   The ſeed of Sorrell drunke in groſſe red wine ſtoppeth the laske and bloudy flix.

CHAP.

CHAP. 85. *Of Biſtort or* **Snake-weed.**

¶ *The Deſcription.*

1 THe great Biſtort hath long leaues much like Patience,but ſmaller,and more wrinkled or crumpled, on the vpper ſide of a darke green,and vnderneath of a blewiſh greene colour much like Woad. The ſtalke is long, ſmooth, and tender, hauing at the top a ſpiked knap or eare ſet full of ſmal whitiſh floures declining to carnation.The root is all in a lumpe, without faſhion ; within of a reddiſh colour like vnto fleſh, in taſte like the kernell of an acorne.

2 The ſmall Biſtort hath leaues about three inches long,and of the bredth of a mans nail;the vpper ſide is of a green colour, and vnderneath of an ouerworne greeniſh colour : amongſt which riſeth vp a ſtalke of the height of a ſpan, full of ioints or knees , bearing at the top ſuch floures as the great Biſtort beareth;which being fallen,the ſeeds appeare of the bigneſſe of a tare,reddiſh of colour,euery ſeed hauing one ſmall green leafe faſtned thereto, with many ſuch leaues thruſt in among the whole bunch of floures and ſeed. The root is tuberous like the other,but ſmaller,and not ſo much crooked.

1 *Biſtorta major.*
Snake-weed.

2 *Biſtorta minor.*
Small Snake-weed.

3 Broad leaued Snake-weed hath many large vneuen leaues,ſmooth and very greene ; among which riſe vp ſmall brittle ſtalks of two hands high, bearing at the top a faire ſpike of floures like to the white Biſtort. The root is knobby or bunched,crookedly turned or writhed this way & that way;whereof it tooke his name *Biſtorta.* ‡ It differs from the firſt only in that the root is more twined in,and the leaues broader and more crumpled. ‡
¶ *The Place.*

1 The great Biſtort growes in moiſt and waterie places,and in the darke ſhadowie woods,being very common in moſt gardens.

2 The

2   The ſmall Biſtort groweth in great aboundance in Weſtmerland, at Cr osby Rauenſwaith, at the head of a Parke belonging to one Mr *Pickering* ; from whence it hath bin diſperſed into many gardens, and alſo ſent vnto me for my garden.

¶ The Time.

They floure in May, and the ſeed is ripe in Iune.

¶ The Names.

Biſtorta is called in Engliſh, Snake-weed : in ſome places, Oiſterloit : in Cheſhire, Paſſions, and Snakeweed, and there vſed for an excellent pot-herbe. It is called *Biſtorta* of his writhed roots, and alſo *Colubrina, Serpentaria, Brittanica, Dracontion Plinij, Dracunculus Dodonai,* and *Limonium Geſneri.*

¶ The Temperature.

Biſtort doth coole and dry in the third degree.

¶ The Vertues.

A   The juice of Biſtort put into the noſe preuaileth much againſt the diſeaſe called *Polypus,* & the biting of ſerpents or any other venomous beaſt, being drunke in wine or the water of Angelica.

B   The root boiled in wine and drunke ſtoppeth the lask and bloudy flix ; it ſtayeth alſo the ouer-much flowing of womens monethly ſickneſſes.

C   The root taken as aforeſaid ſtaieth vomiting, and healeth the inflammation and ſoreneſſe of the mouth and throat : it likewiſe faſtneth looſe teeth, being holden in the mouth for a certaine ſpace, and at ſundry times.

CHAP. 86.   *Of Scuruy-graſſe or Spoonwort.*

¶ The Deſcription.

1   ROund leaued Scuruy-graſſe is a low or baſe herb : it bringeth forth leaues vpon ſmall ſtems or foot-ſtalks of a mean length, comming immediately from the root, very many in number, of a ſhining green colour, ſomwhat broad, thicke, hollow like a little ſpoon, but of no great depth, vneuen or cornered about the edges : among which leaues ſpring vp ſmall ſtalkes of a ſpan high, whereon grow many little white floures ; after which comes the ſeed, ſmall and reddiſh, contained in little round pouches or ſeed-veſſels : the roots be ſmall, white, and thred-die: the whole plant is of a hot and ſpicy taſte.

2   The common Scuruy-graſſe or Spoonwort hath leaues ſomewhat like a ſpoon, hollow in the middle, but altogether vnlike the former : the leaues hereof are bluntly toothed about the edges, ſharpe pointed and ſomewhat long : the ſtalks riſe vp among the leaues , of the length of halfe a foot, whereon do grow white floures with ſome yellowneſſe in the middle : which being paſt, there ſucceed ſmall ſeed-veſſels like vnto a pouch, not vnlike to thoſe of Shepheards purſe, green at the firſt, next yellowiſh, and laſtly when they be ripe of a browne colour, or like a filberd nut. The root is ſmall and tender, compact of a number of threddy ſtrings very thicke thruſt together in manner of a little turſe.

¶ The Place.

The firſt groweth by the ſea ſide at Hull, at Boſton, and Lynne, and in many other places of Lincolnſhire neere vnto the ſea, as in Whapload and Holbeck marſhes in Holland in the ſame county. It hath bin found of late growing many miles from the ſea ſide, vpon a great hil in Lancaſhire called Ingleborough hill ; which may ſeem ſtrange to thoſe that do not know that it will bee content with any ſoile, place, or clyme whatſoeuer. For proofe whereof, my ſelfe haue ſowne the ſeeds of it in my garden, and giuen them vnto others, with whom they floure, flouriſh, and bring forth their ſeed as naturally as by the ſea ſide, and likewiſe retain the ſame hot ſpicy taſt : which proueth that they refuſe no culture, contrary to many other ſea plants.

The ſecond, which is our common Scuruy-graſſe, groweth in diuers places vpon the brimmes of the famous riuer Thames, as at Woolwich, Erith, Greenhithe, Graueſend, as wel on the Eſſex ſhore as the Kentiſh ; at Portſmouth, Briſtow, and many other places alongſt the Weſtern coaſt : but toward the North I haue not heard that any of this kind hath growne.

¶ *The Time.*

It floureth and flourisheth in May. The seed is ripe in Iune.

1 *Cochlearia rotundifolia.*
Round leafed Scuruie-grasse.

2 *Cochlearia Britannica.*
Common English Scuruie-grasse.

¶ *The Names.*

† We are not ignorant that in low Germany, this hath seemed to some of the best learned to be the true *Britannica*, and namely to those next the Ocean in Friesland and Holland. The Germanes call it **Leffelkraut**: that is, *Cochlearia* or Spoonewort, by reason of the compassed roundnes and hallownesse of the leaues, like a spoone, and haue thought it to be *Plinies Britannica*, because they finde it in the same place growing, and endued with the same qualities. Which excellent plant *Cæsars* souldiers (when they remooued their camps beyond the Rhene) found to preuaile (as the Frisians had taught it them) against that plague and hurtfull disease of the teeth, gums, and sinewes, called the Scuruie, being a depriuation of all good bloud and moisture, in the whole body, called *Scorbutum*; in English, the Scuruie, and Skyrby, a disease happening at the sea among Fishermen, and fresh-water souldiers, and such as delight to sit still without labour and exercise of their bodies; and especially aboue the rest of the causes, when they make not cleane their bisket bread from the floure or mealinesse that is vpon the same, which doth spoile many. But sith this agrees not with *Plinies* description, and that there be many other water plants, as *Nasturtium*, *Sium*, *Cardamine*, and such others, like in taste, and not vnlike in proportion and vertues, which are remedies against the diseases aforesaid, there can be no certaine argument drawne therefrom to prooue it to be *Britannica*. For their leaues at their first comming forth are somewhat long like *Pyrola* or Adders tongue, soone after somewhat thicker, and hollow like a nauell, after the manner of Sundew, but in greatnesse like *Soldanella*, in the compasse somewhat cornered, in fashion somewhat like a spoone: the floures white, and in shape like the Cuckow floures: the seed reddish, like the seed of *Thlaspi*, which is not to be seene in *Britannica*, which is rather holden to be Bistort or Garden Patience, than Scuruie grasse. In English it is called, Spoonewort, Scruby grasse and Scuruie grasse.

¶ *The Temperature.*

Scuruie grasse is euidently hot and dry, very like in taste and quality to the garden Cresses, of an aromaticke or spicie taste.

¶ *The*

¶ *The Vertues.*

A   The juice of Spoonewort giuen to drinke in Ale or Beere, is a singular medicine against the corrupt and rotten vlcers, and stench of the mouth : it perfectly cureth the disease called of *Hippocrates*, *Voluulus Hematites* : of *Pliny*, *Stomacace* : of *Marcellus*, *Oscedo* : and of the later writers, *Scorbutum* : of the Hollanders and Frisians, Scuerbuyck : in English, the Scuruie : either giuing the juice in drinke as aforesaid, or putting six great handsfuls to steepe, with long pepper, graines, annise-seed, and liquorice, of each one ounce, the spices being braied, and the herbe bruised with your hands, and so put into a pot, such as is before mentioned in the chapter of bastard Rubarbe, and vsed in like manner ; or boiled in milke or wine and drunke for certaine daies together it worketh the like effect.

B   The juice drunke once in a day fasting with any liquor, ale, beere, or wine, doth cause the foresaid medicine more speedily to worke his effect in curing this filthy, lothsome, heauy, and dull disease, which is very troublesome, and of long continuance. The gums are loosed, swolne, and exulcerate ; the mouth greeuously stinking ; the thighes and legs are withall very often full of blew spots, not much vnlike those that come of bruises : the face and the rest of the body is oftentimes of a pale colour : and the feet are swolne as in a dropsie.

C   There is a disease (saith *Olaus magnus* in his historie of the Northerne regions) haunting the campes, which vexes them that are besieged and pinned vp : and it seemeth to come by eating of salt meates, which is increased and cherished with the cold vapors of the stone walls. The Germanes call this disease (as we haue said) Scorbuck; the symptome or passion which hapneth to the mouth, is called of *Pliny* ſτωραχαλ : *Stomacace* : and that which belongeth to the thighes ωιλιτιρις: *Marcellus* an old writer nameth the infirmitie of the mouth *Oscedo* : which disease commeth of a grosse cold and tough bloud, such as melancholy juice is, not by adustion, but of such bloud as is the feculent or drossie part thereof : which is gathered in the body by ill diet, slothfulnesse to worke, laisinesse (as we terme it) much sleepe and rest on ship-boord, and not looking to make cleane the bisquet from the mealinesse, and vncleane keeping their bodies, which are the causes of this disease called the scuruie or scyrby; which disease doth not onely touch the outer parts, but the inward also : for the liuer oftentimes, but most commonly the spleene, is filled with this kinde of thicke, cold and tough juice, and is swolne by reason that the substance thereof is slacke, spungie and porous, very apt to receiue such kinde of thicke and cold humors. Which thing also *Hippocrates* hath written of in the second booke of his Prorrhetikes : their gums (saith he) are infected, and their mouthes stinke that haue great spleenes or milts : and whosoeuer haue great milts and vse not to bleed, can hardly be cured of this maladie, especially of the vlcers in the legs, and blacke spots. The same is affirmed by *Paulus Ægineta* in his third booke, 49. chapter, where you may easily see the difference betweene this disease and the blacke jaunders, which many times are so confounded together, that the distinction or difference is hard to be knowne, but by the expert chirurgion:who oftentimes seruing in the ships, as well her Maiesties as merchants, are greatly pestered with the curing thereof : it shall be requisite to carry with them the herbe dried : the water distilled, and the juice put into a bottle with a narrow mouth, full almost to the necke, and the rest filled vp with oile oliue, to keep it from putrifaction : the which preparations discreetly vsed, will stand them in great stead for the disease aforesaid.

D   The herbe stamped and laid vpon spots and blemishes of the face, will take them away within six houres, but the place must be washed after with water wherein bran hath been sodden.

CHAP. 87.   *Of Twayblade, or herbe Bifoile.*

¶ *The Description.*

1   HErbe Byfoile hath many small fibres or thready strings, fastened vnto a small knot or root, from which riseth vp a slender stem or stalke, tender, fat, and full of juice ; in the middle whereof are placed in comely order two broad leaues, ribbed and chamfered, in shape like the leaues of Plantaine : vpon the top of the stalke groweth a slender greenish spike made of many small floures, each little floure resembling a gnat, or little gosling newly hatched, very like those of the third sort of Serapias stones.

2   *Ophris Trifolia*, or Trefoile Twaiblade, hath roots, tender stalkes, and a bush of floures like the precedent ; but differeth in that, that this plant hath three leaues which doe clip or embrace the

stalke

ſtalke about ; and the other hath but two, and neuer more, wherein eſpecially conſiſteth the diffe-rence : although in truth I thinke it a degenerate kinde, and hath gotten a third leafe *per accidens,* as doth ſometimes chance vnto the Adders Tongue, as ſhall be declared in the Chapter that fol-loweth.

‡ 3   This kinde of Twaiblade, firſt deſcribed in the laſt edition of *Dodonæus,* hath leaues, floures, and ſtalkes like to the ordinary ; but at the bottome of the ſtalke aboue the fibrous roots it hath a bulbe greeniſh within, and couered with two or three skins: it groweth in moiſt and wet low places of Holland. ‡

1 *Ophris bifolia.*
Twaiblade.

‡ 3 *Ophris bifolia bulboſa.*
Bulbous Twaiblade.

¶ *The Place.*

The firſt groweth in moiſt medowes, fenny grounds, and ſhadowie places. I haue found it in many places, as at South-fleet in Kent, in a Wood of Maſter *Sidleys* by Long-field Downes, in a Wood by London called Hampſtead Wood, in the fields by High-gate, in the Woods by Ouen-den neere to Clare in Eſſex, and in the Woods by Dunmow in Eſſex. The ſecond ſort is ſeldome ſeene.

¶ *The Time.*

They floure in May and Iune.

¶ *The Names.*

It is called of the later Herbariſts, *Bifolium,* and *Ophris.*

¶ *The Nature and Vertues.*

Theſe are reported of the Herbariſts of our time to be good for greene wounds, burſtings, and   A ruptures ; whereof I haue in my vnguents and balſams for greene wounds had great experience, and good ſucceſſe.

Chap. 88. *Of Adders-Tongue.*

¶ *The Deſcription.*

1   Ophiogloſſon, or *Lingua ſerpentis* (called in Engliſh Adders-Tongue; of ſome Adders-Graſſe,though vnproperly)riſeth forth of the ground, hauing one leaſe and no more, fat or oleous in ſubſtance,of a finger long, and very like the young and tender leaues of Marigolds : from the bottome of which leaſe ſpringeth out a ſmall and tender ſtalke one finger and a halfe long,on the end whereof doth grow a long ſmall tongue not vnlike the tongue of a ſerpent,whereof it tooke the name.

2   I haue ſeene another like the former in root,ſtalke,and leaſe ; and differeth,in that this plant hath two, and ſometimes more crooked tongues, yet of the ſame faſhion,which if my iudgement faile not chanceth *per accidens,* euen as we ſee children borne with two thumbes vpon one hand: which moueth me ſo to think, for that in gathering twentie buſhels of the leaues a man ſhall hardly finde one of this faſhion.

1 *Ophiogloſſon.*
Adders-Tongue.

‡ 2 *Ophiogloſſon abortiuum.*
Miſ-ſhapen Adders Tongue.

¶ *The Place.*

Adders-Tongue groweth in moiſt medowes throughout moſt parts of England,;as in a Meadow neere the preaching Spittle adioyning to London ; in the Mantels by London,in the medowes by Cole-brooke,in the fields in Waltham Forreſt,and many other places.

¶ *The Time.*

They are to be found in Aprill and May ; but in Iune they are quite vaniſhed and gone.

¶ *The Names.*

Ophiogloſſum is called in ſhops *Lingua ſerpentis,Linquace,* and *Lingua lace :* it is alſo called *Lancea Chriſti Enephyllon,*and *Lingua vulneraria:* in Engliſh,Adders tongue, or Serpents tongue : in Dutch, ſNatertonguen: of the Germanes, ſater zungelin,

¶ *The*

¶ *The Nature.*

Adders-tongue is dry in the third degree.

¶ *The Vertues.*

The leaues of Adders tongue stamped in a stone morter, and boyled in Oile Oliue vnto the con-   A
sumption of the juyce, and vntill the herbes be dry and parched, and then strained, will yeeld a most
excellent greene oyle, or rather a balsame for greene wounds, comparable to oile of S *Iohns* wort,
if it do not farre surpasse it by many degrees: whose beautie is such that very many Artists haue
thought the same to be mixed with Verdigrease.

CHAP. 89.

Of One-berry, or Herbe True-loue, and Moone-wort.

1 *Herba Paris.*
One-berry, or Herbe True-loue.

2 *Lunaria minor.*
Small Moone-wort.

¶ *The Description.*

1   HErbe Paris riseth vp with one smal tender stalke two hands high, at the very top whereof come forth foure leaues directly set one against another in manner of a Burgundi-
an Crosse or True-loue knot: for which cause among the Antients it hath been called Herbe True-loue. In the midst of the said leafe comes forth a starre-like floure of an herby or
grassie colour, out of the middest whereof there ariseth vp a blackish browne berry: the root is long
and tender, creeping vnder the earth, and dispersing it selfe hither and thither.

2   The small Lunary springeth forth of the ground with one leafe like Adders-tongue, jagged
or cut on both sides into fiue or six deepe cuts or notches, not much vnlike the leaues of *Scolopen-*
dria, or *Ceterach*, of a greene colour; whereupon doth grow a small naked stem of a finger long, bearing at the top many little seeds clustering together; which being gathered and laid in a platter or
such like thing for the space of three weekes, there will fall from the same a fine dust or meale of a
whitish colour, which is the seed if it bring forth any. The root is slender, and compact of many
small threddy strings.

‡ In England (ſaith *Camerarius*) there growes a certaine kinde of *Lunaria*, which hath many leaues, and ſometimes alſo ſundry branches ; which therefore I haue cauſed to be delineated, that other Herbariſts might alſo take notice hereof. Thus much *Camerarius, Epit.Mat.p.644*.where he giues an elegant figure of a variety hauing more leaues and branches than the ordinary, otherwiſe not differing from it. ‡

3   Beſides this varietie there is another kinde ſet forth by *Cluſius*;whoſe figure and deſcription I thinke good here to ſet downe. This hath a root conſiſting of many fibres ſomewhat thicker than thoſe of the common kinde:from which ariſe one or two winged leaues, that is, many leaues ſet to one ſtalke ; and theſe arelike the leaues of the other *Lunaria*, but that they are longer,thicker, and more diuided,and of a yellowiſh greene colour. Amongſt theſe leaues there comes vp a ſtalke fat and juycie, bearing a greater tuft of floures or ſeeds (for I know not whether to call them) than the ordinary,but otherwiſe very like thereto. It groweth in the mountaines of Sileſia, and in ſome places of Auſtria. ‡

‡ 3 *Lunaria minor Ramoſa.*
   Small branched Moon-wort.

¶ *The Place.*

Herba Paris groweth plentifully in all theſe places following ; that is to ſay, in Chalkney wood neere to wakes Coulne, ſeuen miles from Colcheſter in Eſſex, and in the wood by Robinhoods well, neere to Nottingham ; in the parſonage orchard at Radwinter in Eſſex,neere to Saffron Walden ; in Blackburne at a place called Merton in Lancaſhire ; in the Moore by Canterbury called the Clapper ; in Dingley wood, ſix miles from Preſton in Aunderneſſe ; in Bocking parke by Braintree in Eſſex;at Heſſet in Lancaſhire,and in Cotting wood in the North of England ; as that excellent painefull and diligent Phyſition Mr Doctor *Turner* of late memorie doth record in his Herbal.

Lunaria or ſmall Moone-wort groweth vpon dry and barren mountains and heaths. I haue found it growing in theſe places following ; that is to ſay, about Bathe in Somerſetſhire in many places, eſpecially at a place called Carey, two miles from Bruton,in the next cloſe vnto the Churchyard ; on Cockes Heath betweene Lowſe and Linton, three miles from Maidſtone in Kent : it groweth alſo in the ruines of an old bricke-kilne by Colcheſter, in the ground of Mr *George Sayer*, called Miles end : it groweth likewiſe vpon the ſide of Blacke-heath,neere vnto the ſtile that leadeth vnto Eltham houſe,about an hundred paces from the ſtile: alſo in Lancaſhire neere vnto a Wood called Faireſt,by Latham : moreouer, in Nottinghamſhire by the Weſtwood at Gringley,and at Weſton in the Ley field by the Weſt ſide of the towne; and in the Biſhops field at Yorke,neere vnto Wakefield, in the cloſe where Sir *George Sauill* his houſe ſtandeth,called the Heath Hall, by the relation of a learned Doctor in Phyſicke called Maſter *Iohn Merſhe* of Cambridge, and many other places.

¶ *The Time.*

Herba Paris floureth in Aprill,and the berry is ripe in the end of May.

Lunaria or ſmall Moone wort is to be ſeene in the moneth of May.

¶ *The Names.*

One-berry is alſo called Herbe True-loue, and Herbe Paris : in Latine, *Herba Paris*, and *Solanum tetraphyllum* by *Geſner* and *Lobel.*

Lunaria minor is called in Engliſh, Small Lunarie, and Moone-wort.

¶ *The Temperature.*

Herbe Paris is exceeding cold ; whereby it repreſſeth the rage and force of poyſon.

Lunaria minor is cold and dry of temperature.

¶ *The*

¶ *The Vertues.*

The berries of Herbe Paris giuen by the ſpace of twentie daies, are excellent good againſt poy- A ſon, or the pouder of the herbe drunke in like manner halfe a ſpoonefull at a time in the morning faſting.

The ſame is miniſtred with great ſucceſſe vnto ſuch as are become peeuiſh, or without vnder- B ſtanding, being miniſtred as is aforeſaid, euery morning by the ſpace of twenty daies, as *Baptiſta Sardus*, and *Matthiolus* haue recorded. Since which time there hath been further experience made therof againſt poyſon, & put in practiſe in the citie of Paris, in Louaine, and at the baths in Heluetia, by the right excellent Herbariſts *Matthias de Lobel*, and *Petrus Pena*, who hauing o'ten read, that it was one of the Aconites, called *Pardalianches*, and ſo by conſequence of a poyſoning quality, they gaue it vnto dogs and lambes, who receiued no hurt by the ſame: wherefore they further proſecuted the experience thereof, and gaue vnto two dogs faſt bound or coupled together, a dram of Arſenicke, and one dram of Mercury ſublimate mixed with fleſh (‡ in the *Aduerſaria* it is but of each halfe a dram, and theſe *pag*.165. you may finde this Hiſtory more largely ſet downe. t) which the dogs would not willingly eate, and therefore they had it crammed downe their throats: vnto one of theſe dogs they gaue this Antidote following in a little red wine, whereby he recouered his former health againe within a few houres: but the other dog which had none of the medicine, died incontinently.

This is the Receit.

R. *vtriuſque Angelicæ (innuit) domeſticam,& ſylueſtrem, Vicetoxici, valerianæ domeſtica, Polipodij querni, radicum Altheæ, & Vrticæ, ana 3.iiij, Corticis Mezerei Germanici, 3.ij, granorum herba Paridis, N.24. foliorum ejuſdem cum toto, Nym.36. Ex maceratis in aceto radicibus & ſiccatis fit omnium pulvis.*

The people in Germany doe vſe the leaues of Herbe Paris in greene wounds, for the which it is C very good, as *Ioachimus Camerarius* reporteth; who likewiſe ſaith, that the pouder of the roots giuen to drinke, doth ſpeedily ceaſe the gripings and paine of the Collicke.

Small Moone-woort is ſingular to heale greene and freſh wounds: it ſtaieth the bloudy flix. It D hath beene vſed among the Alchymiſts and witches to doe wonders withall, who ſay, that it will looſe lockes, and make them to fall from the feet of horſes that graſe where it doth grow, and hath beene called of them *Martagon*, whereas in truth they are all but drowſie dreames and illuſions; but it is ſingular for wounds as aforeſaid.

Cʜᴀᴘ. 90. *Of Winter-Greene.*

¶ *The Deſcription.*

1 Pʏʀᴏʟᴀ hath many tender and very greene leaues, almoſt like the leaues of Beet, but rather in my opinion like to the leaues of a Peare-tree, whereof it tooke his name *Pyrola*, for that it is *Pyriformis*. Among theſe leaues commeth vp a ſtalke garniſhed with prettie white floures, of a very pleaſant ſweet ſmell, like *Lillium Conuallium*, or the Lillie of the Valley. The root is ſmall and threddy, creeping far abroad vnder the ground.

‡ 2 This differs from the laſt deſcribed in the ſlenderneſſe of the ſtalkes, and ſmalneſſe of the leaues and floures: for the leaues of this are not ſo thicke, and ſubſtantiall, but very thinne, ſharpe pointed, and very finely ſnipt about the edges, blacker, and reſembling a Peare-tree leaſe. The floures are like thoſe of the former, yet ſmaller and more in number: to which ſucceed fiue cornered ſeed veſſels with a long pointell as in the precedent: the root alſo creepes no leſſe than that of the former, and here and there puts vp new ſtalkes vnder the moſſe. It growes vpon the Auſtrian and Styrian Alpes, and floures in Iune and Iuly.

3 This is an elegant plant, and ſometimes becomes ſhrubby, for the new and ſhort branches growing vp each yeere, doe remaine firme and greene for ſome yeeres, and grow ſtreight vp, vntill at length borne downe by their owne weight they fall downe and hide themſelues in the moſſe. It hath commonly at each place where new branches grow forth, two, three, or foure thicke, very greene and ſhining leaues, almoſt in forme and magnitude like to the leaues of *Laureola*, yet ſnipt about the edges, of a very drying taſte, and then bitteriſh. From among theſe leaues at the Spring of the yeere new branches ſhoot vp, hauing ſmall leaues like ſcailes vpon them, and at their tops grow

1 *Pyrola.*
Winter-Greene.

‡ 2 *Pyrola 2 tenerior Cluſ.*
The ſmaller Winter-Greene.

‡ 3 *Pyrola 3 fruticans Cluſ.*
Shrubby Winter-Greehe.

‡ 4 *Pyrola 4 minima Cluſ.*
Round leaued Winter-Greene.

5 *Monophyllon.*
One Blade.

grow floures like to thoſe of the firſt deſcribed, yet ſomewhat larger, of a whitiſh purple colour; which fading, are ſucceeded by fiue corncred ſeed veſſels containing a very ſmall ſeed; the roots are long and creeping. It growes a little from Vienna in Auſtria in the woods of Entze-ſtorf, and in diuers places of Bohemia and Sileſia.

4 This from creeping roots ſends vp ſhort ſtalkes, ſet at certaine ſpaces with ſmall, round, and thinne leaues, alſo ſnipt about the edges, amongſt which vpon a naked ſtem, growes a floure of pretty bignes, conſiſting of fiue white ſharpiſh pointed leaues with ten threds, and a long pointell in the midſt. The ſeed is contained in ſuch heads as the former, and it is very ſmall. This growes in the ſhadowie places of the Alpes of Sneberge, Hochbergerin, Durren-ſtaine, towards the roots of theſe great mountaines. *Cluſ.* ‡

5 *Monophyllon,* or *Vnifolium,* hath a leafe not much vnlike the greateſt leafe o Iuie. with many ribs or ſinewes lik the Plantaine leafe; which ſingle leafe doth alwaies ſpring forth of the earth alone, but when the ſtalke riſeth vp, it bringeth vpon his ſides two leaues, in faſhion like the former; at the top of which ſlender ſtalke come forth fiue ſmall floures like *Pyrola*; which being vaded, there ſucceed ſmall red berries. The root is ſmall, tender, and creeping farre abroad vnder the vpper face of the earth.

¶ *The Place.*

1 *Pyrola* groweth in Lanſdale, and Crauen, in the North part of England, eſpecially in a cloſe called Crag-cloſe.

2 *Monophyllon* groweth in Lancaſhire in Dingley wood, ſix miles from Preſton in Aunderneſſe; and in Harwood neere to Blackeburne likewiſe.

¶ *The Time.*

1 *Pyrola* floureth in Iune and Iuly, and groweth Winter and Summer.

2 *Monophyllon* floureth in May, and the fruit is ripe in September.

¶ *The Names.*

1 *Pyrola* is called in Engliſh Winter-greene: it hath beene called *Limonium* of diuers, but vntruly.

2 *Monophyllon,* according to the etymologie of the word, is called in Latine *Vnifolium*: in Engliſh, One-blade, or One-leafe.

¶ *The Nature.*

1 *Pyrola* is cold in the ſecond degree, and dry in the third.

2 *Monophyllon* is hot and dry of complexion.

¶ *The Vertues.*

Pyrola is a moſt ſingular wound-herbe, either giuen inwardly, or applied outwardly: the leaues   A whereof ſtamped and ſtrained, and the juice made vnto an vnguent, or healing ſalue, with wax oyle, and turpentine, doe cure wounds, vlcers and fiſtulaes, that are mundified from the callous and tough matter, which keepeth the ſame from healing.

The decoction hereof made with wine, is commended to cloſe vp and heale wounds of the en-   B trailes, and inward parts: it is alſo good for vlcers of the kidneies, eſpecially made with water and the roots of Comfrey added thereto.

The leaues of *Monophyllon* or *Vnifolium,* are of the ſame force in wounds with *Pyrola,* eſpecially in   C wounds among the nerues and ſinewes. Moreouer, it is eſteemed of ſome late writers a moſt per-fect medicine againſt the peſtilence, and all poyſons, if a dram of the root be giuen in vinegre mix-ed with wine or water, and the ſicke go to bed and ſweat vpon it.

C H A P.

CHAP. 91.   *Of Lilly in the valley, or May Lilly.*

1 *Lilium convallium.*
Conuall Lillies.

2 *Lilium convallium floribus ſuaue-rubentibus.*
Red Conuall Lillies.

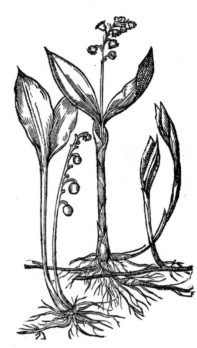

¶ *The Deſcription.*

1   THe Conuall Lilly, or Lilly of the Vally, hath many leaues like the ſmalleſt leaues of Water Plantaine; among which riſeth vp a naked ſtalke halfe a foot high, garniſhed with many white floures like little bels, with blunt and turned edges, of a ſtrong ſauour, yet pleaſant enough; which being paſt, there come ſmall red berries, much like the berries of *Aſparagus*, wherein the ſeed is contained. The root is ſmall and ſlender, creeping far abroad in the ground.

2   The ſecond kinde of May Lillies is like the former in euery reſpect; and herein varieth or differeth, in that this kinde hath reddiſh floures, and is thought to haue the ſweeter ſmell.

¶ *The Place.*

1   The firſt groweth on Hampſted heath, foure miles from London, in great abundance: neere to Lee in Eſſex, and vpon Buſhie heath, thirteene miles from London, and many other places.

2   The other kinde with the red floure is a ſtranger in England: howbeit I haue the ſame growing in my garden.

¶ *The Time.*

They floure in May, and their fruit is ripe in September.

¶ *The Names.*

The Latines haue named it *Lilium Conuallium* : *Geſner* doth thinke it to be *Callionymum* : in the Germane tongue, 𝔐eyen blumlen : the low Dutch, 𝔐eyen bloemkens : in French, *Muguet* : yet there is likewiſe another herbe which they call *Muguet*, commonly named in Engliſh, Woodroof. It is called in Engliſh, Lilly of the Valley, or the Conuall Lillie, and May Lillies, and in ſome places Liriconſancie.

¶ *The Nature.*

They are hot and dry of complexion.

¶ *The*

¶ *The Vertues.*

The floures of the Valley Lillie distilled with wine, and drunke the quantitie of a spoonefull, re- A
store speech vnto those that haue the dumb palsie and that are falne into the Apoplexie, and are
good against the gout, and comfort the heart.

The water aforesaid doth strengthen the memory that is weakened and diminished; it helpeth B
also the inflammations of the eies, being dropped thereinto.

The floures of May Lillies put into a glasse, and set in a hill of ants, close stopped for the space C
of a moneth, and then taken out, therein you shall finde a liquor that appeaseth the paine and griefe
of the gout, being outwardly applied; which is commended to be most excellent.

Chap. 92.   *Of Sea Lauander.*

1 *Limonium.*
Sea Lauander.

2 *Limonium paruum.*
Rocke Lauander.

¶ *The Description.*

1 THere hath beene among writers
from time to time great contention about this plant *Limo-
nium,* no one Author agreeing with another:
for some haue called this herbe *Limonium*;
some another herbe by this name; and some
in remoouing the rocke, haue mired themselues in the mud, as *Matthiolus*, who described two kindes, but made no distinction of
them, nor yet expressed which was the true
Limonium; but as a man herein ignorant, he
speakes not a word of them. Now then to
leaue controuersies and cauilling, the true
Limonium is that which hath faire leaues,
like the Limon or Orenge tree, but of a darke
greene colour, somewhat fatter, and a little
crumpled: amongst which leaues riseth vp an hard and brittle naked stalke of a foot high, diuided
at the top into sundry other small branches, which grow for the most part vpon one side, full of little blewish floures, in shew like Lauander, with long red seed, and a thicke root like vnto the small
Docke.

2   There is a kinde of *Limonium* like the first in each respect, but lesser, which groweth vpon
rockes and chalkie cliffes.

‡ 3   Besides these two here described, there is another elegant plant by *Clusius* and others
referred to this kindred: the description thereof is thus; from a long slender root come forth long
greene leaues lying spread vpon the ground, being also deepely sinuated on both sides, and somewhat roughish. Amongst these leaues grow vp the stalkes welted with slender indented skinnes,
and towards their tops they are diuided into sundry branches after the manner of the ordinarie
one; but these branches are also winged, and at the tops they carry floures some foure or fiue
clu-

clustering together, consisting of one thin crispe or crumpled leafe of a light blew colour (which
continues long, if you gather them in their perfect vigour, and so dry them) and in the middest of
this blew comes vp little white floures, consisting of fiue little round leaues with some white threds
in their middles. This plant was first obserued by *Rouwolfius* at Ioppa in Syria: but it growes also
vpon the coasts of Barbarie, and at Malacca and Cadiz in Spaine: I haue seene it growing with
many other rare plants, in the Garden of my kinde friend Master *Iohn Tradescant* at South-Lam-
beth.

‡ 13 *Limonium folio sinuato.*
Sea-Lauander with the indented leafe.

‡ 4 *Limonio congener, Cluſ.*
Hollow leaued Sea-Lauander.

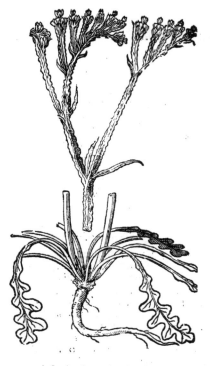

4   *Clusius* in the end of his fourth Booke *Historiæ Plantarum,* sets forth this, and saith, receiued this figure with one dried leafe of the plant sent him from Paris from *Claude Gonier* the Apothecarie of that citie, who receiued it (as you see it here exprest) from Lisbone. Now *Cluſius* describes the leafe, that it was hard, and as if it had been a piece of leather, open on the vpper side,
and distinguished with many large purple veines on the inside, &c. for the rest of his description
was onely taken from the figure (as he himselfe saith) which I hold impertinent to set downe, seeing I here giue you the same figure, which by no meanes I could omit, for the strangenesse thereof, but hope that some or other that trauell into forraine parts may finde this elegant plant, and
know it by this small expression, and bring it home with them, that so we may come to a perfecter
knowledge thereof. ‡

¶ *The Place.*

1   The first groweth in great plenty vpon the walls of the fort against Grauefend: but abun-
dantly on the bankes of the Riuer below the same towne, as also below the Kings Store-house at
Chattam: and fast by the Kings Ferrey going into the Isle of Shepey: in the salt marshes by Lee in
Essex: in the Marsh by Harwich, and many other places.

2 The

The ſmall kind I could neuer find in any other place but vpon the chalky cliffe going from the towne of Margate downe to the ſea ſide, vpon the left hand.

¶ *The Time.*

They floure in Iune and Iuly.

¶ *The Names.*

It ſhall be needleſſe to trouble you with any other Latine name than is expreſt in their titles : The people neere the ſea ſide where it growes do call it Marſh Lauander, and ſea Lauander.

‡ This cannot be the *Limonium* of *Dioſcorides*, for the leaues are not longer than a Beet, nor the ſtalke ſo tall as that of a Lilly ; but you ſhall find more hereafter concerning this, in the chapter of water Plantain. I can not better refer this to any plant deſcribed by the Antients, tha[n] *o Britannica* deſcribed by *Dioſcorides, lib. 4. cap. 2.* ‡

¶ *The Temperature.*

The ſeed of *Limonium* is very aſtringent or binding.

¶ *The Vertues.*

The ſeed beaten into pouder and drunk in Wine, helpeth the Collique, Strangurie, and Dyſen- A terie.

The ſeed taken as aforeſaid, ſtaieth the ouermuch flowing of womens terms, and all other fluxes B of bloud.

CHAP. 93.   *Of Serapia's Turbith, or ſea Star-wort.*

1 *Tripolium vulgare majus.*
Great ſea Star-wort.

‡ 2 *Tripolium vulgare minus.*
Small ſea Star-wort.

¶ *The Deſcription.*

1   THe firſt kind of *Tripolium* hath long and large leaues ſomewhat hollow or furrowed, of a ſhining green colour declining to blewneſſe, like the leaues of Woad : among which riſeth vp a ſtalke of two cubits high and more, which toward the top is diuided into many ſmall branches garniſhed with many floures like Camomill, yellow in the middle, ſet about

or bordered with small blewish leaues like a pale, as in the floures of Camomill; which grow into a whitish rough downe that flieth away with the wind. The root is long and threddy.

     2   There is another kinde of *Tripolium* like the first, but much smaller, wherein consisteth the difference.

¶ *The Place.*

These herbs grow plentifully alongst the English coasts in many places, as by the fort against Grauesend, in the Ile of Shepey in sundry places, in a marsh which is vnder the town wals of Harwich, in the marsh by Lee in Essex, in a marsh which is between the Ile of Shepey and Sandwich, especially where it ebbeth and floweth: being brought into gardens it flourisheth a long time, but there it waxeth huge, great, and ranke, and changeth the great roots into strings.

¶ *The Time.*

These herbs do floure in May and Iune.

¶ *The Names.*

It is reported by men of great fame and learning, That this plant was called *Tripolium* because it doth change the colour of his floures thrice in a day. This rumor we may beleeue as true, for that we see and perceiue things of as great or greater wonder to proceed out of the earth. This herbe I planted in my garden, whither in his season I did repaire to finde out the truth hereof, but I could not espy any such variablenesse herein: yet thus much I may say, that as the heate of the sun doth change the colour of diuers floures, so it fell out with this, which in the morning was very faire, but afterward of a pale or wan colour. Which proueth that to be but a fable which *Dioscorides* saith is reported by some, that in one day it changeth the colour of his floures thrice; that is to say, in the morning it is white, at noone purple, and in the euening, φοινιχε, or crimson. But it is not vntrue, that there may be found three colours of the floures in one day, by reason that the floures are not all perfected together, (as before I partly touched) but one after another by little and little. And there may easily be obserued three colours in them, which is to be vnderstood of them that are beginning to floure, that are perfectly floured, and those that are falling away. For they that are blowing and be not wide open and perfect are of a purplish colour, and those that are perfect and wide open of a whitish blew, and such as haue fallen away haue a white down: which changing hapneth vnto sundry other plants. This herbe is called of *Serapio*, *Turbith*: women that dwell by the sea side call it in English, blew Daisies, or blew Camomill; & about Harwich it is called Hogs beans, for that the swine do greatly desire to feed thereon, as also for that the knobs about the roots doe somewhat resemble the garden bean. It is called in Greeke, τριπολιον: and of diuers others, ψυχη. It may be fitly called *Aster marinus* or *Amellus marinus*: in English, Sea Starwort, *Serapio's* Turbith: of some, blew Daisies. The Arabian *Serapio* calls Sea Starwort, Turbith; and after him, *Auicen*: yet *Actuarius* the Grecian thinketh, that Turbith is the root of *Alypum*. *Mesues* iudged it to be the root of an herb like Fenell. The historie of Turbith of the shops shall be discoursed vpon in his proper place.

¶ *The Temperature.*

Tripolium is hot in the third degree, as *Galen* saith.

¶ *The Vertues.*

A   The root of *Tripolium* taken in wine by the quantitie of two drams driueth forth by siege waterish and grosse humors; for which cause it is often giuen to them that haue the dropsie.

B   It is an excellent herb against poyson, and comparable with *Pyrola*, if not of greater efficacie in healing of wounds either outward or inward.

C H A P. 94   *Of Turbith of Antioch.*

¶ *The Description.*

Garcias a Portugall Physition saith, That Turbith is a plant hauing a root which is neither great nor long: the stalke is of two spannes long, sometimes much longer, a finger thicke, which creepeth in the ground like Ivye, and bringeth forth leaues like those of the Marish Mallow. The floures be also like those of the Mallow, of a reddish white colour. The lower part of the stalke only, which is next vnto the root and gummy, is that which is profitable in medicine, and is the same that is vsed in shops: they chuse that for the best which is hollow, and round like a reed, brittle, and with a smooth bark, as also that whereunto doth cleaue a congealed gum, which is said to be *gummosum*, or gummy, and somewhat white. But as *Garcias* saith, it is not alwayes

gummy

gummy of his own nature; but the Indians becauſe they ſee that our merchants note the beſt Turbith by the gumminesſe, are wont before they gather the ſame, either to wryth or els lightly bruiſe them, that the ſap or liquor may iſſue out; which root being once hardned, they picke out from the reſt to ſell at a greater price. It is likewiſe made white, as the ſame Author ſheweth, being dried in the Sun; for if it be dried in the ſhadow it waxeth blacke, which notwithſtanding may be as good as the white which is dried in the Sun.

Turbith Alex.andrinum officinarum.
Turpetum, or **Turbith** of the ſhops.

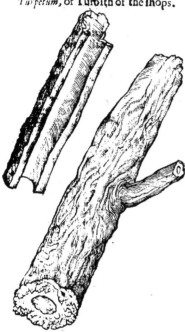

¶ *The Place.*

It groweth by the ſea ſide, but yet not ſo neere that the waſh or water of the ſea may come to it, but neere about, and that for two or three miles in vntilled grounds, rather moiſt than dry. It is found in Cambaya, Surrate, in the Iſle Dion, Bazaim, and the places neere adioyning; alſo in Guzarate, where it groweth plentifully, from whence great aboundance of it is brought into Perſia, Arabia, Aſia the leſſe, and alſo into Portingall and other parts of Europe; but that is preferred which groweth in Cambaia.

¶ *The Names.*

It is called of the Arabians, Perſians, & Turks, *Turbith*: and in Guzarata, *Barcaman*: in the Prouince Canara, in which is the city Goa, *Tiguar*: likewiſe in Europe the Learned call it diuerſly, according to their ſeuerall fancies, which hath bred ſundry controuerſies, as it hath fallen out as well in Hermodactyls as in Turbith; the vſe and poſſeſſion of which we cannot ſeem to want: but which plant is the true Turbith wee haue great cauſe to doubt. Some haue thought our *Tripolium marinum*, deſcribed in the former chapter, to bee the true Turbith: others haue ſuppoſed it to bee one of the Tithymales, but which kind they know not. *Guillandinus* ſaith, that the root of *Tithymalus myrſinitis* is the true Turbith: which cauſed Lobel and Pena to plucke vp by the roots all the kinds of Tithymales and dry them very curiouſly; which when they had beheld and throughly tried, they found it nothing ſo. The Arabians and halfe-Moores that dwell in the Eaſt parts haue giuen diuers names to this plant; and as their words are diuers, ſo haue they diuers ſignifications: but this name Turbith they ſeem to interpret to be any milky root which doth ſtrongly purge flegme, as this root doth. So that as men haue thought good, pleaſing themſelues, they haue many and diuers conſtructions, which hath troubled many excellent learned men to know what root is the true Turbith. But briefly to ſet downe mine opinion, not varying from the iudgement of men of great experience; I thinke aſſuredly, that the root of Scammonie of Antioch is the true and vndoubted Turbith: one reaſon eſpecially that moueth me ſo to thinke is, for that I haue taken vp the roots of Scammony which grew in my garden, and compared them with the roots of Turbith, betweene which I found little or no difference at all.

‡ Through all Spain (as *Cluſius* in his notes vpon *Garcias* teſtifies) they vſe the roots of *Thapſia* for Turbith, which alſo hath bin brought hither, and I keep ſome of them by me, but they purge little or nothing at all being dry, though it may be the green root or iuyce may haue ſome purging facultie. ‡

¶ *The Temperature and Vertues.*

The Indian Phyſitions vſe it to purge flegme, to which if there be no feuer they adde Ginger; A otherwiſe they giue it without in the broth of a chicken, and ſometimes in faire water.

Meſues writeth, that Turbith is hot in the third degree, and that it voideth thicke tough flegme B out of the ſtomack, cheſt, ſinues, and out of the furthermoſt parts of the body; but (as he ſaith) it is ſlow in working, and troubleth and ouerturneth the ſtomacke, and therefore Ginger, Maſtick, and other ſpices are to be mixed with it; alſo oile of ſweet Almonds, or Almonds themſelues, or ſugar, leſt the body with the vſe hereof ſhould pine and fall away. Others temper it with Dates, ſweet

Almonds and certaine other things, making thereof a composition that the Apothecaries call an Electuarie) which is named ⲁⲣⲟⲣⲓⲧⲓⲕⲟⲛ: common in shops, and in continual vse among expert Physitions.

C   There is giuen at one time of this Turbith one dram (more or lesse) two at the most: but in the decoction or in the infusion three or foure.

CHAP. 95. *Of Arrow-head, or Water-Archer.*

1 *Sagittaria maior.*
Great Arrow-head.

2 *Sagittaria minor.*
Small Arrow-head.

¶ *The Description.*

1   THe first kinde of water-Archer or Arrow-head hath large and long leaues in shape like a bearded broad Arrow-head. Among which riseth vp a fat and thicke stalke two or three foot long, hauing at the top many pretty white floures declining to a light carnation, compact of three small leaues: which being past, there come after great rough knops or burres wherein is the seed. The root consisteth of many strings.

2   The second is like the first, and differs in that this kinde hath smaller leaues and floures, and greater burres and roots.

3   The third kinde of Arrow-head hath leaues in shape like the broad Arrow-head, standing vpon the ends of tender foot-stalks a cubit long: among which riseth long naked smooth stalks of a greenish colour, from the middle whereof to the top grow floures like to the precedent. The root is small and threddy.

¶ *The Place.*

These herbs grow in the waterie ditches by S. *Georges* field neere London, in the tower ditch at London, in the ditches neere the walls of Oxford, by Chelmsford in Essex, and in many other places, as namely in the ditch neere the place of execution called S. *Thomas Waterings,* not farre from London.

¶ *The Time.*

They floure in May and Iune.

                                                            ¶ *The*

¶ *The Names.*

Sagittaria may be called in Engliſh, water Archer, or Arrow-head. ‡ Some would haue it the *Phleum* of *Theophraſtus* ; and it is the *Piſtana Magonis*, and *Sagitta* of *Pliny, lib.* 21. *cap.* 17. ‡

¶ *The Nature and Vertues.*

I find nothing extant in writing either concerning their vertues or temperament, but doubtles they are cold and dry in qualitie, and are like Plantain in facultie and temperament.

CHAP. 96. *Of water Plantaine.*

1 *Plantago aquatica maior.*
Great Water Plantain.

‡ 2 *Plantago aquatica minor ſtellata.*
Starry headed ſmall water Plantaine.

3 *Plantago aquatica humilis.*   **Dwarfe water Plantain.**

¶ *The Deſcription.*

1   THe firſt kinde of Water Plantaine hath faire great large leaues like the land Plantaine, but ſmoother, and full of ribs or ſinewes : amongſt which riſeth vp a tall ſtemme, foure foot high, diuiding it ſelfe into many ſlender branches, garniſhed with infinite ſmal white ſloures,

Mm 3   which

which being paſt, there appeare triangle huskes or buttons wherein is the ſeed. The root is as it were a great tuft of threds or thrums.

‡ 2 This plant in his roots and leaues is like the laſt deſcribed,as alſo in the ſtalk,but much leſſe in each of them,the ſtalk being about ſome foot high ; at the top whereof ſtand many pretty ſtar-like skinny ſeed-veſſels,containing a yellowiſh ſeed. ‡

3 The third kinde hath long little narrow leaues much like the Plantaine called Ribwort : among which riſe vp ſmall and feeble ſtalkes branched at the top, whereon are placed white floures conſiſting of three ſlender leaues ; which being fallen , there come to your view round knobs or rough burs : the root is threddy.

¶ *The Place.*

1 This herb growes about the brinks of riuers,ponds,and ditches almoſt euery where.

‡ 2 3 Theſe are more rare. I found the ſecond a little beyond Ilford, in the way to Rumford,and Mr *Goodyer* found it alſo growing vpon Hounſlow heath. I found the third in the companie of Mr *Will. Broad* and Mr *Leonard Buckner*,in a ditch on this ſide Margate in the Iſle of Tenet. ‡

¶ *The Time.*

They floure from Iune till Auguſt.

¶ *The Names.*

The firſt is called *Plantago aquatica*,that is,water Plantain. ‡ The ſecond *Lobel* calls *Aliſma puſillum anguſtifolium muricatum* : and in the *Hiſt.Lugd.* it is called *Damaſonium ſtellatum.* ‡

The third is named *Plantago aquatica humilis*, the low water Plantaine.

‡ I thinke it fit here to reſtore this plant to his antient dignitie, that is, his names and titles wherewith he was antiently dignified by *Dioſcorides* and *Pliny.* The former whereof cals it by ſundry names,and al very ſignificant and proper,as λειμώνιον, πoταμογείτον, πνευρώδεσ, λόγχητε : thus many are Greek and therefore ought not to be reiected,as they haue been by ſome without either reaſon or authoritie. For the barbarous names we can ſay nothing : now it is ſaid to be called *Limonium*, becauſe ὁ λειμώσι φύεται : it growes in wet or ouerflown medowes : it is called *Neuroides*,becauſe the leaf is compoſed of diuers ſtrings or fibers running from one end thereof to the other,as in Plantaine ; which therfore by *Dioſcorides* it is termed for the ſame reaſon πολύνευρον. Alſo it may be as fitly termed *Lonchitis*,for the ſimilitude which the leaſe hath to the top or head of a lance,which λόγχη properly ſignifies,as that other plant deſcribed by *Dioſcorides,lib.3.cap.161.* for that the ſeed (a leſſe eminent part) reſembles the ſame thing.And for *Potamogeiton*,which ſignifies a neighbor to the riuer or water,I thinke it loues the water as well,and is as neere a neighbour to it as that which takes its name from thence,and is deſcribed by *Dioſcorides,lib.4.101.* Now to come to *Pliny,lib.20.cap.8.*he cals it *Beta ſylueſtris,Limonion,*and *Neuroides* : the two later names are out of *Dioſcorides,*and I ſhall ſhew you where alſo you ſhall finde the former in him. Thus much I thinke might ſerue for the vindication of my aſſertion, for I dare boldly affirm,that no late writer can fit all theſe names to any other plant;and that makes me more to wonder,that all our late Herbariſts,as *Matthiolus,Dodonæus,Fuchſius,Caſalpinus,Daleſchampius,* but aboue all,*Pena* and *Lobel*(who, *Aduerſ.pag.126.*cal it to queſtion) ſhould not allow this plant to be *Limonium,*eſpecially ſeeing that *Anguillara* had before or in their time aſſerted it ſo to be : but whether he gaue any reaſons or no for his Aſſertion,I cannot tell, becauſe I could neuer by any meanes get his opinions,but onely find by *Bauhines Pinax*,that ſuch was his opinion hereof. But to return from whence I digreſt : I will giue you *Dioſcorides* his deſcription,with a briefe explanation thereof,and ſo deſiſt. It is thus; It hath leaues like a Beet,thinner and larger,ten or more ; a ſtalke ſlender, ſtraight, and as tall as that of a Lilly , and full of ſeeds of an aſtringent taſte. The leaues of this you ſee are larger than thoſe of a Beet,and thin,and as I formerly told you in the names,neruous;which to be ſo may be plainly gathered by *Dioſcorides* his words in the deſcription of white Hellebore, whoſe leaues he compares to the leaues of Plantain and the wild Beet : now there is no wilde Beet mentioned by any of the Antients,but only this by *Pliny,* in the place formerly quoted ; nor no leaſe more fit to compare thoſe of Hellebore to than thoſe of water Plantaine, eſpecially for the nerues and fibers that run alongſt the leaues : the ſtalke alſo of this is but ſlender,conſidering the height, and it grows ſtraight, and as high as that of a Lilly,with the top plentifully ſtored with aſtringent ſeed : So that no one note is wanting in this, nor ſcarce any to be found in the other plants that many haue of late ſet forth for *Limonium.* ‡

¶ *The Temperature.*

Water Plantain is cold and dry of temperature.

¶ *The*

¶ *The Vertues.*

The leaues of water Plantain, as some Authors report, are good to be laid vpon the legs of such A as are troubled with the dropsie, and hath the same propertie that the land Plantain hath.

‡ *Dioscorides* and *Galen* commend the seed hereof giuen in wine, against fluxes, dysenteries, the B spitting of bloud, and ouermuch flowing of womens terms.

Pliny saith, the leaues are good against burns. ‡   C

Chap. 97. *Of land Plantaine.*

1 *Plantago latifolium.*
Broad leaued Plantaine.

2 *Plantago incana.*
Hoary Plantaine.

¶ *The Description.*

1   AS the Greeks haue called some kinds of herbs Serpents tongue, Dogs tongue, and Ox tongue ; so haue they termed a kinde of Plantain *Arnoglosson*, which is as if you should say Lambs tongue, well known to all, by reason of the great commoditie and plenty of it growing euery where ; and therefore it is needlesse to spend time about them. The greatnes and fashion of the leaues hath been the cause of the varieties and diuersities of their names.

2   The second is like the first, and differeth in that, that this Plantaine hath greater but shorter spikes or knaps ; and the leaues are of an hoary or ouerworne green colour : the stalks are likewise hoary and hairy.

3   The small Plantain hath many tender leaues ribbed like vnto the great Plantain, and is very like in each respect vnto it, sauing that it is altogether lesser.

4   The spiked rose Plantaine hath very few leaues, narrower than those of the second kinde of Plantain, sharper at the ends, and further growing one from another. It beareth a very double floure vpon a short stem like a rose, of a greenish colour tending to yellownesse. The seed groweth vpon a spiky tuft aboue the highest part of the plant; notwithstanding it is but very low in respect of the other Plantains aboue mentioned.

4 *Plantago Roſea ſpicata.*
Spiked Roſe Plantaine.

5 *Plantago Roſea exotica.*
Strange Roſe Plantaine.

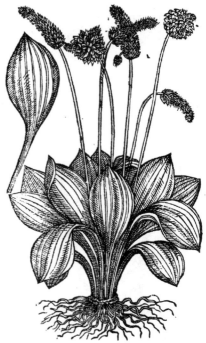

‡ 6 *Plantago panniculis ſparſis.*
Plantain with ſpoky tufts.

5   The fift kinde of Plantain hath beene a
ſtranger in England and elſwhere vntil the impreſſion hereof. The cauſe why I ſay ſo is, the
want of conſideration of the beauty which is
in this plant,wherein it excelleth all the other.
Moreouer, becauſe it hath not been written of
or recorded before this preſent time , though
plants of leſſer moment haue bin very curiouſly ſet forth. This plant hath leaues like vnto
them of the former, and more orderly ſpread
vpon the ground like a Roſe : amongſt which
riſe vp many ſmall ſtalkes like the other Plantains,hauing at the top of euery one a fine double roſe altogether vnlike the former,of an ho-
rie or ruſty green colour.
    ‡ I take this ſet forth by our Author to be
the ſame with that which *Cluſius* receiued from
Iames Garret the yonger of London ; and therefore I giue you the figure thereof in this place,
together with this addition to the hiſtorie out of *Cluſius* : That ſome of the heads are like thoſe of
the former Roſe Plantaine ; other ſome are ſpike-faſhion,and ſome haue a ſpike growing as it were
out of the midſt of the Roſe,and ſome heads are otherwiſe ſhaped ; alſo the whole plant is more
hoary than the common Roſe Plantaine.
    6   This Plantain muſt not here be forgot,though it be ſomwhat hard to be found : his leaues,
roots,and ſtalkes are like thoſe of the ordinarie, but in ſtead of a compact ſpike it hath one much
diuided after the manner as you ſee it heere expreſſed in the figure, and the colour thereof is greeniſh. ‡

¶ The

¶ *The Place.*

The greater Plantains grow almoſt euery where.

The leſſer Plantain is found on the ſea coaſts and banks of great riuers, which are ſomtimes waſhed with brackiſh water.

‡ The Roſe-Plantaines grow with vs in gardens ; and the ſixt with ſpoky tufts grows in ſome
places in the Iſle of Tenet, where I firſt found it, being in company with Mr *Tho. Hicks*, Mr *Leonard
Buckner*, and other London Apothecaries, *Anno* 1632. ‡

¶ *The Time.*

They are to be ſeen from Aprill vnto the end of September.

¶ *The Names.*

Plantain is called in Latine *Plantago* ; and in Greek, ἀρνόγλωσσον , and *Arnogloſſa*, that is to ſay, Lambs
tongue : the Apothecaries keep the Latine name : in Italian, *Piantagine*, and *Plantagine* : in Spaniſh,
Lhantem : the Germanes, Wegrich : in low-Dutch, Wechbre: in Engliſh, Plantain, and Weybred:
in French, *Plantain.*

¶ *The Temperature.*

Plantain (as *Galen* ſaith) is of a mixt temperature ; for it hath in it a certaine waterie coldneſſe,
with a little harſhneſſe, earthy, dry and cold ; therefore they are cold and dry in the ſecond degree.
To be briefe, they are dry without biting, and cold without benumming. The root is of like temperature; but drier, and not ſo cold. The ſeed is of ſubtill parts, and of temperature leſſe cold.

¶ *The Vertues.*

Plantain is good for vlcefs that are of hard curation, for fluxes, iſſues, rheumes, and rottenneſſe,   A
and for the bloudy flix : it ſtayeth bleeding, it heales vp hollow ſores and vlcers as well old as new.
Of all the Plantains the greateſt is the beſt, and excelleth the reſt in facultie and vertue.

The juice or decoction of Plantain drunken ſtoppeth the bloudy flix and all other fluxes of the   B
belly, ſtoppeth the piſſing of bloud, ſpitting of bloud, and all other iſſues of bloud in man or woman, and the deſire to vomit.

Plantain leaues ſtamped and made into a tanſie, with the yelks of egges, ſtayeth the inordinate   C
flux of the terms, although it haue continued many yeares.

The root of Plantaine with the ſeed boiled in white wine and drunke, openeth the conduits or   D
paſſages of the liuer and kidnies, cures the jaundice, and vlceration of the kidnies and bladder.

The juice dropped in the eies cooles the heate and inflammation thereof. I find in antient wri-   E
ters many good-morrowes, which I thinke not meet to bring into your memorie againe ; as, That
three roots will cure one griefe, foure another diſeaſe, ſix hanged about the necke are good for another malady, &c. all which are but ridiculous toyes.

The leaues are ſingular good to make a water to waſh a ſore throat or mouth, or the priuy parts   F
of a man or woman.

The leaues of Plantaine ſtamped and put into oile oliue, and ſet in the hot ſun for a moneth to-   G
gether, and after boiled in a kettle of ſeething water (which we call *Balneum Mariæ*) and then ſtrained, preuaile againſt the pains in the eares, the yard, or matrix, (being dropped into the eares , or
caſt with a ſyringe into the other parts before rehearſed) or the paines of the fundament ; prooued
by a learned gentleman Mr *Godowrus* Sergeant Surgeon to the Queens Majeſtie.

CHAP. 98. *Of Ribwort.*

¶ *The Deſcription.*

1   Ibwort or ſmall Plantaine hath many leaues flat ſpred vpon the ground, narrow, ſharpe
pointed, and ribbed for the moſt part with fiue nerues or ſinues, and therefore it was
called *Quinque-neruia*: in the middle of which leaues riſeth vp a nerued or creſted ſtalk
bearing at the top a darke or duskiſh knap , ſet with a few ſuch white floures as are the floures of
wheat. The root and other parts are like the other Plantains.

‡ There is another kind of this Ribwort, which differs not from the laſt mentioned in any thing
but the ſmalneſſe thereof. ‡

2   Roſe Rib-wort hath many broad and long leaues of a darke greene colour, ſharpe pointed,
and ribbed wrth fiue nerues or ſinewes like the common Rib-wort : amongſt which riſe vp naked
ſtalkes, furrowed, chamfered, or creſted with certaine ſharpe edges : at the top whereof groweth a
great and large tuft of ſuch leaues as thoſe are that grow next the ground, making one entire tuft
or

crumbel,in ſhape reſembling a Roſe(whereof I thought good to giue it his ſirname Roſe)which is
from his floure.

‡ This alſo I thinke differs not from that of *Cluſius*,wherefore I giue his figure in the place of
that ſet forth by our Author. ‡

1 *Plantago quinqueneruia.*   2 *Plantago quinqueneruia roſea.*
Ribwort Plantaine.   Roſe Ribwort.

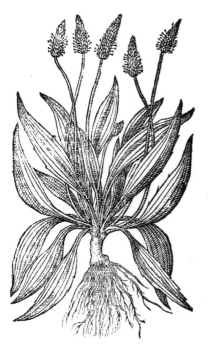

¶ *The Place.*

Ribwort groweth almoſt euerie where in the borders of path-wayes and fertile fields.

Roſe Ribwort is not very common in any place, notwithſtanding it groweth in my garden, and
wilde alſo in the North parts of England, as alſo in a field neere London by a village called Hogſdon, found by a learned merchant of London Mr *Iames Cole*, a louer of plants, and very ſkilfull in
the knowledge of them.

¶ *The Time.*

They floure and flouriſh when the other Plantains do.

¶ *The Names.*

Ribwort is called in Greeke, Ἀρνόγλωσσον μικρὸν : and of ſome, πινπίνυψι : in Latine,*Plantago minor, Quinqueneruia,*and *Lanceola,*or *Lanceolata :* in high-Dutch, **Spitziger wegrich**: in French, *Lanceole :*
in low-Dutch, **Hondts ribbe ;** that is to ſay in Latine, *Coſta Canina,* or Dogs rib : in Engliſh, Ribwort,or Rib Plantaine.

The ſecond I haue thought meet to call Roſe Ribwort in Engliſh, and *Quinquenervia roſea* in
Latine.

¶ *The Temperature.*

Ribwort is cold and dry in the ſecond degree,as are the Plantaines.

¶ *The Vertues.*

The vertues are referred to the kinds of Plantains.

CHAP.

CHAP. **99.** *Of Sea Plantaines.*

1 *Holoſteum Salamanticum.*
Flouring ſea Plantaine.

2 *Holoſteum parvum.*
Small Sea Plantaine.

3 *Plantago marina.*
Sea Plantaine.

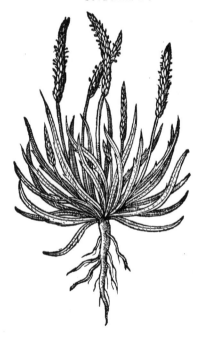

¶ *The Deſcription.*

    1   **C**Arolus *Cluſius* that excellent Herbariſt hath referred theſe two ſorts of *Holoſteum* to the kindes of Sea Plantain. The firſt hath long leaues like the common Rib wort,but narrower,couered with ſome hairineſſe or woollineſſe:among which there riſeth vp a ſtaik bearing at the top a ſpike like the kindes of Plantaine : the root is long and wooddy.This flours in Aprill or May.

    2   The ſecond is like the former,but ſmaller,and not ſo gray or hoary : the flours are like to *Coronopus* or the leſſer Ribwort.This floures at the ſame time as the former.

                                                     3 The

⸪ ₃   The third kinde,which is the ſea Plantaine,hath ſmall narrow leaues like Bucks-horne, but
without any manifeſt inciſure,cuttings,or notches vpon the one ſide : among which riſeth vp a ſpikie ſtalke like the common kinde,but ſmaller.

‡ 4 *Holoſteum, ſiue Leontopodium Creticum.*   ‡ 5 *Holoſteum,ſiue Leontopod.Cret.alterum.*
     Candy Lions foot.   The other Candy Lions foot.

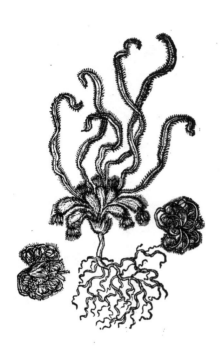 

‡   4   Theſe two following plants are by *Cluſius* and *Bauhine* referred to this Tribe ; wherefore
I thinke it fitting to place them here. The former of them,from a reddiſh and as it were ſcaly root
growing leſſe by little and little,and diuided into fibers,ſends forth many leaues,narrow, hoary, an
handfull long,and hauing three nerues or ribs running alongſt each of them : amongſt theſe come
forth diuers footſtalks couered with a ſoft reddiſh down, and being ſome two or three inches long,
hauing heads ſomewhat thicke and reddiſh : the floures are whitiſh,with a blackiſh middle,which
makes it ſeem as if it were perforated or holed. Now when the plant growes old and withers, the
ſtalks becomming more thicke and ſtiffe bend downe their heads toward the root,ſo that in ſome
ſort they reſemble the foot of a Lion.
    5   This plant which is figured in the vpper place (for I take the lower to be an exacter figure
of the laſt deſcribed) hath leaues like to the ſmall ſea Plantain,but tenderer,and ſtanding vpright;
and amongſt theſe on little foot-ſtalks grow heads like thoſe of *Pſyllium*,but prettier,and of a whitiſh red colour. ‡

¶ *The Place.*

The two firſt grow in moſt of the Spaniſh dominions. *Carolus Cluſius* writeth,that he neuer ſaw
greater or whiter than neere to Valentia a city of Spain,by the highwaies. Since they haue beene
found at Baſtable in the Iſle of Wight,and in the Iſles of Gernſey and Iarſey.
    The third growes neere vnto the ſea in all places of England where I haue trauelled, eſpecially
by the forts on both ſides the water at Graueſend,at Erith neere London,at Lee in Eſſex,at Rie in
Kent, at Weſt-Cheſter,and at Briſtow.
    ‡   The fourth and fifth grow in Candy,from whence they haue beene ſent to Padua and many
other places. ‡

                                               ¶ *The*

¶ *The Names.*

Holosteum, is also called by *Dodonæus, Plantago angustifolia albida,* or *Plantago Hispaniensis* : in English, Spanish hairy small Plantaine, or flouring sea Plantaine.

‡ The fourth is called by *Clusius, Leontopodium Creticum* : by some it hath beene thought to be *Catanance* of *Dioscorides* : the which *Honorius Bellus* will not allow of: *Bauhine* calls it *Holosteum, siue Leontopodium Creticum.*

The fifth is *Leontopodium Creticum alterum* of *Clusius* ; the *Habbures* of *Camerarius* ; and the *Holosteum Creticum alterum* of *Bauhine.* ‡

¶ *The Temperature and Vertues.*

Galen saith, That *Holosteum* is of a binding and drying facultie.

Galen, Dioscorides, and *Pliny* haue prooued it to be such an excellent wound herbe, that it presently closeth or shutteth vp a wound, though it be very great and large : and by the same authority I speake it, that if it be put into a pot where many pieces of flesh are boyling, it will soder them together.

These herbes haue the same faculties and vertues that the other Plantains haue, and are thought to be the best of all the kindes.

† That which was formerly in the fourth place of this chapter, vnder the name of *Holosteum Petraum,* you shall finde hereafter vnder the title of *Muscus Corniculatus,* for vnder that name our Author also gaue another figure thereof, with a description ; and I iudge it more fitly placed there than here amongst the Plantaines.

CHAP. 100.　*Of Sea Buck-horne Plantaines.*

1 *Coronopus.*
Sea Buck-horne.

2 *Coronopus, siue Serpentina minor.*
Small Sea Buck-horne.

¶ *The Description.*

1　THe new Writers following as it were by tradition those that haue written long agone, haue been content to heare themselues speake and set downe certainties by vncertain speeches ; which haue wrought such confusion and corruption of writings, that so many Writers, so many seueral opinions, as may most euidently appeare in these plants and in others: And my selfe am content rather to suffer this scar to passe, than by correcting the error, to renew the old wound. But for mine owne opinion thus I thinke, the plant which is reckoned for a kind of *Coronopus* is doubtlesse a kinde of *Holosteum* : my reason is, because it hath grassie leaues, or rather leaues like *Vetonica sylnestris* or wilde Pinks, a root like those of *Garyophyllata* or Auens, and the spikie eare of *Holosteum* or Sea Plantaine : which are certaine arguments that these writers haue neuer seene the Plant, but only the picture thereof, and so haue set downe their opinions by heare-say.

　　　　This

This plant likewise hath beene altogether vnknowne vnto the old Writers. It groweth most plentifully vpon the cliffes and rocks and the tops of the barren mountaines of Auergne in France, and in many places of Italy.

2   The second sort of wilde sea Plantaine or *Serpentina* differeth not from the former but onely in quantitie and slendernesse of his stalkes, and the smallnesse of his leaues, which exceed not the height of two inches. It groweth on the hills and rockes neere the washings of the sea at Massilia in great plenty almost euery where among the *Tragacanthum*, hauing a most thicke and spreading cluster of leaues after the manner of *Sedum minimum saxeum montanum*, somwhat like *Pinaster*, or the wilde Pine, as well in manner of growing, as stiffenesse, and great increase of his slender branches. It hath the small seed of Plantaine, or *Serpentina vulgaris*, contained within his spikie eares. The root is somewhat long, wooddy, and thicke, in taste somewhat hot and aromaticall.

3 *Coronopus, siue Serpentina minima.*
Small Buck-horne Plantaine.

4 *Cauda Muris.*
Mouse-taile.

3   This small sea plant is likewise one of the kindes of sea Plantaine, participating as well of Buck-horne as of *Holostium*, being as it were a degenerate kinde of sea Plantaine. It hath many grassie leaues very like vnto the herbe Thrift, but much smaller; among which come forth little tender foot-stalkes, whereon doe grow small spikie knobs like those of sea Plantaine. The root is tough and threddy.

4   Mouse-taile or *Cauda muris* resembleth the last kinde of wilde *Coronopus* or sea Plantaine, in small spikie knobs, leaues, and stalkes, that I know no reason to the contrary, but that I may as well place this small herbe among the kindes of *Coronopus* or Bucks-horne, as other writers haue placed kindes of *Holostium* in the same section: and if that be pardonable in them, I trust this may be tolerable in me, considering that without controuersie this little and base herbe is a kinde of *Holostium*, hauing many small grassie leaues spred on the ground, an inch long or somewhat more: among which doe rise small tender naked stalkes of two inches long, bearing at the top a little blackish torch or spikie knob in shape like that of the Plantains, resembling very notably the taile of a Mouse, whereof it tooke his name. The root is small and threddy.

¶ *The Place.*

The first and second of these plants are strangers in England; notwithstanding I haue heard say that they grow vpon the rocks in Silley, Garnsey, and the Isle of man.

Mouse-taile groweth vpon a barren ditch banke neere vnto a gate leading into a pasture on the right hand of the way, as ye go from London to a Village called Hampstead; in a field as you goe from Edmonton (a village neer London) vnto a house thereby called Pims, by the foot-paths sides; in Woodford Row in Waltham Forrest, and in the Orchard belonging to Mr *Francis Whetstone* in Essex, and in other places.

¶ *The*

¶ *The Time.*

They floure and flouriſh in May and Iune.

¶ *The Names.*

Matthiolus writeth, That the people of Goritia doe commonly call theſe two former plants *Serpentaria* and *Serpentina*; but vnproperly, for that there be other plants which may better be called *Serpentina* than theſe two: we may call them in Engliſh, wild ſea Plantaine, whereof doubtleſſe they are kindes.

Mouſe-taile is called in Latine, *Cauda muris*, and *Cauda murina* : in Greeke, μυόσυρος. or μυόσυρος. *Myoſures* is called of the French-men, *Queue de ſouris* : in Engliſh, Bloud-ſtrange, and Mouſe-taile.

¶ *The Temperature.*

Coronopus is cold and dry much like vnto the Plantaine. Mouſe-taile is cold and ſomething drying, with a kinde of aſtriction or binding quality.

¶ *The Vertues.*

Their faculties in working are referred vnto the Plantaines and Harts-horne.

Chap. 101.
Of Bucke-horne Plantaines, or Harts-horne.

1 *Cornu Ceruinum.*
Harts-horne,

2 *Coronopus Ruellij.*
Swines Creſſes, or Bucks-horne.

¶ *The Deſcription.*

1   BVcks-horne or Harts-horne hath long narrow hoary leaues, cut on both the ſides with three or foure ſhort ſtarts or knags, reſembling the branches of an harts horne, ſpreading it ſelfe on the ground like a ſtar : from the middle whereof ſpring vp ſmall round naked hairy ſtalkes; at the top whereof do grow little knops or ſpikie torches like thoſe of the ſmall Plantaines. The root is ſlender and threddy.

2 *Ruellius* Bucks-horne or Swines Cresses hath many small and weake stragling branches, trailing here and there vpon the ground, set with many small cut or jagged leaues, somewhat like the former, but smaller, and nothing at all hairy as is the other. The floures grow among the leaues, in small rough clusters, of a whitish colour: which being past, there come in place little flat pouches broad and rough, in which the seed is contained. The root is white, threddy, and in taste like the garden Cresses.

¶ *The Place.*

They grow in barren plaines, and vntilled places, and sandy grounds; as in Touthill field neere vnto Westminster, at Waltham twelue miles from London, and vpon Blackeheath also neere London.

¶ *The Time.*

They floure and flourish when the Plantaines doe, whereof these same haue beene taken to be kindes.

¶ *The Names.*

Bucks-horne is called in Latine *Cornu Ceruinum*, or Harts-horne: diuers name it *Herba stella*: or *Stellaria*, although there be another herbe so called: in low Dutch, **Hertzhoozen**: in Spanish, *Guiabella*: in French, *Corne de Cerf*: It is thought to be *Dioscorides* his κερατοφυτον, which doth signifie *cornicis pedem*, a Crowes foot. It is called also by certaine bastard names, as *Harenaria, Sanguinaria*: and of many, Herbe Iuy, or herbe Eue.

¶ *The Temperature.*

Bucks-horne is like in temperature to the common Plantaine, in that it bindeth, cooleth, and drieth.

¶ *The Vertues.*

A   The leaues of Buckes-horne boyled in drinke, and giuen morning and euening for certaine daies together, helpe most wonderfully those that haue sore eies, watery or blasted, and most of the griefes that happen vnto the eyes; experimented by a learned Physition of Colchester called Master *Duke*; and the like by an excellent Apothecary of the same Towne called Master *Buckstone*.

B   The leaues and roots stamped with Bay salt, and tied to the wrests of the armes, take away fits of the Ague: and it is reported to worke the like effect being hanged about the necke of the Patient in a certaine number; as vnto men nine plants, roots and all; and vnto women and children seuen.

CHAP. 102.   *Of Saracens Confound.*

¶ *The Description.*

SAracens Confound hath many long narrow leaues cut or slightly snipt about the edges: among which rise vp faire browne hollow stalkes of the height of foure cubits; along which euen from the bottome to the top it is set with long and pretty large leaues like them of the Peach tree: at the top of the stalkes grow faire starre-like yellow floures, which turne into downe, and are carried away with the winde. The root is very fibrous or threddy.

¶ *The Place.*

Saracens Confound groweth by a wood as ye ride from great Dunmow in Essex, vnto a place called Clare in the said country; from whence I brought some plants into my garden.

‡ I formerly in the twenty fourth Chapter of this second booke told you what plant our Author took for Saracens Confound, and (as I haue beene credibly informed) kept in his Garden for it. Now the true *Solidago* here described and figured was found *Anno* 1632, by my kinde Friends Mr *George Bowles* and Mr *William Coot*, in Shropshire in Wales, in the hedge in the way as one goeth from Dudson in the parish of Cherbery to Guarthlow. ‡

¶ *The Time.*

It floureth in Iuly, and the seed is ripe in August.

¶ *The Names.*

Saracens Confound is called in Latine *Solidago Saracenica*, or Saracens Comfrey, and *Consolida Saracenica* in Dutch, **Heidbuisch woundkraut**: of some, *Herba fortis*: in English, Saracens Confound, or Saracens Wound-wort.

¶ *The*

† *Solidago Saracenica.*
Saracens Confound.

¶ *The Nature.*

Saracens Confound is dry in the third degree, with some manifest heate.

¶ *The Vertues.*

Saracens Confound is not inferiour to A any of the wound-herbes whatsoeuer, being inwardly ministred, or outwardly applied in ointments or oyles. With it I cured Master *Cartwright* a Gentleman of Grayes Inne, who was grieuously wounded into the lungs, and that by Gods permission in short space.

The leaues boiled in water and drunke, B restraine and stay the wasting of the liuer, take away the opilation and stopping of the same, and profit against the iaundice and Feuers of long continuace.

The decoction of the leaues made in. C water is excellent against the sorenesse of the throat, if it be therewith gargarised : it increaseth also the vertue and force of loti-on or washing waters, appropriat for priuy maimes, sore mouthes, and such like, if it be mixed therewith.

† The figure that was formerly in this place was of *Consolida palustris* of *Tabernamontanus*; and the true figure belonging to this h ito y was in the next chapter saue one, vnder the title of *Herba Doria Lobelii.*

Chap. 103. *Of Golden Rod.*

¶ *The Description.*

1 GOlden Rod hath long broad leaues somewhat hoary and sharpe pointed; among which rise vp browne stalkes two foot high, diuiding themselues toward the top into sun-dry branches, charged or loden with small yellow floures; which when they be ripe turn into downe which is carried away with the winde. The root is thready and browne of colour. ‡ *Lobel* makes this with vnsnipt leaues to be that of *Arnoldus de villa noua*. ‡

2 The second sort of Golden Rod hath small thin leaues broader than those of the first descri-bed, smooth, with some few cuts or nickes about the edges, and sharpe pointed, of a hot and harsh taste in the throat being chewed; which leaues are set vpon a faire reddish stalke. It tooke his name from the floures which grow at the top of a gold yellow colour : which floures turne into Downe, which is carryed away with the winde as is the former. The root is small, compact of many strings or threds.

¶ *The Place.*

They both grow plentifully in Hampstead Wood, neere vnto the gate that leadeth out of the wood vnto a Village called Kentish towne, not far from London; in a wood by Rayleigh in Essex, hard by a Gentlemans house called Mr *Leonard*, dwelling vpon Dawes heath; in Southfleet and in Swainescombe wood also, neere vnto Grauesend.

¶ *The Place.*

They floure and flourish in the end of August.

¶ *The Names.*

It is called in English, Golden Rod : in Latine, *Virga aurea*, because the branches are like a Gol-den Rod : in Dutch, **Gulden roede** : in French, *Verge d'or.*

1 *Virga aurea.*
Golden Rod.

2 *Virga aurea Arnoldi Villanouani.*
Arnold of the new towne his Golden Rod

¶ *The Temperature.*

Golden Rod is hot and dry in the ſecond degree : it clenſeth, with a certaine aſtriction or binding quality.

¶ *The Vertues.*

A   Golden Rod prouoketh vrine, waſteth away the ſtones in the kidnies, and expelleth them, and
withall bringeth downe tough and raw flegmatick humors ſticking in the vrine veſſels, which now
and then do hinder the comming away of the ſtones, and cauſeth the grauell or ſand which is brittle to be gathered together into one ſtone. And therefore *Arnoldus Villanouanus* by good reaſon
hath commended it againſt the ſtone and paine of the kidnies.

B   It is of the number of thoſe plants that ſerue for wound-drinkes, and is reported that it can fully
performe all thoſe things that Saracens confound can; and in my practiſe ſhall be placed in the for-
moſt ranke.

C   *Arnoldus* writeth, That the diſtilled water drunke with wine for ſome few daies together, worketh
the ſame effect, that is, for the ſtone and grauell in the kidnies.

D   It is extolled aboue all other herbes for the ſtopping of bloud in ſanguinolent vlcers and bleeding wounds; and hath in times paſt beene had in great eſtimation and regard than in theſe daies :
for in my remembrance I haue known the dry herbe which came from beyond the ſea ſold in Bucklers bury in London for halfe a crowne an ounce. But ſince it was found in Hampſtead wood, euen
as it were at our townes end, no man will giue halfe a crowne for an hundred weight of it : which
plainely ſetteth forth our inconſtancie and ſudden mutabilitie, eſteeming no longer of any thing,
how pretious ſoeuer it be, than whileſt it is ſtrange and rare. This verifieth our Engliſh prouerbe,
Far fetcht and deare bought is beſt for Ladies. Yet it may be more truely ſaid of phantaſticall
Phyſitions, who when they haue found an approued medicine and perfect remedy neere home
againſt any diſeaſe; yet not content therewith, they will ſeeke for a new farther off, and by that
meanes many times hurt more than they helpe. Thus much I haue ſpoken to bring theſe new fan-
gled fellowes backe againe to eſteeme better of this admirable plant than they haue done, which
no doubt haue the ſame vertue now that then it had, although it growes ſo neere our owne homes
in neuer ſo great quantity

CHAP.

Chap. 104.　Of Captaine Andreas Dorias his Wound-woort.

† Herba Doria Lobelij.
Dorias Wound-woort.

¶ The Description.

THis plant hath long and large thicke and fat leaues, sharpe pointed, of a blewish greene like vnto Woad, which being broken with the hands hath a pretty spicie smell. Among these leaues riseth vp a stalke of the height of a tall man, diuided at the top into many other branches, whereupon grow small yellowish floures, which turneth into downe that flyeth away with the winde. The root is thicke almost like Helleborus albus.

Of which kinde there is another like the former, but that the leaues are rougher, somewhat bluntly indented at the edges, and not so fat and grosse.

‡ Herba Doria altera.

This herbe growes vp with a greene round brittle stalke, very much chamfered, sinewed, or furrowed, about foure or fiue foot high, full of white pith like that of Elder, and sendeth forth small branches : the leaues grow on the stalke out of order, and are smooth, sharpe pointed, in shape like those of Herba Doria, but much shorter and narrower, the broadest and longest seldome being aboue ten or eleuen inches long, and scarce two inches broad, and are more finely and smally nickt or indented about the edges, their smell being nothing pleasant, but rather when together with the stalke they are broken and rubbed yeeld forth a smell hauing a small touch of the smell of Hemlocke. Out of the bosomes of these leaues spring other smaller leaues or branches. The floures are many, and grow on small branches at the tops of the stalkes like those of Herba Doria, but more like those of Iacobaa, of a yellow colour, as well the middle button, as the small leaues that stand round about, euery floure hauing commonly eight of those small leaues. Which being past the button turneth into downe and containeth very small long seeds which fly away with the wine. The root is nothing else but an infinite of small strings which most hurtfully spread in the ground, and by their infinite increasing destroy and sterue other herbes that grow neere it. Its naturall place of growing I know not; for I had it from M.ʳ Iohn Coys, and yet keepe it growing in my garden. Iohn Goodyer. ‡

¶ The Place.

These plants grow naturally about the borders or brinkes of riuers neere to Narbone in France, from whence they were brought into England, and are contented to be made denizons in my Garden, where they flourish to the height aforesaid.

¶ The Time.

They floured in my Garden about the twelfth of Iune.

¶ The Nature.

The roots are sweet in smell, and hot in the third degree.

¶ The Vertues.

Two drams of the roots of Herba Doria boiled in wine and giuen to drinke, draw downe waterish　A humors, and prouoke vrine.

The same is with good successe vsed in medicines that expell poyson.　　　　　　　　　　　　B

‡ All

‡ All theſe plants mentioned in the three laſt Chapters, to wit, *Solidago, Virga aurea* and this *Herba Dorea,* are by *Bauhine* fitly comprehended vnder the title of *Virga aurea* ; becauſe they are much alike in ſhape, and for that they are all of the ſame facultie in medicine.

† The figure that was here was of *Solidago Saracenica.*

Chap. 105.   *Of Felwoort, or Baldmoney.*

¶ *The Kindes.*

THere be diuers ſorts of Gentians or Felwoorts, whereof ſome be of our owne countrey ; others more ſtrange and brought further off: and alſo ſome not before this time remembred, either of the antient or later writers, as ſhall be ſet forth in this preſent chapter.

¶ *The Deſcription.*

1   THe firſt kinde of Felwoot hath great large leaues, not valike to thoſe of Plantaine, very well reſembling the leaues of the white Hellebore: among which riſeth vp a round hollow ſtalke as thicke as a mans thumbe, full of joints or knees, with two leaues at each of them, and towards the top euery joynt or knot is ſet round about with ſmall yellow ſtarre-like floures, like a coronet or garland : at the bottome of the plant next the ground the leaues do ſpread themſelues abroad, embracing or clipping the ſtalke in that place round about, ſet together by couples one oppoſite againſt another. The ſeed is ſmall, browne, flat, and ſmooth like the ſeeds of the Stocke Gillo-floure. The root is a finger thicke. The whole Plant is of a bitter taſte.

    1 *Gentiana major.*   ‡ 2 *Gentiana major purpurea,* 1 .*Cluſij.*
     Great Felwoort.   Great Purple Felwoort.

3 *Gentiana major ij cæruleo flore Cluſij.*
Blew floured Felwoort.

4 *Gentiana minor Cruciata.*
Croſſewoort Gentian.

5 *Gentiana Pennei minor.*
Spotted Gentian of Dr *Pennie.

‡ 2   **This** deſcribed by *Cluſius,* hath leaues and ſtalkes like the precedent; theſe ſtalkes are ſome cubit and halfe or two cubits high, and towards the toppes they are ingirt with two or three coronets of faire purple floures, which are not ſtarre-faſhioned, like thoſe of the former, but long and hollow, diuided as it were into ſome fiue or ſix parts or leaues, which towards the bottome on the inſide are ſpotted with deepe purple ſpots : theſe floures are without ſmell, and haue ſo many chiues as they haue jagges, and theſe chiues compaſſe the head, which is parted into two cells, and containes ſtore of a ſmooth, chaffie, reddiſh ſeed. The root is large, yellow on the outſide, and white within, very bitter, and it ſends forth euery yeare new ſhoots. It growes in diuers places of the Alps , it floures in Auguſt, and the ſeeds are ripe in September. ‡

3   *Carolus Cluſius* alſo ſetteth forth another ſort of a great Gentian, riſing forth of the ground with a ſtiffe firme or ſolide ſtalke, ſet with leaues like vnto *Aſclepias,* by couples one oppoſite againſt another, euen from the bottome to the top in certaine diſtances : from the boſome of the
leaues

leaues there ſhoot forth ſet vpon ſlender foot-ſtalkes certaine long hollow floures like bels, the mouth whereof endeth in fiue ſharpe corners. The whole floure changeth many times his colour according to the ſoile and climate ; now and then purple or blew, ſometimes whitiſh, and often of an aſh colour. The root and ſeed is like the precedent.

4   Croſſe-woort Gentian hath many ribbed leaues ſpred vpon the ground, like vnto the leaues of Sopewoort, but of a blacker greene colour : among which riſe vp weake joynted ſtalkes trailing or leaning toward the ground. The floures grow at the top in bundles thicke thruſt together, like thoſe of ſweet Williams, of a light blew colour. The root is thicke, and creepeth in the ground far abroad, whereby it greatly increaſeth.

5   *Carolus Cluſius* hath ſet forth in his Pannonicke hiſtorie a kinde of Gentian, which he receiued from Mr *Thomas Pennie* of London, Dr in Phyſicke, of famous memory, and a ſecond *Dioſcorides* for his ſingular knowledge in Plants : which *Tabernamontanus* hath ſet forth in his Dutch booke for the ſeuenth of *Cluſius*, wherein he greatly deceiued himſelfe, and hath with a falſe deſcription wronged others.

This twelfth ſort or kinde of Gentian after *Cluſius*, hath a round ſtiffe ſtalke, firme and ſollide, ſomewhat reddiſh at the bottome, jointed or kneed like vnto Croſſewoort Gentian. The leaues are broad, ſmooth, full of ribbes or ſinewes, ſet about the ſtalkes by couples, one oppoſite againſt another. The floures grow vpon ſmall tender ſtalkes, compact of fiue ſlender blewiſh leaues, ſpotted very curiouſly with many blacke ſpots and little lines ; hauing in the middle fiue yellow chiues. Thee ſeed is ſmall like ſand : the root is little, garniſhed with a few ſtrings of a yellowiſh colour.

¶ The Place.

Gentian groweth in ſhadowie woods, and the mountaines of Italy, Sclauonia, Germany, France, and Burgundie ; from whence Mr *Iſaac de Laune* a learned Phyſitian ſent me plants for the increaſe of my garden. Croſſewoort Gentian groweth in a paſture at the Weſt end of little Rayne in Eſſex on the North ſide of the way leading from Braintree to Much-Dunmow ; and in the horſe way by the ſame cloſe.

¶ The Time.

They floure and flouriſh in Auguſt, and the ſeed is ripe in September.

¶ The Names.

Gentius King of Illyria was the firſt finder of this herbe, and the firſt that vſed it in medicine, for which cauſe it was called Gentian after his owne name : in Greeke, γεντιανή : which name alſo the Apothecaries retaine vnto this day, and call it *Gentiana* : it is named in Engliſh, Felwoort, Gentian, Bitterwoort ; Baldmoyne, and Baldmoney.

  1   This by moſt Writers is called *Gentiana*, and *Gentiana major Lutea*.

  2   *Geſner* cals this *Gentiana punicea* ; *Cluſius*, *Gentiana major flore purpureo*.

  3   This is *Gentiana folijs hirundinariæ* of *Geſner* : and *Gentiana Aſclepiadis folio* of *Cluſius*.

  4   This *Cruciata*, or *Gentiana Cruciata*, of *Tragus*, *Fuchſius*, *Dodonæus*, *Geſner* and others : it is the *Gentiana minor* of *Matthiolus*.

  5   *Cluſius* calls this, *Gentiana major pallida punctis diſtincta*.

¶ The Temperature.

The root of Felwoort is hot, as *Dioſcorides* ſaith, clenſing or ſcouring : diuers copies haue, that it is likewiſe binding, and of a bitter taſte.

¶ The Vertues.

A   It is excellent good as *Galen* ſaith, when there is need of attenuating, purging, clenſing, and remouing of obſtructions, which quality it taketh of his extreme bitterneſſe.

B   It is reported to be good for thoſe that are troubled with crampes and convulſions ; for ſuch as are burſt, or haue fallen from ſome high place : for ſuch as haue euill liuers and bad ſtomackes. It is put into Counterpoyſons, as into the compoſition named *Theriaca diateſſaron* : which *Ætius* calleth *Myſterium*, a myſtery or hid ſecret.

C   This is of ſuch force and vertue, ſaith *Pliny*, that it helpeth cattell which are not onely troubled with the cough, but are alſo broken winded.

D   The root of Gentian giuen in pouder the quantitie of a dramme, with a little pepper and herbe Grace mixed therewith, is profitable for them that are bitten or ſtung with any manner of venomous beaſt or mad dog : or for any that hath taken poyſon.

E   The decoction drunke is good againſt the ſtoppings of the liuer, and cruditie of the ſtomacke, helpeth digeſtion, diſſolueth and ſcattereth congealed bloud, and is good againſt all cold diſeaſes of the inward parts.

CHAP.

CHAP. 106.   *Of English Felwoort.*

¶ *The Description.*

Hollow leafed Felwoort or English Gentian hath many long tough roots, difperfed hither and thither within the vpper cruft of the earth ; from which immediatly rifeth a fat thicke ftalke, jointed or kneed by certaine diftances, fet at euery knot with one leafe, and fometimes moe, keeping no certaine number: which leaues do at the firft inclofe the ftalkes round about, being one whole and entire leafe without any incifure at all, as it were a hollow trunke ; which after it is growne to his fulneffe, breaketh in one fide or other, and becommeth a flat ribbed leafe, like vnto the great Gentian or Plantaine. The floures come forth of the bofome of the vpper leaues, fet vpon tender foot-ftalkes, in fhape like thofe of the fmall Bindeweed, or rather the floures of Sopewoort, of a whitifh colour, wafht about the brims with a little light carnation. Then followeth the feed, which as yet I haue not obferued.

Gentiana concaua.
Hollow Felwoort.

¶ *The Place.*

I found this ftrange kind of Gentian in a fmall groue of a wood called the Spinie, neere vnto a fmall village in Northampton fhire called Lichbarrow : elfewhere I haue not heard of it.

¶ *The Time.*
It fpringeth forth of the ground in Aprill, and bringeth forth his floures and feed in the end of Auguft.

¶ *The Names.*
I haue thought good to giue vnto this plant, in Englifh, the name Gentian, being doubtleffe a kind thereof. The which hath not been fet forth, nor remembred by any that haue written of plants vntill this time. In Latine we may call it *Gentiana concaua,* of the hollow leaues. It may be called alfo hollow leaued Felwoort.

¶ *The Temperature and Vertues.*

Of the faculties of this plant as yet I can fay nothing, referring it vnto the other Gentians, vntill time fhall difclofe that which yet is fecret and vnknowne.

‡ *Bauhine* receiued this plant with the figure thereof from Doctor *Lifter* one of his Majefties Phyfitions, and he referres it vnto *Saponaria,* calling it *Saponaria concaua Anglica* ; and (as farre as I can conjecture) hath a good defcription thereof in his *Prodrom. pag.* 103. Now both by our Authour and *Bauhines* defcription, I gather, that the root in this Figure is not rightly expreffed, for that it fhould be long, thicke, and creeping, with few fibers adhering thereunto ; when as this figure expreffeth an annuall wooddy root. But not hauing as yet feene the plant, I can affirme nothing of certaintie. ‡

‡ Chap. 107   *Of Baſtard Felwoort.*

¶ *The Deſcription.*

‡ **O**Vr Authour in this Chapter ſo confounded all,that I knew not well how,handſomely
to ſet all right;for his deſcriptions they were ſo barren, that little might be gathered
by them,& the figures agreed with their titles,but the place contradicts al;for the firſt
figured is found in England,and the ſecond is not that euer I could learne : alſo the ſecond floures
in the ſpring,according to *Cluſius* and all others that haue written thereof,and alſo by our Authors
owne title,truely put ouer the figure : yet he ſaid they both floure and flouriſh from Auguſt to the
end of September,Theſe things conſidered,I thought it fitter both for the Readers benefit and my
owne credit to giue you this chapter wholly new with additions,rather than mangled and confuſed,
as otherwiſe of neceſſitie it muſt haue beene. ‡

1   This elegant *Gentianella* hath a ſmall yellowiſh creeping root,from which ariſe many greene
ſmooth thicke hard and ſharpe pointed leaues like thoſe of the broad leaued Myrtle,yet larger,and
hauing the veines running alongſt the leaues as in Plantaine. Amongſt the leaues come vp ſhort
ſtalkes,bearing very large floures one vpon a ſtalke ; and theſe floures are hollow like a Bel-floure,
and end in fiue ſharpe points with two little eares betweene each diuiſion,and their colour is an ex-
quiſite blew. After the floure is paſt there followes a ſharpe pointed longiſh veſſell, which opening it ſelfe into two equall parts,ſhewes a ſmall creſted darke coloured ſeed.

‡ 1 *Gentianella verna major.*   2 *Gentianella Alpina verna.*
Spring large floured Gentian.   Alpes Felwoort of the Spring time.

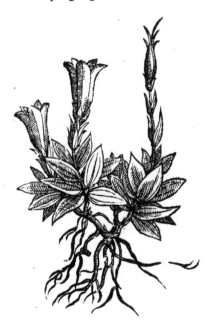

2   This ſecond riſes vp with a ſingle ſlender
and purpliſh ſtalke,ſet at certaine ſpaces with ſix
or eight little ribbed leaues, ſtanding by couples
one againſt another.At the top ſtands a cup, out
whereof comes one long floure without ſmell,
and as it were diuided at the top into fiue parts ;
and it is of ſo elegant a colour, that it ſeemes to
exceed blewneſſe it ſelfe ; each of the foldes or
little leaues of the floure hath a whitiſh line at
the ſide, and other fiue as it were pointed leaues
or appendices ſet between them:and in the mid-
deſt of the floure are certaine pale coloured chiues : a longiſh ſharpe pointed veſſell ſucceeds the
floure which containes a ſmall hard round ſeed. The root is ſmall,yellowiſh and creeping,putting
vp here and there ſtalkes bearing floures, and in other places onely leaues lying orderly ſpread vpon the ground.

3 *Gentianella fugax minor.*
Baſtard or Dwarfe Felwoort.

3　Beſides theſe two whoſe roots laſt long and increaſe euery yeare, there are diuers other Dwarfe or baſtard Gentians which are annuall, and wholly periſh euery yeare aſſoone as they haue perfected their ſeed; and therefore by *Cluſius* they are fitly called *Gentianæ fugaces*. Of theſe I haue onely obſerued two kindes (or rather varieties) in this Kingdome, which I will here deſcribe vnto you. The firſt of theſe, which is the leſſer, and whoſe figure wee here giue you, is a proper plant ſome two or three inches high, diuided immediately from the root into three, foure, or more branches, ſet at certaine ſpaces with little longiſh leaues, being broadeſt at the ſetting on, and ſo growing narrower or ſharper pointed. The tops of theſe ſtalkes are beautified with long, hollow, and pretty large floures, conſidering the magnitude of the plant, and theſe floures are of a darke purpliſh colour, and at their tops diuided into fiue parts. The root is yellowiſh, ſmall, and wooddy. The ſeed which is ſmall and round is contained in longiſh veſſels. The ſtalkes and leaues are commonly of a darke greene, or elſe of a browniſh colour.

4　This from a root like, yet a little larger than the former, ſends vp a pretty ſtiffe round ſtalke of ſome ſpan high, which at certaine ſpaces is ſet with ſuch leaues as the laſt deſcribed, but larger : and out of the boſomes of theſe leaues from the bottome to the top of the ſtalke come forth little foot-ſtalkes, which vſually carry three floures a piece ; two ſet one againſt another, and the third vpon a ſtalke ſomewhat higher; and ſometimes there comes forth a ſingle floure at the root of theſe foot-ſtalkes. The floures in their ſhape, magnitude and colour, are like thoſe of the laſt mentioned, and alſo the ſeed and ſeed veſſels. The manner of growing of this is very well preſented by the figure of the third Gentian, formerly deſcribed in the chapter laſt ſaue one aforegoing.

¶ *The Place.*

1　2　Theſe grow not wilde in England that I know of, but the former is to be found in moſt of our choiſe gardens. As with Mr *Parkinſon*, Mr *Tradeſcant*, and Mr *Tuggye*, &c.

3　4　Theſe are found in diuerſe places, as in the Chalke-dale at Dartford in Kent, and according to our Authour (for I know hee meant theſe) in Waterdowne Forreſt in Suſſex, in the way that leadeth from Charlewoods lodge, vnto the houſe of the Lord of Abergauenie, called Eridge houſe, by a brooke ſide there, eſpecially vpon a Heath by Colbrooke neere London : on the Plaine of Saliſbury, hard by the turning from the ſaid Plain, vnto the right Honourable the Lord of Pembrooks houſe at Wilton, and vpon a Chalkie banke in the highway betweene Saint Albons and Goramberrie.

¶ *The Time.*

1　2　Theſe two floure in Aprill and May. The other from Auguſt vnto the end of October.

¶ *The Names.*

1　This is the *Gentiana* 4. of *Tragus*. The *Gentianella Alpina* of *Geſner*. *Gentianella campanulæ flore* and *Heluetica* of *Lobel* ; the *Gentiana* 5. or *Gentianella major verna* of *Cluſius*.

2　*Geſner* called this *Calathiana verna*: *Lobel*, *Gentianella Alpina* : and *Cluſius*, *Gentiana* 6. and *Gentianella minor verna*.

3　This is the *Calathiana vera* of *Daleſchampius* : and the *Gentiana fugax* 5. or *Gentiana* 11. *minima* of *Cluſius*.

4　I take this to be *Cluſius* his *Gentiana fugax* 4. or *Gentiana* 10. We may call this in Engliſh, Small Autumne Gentian.

¶ *The Temperature and Vertues.*

Theſe by their taſte and forme ſhould be much like to the greater Gentians in the operation and working, yet not altogether ſo effectuall.

　　　　Chap.

CHAP. 108.   *Of Calathian Violet, or Autumne Bell-floure.*

¶ *The Description.*

AMong the number of the base Gentians there is a smal plant,which is late before it commeth vp,hauing stalkes a span high, and sometimes higher, narrow leaues like vnto Time, set by couples about the stalkes by certaine distances : long narrow floures growing at the top of the stalkes, like a cup called a Beaker,wide at the top, and narrower toward the bottome,of a deepe blew colour tending to purple,with certaine white threds or chiues in the bottome : the floure at the mouth or brim is fiue cornered before it be opened,but when it is opened it appeareth with fiue clifts or pleats. The whole plant is of a bitter taste, which plainely sheweth it to be a kinde of wilde Gentian.The root is small, and perisheth when it hath perfected his seed,and recouereth it selfe by falling of the same.

Pneumonanthe.
Calathian Violet.

¶ *The Place.*

It is found sometimes in Medowes,oftentimes in vntilled places. It groweth vpon Long-field downes in Kent, neere vnto a village called Longfield by Grauesend, vpon the Chalkie cliffes neere Greene-Hythe and Cobham in Kent,and many other places. It likewise groweth as you ride from Sugar-loafe hil vnto Bathe,in the West-country.
‡ This plant I neuer found but once, and that was on a wet Moorish ground in Lincolnshite,two or three miles on this side Caster,and as I remember,the place is called Netleton Moore. Now I suspect that our Author knew it not ; first,because he describes it with leaues like vnto Time, when as this hath long narrow leaues more like to Hyssop or Rosemary. Secondly, for that hee saith the root is small and perisheth when as it hath perfected the seed:whereas this hath a liuing, stringie and creeping root. Besides, this seldome or neuer growes on chalkie cliffes, but on wet Moorish grounds and Heaths : wherefore I suspect out Author tooke the small Autumne Gentian (described by me in the fourth place of the last Chapter) for this here treated of. ‡

¶ *The Time.*
The gallant floures hereof bee in their brauery about the end of August,and in September.

¶ *The Names.*
‡ This is thought to be *Viola Calathiana* of *Ruellius*, yet not that of *Pliny* ; and those that desire to know more of this may haue recourse to the twelfth chapter of the first booke of the 2. *Pemp.* of *Dodon.* his Latine Herball, whence our Author tooke those words that was formerly in this place, though he did not well vnderstand nor expresse them ‡. it is called *Viola Autumnalis*, or Autumne Violet,and seemeth to be the same that *Valerius Cordus* doth call *Pneumonanthe*, which he saith is named in the Germane tongue,**Lungen blumen**, or Lung-floure:in English,Autumne Bel-floures, Calathian Violets,and of some,Haruest-bels.

¶ *The Temperature.*
This wilde Felwoort or Violet is in Temperature hot, somewhat like in faculty to Gentian, whereof it is a kinde, but far weaker in operation.

¶ *The Vertues.*
The later Physitions hold it to be effectuall against pestilent diseases, and the bitings and stingings of venomous beasts.

CHAP.

Chap. 109.   *Of Venus Looking-glasse.*

¶ *The Description.*

1   BEsides the former Bel-floures there is likewise a certaine other which is low and little, the stalks whereof are tender, two spannes long, diuided into many branches most commonly lying vpon the ground: the leaues about the stalks are little, sleightly nicked in the edges: the floures are small, of a bright purple colour tending to blewnesse, very beautiful, with wide mouthes like broad bells, hauing a white chiue or thred in the middle. The flours in the day time are wide open, and about the setting of the sun are shut vp and closed fast together in fiue corners, as they are before their first opening, and as the other Bell-floures are. The roots be very slender and perish when they haue perfected their seed.

2   There is another, which from a small and wooddy root sendeth vp a straight stalk, sometimes but two or three inches, yet otherwile a foot high, when as it lights into good ground. This stalke is crested and hollow, hauing little longish leaues crumpled or sinuated about the edges set thereon: and out of the bosome of those leaues, towards the top of the stalke and somtimes lower, come little branches bearing little winged cods; at the tops of which in the midst of fiue little greene leaues stand small purple floures of little or no beauty: which being past, the cods become much larger, and containe in them a small yellowish seed, and they still retaine at their tops the fiue longish green leaues that incompassed the floure. The plant is an Annual like as the former.‡

1 *Speculum Veneris.*
Venus Looking-glasse.

‡ 2 *Speculum Veneris minus.*
Codded Corn Violet.

¶ *The Place.*

It groweth in ploughed fields among the corne in a plentifull and fruitfull soile. I found it in a field among the corn by Greenhithe, as I went from thence toward Dartford in Kent, and in many other places thereabout, but not elsewhere; from whence I brought of the seeds for my garden, where they come vp of themselues from yeare to yeare by falling of the seed.

‡   That which is here figured and described in the first place I neuer found growing in Eng-

land,

land, I haue seene onely some branches of it brought from Leiden by my friend M*r* *William Parker*. The other of my description I haue diuers times found growing among the corne in Chelsey field, and also haue had it brought me from other places by M*r* *George Bowles* and M*r* *Leonard Buckner*. ‡

¶ *The Time.*

It floureth in Iune and Iuly, and the seed is ripe in the end of August.

¶ *The Names.*

It is called *Campanula Aruensis*, and of some *Onobrychis*, but vnproperly : of other, *Caryophyllus Segetum*, or Corn Gillofloure, or Corn Pink, and *Speculum Veneris*, or Ladies Glasse : the Brabanders in their tongue call it 𝕴𝕽𝖔𝖙𝖊𝖓 𝕾𝖕𝖎𝖊𝖌𝖊𝖑.

‡ *Tabernamontanus* hath two figures thereof, the one vnder the name of *Viola aruensis*, and the other by the title of *Viola Pentagonia*, because the floure hath fiue folds or corners. 2 This of my description is not mentioned by any Author, wherefore I am content to follow that name which is giuen to the former, and terme it in Latine *Speculum Veneris minus* ; and from the colour of the floure and codded seed-vessell to call it Codded Corn Violet. ‡

¶ *The Temperature and Vertues.*

We haue not found any thing written either of his vertue or temperature, of the antient or later Writers.

C H A P. 110.   *Of Neesing root or Neesewort.*

1 *Helleborus præcox.*
White Hellebor.

2 *Helleborus albus præcox.*
Timely white Hellebor.

¶ *The Description.*

1   THe first kinde of white Hellebor hath leaues like vnto great Gentian, but much broader, and not vnlike the leaues of the great Plantaine, folded into pleits like a garment plaited to be laid vp in a chest : amongst these leaues riseth vp a stalke a cubit long, see
toward

towards the top full of little star-like floures of an herby green colour tending to whitenes:which being past,there come small husks containing the seed. The root is great and thicke, with many small threds hanging thereat.

2   The second kind is very like the first,and differeth in that, that this hath blacke and reddish floures,and comes to flouring before the other kind,and seldom in my garden commeth to seed.

The white Hellebor groweth on the Alps and such like mountains where Gentian growes. It was reported vnto me by the Bishop of Norwich,That white Hellebor groweth in a wood of his owne neere to his house at Norwich. Some say likewise that it doth grow vpon the mountaines of Wales. I speake this vpon report,yet I thinke it may be true. Howbeit I dare assure you that they grow in my garden at London,where the first kinde floureth and seedeth very well.

¶ The Time.
The first floureth in Iune,and the second in May.

¶ The Names.
Neesewort is called in Greeke , ἐλλέβορε λευκὸς : in Latine, Veratrum album,Helleborus albus,and Sanguis Herculeus. The Germans call it **Weiß nieswurt** : the low-Dutch, **Reißwoztel** : the Italians, Elleboro bianco : the Spaniards,Verde gambre blanco : the French,Ellebore blanche : and we of England call it white Hellebor,Neesewort,Lingwort,and the root Neesing pouder.

. ¶ The Temperature.
The root of white Hellebor is hot and dry in the third degree.

¶ The Vertues.
The root of white Hellebor procureth vomit mightily,wherein consisteth his chiefe vertue,and   A by that means voideth all superfluous slime and naughty humors. It is good against the falling sicknesse,phrensies,sciatica,dropsies,poison,and against all cold diseases that be of hard curation, and will not yeeld to any gentle medicine.

This strong medicine made of white Hellebor, ought not to bee giuen inwardly vnto delicate   B bodies without great correction;but it may be more safely giuen vnto countrey people which feed grosly,and haue hard tough and strong bodies.

The root of Hellebor cut in small pieces, such as may aptly and conueniently be conueyed in-   C to fistula's,doth mundifie them,and take away the callous matter which hinders curation : and after they may be healed vp with some incarnatiue vnguent fit for the purpose. ‡ This facultie by Dioscorides is attributed to the blacke Hellebor,and not to this. ‡

The pouder drawne vp into the nose causeth sneesing,and purgeth the brain from grosse and sli-   D mie humors.

The root giuen to drinke in the weight of two pence,taketh away the fits of agues, killeth mice   E and rats,being made vp with hony and floure of wheat. Pliny addeth,that it is a medicine against the Lowsie euill.

CHAP. III.   Of Wilde white Hellebor.

¶ The Description.

1   HElleborine is like vnto white Hellebor, and for that cause wee haue giuen it the name Helleborine. It hath a straight stalke of a foot high,set from the bottome to the tuft of floures with faire leaues, ribbed and chamfered like those of white Hellebor, but nothing neere so large,of a darke green colour. The floures be orderly placed from the middle to the top of the stalke,hollow within,and white of colour,straked here and there with a dash of purple,in shape like the floures of Satyrion. The seed is small like dust or motes in the sun. The root is small,full of juice,and bitter in taste.

2   The second is like vnto the first,but altogether greater : the floures white without any mixture at all,wherein consisteth the difference.

3   The third kinde of Helleborine,being the sixt of Clusius,hath leaues like the first described,but

ſmaller and narrower. The ſtalke riſeth vp to the height of two ſpans ; at the top wherof grow faire ſhining purple colored flours, conſiſting of ſix little leaues, within or among which lies hid things like ſmall helmets. The plant in proportion is like the other of this kinde. The root is ſmall, and creepeth in the ground.

1 *Helleborine*.   2 *Helleborine anguſtifolia 6. Cluſij*.
Wilde white Hellebor.   Narrow leafed wilde Neeſewort.

¶ *The Place.*

They be found in dankiſh and ſhadowie places : the firſt was found growing in the Woods by Digs wel paſtures, half a mile from Welwen in Hartfordſhire : it grows in a wood fiue miles from London, neere a bridge called Lockbridge ; by Robinhoods wel, where my friend Mr *Stephen Bredwel* a learned Phyſition found the ſame : in the woods by Dunmow in Eſſex, by Southfleet in Kent in a little groue of Iuniper, and in a wood by Clare in Eſſex.

¶ *The Time.*

They floure in May and Iune, and perfeċt their ſeed in Auguſt.

¶ *The Names.*

The likeneſſe it hath with white Hellebor, doth ſhew it may not vnproperly be named *Helleborine*, or white Hellebor : which is alſo called of *Dioſcorides* and *Pliny*, *Ꝟꝓꝓꝓꝓꝓꝓꝓꝓ*, or *Epipaċtis*, but whence that name came it is not apparant : it is alſo named *ꝓꝓꝓꝓ*.

¶ *The Temperature.*

They are thought to be hot and dry of nature.

¶ *The Vertues.*

A   The faculties of theſe wilde Hellebors are referred vnto the white Neeſewort, whereof they are kindes.

B   It is reported, that the decoċtion of wilde Hellebor drunk, opens the ſtoppings of the liuer, and helpeth any imperfeċtions of the ſame.

CHAP.

CHAP. 112. *Of our Ladies Slipper.*

¶ *The Description.*

1   OVr Ladies Shoo or Slipper hath a thicke knobbed root, with certain markes or notes vpon the same, such as the roots of Solomons Seale haue, but much lesser, creeping within the vpper crust of the earth; from which riseth vp a stiffe and hairy stalke a foot high, set by certaine spaces with faire broad leaues, ribbed with the like sinues or nerues as those of the Plantain. At the top of the stalke groweth one single floure, seldome two, fashioned on the one side like an egge; on the other side it is open, empty, and hollow, and of the form of a shoo or slipper, whereof it tooke his name; of a yellow colour on the outside, and of a shining deep yellow on the inside. The middle part is compassed about with foure leaues of a bright purple colour, often of a light red or obscure crimson, and sometimes yellow as in the middle part, which in shape is like an egge as aforesaid.

† 2   This other differs not from the former vnlesse in the colour of the floure, which in this hath the foure long leaues white, and the hollow leafe or slipper of a purple colour. ‡

1 *Calceolus Mariæ.*   ‡ 2 *Calceolus Mariæ alter.*
Our Ladies Slipper.   The other Ladies Slipper.

¶ *The Place.*

Ladies Slipper groweth vpon the mountains of Germany, Hungary, and Poland. I haue a plant thereof in my garden, which I receiued from Mr *Garret* Apothecarie, my very good friend.

‡   It is also reported to grow in the North parts of this kingdome. I saw it in floure with Mr *Tradescant.*

¶ *The Time.*

It floureth about the midst of Iune.

¶ *The Names.*

It is commonly called *Calceolus D. Mariæ*, and *Marianus* : of some, *Calceolus Sacerdotis* : of some, *Alisma*, but vnproperly : in English, our Ladies shoo or slipper : in the Germane tongue, 𝔓𝔣𝔞𝔣𝔣𝔢𝔫 𝔖𝔠𝔥𝔲𝔢𝔱𝔥, 𝔓𝔞𝔭𝔢𝔫 𝔰𝔠𝔬𝔢𝔲 : and of some, *Damasonium nothum*.

¶ The

¶ *The Nature and Vertues.*

Touching the faculties of our Ladies ſhoo we haue nothing to write, it beeing not ſufficiently known to the old writers, no nor to the new.

C H A P. 113.   *Of Sopewort.*

¶ *The Deſcription.*

THe ſtalks of Sopewort are ſlipperie, ſlender, round, jointed, a cubit high or higher: the leaues are broad, ſet with veins very like broad leaued Plantain, yet leſſer, ſtanding out of euery joint by couples for the moſt part, and eſpecially thoſe that are the neereſt the roots bowing back-ward. The floures in the top of the ſtalkes and about the vppermoſt joints are many, well ſmelling, ſometimes of a beautifull red colour like a roſe; otherwhile of a light purple or white, which grow out of long cups conſiſting of fiue leaues, in the middle of which are certaine little threds.   The roots are thicke, long, creeping aſlope, hauing certain ſtrings hanging out of them like the roots of blacke Hellebor, and if they haue once taken good and ſure rooting in any ground, it is impoſſible to deſtroy them.

‡ There is kept in ſome of our gardens a varietie of this, which differs from it, in that the flours are double, and ſomewhat larger: in other reſpects it is altogether like the precedent. ‡

Saponaria.
Sopewort or Bruiſewort.

A

¶ *The Place.*

It is planted in gardens for the flours ſake, to the decking vp of houſes, for the which purpoſe it chie-fly ſerueth. It groweth wilde of it ſelfe neere to ri-uers and running brooks in ſunny places.

¶ *The Time.*

It floureth in Iune and Iuly.

¶ *The Names.*

It is commonly called *Saponaria*, of the great ſcouring qualitie that the leaues haue; for they yeeld out of themſelues a certaine juice when they are bruiſed, which ſcoureth almoſt as well as Sope; although *Ruellius* deſcribe a certaine other Sope-wort. Of ſome it is *Aliſma*, or *Damaſonium*: of others *Saponaria Gentiana*, whereof doubtleſſe it is a kinde: in Engliſh it is called Sopewort, & of ſome, Bruiſ-wort.

¶ *The Nature and Vertues.*

It is hot & dry, and not a little ſcouring withal, hauing no vſe in phyſick ſet downe by any Author of credit.

‡ Although our Authour and ſuch as before him haue written of plants, were ignorant of the fa-cultie of this herbe, yet hath the induſtry of ſome later writers found out the vertue thereof: and *Sep-talius* reports, that it was one *Zapata* a Spaniſh Em-perick. Since whoſe time it hath bin written of by

Rudius, lib. 5. de morbis occult. & venenat. cap. 18. And by *Cæſar Claudinus, de ingreſſu ad infirmos, pag. 411. & pag. 417.* But principally by *Ludouicus Septalius, Animaduerſ. med. lib. 7. num. 214.* where treating of decoctions in vſe againſt the French poxes, he mentions the ſingular effect of this herb againſt that filthy diſeaſe. His words are theſe: I muſt not in this place omit the vſe of another A-lexipharmicall decoction, being very effectuall and vſefull for the poorer ſort; namely that which is made of Sopewort, an herbe common and knowne to all. Moreouer, I haue ſometimes vſed it with happy ſucceſſe in the moſt contumacious diſeaſe: but it is of ſomewhat an vngratefull taſte,

and

and therefore it muſt be reſerued for the poorer ſort. The decoction is thus made : R. *Saponariæ viridⁱ. M. 2. infundantur per noctem in lib. vij. aquæ mox excoquantur ad cocturam Saponariæ: deinde libra vna cum dimidia aqua cum herba iam cocta excoletur cum expreßione, quæ reſeruetur pro potione matutina ad ſudores proliciendos ſumendo ℥ vij. aut viij. quod vero ſupereſt dulcoretur cum paſſulis aut ſaccaro pro potu cum cibis : æſtate & bilioſis naturis addi poterit aut Sonchi, aut Cymbalariæ M. j. Valet & pro mulieribus ad menſtrua alba abſumenda cum M. ſſ. Cymbalariæ, & addito tantundem Philipendulæ.* Thus much *Septalius,* who ſaith he had vſed it *ſæpè ac ſapius,* often and often againe.

B

Some haue commended it to be very good to be applied to greene wounds, to hinder inflammation, and ſpeedily to heale them.

Chap. 114.   Of Arſmart or water Pepper.

¶ The Deſcription.

1   ARſmart bringeth forth ſtalks a cubit high, round, ſmooth, jointed or kneed, diuiding themſelues into ſundry branches, whereon grow leaues like thoſe of the Peach or of the Sallow tree. The floures grow in cluſters vpon long ſtems, out of the boſom of the branches and leaues, and likewiſe vpon the ſtalks themſelues, of a white colour tending to a bright purple : after which come forth little ſeeds ſomwhat broad, of a reddiſh yellow, and ſomtime blackiſh, of an hot and biting taſte, as is all the reſt of the Plant, and like vnto pepper, whereof it tooke his name, yet hath it no ſmell at all.

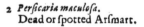

1 *Hydropiper.*
  Arſmart.

2 *Perſicaria maculoſa.*
  Dead or ſpotted Arſmart.

2   Dead Arſmart is like vnto the precedent in ſtalks, cluſtering floures, roots and ſeed, & differeth in that, that this plant hath certain ſpots or marks vpon the leaues, in faſhion of a halfe moon, of a dark blackiſh colour. The whole plant hath no ſharp or biting taſte, as the other hath but as it were a little ſoure ſmacke vpon the tongue. The root is likewiſe full of ſtrings or threds, creeping vp and downe in the ground.

‡ 3 This

‡ 3   This in roots,leaues,and manner of growing is very like the firſt deſcribed,but leſſer by much in all theſe parts: the floures alſo are of a whitiſh and ſometimes of a purpliſh colour: it growes.in barren grauelly and wet places.

4   I haue thought good to omit the impertinent deſcription of our Author fitted to this plant, and to giue one ſomewhat more to the purpoſe: the ſtalkes of this are ſome two foot high, tender, green,and ſometimes purpliſh,hollow,ſmooth,ſucculent and tranſparent, with large and eminent joints,from whence proceed leaues like thoſe of French Mercurie,a little bigger, and broader toward their ſtalkes, and thereabout alſo cut in with deeper notches: from the boſomes of each of theſe leaues come forth long ſtalks hanging downward,and diuided into three or foure branches; vpon which hang floures yellow,and much gaping, with crooked ſpurs or heeles, and ſpotted alſo with red or ſanguine ſpots: after theſe are paſt ſucceed the cods,which contain the ſeed,and they are commonly two inches long,ſlender,knotted, and of a whitiſh greene colour, creſted with greeniſh lines;and as ſoon as the ſeed begins to be ripe,they are ſo impatient,that they wil by no means be touched, but preſently the ſeed will fly out of them into your face. And this is the cauſe that *Lobel* and others haue called this Plant *Noli me tangere*;as for the like reaſon ſome of late haue impoſed the ſame name vpon the *Sium minimum* of *Alpinus*, formerly deſcribed by me in the ſeuenth place of the eighteenth chapter of this booke, *pag.260.* ‡

‡ 3   *Perſicaria puſilla repens.*
Small creeping Arſmart.

4   *Perſicaria ſiliquoſa.*
Codded Arſmart.

¶ *The Place and Time.*

They grow very common almoſt euery where in moiſt and wateriſh plaſhes, and neere vnto the brims of riuers,ditches,and running brooks. They floure from Iune to Auguſt.

‡   The codded or impatient Arſmart was firſt found to grow in this kingdome by the induſtry of my good friend Mr *George Bowles*,who found it at theſe places; firſt in Shropſhire, on the banks of the riuer Kemlet at Marington in the pariſh of Cherberry,vnder a gentlemans houſe called Mr *LLoyd*: but eſpecially at Guerndee in the pariſh of Cherſtock, halfe a mile from the foreſaid riuer,amongſt great Alder trees in the highway.

¶ *The Names.*

1   Arſmart is called in Greeke ὑδρπίδι : of the Latines, *Hydropiper*,or *Piper aquaticum*, or *Aquatile*, or water Pepper: in high-Dutch,𝔚𝔞ſſ𝔢𝔯 𝔓𝔣𝔢ſſ𝔢𝔯: in low-Dutch, 𝔚𝔞𝔱𝔢𝔯 𝔓𝔢𝔭𝔢𝔯: in French, *Curage,*

Curage, or *Culrage*: in Spanish, *Pimenta aquatica*: in English, **Water-Pepper**, Culrage, and Arfe-smart, according to the operation and effect when it is vsed in the abstersion of that part.

2   Dead Arfmart is called *Persicaria*, or Peach-wort, of the likenesse that the leaues haue with those of the Peach-tree. It hath beene called *Plumbago* of the leaden coloured markes which are seene vpon it : but *Pliny* would haue *Plumbago* not to be so called of the colour, but rather of the effect, by reason that it helpeth the infirmitie of the eies called *Plumbum*. Yet there is another *Plumbago* which is rather thought to be that of *Plinies* description, as shal be shewed in his proper place. In English we may call it Peach-wort, and dead Arfmart, because it doth not bite those places as the other doth.

‡ 3   This is by *Lobel* set forth, and called *Persicaria pusilla repens*: of *Tabernamontanus*, *Persicaria pumila*.

4   No plant I thinke hath found more variety of names than this : for *Tragus* calls it *Mercurialis syluestris altera* ; and he also calls it *Esula* : *Leonicerus* calls it *Tithymalus syluestris* : *Gesner*, *Camerarius*, and others, *Noli me tangere* : *Dodonæus*, *Impatiens herba* : *Cæsalpinus*, *Catanance altera* : in the *Hist. Lugd.* (where it is some three times ouer) it is called besides the names giuen it by others, *Chrysæa* : *Lobel*, *Thalius*, and others call it *Persicaria siliquosa* : yet none of these well pleasing *Columna*, hee hath accuratelv described and figured it by the name of *Balsamita altera* : and since him *Bauhine* hath namet it *Balsamina lutea* : yet both these and most of the other keepe the title of *Noli me tangere*. ‡

¶ *The Temperature.*

Arfmart is hot and dry, yet not so hot as Pepper, according to *Galen*.

Dead Arfmart is of temperature cold, and something dry.

¶ *The Vertues.*

The leaues and feed of Arfmart doe waste and consume all cold swellings, dissolue and scatter congealed bloud that commeth of bruisings or stripes.

The same bruifed and bound vpon an impostume in the ioints of the fingers (called among the vulgar fort a fellon or vncome) for the space of an houre, taketh away the paine : but (faith the Author) it must be first buried vnder a stone before it be applied ; which doth somewhat discredit the medicine,

The leaues rubbed vpon a tyred Iades backe, and a good handfull or two laid vnder the saddle, and the same set on againe, wonderfully refresh the wearied horse, and cause him to trauell much the better.

It is reported that Dead Arfmart is good against inflammations and hot swellings, being applied in the beginning : and for greene wounds, if it be stamped and boyled with oyle Oliue, wax and Turpentine.

‡ The faculties of the fourth are not yet knowne. *Lobel* saith it hath a venenate quality : and *Tragus* saith a vomitorie : yet neither of them seems to affirme any thing of certainty, but rather by heare-say.

Chap. 115.   *Of Bell-floures.*

¶ *The Description.*

1   COuentry-Bells haue broad leaues rough and hairy, not vnlike to those of the Golden Buglosse, of a swart greene colour : among which do rise vp stiffe hairie stalkes the second yeare after the sowing of the seed : which stalkes diuide themselues into sundry branches, whereupon grow many faire and pleasant bell-floures, long, hollow, and cut on the brim with fiue sleight gashes, ending in fiue corners toward night, when the floure shutteth it selfe vp, as doe most of the Bell-floures : in the middle of the floures be three or foure whitish chiues, as also much downie haire, such as is in the eares of a Dog, or such like beast. The whole floure is of a blew purple colour: which being past, there succeed great square or cornered seed-vessels, diuided on the inside into diuers cels or chambers, wherein do lie scatteringly many small browne flat seeds. The root is long and great like a Parsenep, garnished with many threddy strings, which perisheth when it hath perfected his seed, which is in the second yeare after his sowing, and recouereth it selfe againe by the falling of the seed.

2   The second agreeth with the first in each respect, as well in leaues, stalkes, or root, and differeth in that, that this plant bringeth forth milke-white floures, and the other not so.

¶ *The*

Viola mariana.   Blew Couentry Bells.

¶ *The Place and Time.*

They grow in woods, mountaines, and dark vallies, and vnder hedges among the buſhes, eſpecially about Couentry, where they grow very plentifully abroad in the fields, and are there called Couentry bells, and of ſome about London, Canterbury bells; but vnproperly, for that there is another kinde of Bell-floure growing in Kent about Canterbury, which may more fitly be called Canterbury Bells, becauſe they grow there more plentifully than in any other countrey. Theſe pleaſant Bell-floures wee haue in our London gardens eſpecially for the beauty of their floure, although they be kinds of Rampions, and the roots eaten as Rampions are.

They floure in Iune, Iuly, and Auguſt; the ſeed waxeth ripe in the mean time; for theſe plants bring not forth their floures all at once, but when one floureth another ſeedeth.

¶ *The Names.*

Couentry bels are called in Latine *Viola mariana*: in Engliſh, *Mercuries* Violets, or Couentry Rapes, and of ſome, Mariets. It hath bin taken to be *Medium*, but vnfitly: of ſome it is called *Rapum ſylveſtre*: which the Greeks call ραπὸν ἄγριον.

¶ *The Nature and Vertues.*

The root is cold and ſomewhat binding, and not vſed in phyſicke, but only for a ſallet root boiled and eaten, with oile, vineger, and pepper.

CHAP. 116.   *Of Throatwort, or Canterbury Bells.*

1 *Trachelium majus.*
Blew Canterbury Bells.

3 *Trachel. majus Belg. ſiue Giganteum.* Gyant Throatwort.

¶ *The Description.*

1   THe first of the Canterbury bells hath rough and hairy brittle stalkes, crested into a certaine squarenesse, diuiding themselues into diuers branches, whereupon do grow very rough sharpe pointed leaues, cut about the edges like the teeth of a sawe; and so like the leaues of nettles, that it is hard to know the one from the other, but by touching them. The floures are hollow, hairy within, and of a perfect blew colour, bell fashion, not vnlike to the Couentry bells. The root is white, thicke, and long lasting. ‡ There is also in some Gardens kept a variety hereof hauing double floures. ‡

2   The white Canterbury bells are so like the precedent, that it is not possible to distinguish them, but by the colour of the floures; which of this plant is a milke white colour, and of the other a blew, which setteth forth the difference.

4 *Trachelium minus.*
Small Canterbury bells.

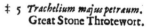
‡ 5 *Trachelium majus petraum.*
Great Stone Throtewort.

‡ Our Author much mistaking in this place (as in many other) did againe figure and describe the third and fourth, and of them made a fifth and sixth, calling the first *Trachelium Giganteum*, and the next, *Viola Calathiana*; yet the figures were such as *Bauhine* could not conjecture what was meant by them, and therefore in his *Pinax*, he saith, *Trachelium Giganteum, & viola Calathiana apud Gerardum, quid?* but the descriptions were better, wherefore I haue omitted the former descriptions, and here giuen you the later. ‡

3   Giants Throatwort hath very large leaues of an ouerworne greene colour, hollowed in the middle like the Moscouites spoone, and very rough, slightly indented about the edges. The stalke is two cubits high, whereon those leaues are set from the bottome to the top; from the bosome of each leafe commeth forth one slender footestalke, whereon doth grow a faire and large floure fashioned like a bell, of a whitish colour tending to purple. The pointed corners of each floure turne themselues backe like a scrole, or the Dalmatian cap; in the middle whereof commeth forth a sharpe stile or clapper of a yellow colour. The root is thicke, with certaine strings annexed thereto.

4   The smaller kinde of Throtewort hath stalkes and leaues very like vnto the great Throte-

woort,

woort, but altogether leſſer, and not ſo hairy : from the boſome of which leaues ſhoot forth very beautifull floures bell faſhion, of a bright purple colour, with a ſmall peſtle or clapper in the middle, and in other reſpects it is like the precedent.

‡ 5   This from a wooddy and wrinkled root of a pale purple colour ſends forth many rough creſted ſtalks of ſome cubit high, which are vnorderly ſet with leaues, long, rough, and ſnipt lightly about their edges, being of a darke colour on the vpper ſide, and of a whitiſh on their vnder part. At the tops of the ſtalkes grow the floures, being many, and thicke thruſt together, white of colour, and diuided into fiue or ſeuen parts, each floure hauing yellowiſh threds, and a pointall in their middles. It floures in Auguſt, and was firſt ſet forth and deſcribed by *Pona* in his deſcription of Mount Baldus. ‡

¶ *The Place.*

The firſt deſcribed and ſometimes the ſecond growes very plentifully in the low Woods and Hedge-rowes of Kent, about Canterbury, Sittingborne, Graueſend, South-fleet, and Greenehyth, eſpecially vnder Cobham Parke-pale in the way leading from Southfleet to Rocheſter, at Eltham about the parke there not farre from Greenewich ; in moſt of the paſtures about Watford and Buſhey, fiſteene miles from London.

‡ 3   The third was kept by our Author in his Garden, as it is alſo at this day preſerued in the Garden of Mr *Parkinſon :* yet in the yeere 1626, I found it in great plenty growing wilde vpon the bankes of the Riuer Ouſe in Yorkeſhire, as I went from Yorke to viſite Selby the place whereas I was borne, being ten miles from thence. ‡

The fourth groweth in the Medow next vnto Ditton ferrie as you goe to Windſore, vpon the chalky hills about Greenehithe in Kent ; and in a field by the high way as you goe from thence to Dartford ; in Henningham Parke in Eſſex ; and in Sion Medow neere to Brandford, eight miles from London.

The fifth groweth on mount Baldus in Italy.

¶ *The Time.*

All the kindes of bell floures doe floure and flouriſh from May vntill the beginning of Auguſt, except the laſt, which is the plant that hath beene taken generally for the Calathian Violet, which floureth in the later end of September ; notwithſtanding the Calathian Violet or Autumne violet is of a moſt bright and pleaſant blew or Azure colour, as thoſe are of this kinde, although this plant ſometimes changeth his colour from blew to whiteneſſe by ſome one accident or other.

¶ *The Names.*

1 2   Throtewoort is called in Latine, *Ceruicaria*, and *Ceruicaria major :* in Greeke : τραχήλιον : of moſt, *Vuularia :* of *Fuchſius, Campanula :* in Dutch, **Halſcrupt :** in Engliſh, Canterburie bells, Haskewoort, Throtewoort, or *Vuula* woort, of the vertue it hath againſt the paine and ſwelling thereof.

‡ 3   This is the *Trachelium majus Belgarum* of *Lobel,* and the ſame (as I before noted) that our Author formerly ſet forth by the name of *Trachelium Giganteum,* ſo that I haue put them as you may ſee, together in the title of the plant.

4   This is the *Trachelium minus* of *Dodonæus, Lobel,* and others : the *Ceruicaria minor* of *Tabernamontanus;* and *Vuularia exigua* of *Tragus :* Our Author gaue this alſo another figure and deſcription by the name of *Viola Calathiana,* not knowing that it was the laſt ſaue one which he had deſcribed by the name of *Trachelium minus.* ‡

¶ *The Temperature.*

Theſe plants are cold and dry, as are moſt of the Bell-floures.

¶ *The Vertues.*

A   The Antients for any thing that we know haue not mentioned, and therefore not ſet downe any thing concerning the vertues of theſe Bell-floures : notwithſtanding wee haue found in the later writers, as alſo of our owne experience, that they are excellent good againſt the inflammation of the throate and *Vuula* or Almonds, and all manner of cankers and vlcerations in the mouth, if the mouth and throat be gargariſed and waſhed with the decoction of them : and they are of all other herbes the chiefe and principall to be put into lotions, or waſhing waters, to inject into the priuie parts of man or woman, being boyled with hony and Allom in water, with ſome white wine.

CHAP.

CHAP. 117.　*Of Peach-bells and Steeple-bells.*

¶ *The Description.*

1　THe Peach-leaued Bell-floure hath a great number of small and long leaues, rising in a great bush out of the ground, like the leaues of the Peach tree: among which riseth vp a stalke two cubits high: alongst the stalke grow many floures like bells, sometime white, and for the most part of a faire blew colour; but the bells are nothing so deepe as they ot the other kindes ; and these are more dilated or spread abroad than any of the rest. The seed is small like Rampions, and the root a tuft of laces or small strings.

2　The second kinde of Bell-floure hath a great number of faire Blewish or Watchet floures, like the other last before mentioned, growing vpon goodly tall stems two cubits and a halfe high, which are garnished from the top of the plant vnto the ground with leaues like Beets, disorderly placed. This whole plant is exceeding full of milke, insomuch as if you do but breake one leafe of the plant, many drops of a milky juice will fall vpon the ground. The root is very great, and full of milke also : likewise the knops wherein the seed should be are empty and void of seed, so that the whole plant is altogether barren, and must be increased with slipping of his root.

1　*Campanula persicifolia.*
Peach-leaued Bell-floure.

2　*Campanula lactescens pyramidalis.*
Steeple milky Bell-floure.

3　The small Bell-floure hath many round leaues very like those of the common field Violet, spred vpon the ground ; among which rise vp small slender stems, disorderly set with many grassie narrow leaues like those of flax. The small stem is diuided at the top into sundry little branches, whereon do grow pretty blew floures bell-fashion. The root is small and thready.

4　The yellow Bell-floure is a very beautifull plant of an handfull high, bearing at the top of his weake and tender stalkes most pleasant floures Bel-fashion, of a faire and bright yellow colour. The leaues and roots are like the precedent, sauing that the leaues that grow next to the ground of this plant are not so round as the former. ‡ Certainly our Author in this place meant to set forth the *Campanula lutea linifolia flore volubili,* described in the *Aduers.pag.* 177. and therefore I haue giuen you the figure thereof. ‡

3 *Campanula rotundifolia.*
Round leaued Bell-floure.

† 4 *Campanula lutea linifoli.*
Yellow Bell-floure.

5 *Campanula minor alba, ſiue purpurea.*
Little white or purple Bell-floure.

5   The litle white Bell-floure is a kind of wilde Rampions, as is that which followeth and alſo the laſt ſaue one before deſcribed. This ſmall plant hath a ſlender root of the bigneſſe of a ſmall ſtraw, with ſome few ſtrings annexed thereto. The leaues are ſomewhat long, ſmooth, and of a perfect greene colour, lying flat vpon the ground : from thence riſe vp ſmall tender ſtalkes, ſet here and there with a few leaues. The floures grow at the top, of a milke white colour.

6   The other ſmall Bel-floure or wilde Rampion differeth not from the precedent but onely in colour of the floures ; for as the others are white, theſe are of a bright purple colour, which ſetteth forth the difference.

‡ 7   Beſides theſe here deſcribed, there is another very ſmall and rare Bell-floure, which hath not beene ſet forth by any but onely by *Bauhine*, in his *Prodromus*, vnder the title of *Campanula Cymbalaria folijs*, and that fitly ; for it hath thinne and ſmall cornered leaues much after the manner of *Cymbalaria*, and theſe are ſet without order on very ſmall weak and tender ſtalkes ſome handfull long ; and at the tops of the branches grow little ſmall and tender Bell-floures of a blew colour. The root, like as the whole plant, is very ſmall and threddy. This pretty plant was firſt diſcouered to grow in England by Maſter *George Bowles*, *Anno* 1632. who found it in Montgomery ſhire, on the dry bankes in the high-way as one rideth from Dolgeogg a Worſhipfull Gentlemans houſe called M^r. *Francis Herbert*, vnto a market towne called Mahuntleth, and in all the way from thence to the ſea ſide. It may be called in Engliſh, The tender Bell-floure.

¶ *The*

¶ *The Place.*

The two first grow in our London Gardens, and not wilde in England.

The rest, except that small one with yellow floures, doe grow wilde in most places of England, especially vpon barren sandy heaths and such like grounds.

¶ *The Time.*

These Bell-floures do flourish from May vnto August.

¶ *The Names.*

Their seuerall titles set forth their names in English and Latine, which is as much as hath been said of them.

¶ *The Nature and Vertues.*

These Bell-floures, especially the foure last mentioned, are cold and dry, and of the Nature of Rampions, whereof they be kindes.

† The figure in the fourth place was of *Rapunculus nemorosus* 3. of *Tabern.* whereof you shall finde mention in the following chapter.

CHAP. 118.   *Of Rampions, or wilde Bell-floures.*

1 *Rapuntium majus.*
Great Rampion.

2 *Rapuntium parvum.*
Small Rampion.

¶ *The Description.*

1　　THe great Rampion being one of the Bell-floures, hath leaues which appeare or come forth at the beginning somewhat large and broad, smooth and plaine, not vnlike to the leaues of the smallest Beet. Among which rise vp stemmes one cubit high, set with such like leaues as those are of the first springing vp, but smaller, bearing at the top of the stalke a great thicke bushie eare full of little long floures closely thrust together like a Fox-taile: which small floures before their opening are like little crooked hornes, and being wide opened they are smal blew-bells, sometimes white, and sometimes purple. The root is white, and as thicke as a mans thumbe.

2 The

2　The ſecond kind being likewiſe one of the bel-floures, and yet a wild kind of Rampion, hath leaues at his firſt comming vp like vnto the Garden Bell-floure. The leaues which ſpring vp afterward for the decking vp of the ſtalke are ſomewhat longer and narrower. The floures grow at the top of tender and brittle ſtalkes like vnto little bells, of a bright blew colour, ſometimes white or purple. The root is ſmall, long, and ſomewhat thicke.

3　This is a wilde Rampion that growes in woods: it hath ſmall leaues ſpread vpon the ground, bluntly indented about the edges: among which riſeth vp a ſtraight ſtem of the height of a cubit, ſet from the bottome to the top with longer and narrower leaues than thoſe next the ground: at the top of the ſtalkes grow ſmall Bell-floures of a watchet blewiſh colour. The root is thicke and tough, with ſome few ſtrings anexed thereto.

‡　There is another variety of this, whoſe figure was formerly by our Author ſet forth in the fourth place of the laſt chapter: it differs from this laſt onely in that the floures and other parts of the plant are leſſer a little than thoſe of the laſt deſcribed. ‡

3 *Rapunculus nemoroſus.*　　　　　　　‡ 4 *Rapunculus Alpinus Corniculatus.*
Wood Rampions.　　　　　　　　　　　Horned Rampions of the Alpes.

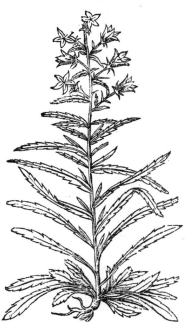 

‡　4　This which growes amongſt the rockes in the higheſt Alpes hath a wooddy and very wrinkled root an handfull and halfe long, from which ariſe many leaues ſet on pretty long ſtalkes, ſomewhat round, and diuided with reaſonable deepe gaſhes, hauing many veines, and being of a darke greene colour: amongſt theſe grow vp little ſtalkes, hauing one leafe about their middles, and three or foure ſet about the floure, being narrower and longer than the bottome leaues. The floures grow as in an vmbell, and are ſhped like that Chymicall veſſell we vſually call a Retort, being big at their bottomes, and ſo becomming ſmaller towards their tops, and hauing many threds in them, whereof one is longer than the reſt, and comes forth in the middle of the floure: it floures in Auguſt. *Pona* was the firſt that deſcribed this, vnder the name of *Trachelium petræum minus.*

5　The roots of this other kinde of horned Rampion grow after an vnuſuàll manner; for firſt or lowermoſt is a root like to that of a Rampion, but ſlenderer, and from the top of that commeth forth as it were another root or two, being ſmalleſt about that place whereas they are faſtned to the vnder root; and all theſe haue ſmall fibres comming from them. The leaues which firſt grow vp are ſmooth, and almoſt like thoſe of a Rampion, yet rounder, and made ſomewhat after the manner of a violet leafe, but nothing ſo big: at the bottome of the ſtalke come forth ſeuen or eight long narrow
leaues

leaues snipt about the edges, and sharpe pointed, and vpon the rest of the stalke grow also three or foure narrow sharpe pointed leaues. The floures which are of a purple colour, at first resemble those of the last described; but afterwards part themselues into fiue slender strings with threds in the middles; which decaying, they are succeeded by little cups ending in fiue little pointels, and containing a small yellow seed. This is described by *Fabius Columna*, vnder the name of *Rapuntium Corniculatum montanum*: And I receiued seeds and roots hereof from Mr *Goodyer*, who found it growing plentifully wilde in the inclosed chalky hilly grounds by Maple-Durham neere Petersfield in Hampshire. In gardens the floures become much longer and fairer.

6   This which is described in *Clusius* his *Cura poster.* by the name of *Pyramidalis*, and was first found and sent to him by *Gregory de Raggio*, a Capuchine Frier, is also of this kindred; wherefore I will giue you a briefe description thereof. The root is white, and long lasting; from which come diuers round hairie and writhen stalkes, about a span long more or lesse. At the top of these stalkes and all amongst the leaues, grow many elegant blew floures, which are succeeded by seed vessels like those of the lesser *Trachelium*, being full of a small seed. The whole plant yeelds milke like as the rest of this kinde, and the leaues as well in shape as hoarinesse on their vnder sides, well resemble those of the second French or Golden Lungwoort of my description. It was first found growing in the chinkes of hard rockes about the mouthes of Caues, in the mountaines of Brescia in Italy, by the foresaid Frier.

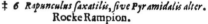

‡ 5 *Rapunculus Corniculatus montanus*.
Mountaine horned Rampions.

‡ 6 *Rapunculus saxatilis, siue Pyramidalis alter*.
Rocke Rampion.

¶ *The Place.*

The first is sowne and set in Gardens, especially because the roots are eaten in Sallads. The second groweth in woods and shadowie places, in fat and clayie soiles.

¶ *The Time.*

They floure in May, Iune, and Iuly.

¶ *The Names.*

‡ Rampions by a generall name are called *Rapuntium* and *Rapunculus*; and the first by reason of the long spokie tuft of floures is called *Rapuntium majus Alopecuri comoso flore*, by *Lobel* and *Pena*: *Rapunculum syluestre*, and *Rapunculus syluestris spicatus* by others. The second which is the

ordinary

ordinary Rampion is called *Rapunculus*, and *Rapuntium minus* ; *Lobel* thinkes it the *Pes Locuſtæ* of *Auicen*;and *Columna* iudges it to be *Erinus* of *Nicander* and *Dioſcorides*. The third is the *Rapunculus nemoroſus ſecundus* of *Tabernamontanus*;and the variety of it is *Rapunc.nemor.tertius*. The names of the reſt are ſhewne in their deſcriptions. ‡

¶ The Temperature.

The roots of theſe are of a cold temperature,and ſomething binding.

¶ The Vertues.

A   The roots are eſpecially vſed in ſallads,being boiled and eaten with oile,vinegre,and pepper.

B   Some affirme, that the decoction of the roots is good for all inflammations of the mouth, and Almonds of the throat,and other diſeaſes happening in the mouth and throte,as the other Throte-woorts.

Cʜᴀᴘ. 119.   Of Wall-floures,or yellow Stocke-Gillofloures.

¶ The Kindes.

‡   THeſe plants which wee terme commonly in Engliſh, Wal-floures and Stocke-Gillo-floures are comprehended vnder one generall name of *Leucoion*,(i)*Viola alba*,White Vi-olet, λευκὸς ſignifying white,and ἴον a Violet,which as ſome would haue it is not from the whiteneſſe of the floure, for that the moſt and moſt vſuall of them are of other colours, but from the whitenes or hoarineſſe of the leaues,which is proper rather to the Stocke Gillouers than to the Wal-floures: I therefore thinke it fit to diſtinguiſh them into *Leucoia folijs viridibus*,that is, Wal-floures ; and *Leucoia folijs incanis*,Stocke-Gillouers.Now theſe againe are diſtinguiſhed into ſeuerall ſpecies,as you may finde by the following Chapters.Moreouer,you muſt remember there is another *Viola alba* or *Leucoion*(which is thought to be that of *Theophraſtus*,and whereof we haue treated in the firſt booke)which is far different from this,and for diſtinction ſake called *Leucoium bulboſum*. ‡

<table>
<tr><td>1 Viola Lutea.
Wal-floure.</td><td>2 Viola lutea multiplex.
Double Wal-floure.</td></tr>
</table>

† 4 *Leucoium syluestre.*
Wilde Wall-floure.

¶ *The Description.*

1   THe stalks of the Wal-floure are ful of greene branches, the leaues are long, narrow, smooth, slippery, of a blackish green color, and lesser than the leaues of stocke Gillofloures. The floures are small, yellow, very sweet of smell, and made of foure little leaues ; which being past, their succeed long slender cods, in which is contained flat reddish seed. The whole plant is shrubby, of a wooddy substance, and can easily endure the cold of Winter.

2   The double Wal-floure hath long leaues greene and smooth, set vpon stiffe branches, of a wooddy substance : whereupon doe grow most pleasant sweet yellow flours very double; which plant is so well knowne to all, that it shall be needlesse to spend much time about the description.

3   Of this double kinde we haue another sort that bringeth his floures open all at once, whereas the other doth floure by degrees, by meanes whereof it is long in flouring.

‡ 4   This plant which was formerly seated in the fourth place of the following chapter, I haue brought to enioy the same place in this, for that by reason of the greenesse of his leaues and other things hee comes nearest to these here described; also I wil describe it anew because the former was almost wholly false: It hath many greene leaues at the top of the root like to those of the Wall-floure, but narrower, and bitter of taste ; among which rise vp one or more stalkes of a foot or more in height, crested and set with carinated leaues. The floures grow at the tops of the stalkes many together, consisting of foure yellow leaues a piece, lesser than those of the ordinary wall-floures; these floures are succeeded by long cods containing a flat seed. The root is long and whitish, with many fibres.

5   Besides these, there is in some Gardens kept another Wall-floure differing from the first in the bignes of the whole plant, but especially of the floure, which is yellow and single, yet very large and beautifull.

6   Also there is another with very greene leaues, and pure white and well smelling floures. ‡

¶ *The Place.*

The first groweth vpon bricke and stone walls, in the corners of churches euery where, as also among rubbish and other stony places.

The double Wall-floure groweth in most gardens of England.

¶ *The Time.*

They floure for the most part all the yeere long, but especially in Winter, whereupon the people in Cheshire do call them Winter-Gillofloures.

¶ *The Names.*

The Wall-floure is called in Greeke ‹‹‹‹‹‹ : in Latine, *Viola lutea,* and *Leucoium luteum :* in the Arabicke tongue, *Keyri :* in Spanish, *Violettas Amarillas :* in Dutch, **Uiolieren** : in French, *Girofflees iaulnes, Violieres des murailles:* in English, Wall-Gillofloure, Wall-floure, yellow stocke Gillofloure, and Winter-Gillofloure.

¶ *The Temperature.*

All the whole shrub of Wall-Gillofloures, as *Galen* saith, is of a cleansing faculty, and of thinne parts.

¶ *The Vertues.*

Dioscorides writeth, that the yellow Wall-floure is most vsed in Physicke, and more than the rest   A of stocke-Gillofloures, whereof this is holden to be a kinde : which hath moued me to preferre it vnto the first place. He saith, that the juyce mixed with some vnctious or oylie thing, and boiled to the forme of a lyniment, helpeth the chops or rifts of the fundament.

The

B   The herbe boiled with white wine, hony, and a little allom, doth cure hot vlcers, and cankers of the mouth.

C   The leaues ſtamped with a little bay ſalt, and bound about the wreſts of the hands, take away the ſhaking fits of the Ague.

D   ‡ A decoction of the floures together with the leaues, is vſed with good ſucceſſe to mollifie Schirrous tumors.

E   The oyle made with theſe is good to be vſed to annoint a Paralyticke, as alſo a gouty part to mitigate paine.

F   Alſo a ſtrong decoction of the floures drunke, moueth the Courſes, and expelleth the dead childe. ‡

CHAP. 120.   *Of Stocke Gillo-floures.*

1 *Leucoium album, ſiue purpureum, ſiue violaceum.*
White, Purple, or Violet coloured Stocke Gillo-floure.

‡ 2 *Leucoium flore multiplici.*
Double Stocke Gillo-floure.

¶ *The Deſcription.*

1   THe ſtalke of the great ſtocke Gillo-floure is two foot high or higher, round, and parted into diuers branches. The leaues are long, white, ſoft, and hauing vpon them as it were a downe like vnto the leaues of willow, but ſofter : the floures conſiſt of foure little leaues growing all along the vpper part of the branches, of a white colour, exceeding ſweet of ſmell: in their places come vp long and narrow cods, in which is contained broad, flat, and round ſeed. The root is of a wooddy ſubſtance, as is the ſtalke alſo.

The purple ſtocke Gillo-floure is like the precedent in each reſpect, ſauing that the floures of this plant are of a pleaſant purple colour, and the others white, which ſetteth forth the difference : of which kinde we haue ſome that beare double floures which are of diuers colours, greatly eſteemed for the beautie of the floures, and pleaſant ſweet ſmell.

This

3 *Leucoium ſpinoſum Creticum.*
Thorny Stocke Gilloſloures.

This kinde of Stocke Gillo floure that beareth floures of the colour of a Violet, that is to ſay of a blew tending to a purple colour, which ſetteth forth the difference betweene this plant and the other ſtocke Gilloſloures, in euery other reſpect is like the precedent.

‡ 2 There were formerly three figures of the ſingle Stockes, which differ in nothing but the colour of their floures; wherefore wee haue made them content with one, and haue giuen (which was formerly wanting) a figure of the double Stock, of which there are many and pretty varieties kept in the Garden of my kinde friend Maſter *Ralph Tuggye* at Weſtminſter, and ſet forth in the bookes of ſuch as purpoſely treat of floures and their varieties.

3   To theſe I thinke it not amiſſe to adde that plant which *Cluſius* hath ſet forth vnder the name of *Leucoium ſpinoſum Creticum.* It growes ſome foot or more high, bringing forth many ſtalkes which are of a grayiſh colour, and armed at the top with many and ſtrong thorny prickles: the leaues which adorne theſe ſtalkes are like thoſe of the ſtocke Gillouer, yet leſſe and ſomewhat hoary; the floures are like thoſe of Mulleine, of a whitiſh yellow colour, with ſome purple threds in their middles; the cods which ſucceed the floures are ſmall and round, containing a little ſeed in them. They vſe, ſaith *Honorius Bellus*, to heat ouens therewith in Candy, where it plentifully growes; and by reaſon of the ſimilitude which the prickles hereof haue with *Stæbe* and the white colour, they call it *Gala Stivida*, or *Galaſtividi*, and not becauſe it yeelds milke, which *Gala* ſignifies. ‡

¶ *The Place.*

1   2   Theſe kindes of Stocke Gilloſloures do grow in moſt Gardens throughout England.

¶ *The Time.*

They floure in the beginning of the Spring, and continue flouring all the Summer long.

¶ *The Names.*

The Stocke Gilloſloure is called in Greeke, Μυκαιον: in Latine, *Viola alba:* in Italian, *Viola bianca:* in Spaniſh, *Violetta blanquas:* in Engliſh, Stocke Gilloſloure, Garnſey Violet, and Caſtle Gilloſloure.

¶ *The Temperature and Vertues.*

They are referred vnto the Wal-floure, although in vertue much inferiour; yet are they not vſed in Phyſicke, except amongſt certaine Empericks and Quackſaluers, about loue and luſt matters, which for modeſtie I omit.   A

Ioachimus Camerarius reporteth, that a conſerue made of the floures of Stocke Gilloſloure, and of- B ten giuen with the diſtilled water thereof, preſerueth from the Apoplexie, and helpeth the palſie.

Chap. 121. *Of Sea Stocke Gillo-floures.*

¶ *The Kindes.*

OF Stocke Gillo-floures that grow neere vnto the Sea there bee diuers and ſundry ſorts, differing as well in leaues as floures, which ſhall bee comprehended in this Chapter next following.

1 *Leucoium marinum flore candido Lobelij.*
White Sea Stocke Gillo-floures.

2 *Leucoium marinum purpureum Lobelij.*
Purple Sea Stocke Gillo-floures,

3 *Leucoium marinum latifolium.*
Broad leafed Sea Stocke Gillo-floure.

¶ *The Description.*

1 THe Sea Stocke Gillo-floure
hath a small wooddy root very thready; from which riseth
vp an hoary white stalk of two foot high,
diuided into diuers small branches, wheron are placed confusedly many narrow
leaues of a soft hoary substance. The
floures grow at the top of the branches,
of a whitish colour, made of foure little
leaues; which beeing past there follow
long cods and seed, like vnto the Garden
stocke Gillo-floure.

‡ 2   The purple stocke Gillofloure
hath a very long tough root, thrusting it
selfe deepe into the ground; from which
rise vp thicke, fat, soft, and hoarie stalkes.
The leaues come forth of the stalkes next
the ground, long, soft, thicke, full of iuice,
couered ouer with a certaine downy hoarinesse, and sinuated somewhat deepe on
both sides, after the manner you may see
exprest in the figure of the fourth described in this Chapter. The stalke is set
here and there with the like leaues, but
lesser. The floures grow at the top of the
stalkes, compact of foure small leaues, of
a light purple colour. The seed is contained in long crooked cods like the garden stocke Gillo-floure.

The figure of *Lobels* which here we
giue you was taken of a dried plant, and
therefore the leaues are not exprest so si-
nuate as they should be. ‡

3 This sea stock Gillofloure hath many
broad leaues spred vpon the ground, somwhat snipt or cut on the edges; amongst
which rise vp smal naked stalkes, bearing
at the top many little floures of a blew
colour tending to a purple. The seed is
in long cods like the others of his kinde.

4 The

4　The great ſea ſtocke Gillofloure hath many broad leaues growing in a great tuft, ſleightly indented about the edges. The floures grow at the top of the ſtalks, of a gold yellow colour. The root is ſmall and ſingle.

5　The ſmall yellow ſea ſtock Gillofloure hath many ſmooth hoary and long leaues ſet vpon a branched ſtalk ; on the top whereof grow pretty ſweet ſmelling yellow floures, bringing his ſeed in little long cods. The root is ſmall and threddy.　‡ The floures of this are ſometimes of a red or purpliſh colour.

4　*Leucoium marinum luteum maius Cluſij & Lobelij.*
　The yellow ſea ſtock Gillofloure.

5　*Leucoium marinum minus Lobel. & Cluſ.*
　Small yellow ſea ſtock Gillofloure.

¶ *The Place.*
Theſe Plants grow neere vnto the ſea ſide about Colcheſter, in the Iſle of Man, neere Preſton in Aunderneſſe, and about Weſtcheſter.

‡ I haue not heard of any of theſe wild on our Coaſts but onely the ſecond, which it may be groweth in theſe places here ſet downe; for it was gathered by Mr *George Bowles* vpon the rocks at Aberdovye in Merioneth ſhire. ‡

¶ *The Time.*
They flouriſh from Aprill to the end of Auguſt.

¶ *The Names.*
There is little to be ſaid as touching the names, more than hath beene touched in their ſeuerall titles.

¶ *The Nature and Vertues.*
There is no vſe of theſe in phyſieke, but they are eſteemed for the beauty of their floures.

CHAP. 122.　*Of Dames Violets or Queens Gillofloures.*

¶ *The Deſcription.*
1　DAmes Violets or Queenes Gillofloures haue great large leaues of a darke green colour, ſomewhat ſnipt about the edges ; among which ſpring vp ſtalks of the height of two

Qp

cubits,

cubits, set with such like leaues : the floures come forth at the tops of the branches, of a faire purple colour, very like those of the stock Gillofloures, of a very sweet smell ; after which come vp long cods, wherein is contained small long blackish seed. The root is slender and threddy.

The Queens white Gillofloures are like the last mentioned, sauing that this plant brings forth faire white floures, and the other purple.

‡ 2 By the industry of some of our Florists, within these two or three years hath bin brought vnto our knowledge a very beautifull kind of these Dame Violets, hauing very faire double white floures : the leaues, stalks, and roots are like to the other plants before described. ‡

1 *Viola Matronalis flore purpureo, siue albo.*
Purple or white Dames Violets.

‡ 3 *Viola Matronalis flore obsoleto.*
Russet Dames Violets.

‡ 3 This plant hath a stalke a cubit high, and is diuided into many branches, vpon which in a confused order grow leaues like those of the Dames violet, yet a little broader and thicker, being first of somewhat an acide, and afterwards of an acride taste ; at the tops of the branches in long cups grow floures like those of the Dames violet, consisting of foure leaues, which stand not faire open, but are twined aside, and are of an ouerworne russet colour, composed as it were of a yellow and browne, with a number of blacke purple veins diuaricated ouer them. Their smell on the day time is little or none, but in the euening very pleasing and sweet. The floures are succeeded by long and here and there swolne cods, which are almost quadrangular, and containe a reddish seed like that of the common kinde. The root is fibrous, and vsually liues not aboue two yeares, for af-ter it hath borne seed it dies ; yet if you cut it downe and keepe it from seeding, it sometimes puts forth shoots whereby it may be increased. I very much suspect that this figure and Description which I here giue you taken out of *Clusius*, is no other plant than that which is kept in some of our gardens, and set forth in the *Hortus Eystettensis* by the name of *Leucoium Melancholicum* : Now I iudge the occasion of this error to haue come from the figure of *Clusius* which we here present you with, for it is in many particulars different from the description ; first in that it expresses not many branches : secondly, in that the leaues are not snipt and diuided : thirdly, in that the floures are not exprest wrested or twined : fourthly, the veines are not rightly exprest in the floure : and lastly, the cods are omitted. Now the *Leucoium melancholicum* hath a hairy stalk diuided into sundry branches of the height formerly mentioned, and the leaues about the middle of the stalke are somwhat sinu-ated or deepely or vnequally cut in ; The shape and colour of the floure is the same with that now described,

‡ 4 *Leucoium melancholicum.*
The melancoly Floure.

described,and the feed-veſſels the ſame,as farre as I can remember : for I muſt confeſſe , I did not in writing take any particular note of them, neither did I euer compare them with this deſcription of *Cluſius* ; onely I tooke ſome yeares agon an exact figure of a branch with the vpper leaues and floures, whereof one is expreſſed as they vſually grow twining backe , and the reſt faire open,the better to ſet forth the veins that are ſet ouer it. There are alſo expreſt a cod or ſeed-veſſell, and one of the leaues that grow about the middle of the ſtalke : all which are agreeable to *Cluſius* deſcription, in myne opinion ; wherefore I only giue you the figure that I then drew,with the title I had it by. ‡

¶ *The Place.*

They are ſowne in gardens for the beauty of their floures.

¶ *The Time.*

They eſpecially floure in May and Iune, the ſecond yeare after they are ſowne.

¶ *The Names.*

Dames Violet is called in Latine, *Viola Matronalis*,and *Viola Hyemalis*,or winter Violets;and *Viola Damaſcena* : it is thought to be the *Heſperis* of *Pliny, lib.21. cap.7.* ſo called, for that it ſmels more,and more pleaſantly in the euening or night,than at any other time. They are called in French, *Violettes de dames,& de domas*,and *Giroſflees des dames*, or *Matrones Violettes* : in Engliſh,Damaske Violets,winter Gilloſloures, Rogues Gilloſloure,and Cloſe Sciences.

¶ *The Temperature.*

The leaues of Dames Violets are ſharpe and hot, very like in taſte and facultie to *Eruca* or Rocket,and ſeemes to be a kinde thereof.

¶ *The Vertues.*

The diſtilled Water of the floures hereof is counted to bee a moſt effectuall thing to procure ſweat.

Chap. 123.

Of white Sattin floure.

¶ *The Deſcription.*

1   Bolbonac or the Sattin floure hath hard and round ſtalks, diuiding themſelues into many other ſmall branches, beſet with leaues like Dames Violets or Queenes Gilloſloures, ſomewhat broad, and ſnipt about the edges, and in faſhion almoſt like Sauce alone, or Iacke by the hedge, but that they are longer and ſharper pointed. The ſtalks are charged or loden with many floures like the common ſtocke Gilloſloure,of a purple colour : which being fallen,the ſeed comes forth,contained in a flat thin cod,with a ſharp point or pricke at one end,in faſhion of the Moon,but ſomewhat blackiſh. This cod is compoſed of three filmes or skins,whereof the two outmoſt are of an ouerworne aſh colour,and the innermoſt or that in the middle,wheron the ſeed doth hang or cleaue, is thinne and cleere ſhining,like a ſhred of white Sattin newly cut from the piece. The whole plant dieth the ſame yeare that it hath borne ſeed,and muſt be ſown yearly. The root is compact of many tuberous parts like Key clogs,or like the great Aſphodill.

The

2 The ſecond kinde of *Bolbonac* or white Sattin hath many great and broad leaues, almoſt like thoſe of the great Bur-Dock: among which riſeth vp a very tal ſtem of the height of foure cubits, ſtiffe, and of a whitiſh green colour, ſet with the like leaues, but ſmaller. The floures grow vpon the ſlender branches, of a purple colour, compaƈt of foure ſmall leaues like thoſe of the ſtocke Gillofloure; after which come thin long cods of the ſame ſubſtance and colour of the former. The root is thicke, whereunto are faſtned an infinite number of long threddy ſtrings: which root dieth not euery yere, as the other doth, but multiplieth it ſelfe as wel by falling of the ſeed, as by new ſhoots of the root.

1 *Viola Lunaris, ſive Bolbonac.*
White Sattin.

2 *Viola Lunaris longioribus ſiliquis.*
Long codded white Sattin.

¶ *The Place.*

Theſe Plants are ſet and ſowne in gardens, notwithſtanding the firſt hath bin found wild in the woods about Pinner and Harrow on the hill 12 miles from London, and in Eſſex likewiſe about Horn-church.

The ſecond groweth about Watford, fifteene miles from London.

¶ *The Time.*

They floure in Aprill the next yeare after they be ſowne.

¶ *The Names.*

They are commonly called *Bolbonac* by a barbarous name: wee had rather call it with *Dodonæus* and *Cluſius*, *Viola latifolia*, and *Viola lunaris*, or as it pleaſeth moſt Herbariſts, *Viola peregrina*: the Brabanders name it 𝔇𝔢𝔫𝔫𝔦𝔫𝔠𝔨 𝔟𝔩𝔬𝔢𝔪𝔢𝔫, of the faſhion of the cods, like after a ſort to a groat or Teſterne; and 𝔓𝔞𝔢ſ𝔠𝔥 𝔟𝔩𝔬𝔢𝔪𝔢𝔫, becauſe it alwayes floureth neere about the feaſt of Eaſter. Moſt of the later Herbariſts doe call it *Lunaria*: Others, *Lunaria Græca*, either of the faſhion of the ſeed, or of the ſiluer brightneſſe that it hath, or of the middle skinne of the cods, when the two outermoſt skins or husks and ſeeds likewiſe are fallen away. We call this herbe in Engliſh, Penny floure, or Mony-floure, Siluer Plate, Prickſong-wort; in Norfolke, Sattin, and white Sattin; and among our
women,

women it is called Honeſtie : it ſeemeth to be the old Herbariſts *Thlaſpi alterum*, or ſecond Treacle muſtard ; and that which *Cratevas* deſcribeth, called of diuers, *Sinapi Perſicum* : for as *Dioſcorides* faith, *Cratevas* maketh mention of a certaine *Thlaſpi* or Treacle muſtard with broad leaues and big roots, and ſuch this Violet hath, which we ſirname *Latifolia* or broad leaued : generally taken of all to be the great *Lunaria* or Moonwort.

¶ *The Temperature and Vertues.*

The ſeed of Bulbonac is of temperature hot and dry, and ſharpe of taſte, like in force to the ſeed A of Treacle muſtard : the roots likewiſe are ſomewhat of a biting qualitie, but not much : they are eaten with ſallads as certaine other roots are.

A certain Heluetian Surgeon compoſed a moſt ſingular vnguent for green wounds, of the leaues B of Bulbonac and Sanicle ſtamped together, adding thereto oile and wax. The ſeed is greatly com-mended againſt the falling ſickneſſe.

Cʜᴀᴘ. 124.   *Of* Galen *and* Dioſcorides *Moonworts or Madworts.*

1 *Alyſſum Galeni.*   † 2 *Alyſſum Dioſcoridis.*
 Galens Madwort.   Dioſcorides Moonwort or Madwort.

¶ *The Deſcription.*

1   THis might be one of the number of the Hore-hounds, but that *Galen* vſed it not for a kinde thereof, but for *Alyſſon* or Madwort : it is like in forme and ſhew to Hore-hound, and alſo in the number of the ſtalks, but the leaues thereof are leſſe, more curled, more hoary & whiter, without any manifeſt ſmell at all. The little coronets or ſpoky whurles that com-paſſe the ſtalks round about are full of ſharp prickles, out of which grow floures of a blewiſh pur-ple colour like thoſe of Horehound : the root is hard, wooddy, and diuerſly parted.

2   I haue one growing in my garden, which is thought to be the true and right Lunary or Moon-wort of *Dioſcorides* deſcription ; hauing his firſt leaues ſomewhat round, and afterward more long, whitiſh, and rough or ſomewhat woolliſh in handling : amongſt which riſe vp rough brittle ſtalkes ſome cubit high, diuided into many branches, whereupon grow many little yellow floures; which

being paſt, there follow flat and rough huskes of a whitiſh colour, in ſhape like little targuets or bucklers, wherein is contained flat ſeed like to the ſeeds of ſtocke Gilloflowres, but bigger. The whole huske is of the ſame ſubſtance, faſhion, and colour that thoſe are of the white Satin.

¶ *The Place.*

Theſe plants are ſowne now and then in gardens, eſpecially for the rareneſſe of them ; the ſeed being brought out of Spain and Italy, from whence I receiued ſome for my garden.

¶ *The Time.*

They floure and flouriſh in May, the ſeed is ripe in Auguſt the ſecond yeare after their ſowing.

¶ *The Names.*

Madwort or Moonwort is called of the Grecians ἄλυσσον or ἄλυσσον : of the Latines, *Alyſſum* : in Engliſh, *Galens* Madwort, and of ſome, Heale-dog : and it hath the name thereof, becauſe it is a preſent remedie for them that are bitten of a mad dog, as *Galen* writeth, *lib. 2. de Antidotis,* in *Antonius Cous* his compoſition deſcribing it in theſe words ; Madwort is an herbe very like to Horehound, but rougher, and more full of prickles about the floures ; it beareth a floure tending to blew.

‡ 2 The ſecond by *Dodonæus, Lobel, Camerarius,* and others, is reputed to be the *Alyſſon* of *Dioſcoridis* ; *Geſner* names it *Lunaria aſpera* ; and *Columna, Leucoium montanum lunatum.* ‡

¶ *The Temperature and Vertues.*

A   *Galen* ſaith it is giuen vnto ſuch as are inraged by the biting of a mad dogge, which thereby are perfectly cured, as is knowne by experience, without any artificiall application or method at all. The which experiment if any ſhall proue, he ſhall find in the working thereof, it is of temperature meanly dry, digeſteth, and ſomthing ſcoureth withall : for this cauſe it taketh away the morphew and ſun-burning, as the ſame Author affirmes.

† That which was formerly figured in the ſecond place, being a kinde of *Sideritis,* I haue here omitted, that I may giue you it more fitly among the reſt of that name and kindred hereafter.

CHAP. 125.   *Of Roſe Campion.*

Lychnis Chalcedonica.
Floure of Conſtantinople.

¶ *The Kindes.*

THere be diuers ſorts of Roſe Campions, ſome of the garden, and others of the field, the which ſhall be diuided into ſeuerall chapters ; and firſt of the Campion of Conſtantinople.

¶ *The Deſcription.*

THe Campion of Conſtantinople hath ſundry vpright ſtalks two cubits high & full of joints, with a certain roughneſſe, and at euery joint two large leaues of a brown green colour. The flours grow at the top like ſweet-Williams or rather like dames Violet, of the colour of red lead or orenge-tawny. The root is ſomewhat ſharpe in taſte.

‡   There are diuers varieties of this, as with white and bluſh coloured floures, as alſo a double kinde with very large double and beautifull floures of a Vermilion colour like as the ſingle one here deſcribed. ‡

¶ *The Place.*

The floure of Conſtantinople is planted in gardens and is very common almoſt euery where.

‡ The white and bluſh ſingle and the double one are more rare, and not to be found but in the gardens of our prime Floriſts. ‡

¶ *The Time.*

It floureth in Iune ond Iuly, the ſecond yeare after it is planted, and many yeares after ; for it conſiſteth of

a root

a root full of life, and endureth long, and can away with the cold of our clymat.

It is called *Conſtantinopolitanus flos*, & *Lychnis Chalcedonica* : of *Aldrouandus*, *Flos Creticus*, or Floure of Candy : of the Germanes, *Flos Hieroſolymitanus*, or Floure of Ieruſalem : in Engliſh, Floure of Conſtantinople : of ſome, Floure of Briſtow, or None-ſuch.

¶ *The Nature and Vertues.*

Floure of Conſtantinople, beſides that grace and beauty that it hath in gardens and garlands, is for ought we know of no vſe, the vertues thereof being not as yet found out.

CHAP. 126. *Of Roſe Campion.*

1 *Lychnis Coronaria rubra.*
Red Roſe Campion.

2 *Lychnis Coronaria alba.*
White Roſe Campion.

¶ *The Deſcription.*

1 THe firſt kind of Roſe Campion hath round ſtalks very knotty and woolly, and at euery knot or joint there do ſtand two woolly ſoft leaues like Mullein, but leſſer & much narrower : the floures grow at the top of the ſtalke, of a perfect red colour ; which beeing paſt, there follow round cods full of blackiſh ſeed. The root is long and threddy.

2 The ſecond Roſe Campion differs not from the precedent in ſtalks, leaues, or faſhion of the floures ; the only difference conſiſts in the colour, for the floures of this plant are of a milke white colour, and the other red.

‡ 3 This alſo in ſtalkes, roots, leaues, and manner of growing differs not from the former, but the floures are much more beautifull, beeing compoſed of ſome three or foure rankes or orders of leaues lying each aboue other. ‡

¶ *The*

‡ 3 *Lychnis Coronaria multiplex.*
Double Rose Campion.

¶ *The Place.*

The Rose Campion groweth plentifully in most gardens.

¶ *The Time.*

They floure from Iune to the end of August.

¶ *The Names.*

The Rose Campion is called in Latine, *Dominarum Rosa, Mariana Rosa, Cæli Rosa, Cæli flos* : of *Dioscorides,* λυχνὶς στεφανωμάτικὴ : that is, *Lychnis Coronaria* or *sativa* : *Gaza* translateth λυχνίς, *Lucernula* ; because the leaues thereof be soft,& fit to make weeks for candles, according to the testimony of *Dioscorides,* it was called *Lychnis* or *Lychnides,* that is, a Torch or such like light, according to the signification of the word, cleere, bright, and light-giuing floures : and therefore they were called the Gardeners Delight, or the Gardeners Eye : in Dutch, **Christes eie** : in French, *Oeillets* and *Oeilets Dieu* : in high-Dutch, **Marien ros-zlin**, and **Himmel roszlin**.

¶ *The Nature.*

The seed of Rose Campion, saith *Galen,* is hot and dry after a sort in the second degree.

¶ *The Vertues.*

The seed drunk in wine is a remedy for them that are stung with a scorpion, as *Dioscorides* testifieth.

Chap. 127.
Of wilde Rose Campion.

¶ *The Description.*

1   THe wild Rose Campion hath many rough broad leaues somewhat hoary and woolly ; among which rise vp long soft and hairy stalks branched into many armes,set with the like leaues,but lesse. The floures grow at the top of the stalks, compact of fiue leaues of a reddish colour : the root is thicke and large,with some threds anexed thereto.

‡   There also growes commonly wilde with vs another of this kind with white floures, as also another that hath them of a light blush color. ‡

2   The sea Rose Campion is a smal herb,set about with many green leaues from the lower part vpward ; which leaues are thicke,somewhat lesser and narrower than the leaues of sea Purslane. It hath many crooked stalks spred vpon the ground a foot long, in the vpper part whereof there is a small white floure in fashion and shape like a little cup or box,after the likenesse of *Behen album* or Spatling Poppy,hauing within the said floure little threds of a blacke colour, in taste salt,yet not vnpleasant.

Mr *Tho.Hesket* reported vnto me,that by the sea side in Lancashire from whence this plant came, there is another sort hereof with red floures.

‡   3   This brings many stalks from one root,round,long,and weaker than those of the first described,lying vsually vpon the ground : the leaues grow by couples at each joint,long,soft, & hairy ; among which alternatly grow the floures,about the bignesse of those of the first described,and of a blush colour ; and they are also succeeded by such seed-vessels,containing a reddish seed. The root is thicke and fibrous,yet commonly outliues not the second yeare.

+ 1 *Lychnis*

† 1 *Lychnis sylvestris rubello flore.*
Red wilde Campion.

2 *Lychnis marina Anglica.*
English Sea Campion.

3 *Lychnis sylvestris hirta 5. Clusij.*
Wilde hairy Campion.

4 *Lychnis sylvestris 8. Clusij.*
Hoary wilde Campion.

5 *Lychnis hirta minima 6. Cluſ.*
Small hairy Campion.

† 6 *Lychnis ſylueſtris incana Lob.*
Ouerworne Campion.

7 *Lychnis caliculis ſtriatis 2. Cluſij.*
Spatling Campion.

† 8 *Lychnis ſylueſtris alba 9. Cluſ.*
White wilde Campion.

4　The fourth kinde of wild Campions hath long and ſlender ſtems, diuiding themſelues into ſundry other branches which are full of joints, hauing many ſmall and narrow leaues proceeding from the ſaid joints, and thoſe of a whitiſh green colour. The floures grow at the top of the ſtalke, of a whitiſh colour on the inner ſide, and purpliſh on the outer ſide, conſiſting of fiue ſmal leaues, euery leafe hauing a cut in the end, which maketh it ſhaped like a forke: the ſeed is like the wilde Poppy ; the root ſomwhat groſſe and thicke, which alſo periſheth the ſecond yeare.

5　The fifth kind of wild Campion hath three or foure ſoft leaues ſomewhat downy, lying flat vpon the ground; among which riſeth vp an hairy aſh-coloured ſtalk diuided into diuers branches, whereupon grow at certain ſpaces, euen in the ſetting together of the ſtalke and branches, ſmal and graſſe-like leaues, hairy, and of an ouerworne dusky colour, as is all the reſt of the plant. The flours grow at the top of the branches, compoſed of fiue ſmall forked leaues of a bright ſhining red colour. The root is ſmall, and of a wooddy ſubſtance.

6　The ſixth kinde of wilde Campion hath many long thicke fat and hoary leaues ſpred vpon the ground, in ſhape & ſubſtance like thoſe of the garden Campion, but of a very duſty ouerworn colour : among which riſe vp ſmall and tender ſtalks ſet at certain diſtances by couples, with ſuch like leaues as the other, but ſmaller. The floures doe grow at the top of the ſtalks in little tufts like thoſe of ſweet-Williams, of a red colour. The root is ſmall, with many thready ſtrings faſtned to it.

‡ 7　This growes ſome cubit high, with ſtalks diſtinguiſhed with ſundry joints, at each where-of are ſet two leaues, green, ſharp pointed, and ſomewhat ſtiffe : the floures grow at the top of the branches, like to thoſe of _Muſcipula_ or Catch-fly, yet ſomewhat bigger, and of a darke red ; which paſt, the ſeed (which is aſh-coloured and ſomewhat large) is contained in great cups or veſſels, co-uered with a hard and very much creſted skin or filme ; whence it is called _Lychnis caliculis ſtriatis_, and not _cauliculis ſtriatis_, as it is falſly printed in _Lobels Icones_, which ſome as fooliſhly haue follow-ed. The root is ſingle and not large, and dies euery yeare.

8　That which our Author figured in this place had great leaues and red floures, which no way ſorted with his deſcription : wherefore I haue in lieu thereof giuen you one out of _Cluſius_, which may fitly carry the title. This at the top of the large fibrous and liuing root ſendeth forth many leaues ſomewhat green, and of ſome fingers length, growing broader by degrees, and at laſt ending againe in a ſharpe point. The ſtalks are ſome cubit high , ſet at each joint with two leaues as it were embracing it with their foot-ſtalks ; which leaues are leſſe and leſſe as they are higher vp, and more ſharp pointed. At the tops of the branches grow the floures, conſiſting of fiue white leaues deeply cut in almoſt to the middle of the floure, and haue two ſharp pointed appendices at the bot-tome of each of them, and fiue chiues or threds come forth of their middles : theſe when they fade contract and twine themſelues vp, and are ſucceeded by thick and ſharp pointed ſeed-veſſels, con-taining a ſmall round aſh-coloured ſeed. I coniecture that the figure of the _Lychnis plumaria_, which was formerly here in the ninth place, out of _Tabern._ might be of this plant as well as of that which _Bauhine_ refers it to, and which you ſhall find mentioned in the end of the Chapter. ‡

¶ _The Place._

They grow of themſelues neere the borders of plowed fields, medowes, and ditch bankes, com-mon in many places. ‡ I haue obſerued none of theſe, the firſt and ſecond excepted, growing wild with vs. ‡

The ſea Campion growes by the ſea ſide in Lancaſhire, at a place called Lytham, fiue miles from Wygan, from whence I had ſeeds ſent me by Mr _Tho. Hesketh,_ who had heard it reported, that in the ſaid place doth grow of the ſame kind ſome with red floures, which are very rare to be ſeen. ‡ This plant (in my laſt Kentiſh ſimpling voiage, 1632, with Mr _Tho. Hicks,_ Mr _Broad,_ &c.) I found growing in great plenty in the low mariſh ground in Tenet, that lieth directly oppoſit to the town of Sandwich. ‡

¶ _The Time._

They floure and flouriſh moſt part of the Summer euen vnto Autumne.

¶ _The Names._

The wilde Campion is called in Greeke Λυχνὶς ἄγρια : in Latine, _Lychnis ſylueſtris :_ in Engliſh, wild Roſe Campion.

¶ _The Temperature._

The temperature of theſe wilde Campions are referred vnto thoſe of the garden.

¶ _The Vertues._

The weight of two drams of the ſeed of wilde Campion beaten to pouder and drunke, purgeth　A

choler

choler by the ftoole, and it is good for them that are ftung or bitten of any venomous beaft.

† The figure that was in the firft place, and was intended for our ordinarie wilde Campion, is that which you fee here in the eighth place ; and thofe that were in the fixth and ei„hth places you fhall hereafter finde with *Mufcipula* or Catch-fly, whereto they are of affinitie. That figure which was in the ninth place out of *Tabern.* vnder the title of *Lychnis plumaria,* as alfo the defcription, I haue omitted as impertinent: for the figure of *Bauhine* himfelfe (who corrected & againe fet forth the Works of *Tabernamontanus*) could not tell what to make thereof ; but queftion, *Quid fit ? an Mufcipula flore mufcofo ?* Which if it be, you fhall find that plant hereafter defcribed, vnder the title of *Sefamoides magnum Salmanticum* : for our Authors defcription is not worth the fpeaking of, being framed only from imagination.

‡ CHAP. 128. *Of diuers other wilde Campions.*

¶ *The Defcription.*

‡ 1   THe firft of thefe which we here giue you is like in leaues, ftalkes, roots, and manner of growing vnto the ordinarie wilde Campion defcribed in the firft place of the precedent Chapter ; but the floures are very double, compofed of a great many red leaues thick packt together, and they are commonly fet in a fhort and broken husk or cod. Now the fimilitude that thefe floures haue to the jagged cloath buttons antiently worne in this Kingdome, gaue occafion to our gentlewomen and other louers of floures in thofe times, to call them Bachelors buttons.

2   This differs not in fhape from the laft defcribed, but only in the colour of the flours, which in this plant are white.

‡ 1 *Lychnis fyl. multiplex purpurea.*
Red Bachelors buttons.

‡ 2 *Lychnis fyl. alba multiplex.*
White Bachelors buttons.

3   Neither in roots, leaues, or ftalks, is there any difference betweene this either degenerate or accidentall varietie of Bachelors buttons, from the two laft mentioned, onely the floures hereof are of a greenifh colour, and fometimes through the middeft of them they fend vp ftalks, bearing alfo tufts of the like double floures.

4   This faith (*Clufius*) hath fibrous roots like to thofe of Primrofes ; out of which come leaues
of

‡ 3 *Lychnis abortiua flore multiplici viridi.*
Degenerate Bachelors Buttons with greene floures.

‡ 4 *Lychnis syl. latifolia Cluf.*
Broad leaued wilde Campion.

‡ 5 *Lychnis montana repens.*
Creeping mountaine Campion.

of a sufficient magnitude, not much vnlike thofe of the great yellow Beares-eare, yet wh'ter, more downy, thicke, and juycie. The next yeare after the fowing thereof it fends vp a ftalke of two or three cubits high, here and there fending forth a vifcous and glutinous juice, which detaines and holds faft flies and fuch infects as do chance to light thereon. At the top of the branches it yeeldeth many floures fet as it were in an vmbel, euen fomtimes an hundred; yet fufficiently fmall, confidering the magnitude of the plant; and each of thefe confifts of fiue little yellowifh greene forked leaues.

5 The ftalkes of this are flender, joynted and creeping like to thofe of the greater Chickweed, and at each joynt grow two leaues like thofe of the myrtle, or of Knot-graffe, yet fomewhat broader. The floures grow in long cups like as thofe of *Saponaria*, and are much leffe, yet of the fame colour. The root is fmall.

¶ *The Place.*

1 2 Thefe are kept in many Gardens of this Kingdome for their beautie, efpecially the firft, which is the more common.

4 This growes naturally in Candy; the fi th by riuelets in the mountainous places of Sauoy.

¶ *The Time.*

Thefe floure in Iune and Iuly with the other wild Campions.

R r ¶ *The*

¶ *The Names.*

2   The first of these is *Lychnis agrestis multiflora* of *Lobel*; and *Ocymoides flore pleno* of *Camerarius*.

2   The second is by *Pena* and *Lobel* also called *Lychnis syluestris multiflora*: it is the *Ocymastrum multiflorum* of *Tabernamontanus*; by which title our Author also had it in the former edition, p.551.

3   *Lobel* hath this by the name of *Lychnis agrestis abortiua multiplici viride flore*.

4   *Clusius* calls this *Lychnis syluestris latifolia*; and he saith he had the seed from *Ioseph de Casa Bona*, by the name of *Muscipula auricula ursi facie*: *Bauhine* hath it by the name of *Lychnis auricula ursi facie*.

5   This (according to *Bauhine*) was set forth by *Matthiolus*, by the name of *Cneoron aliud Theophrasti*: it is the *Ocimoides repens polygonifolia flore Saponaria*, in the *Aduersaria*: and *Saponaria minor Daleschampij*, in the *Hist. Lugd.* It is also *Ocimoides Alpinum*, of *Gesner*; and *Ocymoides repens*, of *Camerarius*.

¶ *The Nature and Vertues.*

The natures and vertues of these, as of many others, lie hid as yet, and so may continue, if chance, or a more curious generation than yet is in being do not finde them out. ‡

CHAP. 129.   *Of Willow-herbe, or Loose-strife.*

1 *Lysimachia lutea.*   ‡ 2 *Lysimachia lutea minor.*
Yellow Willow-herbe.   Small yellow Willow-herbe.

¶ *The Description.*

1   THe first kinde of Willow-herbe hath long and narrow leaues of a grayish greene colour, in shape like the Willow or Sallow leaues, standing three or foure one against another at seuerall distances round about the stalke; which toward the top diuideth it selfe into many other branches, on the tops whereof grow tufts of faire yellow floures, consisting of fiue leaues apiece, without smell: which being past, there commeth forth seed like Coriander. The root is long and slender.

‡ 2 This

‡ 2 This leſſer of *Claſius* his deſcription hath a ſtalke a cubit high, and ſometimes higher, firme, hard, and downy ; about which at certaine diſtances grow commonly foure leaues together, yet ſometimes but three, and they are ſoft and ſomewhat downy, leſſer than thoſe of the former, being firſt of an acide taſte, and then of an acride ; and they are vſually marked on their lower ſides with blacke ſpots. About the top of the ſtalke, out of the boſomes of each leafe come forth little branches bearing ſome few floures, or elſe foot-ſtalkes carying ſingle floures, which is more vnuall towards the top of the ſtalke. The floures are yellow, with ſomewhat a ſtrong ſmell, conſiſting of fiue ſharpe pointed yellow leaues, with ſo many yellow thredl in their middle. The root is joynted, or creeping here and there, putting vp new ſhouts.

‡ 3 *Lyſimachia lutea flore globoſo.*
Yellow Willow-herb with bunched floures.

‡ 4 *Lyſimachia lutea Virginiana.*
Tree Primroſe.

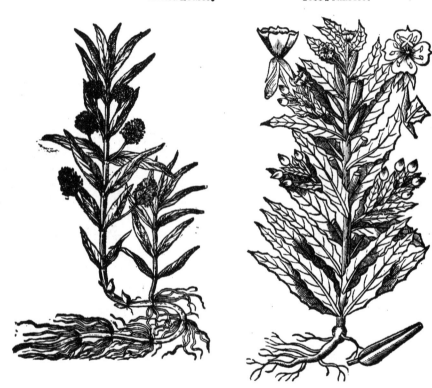

3 This alſo may fitly be referred to the former. The ſtalke is a cubit high, ſtraight, and as it were jointed, naked oft times below by the falling away of the leaues ; but from the middle to the top ſet with two leaues at a joynt, like thoſe of the former ; and out of their boſoms on ſhort ſtalks grow round tufts of ſmall yellow floures as in bunches : the root which creeps ſends forth many ſmall fibres at each joynt. This was ſet forth by *Lobel* vnder the title of *Lyſimachia lutea altera*, or *Lyſimachia ſalicaria : Dodonæus* hath it by the name of *Lyſimachiam aquatile :* and *Cluſius* calls it *Lyſimachia lutea tertia, ſiue minor.*

4 This Virginian hath beene deſcribed and figured onely by *Proſper Alpinus*, vnder the title of *Hyoſcyamus Virginianus :* and by Mr. *Parkinſon*, by the name of *Lyſimachia lutea ſiliquoſa Virginiana :* Alſo *Bauhine* in the Appendix of his *Pinax* hath a large deſcription thereof, by the name of *Lyſimachia lutea corniculata.* The root hereof is longiſh, white, about the thickeneſſe of ones thumbe, from whence growes vp a tall ſtalke diuided into many branches of an ouerworne colour, and a little hairie : the leaues are like thoſe of the former, but ſomewhat ſinuated alongſt their edges, and hauing their middle veine of a whitiſh colour : toward the tops of the branches amongſt the leaues come vp pretty thicke cods, which growing ſmaller on their tops ſuſtaine pretty large yellow floures conſiſting of foure leaues, with a peſtle in the middle, vpon which ſtand foure yellowiſh thrums

in

in faſhion of a croſſe ; and there are alſo eight threds with their pointals in the middle of them. Theſe floures haue ſomewhat the ſmell of a Primroſe (whence M.ʳ *Parkinſon* gaue it the Engliſh name, which I haue alſo here giuen you :) after the floures are fallen, the cods grow to be ſome two inches long, being thicker below, and ſharper at the top, and ſomewhat twined, which in fine open themſelues into foure parts to ſhatter their ſeed, which is blacke and ſmall ; and ſowne, it growes not the firſt yeare into a ſtalke, but ſends vp many large leaues lying handſomely one vpon another Roſe-faſhion. It floures in Iune, and ripens the ſeed in Auguſt. ‡

    5   The ſecond kinde of Willow herbe in ſtalkes and leaues is like the firſt, but that the leaues are longer, narrower, and greener. The floures grow along the ſtalke toward the top, ſpike-faſhion, of a faire purple colour : which being withered turne into downe, which is carried away with the winde.

    5 *Lyſimachia purpurea ſpicata.*   6 *Lyſimachia ſiliquoſa.*
    Spiked Willow-herbe.   Codded Willow-herbe.

    6   This *Lyſimachia* hath leaues and ſtalkes like vnto the former. The floure groweth at the top of the ſtalke, comming out at the end of a ſmall long cod, of a purple colour, in ſhape like a ſtocke Gilloſloure, and is called of many *Filius ante Patrem* (that is, the Sonne before the Father) becauſe that the cod commeth forth firſt, hauing ſeeds therein, before the floure doth ſhew it ſelfe abroad. ‡ The leaues of this are more ſoft, large, and hairy than any of the former : they are alſo ſnipt about the edges, and the floure is large, wherein it differs from the twelfth, hereafter deſcribed ; and from the eleuenth in the hairineſſe of the leaues, and largenſſe of the floures alſo, as you ſhall finde hereafter. ‡

    7   This being thought by ſome to be a baſtard kinde, is (as I do eſteeme it) of all the reſt moſt goodly and ſtately plant, hauing leaues like the greateſt Willow or Ozier. The branches come out of the ground in great numbers, growing to the height of ſix foot, garniſhed with braue floures of great beauty, conſiſting of foure leaues a piece, of an orient purple colour, hauing ſome threds in the middle of a yellow colour. The cod is long like the laſt ſpoken of, and full of downy matter, which flyeth away with the winde when the cod is opened.

    ‡ 8   This alſo, which is the *Chamænerion* of *Geſner*, as alſo his *Epilobion, quaſi ϊον ἐπι λοβω, a Vio-let or floure vpon a cod*, may iuſtly challenge the next place. *Dodonæus* calls it *Pſeudolyſimachium purpureum*

† 7 *Chamænerion.*
Roſe bay Willow-herbe.

‡ 8 *Chamænerion alterum anguſtifolium.*
Narrow leaued Willow-Houre.

‡ 9 *Lyſimachia cærulea.*
Blew Looſe-ſtrife.

‡ 10 *Lyſimachia galericulata.*
Hooded Looſe-ſtrife.

purpureum minus: and it is in the *Hiſtor. Lugdun.* vnder the name of *Linaria rubra.* It groweth vp with ſtalkes ſome foot high, ſet with many narrow leaues like thoſe of Toad-flax, of a grayiſh colour, and the ſtalke is parted into diuers branches, which at their tops vpon long cods carry purple floures conſiſting of foure leaues a piece. The root is long, yellowiſh, and wooddy. ‡

9   There is another baſtard Looſe-ſtrife or Willow herbe, hauing ſtalkes like the other of his kinde, whereon are placed long leaues ſnipt about the edges, in ſhape like the great *Veronica* or herbe *Fluellen.* The floures grow along the ſtalkes, ſpike-faſhion, of a blew colour ; after which ſucceed ſmall cods or pouches. The root is ſmall and fibrous : it may be called *Lyſimachia cærulea,* or blew Willow-herbe.

10   We haue likewiſe another Willow-herbe that groweth neere vnto the bankes of riuers and water-courſes. This I haue found in a watery lane leading from the Lord Treaſurer his houſe called Theobalds, vnto the backeſide of his ſlaughter-houſe, and in other places, as ſhall be decla-red hereafter. Which *Lobel* hath called *Lyſimachia galericulata,* or hooded Willow-herbe. It hath many ſmall tender ſtalkes trailing vpon the ground, beſet with diuers leaues ſomewhat ſnipt about the edges, of a deepe greene colour, like to the leaues of *Scordium* or water Germander : among which are placed ſundry ſmall Bell-floures faſhioned like a little hood, in ſhape reſembling thoſe of Ale-hoofe. The root is ſmall and fibrous, diſperſing it ſelfe vnder the earth farre abroad, whereby it greatly increaſeth.

11 *Lyſimachia campeſtris.*
Wilde Willow-herbe.

‡ 13 *Lyſimachia purpurea minor Cluſ.*
Small purple Willow-herbe.

11   The wilde Willow-Herbe hath fraile and very brittle ſtalkes, ſlender, commonly about the height of a cubit, and ſometimes higher, whereupon do grow ſharpe pointed leaues ſomewhat ſnipt about the edges, and ſet together by couples. There come forth at the firſt long ſlender cods, wherein is contained ſmall ſeed, wrapped in a cottony or downy wooll, which is carried away with the winde when the ſeed is ripe : at the end of which commeth forth a ſmall floure of a purpliſh colour ; whereupon it was called *Filius ante Patrem,* becauſe the floure doth not appeare vntil the cod be filled with his ſeed. But there is another Sonne before the Father, as hath beene declared in the Chapter of Medow-Saffron. The root is ſmall and threddy. ‡ This differeth from the ſixth onely in that the leaues are leſſe, and leſſe hairy, and the floure is ſmaller. ‡

12 The

12   The Wood Willow-Herbe hath a ſlender ſtalke diuided into other ſmaller branches, whereon are ſet long leaues rough and ſharpe pointed, of an ouerworne greene colour. The floures grow at the tops of the branches, conſiſting of foure or fiue ſmall leaues, of a pale purpliſh colour tending to whiteneſſe: after which come long cods, wherein are little ſeeds wrapped in a certaine white Downe that is carried away with the winde. The root is thready. ‡ This differs from the ſixth in that it hath leſſer floures. There is alſo a leſſer ſort of this hairie *Lyſimachia* with ſmall floures.

There are two more varieties of theſe codded Willow-herbes; the one of which is of a middle growth, ſomewhat like to that which is deſcribed in the eleuenth place, but leſſe, with the leaues alſo ſnipped about the edges, ſmooth, and not hairy: and it may fitly be called *Lyſimachia ſiliquoſa glabra media*, or *minor*, The leſſe ſmooth-leaued Willow-herbe. The other is alſo ſmooth leaued, but they are leſſer and narrower: wherefore it may in Latine be termed, *Lyſimachia ſiliquoſa glabra minor anguſtifolia*: in Engliſh, The leſſer ſmooth and narrow leaued Willow-herbe.

‡   13   This leſſer purple Looſe-ſtrife of *Cluſius*, hath ſtalkes ſeldome exceeding the height of a cubit, they are alſo ſlender, weake and quadrangular, towards the top, diuided into branches growing one againſt another, the leaues are leſſe and narrower than the common purple kinde, and growing by couples, vnleſſe at the top of the ſtalkes and branches, whereas they keepe no certaine order; and amongſt theſe come here and there cornered cups containing floures compoſed of ſix little red leaues with threds in their middles. The root is hard, wooddy, and not creeping, as in others of this kinde, yet it endures all the yeere, and ſends forth new ſhoots. It floures in Iune and Iuly, and was found by *Cluſius* in diuers wet medowes in Auſtria. ‡

¶ *The Place.*

The firſt yellow *Lyſimachia* groweth plentifully in moiſt Medowes, eſpecially along the Medowes as you goe from Lambeth to Batterſey neere London, and in many other places throughout England.

‡   The ſecond and third I haue not yet ſeene.

The fourth groweth in many Gardens. ‡

The fifth groweth in places of greater moiſture, yea almoſt in the running ſtreames and ſtanding waters, or hard by them. It groweth vnder the biſhops houſe wall at Lambeth, neere the water of Thames, and in moiſt ditches in moſt places of England.

The ſixth groweth neere the waters (and in the waters) in all places for the moſt part.

The ſeuenth groweth in Yorkeſhire in a place called the Hooke, neere vnto a cloſe called a Cow paſture, from whence I had theſe plants, which do grow in my Garden very goodly to behold, for the decking vp of Houſes and Gardens.

‡   The eighth I haue not yet found growing.

The ninth growes wilde in ſome places of this Kingdome, but I haue ſeene it only in Gardens.

The tenth growes by the ponds and waters ſides in Saint Iames his Parke, in Tuthill fields and many other places. ‡

The eleuenth groweth hard by the Thames, as you goe from a place called the Diuels Neckerchiefe to Redreffe, neere vnto a ſtile that ſtandeth in your way vpon the Thames banke, among the plankes that do hold vp the ſame banke. It groweth alſo in a ditch ſide not farre from the place of execution, called Saint Thomas Waterings.

‡   The other varieties of this grow in wet places, about ditches, and in woods and ſuch like moiſt grounds. ‡

¶ *The Time.*

Theſe herbes floure in Iune and Iuly, and oftentimes vntill Auguſt.

¶ *The Names.*

Lyſimachia, as *Dioſcorides* and *Pliny* write, tooke his name of a ſpeciall vertue that it hath in appeaſing the ſtrife and vnrulineſſe which falleth out among oxen at the plough, if it be put about their yokes: but it rather retaineth and keepeth the name *Lyſimachia*, of King *Lyſimachus* the Sonne of *Agathocles*, the firſt finder out of the nature and vertues of this herbe, as *Pliny* ſaith in his 25.book chap.7. which retaineth the name of him vnto this day; and was made famous by *Eraſiſtratus*. *Ruellius* writeth, that it is called in French, *Cornelle* and *Corneola*: in Greeke, λυσιμάχιον: of the Latines, *Lyſimachium*: of *Pliny*, *Lyſimachia*: of the later Writers, *Salicaria*: in high Dutch, **Wederick**: in Engliſh, Willow-herbe, and Looſe ſtrife.

Chamænerium is called of *Geſner*, *Epilobion*: in Engliſh, Bay Willow, or bay yellow herbe.

‡ The

‡ The names of such as I haue added haue beene sufficiently set forth in their Titles and Histories. ‡

¶ *The Nature.*

The yellow *Lysimachia*, which is the chiefe and best for Physicke vses, is cold and dry, and very astringent.

¶ *The Vertues.*

A   The juyce, according to *Dioscorides*, is good against the bloudy flix, being taken either by potion or Clyster.

B   It is excellent good for greene wounds, and stancheth the bloud: being also put into the nosthrils, it stoppeth the bleeding at the nose.

C   The smoke of the burned herbe driueth away serpents, and killeth flies and gnats in a house; which *Pliny* speaketh of in his 23.booke, chap.8. Snakes, faith he, craull away at the smell of Loofstrife. The same Author affirmeth in his 26.booke, last chap. that it dieth haire yellow, which is not very vnlike to be done by reason the floures are yellow.

D   The others haue not beene experimented, wherefore vntill some matter worthy the noting doth offer it selfe to our consideration, I will omit further to discourse hereof.

E   The iuyce of yellow *Lysimachia* taken inwardly, stoppeth all flux of bloud, and the Dysenteria or bloudy flix.

F   The juyce put into the nose, stoppeth the bleeding of the same, and the bleeding of wounds; and mightily cooleth and healeth them, being made into an vnguent or salue.

G   The same taken in a mother suppositorie of wooll or cotton, bound vp with threds (as the manner thereof is, well knowne to women) staieth the inordinate flux or ouermuch flowing of womens termes.

H   It is reported, that the fume or smoke of the herbe burned, doth driue away flies and gnats, and all manner of venomous beasts.

CHAP. 130. *Of Barren-woort.*

Epimedium.
Barren-Woort.

¶ *The Description.*

THis rare and strange plant was sent to me from the French Kings Herbarist *Robinus*, dwelling in Paris at the signe of the blacke head, in the streete called *Du bout du monde*, in English, The end of the world. This herbe I planted in my Garden, and in the beginning of May it came forth of the ground, with small, hard, and woody crooked stalkes: wherupon grow rough and sharpe pointed leaues, almost like *Alliaria*, that is to say, Sauce alone, or Iacke by the hedge. *Lobel* and *Dodon.* say, that the leaues are somewhat like Iuie; but in my iudgement they are rather like *Alliaria*, somwhat snipt about the edges, and turning themselues flat vpright, as a man turneth his hand vpwards when hee receiueth money. Vpon the same stalkes come forth small floures consisting of foure leaues, whose outsides are purple, the edges on the inner side red, the bottome yellow, & the middle part of a bright red colour, and the whole floure somewhat hollow. The root is small, and creepeth almost vpon the vppermost face of the earth. It beareth his seed in very small cods like Saracens Consound, (‡ to wit that of our Authour formerly

merly deſcribed,*pag.274.*‡) but ſhorter : which came not to ripeneſſe in my Garden, by reaſon that it was dried away with the extreme and vnaccuſtomed heat of the Sun, which happened in the yeare 1590. ſince which time from yeare to yeare it bringeth ſeed to perfection. Further *Dioſcorides* and *Pliny* do report, that it is without floure or ſeed.

¶ *The Place.*

† It groweth in the moiſt medowes of Italy about Bononia and Vincentia : it groweth in the Garden of my friend M' *Iohn Milion* in Old-ſtreet, and ſome other Gardens about towne.

¶ *The Time.*

It floureth in Aprill and May, when it hath taken faſt hold and ſetled it ſelfe in the earth a yeare before.

¶ *The Names.*

It is called *Epimedium* : I haue thought good to call it Barrenwoort in Engliſh ; not becauſe that *Dioſcorides* ſaith it is barren both of floures and ſeeds, but becauſe (as ſome Authors affirme) being drunke it is an enemy to conception.

¶ *The Temperature and Vertues.*

Galen affirmeth that it is moderately cold, with a watery moiſture : we haue as yet no vſe hereof in Phyſicke.

---

‡ Cʜ ᴀ ᴘ . 131.   *Of Fleabane.*

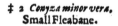

‡ THe ſmalneſſe of the number of theſe plants here formerly mentioned, the confuſion notwithſtanding in the figures, their nominations and hiſtorie, not one agreeing with another, hath cauſed mee wholly to omit the deſcriptions of our Author, and to giue new, agreeable to the figures ; together with an addition of diuers other plants belonging to this kindred. Beſides, there is one thing I muſt aduertiſe you of, which is, that our Author in the firſt place deſcribed the *Baccharis Monſpelienſium* of *Lobel*, or *Conyza maior* of *Matthiolus*, and it is that which grows in Kent and Eſſex on chalkie hils ; yet he gaue no figure of it, but as it were forgetting what he had done, allotted it in a particular chapter afterwards, where alſo another figure was put for it ; but there you ſhall now finde it, though I muſt confeſſe that this is as fit or a fitter place for it, but I follow the courſe of my Author, whoſe matter, not method, I indeauour to amend.

¶ *The*

¶ *The Deſcription.*

1   THis great Fleawoort or Fleabane, from a thick long liuing fibrous root ſends forth many ſtalkes of ſome yard high or more; hard, wooddy, rough, fat, and of an ouerworne colour: the leaues are many, without order, and alternately embrace the ſtalkes, twice as big as thoſe of the Oliue tree, rough and fat, being as it were beſmeared with a gummineſſe or fattineſſe, and of a yellowiſh greene colour: the floures grow after a ſort ſpoke faſhion, ſtanding at the ends of footſtalkes comming out of the boſomes of the leaues, and they are yellow and round almoſt like to Groundſwell, and fly away in downe like as they do; the ſeed is ſmall and aſh coloured. The whole plant is fatty and glutinous, with a ſtrong, yet not altogether vnpleaſant ſmell. This growes not that I know of in theſe cold Countries, vnleſſe ſowne in Gardens. *Cluſius* found it by Lisbone, and in diuers places of Spaine. He, as alſo *Dodonæus, Lobel,* and others, call this *Conyza maior,* and it is thought to be the *Conyza mas* of *Theophraſtus,* and *Conyza maior* of *Dioſcorides.*

2   The leſſer ſeldome ſends vp more than one ſtalke, and that of a cubit high, yet vſually not ſo much: it is diuided into little branches, and alſo rough and glutinous as the precedent, but more greene. The leaues are three times leſſe than thoſe of the former, ſomewhat ſhaped like thoſe of Toad-flax, yet hairie and vnctious: the tops of the branches as in the bigger, carrie leſſe, and leſſe ſhining and ſightly floures, vaniſhing in like ſort into downe. The root is ſingle and annuall, and the whole plant more ſmelling than the former. This is iudged the *Conyza fæmina* of *Theophraſtus;* and *Con. minor* of *Dioſcorides;* it is the *Con. minor* of *Geſner, Lobel, Cluſius* and others. It growes in diuers parts of Spaine and Prouince in France, but not here, vnleſſe in Gardens.

† 3 *Conyza media.*   † 4 *Conyza minima.*
Middle Fleawoort.   Dwarfe Fleabane.

3   The root of this middle kinde is pretty large and fibrous, from whence ariſeth a branched ſtalke of ſome cubit high, engirt at certaine ſpaces with thicke, rough, grayiſh greene leaues: at the tops of the branches grow pretty faire yellow floures of the bigneſſe of a little Marigold; which fading turne to downe, and are carried away with the winde. This floures in Iuly and Auguſt, and may be found growing in moſt places about riuers and pond ſides, as in S. Iames his Parke, Tuthill fields, &c. This is *Conyza media* of *Matthiolus, Dodonæus* and others. Some haue referred it vnto the

Mints,as *Fuchſius*,who makes it *Calamintha 3.genus*; and *Lonicerus*,who calls it *Mentha lutea*. In Cheape-ſide the herbe-women call it Herbe Chriſtopher,and ſell it to Empericks, who with it(as they ſay)make medicines for the eies,but againſt what affeᵭ of them,or with what ſucceſſe I know not.

4 In like places,or rather ſuch as are plaſhic in winter this may be plentifully found growing. The roots are ſmall and fibrous; from whence ariſeth a branched ſtalke ſome foot high, ſet with ſmall longiſh leaues ſomewhat roundiſh pointed,ſoft alſo and woolly, with a ſmell not altogether vnpleaſant,like as the laſt deſcribed: the floures are compaſſed with many yellowiſh threds like to the middle part of Camomill floures,or thoſe of Tanſey: and as the former, turne into downe, and are carried away with the winde; it floures in Iuly and Auguſt. This is the *Conyza minor* of *Tragus*, *Matthiolus*,and others: *Lobel* and *Dodon*.call it *Conyza minima*.

5 This cut leaued Fleabane hath ſmall fibrous roots,from which ariſe thicke, creſted,and hollow ſtalkes,diuided towards the tops into ſundry branches: the leaues that encompaſſe the ſtalke are gaſhed,or els deeply ſinuated on the edges:the floures are ſtarre faſhion and yellow,and alſo fly away in downe; the whole plant is couered ouer with a ſoft and tender downe, and hath ſomewhat the ſmell of hony. This is a variety of the third, and is called by *Dodon*.*Conyza mediæ ſpecies altera*. *Lobel* names it *Conyza helenitis folÿs laciniatis*.

‡ 5 *Conyza folÿs laciniatis*.
Great jagged leaued Fleabane.

‡ 6 *Conyza paluſtris ſerratifolia*.
Water ſnipt Fleabane.

6 The figure which you haue in this ſixth place was formerly vnfitly giuen by our Authour for *Solidago Saracenica*; it hath a large root which ſends forth many fibres, and a creſted hollow ſtalke ſome two cubits or more high, which is vnorderly ſet,with long, yet narrow ſnipt leaues ſomewhat hairy and ſharpe pointed: the top is diuided into branches, which beare pretty large yellow floures, made after the manner of thoſe of Ragwort, and like as they, are alſo carried away with the winde. This *Thalius* calls *Conyza maxima ſerratifolia*. It is the *Lingua major* of *Daleſchampius*,and the *Conſolida paluſtris* of *Tabernamontanus*. It groweth neere water ſides,and floures towards the later end of Summer: I haue not yet heard that it doth grow wilde amongſt vs.

7 The

‡ 7 *Conyza Auſtriaca Cluſij.*
Auſtrian Fleabane.

‡ 8 *Conyza incana.*
Hoary Fleabane.

‡ 9 *Conyza Alpina piloſiſſima.*
Hairie Fleabane of the Alpes.

† 10 *Conyza Cærulea acris.*
Blew floured Fleabane.

The stalkes of this are about a foot high, straight, stiffe, hard, and couered with a whitish Downe: the leaues at the root grow vpon long stalks, and are soft and hairy; but those which are higher vp haue a short or else no stalke at all; rubbed, they yeeld no vnpleasant smell, and tasted they are somwhat bitter and acride. The floures that grow vpon the tops of the branches are large, and fashioned like those of Elecampane, and are of the same yellow colour: the root is long, slender, and blackish, creeping, and putting vp new stalks; it hath many white fibres, and a resinous smell. *Clusius* found it growing on dry hilly places in Austria, and calls it *Conyza 3. Austriaca.*

8   This, which *Lobel* sets forth vnder the title of *Conyza helenitis mellita incana*, I take to bee the same plant that I first figured and described out of *Clusius*, only the root is better exprest in *Clusius* his figure; otherwise by the figures I cannot finde any difference, though *Bauhine* reckon it vp in his *Pinax* as differing therefrom. Hee calls it *Helenitis*, because the floures and leaues haue some semblance of Elecampane; and *Melitta*, for that they smell somewhat like hony.

9   This also seems not much to differ from the last mentioned, but onely in the hairines of the leaues and stalks, and that the floures are smaller. *Lobel* calls it *Conyza Helenitis pilosa*. These last grow vpon mountaines, but none of them with vs in England that I can yet heare of.

10   This hath a small fibrous and yellow root of a very hot and biting taste, which sends vp diuers longish leaues about the head thereof: the stalke is some foot and halfe high, and set alternately with twined longish narrow and somewhat rough leaues of an ouerworne green colour: the top of the stalk and branches are adorned with floures set in longish scaly heads like those of *Hieracium*; the outer little leaues are of a feint blew colour, and the inner threds are yellow. It floures in August, and the floures quickly turn into downe, and are carried away with the wind. It growes in many chalky hils, and I first obserued it in the companie of Mr *George Bowles*, Mr *Iohn Bugs*, and others, close by Farmingham in Kent; and the last yeare Mr *William Broad* found it growing at the Blockhouse at Grauesend. *Tragus* calls it *Tinctorius flos alter: Dodonaus*, because the floure quickly turnes to down, makes it *Erigeron quartum*: and *Gesner*, for that the root is hot, and drawes rheum like Pellitorie of Spain, which therefore is vsed against the tooth-ache, names it *Dentelaria:* he also calls it *Conyza muralis*, and *Conyzoides cærulea*. *Tabernamontanus* also calls it *Conyza cærulea* and lastly, *Fabius Columna* hath it by the name of *Amellus montanus*, to which kinde it may in myne opinion bee as fitly referred as to these *Conyza's.* Our Authour had the figure hereof in the third place in this Chapter.

¶ *The Place, Time, and Names.*

All these haue bin sufficiently shewne in their particular titles and descriptions. ‡

¶ *The Temperature.*

Conyza is hot and dry in the third degree.

¶ *The Vertues.*

The leaues and floures be good against the strangurie, the jaundice, and the gnawing or griping A of the belly.

The same taken with Vineger help the Epilepsie or falling sicknesse.   B

If women sit ouer the decoction thereof it greatly easeth their paines of the Mother.   C

The herb burned where flies, gnats, fleas, or any venomous things are, doth driue them away.   D

‡ The first was formerly of *Conyza media:* the second was of *Conyza minima*; and the third of *Conyza cærulea acris.*

Chap. 132.

Of Starre-wort.

¶ *The Description.*

1   THe first kind of *Aster* or *Inguinalis* hath large broad leaues like *Verbascum Salvifolium* or the great *Conyza*; amongst which riseth vp a stalke foure or fiue handfulls high, hard, rough, and hairy, beset with leaues like Rose Campions, of a darke greene colour. At the top of the said stalks come forth floures of a shining and glistering golden colour; and vnderneath about these floures grow fiue or six long leaues, sharp pointed, and rough, not much in shape

vnlike

vnlike the fiſh called *Stella marina :* the floures turn into down,and are carried away with the wind. The root is fibrous,of a biting and ſharp taſte.

‡ 2   The ſecond,called Italian Starwort, hath leaues not much vnlike Marigolds, but of a parke green colour,rough,and ſomwhat round at the vpper end : the ſtalks are many, & grow ſome cubit high, and at their tops are diuided into ſundry branches which beare faire blewiſh purple floures,yellow in their middles,& ſhaped like Marigolds, and almoſt of the ſame bignes,whence ſome haue called them blew Marigolds. ‡

3   The third kind hath leaues ſo like the Italian Star-wort,that a man can ſcarcely at the ſudden diſtinguiſh the one from the other. The ſingle ſtalk is a cubit long,vpright and ſlender;on the top whereof grow faire yellow floures like thoſe of *Enula Campana,* and they fly away in down:the root is ſmall and threddy.

4   The fourth kinde in talneſſe and floure is not much vnlike that laſt before ſpecified, but in ſtalke and leaues more hairy,and longer,ſomewhat like our ſmall Hounds-tongue, and the roots are leſſe fibrous or threddy than the former.

5   There is another ſort that hath a browne ſtalke,with leaues like the ſmall *Conyza :* the flours are of a darke yellow,which turne into down that flieth away with the wind like *Conyza.* The root is full of threds or ſtrings.

6   There is alſo another that hath leaues like the great Campion, ſomewhat hairie : amongſt which come vp crooked crambling ſtalks leaning lamely many wayes. Whereupon do grow faire yellow floures ſtar-faſhion; which paſt, the cups become ſo hard that they will ſcarce be broken with ones nails to take forth the ſeed:the root is long and ſtraight as a finger,with ſome few ſtrings anexed to the vppermoſt part thereof. It groweth wilde in ſome parts of Spain.

1 *After Atticus.*
Starwort.

‡ 2 *After Italorum.*
Italian Starwort.

7   There groweth another kinde of Starre-woort which hath many leaues like Scabious, but thinner,and of a more greene colour, couered with a woolly hairineſſe, ſharpe and bitter in taſte ; amongſt which ſpringeth vp a round ſtalke more than a cubit high ; often growing vnto a reddiſh

diſh colour, ſet with the like leaues, but ſmaller and ſharper pointed, diuiding it ſelfe toward the top into ſome few branches, whereon grow large yellow floures like *Doronicum* or *Sonchus*. The root is thicke and crooked. ‡ This is *Aſter Pannonicus maior, ſiue tertius* of *Cluſ.* and his *Auſtriacus primus.*

8　We haue ſeene growing vpon wild mountains another ſort, hauing leaues much leſſe than the former, ſomewhat like to the leaues of Willow, of a faire green colour, which doe adorne and deck vp the ſtalke euen to the top; whereupon do grow yellow floures ſtar-faſhion, like vnto the former. The root is ſmall and tender, creeping far abroad, whereby it mightily encreaſeth. ‡ This is *Aſter Pannonicus ſalignis folijs, ſiue Aſter 4. Auſtriacus 2.* of *Cluſius*. It is *Bubonium luteum* of *Tabern.* and our Author gaue the figure hereof for *Aſter Italorum*. ‡

9　*Cluſius* hath ſet forth a kinde that hath an vpright ſtalke ſomewhat hairy, two cubits high, beſet with leaues ſomewhat woolly, like to thoſe of the Sallow, hauing at the top of the ſtalke faire yellow floures like *Enula Campana*, which turn into down that is caried away with the winde. The root is thicke, with ſome haires or threds faſtned thereto. ‡ This is *Aſter lanuginoſo folio, ſiue 3.* of *Cluſius*. Our Author gaue the figure hereof vnder the title of *Aſter hirſutus*. It is *Aſter flore luteo* of *Taber.* ‡

10　He hath likewiſe deſcribed another ſort, that hath leaues, ſtalks, floures, and roots like the ninth, but neuer groweth to the height of one cubit. ‡ It bringeth forth many ſtalkes, and the leaues that grow diſorderly vpon them are narrower, blacker, harder and ſharper pointed than the former, not vnlike thoſe of the common *Ptarmica*, yet not ſnipt about the edges: The floures are yellow, and like thoſe of the laſt deſcribed, but leſſe. This is the *Aſter anguſtifolius ſiue ſextus* of *Cluſius*. ‡

11　There is likewiſe ſet forth in his Pannonicke Obſeruations, a kind of *Aſter* that hath many ſmall hairy leaues like the common great Daiſie; among which riſeth vp an hairy ſtalke of a foot high, hauing at the top faire blew floures inclining to purple, with their middle yellow, which turn in the time of ſeeding into a woolly down that flieth away with the wind. The whole plant hath a drying binding and bitter taſte. The root is threddy like the common Daiſie or that of the Scabious. ‡ This is *Aſter Alpinus cæruleo flore, ſiue 7.* of *Cluſius*. ‡

‡ *Aſter mont. flore albo.　4 Aſter hirſutus.*
Mountain Starwort.　　Hairy Starwort.

‡ *5 Aſter Conyzoides Geſneri.*
Fleabane Starwort.

‡ 6 *Aſter luteus ſupinus Cluſij.*
Creeping Starwort.

‡ 7 *Aſter luteus folijs ſucciſa.*
Scabious leafed Star-wort.

‡ 8 *Aſter Salicis folio.*
Willow-leaued Starwort.

‡ 9 *Aſter Auſtriacus 5. Cluſ.*
Sallow leaued Starwort.

‡ 12　There are kept in the gardens of Mr *Tradescant*, Mr *Tuggye*, and others, two Starworts differing much from all these formerly mentioned : the first of them is to be esteemed for that it floures in October and Nouember,when as few other floures are to be found : the root is large and liuing,which sends vp many small stalks some two cubits high, wooddy, slender, and not hollow, and towards the top they are diuided into aboundance of small twiggy branches : the leaues that grow alternatly vpon the stalks,are long,narrow,and sharpe pointed,hauing foure or six scarce discernable nicks on their edges : the floures which plentifully grow on smal branches much after the manner of *Virga aurea*, consist of twelue white leaues set in a ring, with many threds in their middles,which being yong are yellow,but becomming elder and larger they are of a reddish color,and at length turne into downe. I haue thought fit to call this plant, not yet described by any that I know of,being reported to be a Virginian,by the name of *Aster Virginianus fruticosus*, Shrubby starwort.

13　This which in gardens floures some moneth before the former, grows not so high,neither are the stalks so straight,but often crooked, yet are they diuided into many branches which beare small blewish floures like those of the former : the leaues are longish and narrow. This also is said to haue come from Canada or Virginia,and it may be called *Aster fruticosus minor*,Small shrubby Starwort. ‡

‡ 10　*After 6. Clusij.*　　　　　‡ 11　*After 7. Clusij.*
Narrow leaued Starwort.　　　Dwarfe Daisie-leaued Starwort.

¶ *The Place.*
The kindes of Starwort grow vpon mountaines and hilly places,and somtimes in woods & medowes lying by riuers sides.

The two first kinds grow vpon Hampsted heath foure miles from London, in Kent vpon Southfleet Downes,and in many other such downy places.

‡　I could neuer yet find nor heare of any of these Star-floures to grow wilde in this kingdom, but haue often seen the Italian Starwort growing in gardens. These two kindes that our Author mentions to grow on Hampsted heath and in Kent,are no other than two *Hieracia* or Hawkeweeds, which are much differing from these. ‡

¶ *The Time.*
They floure from Iuly to the end of August.

　　　　　　　¶ *The*

¶ *The Names.*

This herb is called in Greeke Ἀστὴρ ἀττικὸς, and also Βουβώνιον : in Latine, *After Atticus*, *Bubonium*, and *Inguinalis* : of some, *Asterion*, *Asteriscon*, and *Hyophthalmon* : in high-Dutch, Megerkraut : in Spanish, *Bobas* : in French, *Estrille*, and *Asper goutte menne* : in English, Starwort, and Sharewort.

¶ *The Temperature.*

It is of a mean temperature in cooling and drying. *Galen* saith it doth moderatly wast and consume, especially while it is yet soft and new gathered.

That with the blew or purple floure is thought to be that which is of *Virgil* called *Flos Amellus* : of which he maketh mention, *lib.4.* of his Georgicks :

Est etiam flos in pratis, cui nomen Amello
Fecere agricolæ : facilis quærentibus herba ;
Namque vno ingentem tollit de cespite sylvam :
Aureus ipse, sed in folijs, quæ plurima circum
Funduntur, violæ sublucet purpura nigra.

In English thus :

In Medes there is a floure *Amello* nam'd,
By him that seeks it easie to be found,
For that it seems by many branches fram'd
Into a little wood : like gold the ground
Thereof appeares ; but leaues that it beset
Shine in the colour of the Violet.

¶ *The Vertues.*

A   The leaues of *After* or *Inguinalis* stamped, and applied vnto botches, impostumes, and venereous bubones, which for the most part happen in *Inguine*, that is, the flanke or share) do mightily maturate and suppurate them, whereof this herb *After* tooke the name *Inguinalis*.

B   It helpeth and preuaileth against the inflammation of the fundament, and the falling forth of the gut called *Saccus ventris*.

C   The floures are good to be giuen vnto children against the squinancie and falling sicknesse.

† The figure which formerly was in the second place vnder the title of *After Atticus*, was of the eighth here described ; also in the third place formerly were these two figures which we here giue you, whereof the former is of *After montanus*, and the later of *After hirsutus* ; and that which was vnder the title of *After hirsutus* in the fourth place, belongs to the ninth description,

---

CHAP. 133. *Of Woad.*

¶ *The Description.*

1   GLastum or garden Woad hath long leaues of a blewish green colour. The stalk growes two cubits high, set about with a great number of such leaues as come vp first, but smaller, branching it selfe at the top into many little twigs, whereupon do grow many small yellow floures : which being past, the seed comes forth like little blackish tongues. The root is white and single.

2   There is a wild kind of Woad very like vnto the former in stalks, leaues, and fashion, sauing that the stalke is tenderer, smaller, and browner, and the leaues and tongues narrower ; otherwise there is no difference betwixt them.

¶ *The Place.*

The tame or garden Woad growes in fertile fields where it is sowne : the wilde kinde groweth where the tame hath bin sowne.

¶ *The Time.*

They floure from Iune to September.

¶ *The Names.*

Woad is called in Greeke Ἰσάτις : in Latine, *Isatis*, and *Glastum*. *Cæsar*, *lib.5.* of the French warres, saith, That all the Brittons do colour themselues with Woad, which giueth a blew colour : which thing also *Pliny*, *lib.22. cap.1.* doth testifie : In France they call it *Glastum*, which is like vnto Plantaine, wherewith the Brittish wiues and their daughters are colored all ouer, and go naked in some kinde of sacrifices. It is likewise called of diuers *Guadum* : of the Italians, *Guado*, a word as it seemeth wrung out of the word *Glastum* : in Spanish and French *Pastel* : in Dutch, weet : in English, Woad, a nd Wade.

¶ *The*

1 *Glaſtum ſativum.*
Garden Woad.

‡ 2 *Glaſtum ſylveſtre.*
Wilde Woad.

¶ *The Temperature.*

Garden Woad is dry without ſharpneſſe : the wilde Woad drieth more, and is more ſharp and biting.

¶ *The Vertues.*

The decoction of Woad drunken is good for ſuch as haue any ſtopping or hardneſſe in the milt   A
or ſpleen, and is alſo good for wounds or vlcers in bodies of a ſtrong conſtitution, as of countrey people, and ſuch as are accuſtomed to great labour and hard courſe fare.

It ſerueth well to dye and colour cloath, profitable to ſome few, and hurtfull to many.   B

CHAP. 134. *Of Cow Baſill.*

¶ *The Deſcription.*

1   THis kinde of wilde Woad hath fat long leaues like *Valeriana rubra Dodonæi,* or *Behen rubrum :* the ſtalke is ſmall and tender, hauing thereupon little purple floures conſiſting of foure leaues ; which being paſt, there come ſquare cornered husks ful of round black ſeed like Coleworts. The whole plant is couered ouer with a clammy ſubſtance like birdlime, ſo that in hot weather the leaues thereof will take flies by the wings (as *Muſcipula* doth) in ſuch manner as they cannot eſcape away.

2   *Ephemerum Matthioli* hath long fat and large leaues like vnto Woad, but much leſſe: among which riſeth vp a round ſtalke a cubit high, diuiding it ſelfe into many branches at the top, which are ſet with many ſmall white floures conſiſting of many leaues ; which being paſt, there follow little round bullets containing the ſeed. The root is ſmall and full of fibres.

¶ *The Place.*

Cow-Baſill groweth in my garden, but *Ephemerum* is a ſtranger as yet in England.

¶ *The Time.*

They floure in May and Iune.

¶ *The*

1 *Vaccaria.*
Cow Baſill.

2 *Ephemerum Matthioli.*
Quick-fading floure.

‡ ¶ *The Names.*

1   Cow-Baſill is by *Cordus* called *Thameenemon* : by ſome, according to *Geſner, Lychnis,* and *Per-foliata rubra. Lobel* termes it *Iſatis ſylueſtris,* and *Vaccaria* : the laſt of which names is retained by moſt late writers.

2   This by *Lobel* is ſaid to be *Ephemerum* of *Matthiolus* : yet I think *Matthiolus* his figure, which was in this place formerly, was but a counterfeit ; and ſo alſo do *Columna* and *Bauhinus* iudge of it ; and *Bauhine* thinks this of *Lobel* to be ſome kinde of *Lyſimachia.*

¶ *The Nature and Vertues.*

I finde not any thing extant concerning the Nature and Vertues of *Vaccaria* or Cow-Baſill.

A   *Ephemerum* (as *Dioſcorides* writeth) boiled in Wine, and the mouth waſhed with the decoction thereof, taketh away the tooth-ache.

Chap. 135.

Of Seſamoides, or baſtard Weld or Woad.

¶ *The Deſcription.*

1   THe great *Seſamoides* hath very long leaues and many, ſlender toward the ſtalk, and broa-der by degrees toward the end, placed confuſedly vpon a thick ſtiffe ſtalke : on the top whereof grow little fooliſh or idle white floures : which being paſt, there follow ſmall ſeeds like vnto Canary ſeed that birds are fed withall. The root is thick, and of a woody ſubſtance.

‡ 2   This leſſer *Seſamoides* of Salamanca, from a long liuing white hard and pretty thicke root ſends vp many little ſtalkes ſet thicke with ſmall leaues like thoſe of Line ; and from the mid-dle to the top of the ſtalke grow many floures, at firſt of a greeniſh purple, and then putting forth yellowiſh threds ; out of the midſt of which appeare as it were foure greene graines, which when the floure is fallen grow into little cods full of a ſmall blackiſh ſeed. It growes in a ſtony ſoile vp-on the hills neere Salamanca, where it floures in May, and ſhortly after perfects his ſeed. ‡

3   Our

1 *Seſamoides Salamanticum magnum.*
Great baſtard Woad.

2 *Seſamoides Salamanticum parvum.*
Small baſtard Woad.

3 *Seſamoides parvum Matthioli.*
Bucks horne gum Succorie.

‡ 3　Our Authour formerly in the chapter
of *Chondrylla* ſpoke (in *Dodonæus* his words) a-
gainſt the making of this plant a *Seſamoides* : for
of this plant were the words of *Dodonæus* ; which
are theſe : Diuers (ſaith he) haue taken the plant
with blew floures to be *Seſamoides parvum*, but
without any reaſon ; for that *Seſamoides* hath borrowed his name from the likeneſſe it hath with
Seſamum.　But this herb is not like to *Seſamum* in
any one point , and therefore I thinke it better
referred vnto the Gum Succories ; for the floures
haue the forme and colour of gum Succorie, and
it yeeldeth the like milky juice.　Our Author it
ſeemes was either forgetfull or ignorant of what
he had ſaid ; for here hee made it one, and deſcribed it meerely by the figure and his own fancie.
Now I following his traét, haue (thongh vnfitly)
put it here, becauſe there was no hiſtorie nor figure of it formerly there, but both heere, though
falſe and vnperfeét. This plant hath a root ſomewhat like that of Goats-beard ; from which ariſe
leaues rough and hairy, diuided or cut in on both
ſides after the manner of Bucks horne, and larger
than they. The ſtalke is ſome foot high, diuided
into branches, which on their tops carry flours of
a faire blew colour like thoſe of Succory, which
ſtand in rough ſcaly heads like thoſe of Knapweed. ‡

¶ *The*

¶ *The Place.*

Theſe grow in rough and ſtony places,but are all ſtrangers in England.

¶ *The Time.*

They floure in May and Iune,and ſhortly after ripen their ſeed.

‡ ¶ *The Names.*

I thinke none of theſe to be the *Seſamoides* of the Antients. **1** The firſt is ſet forth by *Cluſius* vnder the name we here giue you: it is the *Muſcipula altera muſcoſo flore* of Lobel: *Viſcago major* of *Camerarius.*

2 This alſo *Cluſius* and *Lobel* haue ſet forth by the ſame name as we giue you them.

3 *Matthiolus,Camerarius,*and others haue ſet this forth for *Seſamoides parvum,* in the *Hiſt.Lugd.* it is called *Catanance quorundam* ; but moſt fitly by *Dodon. Chondryllæ ſpecies tertia,* The third kind of Gum Succorie. ‡

¶ *The Temperature.*

Galen affirmeth,That the ſeed contains in it ſelfe a bitter quality,and ſaith that it heateth,breaketh,and ſcoureth.

¶ *The Vertues.*

A   *Dioſcorides* affirmeth, that the weight of an halfepenny of the ſeed drunke with Mead or honied water,purgeth flegme and choler by the ſtoole.

B   The ſame being applied,doth waſte hard knots and ſwellings.

† That which here formerly enioyed the third place,by the title of *Seſamoides maius Scaligeri,* was no other then the plant that is hereafter deſcribed, by the name of *Turton-ratre Gallo-prouincia,* where you may finde both the figure and deſcription.

Cʜᴀᴘ. 136. *Of Dyers Weed.*

Lutcola. Dyers weed,or Yellow weed.

¶ *The Deſcription.*

DYers weed hath long narrow and greeniſh yellow leaues not much vnlike to Woad, but a great deale ſmaller and narrower;from among which commeth yp a ſtalke two cubits high, beſet with little narrow leaues:euen to the top of the ſtalk come forth ſmall pale yellow floures,cloſely cluſtering together one within another,which doe turne into ſmall buttons, cut as it were croſſe-wiſe, wherein the ſeed is contained. The root is very long and ſingle.

¶ *The Place.*

Dyers weed groweth of it ſelfe in moiſt barren and vntilled places, in and about villages almoſt euerie where.

¶ *The Time.*

This herb flouriſheth in Iune and Iuly.

¶ *The Names.*

Pliny, lib.33.cap.5. maketh mention by the way of this herb,and calleth it *Lutea* : and *Vitruvius,li.7. Lutum* : it is the *Antirrhinum* of *Tragus* : and *Pſeudoſtruthium* of *Matthiolus. Virgil* in his Bucolicks, Eclog 4. calls it alſo *Lutum* : in Engliſh, Weld, or Dyers weed.

¶ *The Nature.*

It is hot and dry of temperature.

‡ ¶ *The Vertues.*

A   The root as alſo the whole herb heates and dries in the third degree : it cuts, attenuates, reſolueth,opens,digeſts. Some alſo commend it againſt the punctures and bites of venomous creatures,

tures, not only outwardly applied to the wound, but also taken inwardly in drinke.

Also it is commended against infection of the plague : some for these reasons term it *Theriaca-* B *ria. Mat.* ‡

Chap. 137.   Of Staues-acre.

Staphis-agria.   Staues-acre.

¶ *The Description.*

S Saues-acre hath straight stalks of a browne colour, with leaues clouen or cut into sundry sections, almost like the leaues of the wilde Vine : the floures grow vpon short stems, fashioned somwhat like vnto our common Monkes hood, of a perfect blew colour : which beeing past, there succeed welted huskes like those of Wolfs-bane, wherein is contained triangular brownish rough seed. The root is of a wooddy substance, and perisheth when it hath perfected his seed.

¶ *The Place.*

It is with great difficultie preserued in our cold countries, albeit in some milde Winters I haue kept it couered ouer with a little Ferne, to defend it from the injurie of the March wind, which doth more harm to plants that come forth of hot countries, than doth the greatest frosts.

¶ *The Time.*

It floureth in Iune, and the seed is ripe the second yeare of his sowing.

¶ *The Names.*

It is called in Greeke ⲥⲧⲁⲫⲓⲥ ⲁⲅⲣⲓⲁ : in Latine, *Herba Pedicularis*, and *Peduncularia*, as *Marcellus* reporteth. *Pliny, lib.26. cap.13.* seems to name it *Vua taminia :* of some, *Pituitaria,* and *Passula montana :* in shops, *Staphisagria :* in Spanish, *Yerua piolente :* in French, *Herbe aux poulx :* in high-Dutch, **Leng kraut:** in low-Dutch, **Lupsruit :** in English, Staues-acre, Louse-wort, and Louse-pouder.

¶ *The Temperature.*

The seeds of Staues-acre are extreame hot, almost in the fourth degree, of a biting and burning qualitie.

¶ *The Vertues.*

Fifteen seeds of Staues-acre taken with honied water will cause one to vomit grosse flegme and A slimy matter, but with great violence ; and therefore those that haue taken them ought to walke without staying, and to drinke honied water, becaufe it bringeth danger of choking and burning the throat, as *Dioscorides* noteth : for which cause they are rejected and not vsed of Physitions, either in prouoking vomit, or else in mixing them with other inward medicines.

The seed mingled with oile or greafe driueth away lice from the head, beard, and all other parts B of the body, and cureth all scuruy itch and manginesse.

The same boiled in vineger and holden in the mouth, asswageth the tooth-ache. C

The same chewed in the mouth draweth forth much moisture from the head, and cleanseth the D brain, especially if a little of the root of Pellitorie of Spain be added thereto.

The same tempered with vineger is good to be rubbed vpon lousie app arell, to destroy & driue E away lice.

The feeds hereof are perillous to be taken inwardly without good aduice, and correction of the F same ; and therefore I aduise the ignorant not to be ouer-bold in medling with it, sith it is so dangerous that many times death ensueth vpon the taking thereof.

Chap.

Chap. 138. *Of Palma Chriſti.*

¶ *The Deſcription.*

1   R Icinus, *Palma Chriſti,* or *Kik,* hath a great round hollow ſtalke fiue cubits high, of a browne colour, died with a blewiſh purple vpon green. The leaues are great and large, parted into ſundry ſections or diuiſions, faſhioned like the leaues of a figtree, but greater, ſpred or wide open like the hand of a man, and hath toward the top a bunch of flours cluſtring together like a bunch of grapes ; whereof the loweſt are of a pale yellow colour, and wither away without bearing any fruit ; and the vppermoſt are reddiſh, bringing forth three cornered huskes which containe the ſeed as big as a kidney bean, of the colour and ſhape of a certaine vermin which haunteth cattell, called a Tik.

2   This *Palma Chriſti* of America growes vp to the height and bigneſſe of a ſmal tree or hedge ſhrub, of a wooddy ſubſtance, whoſe fruit is expreſſed by the figure, being of the bignes of a great bean, ſomewhat long, and of a blackiſh colour, rough and ſcaly.

1 *Ricinus.*
Palma Chriſti.

2 *Ricinus Americanus.*
Palma Chriſti of America.

¶ *The Place.*

The firſt kinde of *Ricinus* or Palma Chriſti groweth in my garden, and in many other gardens likewiſe.

¶ *The Time.*
Ricinus or Kik is ſowne in Aprill, and the ſeed is ripe in the end of Auguſt.

The Name, and cauſe thereof.
Ricinus (whereof mention is made in the fourth chapter and ſixt verſe of the prophecie of *Ionas*)
was

was called of the Talmudists, כ *Kik*, for in the Talmud we reade thus, כ ושמען ולא *Velo beschemen Kik:* that is, in English, And not with the oyle of *Kik*: which oile is called in the Arabian tongue, *Alkerna*, as *Rabbi Samuel* the sonne of *Hophni* testifieth. Moreouer a certaine Rabbine moueth a question, saying, what is *Kik?* Hereunto *Resch Lachish* maketh answer in Ghemara, saying, Kik is nothing else but *Ionas* his Kikaijon. And that this is true, it appeareth by that name κικι: which the Antient Greeke Physitions, and the Ægiptians vsed; which Greeke word commeth of the Hebrew word *Kik.* Hereby it appeareth, that the old writers long agoe called this plant by the true and proper name. But the old Latine writers knew it by the name *Cucurbita*, which euidently is manifested by an history which Saint *Augustine* recordeth in his Epistle to Saint *Ierome*, where in effect he writeth thus ; That name *Kikaijon* is of small moment, yet so small matter caused a great tumult in Africa. For on a time a certaine Bishop hauing an occasion to intreat of this which is mentioned in the fourth chapter of *Ionas* his prophesie (in a collation or sermon, which he made in his cathedrall church or place of assembly) said, that this plant was called *Cucurbita*, a Gourd, because it encreased to so great a quantitie, in so short a space, or else (saith he) it is called *Hedera.* Vpon the nouelty and vntruth of this his doctrine, the people were greatly offended, and thereof suddenly arose a tumult and hurly burly , so that the Bishop was enforced to goe to the Iewes, to aske their iudgement as touching the name of this plant. And when he had receiued of them the true name, which was *Kikaijon*, he made his open recantation, and confessed his errour, and was iustly accused for a falsifier of the holy Scripture. ‡ The Greekes called this plant also κροτων : i. *Ricinus*, by reason of the similitude that the seed hath with that insect, to wit, a Tik. ‡

¶ *The Nature.*

The seed of Palma Christi, or rather *Kik*, is hot and dry in the third degree.

¶ *The Vertues.*

Ricinus his seed taken inwardly, openeth the belly, and causeth vomit, drawing slimy flegme and choler from the places possessed therewith.   **A**

The broth of the meat supped vp, wherein the seed hath beene sodden, is good for the colicke and the gout, and against the paine in the hips called *Sciatica*: it preuaileth also against the iaundise and dropsie.   **B**

The oyle that is made or drawne from the seed is called *Oleum Cicinum* : in shops it is called, *Oleum de Cherua* : it heateth and drieth, as was said before, and is good to anoint and rub all rough hardnesse and scuruinesse gotten by itch.   **C**

This oyle, as *Rabbi David Chimchi* writeth, is good against extreme coldnesse of the body.   **D**

CHAP. 139. *Of Spurge.*

¶ *The Description.*

1   THe first kinde of Sea Spurge riseth forth of the sands, or baich of the sea, with sundry reddish stems or stalkes growing vpon one single root, of a wooddy substance : and the stalkes are beset with small, fat, and narrow leaues like vnto the leaues of Flax. The floures are yellowish, and grow out of little dishes or Saucers like the common kinde of Spurge. After the floures come triangle seeds, as in the other Tithymales.

2   The second kinde (called *Helioscopius*, or *Solisequium* : and in English, according to his Greeke name, Sunne Spurge, or time Tithymale, of turning or keeping time with the Sunne) hath sundry reddish stalkes of a foot high: the leaues are like vnto Purslane, not so great nor thicke, but snipt about the edges : the floures are yellowish, and growing in little platters.

3   The third kinde hath thicke, fat, and slender branches trailing vpon the ground, beset with leaues like Knee-holme, or the great Myrtle tree. The seed and floures are like vnto the other of his kinde.

4   The fourth is like the last before mentioned, but it is altogether lesser, and the leaues are narrower ; it groweth more vpright, otherwaies alike.

5   Cyprus Tithymale hath round reddish stalkes a foot high, long and narrow like those of Flax, and growing bushie, thicke together like as those of the Cyprus tree. The floures, feed, and root, are like the former, sometimes yellow, oftentimes red.

6   The sixt is like the former, in floures, stalkes, roots, and seeds, and differeth in that, this kinde hath leaues narrower, and much smaller, growing after the fashion of those of the Pine tree, otherwise it like.

7   There is another kinde that groweth to the height of a man , the stalke is like the last

1 *Tithymalus paralius.*
Sea Spurge.

2 *Tithymalus Helioſcopius.*
Sunne Spurge.

3 *Tithymalus Myrtifolius latifolius.*
Broad leaued Myrtle Spurge.

4 *Tithymalus Myrſinitis anguſtifolius.*
Narrow leaued Myrtle Spurge.

5 *Tithymalus Cupreßinus.*
Cypreſſe Spurge.

6 *Tithymalus Pineus.*
Pine Spurge.

† 7 *Tithymalus Myrſinites arboreſcens.*
Tree Myrtle Spurge.

† 8 *Tithymalus Characias Monſpell.*
Sweet wood Spurge.

† 9 *Tithymalus Characias amygdaloides.*
Vnſauorie Wood-ſpurge.

‡ 10 *Tithymalus Characias anguſtifolius.*
Narrow leaued Wood-ſpurge.

‡ 11 *Tithymalus Characias ſerratifolius.*
Cut leaued Wood-ſpurge.

12 *Tithymalus platyphyllos.*
Broad leaued Spurge.

mentioned, but diuided into sundry branches a finger thicke, and somewhat hairy, not red as the others, but white: the leaues be long and narrow, whitish and a little downy: the floures are yellow, but in the other point like to the rest of this kinde.

8   The eighth kinde riseth vp with one round reddish stalke two cubits high, set about with long, thin, and broad leaues like the leaues of the Almond tree: the floures come forth at the top like the others, and of a yellow colour. The seed and root resemble the other of his kinde.

9   The ninth (which is the common kinde growing in most woods) is like the former, but his leaues be shorter and lesse, yet like to the leaues of an Almond tree: the floures are also yellow, and the seed contained in three cornered seed-vessels.

‡ 10   This fourth kinde of *Tithymalus Characias*, or Valley Tithymale (for so the name imports) hath long, yet somewhat narrower leaues than the former, whitish also, yet not hoary; the vmbels or tufts of floures are of a greenish yellow, which before they be opened doe represent the vmbels or tufts of floures are of a greenish yellow, which before they be opened doe represent the shape of a longish fruit, as an Almond, yet in colour it is like the rest of the leaues: the floures and seeds are like those of the former, and the root descends deepe into the ground.

11   The fifth *Characias* hath also long leaues sharpe pointed, and broader at their setting on, and of a light greene colour, and snipt or cut about the edges like the teeth of a saw. The vmbels are smaller, yet carry such floures and seeds as the former. ‡

12   This kinde hath great broad leaues like the young leaues of Woad, set round about a stalk of a foot high, in good order: on the top whereof grow the floures in small platters like the common kinde, of a yellow colour declining to purple. The whole plant is full of milke, as are all the rest before specified.

‡ 13   *Tithymalus Dendroides ex Cod. Casareo.*   14   *Esula maior Germanica.*
       Great Tree Tithymale.   Quacksaluers Turbith.

13   There is another kinde of Tithymale, whose figure was taken forth of a Manuscript of the Emperors by *Dodoneus*, that hath a stalke of the bignesse of a mans thigh, growing like a tree vnto the height of two tall men, diuiding it selfe into sundry armes or branches toward the top, of a red colour. The leaues are small and tender, much like vnto the leaues of *Myrtus*: the seed is like vnto that of wood Tithymale, or *Characias*, according to the authority of *Peter Bellone*.

14   There is a kinde of Tithymale called *Esula maior*, which *Martinus Rulandus* had in great

veneration, as by his extraction which he vſed for many infirmities may and doth appeare at large, in his booke entituled *Centuria curationum Empiricarum*, dedicated vnto the duke of Bauaria. This plant of *Rulandus* hath very great and many roots couered ouer with a thicke barke, plaited as it were with many ſurculous ſprigs; from which ariſe ſundry ſtrong and large ſtemmes of a fingers thickeneſſe, in height two cubits, beſet with many pretty large and long leaues like *Lathyris*, but that they are not ſo thicke : the ſeed and floure are not vnlike the other Tithymales.

15   This is like the fifth, ſaue that it hath ſmaller and more feeble branches; and the whole plant is altogether leſſer, growing but a ſpan or ſome foot high; and the floures are of a red or els a greene colour.

16   There is another rare and ſtrange kinde of *Eſula*, in alliance and likeneſſe neere vnto *Eſula minor*, that is the ſmall *Eſula* or *Pityuſa* vſed among the Phyſitions and Apothecaries of Venice as a kind of *Eſula*, in the confection of their *Bened:cta* and Catharticke pills, in ſtead of the true *Eſula* : It yeeldeth a fungous, rough, and browne ſtalke two cubits high, diuiding it ſelfe into ſundry branches, furniſhed with ſtiffe and ſat leaues like Liquorice, growing together by couples. The floures are pendulous, hanging downe their heads like ſmall bells, of a purple colour, and within they are of a darke colour like *Ariſtolochia rotunda*.

† 15 *Eſula minor, ſeu Pityuſa.*   ‡ 16 *Eſula Veneta maritima.*
Small Eſula.   Venetian Sea-Spurge.

‡ 17   There growes in many chalkie grounds and ſuch dry hilly places, among corne, a ſmall Spurge which ſeldome growes to two handfuls high; the root is ſmall, and ſuch alſo are the ſtalkes and leaues, which grow pretty thicke thereon; which oft times are not ſharpe, but flat pointed : the ſeed-veſſels and floures are very ſmall, yet faſhioned like thoſe of the other Tithymales. It is to be found in corne fields in Iuly and Auguſt. ‡

18   The bigger *Cataputia* or the common garden Spurge is beſt knowne of all the reſt, and moſt vſed; wherefore I will not ſpend time about his deſcription.

The ſmall kinde of *Cataputia* is like vnto the former, but leſſer, whereby it may eaſily be diſtinguiſhed; being ſo well knowne vnto all, that I ſhall not need to deſcribe it.

‡ Theſe two (I meane the bigger and leſſer *Cataputia* of our Author) differ not but by reaſon of their age, and the fertileneſſe and barrenneſſe of the ſoile, whence the leaues are ſometimes broader, and otherwhiles narrower. ‡

‡ 17 *Eſula exigua Tragi.*
Dwarfe Eſula.

18 *Lathyris ſeu Cataputia minor.*
Garden Spurge.

19 *Peplus, ſive Eſula rotunda.*
Pettie Spurge.

20 *Peplis.*
Iſope Spurge.

21 *Chamæſyce.*
Spurge Time.

22 *Apios vera.*
Knobbed Spurge.

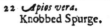

‡ 23 *Apios radice oblonga.*
Long knottie rooted Spurge.

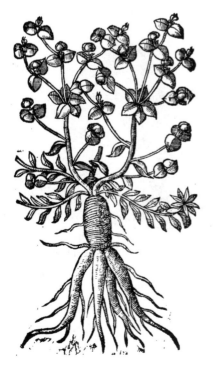

19   The nineteenth kind called *Peplus*, hath a ſmall, and fibrous root, bringing forth many fruitfull branches two handfuls long, but little and tender, with leaues like the Sun Tithymale, but rounder and much ſmaller : it hath alſo ſmall yellow floures : which being paſt there appeareth a ſlender pouchet, three cornered like the other Tithymales, hauing within it a very medullous whitiſh ſeed like Poppie : the whole plant yeeldeth a milky juyce, which argueth it to be a kinde of Tithymale.

20   As in name ſo in ſhape this twentieth reſembleth *Peplus*, and commeth in likelihood neerer the ſignification of *Peplum*, or *Flammeolum* than the other; therefore *Dioſcorides* affirmeth it to be *Thamnos amphilaphes*, for that it bringeth forth a greater plenty of branches, more cloſely knit and wound together, with ſhining twiſts and claſpers an handfull and a halfe long. The leaues are leſſer than thoſe of *Peplus*, of an indifferent likeneſſe and reſemblance betweene *Chamæſyce* and wilde Purſlane. The ſeed is great and like that of *Peplus* : the root is ſmall and ſingle.

21   The one and twentieth kinde may be eaſily knowne from the two laſt before mentioned, although they be very like. It

hath

hath many branches and leaues creeping on the ground, of a pale greene colour, not vnlike to *Herniaria*, but giuing milke as all the other Tithymales doe, bearing the like seed, pouch, and floures, but smaller in each respect.

22   The two and twentieth kinde of Tithymale hath a round root like a small Turnep, as euery Author doth report : yet my selfe haue the same plant in my garden which doth greatly increase, of which I haue giuen diuers vnto my friends, whereby I haue often viewed the roots, which do appeare vnto me somewhat tuberous, and therein nothing answering the descriptions which *Dioscorides, Pena*, and others haue expressed and set forth. This argueth, that either they were deceiued, and described the same by hearesay, or else the plant doth degenerate being brought from his natiue soile. The leaues are set all alongst a small rib like *Fraxinella*, somewhat round, greene aboue, and reddish vnderneath. The seed groweth among the leaues like the seed of *Peplus*. The whole plant is full of milke like the other Tithymales.

‡   Our Author here wrongfully taxes other Writers of plants, and *Dioscorides* and *Pena* by name, which shewes that hee either neuer read, or else vnderstood not what they writ, for neither of them (nor any other that I know of) resembles the root of this to a Turnep, but say it hath a tuberous peare fashioned root, &c. as you may see in *Diosc. lib.4.cap.177.* and in the *Aduersaria, pag.204.* The leaues also grow not by couples one against another, as in *Fraxinella*, but rather alternately, or else without any certaine order, as in other Tithymales.

23   This, saith *Clusius*, hath also a tuberous root, but not peare fashioned like as the former, but almost euery where of an equall thickenesse ; being about an inch and sometimes two inches long, and the lower part thereof is diuided into foure other roots, or thicke fibres, growing smaller by little and little, and sending forth some few fibres : it is blacke without, and white within, and full of a milky juyce : the stalkes are short and weake, set with little leaues like those of the former : the floures are of a yellowish red colour, and the seed is contained in such vessels as the other Tithymales. This is *Tithymalus tuberosus*, or *Ischas altera* of *Clusius*.

¶ *The Place.*

The first kinde of Spurge groweth by the sea side vpon the rowling Sand and Baich, as at Lee in Essex, at Lang-tree point right against Harwich, at Whitstable in Kent, and in many other places.

The second groweth in grounds that lie waste, and in barren earable soile, almost euery where.

The third and fourth, as also the fourteenth and eighteenth, grow in Gardens, but not wilde in England.

The ninth Spurge called *Characias* groweth in most Woods of England that are drie and warme.

The eighteenth and nineteenth grow in salt marshes neere the sea, as in the Isle of Thanet by the sea side, betweene Reculuers and Margate in great plenty.

¶ *The Time.*

These plants floure from Iune to the end of Iuly.

¶ *The Names.*

Sea Spurge is called in Latine *Tithymalus paralius* : in Spanish, *Leche tresua* : in high Dutch, **Wolfet milch** : that is to say, *Lupinum lac*, or Wolfes milke. Wood Spurge is called *Tithymalus characias*. The first is called in English, Sea Spurge, or sea Wartwoort. The second, Sun Spurge ; the third and fourth, Mirtle Spurge : the fifth Cypresse Spurge ; or among Women, Welcome to our house ; the sixth, Pine Spurge ; the seuenth, Shrub Spurge, and tree Mirtle Spurge ; the eighth and ninth, Wood Spurge ; the twelfth, Broad leafed Spurge ; the thirteenth, Great Tree Spurge ; the fourteenth and fifteenth, Quacksaluers Spurge; the sixteenth, Venice Spurge, the seuenteenth, Dwarfe Spurge ; the eighteenth, Common Spurge ; the nineteenth and twentieth, Pettie Spurge; the one and twentieth, Spurge Time ; the two and twentieth, True *Apios* or the knobbed Spurge.

¶ *The Temperature.*

All the kinds of Tithymales or Spurges are hot and dry almost in the fourth degree, of a sharpe and biting quality, fretting or consuming. First the milke and sap is in speciall vse, then the fruit and leaues, but the root is of least strength. The strongest kinde of Tithymale, and of greatest force is that of the sea.

⸪ Some write by report of others, that it enflameth exceedingly, but my selfe speak by experience ; for walking along the sea coast at Lee in Essex, with a Gentleman called M*r Rich*, dwelling in the same towne, I tooke but one drop of it into my mouth ; which neuerthelesse did so inflame and swell in my throte that I hardly escaped with my life. And in like case was the Gentleman, which caused vs to take our horses, and poste for our liues vnto the next farme house to drinke some milke to quench the extremitie of our heat, which then ceased.

¶ *The*

¶ *The Vertues.*

A   The juyce of Tithymale, I do not meane ſea Tithymale, is a ſtrong medicine to open the belly, and cauſing vomit, bringeth vp tough flegme and cholericke humours. Like vertue is in the ſeed and root, which is good for ſuch as fall into the dropſie, being miniſtred with diſcretion and good aduice of ſome excellent Phyſition, and prepared with his Correctories by ſome honeſt Apothecarie.

B   The juyce mixed with hony, cauſeth haire to fall from that place which is anointed therewith, if it be done in the Sun.

C   The juyce or milke is good to ſtop hollow teeth, being put into them warily, ſo that you touch neither the gums, nor any of the other teeth in the mouth with the ſaid medicine.

D   The ſame cureth all roughneſſe of the skin, mangineſſe, leprie, ſcurfe, and running ſcabs, and the white ſcurfe of the head. It taketh away all manner of warts, knobs, and the hard callouſneſſe of Fiſtulaes, hot ſwellings and Carbuncles.

E   It killeth fiſh, being mixed with any thing that they will eat.

F   Theſe herbes by mine aduiſe would not be receiued into the body, conſidering that there be ſo many other good and wholeſome potions to be made with other herbes, that may be taken without perill.

† The ſiuenth figure was formerly of *Tithymalus myrſinites 3-anguſtifolius* of *Tabernamontanus.* The 8. and 9. were both of the ſame plant : the 12. was the figure of the *Eſula exigua Tragi*, whoſe hiſtory I haue giuen you in the 17. place.

C H A P. 140.   *Of Herbe Terrible.*

1 *Alypum montis Ceti.*   1 *Tarton-Raire Gallo-Prouinciæ.*
Herbe Terrible.   Gutwoort.

¶ *The Deſcription.*

1   HErbe Terrible is a ſmall ſhrub two or three cubits high, branched with many ſmall twigs, hauing a thinne rinde, firſt browne, then purple, with many little and thinne leaues like Myrtle. The floures are rough like the middle of Scabious floures, of a blew purple colour. The root is two fingers thicke, browne of colour, and of a wooddy ſubſtance: the whole plant very bitter, and of an vnpleaſant taſte like *Chamelæa*, yet ſomewhat ſtronger.

2   Tartonraire, called in Engliſh Gutwoort, groweth by the ſea, and is Catharticall, and a ſtranger with vs. In the mother tongue of the Maſſilians, it is called Tartonraire, of that abundant

and

and vnbridled faculty of purging, which many times do cause *Dysenteria*, and such like immode-
rate fluxes, especially when one not skilfull in the vse thereof shall administer the pouder of the
leaues mixed with any liquor. This plant groweth in manner of a shrub, like *Chamelæa*, and bringeth forth many small, tough, and pliant twigs, set about with a thin and cottony hairinesse, and
hath many leaues of a glistering siluer color, growing from the lowest part euen to the top, altogether like *Alypum* before mentioned: and vpon these tough and thicke branches (if my memory faile
not) do grow small floures, first white, afterward of a pale yellow : the seed is of a russet colour : the
root hard and wooddy, not very hot in the mouth, leauing vpon the tongue some of his inbred heat
and taste, somewhat resembling common Turbith, and altogether without milke.

¶ *The Place.*

These plants do grow vpon the mountaines in France, and other places in the grauelly grounds,
and are as yet strangers in England.

¶ *The Time.*

They flourish in August and September. ‡ The first *Clusius* found flouring in diuerse parts of
Spaine, in February and March ; and I conjecture the other floures about the same time, yet I can
finde nothing said thereof in such as haue deliuered the history of it. ‡

¶ *The Names.*

There are not any other names appropriate to these plants more than are set forth in the titles.
‡ The first of these is the *Alypum montis Ceti*, and *Herba terribilis* of *Lobel*; *Clus.* cals it, *Hippoglossum
Valentinum*, and in *Hist. Lugd.* it is named *Alypum Penæ*, and *Empetrum Phacoides*. The second is the
Tartonraire Galloprouinciæ Massiliensium, in the *Aduersaria*; *Sesamoides majus multorum* of *Dalec.* and the
Sesamoides maius Scaligeri of *Tabern.* by which title our Author also gaue his figure, in the 397. pag.
of the former edition. ‡

¶ *The Temperature and Vertues.*

There is nothing either of their nature or vertues, more than is set forth in the Descriptions.
‡ Both these plants haue a strong purging faculty like as the Tithymales ; but the later is far
more powerfull, and comes neere to the quality of *Mezereon* ; wherefore the vse of it is dangerous,
by reason of the violence and great heat thereof. ‡

Chap. 141.   *Of Herbe Aloe, or Sea Houseleeke.*

‡ 1 *Aloe vulgaris, siue Sempervivum marinum.*   2 *Aloe folio mucronato.*
Common *Aloe*, or Sea-Housleeke.   Prickly herbe Aloe, or Sea-Housleeke.

¶ *The Deſcription.*

1 HErbe Aloe hath hath leaues like thoſe of ſea Onion, very long, broad, ſmooth, thicke, bending backewards, notched in the edges, ſet with certaine little blunt prickles, full of tough and clammie juyce like the leaues of Houſleeke. The ſtalke, as *Dioſcorides* ſaith, is like to the ſtalke of Affodill: the floure is whitiſh; the ſeed like that of Affodill, the root is ſingle, of the faſhion of a thicke pile thruſt into the ground. The whole herbe is extreame bitter, ſo is the juyce alſo that is gathered thereof.

† 2 There is another herbe Aloe that groweth likewiſe in diuers prouinces of America; the leaues are two cubits long, alſo thicker, broader, greater, and ſharper pointed than the former, and it hath on the edges far harder prickles. The ſtalke is three cubits high, and a finger thicke, the which in long cups beares violet coloured floures. †

¶ *The Place.*

This plant groweth very plentifully in India, and in Arabia, Cœloſyria, and Ægipt, from whence the juyce put into skins is brought into Europe. It groweth alſo, as *Dioſcorides* writeth, in Aſia, on the ſea coaſts, and in Andros, but not very fit for juyce to be drawn out. It is likewiſe found it Apulia, and in diuers places of Granado and Andaluſia, in Spaine, but not farre from the ſea: the juyce of this is alſo vnprofitable.

¶ *The Time.*

The herbe is alwaies greene, and likewiſe ſendeth forth branches, though it remaine out of the earth, eſpecially if the root be couered with lome, and now and then watered: for ſo being hanged on the feelings and vpper poſts of dining roomes, it doth not onely continue a long time greene, but it alſo groweth and bringeth forth new leaues: for it muſt haue a warme place in Winter time, by reaſon it pineth away if it be frozen.

¶ *The Names.*

The herbe is called in Greeke ελΛη: in Latine, and in ſhops alſo, *Aloë*: and ſo is likewiſe the juice. The plant alſo is named ἀφαιζον, ἰνζγαον, θυμπι, γεραμελίερε: but they are baſtard words: it is called ἀφαιζον becauſe it liueth not onely in the earth, but alſo out of the earth. It is named in French, *Poroquet*: in Spaniſh, *Azenar*, and *Yerua banoſa*: in Engliſh, *Aloes*, herbe *Aloes*, Sea Houſeleeke, Sea Aigrene.

The herbe is called of the later Herbariſts oftentimes *Sempervivum*, and *Sempervivum Marinum*, becauſe it laſteth long, after the manner of Houſe-leeke. It ſeemeth alſo that *Columella* in his tenth booke nameth it *Sedum*, where he ſetteth downe remedies againſt the canker-wormes in trees.

> *Profuit & plantis latices infundere amaros*
> *Marrubÿ, multoque Sedi contingere ſucco.*

In Engliſh thus:

> Liquors of Horehound profit much b'ing pour'd on trees:
> The ſame effect Sea Houſleeke works as well as theſe.

For hee reciteth the juyce of *Sedum* or Houſleeke among the bitter juices, and there is none of the Houſleekes bitter but this.

¶ *The Temperature.*

Aloë, that is to ſay, the juyce which is vſed in Phyſicke, is good for many things. It is hot, and that in the firſt or ſecond degree, but dry in the third, extreme bitter, yet without biting. It is alſo of an emplaiſticke or clammie quality, and ſomething binding, externally applied.

¶ *The Vertues.*

A It purgeth the belly, and is withall a wholeſome and conuenient medicine for the ſtomacke, if any at all be wholeſome. For as *Paulus Ægineta* writeth, when all purging medicines are hurtfull to the ſtomacke, *Aloës* onely is comfortable. And it purgeth more effectually if it be not waſhed: and if it be, it then ſtrengtheneth the ſtomacke the more.

B It bringeth forth choler, but eſpecially it purgeth ſuch excrements as be in the ſtomacke, the firſt veines, and in the neereſt paſſages. For it is of the number of thoſe medicines, which the Græcians call ιαιεαφκκυα, of the voiding away of the ordure; and of ſuch whoſe purging force paſſeth not far beyond the ſtomacke. Furthermore, *Aloës* is on enemie to all kindes of putrefactions; and defendeth the body from all manner of corruption. It alſo preſerueth dead carkaſes from putrifying;
                                 it

it killeth and purgeth away all mann ro wormes of the belly. It is good againſt a ſtinking breath proceeding from the imperfeċtion of the ſtomacke : it openeth the piles or hemorrhoides of the fundament ; and being taken in a ſmall quantity, it bringeth downe a monthly courſe : it is thought to be good and profitable againſt obſtruċtions and ſtoppings in the reſt of the intrals. Yet ſome there be who thinke, that it is not conuenient for the liuer.

One dramme thereof giuen, is ſufficient to purge. Now and then halfe a dramme or little more   C is enough.

It healeth vp greene wounds and deepe ſores, clenſeth vlcers, and cureth ſuch ſores as are hardly   D to be helped, eſpecially in the fundament and ſecret parts. It is with good ſucceſſe mixed with aequis, or medicines which ſtanch bleeding, and with plaiſters that be applied to bloudy wounds ; for it helpeth them by reaſon of his emplaiſticke qualitie and ſubſtance. It is profitably put into medicines for the eies, foraſmuch as it clenſeth and drieth without biting.

Dioſcorides ſaith, that it muſt be torrified or parched at the fire, in a cleane and red hot veſſell,   E and continually ſtirred with a *Spatula*, or Iron Ladle, till it bee torrified in all the parts alike : and that it muſt alſo bee waſhed, to the end that the vnprofitable and ſandie droſſe may ſinke downe vnto the bottome, and that which is ſmooth and moſt perfeċt be taken and reſerued.

The ſame Author alſo teacheth, that mixed with honie it taketh away blacke and blew ſpots,   F which come of ſtripes : that it helps the inward ruggednes of the eye-lids, and itching in the corners of the eyes : it remedieth the head-ache, if the temples and forehead be anointed therewith, being mixed with vineger and oile of Roſes : being tempered with wine, it ſtaieth the falling off of the haire, if the head be waſhed therewith : and mixed with wine and honie, it is a remedie for the ſwelling of the Vvula, and ſwelling of the Almonds of the throat, for the gums and all vlcers of the mouth.

The juice of this herbe *Aloë* (whereof is made that excellent and moſt familiar purger, called   G *Aloë Succotrina*) the beſt is that which is cleare and ſhining, of a browne yellowiſh colour : it openeth the bellie, purging cold, flegmaticke, and cholericke humours, eſpecially in thoſe bodies that are ſurcharged with ſurfetting, either of meat or drinke, and whoſe bodies are fully repleat with humours, fairing daintily, and wanting exerciſe. This *Aloës* I ſay, taken in a ſmall quantitie after ſupper (or rather before) in a ſtewed prune, or in water the quantitie of two drammes in the morning, is a moſt ſoueraigne medicine to comfort the ſtomacke, and to cleanſe and driue foorth all ſuperfluous humours. Some vſe to mixe the ſame with Cinnamon, Ginger, and Mace, for the purpoſe aboueſaid ; and for the Iaundies, ſpitting of bloud, and all extraordinarie iſſues of bloud.

The ſame vſed in vlcers, eſpecially thoſe of the ſecret parts or fundament, or made into pouder, and ſtrawed on freſh wounds, ſtaieth the bloud and healeth the ſame, as thoſe vlcers before ſpoken of.

The ſame taken inwardly cauſeth the Hemorrhoids to bleed, and beeing laid thereon it cauſeth   I them to ceaſe bleeding.

Chap. 142. *Of Houſleeke or Sengreene.*

¶ *The Kindes.*

SEngreene, as *Dioſcorides* writeth, is of three ſorts, the one is great, the other ſmall, and the third is that which is called *Illecebra*, biting Stone-crop, or Wall-pepper.

¶ *The Deſcription.*

1   THe great Sengreene, which in Latine is commonly called *Iouis barba*, Iupiters beard, bringeth forth leaues hard adioyning to the ground and root, thicke, fat, full of tough juice, ſharp pointed, growing cloſe and hard together, ſet in a circle in faſhion of an eye, and bringing forth verie many ſuch circles, ſpreading it ſelfe out all abroad : it oftentimes alſo ſendeth forth ſmall ſtrings, by which it ſpreadeth farther, and maketh new circles; there riſeth vp oftentimes in the middle of theſe an vpright ſtalke about a foot high couered with leaues growing leſſe and leſſe toward the points, parted at the top into certaine wings or branches, about which are floures orderly placed, of a darke purpliſh colour : the root is all of ſtrings.

2   There is also another great Housleek or Sengreene (syrnamed tree Housleeke) that bringeth forth a stalke a cubit high, somtimes higher, and often two; which is thick, hard, wooddy, tough, and that can hardly be broken, parted into diuers branches, and couered with a thick grosse bark, which in the lower part reserueth certaine prints or impressed markes of the leaues that are fallen away. The leaues are fat, well bodied, full of juice, an inch long and somewhat more, like little tongues, very curiously minced in the edges, standing vpon the tops of the braunches, hauing in them the shape of an eye. The floures grow out of the branches, which are diuided into many springs; which floures are slender, yellow, and spred like a star; in their places commeth vp very fine seed, the springs withering away: the root is parted into many off-springs. This plant is alwaies greene, neither is it hurt by the cold in winter, growing in his natiue soile; whereupon it is named *ἀειζωον*, and *Sempervivum*, or Sengreene.

<table>
<tr><td>1 Sempervivum majus.
Great Housleeke.</td><td>‡ 2 Sedum majus arborescens.
Tree Housleeke.</td></tr>
</table>

3   There is also another of this kinde, the circles whereof are answerable in bignesse to those of the former, but with lesser leaues, moe in number, and closely set, hauing standing on the edges very fine haires as it were like soft prickles. This is somewhat of a deeper greene: the stalke is shorter, and the floures are of a pale yellow. ‡ This is the third of *Dodonæus* description, *Pempt. 1. lib. 5. cap. 8.* ‡

4   There is likewise a third to be referred hereunto: the leaues hereof be of a whitish green, and are very curiously nicked round about. ‡ The floure is great, consisting of six white leaues. This is that described by *Dodonæus* in the fourth place; being the *Cotyledon altera secunda* of *Clusius.* ‡

5   There is also a fourth, the circles wherof are lesser, the leaues sharp pointed, very closely set, of a darke red colour on the top, and hairy in the edges: the floures on the sprigs are of a gallant purple colour. ‡ This is the fift of *Dodonæus*, and the *Cotyledon altera tertia* of *Clusius.* ‡

¶ *The Place.*

1   The great Sengreene is well knowne not onely in Italy, but also in France, Germany, Bohemia, and the Low-Countries. It groweth vpon stones in mountaines, vpon old walls, and auntient Buildings, especially vpon the tops of houses. The forme hereof doth differ according to the nature of the soile; for in some places the leaues are narrower and lesser, but moe in number, and haue one only circle: in some they are fewer, thicker, and also broader: they are greene, and of a deeper
greene

‡ 5 *Sedum maius anguſtifolium.*
Great narrow leaued Houſleek.

green in ſome places, and in others of a lighter green; for thoſe which we haue deſcribed grow not in one place, but in diuers and ſundry.

2   Great Sengreene is found growing of it ſelfe on the tops of houſes, old walls, and ſuch like places, in very many prouinces of the Eaſt, and of Greece, and alſo in the Iſlands of the Mediterranean ſea, as in Creet, now called Candy, Rhodes, Zant, and others : neither is Spain without it ; for (as *Cluſius* witneſſeth) it groweth in many places of Portingall ; otherwiſe it is cheriſhed in pots. In cold countries and ſuch as lie Northward, as in both the Germanies, it neither groweth of it ſelfe, nor yet laſteth long , though it be carefully planted, and diligently looked vnto, but through the extremitie of the weather and the ouermuch cold of winter it periſheth.

¶ *The Time.*

The ſtalk of the firſt doth at length floure after the Summer ſolſtice, which is in Iune about S. *Barnabies* day, and now and then in the month of Auguſt : but in Aprill, that is to ſay, after the Æquinoctiall of the Spring , which is about a moneth after the ſpring is begun, there grow out of this among the leaues ſmall ſtrings, which are the groundwork of the circles ; by which beeing at length full grown, it ſpreads it ſelfe into verie many circles.

2   Houſleek that growes like a tree floureth in Portugall preſently after the winter ſolſtice, being in December about S. *Lucies* day.

¶ *The Names.*

The firſt is commonly called *Iouis barba*, or *Iupiters* beard, and alſo *Sedum maius vulgare*: the Germanes call it Ⱨⱥⱨⱨⱬⱳⱳⱪⱪⱬ, Ᵹⱥⱥⱬ Ⲥⱺⱨⱶⱶⱬⱥⱥⱬ : the Low-Dutch, Ⲥⱺⱨⱶⱶⱬⱥⱥⱨⱬ : the Hollanders, Ⱨⱨⱴⱥⱥⱺⱥⱨ : the Frenchmen *Ioubarbe* : the Italians, *Sempreuiuo maggiore* : the Spaniards, *Siempreuiua, yerua pentera* : the Engliſhmen, Houſleeke, Sengreen, and Aygreen : of ſome, Iupiters Eye, Bullocks eye, and *Iupiters* beard : of the Bohemians, *Netreske*. Many take it to be *Cotyledon altera Dioſcoridis* ; but we had rather haue it one of the Sengreens, for it is continually green, and alwaies flouriſheth, and is hardly hurt by the extremitie of Winter.

The other without doubt is *Dioſcorides* his ἀείζωον μέγα ; that is, *Sempervivum magnum*, or *Sedum maius*, great Houſleeke or Sengreen : *Apuleius* calleth it *Vitalis*, and *Semperflorium* : it is alſo named ζωόφθαλμον, ζιζοειδές, αείζωον.

¶ *The Temperature.*

The great Houſleeks are cold in the third degree ; they are alſo dry, but not much, by reaſon of the waterie eſſence that is in them.

¶ *The Vertues.*

They are good againſt S. *Anthonies* fire, the ſhingles, and other creeping vlcers and inflammations, as *Galen* ſaith, that proceed of rheumes and fluxes ; and as *Dioſcorides* teacheth, againſt the inflammation or fiery heate in the eies: the leaues, ſaith *Pliny*, being applied, and the juice laid on, are a remedie for rheumatick and watering eies.   A

They take away the fire of burnings and ſcaldings, and being applied with barly meale dried, do take away the paine of the gout.   B

Dioſcorides teacheth, that they are giuen to them that are troubled with a hot laske: that they likewiſe driue forth wormes of the belly, being drunke with wine.   C

The juice put vp in a peſſary do ſtay the fluxes in women, proceeding of a hot cauſe ; the leaues held in the mouth do quench the thirſt in hot burning feuers.   D

The juice mixed with barly meale and vineger preuaileth againſt S. *Anthonies* fire, all hot burning and fretting vlcers, and againſt ſcaldings, burnings, and hot inflammations, and alſo the gout comming of an hot cauſe.   E

The

F   The juice of Housleeke, garden Nightshade, and the buds of Poplar boiled in *Axungia porci*, or hogs grease, make the most singular Populeon that euer was vsed in Surgerie.

G   The juice hereof taketh away cornes from the toes and feet, if they be washed and bathed therewith, and euery day and night as it were emplaistered with the skin of the same Housleeke, which certainly taketh them away without incision or such like, as hath been experimented by my verie good friend Mr *Nicolas Belson*, a man painfull and curious in searching forth the secrets of nature.

H   The decoction of Housleek or the juice thereof drunke, is good against the bloudy flixe, and cooleth the inflammation of the eyes, being dropped thereinto, and the herb bruised and layd vpon them.

Chap. 143.
Of the lesser Housleekes or Prick-madams.

1 *Sedum minus hematoides.*
Prick-madam.

2 *Sedum minus Officinarum.*
White floured Prick-madam.

¶ *The Description.*

1   THe first of these is a very little hearbe creeping vpon the ground with many slender stalks, which are compassed about with a great number of leaues that are thick, full of joints, little, long, sharp pointed, inclining to a green blew. There rise vp amongst these little stalks an handfull high, bringing forth at the top as it were a shadowie tuft, and therein fine yellow floures: the root is full of strings.

2   The other little Sengreene is also a small herbe, bringing forth many slender stalkes, seldome aboue a span high; on the tops whereof stand little floures like those of the other, in small

loose

‡ 3 *Sedum minus æstivum:*
Small Summer Sengreen.

‡ 4 *Sedum minus flore amplo.*
Small large floured Sengreen.

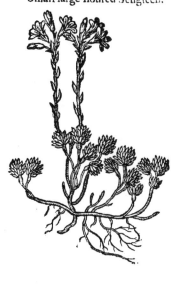

7 *Sedum medium teretifolium.*
Small Prick-madam.

‡ 6 *Aizoon Scorpioides.*
Scorpion Sengreen.

loofe tufts, but they are white & fomthing leſſer : the leaues about the ſtalkes are few and little,but long,blunt,and round,bigger than wheaten cornes, fomething leſſer than the kernels of Pine apples, otherwiſe not vnlike ; which oftentimes are fomthing red ſtalks and all : the root creepeth vpon the ſuperficiall or vppermoſt part of the earth, ſending downe ſlender threds.

‡ 7 *Sedum Portlandicum*.
Portland Sea-green.

3   This is a ſmall kinde of Stonecrop, which hath little narrow leaues,thick ſharp pointed and tender ſtalks ful of fatty juice; on the top wherof grow fmal yellow flours ſtar-faſhion. The root is ſmall,and running by the ground.

4   There is likewiſe another Stonecrop called Frog Stonecrop, which hath little tufts of leaues riſing from ſmall and threddy roots, creeping vpon the ground like to *Kali* or Frog-graſſe : from the which tufts of leaues riſeth a ſlender ſtalke ſet with a few ſuch like leaues,hauing at the top pretty large yellow floures,the ſmalneſſe of the plant being conſidered.

‡ 8 *Sedum petræum*.
Small rocke Sengreene.

‡ 5   This is like that which is deſcribed in the ſecond place,but that the ſtalkes are leſſer, and not ſo tall, and the floures of this are ſtar-faſhioned and of a golden yellow colour. ‡

8   There is another Stonecrop or Prickmadam called *Aizoon Scorpioides* , which is altogether like the great kinde of Stonecrop,and differeth in that, that this kind of Stonecrop or Prickmadam hath his tuft of yellow floures turning againe,not much vnlike the taile of a ſcorpion, reſembling *Myoſitis ſcorpioides* , and the leaues fomewhat thicker and cloſer thruſt together:the root is ſmall and tender.

7   There is a plant called *Sedum Portlandicum*,or Portland Stonecrop,of the Engliſh Iſland called Portland, lying in the South coaſt,hauing goodly branches,and a rough rinde.The leaues imitate *Laureola*,growing amongſt the Tithymales,but thicker,ſhorter,more fat and tender. The ſtalke is of a wooddy ſubſtance like *Laureola*, participating of the kindes of *Craſſula, Sempervivum*, and the Tithymales,whereof we thinke it to be a kinde : yet not daring to deliuer any vncertaine ſentence, it ſhall be leſſe prejudiciall to the truth, to account it as a ſhrub degenerating from both kindes.

‡   *Pena* and *Lobel*,who firſt ſet this forth, knew not very well what they ſhould ſay thereof; nor any ſince them : wherefore I haue onely giuen you their figure put to our Authours deſcription. ‡

8   There is a plant which hath receiued his name *Sedum petræum*, becauſe it doth for the moſt part grow vpon the rockes, mountaines, and ſuch like ſtony places, hauing very ſmall leaues comming forth of the ground in tufts like *Pſeudo-Moly*,that is,our common herb called Thrift : among the leaues come forth ſlender ſtalk, an handfull high, loden with ſmall yellow floures like vnto the common Prick-madam : after which come little thick ſharp pointed cods,which contain the ſeed, which is ſmall,flat,and yellowiſh.

¶ *The*

¶ *The Place.*

The former of these groweth in gardens in the Low-Countries : in other places vpon stone wals and tops of houses in England almost euerie where.

The other growes about rubbish in the borders of fields, and in other places that lie open to the Sun.

¶ *The Time.*

They floure in the Summer moneths.

¶ *The Names.*

The lesser kinde is called in Greeke ἀείζωον μικρὸν : in Latine, *Sedum*, and *Sempervivum minus* : of the Germanes, **Klepn Donderbaer**, and **Klepn Hauszwurtz** : of the Italians, *Sempervivo minore* : of the Frenchmen, *Tricque-madame* : of the English, Prick-madam, dwarfe Housleeke, and small Sengreene.

The second kind is named in shops *Crassula minor* : and they doe syrname it *Minor*, for the difference between it and the other *Crassula*, which is a kind of *Orchis* : it is also called *Vermicularis* : in Italian, *Pignola, Cranellosa*, and *Grasella* : in low-Dutch, **Blader loosen** : in English, wild Prickmadam, Great Stone-crop, or Worm-grasse.   ‡ That which is vulgarly known and called by the name of Stonecrop is the *Illecebra* described in the following chapter, & such as grow commonly with vs of these small Housleeks mentioned in this chapter are generally named Prickmadams : but our Author hath confounded them in this and the next chapter ; which I would not alter, thinking it sufficient to giue you notice thereof.

¶ *The Temperature and Vertues.*

All these small Sengreens are of a cooling nature like to the great ones, and are good for those   A things that the others be. The former of these is vsed in many places in sallads, in which it hath a fine rellish, and a pldasant taste, and is good for the heart-burne.

Chap. 144.

‡ *Of diuers other small Sengreenes.*

¶ *The Description.*

‡ 1   THe stalke of this small water Sengreene is some span long, reddish, succulent, and weake ; the leaues are longish, a little rough, and ful of juice : the floures grow vpon the tops of the stalks, consisting of six purple or else flesh-colored leaues, which are succeeded by as many little cods containing a small seed : the root is small and thready, & the whole plant hath an insipide or waterish taste.   This was found by *Clusius* in some waterie places of Germany about the end of Iune, and he calls it *Sedum minus 3. siue palustre.*

2   This second from small fibrous and creeping roots sends vp sundry little stalkes set with leaues like those of the ordinary Prick-madam, yet lesse, thick, and flatter, and of a more astringent taste : the floures, which are pretty large, grow at the tops of the branches, and consist of fiue pale yellowish leaues. It growes in diuers places of the Alps, and floures about the end of Iuly, and in August.   This is the *Sedum minus 6.* or *Alpinum 1.* of *Clusius.*

3   This hath small little and thicke leaues, lying bedded, or compact close together, and are of an Ash-colour inclining to blew : the stalkes are some two inches long, slender, and almost naked ; vpon which grow commonly some three floures consisting of fiue white leaues apiece, with some yellow threds in the middle. This mightily encreases, and will mat and couer the ground for a good space together. It floures in August, and growes vpon the craggy places of the Alps. *Clusius* cals it *Sedum minus nonum, siue Alpinum 3.*

3   The leaues of this are somwhat larger and longer, yet thick, and somwhat hairy about their edges ; at first also of an acide taste, but afterwards bitterish and hot : it also sends forth shoots, and in the midst of the leaues it puts forth stalkes some two inches high, which at the top as in an vmbel carry some six little floures consisting of fiue leaues apiece, hauing their bottomes of a yellowish colour. It is found in the like places, and floures at the same time as the former. *Clusius* maketh it his *Sedum minus 10. Alpinum 4.* and in the *Hist. Lugd.* it is called *Iasme montana.*

5 For

‡ 1 *Sedum minus paluſtre.*
Small water Sengreen,

‡ 2 *Sedum Alpinum* 1. *Cluſij.*
Small Sengreen of the Alps.

‡ 3 *Sedum Alpinum* 3. *Cluſij.*
White Sengreen of the Alps.

‡ 4 *Sedum Alpinum* 4. *Cluſij.*
Small Sengreen of the Alps.

‡ 5 *Sedum petraum Bupleuri folio.*
Long leaued rock Sengreen.

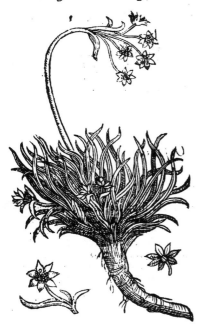

5 For theſe foure laſt deſcribed we are beholden to *Cluſius*, and for this fifth to *Pona*, who thus deſcribes it : it hath one thicke and large root with few or no fibres, but ſome knots bunching out here and there : it is couered with a thicke barke, and is of a blackiſh red colour on the out ſide : the leaues are many, long and narrow, lying ſpred vpon the ground : the ſtalke growes ſome foot high, and is round and naked, and at the top carries floures conſiſting of ſeuen ſharpe pointed pale yellow leaues ; which are ſucceeded by ſeeds like thoſe of *Bupleurum*, and of a ſtrong ſmel : it floures about the middle of July, and the ſeed is ripe about the middle of Auguſt. *Pona*, who firſt obſerued this growing vpon mount Baldus in Italy, ſets it forth by the name of *Sedum petraum Bupleuri folio.* *Bauhine* hath it by the name of *Perfoliata Alpina Gramineo folio*, and *Bupleuron anguſtifolium Alpinum.*

¶ *The Nature and Vertues.*

The three firſt deſcribed without doubt are cold, and partake in vertues with the other ſmal Sengreens : but the two laſt are rather of an hot and attenuating faculty. None of them are commonly known or vſed in phyſicke. ‡

Vermicularis, ſiue Illecebra minor acris.
Wall Pepper, or Stone-crop.

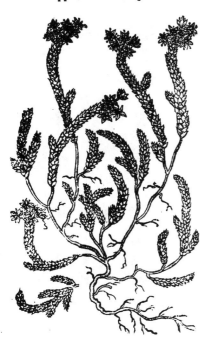

Chap. 115.
Of Stonecrop, called Wall Pepper.

¶ *The Deſcription.*

THis is a low and little herbe, the ſtalkes be ſlender & ſhort; the leaues about theſe ſtand very thick, and ſmall in growth, ful bodied, ſharp pointed , and full of juice : The floures ſtand at the top, and are maruellous little ; of colour yellow, and of a ſharpe biting taſte : the root is nothing but ſtrings.

¶ *The Place.*

It groweth euery where in ſtony and dry places, and in chinks and crannies of old walls, and on the tops of houſes : it is alwaies greene, and therefore it is very fitly placed amongſt the Sengreenes.

¶ *The Time.*

It floureth in the Summer moneths.

¶ *The Names.*

This is *Tertium ſempervjuum Dioſcoridis*, or *Dioſcorides* his third Sengreen, which he ſaith is called of the Grecians, ειδεσꝑι αꝗμα : and παιϙιν : and of the Romans *Illecebra.* *Pliny* alſo witneſſeth, that the Latines name it *Illecebra* : yet there is another αιεξρꝑι αꝗμα : and another παιϙιν the Germans call this herb **Mautpfeffer**, & **Katzen treutle** : the Frenchmen, *Pain d'oiſeau* : the Low-Dutch, **Muer**

𝕸uer 𝕻epper: the Englifhmen, Stonecrop, and Stonehore, little Stonecrop, Pricket, Mouf-tail, Wall Pepper, Country Pepper, and Iack of the Butterie.

¶ *The Temperature.*

This little herb is fharpe and biting, and very hot. Being outwardly applied it raifeth blifters, and at length exulcerateth.

¶ *The Vertues.*

A   It wafteth away hard kernels and the Kings euill, if it be laid vnto them, as *Diofcorides* writes.

B   The juice hereof extracted or drawne forth, and taken with vineger or other liquor, procures vomit, and brings vp groffe and flegmatick humors, and alfo cholerick, and doth thereby often times cure the Quartan Ague and other Agues of long continuance: and giuen in this manner it is a remedie againft poifons inwardly taken.

Chap. 146. *Of Orpine.*

¶ *The Defcription.*

1   THe Spanifh Orpyne fends forth round ftalkes, thicke, flipperie, hauing as it were little joints fomwhat red now and then about the root: the leaues in like manner be thicke, fmooth, groffe, full of tough juice, fometimes fleightly nicked in the edges, broader leafed, and greater than thofe of Purflane, otherwife not much vnlike, which by couples are fet oppofit one againft another vpon euerie joint, couering the ftalke in order by two and two: the floures in the round tufts are of a pale yellow: the root groweth ful of bumps like vnto long kernels, waxing fharp toward the point: thefe kernels be white, and hauing ftrings growing forth of them.

1 *Craffula major Hifpanica.*
Spanifh Orpyne.

2 *Craffula fiue faba inuerfa.*
Common Orpyne.

2   The fecond, which is our common Orpyne, doth likewife rife vp with very many round ftalkes that are fmooth, but not jointed at all: the leaues are groffe or corpulent, thicke, broad,

and

and oftentimes ſomewhat nicked in the edges , leſſer than thoſe of the former, placed out of order. The floures be either red or yellow,or elſe whitiſh:the root is white,wel bodied and full of kernels. This plant is very full of life:the ſtalks ſet only in clay continue greene a long time, and if they be now and then watered they alſo grow. Wee haue a wilde kinde of Orpyne growing in corne fields and ſhadowy woods in moſt places of England, in each reſpect like that of the garden, ſauing that it is altogether leſſer.

¶ The Place.

They proſper beſt in ſhadowie and ſtony places,in old wals made of lome or ſtone. Oribaſius ſaith That they grow in Vineyards and tilled places.The firſt groweth in gardens;the other euerie where: the firſt is much found in Spaine and Hungarie;neither is Germany without it;for it groweth vpon the bankes of the riuer of Rhene neere the Vineyardes,in rough and ſtony places, nothing at all differing from that which is found in Spaine.

The ſecond groweth plentifully both in Germany,France,Bohemia,England, and in other countries among Vines,in old lomy daubed and ſtony walls.

¶ The Time.

The Orpynes floure about Auguſt or before.

¶ The Names.

The firſt is that which is called of the Grecians ꝟꝟ and ꝟꝟꝟ of the Latines, Telephium, and Sempervivum ſylveſtre,and Illecebra: but Illecebra by reaſon of his ſharpe and biting quality differeth much from it,as we haue declared in the former chapter. Some there be that name it ꝟꝟ, or Portulaca ſylveſtris : yet there is another Portulaca ſylveſtris,or wild Purſlane,like to that which groweth in gardens,but leſſer:we may call this in Engliſh,Spaniſh Orpyne,Orpyne of Hungarie, or ioynted Orpyne.

The ſecond kind of Orpyne is called in ſhops Craſſula,and Craſſula Fabaria,and Craſſula major,that it may differ from that which is deſcribed in the chapter of little Houſleeke:it is named alſo Fabaria. in high-Dutch,Wandkraut,Knauenkraut,Fottzwang,& Fottzwepn:in Italian Fabagraſſa: in French,Ioubarbe des vignes,Feue eſpeſſe : in Low-Dutch,Smer woztele,and Hemel Sluetel : in Engliſh,Orpyne:alſo Liblong or Liue-long.

¶ The Temperature.

The Orpyns be cold and dry,and of thin or ſubtile parts.

¶ The Vertues.

Dioſcorides ſaith,That being laid on with vinegre,it taketh away the white morphew:Galen ſaith A the black alſo;which thing it doth by reaſon of the ſcouring or clenſing quality that it hath:whereupon Galen attributeth vnto it an hot facultie,though the taſte ſheweth the contrarie : which aforeſaid ſcouring facultie declareth, That the other two alſo be likewiſe cold. But cold things may as well clenſe,if drineſſe of temperature and thinneſſe of eſſence be ioined together in them.

CHAP. 147.   Of the ſmaller Orpyns.

¶ The Deſcription.

1   The Orpyn with purple floures is lower and leſſer than the common Orpyn:the ſtalks be ſlenderer, and for the moſt part lie along vpon the ground. The leaues are alſo thinner and longer,&of a more blew greene,yet well bodied,ſtanding thicker below than aboue, confuſedly ſet together without order : the floures in the tufts at the tops of the ſtalkes be of a pale blew tending to purple.The roots be not ſet with lumpes or knobbed kernels,but with a multitude of hairy ſtrings.

2   This ſecond Orpyn,as it is known to few,ſo hath it found no name,but that ſome Herbariſts doe call it Telephium ſempervivum or virens : for the ſtalkes of the other do wither in winter,the root remaineth greene;but the ſtalkes and leaues of this endure alſo the ſharpneſſe of winter;and therefore we may call it in Engliſh,Orpyn euerlaſting,or neuer-dying Orpyn. This hath leſſer and rounder leaues than any of the former:the floures are red, and the root fibrous.

‡   3 Cluſius receiued the ſeeds of this from Ferranto Imperato of Naples , vnder the name of
Telephium

1 *Telephium floribus purpureis.*
Purple Orpyn.

2 *Telephium semper-virens.*
Neuer-dying Orpyn.

‡ 3 *Telephium legitimum Imperati.* Creeping Orpyn. ‡

Telephium legitimum ; and he hath thus giuen vs the history thereof: It produces from the top of the root many branches spred vpon the ground, which are about a foot long, set with many leaues, especially such as are not come to floure; for the other haue fewer: these leaues are smaller, lesse thick also and succulent than those of the former kindes, neither are they so brittle: their colour is green, inclining a little to blew : the tops of the branches are plentifully stored with little floures growing thick together and composed of fiue little white leaues a peece : which fading, there succeed cornered seed vessels full of a brownish seed. The root is sometimes as thicke as ones little finger, tough, white, diuided into some branches, and liuing many yeares. ‡

¶ *The Place, Time, Names, Temperature, and Vertues.*

The first growes not in England. The second flourishes in my garden. ‡ The third is a stranger with vs. ‡ They floure when the common Orpyn doth. Their names are specified in their seuerall descriptions : and their temperature and faculties in working are referred to the common Orpyn.

CHAP.

CHAP. 148.   Of Purſlane.

¶ *The Deſcription.*

1   THe ſtalkes of the great Purſlane be round, thicke, ſomewhat red, full of juyce, ſmooth, glittering, and parted into certaine branches trailing vpon the 'ground : the leaues be an inch long, ſomething broad, thicke, fat, glib, ſomewhat greene, whiter on the ne-ther ſide : the floures are little, of a faint yellow, and grow out at the bottome of the leaues. After them ſpringeth vp a little huske of a greene colour, of the bigneſſe almoſt of halfe a barly corne, in which is ſmall blacke ſeed : the root hath many ſtrings.

1 *Portulaca domeſtica.*
Garden Purſlane.

2 *Portulaca ſylueſtris.*
Wilde Purſlane.

2   The other is leſſer and hath like ſtalkes, but ſmaller, and it ſpreadeth on the ground : the leaues be like the former in faſhion, ſmoothneſſe, and thickeneſſe, but farre leſſer.

¶ *The Place.*

The former is fitly ſowne in gardens, and in the waies and allies thereof, being digged and dun-ged ; it delighteth to grow in a fruitfull and fat ſoile not dry.

The other commeth vp of his owne accord in allies of gardens and vineyards, and oftentimes vpon rockes : this alſo is delighted with watery places : being once ſowne, if it be let alone till the ſeed be ripe it doth eaſily ſpring vp afreſh for certaine yeeres after.

¶ *The Time.*

It may be ſowne in March or Aprill ; it flouriſheth and is greene in Iune, and afterwards euen vntill Winter.

Purſlane is called in Greeke, *ἀνδράχνη* : in Latine, *Portulaca* : in high Dutch, **Burkelkraut** : in French, *Poupier* : in Italian, *Prochaccia* : in Spaniſh, *Verdolagas* : in Engliſh, Purſlane, and Porcelane.

¶ *The Temperature.*

Purslane is cold,and that in the third degree,and moist in the second : but wilde Purslane is not so moist.

¶ *The Vertues.*

A   Raw Purslane is much vsed in sallades,with oyle, salt and vinegre : it cooleth an hot stomacke, and prouoketh appetite ; but the nourishment which commeth thereof is little, bad, cold, grosse, and moist : being chewed it is good for teeth that are set on edge or astonied ; the juyce doth the same being held in the mouth,and also the distilled water.

B   Purslane is likewise commended against wormes in young children, and is singular good especially if they be feuerish withall, for it both allaies the ouermuch heate, and killeth the wormes : which thing is done through the saltnes mixed therewith, which is not only an enemy to wormes, but also to putrifaction.

C   The leaues of Purslane either raw , or boiled, and eaten as sallades, are good for those that haue great heate in their stomackes and inward parts, and doe coole and temper the inflamed bloud.

D   The same taken in like manner is good for the bladder and kidnies, and allaieth the outragious lust of the body : the juyce also hath the same vertue.

E   The juyce of Purslane stoppeth the bloudy flix, the flux of the hemorrhoides,monethly termes, spitting of bloud,and all other fluxes whatsoeuer.

F   The same throwne vp with a mother syringe,cureth the inflammations,frettings,and vlcerations of the matrix ; and put into the fundament with a clister pipe, helpeth the vlcerations and flux of the guts.

G   The leaues eaten raw, take away the paine of the teeth, and fasten them ; and are good for teeth that are set on edge,with eating of sharpe or sowre things.

H   The seed being taken,killeth and driueth forth wormes, and stoppeth the laske.

Chap. 149.

Of sea Purslane, and of the shrubby Sengreenes.

¶ *The Description.*

1   SEa Purslane is not a herbe as Garden Purslane, but a little shrub : the stalkes whereof be hard and wooddy : the leaues fat, full of substance, like in forme to common Purslane, but much whiter and harder : the mossie purple floures stand round about the vpper parts of the stalkes,as doe almost those of Blyte,or of Orach:neither is the seed vnlike,being broad and flat : the root is wooddy, long lasting, as is also the plant, which beareth out the Winter with the losse of a few leaues.

† 2   There is another sea Purslane or *Halimus*, or after *Dodonæus*, *Portulaca marina*,which hath leaues like the former,but not altogether so white,yet are they somewhat longer and narrower, not much vnlike the leaues of the Oliue tree. The slender branches are not aboue a cubit or cubit and, halfe long,and commonly lie spred vpon the ground,and the floures are of a deepe ouerworne herby colour,and after them follow seeds like those of the former,but smaller.

‡ 3   Our ordinary *Halimus* or sea Purslane hath small branches some foot or betterlong, lying commonly spred vpon the ground, of an ouerworne grayish colour, and sometimes purple;the leaues are like those of the last mentioned, but more fat and thicke, yet lesse hoary. The floures grow on the tops of the branches,of an herby purple colour, which is succeeded by small seeds like to that of the second kinde. ‡

4   There is found another wilde sea Purslane,whereof I haue thought good to make mention ; which doth resemble the kindes of Aizoons. The first kinde groweth vpright, with a trunke like a small tree or shrub, hauing many vpright wooddy branches, of an ash colour, with many thicke, darke,greene leaues like the small Stone-crop,called *Vermicularis* : the floures are of an herby yellowish greene colour : the root is very hard and fibrous : the whole plant is of a salt tang taste, and the juyce like that of Kali.

5   There is another kinde like the former,and differeth in that, this strange plant is greater, the leaues more sharpe and narrower,and the whole plant more wooddy, and commeth neere to the forme of a tree. The floures are of a greenish colour.

¶ *The*

‡ 1 *Halimus latifolius.*
Tree Sea Purſlane.

‡ 2 *Halimus anguſtifolius procumbens.*
Creeping Sea Purſlane.

† 3 *Halimus vulgaris, ſiue Portulaca marina.*
Common Sea Purſlane.

‡ 4 *Vermicularis frutex minor.*
The leſſer ſhrubby Sengreene.

‡ 5 *Vermicularis frutex major.*
The greater Tree Stone-crop.

¶ *The Place.*

‡ The firſt and ſecond grow vpon the Sea coaſts of Spaine and other hot countries ‡: and the third groweth in the ſalt mariſhes neere the ſea ſide, as you paſſe ouer the Kings ferrey vnto the Iſle of Shepey, going to Sherland houſe (belonging ſometime vnto the Lord *Cheiny,* and in the yeare 1590, vnto the Worſhipfull Sr. *Edward Hobby*) faſt by the ditches ſides of the ſame mariſh : it groweth plentifully in the Iſle of Thanet as you go from Margate to Sandwich, and in many other places along the coaſt. The other ſorts grow vpon bankes and heapes of ſand on the Sea coaſts of Zeeland, Flanders, Holland, and in like places in other countries, as beſides the Iſle of Purbecke in England ; and on Rauen-ſpurne in Holderneſſe, as I my ſelfe haue ſeene.

¶ *The Time.*
Theſe flouriſh and floure eſpecially in Iuly and Auguſt.

¶ *The Names.*
Sea Purſlane is called *Portulaca Marina* : In Greeke, ἅλιμοϲ: it is alſo called in Latine, *Halimus* : in Dutch, **Zee Poꝛceleſine:** in Engliſh, Sea Purſlane.
The baſtard ground Pine is called of ſome, *Chamepitys vermiculata* : in Engliſh, Sea ground Pine : ‡ or more fitly, Tree Ston-crop, or Pricket, or Shrubby Sengreene. ‡

¶ *The Temperature.*
Sea Purſlane is (as *Galen* ſaith) of vnlike parts, but the greater part thereof is hot in a meane, with a moiſture vnconcoĉted, and ſomewhat windie.

¶ *The Vertues.*

A   The leaues (ſaith *Dioſcorides*) are boyled to be eaten : a dram weight of the root being drunke with meade or honied water, is good againſt cramps and drawings awrie of ſinewes, burſtings, and gnawings of the belly : it alſo cauſeth nurſes to haue ſtore of milke. The leaues be in the Low-countries preſerued in ſalt or pickle as capers are, to be ſerued and eaten at mens tables in ſtead of them, and that without any miſlike of taſte, to which it is pleaſant. *Galen* doth alſo report, that the young and tender buds are wont in Cilicia to be eaten, and alſo laid vp in ſtore for vſe.

B   ‡ *Cluſius* ſaith, That the learned Portugal Knight *Damianus a Goes* aſſured him, That the leaues of the firſt deſcribed boyled with bran, and ſo applied, mitigate the paine of the Gout proceeding of an hot cauſe. ‡

† The figure that was formerly giuen by our Author with the title of *Portulaca marina,* and which is ſet forth by *Tabern.* vnder the ſame name, is either of none of theſe plants, or elſe it is imperfeĉt. *Bauhine* knowes not what to make of it, but queſtions it, *Quid ſit ?*

CHAP. 150.   *Of Herbe-Iuy, or Ground-Pine.*

¶ *The Deſcription.*

1   THe common kinde of *Chamæpitys* or Ground-Pine is a ſmal herbe and very tender, creeping vpon the ground, hauing ſmall and crooked branches trailing about. The leaues be ſmall, narrow and hairy, in fauor like the Firre or Pine tree; but if my ſence of ſmelling be perfeĉt, me thinkes it is rather like vnto the ſmell of hempe. The floures be little, of a pale yellow colour, and ſometimes white : the root is ſmall and ſingle, and of a wooddy ſubſtance.

†   2   The ſecond hath pretty ſtrong foure ſquare ioynted ſtalkes, browne and hairy ; from which grow pretty large hairy leaues much clouen or cut : the floures are of a purple colour, and grow about the ſtalks in roundles like the dead Nettle : the ſeed is black and round, and the whole plant ſauouring like the former : ‡ which ſheweth this to be fitly referred to the *Chamæpitys,* and not to be well called *Chamædrys fœmina,* or iagged Germander, as ſome haue named it. ‡

1 *Chamæpitys mas.*
The male ground-Pine.

2 *Chamæpitys fæmina.*
The female ground-Pine.

3 *Chamæpitys 3. Dodon.*
Small Ground-Pine.

4 *Iua muscata Monspeliaca.*
French herbe-Iuy or Ground-Pine.

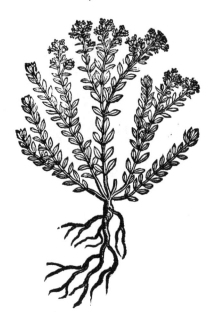

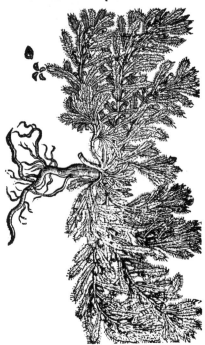

3   This kinde of Herbe-Iuy,growing for the moſt part about Montpelier in France, is the leaſt of all his kind,hauing ſmall white and yellow floures, in ſmell and proportion like vnto the others, but much ſmaller.

† 4   There is a wilde or baſtard kinde of *Chamæpitys*, or ground-Pine, that hath leaues ſome-what like vnto the ſecond kinde, but not jagged in that manner, but onely ſnipt about the edges. The root is ſomewhat bigger,wooddy,whitiſh,and bitter,and like vnto the root of Succorie. All this herbe is very rough,and hath a ſtrong vnpleaſant ſmell,not like that of the ground-Pines.

† 5   There is another kind that hath many ſmall and tender branches beſet with little leaues for the moſt part three together, almoſt like the leaues of the ordinary ground-Pine : at the top of which branches grow ſlender white floures ; which being turned vpſide downe, or the lower part vpward, doe ſomewhat reſemble the floures of *Lamium :* the ſeeds grow commonly foure together in a cup,and are ſomewhat big and round : the root is thicke, whitiſh, and long laſting.

‡ 5 *Chamæpitys ſpuria altera Dodon.*   ‡ 6 *Chamæpitys Auſtriaca.*
   Baſtard Ground-Pine.   Auſtrian Ground-pine.

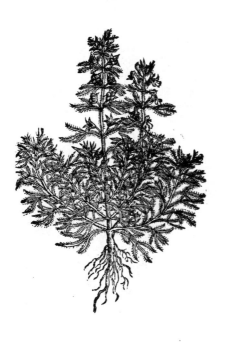 

6   There groweth in Auſtria a kinde of *Chamæpitys*,which is a moſt braue and rare plant, and of great beautie,yet not once remembred either of the ancient or new Writers, vntill of late that fa-mous *Carolus Cluſius* had ſet it forth in his Pannonicke Obſeruations ; who for his ſingular skil and induſtrie hath woon the garland from all that haue written before his time. This rare and ſtrange plant I haue in my garden, growing with many ſquare ſtalkes of halfe a foot high, beſet euen from the bottome to the top with leaues ſo like our common Roſemary, that it is hard for him which doth not know it exactly to finde the difference;being greene aboue,and ſomewhat hairie and hoa-ry vnderneath : among which come forth round about the ſtalkes (after the manner of roundles or coronets)certain ſmall cups or chalices of a reddiſh colour ; out of which come the floures like vnto Archangell in ſhape, but of a moſt excellent and ſtately mixed colour, the outſide purple de-clining to blewneſſe,and ſometimes of a violet colour.The floure gapeth like the mouth of a beaſt, and hath as it were a white tongue ; the lower and vpper jawes are white likewiſe, ſpotted with ma-ny bloudy ſpots : which being paſt, the ſeeds appeare very long, of a ſhining blacke colour, ſet in order in the ſmall huskes as the *Chamæpitys ſpuria*. The root is blacke and hard,with many hairy ſtrings faſtned thereto.

¶ *The*

¶ *The Place.*

These kindes of *Chamæpitys* (except the last) grow very plentifully in Kent, especially about Grauesend, Cobham, Southfleet, Horton, Dartford, and Sutton, and not in any other shire in England that I euer could finde.

‡ None of these except the first, for any thing I know, or can learne, grow wilde in England; the second I haue often seene in Gardens. ‡

¶ *The Time.*

They floure in Iune, and often in August.

¶ *The Names.*

Ground Pine is called in Greeke, χαμαίπιτυς: in Latine, *Ibiga, Aiuga*, and *Abiga:* in shops, *Iua Arthritica* and *Iua moschata:* in Italian, *Iua:* in Spanish, *Chamæpiteos:* in high Dutch, **Bergiß mich nicht:** in low Dutch, **Velt Cipzes:** in French, *Iue moschate:* in English, Herbe Iuie, Forget me not, Ground Pine, and field Cypresse.

‡ 1 The first of these is the *Chamæpitys prima*, of *Matthiolus*, *Dodonæus* and others, and is that which is commonly vsed in shops and in Physicke.

2 This *Matthiolus* cals *Chamædrys altera*: Lobel, *Chamædrys Liciniatis folijs*: *Lonicerus, Traxago vera*; *Tabernamontanus, Iua moscata*; and *Dodon.* (whom in this chapter we chiefly follow) *Chamæpitys altera*.

3 Thirdly, this is the *Chamæpitys* 1. of *Fuchsius* and others; the *Chamæpitys* 1. *Dioscoridis odoratior* of Lobel; and the *Chamæpitys* 3. of *Matthiolus* and *Dodonæus*.

4 *Gesner* cals this *Chamæpitys species Monspelij*: *Clusius* and *Dodon*: *Anthyllis altera;* and Lobel, *Anthyllis Chamepityides minor;* and *Tabern. Iua Moschata Monspeliensium*.

5 This is *Chamæpitys adulterina* of Lobel: *Pseudochamæpitys* and *Aiuga adulterina* of *Clusius*: and *Chamæpitys spuria altera* of *Dodon*.

6 This is *Chamæpitys Austriaca* of *Clusius*; and *Chamæpitys cærulea* of *Camerarius*. ‡

¶ *The Temperature.*

These herbes are hot in the second degree, and dry in the third.

¶ *The Vertues.*

The leaues of *Chamæpitys* tunned vp in Ale, or infused in wine, or sodden with hony, and drunke **A** by the space of eight or ten daies, cure the iaundise, the Sciatica, the stoppings of the liuer, the difficultie of making water, the stoppings of the spleene, and cause women to haue their naturall sicknesse.

Chamæpitys stamped greene with hony cureth wounds, malignant and rebellious vlcers, and dis- **B** solueth the hardnesse of womens brests or paps, and profitably helpeth against poyson, or biting of any venomous beast.

The decoction drunke, dissolueth congealed bloud, and drunke with vinegre, driueth forth the **C** dead childe.

It clenseth the intrals: it helpeth the infirmities of the liuer and kidnies; it cureth the yellow **D** iaundise being drunke in wine: it bringeth downe the desired sickenesse, and prouoketh vrine: being boiled in Mead or honied water and drunke, it helpeth the Sciatica in forty daies. The people of Heraclea in Pontus do vse it against Wolfes bane in stead of counterpoyson.

The pouder hereof taken in pils with a fig, mollifieth the belly: it wasteth away the hardnesse **E** of the paps: it healeth wounds, it cureth purrified vlcers being applied with hony: and these things the first ground Pine doth performe, so do the other two: but not so effectually, as witnesseth *Dioscorides*.

Clusius of whom mention was made, hath not said any thing of the Vertues of *Chamæpitys Au-* **F** *striaca*; but verily I thinke it better by many degrees for the purposes aforesaid: my conjecture I take from the taste, smell, and comely proportion of this Herbe, which is more pleasing and familiar vnto the nature of man, than those which wee haue plentifully in our owne Country growing.

CHAP. 151. *Of Nauelwoort, or Penniwoort of the Wall.*

¶ *The Description.*

1 THe great Nauelwoort hath round and thicke leaues, somewhat bluntly indented about the edges, and somewhat hollow in the midst on the vpper part, hauing a short tender
stemme

ftemme faftened to the middeft of the leafe, on the lower fide vnderneath the ftalke, whereon the floures do grow, is fmall and hollow, an handfull high and more, befet with many fmall floures of an ouerworne incarnate colour. The root is fmall like an oliue, of a white colour.

‡ The root is not well expreft in the figure, for it fhould haue been more vnequall or tuberous, with the fibres not at the bottome, but top thereof. ‡

2 The fecond kinde of Wall Penniwort or Nauelwoort hath broad thicke leaues fomewhat deepely indented about the edges : and are not fo round as the leaues of the former, but fomewhat long towards the fetting on, fpred vpon the ground in manner of a tuft, fet about the tender ftalke, like to Sengreene or Houfleeke ; among which rifeth vp a tender ftalke whereon doe grow the like leaues. The floures ftand on the top confifting of fiue fmall leaues of a whitifh colour, with redde fpots in them. The root is fmall and threddy. ‡ This by fome is called *Sedum Serratum*. ‡

1 *Vmbilicus Veneris.*
Wall Penniwoort.

‡ 2 *Vmbilicus Ven. fiue Cotyledon altera.*
Iagged or Rofe Penniwoort.

 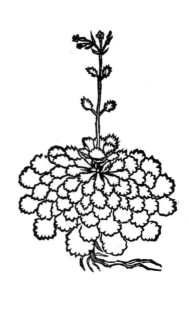

‡ 3 This third kinde hath long thicke narrow leaues, very finely fnipt or nickt on the edges, which lie fpred very orderly vpon the ground ; and in the midft of them rifeth vp a ftalke fome foot high, which beares at the top thereof vpon three or foure little branches, diuers white floures con-fifting of fiue leaues apiece.

4 The leaues of this are long and thicke, yet not fo finely fnipt about the edges, nor fo narrow as thofe of the former : the ftalke is a foot high, fet here and there with fomewhat fhorter and roun-der leaues than thofe below ; and toward the top thereof, out of the bofomes of thefe leaues come fundry little foot-ftalkes, bearing on their tops pretty large floures of colour white, and fpotted with red fpots. The roots are fmall, and here and there put vp new tufts of leaues, like as the com-mon Houfleeke. ‡

5 There is a kinde of Nauelwoort that groweth in watery places, which is called of the huf-bandman Sheeps bane, becaufe it killeth fheepe that do eat thereof : it is not much vnlike the pre-cedent, but the round edges of the leaues are not fo euen as the other ; and this creepeth vpon the ground, and the other vpon the ftone walls.

‡ 6 Becaufe fome in Italy haue vfed this for *Vmbilicus Veneris*, and otherfome haue fo called it, I thought it not amiffe to follow *Matthiolus*, and giue you the hiftory thereof in this place, rather than to omit it, or giue it in another which may be perhaps as vnfit, for indeed I cannot fitly ranke
it

3 *Vmbilicus Veneris minor.*
Small Nauelwoort.

‡ 4 *Cotyledon minor montana altera.*
The other ſmall mountaine Nauelwoort.

5 *Cotyledon paluſtris.*
Water Penniwoort.

‡ 6 *Cymbalaria Italica.*
Italian Baſtard Nauelwoort.

it with any other plant. *Bauhine* ſets it betweene *Hedera Terreſtris,* and *Naſturtium Indicum:* and *Co-lumna* refers it to the *Linaria's,* but I muſt confeſſe I cannot referre it to any ; wherefore I thinke it as proper to giue it here as in any other place. The branches of this are many, long, ſlender, and cree-ping, vpon which grow without any certaine order many little ſmooth thicke leaues faſhioned like thoſe of Iuie, and faſtened to ſtalkes of ſome inch long: and together with theſe ſtalkes come forth others of the ſame length, that carry ſpur-faſhioned floures, of the ſhape and bigneſſe of thoſe of the female Fluellen: their outſide is purple, their inſide blew, with a ſpot of yellow in the opening. The root is ſmall, creeping and thriddy. It floures toward the end of Summer, and growes wilde vpon walls in Italy, but in gardens with vs. *Matthiolus* calls it *Cymbalaria* (to which *Lobel* addes) *Italica Hederacio folio: Lonicerus* termes it *Vmbilicus Veneris Officinarum:* and laſtly *Columna* calls it *Linariahederæ folio.* ‡

¶ The Place.

The firſt kind of Penniwoort groweth plentifully in Northampton vpon euery ſtone wall about the towne, at Briſtow, Bathe, Wells, and moſt places of the Weſt countrie vpon ſtone walls. It groweth vpon Weſtminſter Abbey, ouer the doore that leadeth from *Chaucers* tombe to the old palace. ‡ In this laſt place it is not now to be found. ‡

The ſecond, third, and fourth grow vpon the Alpes neere Piedmont, and Bauier, and vpon the mountaines of Germany: I found the third growing vpon Bieſton Caſtle in Cheſhire.

‡ The fifth growes vpon the Bogges vpon Hampſtead Heath, and many ſuch rotten grounds in other places. ‡

¶ The Time.

They are greene and flouriſh eſpecially in Winter: They floure alſo in the beginning of Sum-mer.

¶ The Names.

Nauelwoort is called in Greeke, κοτυληδών: in Latine, *Vmbilicus Veneris,* and *Acetabulum* of diuers, *Herba Coxendicum: Iacobus Manlius* nameth it, *Scatum Cœli,* and *Scatellum:* in Dutch, 𝕹𝖆𝖚𝖊𝖑𝖈𝖗𝖚𝖕𝖙 : in Italian, *Cupertoiule:* in French, *Eſcuelles:* in Spaniſh, *Capadella:* of ſome, *Hortus Veneris,* or Venus garden, and *Terra vmbilicus,* or the Nauel of the earth: in Engliſh, Penniwoort, Wall-Penniwoort, Ladies Nauell, Hipwoort and Kidney-woort.

Water Penniwoort is called in Latine, *Cotyledon paluſtris:* in Engliſh, Sheepe-killing Penni-graſſe, Penny-rot, and in the North Countrey, White-rot: for there is alſo Red-rot, which is *Roſa ſolis:* in Northfolke it is called, Flowkwoort. ‡ *Columna* and *Bauhine* fitly refer this to the *Ranun-culi,* or Crowfeet; for it hath no affinitie at all with the Cotyledons (but onely in the roundneſſe of the leafe) the former of them calls it, *Ranunculus aquaticus vmbilicato folio,* and the later, *Ranunculus aquat. Cotyledonis folio.* ‡

¶ The Temperature.

Nauelwoort is of a moiſt ſubſtance and ſomewhat cold, and of a certaine obſcure binding qua-litie: it cooleth, repelleth, or driueth backe, ſcoureth and conſumeth, or waſteth away, as *Galen* te-ſtifieth.

‡ The Water Pennywoort is of an hot and vlcerating quality, like to the Crowfeet, whereof it is a kinde. The baſtard Italian Nauelwoort ſeemes to partake with the true in cold and moi-ſture. ‡

¶ The Vertues.

A   The juyce of Wall Pennywoort is a ſingular remedy againſt all inflammations and hot tumors, as Eryſipelas, Saint Anthonies fire and ſuch like: and is good for kibed heeles, being bathed there-with, and one or more of the leaues laid vpon the heele.

B   The leaues and roots eaten doe breake the ſtone, prouoke vrine, and preuaile much againſt the dropſie.

C   The ignorant Apothecaries doe vſe the Water Pennywort in ſtead of this of the wall, which they cannot doe without great error, and much danger to the patient: for husbandmen know well, that it is noiſome vnto Sheepe, and other cattell that feed thereon, and for the moſt part bringeth death vnto them, much more to men by a ſtronger reaſon.

CHAP. 152. *Of Sea Pennywoort.*

1 *Androsace Matthioli.*
Sea Nauel-woort.

2 *Androsace annua spuria.*
One Sommers Nauell-woort.

¶ *The Description.*

1 THe Sea Nauel-woort hath many round thicke leaues like vnto little saucers, set vpon small and tender stalks, bright, shining, and smooth, of two inches long, for the most part growing vpon the furrowed shells of cockles or the like, euery small stem bearing vpon the end or point, one little buckler and no more, resembling a nauell: the stalke and leafe set together in the middle of the same. Whereupon the Herbarists of Montpelier haue called it *Vmbilicus Marinus*, or sea Nauell. The leaues and stalkes of this plant, whilest they are yet in the water, are of a pale ash colour, but being taken forth, they presently wax white, as Sea Mosse, called *Corallina*, or the shell of a Cockle. It is thought to be barren of seed, and is in taste saltish.

2 The second *Androsace* hath little smooth leaues, spred vpon the ground like vnto the leaues of small Chickweed or Henbit, whereof doubtles it is a kind: among which riseth vp a slender stem, hauing at the top certaine little chaffie floures of a purplish colour. The seed is contained in small scaly husks, of a reddish colour, and a bitter taste. The whole plant perisheth when it hath perfected his seed, and must be sowne againe the next yere: which plant was giuen to *Matthiolus* by *Cortusus*, who (as he affirmeth) receiued it from Syria; but I thinke hee said so to make *Matthiolus* more joyfull: but surely I surmise hee picked it out of one old wal or other, where it doth grow euen as the small Chickweed, or Nailewoort of the wall do.

‡ The figure that was here was that vnperfect one of *Matthiolus*; and the description of our Author was framed by it, vnlesse the last part thereof, which was taken out of the *Aduersaria pag. 166.* to amend both these, wee here present you with the true figure and description, taken out of the workes of the iudicious and, painefull Herbarist *Carolus Clusius.* It hath (saith hee) many leaues lying flat vpon the ground, like to those of Plantaine, but lesser, and of a pale greene colour, and toothed about the edges, soft also and juycie, and of somewhat a biting taste. Amongst these leaues rise vp fiue or six stalkes of an handfull high, commonly of a greene, yet sometimes of a purple colour, naked and somwhat hairy, which at their tops carry in a circle fiue roundish leaues also a little toothed and hairy; from the midst of which arise fiue or more foot-stalkes, each bearing a greenish rough or hairie cup parted also into fiue little leaues or jags, in the midst of which stands a little white floure parted also into fiue; after which succeed pretty large seed vessels which

which containe an vnequall red seed like that of Primroses, but bigger : the root is single and slender, and dies as soone as the seed is perfected. It growes naturally in diuers places of Austria, and amongst the corne about the Bathes of Baden ; whereas it floures in Aprill, and ripens the seed in May and Iune. ‡

¶ The Place.

Androsace will not grow any where but in water:great store of it is about Frontignan by Montpellier in Languedoc, where euery fisher-man doth know it.

The second groweth vpon old stone and mud walls : notwithstanding I haue (the more to grace *Matthiolus* great jewell) planted it in my garden.

¶ The Time.

The bastard *Androsace* floureth in Iuly, and the seed is ripe in August.

¶ The Names.

Androsace is of some called *Vmbilicus marinus*, or sea Nauell.

‡ The second is knowne and called by the name of *Androsace altera Matthioli*. ‡

¶ The Temperature.

The sea Nauell is of a diureticke qualitie, and more dry than *Galen* thought it to be, and lesse hot than others haue deemed it : there can no moisture be found in it.

¶ The Vertues.

A   Sea Nauel-woort prouoketh vrine, and digesteth the filthinesse and sliminesse gathered in the joynts.

B   Two drams of it, as *Dioscorides* saith, drunke in wine, bring downe great store of vrine out of their bodies that haue the dropsie, and it maketh a good plaister to ease the paine of the gout.

CHAP. 153. *Of Rose-root, or Rosewoort.*

Rhodia Radix.
Roose-root.

¶ The Description.

ROsewoort hath many small, thicke, and fat stems, growing from a thicke and knobby root:the vpper end of it for the most part standeth on the ground, and is there of a purplish colour, bunched & knobbed like the root of Orpin, with many hairy strings hanging therat, of a pleasant smel when it is broken, like the damaske rose, whereof it tooke his name. The leaues are set round about the stalks, euen from the bottome to the top, like those of the field Orpin, but narrower, and more snipt about the edges. The floures grow at the top, of a faint yellow colour.

¶ The Place.

It groweth very plentifully in the North part of England, especially in a place called Ingleborough Fels, neere vnto the brookes sides, and not elsewhere that I can as yet finde out, from whence I haue had plants for my garden.

¶ The Time.

It floureth and flourisheth in Iune, and the seed is ripe in August.

¶ The Names.

Some haue thought it hath taken the name *Rhodia* of the Island in the Mediterranean sea, called Rhodes:but doubtlesse it took his name *Rhodia radix*, of the root which smelleth like a rose : in English, Rose-root, and Rose-woort.

¶ The Vertues.

There is little extant in writing of the faculties of Rosewoort : but this I haue found, that if the root be stamped with oile of Roses and laid to the temples of the head, it easeth the paine of the head.

CHAP.

CHAP. 154.   Of Sampier.

1 *Crithmum marinum.*
Rocke Sampier.

2 *Crithmum spinosum.*
Thorny Sampier.

3 *Crithmum chrysanthemum.*
Golden Sampier.

¶ *The Description.*

1 ROcke Sampier hath many fat and thicke leaues somwhat like those of the lesser Purslane, of a spicie taste, with a certain saltnesse ; amongst which rises vp a stalk diuided into many smal spraies or sprigs, on the top whereof grow spoky tufts of white floures, like the tufts of Fennell or Dill ; after that comes the seed, like the seed of Fenell, but greater : the root is thicke and knobby, beeing of smell delightfull and pleasant.

2 The second Sampier, called *Pastinaca marina,* or sea Parsnep, hath long fat leaues very much jagged or cut euen to the middle rib, sharp or prickely pointed, which are set vpon large fat jointed stalks ; on the top wherof do grow tufts of whitish or else reddish floures. The seed is wrapped in thorny husks: the root is thicke and long, not vnlike to the Parsenep, very good and wholsome to be eaten.

3 Golden Sampier bringeth forth many stalks from one root, compassed about with a multitude of long fat leaues, set together by equal distances : at the top whereof come yellow floures. The seed is like those of the rock Sampier.

Yy   ¶ *The*

¶ *The Place.*

Rocke Sampier growes on the rocky clifts at Douer, Winchelfey, by Rie, about Southampton, the Ifle of Wight, and moft rocks about the Weft and North parts of England.

The fecond groweth neere the fea vpon the fands and Baych betweene Wnitftable and the Ifle of Tenet, by Sandwich, and by the fea neere Weft-chefter.

The third growes in the myrie marfh in the Ifle of Shepey, as you go from the Kings Ferrey to Sherland houfe.

¶ *The Time.*

Rocke Sampier flourifheth in May and Iune, and muft be gathered to be kept in pickle in the beginning of Auguft.

¶ *The Names.*

Rock Sampier is called in Greeke κρίθμον : in Latine *Crithmum*, and of diuers *Bati* : in fome fhops, *Creta marina*: of *Petrus Crefcentius*, *Cretamum*, and *Rincum marinum* : in high-Dutch, Meerfenchel: which is in Latine *Fœniculum marinum*, or fea Fennell : in Italian, *Fenocchio marino*, *Herba di San Pietro* ; and hereupon diuers name it *Sampetra*: in Spanifh, *Perexil de la mer*, *Hinoio marino*, *Fenol marin* : in Englifh, Sampier, rock Sampier, and of fome, Creftmarine : and thefe be the names of the Sampier generally eaten in fallads.

The other two be alfo *Crithma* or Sampiers, but moft of the later Writers would draw them to fome other plant ; for one calls the fecond *Paftinaca marina*, or fea Parfnep ; and the third, *After Atticus marinus* ; and *Lobel* names it *Chryfanthemum littoreum* : but wee had rather entertaine them as *Matthiolus* doth, among the kindes of *Crithmum* or Sampier.

¶ *The Temperature.*

Sampier doth dry, warm, and fcoure, as *Galen* faith.

¶ *The Vertues.*

A The leaues, feeds, and roots, as *Diofcorides* faith, boiled in wine and drunk, prouoke vrine and womens ficknefſe, and preuaile againft the jaundice.

B The leaues kept in pickle, and eaten in fallads with oile and vineger, is a pleaſant fauce for meat, wholfome for the ftoppings of the liuer, milt, kidnies, and bladder : it prouoketh vrine gently : it openeth the ftoppings of the intrals, and ftirreth vp an appetite to meat.

C It is the pleafanteft fauce, moft familiar, and beft agreeing with mans body, both for digeftion of meats, breaking of the ftone, and voiding of grauell in the reins and bladder.

Chap. 155.

Of Glaffe Saltwort.

¶ *The Defcription.*

1 Glaffewort hath many groffe thicke and round ftalks a foot high, full of fat and thicke fprigs, fet with many knots or joints, without any leaues at all, of a reddifh greene colour : the whole plant refembles a branch of Corall : the root is very fmall and fingle.

2 There is another kind of Saltwort which hath been taken among the antient Herbarifts for a kinde of Sampier. It hath a little tender ftalke a cubit high, diuided into many fmall branches, fet full of little thicke leaues very narrow, fomewhat long and fharp pointed, yet not pricking ; amongft which come forth fmall feed wrapped in a crooked huske, turning round like a crooked perwinkle : the ftalks are of a reddifh colour : the whole plant is of a falt and biting taft : the root is fmall and threddy.

† 3 There is likewife another kinde of *Kali*, whereof *Lobel* maketh mention vnder the name of *Kali minus*, which is like to the laft before remembred, but altogether leffe, ‡ hauing many flender weake branches lying commonly fpred vpon the ground, and fet with many fmall round long fharp pointed leaues of a whitifh green colour: the feed is fmall and fhining, not much vnlike that of Sorrell : the root is flender with many fibres : the whole plant hath a faltifh tafte like as the former. *Dodon.* calls this *Kali album.*

¶ *The*

1 Salicornia, ſiue Kali geniculatum.
Glaſſewort, Saltwort, or Sea-graſſe.

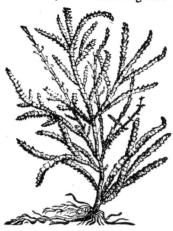

‡ *2 Kali majus ſemine cochleato.*
Snaile Glaſſewort.

‡ *3 Kali minus.*
Small Glaſſewort.

¶ *The Place.*

Theſe plants are to be found in ſalt marſhes almoſt euerie where.

‡ The ſecond excepted,which groweth not here,but vpon the coaſts of the Mediterranean ſea. ‡

¶ *The Time.*

They floure and flouriſh in the Summer moneths.

¶ *The Names.*

Saltwort is called of the Arabians, *Kali*, and *Alkali. Avicen, cap.* 724. deſcribeth them vnder the name of *Vſnen*, which differs from *Vſnee*; for *Vſnee* is that which the Grecians call *ϲρύον*: and the Latines *Muſcus*, or Moſſe : of ſome,as *Baptiſta Montanus*, it hath been iudged to be *Empetron*.

The axen or aſhes hereof, are named of *Matthiolus, Syluaticus Soda*: of moſt,*Sal Alkali* : diuers call it *Alumen catinum*. Others make this kinde of difference betweene *Sal Kali* and *Alumen catinum*, That *Alumen catinum* is the aſhes it ſelf,and that the ſalt that is made of the aſhes is *Sal Alkali*.

Stones are beaten to pouder and mixed with aſhes, which beeing melted together, become the matter whereof glaſſe is made. Which while it is made red hot in the furnace,and is melted, becomming liquid and fit to worke vpon,doth yeeld as it were a fat floting aloft;which when it is cold waxeth as hard as a ſtone,yet is it brittle and quickly broken. This is commonly called *Axunzia vitri* : in Engliſh, Sandeuer : in French,*Suin de Voirre* : in Italian,*Fior de Criſtallo, 1.* Floure of Cryſtal. The herb is alſo called of diuers, *Cali articulatum*, or jointed Glaſſewort : in Engliſh,Crab-graſſe, and Frog-graſſe.

¶ *The Temperature.*

Glaſſewort is hot and dry : the aſhes are both drier and hotter , and that euen to the fourth degree : the aſhes haue a cauſticke or burning facultie.

¶ *The Vertues.*

A   A little quantitie of the herb taken inwardly doth not only mightily prouoke vrine,but in like ſort caſteth forth the dead childe. It draweth forth by ſiege waterie humors,and purgeth away the dropſie.

B   A great quantitie taken is miſchieuous and deadly : the ſmel and ſmoke alſo of this herb being burnt driues away ſerpents.

C   The aſhes are likewiſe tempered with thoſe medicines that ſerue to take away ſcabs and filth of the skin. It eaſily conſumeth proud and ſuperfluous fleſh that groweth in poiſonſome vlcers,as *Auicen* and *Serapio* report.

D   We reade in the copies of *Serapio,* That *Kali* is a tree ſo great that a man may ſtand vnder the ſhadow thereof : but it is very like that this errour proceeds rather from the interpreter, than from the Author himſelfe.

E   The floure of Cryſtall,or as they commonly terme it Sandeuer, doth wonderfully dry : it eaſily taketh away ſcabs and mangineſſe, if the foule parts be waſhed and bathed with the water wherein it is boiled.

CHAP. 156.   *Of Thorow-Wax.*

1 *Perfoliata vulgaris.* Common Thorow-wax.	2 *Perfoliata ſiliquoſa.* Codded Thorow-wax.

¶ *The Deſcription.*

1　THorow-wax or Thorow leafe, hath a round, ſlender, and brittle ſtalke, diuided into many ſmal branches, which paſſe or go thorow the leaues, as though they had been drawn or thruſt thorow, and to make it more plain, euery branch grows thorow euery leaf, making them like hollow cups or ſaucers. The ſeed groweth in ſpoky tufts or rundles like Dill, long and blackiſh. The floures are of a faint yellow colour. The root is ſingle, white and threddy.

2　Codded Thorow-wax reckoned by *Dodonæus* among the Braſſickes or Colewoorts, hee making it a kind thereof, and calling it *Braſſica ſylveſtris perfoliata* : though in mine opinion without reaſon, ſith it hath neither ſhape, affinitie, nor likeneſſe with any of the Colewoorts, but altogether moſt vnlike, reſembling very well the common Thorow-wax ; whereunto I rather refer it. It hath ſmall, tender, and brittle ſtalkes two foot high, bearing leaues, which wrap and incloſe themſelues round about, although they do not run thorow as the other do, yet they grow in ſuch manner, that vpon the ſudden view thereof, they ſeeme to paſſe thorow as the other: vpon the ſmall branches do grow little white floures ; which being paſt, there ſucceed ſlender and long cods like thoſe of Turneps or Nauewes, whoſe leaues and cods do ſomewhat reſemble the ſame, from whence it hath the name *Napifolia*, that is, Thorow-wax with leaues like vnto the Nauew. The root is long and ſingle, and dieth when it hath brought forth his ſeed.

There is a wilde kinde hereof growing in Kent in many places among the corn, like the former in each reſpect, but altogether leſſe : the which no doubt brought into the garden would proue the very ſame.

¶ *The Place.*

‡　The firſt deſcribed growes plentifully in many places about Kent, and between Farningham and Ainsford it growes in ſuch quantitie (as I haue bin informed by Mr Bowles) in the corn fields on the tops of the hills, that it may well be termed the infirmitie of them.

The later growes not wilde with vs, that euer I could finde ; though *Lobel* ſeemes to affirme the contrarie. ‡

They grow in the gardens of Herbariſts, and in my garden likewiſe.

¶ *The Time.*

They floure in May and Iune, and their ſeed is ripe in Auguſt.

¶ *The Names.*

1　It hath been called from the beginning *Perfoliata*, becauſe the ſtalk doth paſſe through the leaſe; following the ſignification of the ſame. We call it in Engliſh, Thorow-wax, and Thorow-leaſe.

‡　2　This by the moſt and beſt part of Writers (though our Author be of another opinion) is very fitly referred to the wilde Coleworts, and called *Braſſica campeſtris* by *Cluſius* and by *Camerarius* : *Braſſica agreſtis* by *Tragus* : yet *Lobel* calls it *Perfoliata Napifolia Anglorum ſiliquoſa*. ‡

¶ *The Temperature.*

Thorow-wax is of a dry complexion.

¶ *The Vertues.*

The decoction of Thorow-wax made of water or wine healeth wounds. The juice is excellent　A for wounds, made either into oile or vnguent.

The greene leaues ſtamped, boiled wath waxe, oile, roſin, and turpentine, make an excellent Vn-　B guent or oile to incarnate, or bring vp fleſh in deep wounds.

CHAP. 157.　*Of Hony-wort.*

¶ *The Deſcription.*

1　CErinthe or Hony-wort riſeth forth of the ground after the ſowing of his ſeed, with two ſmall leaues like thoſe of Baſill; between the which leaues commeth forth a thick fat ſmooth tender and brittle ſtalke ful of juice, that diuides it ſelfe into many other branches, which alſo are diuided into ſundry other armes or branches likewiſe, crambling or leaning toward the ground, being not able without props to ſuſtain it ſelfe, by reaſon of the great weight of

　　　　　　　leaues,

1 *Cerinthe maior.*
Great Honywort.

‡ 2 *Cerinthe aſperior flore flauo.*
Rough Honywort.

3 *Cerinthe minor.*
Small Honywort.

leaues, branches, and much juice the whole plant is ſurcharged with,vpon which branches are placed many thicke rough leaues,ſet with very ſharp prickles like the rough skinne of a Thornback,of a blewiſh green colour, ſpotted very notably with white ſtrakes and ſpots,like thoſe leaues of the true *Pulmonaria* or Cow-ſlips of Ieruſalem, and in ſhape like thoſe of the codded Thorow-wax,which leaues do clip or embrace the ſtalke round about: from the boſome whereof come forth ſmall cluſters of yellow floures, with a hoope or band of bright purple round about the middeſt of the yellow floure. The floure is hollow, faſhioned like a little box,of the taſt of hony when it is ſucked, in the hollownes wherof are many ſmal chiues or threads; which beeing paſt, there ſucceed round blacke ſeed contained in ſoft skinny husks. The root periſheth at the firſt approch of winter. ‡ This varieth in the colour of the floures,which are yellow or purple, and ſome-times of both mixt together. ‡

‡ 2 The leaues of this other great Honywort of *Cluſius* deſcription,are ſhaped like thoſe laſt deſcribed,but are narrower at their ſetting on, and rougher,the floures alſo are yellow, but in ſhape & magnitude like the former,as it is al-ſo in the ſeeds & all other parts thereof. ‡

3 This

3   This other Cerinth or Honywort hath small long and slender branches, reeling this way and that way as not able to sustain it self, very brittle, beset with leaues not much vnlike the precedent, but lesser, neither so rough nor spotted, of a blewish green colour. The floures be small, hollow, and yellow : the seed is small, round, and as black as Ieat : the root is white, with some fibres, the which dieth as soon as the former. There is a taste as it were of new wax in the floures or leaues chewed, as the name doth seem to import.

¶ The Place.

These plants do not grow wilde in England, yet I haue them in my garden : the seed whereof I receiued from right honorable good friend the Lord *Zouch.*

¶ The Time.

They floure from May to August, and perish at the first approch of winter, and must be sowne a-gain the next spring.

‡ ¶ The Names.

1   The first of these by *Gesner* is called *Cynoglossa montana,* and *Cerinthe : Dodonaus* cals it *Maru herba : Lobel,* and others, *Cerinthe major.*

2   The second is *Cerinthe quorundam major flauo flore* of *Clusius.*

3   The third by *Dodonaus* is called *Maru herba minor ;* and by *Clusius, Cerinthe quorundam minor flauo flore : Lobel* also cals it *Cerinthe minor.*

¶ The Nature and Vertues.

Pliny & *Auicen* seem to agree, that these herbs are of a cold complexion ; notwithstanding there is not any experiment of their vertues worth the writing.

CHAP. 158.   Of S. Iohns Wort.

1 Hypericum.   2 Hypericum Syriacum.
S. Iohns wort.   Rue S. Iohns wort.

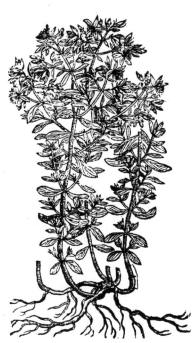
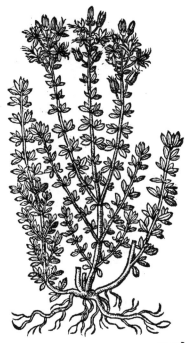

¶ The

¶ *The Description.*

1   SAint Iohns wort hath brownish stalks beset with many small and narrow leaues,which if you behold betwixt your eies and the light,do appeare as it were bored or thrust thorow in an infinite number of places with pinnes points. The branches diuide themselues into sundry final twigs,at the top whereof grow many yellow floures,which with the leaues bruised do yeeld a reddish juice of the colour of bloud, The seed is contained in little sharp pointed huskes blacke of colour,and smelling like Rosin. The root is long,yellow,and of a wooddy substance.

2   The second kind of S.Iohns wort is named *Syriacum*,of those that haue not seen the fruitful and plentifull fields of England,wherein it groweth aboundantly, hauing small leaues almost like Rue or Herb-Grace : wherein *Dodonæus* hath failed,intituling the true *Androsæmum* by the name of *Ruta syluestris* ; whereas indeed it is no more like Rue than an Apple to an Oister. This plant is altogether like the precedent,but smaller,wherein consisteth the difference. ‡ It had beene fitter for our Author to haue giuen vs a better and perfecter description of this plant(which as hee saith growes so abundantly with vs)than so absurdly to cauil with *Dodonæus*,for calling(as he saith)the true *Androsæmum*, *Ruta syluestris :* for if that be the true *Androsæmum* which *Dodonæus* made mention by the foresaid name,why did not our Authour figure and describe it in the next chapter saue one,for *Androsæmum*,but followed *Dodonæus* in figuring and describing Tutsan for it ? See more hereof in the chapter of Tutsan. I cannot say I haue seen this plant ; but *Lobel* the Author & setter forth thereof thus briefly describes it: The leaues are foure times lesse than those of ours,which grow thicke together as in rundles,vpon stalks being a cubit high. The floures are yellow,and like those of our common kinde. ‡

3   Woolsy S.Iohns wort hath many small weake branches trailing vpon the ground,beset with many little leaues couered ouer with a certain soft kinde of downinesse : among which come forth weake and tender branches charged with small pale yellow floures. The seeds and roots are like to the true S.Iohns wort.

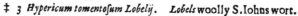

‡ *3 Hypericum tomentosum Lobelij.*   *Lobels* woolly S.Iohns wort.

‡   The figure that our Author gaue was of that which I here giue you the second in the third place,vnder the title of *Hyper.toment.Clusij :* for *Clusius* saith it was his,and blames *Lobel* for making it all one with that he found about Montpelier:whose figure also I giue you first in the third place that you may see what difference you can obserue by them : for *Clusius* saith, *Lobels* is but an handfull high ; yet tells he not vs how high his growes,neither instances how they differ, neither can I gather it by *Lobels* description : but I conjecture it thus ; That of *Clusius* his description is taller, more white and hairy,and hath the floures growing along little foot-stalks,and not in maner of an vmbel,as in the other.

‡ 4   Besides these two creeping hoary S.Iohns Worts here described,there is another small kinde which is called by *Dodonæus*, *Hypericum minus* ; and by *Lobel*, *Hypericum minimum supinum Septentrionale.* It growes some handfull or more high,with weake and slender branches set with leaues like those of the ordinarie kinde,but lesse : the floures are also like those of the first described, but fewer in number,and lesse. It is to be found in dry and barren grounds,and flours at the same time as the former.

5   I haue obserued growing in S.Iohns Wood and other places,that kinde of S.Iohns Wort which

which by *Tragus* is called *Hypericum pulchrum* ; and both by him and *Lonicerus* is thought to be *Dioscorides* his *Androsæmum* ; the which we in English may for distinctions sake call vpright S.Iohns wort. It hath roots like those of the ordinarie kinde,from which arise streight slender stalkes some cubit high,set at equall spaces with pretty smooth leaues,broad,and almost incompassing the stalk at their setting on,being sometimes of a green, otherwhiles of a reddish colour : towards the top they are parted into some few branches,which beare such yellow floures as the common kind,but somewhat smaller. It floures about the same time as the former,or a little after. ‡

3 *Hypericum tomentosum Clusij.*
Woolly S.Iohns wort of *Clusius*.

‡ 4 *Hypericum supinum glabrum.*
Small creeping S.Iohns wort.

¶ *The Place.*
They grow very plentifully in pastures in euery countrie.
¶ *The Time.*
They floure and flourish for the most part in Iuly and August.
¶ *The Names.*
S.Iohns wort is called in Greeke ὑπέρικον : in Latine *Hypericum*: in shops,*Perforata* : of diuers,*Fuga dæmonum* : in Dutch, **San Iohans kraut**: in Italian,*Hyperico* : in Spanish,*Caraconzillo* : in French *Mille Pertuys* : in English,S.Iohns wort,or S.Iohns grasse.
¶ *The Temperature.*
S.Iohns wort,as *Galen* teacheth,is hot and dry,being of substance thin.
¶ *The Vertues.*

A S.Iohns wort with his floures and seed boiled and drunke, prouoketh vrine, and is right good against the stone in the bladder,and stoppeth the laske. The leaues stamped are good to be laid vpon burnings,scaldings,and all wounds,and also for rotten and filthy vlcers.

B The leaues,floures,and seeds stamped,and put into a glasse with oile oliue,and set in the hot sun for certain weeks together,and then strained from those herbs,and the like quantitie of new put in and sunned in like manner,doth make an oile of the colour of bloud,which is a most pretious remedie for deep wounds and those that are thorow the body,for the sinues that are prickt,or any wound made with a venomed weapon. I am accustomed to make a compound oile hereof, the making of which you shall receiue at my hands,because I know that in the world there is not a better,no not the naturall Balsam it selfe ; for I dare vndertake to cure any such wound as absolutely in each respect,if not sooner and better,as any man shall or may with naturall Balsam.

C Take white wine two pintes,oile oliue foure pounds,oile of Turpentine two pounds,the leaues, floures,and seeds of S.Iohns wort of each two great handfulls gently bruised ; put them all together into a great double glasse,and set it in the Sun eight or ten daies ; then boile them in the same glasse *per Balneum Mariæ*,that is,in a kettle of water, with some straw in the bottome, wherein the glasse must stand to boile : which done,strain the liquor from the herbs, and do as you did before, putting in the like quantitie of herbs,floures,and seeds, but not any more wine. Thus haue you a great secret for the purposes aforesaid.

Dioscorides

Dioscorides saith, That the seed drunke for the space of forty daies together cureth the Sciatica, and all aches that happen in the hips.

The same Author saith, That being taken in wine it takes away tertian and quartan Agues.

CHAP. 159.

Of S. Peters wort or square S. Iohns grasse.

1 *Ascyron.*
S. Peters wort.

¶ *The Description.*

1   SAint Peters wort groweth to the height of a cubit and a halfe, hauing a straight vpright stalk somwhat brown, set by couples at certaine distances, with leaues much like those of S. Iohns Wort, but greater, rougher, and rounder pointed: from the bosom of which leaues come forth many small leaues, the which are not bored through, as those of S. Iohns Wort are; yet somtimes there be some few so bored through. The floures grow at the top of the branches of a yellow colour: the leaues and floures when they are bruised do yeeld forth a bloudy juice as doth S. Iohns wort, whereof this is a kind. The root is tough, and of a wooddy substance.

‡ 2 Vpon diuers boggy grounds of this kingdome is to be found growing that S. Peters Wort which *Clusius* describes in his *Auctarium* by the name of *Ascyrum supinum* ιωνε. This sends forth diuers round hairy creeping stalks, which heere and there put vp new fibres or roots, and these are set at certain spaces with very round and hairy leaues of a whitish colour, two at a joint, and on the tops of these stalks grow a few small yellow floures which consist of fiue leaues apiece. These stalks seldome send forth branches, vnlesse it be one or two at the tops: it may well be called in English, Round leaued S. Peters wort. ‡

¶ *The Place.*

S. Peters wort or S. Iohns grasse groweth plentifully in the North parts of England, especially in Landsdale and Crauen: I haue found it in many places of Kent, especially in a copse by Mr. *Sidleys* house neere South-fleet.

¶ *The Time.*

It floureth and flourisheth when S. Iohns wort doth.

¶ *The Names.*

It is called in Greeke ἀσκυρον: the Latines haue no other name but this Greek name *Ascyron*. It is called of some *Androsæmum: Galen* makes it both a kind of Tutsan and S. Iohns wort, & saith it is named *Ascyron*, or *Ascyroides*: in English, S. Peters wort, Square or great S. Iohns grasse, and of some Hardhay. Few know it from S. Iohns wort.

¶ *The Temperature.*

This herb is of temperature hot and dry.

¶ *The Vertues.*

A   It is endued with the same vertues that S. Iohns wort is. The seed, saith *Dioscorides* and *Galen*, being drunk in foure ounces and a halfe of Meade, doth plentifully purge by siege cholericke excrements.

CHAP.

CHAP. 160.   Of Tutſan or Parke leaues.

¶ *The Deſcription.*

1   THe ſtalks of Tutſan be ſtraight, round, chamfered or creſted, hard and wooddy, beeing for the moſt part two foot high. The leaues are three or foure times bigger than thoſe of S. Iohns wort, which be at the firſt green; afterwards and in the end of Summer of a darke red colour: out of which is preſſed a juice not like blacke bloud, but Claret or Gaſcoigne wine. The floures are yellow, and greater than thoſe of S. Peters wort; after which riſeth vp a little round head or berry, firſt green, afterwards red, laſt of all blacke, wherein is contained yellowiſh red ſeed: the root is hard, wooddy, and of long continuance.

‡   2   This, which *Dodonæus* did not vnfitly call *Ruta ſylveſtris Hypericoides*, and which others haue ſet forth for *Androſæmum*, and our Author the laſt chapter ſaue one affirmed to be the true *Androſæmum*, (though here it ſeemes he had either altered his mind, or forgot what he formerly wrot) may fitly ſtand in competition with the laſt deſcribed, which may paſſe in the firſt place for the *Androſæmum* of the Antients; for *adhuc ſub judice lis eſt*. I will not here inſiſt vpon the point of controuerſie, but giue you the deſcription of the plant, which is this: It ſends vp round ſlender reddiſh ſtalks ſome two cubits high, ſet with fewer yet bigger leaues than the ordinarie S. Iohns wort, and theſe alſo more hairy: the floures and ſeeds are like thoſe of the common S. Iohns wort, but ſomewhat larger: it growes in ſome mountainous wooddy places; and in the *Aduerſaria* it is called *Androſamum excellentius, ſeu magnum*: and by *Dodonæus* (as we but now noted) *Ruta ſylveſtris Hypericoides*, thinking it to be the *Ruta ſylueſtris* which is deſcribed by *Dioſcorides, lib.3. cap.48.* in the old Greeke edition of *Manutius*, κᾳ. υμς. And in that of *Marcellus Virgilius* his interpretation, in the chapter and booke but now mentioned; but rejected amongſt the *Nothi* in the Paris edition, *Anno* 1549. You may finde the deſcription alſo in *Dodonæus, Pempt. prima, lib.3. cap.25.* whither I refer the Curious, being loth here to inſiſt farther vpon it.   ‡

1 *Clymenon Italorum.*
Tutſan or Parke leaues.

‡ 2 *Androſamum Hypericoides.*
Tutſan S. Iohns wort.

¶ *The Place*.

Tutſan groweth in woods and by hedges,eſpecially in Hampſted wood, where the Golden Rod doth grow ; in a wood by Railie in Eſſex,and many other places.

¶ *The Time*.

It floures in Iune and Auguſt: the ſeed in the mean time waxeth ripe.　The leaues become red in Autumne,at which time is very eaſily preſſed forth his winy iuice.

¶ *The Names*.

It is called in Greeke *αἰδροταίμων* : and the Latines alſo *Androſamon* : it is likewiſe called *Dionyſia*, as *Galen* witneſſeth. They are farre from the truth that take it to be *Clymenum*, and it is needleſſe to finde fault with their error. It is alſo called *Siciliana*,and *Herba Siciliana* : in Engliſh, Tutſan, and Parke leaues.

¶ *The Temperature*.

The faculties are ſuch as S.Peters wort,which doth ſufficiently declare it to be hot and dry.

¶ *The Vertues*.

A　　The ſeed hereof beaten to pouder,and drunke to the weight of two drams,doth purge cholerick excrements,as *Dioſcorides* writeth ; and is a ſingular remedie for the Sciatica,prouided that the patient drinke water for a day or two after purging.

B　　The herbe cureth burnings , and applied vpon new wounds it ſtancheth the bloud and healeth them.

C　　The leaues laid vpon broken ſhins and ſcabbed legs heale them,and many other hurts & griefs ; whereof it tooke his name Tout-ſaine or Tutſane,of healing all things.

---

‡ Cʜᴀᴘ. 161.　*Of baſtard S.Iohns wort.*

‡ 1 *Coris Matthioli.*
Matthiolus his baſtard S.Iohns wort.

‡ 2 *Coris cærulea Monſpeliaca.*
French baſtard S.Iohns wort.

‡　THe diligence of theſe later times hath been ſuch to finde out the *Materia Medica* of the Antients,that there is ſcarce any Plant deſcribed by them,but by ſome or other of late there haue been two or more ſeuerall plants referred thereto : and thus it hath happened vnto that
which

which *Dioſcorides lib.3.cap.*174. hath ſet forth by the name of *Coris*; and preſently deſcribes after the kindes of *Hypericon*, and that with theſe words; ἡ Ν η ὦτι ὑἀφιαμχαχόμ.: Some alſo call this *Hypericon*; to which *Matthiolus* and others haue fitted a plant, which is indeed a kinde of *Hypericon*, as you may perceiue by the figure and deſcription which I giue you in the firſt place. Some (as *Heſychius*) re-ferre it to *Chamæpitys*, (and indeed by *Dioſcorides* it is placed betweene *Androſæmon* and *Chamæpitys*) and to this that which is deſcribed by *Pena* and *Lobel* in the *Aduerſ.* and by *Cluſius* in his Hiſtorie, may fitly be referred : this I giue you in the ſecond place.

¶ *The Deſcription.*

1　THe firſt hath a wooddy thicke and long laſting root, which ſendeth vp many branches ſome foot or more high, and it is ſet at certaine ſpaces with round leaues like thoſe of the ſmall Glaſſe-wort or Sea-Spurry, but ſhorter : the tops of the ſtalkes are diuided into ſundry branches, which carry floures like thoſe of S. Iohns woort, of a whitiſh red colour, with threds in their middles hauing little yellow pendants. It growes in Italy and other hot countries, in places not far from the ſea ſide. This is thought to be the true *Coris*, by *Matthiolus*, *Geſner*, *Lonice-rus*, *Lacuna*, *Bellus*, *Pena*, and others.

2　This from a thicke root red on the outſide ſendeth vp ſundry ſtalkes, ſome but an handfull, other ſome a foot or more long, ſtiffe, round, purpliſh, ſet thicke with leaues like thoſe of Heath, but thicker, more ſucculent and bitter, which ſometimes grow orderly, and otherwhiles out of or-der. The ſpikes or heads grow on the tops of the branches, conſiſting of a number of little cups, diuided into fiue ſharpe points, and marked with a blacke ſpot in each diuiſion : out of theſe cups comes a floure of a blew purple colour, of a moſt elegant and not fading colour; and it is compo-ſed of foure little bifide leaues, whereof the two vppermoſt are the larger : the ſeed, which is round and blackiſh, is contained in ſeed veſſels hauing points ſomewhat ſharpe or prickly. It floures in Aprill or May, and is to be found growing in many places of Spaine, as alſo about Mompelier in France; whence *Pena* and *Lobel* called it *Coris Monſpeliaca*; and *Cluſius*, *Coris quorundam Gallorum & Hiſpanorum.*

¶ *The Temperature.*

Theſe plants ſeeme to be hot in the ſecond or third degree.

¶ *The Vertues.*

Dioſcorides ſaith, That the ſeeds of *Coris* drunke mooue the courſes and vrine, are good againſt the　A biting of the Spider *Phalangium*, and the Sciatica; and drunke in Wine, againſt that kinde of Convulſions which the Greekes call *Opiſthotonos*, (which is when the boby is drawne backwards) as alſo againſt the cold fits in Agues. It is alſo good annointed with oyle, againſt the aforeſaid Convulſions. ‡

Cʜᴀᴘ. 153. *Of the great Centorie.*

¶ *The Deſcription.*

1　THe great Centorie bringeth forth round ſmooth ſtalkes three cubits high : the leaues are long, diuided as it were into many parcels like to thoſe of the Walnut tree, and of an ouerworne grayiſh colour, ſomewhat ſnipt about the edges like the teeth of a ſaw. The floures grow at the top of the ſtalkes in ſcaly knaps like the great Knapweed, the middle thrums whereof are of a light blew or sky colour : when the ſeed is ripe the whole knap or head turneth into a downy ſubſtance like the head of an Artichoke, wherein is found a long ſmooth ſeed, bearded at one end like thoſe of Baſtard Saffron, called *Cartamus*, or the ſeed of *Carduus Bene-dictus*. The root is great, long, blacke on the outſide, and of a ſanguine colour on the inſide, ſome-what ſweet in taſte, and biting the tongue.

2　There is likewiſe another ſort, hauing great and large leaues like thoſe of the water Docke, ſomewhat ſnipt or toothed about the edges. The ſtalke is ſhorter than the other, but the root is more oleous or fuller of juyce, otherwiſe like. The floure is of a pale yellow purpliſh colour, and the ſeed is like that of the former.

1 *Centaurium magnum.*
Great Centorie.

‡ 2 *Centaurium majus alterum.*
Whole leaued great Centorie.

¶ *The Place.*

The great Centorie ioyeth in a fat and fruitfull ſoile, and in Sunny bankes full of Graſſe and herbes. It groweth very plentifully, ſaith *Dioſcorides*, in Lycia, Peloponneſus, Arcadia, and Morea: and it is alſo to be found vpon Baldus a Mountaine in the territories of Verona, and likewiſe in my Garden.

¶ *The Time.*

It floureth in Summer, and the roots may be gathered in Autumne.

¶ *The Names.*

It is called in Greeke, Κεντκύριον τὸ μέγα : of *Theophraſtus* alſo *Centauris* · in diuers ſhops falſly *Rha Ponticum*: for *Rha Ponticum* is *Rha* growing in the countries of Pontus ; a plant differing from great Centorie. *Theophraſtus* and *Pliny* ſet downe among the kindes of *Panaces* or All-heales, this great Centorie, and alſo the leſſer, whereof we will write in the next chapter following. *Pliny* reciting the words of *Theophraſtus*, doth in his twenty fifth booke, and fourth chapter write, that they were found out by *Chiron* the Centaure, and ſurnamed *Centauria*. Alſo affirming the ſame thing in his ſixth chapter (where he more largely expoundeth both the Centauries) hee repeateth them to be found out by *Chiron*: and thereupon he addeth, that both of them are named *Chironia*. Of ſome it is reported, That the ſaid *Chiron* was cured therewith of a wound in his foot, that was made with an arrow that fell vpon it when he was entertaining *Hercules* into his houſe ; whereupon it was called *Chironium*: or of the curing of the wounds of his ſouldiers, for the which purpoſe it is moſt excellent.

¶ *The Temperature.*

It is hot and dry in the third degree. *Galen* ſaith, by the taſte of the root it ſheweth contrarie qualities, ſo in the vſe it performeth contrary effects.

¶ *The Vertues.*

A   The root taken in the quantitie of two drams is good for them that be burſten, or ſpit bloud; againſt the crampe and ſhrinking of ſinewes, the ſhortneſſe of winde or difficulty of breathing, the cough and gripings of the belly.

B   There is not any part of the herbe but it rather worketh miracles than ordinary cures in greene wounds ; for it ioyneth together the lips of ſimple wounds in the fleſh, according to the firſt intention, that is, glewing the lips together, not drawing to the place any matter at all.

The

☞ The root of this plant (ſaith *Dioſcorides*) is a remedy for ruptures, convulſion, and cramps, taken　C
in the weight of two drams, to be giuen with wine to thoſe that are without a feuer, and vnto thoſe
that haue, with water.

Galen ſaith, that the juyce of the leaues thereof performeth thoſe things that the root doth ;　D
which is alſo vſed in ſtead of *Lycium*, a kinde of hard juyce of a ſharpe taſte.

Cʜᴀᴘ. 142. *Of Small Centorie.*

¶ *The Deſcription.*

1　THe leſſer Centorie is a little herbe : it groweth vp with a cornered ſtalke halfe a foot
high, with leaues in forme and bigneſſe of S. Iohns wort: the floures grow at the top in
a ſpoky buſh or rundle, of a red colour tending to purple ; which in the day time and
after the Sun is vp do open themſelues, but towards euening ſhut vp againe : after them come forth
ſmall ſeed-veſſels, of the ſhape of wheat cornes ; in which are contained very little ſeeds. The root
is ſlender, hard, and ſoone fading.

2　The yellow Centory hath leaues, ſtalkes, and ſeed like the other, and is in each reſpect as
like, ſauing that the floures hereof are of a perfect yellow colour, which ſetteth forth the difference.

‡　This is of two ſorts ; the one with broad leaues through which the ſtalkes paſſe ; and the
other hath narrow leaues like thoſe of the common Centorie. ‡

1 *Centaurium parvum.*
Small Centorie.

2 *Centaurium parvum luteum Lobelij.*
Yellow Centorie.

¶ *The Place.*

1　The firſt is growing in great plenty throughout all England, in moſt paſtures and graſſie
fields.

2　The yellow doth grow vpon the chalkie cliffes of Greenehithe in Kent, and ſuch like places.

¶ *The Time.*

They are to be gathered in their flouring time, that is in Iuly and Auguſt: of ſome that gather them ſuperſtitiouſly they are gathered betweene the two Lady daies.

¶ *The Names.*

The Greekes call this, ꝃⲉⲛⲧⲁⲩⲣⲓⲟⲛ ⲙⲓⲕⲣⲟⲛ: in Latine it is called *Centaurium minus* ; yet *Pliny* nameth it *Libadion*, and by reaſon of his great bitterneſſe, *Fel terra*. The Italians in Hetruria call it, *Biondella* : in Spaniſh, *Centoria* : in low Dutch, **Centozype** : in Engliſh, ſmall, little, or common Centorie : in French, *Centoire*.

¶ *The Temperature.*

The ſmall Centorie is of a bitter quality, and of temperature hot and dry in the ſecond degree; and the yellow Centorie is hot and dry in the third degree.

¶ *The Vertues.*

A   Being boyled in water and drunke it openeth the ſtoppings of the liuer, gall, and ſpleene, it helpeth the yellow jaundiſe, and likewiſe long and lingering agues : it killeth the wormes in the belly ; to be briefe, it clenſeth, ſcoureth, and maketh thinne humours that are thicke, and doth effectually performe whatſoeuer biting things can.

B   *Dioſcorides*, and *Galen* after him report, that the decoction draweth downe by ſiege choler and thicke humors, and helpeth the Sciatica ; but though we haue vſed this often and luckily, yet could we not perceiue euidently that it purges by the ſtoole any thing at all, and yet it hath performed the effects aforeſaid.

C   This Centorie being ſtamped and laid on whileſt it is freſh and greene, doth heale and cloſe vp greene wounds, cleanſeth old vlcers, and perfectly cureth them.

D   The juyce is good in medicines for the eies ; mixed with hony it cleanſeth away ſuch things as hinder the ſight ; and being drunke it hath a peculiar vertue againſt the infirmities of the ſinues, as *Dioſcorides* teacheth.

E   The Italian Phyſitians do giue the pouder of the leaues of yellow Centorie once in three daies in the quantity of a dram, with anniſe or caraway ſeeds, in wine or other liquor, which preuaileth againſt the dropſie and greene ſickeneſſe. Of the red floured, *Ioannes Poſtius* hath thus written :

Flos mihi ſuaue rubet, ſed ineſt quoque ſuccus amarus,
Qui juuat obſeſſum bile, aperitque jecur.

My floure is ſweet in ſmell, bitter my juyce in taſte,
Which purge choler, and helps liuer, that elſe would waſte.

C H A P. 164.   *Of Calues ſnout, or Snapdragon.*

¶ *The Deſcription.*

1   THe purple Snapdragon hath great and brittle ſtalks, which diuideth it ſelfe into many fragile branches, whereupon do grow long leaues ſharpe pointed, very greene, like vnto thoſe of wilde flax, but much greater, ſet by couples one oppoſite againſt another. The floures grow at the top of the ſtalkes, of a purple colour, faſhioned like a frogs mouth, or rather a dragons mouth, from whence the women haue taken the name Snapdragon. The ſeed is blacke, contained in round huskes faſhioned like a calues ſnout, (whereupon ſome haue called it Calues ſnout) or in mine opinion it is more like vnto the bones of a ſheeps head that hath beene long in the water, or the fleſh conſumed cleane away.

2   The ſecond agreeth with the precedent in euery part, except in the colour of the floures, for this plant bringeth forth white floures, and the other purple, wherein conſiſts the difference.

3   The yellow Snapdragon hath a long thicke wooddy root, with certain ſtrings faſtned thereto; from which riſeth vp a brittle ſtalke of two cubits and a halfe high, diuided from the bottome to the top into diuers branches, whereupon do grow long greene leaues like thoſe of the former, but greater and longer. The floures grow at the top of the maine branches, of a pleaſant yellow colour, in ſhape like vnto the precedent.

4   The ſmall or wilde Snapdragon differeth not from the others but in ſtature : the leaues are leſſer and narrower : the floures purple, but altogether ſmaller : the heads or ſeed-veſſels are alſo like thoſe of the former.

‡   5   There is another kinde hereof which hath many ſlender branches lying oftentimes vpon the ground : the leaues are much ſmaller than theſe of the laſt deſcribed : the floures and ſeed-veſſels are alſo like, but leſſer, and herein conſiſts the onely difference. ‡

¶ *The*

1 2 *Antirrhinum purpureum ſiue album.*
Purple or white floured Snapdragon.

3 *Antirrhinum luteum.*
Yellow Snapdragon.

4 *Antirrhinum minus.*
Small Snapdragon.

‡ 5 *Antirrhinum minimum repens.*
Small creeping Snapdragon.

¶ *The Place.*

The three first grow in most gardens; but the yellow kinde groweth not common, except in the gardens of curious Herbarists.

‡ The fourth and fifth grow wilde among corne in diuers places. ‡

¶ *The Time.*

That which hath continued the whole Winter doth floure in May, and the rest of Summer afterwards; and that which is planted later, and in the end of Summer, floureth in the Spring of the following yeare: they do hardly endure the injurie of our cold Winter.

¶ *The Names.*

Snapdragon is called in Greeke, *ἀντίρινον*: in Latine also, *Antirrhinum*: of *Apuleius*, *Canis cerebrum*, *Herba Simiana*, *Venusta minor*, *Opalis grata*, and *Orontium*: it is thought to be *Leo herba*, which *Columella, lib.* 10. reckons among the floures: yet *Gesner* hath thought that this *Leo* is *Columbine*, which for the same cause he hath called *Leontostomium*: but this name seemeth to vs to agree better with Calues snout than with columbine; for the gaping floure of Calues snout is more like to Lyons snap than the floure of Columbine: it is called in Dutch, **Oȝant:** in Spanish, *Cabeza de ternera*: in English, Calues snout, Snapdragon, and Lyons snap: in French, *Teste de chien*, and *Teste de Veau*.

¶ *The Temperature.*

They are hot and dry, and of subtill parts.

¶ *The Vertues.*

A   The seed of Snapdragon (as *Galen* saith) is good for nothing in the vse of Physicke; and the herb it selfe is of like faculty with *Bubonium* or Star-wort, but not so effectuall.

B   They report (saith *Dioscorides*) that the herbe being hanged about one preserueth a man from being bewitched, and that it maketh a may gracious in the sight of people.

C   *Apuleius* writeth, that the distilled water, or the decoction of the herbe and root made in water, is a speedy remedy for the watering of eies proceeding of a hot cause, if they be bathed therewith.

Chap. 165.   *Of Tode-flax.*

1 *Linaria vulgaris lutea.*
Great Tode-flax.

2 *Linaria purpurea odorata.*
Sweet purple Tode-flax.

¶ *The Deſcription.*

1   Inaria being a kinde of *Antirrhinum*, hath ſmall, ſlender, blackiſh ſtalkes; from which do grow many long narrow leaues like flax. The floures be yellow, with a ſpur hanging at the ſame like vnto a Larkes ſpur, hauing a mouth like vnto a frogs mouth, euen ſuch as is to be ſeene in the common Snapdragon; the whole plant before it come to floure ſo much reſembleth *Eſula minor*, that the one is hardly knowne from the other, but by this old verſe:

Eſula lacteſcit, ſive lacte Linaria creſcit.

‡ *Eſula* with milke doth flow,
   Toad-flax without milke doth grow. ‡

2   The ſecond kinde of Tode-flax hath leaues like vnto *Bellis maior*, or the great Daſie, but not ſo broad, and ſomewhat jagged about the edges. The ſtalke is ſmall and tender, of a cubit high, beſet with many purple floures like vnto the former in ſhape. The root is long, with many threds hanging thereat, the floures are of a reaſonable ſweet ſauour.

3   The third, being likewiſe a kinde of Tode-flax, hath ſmall and narrow leaues like vnto the firſt kinde of *Linaria*: the ſtalke is a cubit high, beſet with floures of a purple colour, in faſhion like *Cinaria*, but that it wanteth the taile or ſpurre at the end of the floure which the other hath. The root is ſmall and threddy.

† 3 *Linaria purpurea altera.*
Variable Tode-flax.

† 4 *Linaria Valentina Cluſ.*
Tode-flax of Valentia.

† 4   *Linaria Valentina* hath leaues like the leſſer Centorie, growing at the bottome of the ſtalke by three and three, but higher vp towards the top, without any certaine order: the ſtalkes are of a foot high; and it is called by *Cluſius, Valentina*, for that it was found by himſelfe in *Agro Valentino*, about Valentia in Spaine, where it beareth yellow floures about the top of the ſtalke like common *Linaria*, but the mouth of the floure is downy, or moſſie, and the taile of a purple colour. It floureth at Valentia in March, and groweth in the medowes there, and hath not as yet been ſeene in theſe Northerne parts.

5   *Oſyris alba* hath great, thicke, and long roots, with ſome threds or ſtrings hanging at the ſame, from which riſe vp many branches very tough and pliant, beſet towards the top with floures not much vnlike the common Toad-flax, but of a pale whitiſh colour, and the inner part of the mouth ſomewhat more wide and open, and the leaues like the common Tode-flax,

† 6 *Oſyris*

† 5 *Oſyris alba, Lob.*
White Tode-flax.

6 *Oſyris purpurocærulea* is a kinde of Tode-flax that hath many ſmall and weake branches, trailing vpon the ground, beſet with many little leaues like flax. The floures grow at the top of the ſtalke like vnto the common kinde, but of a purple colour declining to blewneſſe.The root is ſmall and threddy.

‡ 7   This hath many ſmall creeping bran-ches ſome handfull or better high,and hath ſuch leaues,floures,and ſeed,as the common kind,but all of them much leſſe,and therein conſiſte th the difference. It growes naturally in the dry fields about Salamanca in Spaine,and floures all Sum-mer long. *Lobel* cals it *Oſyris flava ſylueſtris* ; and *Cluſius*,*Linaria Hiſpanica.*

8   The branches of this eight kind are ſpred vpon the ground,and of the length of thoſe of the laſt deſcribed : the leaues are leſſer than thoſe of the common Tode-flax,thicke, juycie, and of a whitiſh greene colour, and they grow not diſor-derly vpon the ſtalks,but at certain ſpaces,ſome-times three,but moſt vſually foure together : the floures in ſhape are like thoſe of the ordinarie kinde, but of a moſt perfect Violet co lour, and the lower lip where it gapes of a golden yellow; the taſte is bitter.After the floures are paſt come veſſels round & thick,which contain a flat black ſeed in two partitions or cels:the root is ſlender, white, and long laſting, and it floures vnto the end of Autumne. It growes naturally vpon the higheſt Alps.*Geſner* cals it, *Linaria Alpina:* and *Cluſius*, *Linaria tertia Styriaca.* ‡

† *6 Oſyris Purpurocærulea repens.*   Purple Tode-flax.

† 9   Foraſmuch as this plant is ſtalked and leafed like common flaxe, and thought by ſome to be *Oſyris* ; the new writers haue called it *Lynoſyris :* it hath ſtalkes very ſtiffe and wooddy, beſet with leaues like the common *Linaria*, with floures at the top of the ſtalkes of a faint ſhining yel-low colour,in forme and ſhape ſomewhat like vnto *Conyza major*. The whole plant groweth to the height of two cubits, and is in taſte ſharpe and clammie, or glutinous,and ſomewhat bitter. The root is compact of many ſtrings,intangled one within another.

† 10   *Guillandinus* calleth this plant *Hyſſopus vmbellifera Dioſcoridis*,that is,*Dioſcorides* his Hy-ſope,

ſope, which beareth a tuft in all points like *Linoſyris*, whereof it is a kinde, not differing from it in ſhew and leaues. The ſtalkes are a cubit high, diuided aboue into many ſmall branches, the tops wherof are garniſhed with tufts of ſmall floures, each little floure being parted into fiue parts with a little thred or peſtle in the middle, ſo that it ſeemes full of many golden haires or thrums. The ſeed is long and blackiſh, and is carried away with the winde. ‡ *Bauhine* in his *Pinax* makes this all one with the former, but vnfitly, eſpecially if you marke the deſcription of their floures which are far vnlike. *Fabius Columna* hath proued this to be the *Chryſocome* deſcribed by *Dioſc. lib.* 4. *cap.* 55. ‡

‡ 7 *Oſyris flaua ſylueſtris.*   Creeping yellow Tode-flax.

‡ 8 *Linaria quadrifolia ſupina.*
Foure leaued creeping Tode-flax.

† 9 *Linoſyris Nuperorum, Lob.*
Golden Star-faſhioned Tode-flax.

10 *Linaria.*

10 _Linaria aurea Tragi._
Golden Tode-flax.

11 _Scoparia ſive Oſyris Græcorum._
Buſhie or Beſome Tode-flax.

† 12 _Paſſerina linaria folio, Lob._
Sparrowes Tode-flax.

† 13 _Paſſerina altera._
Sparrow-tongue.

‡ 14 *Linaria adulterina.*
Baſtard Tode-flax.

† 11 *Scoparia*, or after *Dodonæus*, *Oſyris*, which the Italians cal *Belvidere*, hath very many ſhoots or ſprigs riſing from one ſmall ſtalk, making the whole plant to reſemble a Cypres tree, the branches grow ſo handſomely : now it growes ſome three foot high, and very thick and buſhie, ſo that in ſome places where it naturally groweth they make beſomes of it, whereof it tooke the name *Scoparia*. The leaues be ſmall and narrow, almoſt like to the leaues of flax. The floures be ſmall, and of an herby colour, growing among the leaues, which keep greene all the Winter. ‡ I neuer knew it here to ripen the ſeed, nor to outliue the firſt froſt. ‡

12 This plant alſo for reſemblance ſake is referred to the Linaries, becauſe his leaues be like *Linaria*. At the top of the ſmall branched ſtalkes doe grow little yellowiſh floures, pale of colour, ſomewhat like the tops of *Chryſocome*. *Iohn Mouton* of Turnay taketh it to be *Chryſocome altera*. And becauſe there hath been no accordance among Writers, it is ſufficient to ſet forth his deſcription with his name *Paſſerina*. ‡ *Bauhine* refers it to the *Gromills*, and cals it, *Lithoſpermum Linariæ folio Monſpeliacum.*

‡ 13 This which *Tabern.* calls *Lingua Paſſerina*, and whoſe figure was giuen by our Authour for the former, hath a ſmall ſingle whitiſh root, from which it ſends vp a ſlender ſtalke ſome cubit and halfe high, naked on the lower part, but diuided into little branches on the vpper, which branches are ſet thicke with little narrow leaues like thoſe of Winter Sauorie or Tyme : amongſt which grow many little longiſh ſeeds of the bigneſſe and taſte of Millet, but ſomewhat hotter and bitterer. The floures conſiſt of foure ſmall yellow leaues. *Tragus* calls this *Paſſerina*; *Dodonæus* makes it, *Lithoſpermum minus* : and *Columna* hath ſet it forth by the name of *Linaria altera botryodes montana.*

14 This which *Cluſius* hath ſet forth by the name of *Anonymos*, or Nameleſſe, is called in the *Hiſt. Lugd. pag.* 1150. *Anthyllis montana* ; and by *Tabern. Linaria adulterina.* It hath many hard pale greene branches of ſome foot high; and vpon theſe without any order grow many hard narrow long leaues like thoſe of flaxe, at firſt of a very tart, and afterwards of a bitteriſh taſte : the tops of the ſtalkes are branched into ſundry foot ſtalkes, which carry little white floures conſiſting of fiue ſmall leaues lying ſtarre-faſhion, with ſome threds in their middles : after which at length come ſingle ſeeds fiue cornered, containing a white pith in a hard filme or skin. The root is white, diuided into ſundry branches, and liues long, euery yeare ſending vp many ſtalkes, and ſometimes creeping like that of Tode-flax. It floures in May, and growes vpon mountainous places of Germany ; Mr *Goodyer* found it growing wilde on the ſide of a chalkie hill in an incloſure on the right hand of the way, as you go from Droxford to Poppie hill in Hampſhire. ‡

¶ *The Place.*

The kindes of Tode-flax grow wilde in many places, as vpon ſtone walls, grauelly grounds, barren medowes, and along by hedges.

‡ I do not remember that I haue ſeene any of theſe growing wilde with vs, vnleſſe the firſt ordinary kinde, which is euery where common. ‡

¶ *The Time.*

They floure from Iune to the end of Auguſt.

¶ *The Names.*

† Tode-flax is called of the Herbariſts of our time, *Linaria*, or Flax-weed, and *Vrinalis*: of ſome, *Oſyris*, in high Dutch, **Lynkraut**, and **Onſer fraumen flaſch**: in low Dutch, **vlaſt Ulas**: in Engliſh, Wild-flax, Tode-flax, and Flax-weed : the eleuenth is called in Italian, *Bel-videre*, or faire in ſight. The ſame plant is alſo called *Scoparia*, and *Herba ſtudioſorum*, becauſe it is a fit thing to make brooms

of,

of, wherewith ſchollers and ſtudents may ſweepe their owne ſtudies and cloſets. The particular names are expreſſed both in Latine and Engliſh in their ſeuerall titles, whereby they may be diſtinguiſhed. ‡ It is thought by moſt, that this *Beluidere,* or *Scoparia,* is the *Oſyris* deſcribed by *Dioſcorides, lib.4.cap.*143. For beſides the notes it hath agreeing with the deſcription, it is at this day by the Greekes called *ἄξυρις.* ‡

¶ The Temperature.

The kindes of Tode-flax are of the ſame temperature with wilde Snap-dragons, whereof they are kindes.

¶ The Vertues.

A   The decoction of Tode-flax taketh away the yellowneſſe and deformitie of the skinne, being waſhed and bathed therewith.

B   The ſame drunken, openeth the ſtoppings of the Liuer and Spleene, and is ſingular good againſt the jaundiſe which is of long continuance.

C   The ſame decoction doth alſo prouoke vrine, in thoſe that piſſe drop after drop, vnſtoppeth the kidnies and bladder.

† The figures in this chapter were moſt of them falſe placed, as thus: The third was of *Linaria Penon.*1. of *Cluſius,* being the *Linaria alba* of *Lobel,* deſcribed in the fifth place. The fourth was of the *Oſyris flauaſil.* of *Lobel,* deſcribed here by mee in the ſeuenth place. The fifth was of *Linaria* 3 *Striaca* of *Cluſius,* which you finde deſcribed by me in the eighth place. The ſixth was of *Linaria aurea minor* of *Taberm* being onely a variety of the *Linaria aurea* ſet forth in the tenth place. The ſeuenth was of the *Linaria Adulterina,* whoſe hiſtorie I haue giuen you in the fourteenth place. That which was formerly vnder the title of *Paſſerina Linaria* in with a hiſtorie fitted thereto in the thirteenth place.

C H A P . 146. Of Garden flaxe.

† *Linum ſatiuum.*
Garden flax.

¶ *The Deſcription.*

FLax riſeth vp with ſlender and round ſtalks The leaues thereof bee long, narrow, and ſharpe pointed : on the tops of the ſprigs are faire blew floures, after which ſpring vp little round knobs or buttons, in which is contained the ſeed, in forme ſomewhat long, ſmooth, glib or ſlipperie, of a darke colour. The roots be ſmall and thready.

¶ *The Place.*

It proſpereth beſt in a fat and fruitfull ſoile, in moiſt and not dry places ; for it requireth, as *Columella* ſaith, a very fat ground, and ſomewhat moiſt. Some, ſaith *Palladius,* do ſow it thicke in a leane ground, and by that meanes the flaxe groweth fine. *Pliny* ſaith that it is to be ſowne in grauelly places, eſpecially in furrowes : *Nec magis feſtinare aliud :* and that it burneth the ground, and maketh it worſe: which thing alſo *Virgil* teſtifieth in his Georgickes.

Vrit lini campum ſeges, vrit Auena.
Vrunt lethæo perfuſa papauera ſomno.

In Engliſh thus :

Flaxe and Otes ſowne conſume
    The moiſture of a fertile field :
The ſame worketh Poppy, whoſe
    Iuyce a deadly ſleepe doth yeeld.

¶ The

¶ *The Time.*

Flaxe is ſowne in the ſpring, it floureth in Iune and Iuly. After it is cut downe (as *Pliny*, *lib.19.cap.1.*ſaith)the ſtalks are put into the water,ſubject to the heat of the Sun,& ſome weight laid on them to be ſteeped therein;the looſenes of the rinde is a ſigne when it is well ſteeped:then is it taken vp and dried in the Sun,and after vſed as moſt huſwiues can tell better than my ſelfe.

¶ *The Names.*

It is called both in Greeke and Latine *also, Linum:* in high Dutch, **Flachß:**in Italian and Spaniſh,*Lino:* in French,*Dulin:* in low Dutch,**Ulag:**in Engliſh,Flax,and Lyne.

¶ *The Temperature and Vertues.*

Galen in his firſt booke of the faculties of nouriſhments ſaith, that diuers vſe the ſeed hereof parched as a ſuſtenance with *Garum*, no otherwiſe than made ſalt.　**A**

They alſo vſe it mixed with hony,ſome likewiſe put it among bread;but it is hurtfull to the ſtomack,and hard of digeſtion,and yeelds to the body but little nouriſhment : but touching the quality which maketh the belly ſoluble,neither will I praiſe or diſpraiſe it;yet that it hath ſome force to prouoke vrine,is more apparant when it is parched : but then it alſo ſtaieth the belly more.　**B**

The ſame author in his books of faculties of ſimple medicines ſaith, that Lineſeed being eaten is windy although it be parched,ſo full it is of ſuperfluous moiſture: & it is alſo after a ſort hot in the firſt degree,and in a meane betweene hot and dry.But how windy the ſeed is,and how full of ſuperfluous moiſture it is in euery part, might very well haue beene perceiued a few yeares ſince as at Middleborough in Zeland,where for want of grain and other corne,moſt of the Citizens were fain to eat bread and cakes made hereof with hony and oile, who were in ſhort time after ſwolne in the belly below the ſhort ribs, faces, and other parts of their bodies in ſuch ſort, that a great number were brought to their graues therby:for theſe ſymptomes or accidents came no otherwiſe than by the ſuperfluous moiſture of the ſeed which cauſeth windineſſe.　**C**

Lineſeed, as *Dioſcorides* hath written,hath the ſame properties that Fenugreek hath : it waſteth away & mollifieth al inflammations or hot ſwellings,as wel inward as outward,if it be boiled with hony,oile,and a little fair water, and made vp with clarified hony; it taketh away blemiſhes of the face,and the Sun burning,being raw and vnboiled;and alſo foule ſpots,if it be mixed with ſalt-peter and figs:it cauſeth rugged and ill fauored nails to fall off,mixed with hony and water-Creſſes.　**D**

It draweth forth of the cheſt corrupted flegme and other filthy humours, if a compoſition with hony be made thereof to licke on,and eaſeth the cough.　**E**

Being taken largely with pepper and hony made into a cake, it ſtirreth vp luſt.　**F**

The oile which is preſſed out of the ſeed, is profitable for many purpoſes in Phyſicke and Surgerie;and is vſed of painters,picture makers,and other artificers.　**G**

It ſoftneth all hard ſwellings ; it ſtretcheth forth the ſinewes that are ſhrunke and drawne together,mitigateth pain,being applied in manner of an ointment.　**H**

Some alſo giue it to drinke to ſuch as are troubled with pain in the ſide and collick;but it muſt be freſh and newly drawne : for if it be old and ranke,it cauſeth aptneſſe to vomit,and withall it ouermuch heateth.　**I**

Lineſeed boiled in water with a little oile, and a quantity of Anniſe ſeed, impoudered and implaiſtred vpon an *angina*,or any ſwelling in the throat,helpeth the ſame.　**K**

It is with good ſucceſſe vſed plaiſterwiſe,boiled in vineger,vpon the diſeaſes called *Coliaca* and *Dyſenteria*,which are bloudy fluxes and paines of the belly.　**L**

The ſeeds ſtamped with the roots of wild Cucumbers,draweth forth ſplinters,thornes, broken bones,or any other thing fixed in any part of the body.　**M**

The decoction is an excellent bath for women to ſit ouer for the inflammation of the ſecret parts,becauſe it ſoftneth the hardneſſe thereof,and eaſeth paine and aking.　**N**

The ſeed of Line and Fenugreek made into pouder,boiled with Mallowes,Violet leaues,Smallage, and Chickweed,vntill the herbs be ſoft ; then ſtamped in a ſtone morter with a little hogges greaſe to the forme of a cataplaſme or pulteſſe,appeaſeth all manner of paine,ſoftneth all cold humors or ſwellings,mollifieth and bringeth to ſuppuration all apoſtumes ; defends wounded members from ſwelling and rankling, and when they be already rankled,it taketh the ſame away,being applied very warme euening and morning.　**O**

† The figure that was formerly in this place, for the ordinary flax was of *Linum ſylueſtre latifolium* 3. of *Cluſius*, which is deſcribed by me in the ſixth place in the enſuing Chapter.

C H A P. 167.   *Of Wilde Flaxe.*

¶ *The Deſcription.*

1   THis wilde kinde of Line or Flaxe hath leaues like thoſe of garden Flaxe, but narrower, growing vpon round bright and ſhining ſprigs, a foot long, and floures like the manured Flax, but of a white colour. The root is tough and ſmall, with ſome fibres annexed thereto. ‡ This is ſometimes found with deep blew floures, with violet colored floures, and ſometimes with white, ſtreaked with purple lines.

1 *Linum ſylueſtre floribus albis.*   2 *Linum ſylueſtre tennifolium.*
Wilde white flaxe.   Thin leaued wilde flaxe.

2   The narrow and thin leafed kind of Line is very like to the common Flax, but in all points leſſer. The floures conſiſt of fiue leaues, which do ſoone fade and fall away, hauing many ſtalkes proceeding from one root, of a cubit high, beſet with ſmall leaues, yea leſſer than thoſe of *Linaria purpurea.*

‡   Our Author in the former edition gaue two figures vnder this one title of *Linum ſylueſtre tenuifolium,* making them the ſecond & third; but the deſcription of the third was of the rough broad leaued wilde Flaxe, whoſe figure therefore we haue put in that place. Now the two whoſe figures were formerly here are but varieties of one ſpecies, and differ thus; the former of them (whoſe figure we haue omitted as impertinent) hath fewer leaues, which therefore ſtand thinner vpon the ſtalke, and the floures are either blew or elſe white. The later, whoſe figure you may finde here ſet forth, hath more leaues, and theſe growing thicker together: the floure is of a light purple, or fleſh colour. ‡

3   There is a kinde of wilde flaxe which hath many hairy branches, riſing vp from a very ſmall root, which doth continue many yeres without ſowing, increaſing by roots into many other plants, with ſtalks amounting to the height of one cubit, beſet with many rough and hairy broad leaues: at the top of the ſtalkes do grow many blew floures, compact of fiue leaues, much greater and fairer than common Line or flaxe; which being paſt, there ſucceed ſmall ſharpe pointed heads full of ſeeds, like Line ſeed, but of a blackiſh ſhining colour.

4 *Chamalinum*

4 *Chamælinum,*of ſome called *Linum ſylveſtre perpuſillum*,and may be called in Engliſh,very low or dwarfe wilde Flax : for this word *Chamæ* ioyned to any Simple,doth ſignifie, that it is a low or dwarfe kinde thereof; being ſcarce an handful high,hath pale and yellow floures : but as it is in all things like vnto flax,ſo the floures,leaues, ſtalkes, and all other parts thereof, are foure times leſſer than *Linum.*

‡ 5　There is alſo growing wilde in this kingdome a ſmall kinde of wilde Flax, which I take to be the *Linocarpos* deſcribed by *Thalius*,and mentioned by *Camerarius*,by the name of *Linum ſylue-ſtre puſillum candicantibus floribus. Anno* 1529, when I firſt found it, in a Iournall written of ſuch plants as I gathered, I ſet this by the name of *Linum ſylveſtre puſillum candidis floribus* ; which my friend Mr *Iohn Goodyer* ſeeing,he told me he had long knowne the plant,and refer'd it to the Lines : but there were ſome which called it in Engliſh Mil-mountain,and vſed it to purge ; and of late he hath ſent me this hiſtorie of it,which you ſhall haue as I receiued it from him.

Linum ſylveſtre Catharticum. Mil-mountaine

It riſeth vp from a ſmall white thready crooked root, ſometimes with one, but moſt commonly with fiue or ſix or more round ſtalks,about a foot or nine inches high,of a brown or reddiſh color, euery ſtalk diuiding it ſelfe neere the top,or from the middle vpward,into many parts or branches of a greener colour than the lower part of the ſtalke : the leaues are ſmall,ſmooth,of colour green, of the bigneſſe of Lentill leaues,and haue in the middle one rib or ſinew,and no more that may be perceiued,and grow alongſt the ſtalke in very good order by couples one oppoſit againſt the other : at the tops of the ſmall branches grow the floures,of a white colour,conſiſting of fiue ſmall leaues apiece,the nailes whereof are yellow : in the inſide are placed ſmall ſhort chiues alſo of a yellow colour ; after which come vp little knobs or buttons, the top whereof when the ſeed is ripe diuides it ſelfe into fiue parts ; wherein is contained ſmall ſmooth flat ſlipperie yellow ſeed:when the ſeed is ripe the herb periſheth : the whole herb is of a bitter taſt,and herby ſmell. It groweth plentiful-ly in the vnmanured incloſures of Hampſhire,on chalky downes,and on Purfleet hils in Eſſex,and in many other places. It riſeth forth of the ground at the beginning of the ſpring,and floureth all the Summer.

‡ 3 *Linum ſylveſtre latifolium.*
Broad leaued wilde Flax.

4 *Chamælinum perpuſillum,*
Dwarfe wilde Flax.

 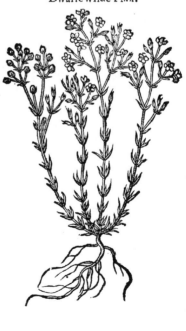

‡ 5 *Linum ſyl. catharticum.*
Mill-mountaine.

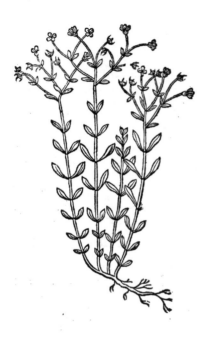

‡ 4 *Linum ſyl. latifolium 3.Cluſ.*
The third broad leaued wild Flax.

‡ 7 *Linum marinum luteum.*
Yellow floured wilde Flax.

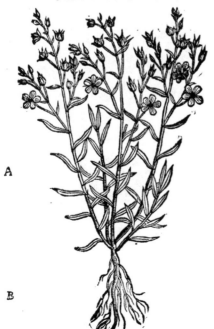

A

B

I came to know this herbe by the name of Mill-mountaine, and his vertue, by this meanes : On the ſecond of October, 1617, going to Mr *Colſons* ſhop an Apothecarie of Wincheſter in Hampſhire, I ſaw this herb lying on his ſtall , which I had ſeene growing long before ; I deſired of him to know the name of it ; he told mee that it was called Mill-mountaine ; and hee alſo told me, That beeing at Doctour *Lake* his houſe at Saint Croſſe a mile from Wincheſter, ſeeing a man of his haue this herb in his hand, he deſired the name : hee told him as before ; and alſo the vſe of it, which is this :

Take a hand full of Mil-mountain, the whole plant, leaues, ſeeds, floures and all, bruiſe it and put it in a ſmal tunne or pipkin of a pinte filled with white Wine, and ſet it in on the embers to infuſe all night, and drinke that Wine in the morning faſting, and hee ſaid it would giue eight or ten ſtooles. This Dr *Lake* was afterwards made Biſhop of Bathe and Wels, who alwaies vſed this herbe for his purge , after the ſaid manner, as his man affirmed, Iuly 20, 1619. *Iohn Goodyer.*

I haue not as yet made trial hereof, but ſince in *Geſn. de Lunarijs, p.* 34. I haue found the

the like or a more purging facultie attributed to this herb, as I thinke (for I cannot refer it to any other) where he would haue it to be *Helleborine* of the Antients. I thinke it not amisse here to set downe his words, becaufe the booke is not commonly to be had, being fet forth *Anno* 1555 : *Ante annos 15. aut circiter, cum Anglicus ex Italia radiens, me salutaret* (Turnerus *is fuerit, vir excellentis tum in re medica, tum alijs plerisque disciplinis doctrina, aut alius quispiam, vix satis memini) inter alias rariorum stirpium icones quas depingendas commodabat, Elleborinem quoque ostendebat pictam, herbulam fruticosam, pluribus ab vna radice cauliculis quinque fere digitorum proceritate erectis, foliolis perexiguis, binis per intervalla (eiusmodi vt ex aspectu genus quoddam Alsine exiguum videretur) vasculis in summo exiguis, rotundis tanquam lini. Hanc ajebat crescere in pratis siccis, vel clivis Montium: inutili radice, subamara, purgare vtrinque & in Anglia vulgo vsurpari a rusticis.* Thus much for Gesner.

6 *Clusius* amongst other wilde Lines or Flaxes hath set forth this, which from a liuing thicke wrythen root sends vp many stalkes almost a cubit high, fomewhat red and stiffe, set with prettie large and thicke leaues not rough and hairy, but smooth and hard: the floures grow plentifully on the tops of the stalkes, being large, and composed of fiue leaues of a faire yellow colour, with fiue threds comming forth of their middles, with as many smaller and shorter haires. The seed is contained in flatter heads than those of the first described, containing a blacke but not shining seed. It floures in Iune and Iuly, and ripens the seed in August. It growes naturally vpon diuers hills in Germany.

7 *Matthiolus* and *Dodoneus* haue vnder the name of *Linum sylvestre*; and *Lobel* by the name of *Linum marinum luteum Narbonense*, set forth another yellow floured wilde flax. This growes with slender stalks some cubit high, set with leaues like those of flaxe, but somewhat lesser, and fewer in number: at the tops of the stalks grow floures smaller than those of the common Lyne, and yellow of colour. It growes naturally vpon the coasts of France that lie towards the Mediterranean sea, but not in England that I haue heard of. ‡

¶ *The Place.*

They grow naturally in grauelly grounds: the first growes in wel manured places, as in gardens and such like soiles. The second groweth by the sea side. The third and fourth grow vpon rockes and cliffes neere the sea side: I haue seen them grow vpon the sea banks by Lee in Essex, & in many places of the Isle of Shepey. They grow also between Quinborow and Sherland house.
‡ I haue not seen any of these growing wilde, but only the fift of my description. ‡

¶ *The Time.*

They floure from May to the end of August.

¶ *The Names.*

Their names are sufficiently exprest in their seuerall titles.

¶ *The Nature and Vertues.*

The faculties of these kindes of wilde flax are referred vnto the manured flax, but they are seldome vsed either in physicke or surgerie.

CHAP. 168. *Of black Saltwort.*

¶ *The Description.*

IN old time, saith the Authors of the *Aduersaria*, this plant was vsed for meat, & receiued among the *Legumina*. It was called *Glaux* by reason of the colour of the leaues, which are of a blewish gray colour, called in Latine *Glaucus color*, such as is in the Sallow leaf: of others it is called *Galax* or *Glax*; and *Engalacton, quasi lactea*, or *Lactifica*, becaufe it is good to encreafe milk in the brests of women, if it be much vsed. *Ruellius* and others haue set downe *Galega, Securidica, Polygala*, & many other plants for the true *Glaux*, which hath bred a confusion. The true *Glaux* of *Dioscorides* hath many small branches, some creeping on the ground, and some standing vpright, tender and smal, be set with many little fat leaues like *Tribulus sylvestris*, or *Herniaria*, growing along the stalks by couples, between which grow small purple floures: which beeing past, there succeed little bullets or seedvessels. The root is very small and thready, and taking hold of the vpper face of the earth, it runnes abroad, by which means it mightily encreafeth.

Glaux exigua maritima.
Blacke Saltwort.

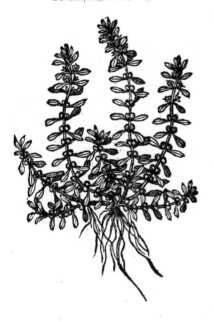

¶ *The Place.*

The true *Glaux* or Milkwort groweth very plentifully in ſalt places & marſhes neer
the ſea, from whence I haue brought it into my garden, where it proſpereth as wel as
in his natiue ſoile. I found it eſpecially betweene Whitſtable and the Iſle of Thanet
in Kent, & by Graueſend in the ſame coun-
tie, by Tilberry Blockhouſe in Eſſex, and in
the Iſle of Shepey, going from Kings ferry
to Sherland houſe.

¶ *The Time.*

It floureth in May, and the ſeed is ripe in
Iune.

¶ *The Names.*

The names haue been ſufficiently ſpoken
of in the deſcription. It ſhall ſuffice to call
it in Engliſh, ſea Milkwort.

¶ *The Nature.*

Paulus Ægineta ſaith it is hot and moiſt
of temperature.

¶ *The Vertues.*

This Milkwort taken with milk, drinke,
or pottage, ingendereth ſtore of milke, and
therefore it is good to be vſed by Nurſes
that want the ſame.

Chap. 169.   Of Milkewort.

¶ *The Deſcription.*

1   THere haue been many plants neerely reſembling *Polygala*, and yet not the ſame indeed,
which doth verifie the Latine ſaying, *Nullum ſimile eſt idem*. This neere reſemblance
doth rather hinder thoſe that haue ſpent much time in the knowledge of Simples, than
increaſe their knowledge : and this alſo hath been an occaſion that many haue imagined a ſundrie
Polygala vnto themſelues, and ſo of other plants. Of which number this whereof I ſpeake is one, ob-
taining this name of the beſt writers and herbariſts of our time, deſcribing it thus : It hath many
thicke ſpreading branches creeping on the ground, bearing leaues like thoſe of *Herniaria*, ſtanding
in rowes like the ſea Milkwort ; among which grow ſmal whorles or crownets of white floures, the
root being exceeding ſmall and thready.

2   The ſecond kinde of *Polygala* is a ſmall herbe with pliant ſlender ſtemmes, of a wooddy ſubſtance, an handfull long, creeping by the ground : the leaues be ſmall and narrow like to Lintels, or
little Hyſſop. The floures grow at the top, of a blew colour, faſhioned like a little bird, with wings,
taile, and body eaſie to be diſcerned by them that do obſerue the ſame: which being paſt, there ſucceed ſmall pouches like thoſe of *Burſa paſtoris*, but leſſer. The root is ſmall and wooddy.

3   This third kinde of *Polygala* or Milkwort, hath leaues and ſtalkes like the laſt before mentioned, and differeth from it only herein, that this kinde hath ſmaller branches, and the leaues are not
ſo thicke thruſt together, and the floures are like the other, but that they be of a red or purple colour.

4   The fourth kinde is like the laſt ſpoken of in euery reſpect, but that it hath white floures, o-
therwiſe it is very like.

5   Purple Milkewort differeth from the others in the colour of the floures , it bringeth forth
moe branches than the precedent, and the floures are of a purple colour, wherin eſpecially conſiſts
the difference.

1 *Polygala repens.*
Creeping Milkwort.

2 *Polygala flore cæruleo.*
Blew Milkewort.

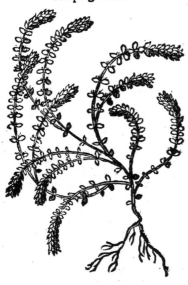

3 *Polygala rubris floribus.*
Red Milkwort.

4 *Polygala albis floribus.*
White Milkewort.

6   The ſixt Milkwort is like vnto the reſt in each reſpect, ſauing that the floures are of an ouer-worne ilfauored colour, which maketh it to differ from all the other of his kinde.

5 *Polygala purpurea.*
Purple Milkwort.

A

B

¶ *The Place.*

Theſe plants or Milke-worts grow commonly in euery wood or fertil paſture whereſoeuer I haue tra-uelled.

¶ *The Time.*

They floure from May to Auguſt.

¶ *The Names.*

Milkwort is called by *Dodonæus, Flos Ambaruualis,* becauſe it doth eſpecially floure in the Croſſe or Gangweeke, or Rogation weeke: of which floures the maidens which vſe in the countries to walke the Proceſſion do make themſelues garlands and Noſe-gaies: in Engliſh wee may call it Croſſe-floure, or Proceſſion floure, Gang-floure, Rogation-floure, and Milkwort, of their vertues in procuring milke in the breſts of nurſes. *Hieronymus Tragus* as alſo *Dioſcori-des* call it *Polygalon.* ‡ *Geſner* calls this *Crucis flos;* and in his Epiſtles he nameth it *Amarella:* it is vulgarly knowne in Cheapeſide to the herbe women by the name of hedge Hyſſop; for they take it for *Gratiola* or hedge Hyſſop, and ſell it to ſuch as are ignorant for the ſame. ‡

¶ *The Vertues.*

Galen, Dioſcorides, and *Theophraſtus* doe account theſe for Milke-woorts, and that they may without errour be vſed for thoſe purpoſes whereunto *Glaux* ſerueth.

‡ I doubt that this is not the *Polygalon* of *Dio-ſcorides:* for *Geſner* affirmes, That an handfull hereof ſteeped all night in wine, and drunk in the morning, faſting, will purge choler effectually by ſtoole without any danger, as he himſelfe had tried. ‡

Chap. 170. *Of Knot-graſſe.*

¶ *The Deſcription.*

1   THe common male Knot-graſſe creepes along vpon the ground, with long ſlender weak branches full of joints or knots, whereof it tooke his name. The leaues grow vpon the weake branches like thoſe of ſmall S. Iohns wort, but longer and narrower. The floures are maruellous little, and grow out of the knots, of an hearby colour; in their places come vp trian-gular ſeed: the root is long, ſlender, and full of ſtrings.

2   The ſecond differeth not from the former, but only that it is altogether leſſer, wherein eſpe-cially conſiſteth the difference. ‡ Becauſe the difference is no otherwiſe I haue thought good to omit the figure.

3   The Authors of the *Aduerſaria* mention another larger Knot-graſſe, which growes in diuers places of the coaſt of the Mediterranian ſea, hauing longer and larger branches and leaues, & thoſe of a white ſhining colour. The ſeeds grow at the joints in chaffie white huſks, and the whole plant is of a ſalt and aſtringent taſte. They call it *Polygonum marinum maximum* ‡

¶ *The Place.*

Theſe Knot-graſſes grow in barren and ſtony places almoſt euery where.

¶ *The Time.*

They are in floure and ſeed all the Summer long.

1 *Polygonum mas vulgare.*
Common Knot-grasse.

¶ *The Names.*

Knot-grasse is called in Greeke, πολυγονον αρρεν : that is to say, *Polygonum mas,*or male Knot-grasse : in Latine,*Seminalis,Sanguinaria* : of *Columella, Sanguinalis* : in shops, *Centumnoda,* and *Corrigiola* : of *Apuleius,Proserpinaca:* in high-Dutch,ꝲoagꝺꝛyt: in low-Dutch,Uerkens gras,& Ꝺuiſent knop: in Italian,*Polygono* : in Spanish,*Carriola:*in French *Renovee:*in Wallon,*Mariolaine de Cure:*in English Knot-grasse, and Swines grasse : in the North, Birds tongue.

¶ *The Temperature.*

Knot-grasse, as *Galen* teacheth, is of a binding qualitie, yet is it cold in the second if not in the beginning of the third degree.

¶ *The Vertues.*

The juice of Knot-grasse is good against the **A** spitting of bloud,the pissing of bloud,and al other issues or fluxes of bloud, as *Brasauolus* reporteth : and *Camerarius* saith he hath cured many with the juice thereof,that haue vomited bloud, giuen in a little stiptick wine. It greatly preuaileth against the Gonorrhæa or running of the reines, and the weakenesse of the back comming by means thereof,being shred and made in a tansie with eggs,and eaten.

The decoction of it cureth the disease afore- **B** said in as ample manner as the juice:or giuen in pouder in a reare egge it helpeth the backe very much.

The herbe boiled in wine and honey cureth the vlcers and inflammations of the secret parts of **C** man or woman,adding thereto a little allom,and the parts washed therewith.

Dioscorides saith that it prouoketh vrine,and helpeth such as do pisse drop after drop, when the **D** vrine is hot and sharp.

It is giuen vnto Swine with good successe,when they are sicke and will not eat their meat:wher- **E** vpon country people do call it Swines grasse,or Swines skir.

Chap. 171.
Of sundry sorts of Knot-grasses.

¶ *The Description.*

1   THe snowy white and least kinde of *Polygonum* or Knot-grasse, called of *Clusius, Paronj-chia Hispanica,*is a strange and worthy plant to behold,handle,and consider,although it be but small.It is seldome aboue a foot long,hauing small branches,thick,rough,hard, and full of joints ; out of which the leaues come forth like smal teeth,lesse than the leaues of *Herniaria,*or *Thymum tenuifolium.* At the top of the stalks stand most delicate floures,framed by nature as it were with fine parchment leaues about them,standing in their singular whitenesse and snowie colour,resembling the perfect white silke, so many in number at the top, and so thicke, that they ouershadow the rest of the plant beneath. The root is slender and of a wooddy substance : the seed is couered as it were with chaffe,and is as small as dust or the motes in the sun.

2   *Anthyllis* of Valentia being likewise a kind of Knot-grasse, hath small leaues like *Glaux exigua,* or rather like *Chamasyce,*set orderly by couples at the joints : among which come floures consisting of foure little whitish purple leaues,and other small leaues like the first, but altogether lesser. The root is small,blacke,and long,and of a wooddy substance.

‡   Our Author, though hee meant to haue giuen vs the figure of Knawel in the third place, as may be perceiued by the title, yet he described it in the fourth, and in the third place went about

to

1 *Polygonum montanum.*   Mountaine Knot-graſſe.

‡ 2 *Anthyllis Valentina Cluſij.*   Valentia Knot-graſſe.

‡ 3 *Polygonum ſerpillifolium.*
Small round leaued Knot-graſſe.

† 4 *Polygonum Selinoides, ſiue Knawel.*
Parſley-Piert.

to deſcribe *Polygonum Serpillifolia* of *Pena*; as may be gathered by the deſcription which ſhould haue ſtood, but that I opportunely receiued a better from my oft mentioned friend M^r.*Goodyer*, which therefore I thought good to impart vnto you.

Polygonum alterum puſillo vermiculato Serpilli foliolo Pena.

This hath many ſmall round ſmooth wooddy branches, ſomwhat reddiſh, trailing on the ground, nine inches or a foot long; wheron by ſmall diſtances on ſhort joints grow tufts of very ſmall ſhort blunt topped ſmooth greene leaues, in a manner round, like thoſe of the ſmalleſt Time, but much ſmaller, and without ſmel, diuiding themſelues at the boſoms of thoſe leaues into ſmall branches; at the tops of which branches grow ſmall floures, one floure on a branch, and no more, conſiſting of foure little round topped leaues apiece of a faint or pale purpliſh colour: I obſerued no ſeed. The root is wooddy, blackiſh without, very bitter, with ſome taſt of heate, and groweth deepe into the ground. The leaues are nothing ſo full of juice as Aizoon. I found it flouring the third day of September, 1621, on the ditch banks at Burſeldon ſerrey by the ſea ſide in Hampſhire. *Io.Goodyer.* ‡

4　Among the Knot-graſſes may well be ſuted this ſmal plant, but lately written of, and not ſo commonly knowne as growing in England, being about an handfull high, and putting out from a fibrous root ſundry ſlender ſtalkes full of little branches and joints: about which grow confuſedly many narrow leaues, for the moſt part of an vnequall quantity, yet here and there two longer than the reſt, and much alike in greatneſſe: at the outmoſt part of the branches and ſtalks (where it hath thickeſt tufts) appeare out of the middeſt of the leaues little floures of an herby colour, which are ſucceeded by ſeed-veſſels ending in fiue ſharp points: the whole plant is of a whitiſh colour. If my memorie faile me not, *Pena* means this herbe where he ſpeaketh of *Saxifr. Angl.* in his *Adver.p.*103. and alſo reporteth that he found this plant by the way ſide as he rode from London to Briſtow, on a litle hill not far from Chipnam: his picture doth very well reſemble the kinde of Knot-graſſe called among the Germanes **Knawel**: and calling it *Saxifraga Anglicana* cauſeth me to thinke, that ſome in the Weſt parts where he found it do call it Saxifrage, as we do call ſundry other herbs, eſpecially if they ſerue for the ſtone. My friend M^r.*Stephen Bredwell*, Practitioner of Phyſicke in thoſe parts, heard of a ſimple man who did much good with a medicine that he made with Parſley Piert againſt the ſtone, which he miniſtred vnto all ſorts of people. This my friend requeſted the poore man to ſhew him the hearbe called Parſley Piert; who frankly promiſed it him, and the next morning brought him an handfull of the herb, and told him the compoſition of his medicin withall, which you ſhal find ſet downe in the vertues, and proued by ſundry of good account to be a ſingular remedie for the ſame.

† 5　*Saxifraga Anglicana alſinefolia.*
Chick-weed Breake-ſtone.

‡ 6　*Saxifraga paluſtris alſinefolia.*
Small water Saxifrage.

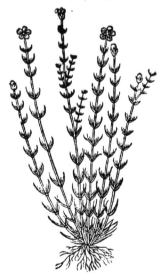

‡ 5 Our

‡ 4　Our Author here in the fourth place deſcribed the Knawel,& he figured it in the ſecond place,vnder the title of *Anthyllis Valentina Cluſij*:for the figure which was in the third place we here giue you in the fifth;and I conjecture it is not of Knawel, but thought to bee of *Saxifraga anglicana* of the *Adverſ*. But the conjecture of *Pena* and *Lobel* being true,who judge their *Saxifraga Anglicana* to be *Synanchice Daleſchampij*,then is it neither *Knawel*, as our Authour had it,nor this which I giue,but a ſmall plant which you ſhal find among the *Rubia's*.Now this plant that I once took here to be *Saxifraga Anglicana* of *Pena* and *Lobel*,is a ſmall littleherbe growing thicke,with very many branches ſome two or three inches high,with ſome ſtalks ſtanding vpright,and other ſome creeping: at each joynt grow two ſhort narrow ſharp pointed greene leaues, out of whoſe boſomes come diuers leſſer leaues:at the tops of the branches vpon prety long ſtalks grow vpon each ſtalk one round whitiſh ſcaly head, conſiſting commonly of foure vnder greeniſh leaues which make the cup, and foure grayiſh or whitiſh leaues which are the floure. Now after theſe come to ſome maturitie they appeare all of a whitiſh colour,and through the thin films of theſe heads appeares the ſeed, which at the firſt view ſeems to be pretty large and blacke ; for it lieth all cluſtering together ; but if you rub it out you ſhall find it as ſmall as ſand,and of a darke reddiſh colour. The taſt of this plant is very hot and piercing like that of Golden rod or our common Saxifrage, and without doubt it is more effectuall to moue vrine than the former Knawel. I haue found it growing in many places about brick and ſtone walls,and vpon chalky barren grounds. I called this in my Iournall *An.*1632. *Saxifraga minor altera floſculis albis ſemine nigro* ; and queſtioned whether it were not *Alſine Saſſifraga anguſtifol.minima mont*.of *Columna*.But now I think it rather(if the number of leaues in the flour did not diſagree) the other which is deſcribed in the next place, of which I ſince that time haue receiued both the figure and deſcription, as alſo a dry plant from Mᵣ.*Goodyer*. He conjectures it may be this plant which I haue here deſcribed, that is ſet forth in the *Hiſt Lugd.pag.*1235.by the name of *Alſine muſcoſa*.

Alſine paluſtris,folijs tenuiſſimis : ſiue Saxifraga paluſtris alſinefolia.

6　This hath a great number of very ſmall graſſe-like leaues , growing from the root, about an inch long,a great deale ſmaller and ſlenderer than ſmall pins ; among which ſpring vp many ſmall ſlender round ſmooth firme branches ſome handfull or handfull and halfe high,from which ſometimes grow a few other ſmaller branches,whereon at certain joints grow leaues like the former,and thoſe ſet by couples with other ſhorter comming forth of their boſomes ; and ſo by degrees they become ſhorter and ſhorter towards the top, ſo that toward the top this plant ſomwhat reſembleth *Thymum durius*. The floures are great for the ſlenderneſſe of the plant, growing at the tops of the branches,each floure conſiſting of 5 ſmall blunt roundiſh topped white floures, with white chiues in the midſt:the ſeed I obſerued not.The root is ſmall,growing in the myre with a few ſtrings:this groweth plentifully on the boggy ground below the red Wel of Wellingborough in Northampton ſhire.This hath not bin deſcribed that I find.I obſerued it at the place aforeſaid *Aug.*12. 1626. *Iohn Goodyer.* ‡

The Place.

†　The firſt and ſecond are ſtrangers in England:the reſt grow in the places mentioned in their deſcriptions.

¶ The Time.

Theſe floure for the moſt part from May to September.

¶ The Names.

That which hath been ſaid of their names in their ſeuerall deſcriptions ſhall ſuffice.

¶ The Temperature.

They are cold in the ſecond degree,and dry in the third,aſtringent and making thick.
‡　Theſe,eſpecially the three laſt,are hot in the ſecond or third degree,and of ſubtil parts;but the Parſley Piert ſeemes not to be ſo hot as the other two. ‡

¶ The Vertues.

A　Here according to my promiſe I haue thought good to inſert this medicin made with Knawel: which herbe is called(as I ſaid before)Parſley Piert, but if I might without offence, it ſhould bee called *Petra pungens*:for that barbarous word Parſley Piert was giuen by ſome ſimple man (‡alſo the other ſauors of ſimplicitie‡) who had not well learned the true terme.The compoſition which followeth muſt be giuen in warme white wine,halfe a dram,two ſcruples,or more,according to the conſtitution of the body which is to receiue it.

The

The leaues of Parſley Piert, Mouſ-eare, of each one ounce when the herbes be dried, bay berries, **E**
Turmericke, Cloues, the ſeeds of the great Burre, the ſeeds in the berries of Hippes, or Briertree,
Fenugreeke, of each one ounce, the ſtone in the oxe gall, the weight of 24. Barley cornes, or halfe a
dram, made together into a moſt fine and ſubtill pouder, taken and drunke in maner aforeſaid, hath
been proued moſt ſingular for the diſeaſe aforeſaid.

‡   The fifth and ſixth are of the ſame faculty, and may be vſed in the like caſes. ‡   **F**

† The figure that formerly was in the ſecond place was of Knawell, and that in the third place of *Polygonum minus Polycarpon of Tabern.*

CHAP. 172.   *Of Rupture wort.*

1 *Herniaria.*   Rupture wort.

‡ 2 *Millegrana minima.*
Dwarfe All-ſeed.

¶ *The Deſcription.*

1   THere is alſo a kinde of Knot-graſſe commonly called in Latine *Herniaria*: in Engliſh, Rupture woort, or Rupture graſſe. It is
a baſe and low creeping herbe, hauing many ſmall ſlender branches trailing vpon the ground, yet very tough,
and full of little knots ſomewhat reddiſh, whereupon
doe grow very many ſmall leaues like thoſe of Time;
among which come forth little yellowiſh flours which
turne into very ſmall ſeed, and great quantitie thereof,
conſidering the ſmalneſſe of the plant, growing thicke
cluſtering together by certain ſpaces. The whole plant
is of a yellowiſh greene colour. The root is very ſlender and ſingle.

2   There is alſo a kinde of *Herniaria*, called *Mille
grana* or All-ſeed, that groweth vpright an handfull
high, with many ſmall and tender branches, ſet with
leaues like the former, but few in number, hauing as it were two ſmall leaues and no more. The
whole plant ſeemeth as it were couered ouer with ſeeds or graines, like the ſeed of Panicke, but
much leſſer. ‡ I haue not ſeene many plants of this, but all that euer I yet ſaw neuer attained to the
height of two inches. ‡

¶ *The Place.*

1   It ioyeth in barren and ſandy grounds, and is likewiſe found in dankiſh places that lie wide
open to the ſun: it doth grow and proſper in my garden exceedingly. ‡ 2 I found this in Kent on a
Heath not farre from Chiſte-hurſt, being in company with Mr *Bowles* and diuers others, in July,
1630. ‡

¶ *The Time.*

It floureth and flouriſheth in May, Iune, Iuly and Auguſt.

¶ *The Names.*

It is called of the later Herbariſts *Herniaria* and *Herniola*; taken from the effect in curing the
diſeaſe *Hernia*: of diuers, *Herba Turca*, and *Empetron*; in French, *Boutonet*: in Engliſh, Rupture wort,
and Burſtwort.

¶ *The Temperature and Vertues.*

A   Rupture wort doth notably drie, and throughly cloſeth vp together and faſteneth. It is reported that being drunke it is ſingular good for Ruptures, and that very many that haue been burſten were reſtored to health by the vſe of this herbe; alſo the pouder hereof taken with wine, doth make a man to piſſe that hath his water ſtopt; it alſo waſteth away the ſtones in the kidnies, and expelleth them.

Cℏᴀᴘ. **173.**   *Of wilde Time.*

1 *Serpillum vulgare.*
Wilde Time.

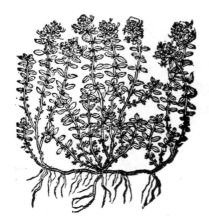

3 *Serpillum majus flore purpureo.*
Great purple wilde Time.

¶ *The Deſcription.*

1   **B**Oth *Dioſcorides* and *Pliny* make two kindes of *Serpillum*, that is, of creeping or wilde Time; whereof the firſt is our common creeping Time, which is ſo well knowne, that it needeth no deſcription; yet this ye ſhall vnderſtand, that it beareth floures of a purple colour, as euery body knoweth. Of which kinde I found another ſort, with floures as white as ſnow, and haue planted it in my garden, where it becommeth an herbe of great beauty.

2   This wilde Time that bringeth forth white floures differeth not from the other, but onely in the colour of the floures, whence it may be called *Serpillum vulgare flore albo.* White floured Wilde Time.

There is another kinde of *Serpillum* which groweth in Gardens, in ſmell and fauour reſembling Marjerome. It hath leaues like Organy, or wilde Marjerome, but ſomewhat whiter, putting forth many ſmall ſtalkes, ſet full of leaues like Rue, but lon...er, narrower, and harder. The floures are of a biting taſte, and pleaſant ſmell. The whole plant groweth vpright, whereas the other creepeth along vpon the earth, catching hold where it growes, & ſpreading it ſelfe far abroad.

3   This great wilde Time creepeth not as the others doe, but ſtandeth vpright, and bringeth forth little ſlender branches full of leaues like thoſe of Rue; yet narrower, longer, and harder. The floures be of a purple colour, and of a twinging biting taſte: it groweth vpon rockes, and is hotter than any of the others.

4   This other great one with white floures differeth not from the precedent, hauing many knaps or heads, of a milke white colour, which ſetteth forth the difference; and it may be called *Serpillum majus flore albo.* Great white floured wild Time.

5   This wilde Time creepeth vpon the ground, ſet with many leaues by couples like thoſe of Marjerome, but leſſer, of the ſame ſmel: the flours are of a reddiſh color. The root is very threddy.

6   Wilde Time of Candy is like vnto the other wild Times, ſauing that his leaues are narrower and longer, and more in number at each joynt. The ſmell is more aromaticall than any of the others, wherein is the difference.

7   There is a kinde of wilde Time growing vpon the mountaines of Italy, called *Serpillum Citratum,*

5 *Serpillum folijs amaraci.*
Marjerome Time.

6 *Serpillum Creticum.*
Wilde Time of Candy.

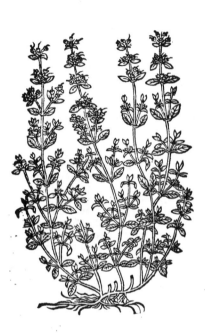

‡ 7 *Serpillum cictatum.*
Limon Time.

‡ 8 *Serpillum hirſutum.*
Hoary wilde Time.

Citratum, that is, hauing the smell of a Pome-Citron, or a Limon, which giueth it the difference from the other wilde times. ‡ It growes in many gardens also, and (as I haue been told) wilde in diuers places of Wales.

8   This (which is the *Serpillum Pannonicum* 3. of *Clusius*) runnes or spreds it selfe far vpon the ground. For though it haue a hard and wooddy root like as the former kindes, yet the branches which lie spread round about here and there take root, which in time become as hard and wooddie as the former. The leaues and stalkes are like those of the last described, but rough and hoarie: the floures also are not vnlike those of the common kinde. The whole plant hath a kinde of resinous smell. It floures in Iune with the rest, and growes vpon the like mountainous places; but whether with vs in England or no I cannot yet affirme any thing of certaintie. ‡

¶ *The Place.*

The first groweth vpon barren hills and vntoiled places: the second groweth in Gardens. The white kinde I found at South fleet in Kent, in a barren field belonging to one Mr *William Swan*.

¶ *The Time.*

They floure from May to the end of Summer.

¶ *The Names.*

Wild Time is called in Latine, *Serpillum*, *à serpendo*, of creeping: in high and low Dutch, **Quendel**, and **wilden Thymus**, and also **Onser Urouwen bedstroo**: in Spanish, *Serpoll*: in Italian, *Serpillo*: in French, *Pillolet*: in English, wilde Time, Puliall mountaine, Pella Mountaine, running Time creeping Time, Mother of Time: in shops it is called *Serpyllum*; yet some call it, *Pulegium montanum*: and it is euery where (saith *Dodonæus*) thought to be the *Serpyllum* of the Antients. Notwithstanding it answereth not so well to the wilde Times as to *Dioscorides* his *Saxifraga*; for if it be diligently compared with the description of both the *Serpilla* and the *Saxifraga*, it shall be found to be little like the wilde Times, but very much like the *Saxifraga*: for (saith *Dioscorides*) *Saxifraga* is an herbe like Time, growing on rockes, where our common wilde Time is oftentimes found.

Ælianus in his ninth booke of his sundry Histories seemeth to number wilde Time among the floures. *Dionysius Iunior* (saith he) comming into the city Locris in Italy, possessed most of the houses of the city, and did strew them with roses, wilde Time, and other such kindes of floures. Yet *Virgil* in the second Eclog of his Bucolicks doth most manifestly testifie, that wilde Time is an herbe, in these words:

> *Thestylis & rapido fessis messoribus astu*
> *Allia, serpillumque, herbas contundit olentes.*

Thestilis for mowers tyr'd with parching heate,
Garlicke, wilde Time, strong smelling herbes doth beate.

Out of which place it may be gathered, that common wilde time is the true and right *Serpillum*, or wilde Time, which the Grecians call ἕρπυλλον. *Marcellus* an old antient Author among the Frenchmen saith it is called *Gilarum*; as *Plinius Valerianus* saith it is called of the same, *Laurio*.

¶ *The Temperature.*

Wilde Time is of temperature hot and dry in the third degree: it is of thin and subtill parts, cutting and much biting.

¶ *The Vertues.*

A   It bringeth downe the desired sickenesse, prouoketh vrine, applied in bathes and fomentations it procureth sweat: being boyled in wine, it helpeth the ague, it easeth the strangurie, it stayeth the hicket, it breaketh the stones in the bladder, it helpeth the Lethargie, frensie, and madnesse, and stayeth the vomiting of bloud.

B   Wilde Time boiled in wine and drunke, is good against the wambling and gripings of the belly, ruptures, convulsions, and inflammations of the liuer.

C   It helpeth against the bitings of any venomous beast, either taken in drinke, or outwardly applied.

D   *Aëtius* writeth, That *Serpillum* infused well in Vinegre, and then sod and mingled with rose water, is a right singular remedy to cure them that haue had a long phrensie or lethargic.

E   *Galen* prescribeth one dram of the juyce to be giuen in vinegre against the vomiting of bloud, and helpeth such as are grieued with the spleene.

CHAP.

CHAP. 174.   *Of Garden Time.*

¶ *The Deſcription.*

1   THe firſt kinde of Time is ſo well knowne that it needeth no deſcription; becauſe there is not any which are ignorant what *Thymum durius* is, I meane our common garden Time.

2   The ſecond kinde of Time with broad leaues hath many wooddy branches riſing from a threddy root, beſet with leaues like *Myrtus*. The floures are ſet in rundles about the ſtalke like Horehound. The whole plant is like the common Time in taſte and ſmell.

1 *Thymum durius.*
Hard Time.

† 2 *Thymum latifolium.*
Great broad leaued Time.

3   Time of Candy is in all reſpects like vnto common Time, but differeth in that, that this kinde hath certaine knobby tufts not much vnlike the ſpikes or knots of *Stæcados*, but much leſſer, beſet with ſlender floures of a purple colour. The whole plant is of a more gracious ſmell than any of the other Times, and of another kinde of taſte, as it were ſauouring like ſpice. The root is brittle, and of a wooddy ſubſtance.

4   Doubtleſſe that kinde of Time whereon *Epithymum* doth grow, and is called for that cauſe *Epithymum*, and vſed in ſhops, is nothing elſe but Dodder that growes vpon Time; and is all one with ours, though *Matthiolus* makes a controuerſie and difference thereof: for *Pena* trauelling ouer the hills in Narbone neere the ſea, hath ſeene not onely the garden Time, but the wilde Time alſo loden and garniſhed with this *Epithymum*. So that by his ſight and mine owne knowledge I am aſſured, that it is not another kinde of Time that beareth *Epithymum*, but is common Time: for I haue often found the ſame in England, not only vpon our Time, but vpon Sauorie, and other herbes alſo: notwithſtanding thus much I may conjecture, that the clymate of thoſe Countries doth yeeld the ſame forth in greater abundance than ours, by reaſon of the intemperance of cold, whereunto our country is ſubject.

¶ *The*

† 3 *Thymum Creticum.*
Time of Candy.

4 *Epithymum Græcorum.*
Laced Time.

¶ *The Place.*

Theſe kindes of Time grow plentifully in England in moſt gardens euery where, except that with broad leaues, and Time of Candy which I haue in my Garden.

¶ *The Time.*

They flouriſh from May vnto September.

¶ *The Names.*

The firſt may be called hard Time, or common Garden Time : the ſecond, broad leaued Time : the third, Time of Candy ; our Engliſh women call it Muske Time : the laſt may be called Dodder Time.

¶ *The Temperature.*

Theſe kindes of Time are hot and dry in the third degree.

¶ *The Vertues.*

A   Time boyled in water and hony drunken, is good againſt the cough and ſhortnes of the breath ; it prouoketh vrine, expelleth the ſecondine or after-birth, and the dead childe, and diſſolues clotted or congealed bloud in the body.

B   The ſame drunke with vinegre and ſalt purgeth flegme : or boyled in Mede or Methegline, it cleanſeth the breaſt, lungs, reines, and matrix, and killeth wormes.

C   Made into pouder, and taken in the weight of three drams with Mede or honied vinegre, called Oxymel, and a little ſalt, it purgeth by ſtoole tough and clammy flegme, ſharpe and cholericke humors, and all corruption of bloud.

D   The ſame taken in like ſort, is good againſt the Sciatica, the paine in the ſide and breſt, againſt the winde in the ſide and belly, and is profitable alſo for ſuch as are fearefull, melancholicke, and troubled in minde.

E   It is good to be giuen vnto thoſe that haue the falling ſickeneſſe to ſmell vnto.

F   *Epithymum,* after *Galen,* is of more effectuall operation in Phyſicke than Time, being hot and dry in the third degree, more mightily cleanſing, heating, drying, and opening than *Cuſcuta,* hauing right good effect to eradicat melancholy, or any other humor in the ſpleen, or other diſeaſe, ſprung by occaſion of the ſpleene.

                                                     It

It helpeth the long continued paine of the head, and beſides his ſingular effects about ſplene. G
ticall matters, it helpeth the lepry, or any diſeaſe of melancholy; all quartaine agues, and ſuch like
griefes proceeding from the ſpleene.

Dioſcorides ſaith, *Epithymum* drunke with honied water, expelleth by ſiege, flegme, and melan- H
choly.

Of his natiue propertie it relieueth them which be melancholicke, ſwolne in the face and other I
parts, if you pound *Epithymum*, and take the fine pouder thereof in the quantitie of foure ſcruples in
the liquor which the Apothecaries call *Paſſum*, or with Oxymell and ſalt, which taketh away all
flatuous humors and ventoſities.

† The ſecond figure was of *Serpillum Citratum* deſcribed in the ſeuenth place of the foregoing chapter; the third was of *Marum Matthioli*, *Tabern.* being the *Tra-gſingonum alterum* of *Lobel.*

CHAP. 175.   *Of Sauorie.*

¶ *The Kindes.*

THere be two kindes of Sauorie, the one that indureth Winter, and is of long continuance: the
other an annuall or yearely plant, that periſheth at the time when it hath perfected his ſeed,
and muſt be ſowne againe the next yeare; which we call Summer Sauorie, or Sauorie of a yeare.
There is likewiſe another, which is a ſtranger in England, called of *Lobel*, *Thymbra S.Iuliani*, deny-
ing it to be the right *Satureia*, or Sauorie: whether that of *Lobel*, or that we haue in our Engliſh
gardens be the true Winter Sauorie, is yet diſputable; for we thinke that of S.Iulians rocke to be
rather a wilde kinde than otherwiſe. ‡ *Pena* and *Lobel* do not deny, but affirme it in theſe words,
Nullus non fatetur Satureiam veram; that is, which none deny to be the true *Satureia* or Sauorie. *Vid.*
Aduerſar.pag.182. ‡

1 *Satureia hortenſis.*
Winter Sauorie.

2 *Satureia hortenſis æſtiua.*
Summer Sauorie.

¶ *The Deſcription.*

1   WInter Sauorie is a plant reſembling Hyſſope, but lower, more tender and brittle : it bringeth forth very many branches, compaſſed on euery ſide with narrow and ſharpe pointed leaues, longer than thoſe of Time ; among which grow the floures from the bottome to the top, out of ſmall huskes, of colour white, tending to a light purple. The root is hard and wooddy, as is the reſt of the plant.

2   Summer Sauorie groweth with a ſlender brittle ſtalke of a foot high, diuided into little branches: the leaues are narrow, leſſer than thoſe of Hyſſope, like the leaues of Winter Sauorie, but thinner ſet vpon the branches. The floures ſtand hard to the branches, of a light purple tending to whiteneſſe. The root is ſmall, full of ſtrings, and periſheth when it hath perfected his ſeed.

3   *Satureia Sancti Iuliani.*
Rocke Sauorie.

‡ 4   *Satureia Cretica.*
Candy Sauorie.

3   This ſmall kinde of Sauorie, which *Lobel* hath ſet forth vnder the title of *Thymbra S.Iuliani,* becauſe it groweth plentifully vpon the rough cliffes of the Tyrrhenian ſea in Italie, called Saint Iulians rocke, hath tender twiggie branches an handfull high, of a wooddie ſubſtance, ſet full of leaues from the bottome to the top, very thicke thruſt together like vnto thoſe of Time, ſauing that they be ſmaller and narrower, bringing forth at the top of the ſprigs a round ſpikie tuft of ſmall purpliſh floures. The whole plant is whitiſh, tending to a bleake colour, and of a very hot and ſharpe taſte, and alſo well ſmelling.

‡   4   This in the opinion of *Honorius Bellus, Cluſius,* and *Pona,* is thought, and not without good reaſon, to be the true *Thymbra,* or *Satureia* of *Dioſcorides* and the Antients, for beſides that it agrees with their deſcription, it is to this day called in Candie θυμ-̔ρι and θρυϛι. *Cluſius* deſcribes it thus : It ſends forth many branches immediately from the root like as Time, and thoſe quadrangular, rough, and of a purpliſh colour: vpon theſe grow alternately little roughiſh leaues much like thoſe of the true Tyme ; and out of their boſomes come little branches ſet with the like, but leſſer leaues. The toppes of the branches are compaſſed with a rundle made of many little leaues, whereout come floures of a fine purple colour, and like the floures of Tyme, being diuided into foure parts, whereof the lower is the broader, and hangs downe : The vpper is alſo broad, but ſhorter, and the other two leſſe. Out of the middle of the floure come fine whitiſh threds, pointed with browne and a forked ſtile. The ſeed is ſmall and blacke like that of Tyme. The root hard and wooddie. It floured with *Cluſius* (who receiued the ſeeds out of Candie from *Honorius Bellus*) in October and Nouember. ‡

¶ *The Place.*

They are ſowne in Gardens, and bring forth their floures the firſt yeare of their ſowing.

¶ *The*

¶ *The Time.*

They floure in Iuly and Auguſt.

¶ *The Names.*

Sauorie is called in Greeke, θύμβρα, neither hath it any other true name in Latine than *Thymbra*. The Interpreters would haue it called *Satureia*, wherein they are repugnant to *Columella* a Latine Writer, who doth ſhew a manifeſt difference betweene *Thymbra* and *Satureia*, in his tenth booke, where he writeth, that Sauorie hath the taſte of Tyme, and of *Thymbra* or the Winter Sauorie.

Et Satureia Thymi referens Thymbræque ſaporem.

† Notwithſtanding this aſſertion of *Columella*, *Pliny lib.9.cap.8.* makes *Satureia*, or Sauorie, to be that *Thymbra* which is called alſo *Cunila*. Sauorie in high Dutch is called **Knuel Saturey**, and **Sabaney** : in low Dutch **Cettlen** : which name as it ſeemeth is drawne of out *Cunila* in italian, *Sauoreggia*: in Spaniſh *Axidrea* and *Sagorida* : in French, *Sarriette* : in Engliſh, Sauorie, Winter Sauorie, and Summer Sauorie.

¶ *The Temperature and Vertues.*

Winter Sauorie is of temperature hot and dry in the third degree, it maketh thinne, cutteth, it A clenſeth the paſſages : to be briefe, it is altogether of like vertue with Tyme.

Summer Sauorie is not full ſo hot as Winter Sauorie, and therefore ſaith *Dioſcorides*, more fit to B be vſed in medicine : it maketh thin and doth maruellouſly preuaile againſt winde : therefore it is with good ſucceſſe boyled and eaten with beanes, peaſon, and other windie pulſes, yea, if it be applied to the belly in a fomentation, it forthwith helpeth the affects of the mother proceeding from winde.

CHAP. 176.   *Of Dodder.*

Cuſcuta ſiue Caſſuthe.
Dodder.

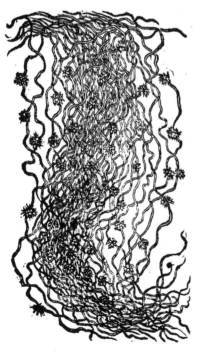

¶ *The Deſcription.*

CVſcuta, or Dodder, is a ſtrange herbe, altogether without leaues or root, like vnto threds very much ſnarled or wrapped together, confuſedly winding it ſelfe about buſhes and hedges, and ſundry kindes of herbes. The threds are ſomewhat redde : vpon which grow here & there little round heads or knops, bringing forth at the firſt ſlender white floures, afterward a ſmall ſeed.

¶ *The Place.*

This herbe groweth vpon ſundry kindes of herbes, as vpon Time, Winter Sauorie, Germander, and ſuch like, taking his name from the herbe whereupon it doth grow, as that vpon Tyme is called *Epithymum*, vpon Line or flax, *Epilinum* : and ſo of others, as *Dodonæus* ſetteth forth at large: yet hath he forgotten one among the reſt, which groweth very plentifully in Summerſetſhire vpon nettles : neither is it the leaſt among many, either in beautie or operation, but comparable to the beſt *Epithymum* : following therefore the example of *Dioſcorides*, I haue thought good to call it *Epiurtica*, or rather, ἐπίκνιδη, and ſo of the reſt according to the herbes whereon they grow.

¶ *The Names.*

The greateſt is called in ſhops euery where *Cuſcuta*: and of diuers becauſe it groweth vpon Flaxe fluxe or Lyne, *Podagra Lini*; the better learned do name it *Caſſutha*, or *Caſſytha*: and *Geſnerus*, ᴧᴙᴣᴙᴒᴙᴨ; the Arabians, *Keſſuth* and *Chaſuth* : in Dutch, **Schozſte**, and **Wzanghe**: in high Dutch, **Filtraut**: in French, *Goute d'Lin*, and *Tigne de Lin* : in Engliſh, Dodder.

The leſſer and ſlenderer which wrappeth it ſelfe vpon Time and Sauorie, is called of *Dioſcorides* ᴣᴙᴅᴤᴙᴜ; the Apothecaries keep the name *Epithymum* : others, among whom is *Actuarius*, name that *Epithymum* which groweth vpon Time only, and that which groweth on Sauorie *Epithymbrum*, and that alſo which hangeth vpon *Stæbe*, they terme *Epiſtæbe*; giuing a peculiar name to euery kinde.

¶ *The Nature.*

The nature of this herbe changeth and altereth, according to the nature and qualitie of the herbs whereupon it groweth : ſo that by ſearching the nature of the plant you may eaſily finde out the temperament of the laces growing vpon the ſame. But more perticularly : it is of temperature ſomewhat more dry than hot, and that in the ſecond degree : it alſo cleanſeth with a certaine aſtrictiue or binding quality, and eſpecially that which is found growing vpon the bramble : for it alſo receiueth a certaine nature from his parents on which it groweth ; for when it groweth vpon the hotter herbes, as Time and Sauory, it becommeth hotter and dryer, and of thinner parts : that which commeth of Broome prouoketh vrine more forcibly, and maketh the belly more ſoluble : and that is moiſter which groweth vpon flax : that which is found vpon the bramble hath ioyned with it, as we haue ſaid, a binding qualitie, which by reaſon of this facultie ioyned with it is good to cure the infirmities of the Liuer and Milt : for ſeeing that it hath both a binding and purging facultie vnited to it, it is moſt ſingular good for the entrals : for *Galen* in his thirteenth booke of the Method of curing, doth at large declare, that ſuch medicines are fitteſt of all for the liuer and Milt.

¶ *The Vertues.*

A   Dodder remooueth the ſtoppings of the liuer and of the milt or ſpleene, it disburdeneth the veines of flegmaticke, cholericke, corrupt and ſuperfluous humors : prouoketh vrine gently, and in a meane openeth the kidnies, cureth the yellow iaundiſe which are ioyned with the ſtopping of the liuer and gall : it is a remedy againſt lingring agues, baſtard and long tertians, quartains alſo, and properly agues in infants and young children, as *Meſues* ſaith in *Serapio* ; who alſo teacheth, that the nature of Dodder is to purge choler by the ſtoole, and that more effectually if it haue Wormewood ioyned with it, but too much vſing of it is hurtfull to the ſtomacke : yet *Auicen* writeth, that it doth not hurt it, but ſtrengtheneth a weake and feeble ſtomacke ; which opinion alſo we do better allow of.

B   *Epithymum*, or the Dodder which groweth vpon Time, is hotter and dryer than the Dodder that groweth vpon flax, that is to ſay euen in the third degree, as *Galen* ſaith. It helpeth all the infirmities of the milt : it is a remedy againſt obſtructions and hard ſwellings. It taketh away old headaches, the falling ſickeneſſe, madneſſe that commeth of melancholy, and eſpecially that which proceedeth from the ſpleene and parts thereabout : it is good for thoſe that haue the French diſeaſe, and ſuch as be troubled with contagious vlcers, the leproſie, and ſcabby euill.

C   It purgeth downewards blacke and Melancholy humors, as *Aëtius*, *Actuarius*, and *Meſue* write, and alſo flegme, as *Dioſcorides* noteth : that likewiſe purgeth by ſtoole which groweth vpon Sauory and Scabious, but more weakely, as *Actuarius* ſaith.

D   *Cuſcuta*, or Dodder that groweth vpon flax, boiled in water or wine and drunke, openeth the ſtoppings of the liuer, the bladder, the gall, the milt, the kidnies and veines, and purgeth both by ſiege and vrine cholericke humours.

E   It is good againſt the ague which hath continued a long time, and againſt the Iaundiſe, I meane that Dodder eſpecially that groweth vpon brambles.

F   *Epiurtica* or Dodder growing vpon nettles, is a moſt ſingular and effectuall medicine to prouoke vrine, and to looſe the obſtructions of the body, and is proued oftentimes in the Weſt parts with good ſucceſſe againſt many maladies.

Chap. 177. *Of Hyſſope.*

¶ *The Deſcription.*

1   Dioſcorides that gaue ſo many rules for the knowledge of ſimples, hath left Hyſſope altogether without deſcription, as being a plant ſo well knowne that it needed none : whoſe example I follow not onely in this plant, but in many others which be common, to auoid tediouſneſſe to the Reader.

a The

1 *Hyſſopus Arabum.*
Hyſſope with blew floures.

2 *Hyſſopum Arabus flore rubro.*
Hyſſope with reddiſh floures.

†∙3 *Hyſſopus albis floribus.*
White floured Hyſſope.

4 *Hyſſopus tenuifolia.*
Thinne leafed Hyſſope.

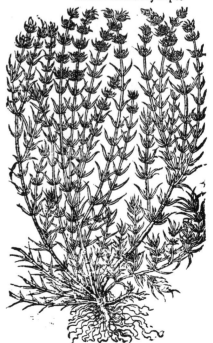

‡ 5 . *Hyſſopus parua anguſtis folijs.*
Dwarfe narrow leaued Hyſſope.

2   The ſecond kinde of Hyſſope is like the former, which is our common Hyſſope, and differeth in that, that this Hyſſope hath his ſmall and ſlender branches decked with faire red floures.

3   The third kinde of Hyſſope hath leaues ſtalkes, branches, ſeed, and root, like the common Hyſſope, and differeth in the floures onely, which are as white as ſnow.

4   This kinde of Hyſſope of all the reſt is of the greateſt beautie ; it hath a wooddy root tough, and full of ſtrings ; from which riſe vp ſmall, tough, and ſlender flexible ſtalkes, whereupon doe grow infinite numbers of ſmall Fennell-like leaues, much reſembling thoſe of the ſmalleſt graſſe, of a pleaſant ſweet ſmel, & aromaticke taſte, like vnto the reſt of the Hyſſops but much ſweeter ; at the tops of the ſtalkes do grow amongſt the leaues ſmal hollow floures, of a blewiſh colour tending to purple. The ſeeds as yet I could neuer obſerue.

‡ 5   This differs from the firſt deſcribed, in that the ſtalkes are weaker and ſhorter, the leaues alſo narrower , and of a darker colour: the floures grow after the ſame manner, and are of the ſame colour as thoſe of the common kinde. ‡

We haue in England in our Gardens another kinde, whoſe picture it ſhall be needleſſe to expreſſe, conſidering that in few words it may be deliuered. It is like vnto the former, but the leaues are ſome of them white, ſome greene, as the other ; and ſome greene and white mixed and ſpotted, very goodly to behold.

Of which kindes we haue in our Gardens moreouer another ſort, whoſe leaues are wonderfully curled, rough, and hairie, growing thicke thruſt together, making as it were a tuft of leaues; in taſte and ſmell, and all other things like vnto the common Hyſſope.

I haue likewiſe in my garden another ſort of Hyſſope, growing to the forme of a ſmall wooddie ſhrub, hauing very faire broad leaues like vnto thoſe of *Nummularia*, or Monywort, but thicker, fuller of juyce, and of a darker greene colour ; in taſte and ſmell like the common Hyſſope.

¶ *The Place.*

All theſe kindes of Hyſſope do grow in my Garden, and in ſome others alſo.

¶ *The Time.*

They floure from Iune to the end of Auguſt.

¶ *The Names.*

Hyſſope is called in Latine, *Hyſſopus :* the which name is likewiſe retained among the Germans, Brabanders, Frenchmen, Italians, and Spaniards. Therefore that ſhall ſuffice which hath beene ſet downe in their ſeuerall titles.

‡   This is by moſt Writers iudged to be Hyſſope vſed by the Arabian Phyſitions, but not that of the Greekes, which is neerer to *Origanum* and Marjerome, as this is to *Satureia* or Sauorie. ‡

¶ *The Temperature and Vertues.*

A   A decoction of Hyſſope made with figs, and gargled in the mouth and throte, ripeneth and breaketh the tumors and impoſthumes of the mouth and throte, and eaſeth the difficultie of ſwallowing, comming by cold rheumes.

B   The ſame made with figges, water, honey, and rue, and drunken, helpeth the inflammation of the lungs, the old cough, and ſhortneſſe of breath, and the obſtructions and ſtoppings of the breaſt.

C   The ſirrup or juyce of Hyſſope taken with the ſirrup of vnegre, purgeth by ſtoole tough and clammie flegme, and driueth forth wormes if it be eaten with figges.

D   The diſtilled water drunke, is good for thoſe diſeaſes before named, but not with that ſpeed and force.

† That figure in the third place was of the *Satureia Romana*, 2. of *Tabernamontanus*.

CHAP. 178.   *Of Hedge Hyſſope.*

¶ *The Deſcription.*

1   HEdge Hyſſop is a low plant or herb about a ſpan long, very like vnto the common Hyſſope, with many ſquare ſtalks or ſlender branches, beſet with leaues ſomewhat larger
than Hyſſope, but very like : the floures grow betwixt the leaues vpon ſhort ſtems, of a
white colour declining to blewneſſe. All the herb is of a moſt bitter taſte like the ſmall Centory.
The root is little and threddy, dilating it ſelfe far abroad ; by which means it multiplieth greatly,
and occupieth much ground where it groweth.

1 *Gratiola.*
Hedge Hyſſope.

‡ 2 *Gratiola anguſtifolia.*   3 *Gratiola latifolia.*
Graſſe Poley.   Broad leaued hedge Hyſſop.

‡   2   Narrow leaued hedge Hyſſop from a ſmall fibrous white root ſends vp a reddiſh round
creſted ſtalke diuided into ſundry branches, which are ſet with leaues like thoſe of Knot-graſſe, of
a pale greene colour, and without any ſtalks : out of the boſome of theſe come floures ſet in long
cups compoſed of foure leaues of a pleaſing blew colour, which are ſucceeded by longiſh ſeed veſſels containing a ſmall dusky ſeed. The whole plant is without ſmell, neither hath it any bitterneſſe or other manifeſt taſte. It varies in leaues, ſometimes broader, and otherwhiles narrower : the
plant growing ſomtimes but an handfull, and otherwhiles a foot high. *Geſner* called this *Gratiola
minor* : and *Camerarius, Hyſſopoides* : *Bauhine* onely hath figured it, and that by the name of *Hyſſopifolia, ſiue Gratiola minor. Cordus* firſt mentioned it, and that by the Dutch name of Graſſe Poly, which
name we may very fitly retain in Engliſh. ‡

3   Broad leaued hedge Hyſſope hath many ſmall and tender branches, foure ſquare, and ſomewhat hollow or furrowed, beſet with leaues by couples one oppoſit againſt the other, like vnto the
former, but ſomewhat ſhorter, and much broader : amongſt which grow the floures of a purple

colour, spotted on the inside with white, and of a brighter purple than the rest of the floure, fashioned like the smallest *Antirrhinum* or least Snapdragon: which being past, there succeed little seed vessels, fashioned like the nut of a crosse-bow, which contain small yellowish seed extreame bitter of taste; the whole plant is likewise bitter, as the common or well knowne *Gratiola*. The root is compact of a great number of whitish strings intangled one within another, which mightily encreaseth or spreadeth abroad.

‡   This plant is only a lesser kinde of the *Lysimachia galericulata* of *Lobel*, which some haue called *Gratiola latifolia*. Our Authors figure was very ill, wherefore I haue endeauored with the helpe of some dried plants and my memorie, to present you with a better expression thereof. ‡

¶ *The Place.*

The first groweth in low and moist places naturally, which I haue planted in my garden.

‡   The second was found growing by my oft mentioned friend Mr *Bowles*, at Dorchester in Oxfordshire, at the backe side of the inclosed grounds on the left hand of the town, if you would ride from thence to Oxford, in the grassy places of the champian corne fields. ‡

The third growes likewise in moist places: I found it growing vpon the bog or marish ground at the farther end of Hampsted heath, and vpon the same heath toward London, neere vnto the head of the springs that were digged for water to be conueied to London, 1590. attempted by that carefull citisen *Iohn Hart* Knight, Lord Major of the City of London; at which time my selfe was in his Lordships company, viewing for my pleasure the same goodly springs.

¶ *The Time.*

The first floureth in May: the second in Iune and Iuly: the third in August.

¶ *The Names in generall.*

Hedge Hyssope is called in Latine *Gratiola*, and *Gratia Dei*, or the Grace of God; notwithstanding there is a kinde of *Geranium* or Storks bill called by the later name: of *Cordus*, *Limnesium*, and *Centauroides*: of *Anguillara* it is thought to be *Dioscorides* his *Papauer Spumeum*, or Spatling Poppy: but some thinke *Papauer spumeum* to be that which we cal *Behen album*: in Dutch it is called **Gods gratie**: in Italian, *Stanea canallo*, because horses hauing eaten thereof wax lean, and languish thereupon: In English, Gratia Dei, and hedge Hyssop. The seed hereof is called *Gelbenech*, which name the Arabians retain to this day.

¶ *The Names in particular.*

‡ 1   *Matthiolus*, *Dodonæus*, and others haue called this *Gratiola*: *Anguillara*, *Gratia Dei*: *Cordus*, *Limnesium*, *Centauroides*; he also thought it (but vnfitly) to be the *Eupatorium* of *Mesue*: *Gesner* thinks it may be *Polemonium palustre amarum* of *Hippocrates*, that write of the diseases of cattell.

2   *Cordus* called this Grasse Poley; *Gesner*, *Gratiola minor*: *Camerarius*, *Hyssopoides*: and *Bauhine*, *Hyssopifolia*.

3   This is not set forth by any but our Author, and it may fitly be named *Lysimachia galericulata minor*, as I haue formerly noted. ‡

¶ *The Temperature.*

Hedge Hyssope is hot and dry of temperature; and the first is only vsed in medicine.

¶ *The Vertues.*

A   Whoso taketh but one scruple of *Gratiola* bruised, shall perceiue euidently his effectuall operation and vertue, in purging mightily, and that in great aboundance, waterish grosse and slimy humors. *Conradus Gesnerus* experimented this, and found it to be true, and so haue I my selfe, and many others.

B   *Gratiola* boiled, and the decoction drunke or eaten with any kinde of meat, in manner of a sallad, openeth the belly, and causeth notable loosenesse, scouring freely, whereby it purgeth grosse flegm and cholericke humors.

C   *Gratiola* or hedge Hyssop boiled in wine and giuen to drinke, helpeth feuers of what sort soeuer, and is most excellent in dropsies and such like diseases proceeding of cold and waterie causes.

D   The extraction giuen with the pouder of Cinnamon, and a little of the juice of Calamint, preuaileth against tertian and quotidian feuers, set downe for most certain by the learned *Ioachimus Camerarius*.

C H A P.

Chap. 179. *Of Lauander Spike.*

¶ *The Description.*

1   Lauander Spike hath many ftiffe branches of a wooddy fubftance, growing vp in the manner of a fhrub, fet with many long hoarie leaues, by couples for the moft part, of a ftrong fmell, and yet pleafant enough to fuch as do loue ftrong fauors. The floures grow at the top of the branches, fpike fafhion, of a blew colour. The root is hard and wooddie.

2   The fecond differeth not from the precedent, but in the colour of the floures: For this plant bringeth milke white floures; and the other blew, wherein efpecially confifteth the difference.

3   We haue in our Englifh gardens a fmall kinde of Lauander, which is altogether leffer than the other, ‡ and the floures are of a more purple colour, and grow in much leffe and fhorter heads, yet haue they a far more gratefull fmell: the leaues are alfo leffe and whiter than thofe of the ordinarie fort. This did, and I thinke yet doth grow in great plenty, in his Majefties priuate garden at White-Hall. And this is called Spike, without addition, and fometimes Lauander Spike: and of this by diftillation is made that vulgarly known and vfed oile which is tearmed *Oleum fpicæ*, or oile of Spike. ‡

1   *Lavandula flore cæruleo.*
    Common Lauander.

2   *Lavandula flore albo.*
    White floured Lauander.

¶ *The Place:*

In Spaine and Languedock in France, moft of the mountaines and defert fieldes, are as it were

3 *Lauendula minor, ſiue Spica.*
Lauander Spike.

couered ouer with Lauander. In theſe cold countries they are planted in gardens.

¶ *The Time.*

They floure and flouriſh in Iune and Iuly.

¶ *The Names.*

Lauander Spike is called in Latine *Lauendula,* and
Spica: in Spaniſh, *Spigo,* and *Languda.* The firſt is the
male , and the ſecond the female. It is thought of
ſome to bee that ſweet herbe *Caſia* , whereof *Virgil*
maketh mention in the ſecond Eclog of his Bucolicks:

> *Tum Caſia atque alijs intexens ſuauibus herbis,*
> *Mollia luteola pingit vacinia caltha.*

And then ſhee'l Spike and ſuch ſweet hearbs infold
And paint the Iacinth with the Marigold.

And likewiſe in the fourth of his Georgickes ,
where he intreateth of chuſing of ſeats and places
for Bees, and for the ordering thereof, he ſaith thus:

> *Hæc circum Caſia virides & olentia late*
> *Serpilla, & grauiter ſpitantis copia Thymbræ*
> *Floreat; &c.* ————————

About them let freſh Lauander and ſtore
Of wilde Time with ſtrong Sauorie to floure.

Yet there is another *Caſia* called in ſhops *Caſia
Lignea,* as alſo *Caſia nigra,* which is named *Caſia fiſtula* ; and another a ſmall ſhrubby plant extant a-
mongthe ſhrubs or hedge buſhes, which ſome think to be the *Caſia Poëtica,* mentioned in the precedent verſes.

¶ *The Temperature.*

Lauander is hot and dry, and that in the third degree, and is of a thin ſubſtance, conſiſting of many airie and ſpirituall parts. Therefore it is good to be giuen any way againſt the cold diſeaſes of
the head, and eſpecially thoſe which haue their original or beginning not of abundance of humors,
but chiefely of a cold quality only.

¶ *The Vertues.*

A   The diſtilled water of Lauander ſmelt vnto, or the temples and forehead bathed therewith, is a
refreſhing to them that haue the Catalepſy, a light migram, and to them that haue the falling ſickneſſe, and that vſe to ſwoune much. But when there is abundance of humours, eſpecially mixt with
bloud, it is not then to be vſed ſafely , neither is the compoſition to be taken which is made of diſtilled wine: in which ſuch kinds of herbes, floures, or ſeeds, and certain ſpices are infuſed or ſteeped, though moſt men do raſhly and at aduenture giue them without making any difference at al.
For by vſing ſuch hot things that fill and ſtuffe the head, both the diſeaſe is made greater, and the
ſick man alſo brought into daunger, eſpecially when letting of bloud, or purging haue not gon before. Thus much by way of admonition, becauſe that euery where ſome vnlearned Phyſitians and
diuers raſh & ouerbold Apothecaries, and other fooliſh women, do by and by giue ſuch compoſitions, and others of the like kind, not only to thoſe that haue the Apoplexy ; but alſo to thoſe that
are taken, or haue the Catuche or Catalepſis with a Feuer; to whom they can giue nothing worſe,
ſeeing thoſe things do very much hurt, and oftentimes bring death it ſelfe.

B   The floures of Lauander picked from the knaps, I meane the blew part and not the buſk, mixed
with Cinnamon, Nutmegs, & Cloues, made into pouder, and giuen to drinke in the diſtilled water
thereof, doth helpe the panting and paſſion of the heart , preuaileth againſt giddineſſe, turning, or
ſwimming of the braine, and members ſubiect to the palſie.

C   Conſerue made of the floures with ſugar, profiteth much againſt the diſeaſes aforeſaid , if the
quantitie of a beane be taken thereof in the morning faſting.

D   It profiteth them much that haue the palſie , if they be waſhed with the diſtilled water of the
floures,

floures,or anointed with the oile made of the floures,and oile oliue, in ſuch maner as oile of Roſes is,which ſhall be expreſſed in the treatiſe of Roſes.

Chap. 180.　*Of French Lauander or Stickadoue.*

¶ *The Temperature.*

1　FRench Lauander hath a body like Lauander,ſhort,and of a woody ſubſtance, but ſlende-rer, beſet with long narrow leaues of a whitiſh colour, leſſer than thoſe of Lauander : it hath in the top buſhy or ſpiky heads,well compact or thruſt together ; out of the which grow forth ſmall purple floures of a pleaſant ſmel.The ſeed is ſmall and blackiſh : the root is hard and wooddy.

2　This jagged Sticadoue hath many ſmall ſtiffe ſtalks of a woody ſubſtance, whereupon doe grow jagged leaues in ſhape like the leaues of Dil,but of an hoary colour : on the top of the ſtalks grow ſpike flours of a blewiſh colour,and like vnto the common Lauander Spike.The root is like-wiſe wooddy.　‡ This by *Cluſius* who firſt deſcribed it, as alſo by *Lobel*, is called *Lauendula multi-fido folio,*or Lauander with the diuided leafe ; the plant more reſembling Lauander than Sticka-doue. ‡

3　There is alſo a certain kind hereof,differing in ſmalneſſe of the leaues only,which are round about the edges nicked or toothed like a ſaw,reſembling thoſe of Lauander cotton:the root is like-wiſe wooddy.

‡ 4　There is alſo another kinde of *Stæchas* which differs from the firſt or ordinarie kinde, in that the tops of the ſtalks are not ſet with leaues almoſt cloſe to the head,as in the common kinde, but are naked and wholly without leaues : alſo at the top of their ſpike or floures (as it were to re-compence their defect below) there grow larger and fairer leaues than in the other ſorts.The other parts of the plant differ not from the common *Stæchas.*

† 1 *Stæchas, ſiue Spica hortulana.*　　　　　　2 *Stæchas multifida.*
Sticadoue and Sticados.　　　　　　　　　　Iagged Sticados.

3 *Stæchas folio ſerrato.*
Toothed Sticadoue.

‡ 4 *Stæchas ſummis cauliculis nudis.*
Naked Sticadoue.

¶ *The Place.*

Theſe herbes grow wilde in Spaine, in Languedoc in France, and the Iſlands called Stœchades ouer againſt Maſſilia : we haue them in our gardens, and keep them with great diligence from the injurie of our cold clymat.

¶ *The Time.*

They are ſowne of ſeed in the end of Aprill, and couered in Winter from the cold, or elſe ſet in pots or tubs with earth, and caried into houſes.

¶ *The Names.*

The Apothecaries call the floure *Stæcados : Dioſcorides,* σ.. *Galen,* ..., by the dipthong *oi* in the firſt ſyllable : in Latine, *Stæchas :* in high-Dutch, **Stichas kraut :** in Spaniſh, *Thomani,* and *Cantueſſo :* in Engliſh, French Lauander, Stecado, Stecadoue, Sticadoue, Caſſidonie, and ſome ſimple people imitating the ſame name do call it Caſt me downe.

¶ *The Temperature.*

French Lauander, ſaith *Galen,* is of temperature compounded of a little cold earthy ſubſtance, by reaſon whereof it bindeth : it is of force to take away obſtructions, to extenuate or make thin, to ſcoure and clenſe, and to ſtrengthen not only all the intrals, but the whole body alſo.

¶ *The Vertues.*

A   *Dioſcorides* teacheth, That the decoction hereof doth help the diſeaſes of the cheſt, and is with good ſucceſſe mixed with counterpoiſons.

B   The later Phyſitions affirm, That *Stæchas,* and eſpecially the floures of it, are moſt effectuall againſt paines of the head, and all diſeaſes thereof proceeding of cold cauſes, and therefore they bee mixed in all compoſitions almoſt which be made againſt head-ache of long continuance, the Apoplexy, the falling ſickneſſe, and ſuch like diſeaſes.

C   The decoction of the heads and floures drunke, openeth the ſtoppings of the liuer, the lungs, the milt, the mother, the bladder, and in one word, all other inward parts, clenſing and driuing forth all euill and corrupt humors, and procuring vrine.

C H A P.

CHAP. 181. *Of Fleawort.*

¶ *The Description.*

1   PSyllium, or the common Fleawort, hath many round and tender branches, set full of long and narrow leaues somewhat hairy. The tops of the stalks are garnished with sundrie round chaffie knops beset with small yellow floures : which beeing ripe containe many little shining seeds, in proportion, colour, and bignesse like vnto fleas.

2   The second kinde of *Psyllium* Fleawort hath long and tough branches, of a woody substance like the precedent, but longer and harder, with leaues resembling the former, but much longer and narrower. The chaffie tuft which containeth the seed is like the other, but more like the eare of *Phalaris*, which is the eare of *Alpisti*, the Canary seed which is meat for birds that come from the Islands of Canarie. The root hereof lasteth all the winter, and likewise keepes his greene leaues ; whereof it tooke this addition of *Sempervirens*.

1 *Psyllium, siue pulicaris herba.*
Fleawort.

2 *Psyllium sempervirens Lobelij.*
Neuer dying Fleawort.

¶ *The Place.*

These plants are not growing in our fields of England, as they do in France and Spaine, yet I haue them growing in my garden.

¶ *The Time.*

They floure in Iune and Iuly.

¶ *The Names.*

Fleawort is called in Greeke ~~~~ : in Latine, *Pulicaria*, and *Herba Pulicaris* : in shops, *Psyllium* : in English, Fleawort ; not becaufe it killeth fleas, but becaufe the feeds are like fleas : of fome, Fleabane, but vnproperly : in Spanish, *Zargatona* : in French, *L'herbe aus pulces* : in Dutch, ꝛuꝑls bloꝑe crupt.

¶ *The Temperature.*

Galen and *Serapio* record, That the feed of *Psyllium* (which is chiefly vfed in medicine) is cold in the second degree, and temperat in moisture and drinesse.

¶ *The*

¶ *The Vertues.*

A   The ſeed of Fleawort boiled in water or infuſed, and the decoction or infuſion drunke, purgeth downwards aduſt and cholericke humors, cooleth the heate of the inward parts, hot feuers, burning agues, and ſuch like diſeaſes proceeding of heate, and quencheth drowth and thirſt.

B   The ſeed ſtamped, and boiled in water to the form of a plaiſter, and applied, takes away all ſwellings of the joints, eſpecially if you boile the ſame with vineger and oile of roſes, and apply it as a-foreſaid.

C   The ſame applied in manner aforeſaid, vnto any burning heate called S. *Anthonies* fire, or any hot and violent impoſtume, aſſwageth the ſame, and bringeth it to ripeneſſe.

D   Some hold, That the herb ſtrewed in the chamber where any fleas be, will driue them away ; for which cauſe it tooke the name Fleawort : but I thinke it is rather becauſe the ſeed doth reſemble a flea ſo much, that it is hard to diſcern the one from the other.

¶ *The Danger.*

Too much Fleawort ſeed taken inwardly is hurtfull to mans nature: ſo that I wiſh you not to follow the minde of *Galen* and *Dioſcorides* in this point, being a medicine rather bringing a malady, than taking away the griefe : remembring the old prouerb, A man may buy gold too deare, and the hony is too deare that is lickt from thorns.

‡ *Dioſcorides* nor *Galen* mention no vſe of this inwardly: but on the contrary, *Dioſcorides, lib. 6.* which treats wholly of the curing and preuenting poiſons, mentions this in the tenth chapter for a poiſon, and there ſets down the ſymptomes which it cauſes, and refers you to the foregoing chapter for the remedies. ‡

Cʜᴀᴘ. 182.   *Of Cloue Gillofloures.*

¶ *The Kindes.*

THere are at this day vnder the name of *Cariophyllus* comprehended diuers and ſundry ſorts of plants, of ſuch various colours, and alſo ſeuerall ſhapes, that a great and large volume would

not

not ſuffice to write of euery one at large in particular ; conſidering how infinite they are, and how
euery yeare euery clymate and country bringeth forth new ſorts, ſuch as haue not heretofore been
written of ; ſome whereof are called Carnations , others Cloue Gilloſloures,ſome Sops in wine,
ſome Pagiants,or Pagion color,Horſe-fleſh,blunket,purple,white,double and ſingle Gilloſloures,
as alſo a Gilloſloure with yellow flours:the which a worſhipful Merchant of London Mr. *Nicholas*
Lete procured from Poland,and gaue me thereof for my garden, which before that time was neuer
ſeen nor heard of in theſe countries. Likewiſe there be ſundry ſorts of Pinks comprehended vnder
the ſame title, which ſhall be deſcribed in a ſeuerall chapter. There be vnder the name of Gilloſloures alſo thoſe ſloures which we call Sweet-Iohns and Sweet-Williams. And firſt of the great
Carnation and Cloue Gilloſloure.

   ‡   There are very many kinds both of Gilloſloures,Pinkes,and the like,which differ very little
in their roots,leaues,ſeeds,or maner of growing,though much in the colour,ſhape,and magnitude
of their ſloures;whereof ſome are of one colour,other ſome of more,and of them ſome are ſtriped
others ſpotted,&c. Now I(holding it a thing not ſo fit for me to inſiſt vpon theſe accidental differences of plants, hauing ſpecifique differences enough to treat of)refer ſuch as are addicted to
theſe commendable and harmeleſſe delights, to ſuruey the late and oft mentioned Worke of my
frend Mr. *Iohn Parkinſon*, who hath accurately and plentifully treated of theſe varieties;and if they
require further ſatisfaction,let them at the time of the yeare repair to the garden of Miſtreſſe *Tug-*
gy (the wife of my late deceaſed friend Mr. *Ralph Tuggy*) in Weſtminſter,which in the excellencie
and varietie of theſe delights exceedeth all that I haue ſeen:as alſo he himſelfe whilſt he liued exceeded moſt,if not all of his time, in his care,induſtry,and skil in raiſing,encreaſing,and preſeruing
of theſe plants and ſome others;whoſe loſſe therefore is the more to be lamented by all thoſe that
are louers of plants.I will only giue you the figures of ſome three or foure more,whereof one is of
the ſingle one,which therefore ſome tearme a Pinke, though in mine opinion vnfitly, for that it is
produced by the ſeed of moſt of the double ones , and is of different colour and ſhape as they are,
varying from them only in the ſingleneſſe of the ſloures. ‡

‡ *Caryophyllus major & minor,rubro & albo variegati:*   ‡ *Caryophyllus purpureus profunde laciniatus:*
    The white Carnation,and Pageant.   **The blew, or deep purple Gilloſloure.**

¶ *The Description.*

1   THe great Carnation Gillo-floure hath a thick round wooddy root, from which riseth vp many strong joynted stalks set with long green leaues by couples : on the top of the stalks do grow very fair floures of an excellent sweet smell, and pleasant Carnation colour, whereof it tooke his name.

2   The Cloue Gillofloure differeth not from the Carnation but in greatnesse as well of the flowres as leaues. The floure is exceeding well knowne, as also the Pinkes and other Gillofloures; wherefore I will not stand long vpon the description.

‡ *Caryophyllus simplex major.*
The single Gillofloure or Pinke.

¶ *The Place.*

These Gillofloures, especially the Carnations, are kept in pots from the extremitie of our cold Winters. The Cloue Gillofloure endureth better the cold, and therefore is planted in gardens.

The Time.

They flourish and floure most part of the Summer.

¶ *The Names.*

The Cloue Gillofloure is called of the later Herbarists *Caryophylleus Flos* , of the smell of cloues wherewith it is possessed: in Italian, *Garofols*: in Spanish, *Clauel* : in French, *Oeilletz* : in low-Dutch, Ginoffelbloemen ‡ in Latine of most, *Ocellus Damascenus, Ocellus Barbaricus*, and *Barbarica*: in English, Carnations, and Cloue Gillofloures. Of some it is called *Vetonica*, and *Herba tunica*. The which *Bernardus Gordonius* hath set downe for *Dioscorides* his *Polemonium*.

That worthy Herbarist and learned Physition of late memorie Mr. Doctor *Turner* maketh *Caryophyllus* to be *Cantabrica*, which *Plin. lib. 23. cap.* 8. writeth to haue been found out in Spaine about *Augustus* time, and that by those of Biscay.

Iohannes Ruellius saith, That the Gillofloure was vnknowne to the old writers: whose iudgement is very good, especially because this herb is not like to that of *Vetonica*, or *Cantabrica*. It is maruell, saith he, that such a famous floure, so pleasant & sweet, should lie hid, and not be made known by the old writers: which may be thought not inferior to the rose in beautie, smell, and varietie.

¶ *The Temperature.*

The Gillofloure with the leaues and roots for the most part are temperat in heate and drinesse.

¶ *The Vertues.*

A   The conserue made of the floures of the Cloue Gillofloure and sugar, is exceeding cordiall, and wonderfully aboue measure doth comfort the heart, being eaten now and then.

B   It preuaileth against hot pestilentiall feuers, expelleth the poison and furie of the disease, and greatly comforteth the sicke, as hath of late been found out by a learned Gentleman of Lee in Essex, called Mr. *Rich.*

CHAP. 183.   *Of Pinks, or wilde Gillofloures.*

¶ *The Description.*

1   THe double purple Pinke hath manie grassie leaues set vpon small joynted stalkes by couples , one opposite against another, whereupon doe grow pleasant double purple floures,

1 *Cariophyllus syluestris simplex.*
Single purple Pinks.

2 *Cariophyllus syluestris simplex suaue rubens.*
Single red Pinks.

3 *Cariophyllus plumarius albus.*
White jagged Pinks.

‡ *Cariophyllus plumarius albus adoratior.*
Large white jagged Pinks.

floures of a moſt fragrant ſmell, **not inferior to the Cloue Gilloſloure : the root is ſmal & woody·**

‡   There is alſo a ſingle one of this kinde, whoſe figure I heere giue you in ſtead of the double one of our Author. ‡

2   The ſingle red Pinke hath likewiſe many ſmall graſſy leaues leſſer than the former. The floures grow at the top of the ſmall ſtalks ſingle, and of a ſweet bright red colour.

3   The white jagged Pinke hath a tough wooddy root, from which riſe immediately many graſſie leaues ſet vpon a ſmall ſtalke full of joints or knees, at euery joint two one againſt another euen to the top ; whereupon do grow faire double purple floures of a ſweet and ſpicy ſmell, conſiſting of fiue leaues, ſomtimes more, cut or deeply jagged on the edges, reſembling a feather : wherupon I gaue it the name *Plumarius*, or feathered Pink. The ſeed is ſoft, blackiſh, and like vnto onion ſeed.

‡   There is another varietie of this, with the leaues ſomewhat larger and greener than the laſt me ntioned : the floures alſo are ſomewhat bigger, more cut in or diuided, and of a much ſweeter ſmell. ‡

4   This purple coloured Pinke is very like the precedent in ſtalks, roots, and leaues : the floures grow at the tops of the branches, leſſer than the laſt deſcribed, and not ſo deeply jagged, of a purple colour tending to blewneſſe, wherein conſiſteth the difference.

There be diuers other ſorts of Pinks, whereof to write particularly were to ſmall purpoſe, conſidering they are all well known to the moſt, if not to all. Therefore theſe few ſhall ſerue at this time for thoſe that we doe keep in our gardens : notwithſtanding I thinke it conuenient to place theſe wilder ſorts in this ſame chapter, conſidering their nature and vertues do agree, and few or none of them be vſed in phyſicke ; beſides their neereneſſe in kindred and neighborhood.

4 *Cariophyllus plumarius purpureus*.   5 *Cariophyllus plumarius ſylueſtris albus*.
Purple jagged Pinks.   White wilde jagged Pinks.

5   This wilde jagged Pinke hath leaues, ſtalks, and floures like vnto the white jagged Pinke of the garden, but altogether leſſe, wherein they eſpecially differ.

6   The purple mountain or wild Pink hath many ſmal graſſy leaues : among which riſe vp ſlender ſtalks ſet with the like leaues but leſſe ; on the top wherof do grow ſmall purple floures, ſpotted finely with white or yellowiſh ſpots, and much leſſe than any of the others before deſcribed.

6 *Caryophyllus montanus purpureus.*
Wilde Purple jagged Pinke.

7 *Caryophyllus montanus Clufij.*
Clufius mountaine Pinke.

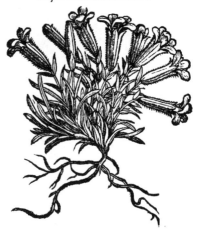

‡ 8 *Caryophyllus pumilio Alpinus.*
Dwarfe Mountaine Pinke.

9 *Caryophyllus cæruleus fiue Aphyllanthos.*
Leafeles Pinke, or ruſhy Pinke.

7. The Mountaine Pinke of *Clufius* defcription hath many large leaues growing in a tuft like vnto thofe of Thrift, and of a bitter tafte : among which rife vp fmall tender foot-ftalkes, rather than ftalkes or ftems themfelues, of the height of two inches ; whereupon do grow fuch leaues as thofe that were next the ground, but leffer, fet by couples one oppofite to another : at the top of each fmall foot-ftalke doth ftand one red floure without fmell, confifting of fiue little leaues fet in a rough hairy huske or hofe fiue cornered, of a greenifh colour tending to purple. The root is tough and thicke, cafting abroad many fhoots, whereby it greatly encreafeth.

‡ 8   This for his ftature may juftly take the next place ; for the ftalke is fome inch high, fet with little fharpe pointed greene graffie leaues : the floures which grow vpon thefe ftalkes are com-pofed of fiue little flefh-coloured leaues a little diuided in their vpper parts : the feed is contained in blacke fhining heads, and it is fmall and reddifh, and fhaped fomewhat like the fafhion of a kid-ney, whereby it comes neerer to the *Lychnides*, than to the *Caryophylli* or Pinkes. The root is long, blacke, and much fpreading, whereby this little plant couers the ground a good fpace together like as a moffe, and makes a curious fhew when the floures are blowne, which is commonly in Iune. It

     floures

10 *Caryophyllus montanus albus.*
White mountaine Pinke.

‡ 11 *Caryophyllus pratenſis.*
Deptford Pinke.

12 *Caryophyllus Virgineus.*
Maiden Pinkes.

‡ 13 *Caryophyllus montanus humilis latifolius.*
Small mountaine broad leaued Pinke.

‡ 14 *Caryophyllus montanus albus.*
White mountaine Pinke.

15 *Caryophyllus Holostius.*
Wilde Sea Pinke.

16 *Caryophyllus Holostius aruensis.*
Broad leafed wilde Pinke.

‡ 17 *Caryophyl. humilis flor. caud. amæno.*
White Campion Pinke.

It growes naturall on diuers places of the Alpes. *Gesner* called it *Muscus floridus : Pona, Ocimoides Muscosus :* and *Clusius,Caryophyllus pumilio Alpinus 9.* ‡

9　This leafe-lesse Pinke (as the Greeke word doth seeme to import) hath many small rushy or benty leaues rising immediately from a tough rushy root : among which rise vp stalkes like vnto rushes,of a span high,without any joynt at all,but smooth and plaine ; on the top whereof groweth a small floure of a blewish or sky colour, consisting of foure little leaues somewhat jagged in the edges,not vnlike those of wilde flax. The whole plant is very bitter, and of a hot taste.

10　The white mountaine Pinke hath a great thicke and wooddy root; from the which immediately rise vp very many small and narrow leaues, finer and lesser than grasse,not vnlike to the smallest rush : among which rise vp little tender stalkes, joynted or kneed by certaine distances, set with the like leaues euen to the top by couples,one opposite against another : at the top whereof grow pretty sweet smelling floures composed of fiue little white leaues. The seed is small and blackish.

11　There is a wilde creeping Pinke which groweth in our pastures neere about London,and in other places,but especially in the great field next to Detford, by the path side as you go from Redriffe to Greenewich ; which hath many small tender leaues shorter than any of the other wilde Pinkes, set vpon little tender stalkes which lie flat vpon the ground, taking hold of the same in sundry places, whereby it greatly encreaseth : whereupon grow little reddish floures. The root is small,tough,and long lasting.

12　This Virgin-like Pinke is like vnto the rest of the Garden Pinkes in stalkes, leaues, and roots. The floures are of a blush colour,whereof it tooke his name, which sheweth the difference from the other.

‡　This whose figure I giue you for that small leaued one that was formerly in this place, hath slender stalkes some spanne high, set with two long narrow hard sharpe pointed leaues at each joynt. The floures (which grow commonly but one on a stalke) consist of fiue little snipt leaues of a light purple colour, rough and deeper coloured about their middles, with two little crooked threds or hornes : the seed is chaffie and blacke : the root long, and creeping : it floures in Aprill and May,and is the *Flos caryophylleus syluestris 1.of Clusius.* ‡

13　*Clusius* mentions also another whose stalkes are some three inches high : the leaues broader,softer,and greener than the former : the floures also that grow vpon the top of the stalkes are larger than the former, and also consist of fiue leaues of a deeper purple than the former, with longer haires finely intermixt with purple and white.

‡　14　This from a hard wooddy root sends vp such stalkes as the former, which are set at the joynts with short narrower and darker greene leaues : the floures are white, sweet-smelling, consisting of fiue much diuided leaues,hauing two threds or hornes in their middle. It floures in May, and it is the *Caryophyllus syluestris quintus* of *Clusius.* ‡

15　This wilde sea Pinke hath diuers small tender weake branches trailing vpon the ground, whereupon are set leaues like those of our smallest garden Pinke, but of an old hoary colour tending to whitenesse,as are most of the sea Plants.The floures grow at the top of the stalkes,in shape like those of Stitch-woort,and of a whitish colour.Neither the seeds nor seed-vessels haue I as yet obserued : the root is tough and single.

16　There is another of these wilde Pinkes which is found growing in ploughed fields, yet in such as are neere vnto the sea : it hath very many leaues spread vpon the ground of a fresh greene colour ; amongst which rise vp tender stalkes of the height of a foot, set with the like leaues by couples at certaine distances. The floures grow at the top many together, in manner of the Sweet-William, of a white or sometimes a light red colour. The root is small, tough and long lasting. ‡ This is a kinde of *Gramen Leucanthemum*, or *Holosteum Ruellij*,described in the 38. Chapter of the first booke.

17　*Clusius* makes this a *Lychnis :* and *Lobel* (whom I here follow) a Pinke,calling it *Caryophyllus minimus humilis alter exoticus flore candido amæno.* This from creeping roots sendeth vp euery yeare many branches some handfull and better high, set with two long narrow greene leaues at each joynt : the floures which grow on the tops of the branches are of a pleasing white colour, composed of fiue jagged leaues without smell. After the floures are gone there succeed round blunt pointed vessels,containing a small blackish flat seed like to that of the other Pinks. This hath a viscous or clammy juyce like as that of the *Muscipula's* or Catch-flies. *Clusius* makes this his *Lychnis syluestris decima.* ‡

¶ *The Place.*

These kindes of Pinkes do grow for the most part in gardens, and likewise many other sorts, the which were ouer long to write of particulary. Those that be wilde do grow vpon mountaines,stony rockes, and desert places. The rest are specified in their descriptions.

¶ *The*

¶ *The Time.*

They floure with the Cloue Gillofloure, and often after.

¶ *The Names.*

The Pinke is called of *Pliny* and *Turner, Cantabrica* and *Statice*: of *Fuchſius* and *Dodonæus, Vetonica altera,* and *Veronica altilis*: of *Lobelius* and *Fuchſius, Superba*: in French, *Gyrofflees, Oeilletz,* and *Violettes herbues*: in Italian, *Garofoli,* and *Garoni*: in Spaniſh, *Clauis*: in Engliſh, Pinkes and ſmall Honeſties.

¶ *The Temperature.*

The temperature of the Pinkes is referred to the Cloue Gillofloures.

¶ *The Vertues.*

Theſe are not vſed in Phyſicke, but eſteemed for their vſe in Garlands and Noſegaies. They are A good to be put into Vinegre, to giue it a pleaſant taſte and gallant colour, as *Ruellius* writeth. *Fuchſius* ſaith, that the roots are commended againſt the infection of the plague; and that the juyce thereof is profitable to waſte away the ſtone, and to driue it forth: and likewiſe to cure them that haue the falling ſickeneſſe.

CHAP. 184.   *Of Sweet Saint Iohns and Sweet Williams.*

1 *Armeria alba.*
White Iohns.

2 *Armeria alba & rubra multiplex.*
Double white and red Iohns.

¶ *The Deſcription.*

1   SWeet Iohns haue round ſtalkes as haue the Gillofloures, (whereof they are a kinde) a cubit high, whereupon doe grow long leaues broader than thoſe of the Gillofloure, of a greene graſſie colour: the floures grow at the top of the ſtalkes, very like vnto Pinkes, of a perfect white colour.

2   The ſecond differeth not from the other but in that, that this plant hath red floures, and the other white.

We haue in our London Gardens a kinde hereof bearing moft fine and pleafant white floures, fpotted very confufedly with reddifh fpots, which fetteth forth the beautie thereof ; and hath bin taken of fome (but not rightly) to be the plant called of the later Writers *Superba Auftriaca*, or the Pride of Auftria. ‡ It is now commonly in moft places called London-Pride. ‡

† We haue likewife of the fame kinde bringing forth moft double floures, and thefe either very white, or elfe of a deepe purple colour.

3   The great Sweet-William hath round joynted ftalkes thicke and fat, fomewhat reddifh about the lower joynts, a cubit high, with long broad and ribbed leaues like as thofe of the Plantaine, of a greene graffie colour. The floures at the top of the ftalkes are very like to the fmall Pinkes, many joyned together in one tuft or fpoky vmbell, of a deepe red colour : the root is thicke and wooddy.

3 *Armeria rubra latifolia.*
Broad leaued Sweet-William.

4 *Armeria fuaue rubens.*
Narrow leaued Sweet-Williams.

4   The narrow leaued Sweet-William groweth vp to the height of two cubits, very well refembling the former, but leffer, and the leaues narrower : the floures are of a bright red colour, with many fmall fharpe pointed graffie leaues ftanding vp among them, wherein efpecially confifteth the difference.

‡ 5   this little fruitfull Pinke (whofe figure our Author formerly gaue in the firft place of the next chapter faue one) hath a fmall whitifh wooddy root, which fends forth little ftalkes fome handfull and better high ; and thefe at each joynt are fet with two thinne narrow little leaues : at the top of each of thefe ftalkes growes a fingle skinny fmooth fhining huske, out of which (as in other Pinkes) growes not one onely floure, but many, one ftill comming out as another withers ; fo that oft times out of one head come feuen, eight, or nine floures one after another, which as they fade leaue behinde them a little pod containing fmall blacke flattifh feed. The floure is of a light red, and very fmall, ftanding with the head fomewhat far out of the hofe or huske. ‡

¶ *The Place.*

Thefe plants are kept and maintained in gardens more for to pleafe the eye, than either the nofe or belly.

¶ *The*

‡ 5 *Armeria prolifera, Lob.*
Childing ſweet Williams.

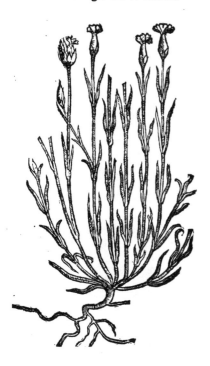

¶ *The Time.*

They flouriſh and bring forth their floures in Aprill and May, ſomewhat before the Gillofloures, and after beare their floures the whole Summer.

¶ *The Names.*

The ſweet Iohn, and alſo the ſweet William are both comprehended vnder one title, that is to ſay, *Armeria*: of ſome *Superba*, and *Caryophyllus ſylueſtris*: of ſome Herbariſts, *Vetonica agreſtis*, or *Sylueſtris*: of ſome, *Herba tunica*: but it doth no more agree herewith than the Cloue Gillo-floure doth with *Vetonica altera*, or *Polemonium*: in French, *Armories*: hereupon *Ruellius* nameth them *Armery Flores*: in Dutch, **Keykens**: as though you ſhould ſay, a bundle or cluſter, for in their vulgar tongue, bundles of floures or noſegaies they call **Keykens**: doubtleſſe they are wild kindes of Gillofloures: In Engliſh the firſt two are called Sweet Iohns; and the two laſt, Sweet Williams, Tolmeiners, and London Tufts.

¶ *The Temperature and Vertues.*

Theſe plants are not vſed either in meat or medicine, but eſteemed for their beauty to decke vp gardens, the boſomes of the beautifull, garlands and crownes for pleaſure.

CHAP. 185. *Of Crow floures, or Wilde Williams.*

¶ *The Deſcription.*

1 BEſides theſe kindes of Pinkes before deſcribed, there is a certaine other kinde, either of the Gillofloures or elſe of the Sweet Williams, altogether and euery where wilde, which of ſome hath beene inſerted amongſt the wilde Campions; of others taken to be the true *Flos Cuculi*. Notwithſtanding I am not of any of their mindes, but doe hold it for neither, but rather a degenerate kinde of wilde Gillofloure. The Cuckow floure I haue comprehended vnder the title of *Siſimbrium*: Engliſhed, Ladies ſmockes; which plant hath been generally taken for *Flos Cuculi*. It hath ſtalks of a ſpan or a foot high, whereupon the leaues do ſtand by couples out of euery joynt; they are ſmall and bluntly pointed, very rough and hairy. The floures are placed on the tops of the ſtalkes, many in one tuft, finely and curiouſly ſnipt in the edges, leſſer than thoſe of Gillofloures, very well reſembling the Sweet William (whereof no doubt it is a kinde) of a light red or skarlet colour.

2 This female Crow-floure differeth not from the male, ſauing that this plant is leſſer, and the floures more finely jagged like the feathered Pinke, whereof it is a kinde.

3 Of this Crow-floure wee haue in our Gardens one that doth not differ from the former of the field, ſauing that the plant of the garden hath many faire red double floures, and thoſe of the field ſingle.

¶ *The Place.*

Theſe grow all about in Medowes and paſtures, and dankiſh places.

¶ *The*

1 *Armoraria pratensis mas.*
The male Crow-floure.

‡ 3 *Armoraria pratensis flore pleno.*
The double Crow-floure.

¶ *The Time.*

They begin to floure in May and end in Iune.

¶ *The Names.*

The Crow-floure is called in Latine, *Armoraria syluestris,* and *Armoracia :* of some, *Flos Cuculi,* but not properly ; it is also called, *Tunix :* of some *Armeria, Armerius flos primus* of *Dodonæus,* and likewise, *Caryophyllus minor syluestris folijs latioribus :* in Dutch, **Crakenbloemkens :** that is to say, *Cornicis floris :* in French, *Cuydrelles :* in English, Crow-floures, wilde Williams, marsh Gillofloures, and Cuckow Gillofloure.

¶ *The Temperature and Vertues.*

These are not vsed either in medicine or in nourishment:but they serue for garlands and crowns, and to decke vp gardens.

CHAP. 146. *Of Catch-flie, or Limewort.*

¶ *The Description.*

1   THis plant, called *Viscaria,* or Lymewort, is likewise of the stocke and kindred of the wilde Gillofloures:notwithstanding *Clusius* hath joined it with the wilde Campions, making it a kinde thereof,but not properly. *Lobel* among the sweet Williams,where-of doubtlesse it is a kinde.It hath many leaues rising immediately from the root like those of the Crowfloure,or wilde sweet William : among which rise vp many reddish stalkes joynted or kneed at certaine spaces,set with leaues by couples one against another : at the top whereof come forth pretty red floures;which being past,there commeth in place small blackish seed. The root is large with manyfibres. The whole plant, as well leaues and stalkes,as also the floures are here and there couered ouer with a most thicke and clammy matter like vnto Bird-lime,which if you take in your

hands,

† 1 *Viſcaria, ſiue Muſcipula.*
Limewort.

2 *Muſcipula Lobelij.*
Catch Flie.

‡ 3 *Muſcipula anguſtifolia.*
Narrow leaued Catch-flie.

hands, the ſlimineſſe is ſuch, that your fingers will ſticke and cleaue together, as if your hand touched Bird-lime: and furthermore, if flies doe light vpon the ſame, they will be ſo intangled with the limineſſe, that they cannot flie away; inſomuch that in ſome hot day or other you ſhal ſee many flies caught by that means. Whereupon I haue called it Catch Flie, or Limewort. ‡ This is *Lychnis ſyl. 3.* of *Cluſius* ; *Viſcago* of *Camerarius*, and *Muſcipula ſiue Viſcaria* of *Lobel.* ‡

2   This plant hath many broad leaues like the great ſweet William, but ſhorter (whereof it is likewiſe a kinde) ſet vpon a ſtiffe and brittle ſtalk; from the boſom of which leaues ſpring forth ſmaller branches, clothed with the like leaues, but much leſſer. The floures grow at the top of the ſtalkes many together tuft faſhion, of a bright red colour. The whole plant is alſo poſſeſſed with the like limineſſe as the other is, but leſſe in quantity. ‡ Thisis *Lychnis ſyl. 1.* of *Cluſius* ; and *Muſcipula ſiue Armoraria altera* of *Lobel* : *Dodonæus* calls it *Armerius flos* 3. in his firſt Edition : but makes it his fourth in the laſt Edition in *Folio.* ‡

‡ 3   There is alſo belonging vnto this kindred another plant which *Cluſius* makes his *Lychnis ſyl.* 4. It comes vp commonly with one ſtalke a foot or more high, of a greene purpliſh

plifh colour, with two long fharpe pointed thicke greene leaues, fet at each joynt : from the middle to the top of the ftalke grow little branches, which vpon pretty long ftalkes carry floures confifting of fiue little round leaues, yet diuided at the tops ; they are of a faire incarnate colour, with a deepe purple ring in their middles, without fmell : after the floures are paft, fucceed skinny and hard heads, fmaller towards the ftalkes, and thicker aboue ; and in thefe are contained very fmall darke red feeds. The root is thicke, and blacke with many fibres, putting vp new fhoots and ftalkes after the firft yeare, and not dying euery yeare like as the two laft defcribed.

¶ *The Place.*

Thefe plants do grow wilde in the fields in the Weft part of England, among the corne : we haue them in our London gardens rather for toies of pleafure, than any vertues they are poffeffed with, that haue as yet beene knowne.

¶ *The Time.*

They floure and flourifh moft part of the Summer.

¶ *The Names.*

Catch Flies hath beene taken for *Behen*, commonly fo called, for the likeneffe that it hath with *Behen rubente flore :* or with *Behen* that hath the red floure, called of fome *Valeriana rubra*, or red Valerian; for it is fomething like vnto it in jointed ftalkes and leaues, but more like in colour : of *Lobel*, *Mufcipula* and *Vifcaria* : of *Dodon. Armerius flos tertius* : of *Clufius*, *Lychnis fylueftris, Silene Theophrafti*, and *Behen rubrum Salamanticum* : in Englifh, Catch Flie, and Limewort.

¶ *The Nature and Vertues.*

The nature and vertues of thefe wilde Williams are referred to the Wilde Pinkes and Gillofloures.

†   Our Author certainely intended in this firft place to figure and defcribe the *Mufcipula* or *Vifcaria* of *Lobel*, but the figure he here gaue in the firft place was of that plant which I haue giuen you in the laft Chapter faue one by the name of *Armeria proiifera Lobeln.* The figure which belonged to this place was in the Chapter of wilde Campions, vnder the title of *Lychnis fyluestris incana.*

CHAP. 167. *Of Thrift, or our Ladies Cufhion.*

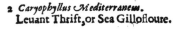

  1 *Caryophyllus marinus minimus Lobelij.*   **2** *Caryophyllus Mediterraneus.*
    Thrift, or Sea Gillofloure.   Leuant Thrift, or Sea Gillofloure.

¶ The Deſcription.

1 THriſt is alſo a kinde of Gilloſloure, by *Dodonæus* reckoned among graſſes, which brings forth leaues in great tufts, thicke thruſt together, ſmaller ſlenderer and ſhorter than graſſe: among which riſe vp ſmall tender ſtalkes of a ſpanne high, naked and without leaues; on the tops whereof ſtand little floures in a ſpokie tuſt, of a white colour tending topurple. The root is long and threddy.

2 The other kind of Thriſt, found vpon the mountaines neer vnto the Leuant or Mediterranean ſea, differeth not from the precedent in leaues, ſtalkes or floures, but yet is altogether greater, and the leaues are broader.

¶ The Place.

1 The firſt is found in the moſt ſalt marſhes in England, as alſo in Gardens, for the bordering vp of beds and bankes, for the which it ſerueth very fitly. The other is a ſtranger in theſe Northerne Regions.

¶ The Time.

They floure from May, till Summer be farre ſpent.

¶ The Names.

Thriſt is called in Latine, *Gramen Polyanthemum*, of the multitude of the floures: of ſome, *Gramen marinum*: of *Lobel*, *Caryophyllus Marinus*: in Engliſh, Thriſt, Sea-graſſe, and our Ladies Cuſhion.

¶ The Temperature and Vertues.

Their vſe in Phyſicke as yet is not knowne, neither doth any ſeeke into the Nature thereof, but eſteeme them only for their beautie and pleaſure.

---

‡ CHAP. 188.

Of the Saxifrage of the Antients, and of the great one of Matthiolus, with that of Pena and Lubel.

‡ THis name *Saxifraga* or Saxifrage, hath of late beene impoſed vpon ſundry plants farre different in their ſhapes, places of growing, & temperature, but all agreeing in this one faculty of expelling or driuing the ſtone out of the Kidnies, though not all by one meane or manner of operation. But becauſe almoſt all of them are deſcribed in their ſir places by our Author, I will not inſiſt vpon them: yet I thinke it not amiſſe a little to enquire, whether any *Saxifraga* were knowne to the Antients; and if knowne, to what kinde it may probably be referred? Of the Antients, *Dioſcorides*, *Paulus Ægineta*, and *Apuleius*, ſeeme to mention one *Saxifraga*, but *Pliny*, *lib. 22. cap. 21.* by the way ſhewes, that ſome called *Adianthum* by the name of *Saxifragum*: but this is nothing to the former; wherefore I will nor inſiſt vpon it, but returne to examine that the other three haue written thereof. *Dioſcorides lib. 4.* betweene the chapters of *Tribulus* and *Limonium*, to wit, in the ſeuenteenth place hath deliuered the Hiſtory of this plant, both in the Greeke Edition of *Aldus Manutius*, as alſo in that of *Marcellus Virgilius*, yet the whole chapter in the Paris Edition, 1549, is rejected and put amongſt the *Notha*. The beginning thereof (againſt which they chiefely except) is thus: Σαξίφραγον, ἡ δὲ σαξίφραγον, ἡ δὲ ἡμετέρα, ἡαιδοι σαξίφραξιφα, (i) Sarxiphagon, alij vero Sarxifrangon, alij vero Empetron, Romani Sarxifranga. The firſt exception of *Marcellus Virgilius* againſt this Chapter is *Peregrina Græcis & aliena vox Saxifraga eſt*, &c. The ſecond is, *Quod multo feliciores in componendis ad certiorem, yei alicuius ſignificationem vocibus Græi, quam Latini*, &c. The third is, *Solam in toto hoc opere primam, & a principio propoſitam audiri Romanam vocem, tamque inopes in appellanda hæc herba fuiſſe Græcos, vt niſi Romana voce eam indicaſſent, nulla ſibi futura eſſet.* Theſe are the arguments which he vſes againſt this Chapter; yet rejects it not, but by this means hath occaſioned others without ſhewing any reaſon to doe it: Now I will ſet downe what my opinion is concerning this matter, and ſo leaue it to the judgement of the learned. I grant *Marcellus*, that *Saxifraga* is a ſtrange and no Greeke word; but the name in the title, and firſt in the Chapter both in his owne Edition and all the Greeke Editions that I haue yet ſeene is Σαξίφραγον, which none, no not he himſelfe can deny to haue a Greeke originall ἀπὸ τὶς σάρκας ἐσθίειν of eating the fleſh: yet becauſe there is no ſuch facultie as this denomination imports attributed thereto by the Authour, therefore he will nor allow it to be ſo. But you muſt note that many names are impoſed by the vulgar, and the reaſon of the name not alwaies explained by thoſe that haue written of them, as in this ſame Author may be ſeene in the

the Chapters of *Catanance,Cynosbatos,Hemerocallis,Crotæogonon*, and diuers others,which are or feeme to be fignificant, and to import fomething by their name ; yet he faith nothing thereof. It may be that which they would expreffe by the name, was, that the herbe had fo piercing a faculty that it would eat into the very flefh. The fecond and third argument both are anfwered,if this firft word be Greeke, as I haue already fhewed it to be, and there are not many words in Greeke that more frequently enter into fuch compofition than ·· : as *Pamphagos* , *Polyphagos; Opfiphagos*, and many other may fhew. Moreouer, it had been obfurd, for *Diofcorides*, or any elfe how fimple foeuer they were, if they had knowne the firft word to haue beene Latine, and *Saxifraga*, to fay againe prefent-ly after,that the Romanes called it *Saxifranga*,or *Saxifraga*,for fo it would be and not *Sarxifranga :* but I feare that the affinitie of founds niore than of fignification hath caufed this confufion, efpe-cially in the middle times betweene vs and *Diofcorides*, when learning was at a very low ebbe. The chiefe reafons that induce me to thinke this chapter worthy to keepe his former place in *Diofcori-des*,are thefe : Firft, the generall confent of all both of Greeke and Latine copies (as *Marcellus* faith) how antient foeuer they be. Secondly, the mention of this herbe,for the fame effect in fome Greeke Authors of a reafonable good antiquitie ; for *Paulus Ægineta* teftifieth, that Σ··········· ···· ·· ··· ·· ····· ·········· . Then *Trallianus* amongft other things in a *Conditum Nephriticum* mentions Σ·········· : but *Nonus* a later Greeke calls it Σ········· : fo that it is euident they knew and vfed fome fim-ple medicine that had both the names of *Sarxiphagon* and *Saxiphragos*, which is the Latine *Saxifra-ga*. Now feeing they had, and knew fuch a fimple medicine, it remaines we enquire after the fhape and figure thereof. *Diofcorides* defcribes it to be a fhrubby plant, growing vpon rockes and craggie places,like vnto *Epithymum:* boyled in wine and drunke, it hath the faculty to helpe the Strangurie and Hicket ; it alfo breakes the ftone in the bladder and prouokes vrine. This word *Epithymum* is not found in moft copies,but a fpace left for fome word or words that were wanting:But *Marcellus* faith, he found it expreft in a booke which was *Omnium vetuftißimus & probatißimus :* and *Hermo-laus Barbarus* faith, *Veterum in* Diofcoride *pi&uram huius herbæ vidi, non plus folijs quam cirris minutis-per ramos ex interuallo conditis,nec frequentibus,in cacumine furculorum flocci feu arentes potius quam flofcu-li,fubrubida radice non fiue fibris.* A figure reafonable well agreeing with this defcription of *Hermolaus*, I lately receiued from my friend Mr *Goodyer*,who writ to me that he had fought to know what *Saxifraga* (to wit,of the Ancients) fhould be : and finding no antient Author that had defcribed it to any purpofe,he fought *Apuleius* ; which word *Apuleius* (faith he) is the printed title : my Manu-fcript acknowledgeth no Author but *Apolienfis Plato*; there is no defcription neither, but the Manu-fcript hath a figure which I haue drawne and fent you, and all that *verbatim* that he hath written of it, I fhould be glad to haue this figure cut and added to your woike, together with his words, be-caufe there hath beene fo little written thereof by the Antients. This his requeft I though fit to performe, and haue (for the better fatisfaction of the Reader) as you fee made a further enquirie thereof : wherefore I will onely adde this,that the plants here defcribed, and the *Alfine Saxifraga* of *Colum.*together with the two Chickweed *Saxifrages* formerly defcribed Chap.171.come neereft of any that I know to the figure and delineation of this of the Antients.

Nomen iftius herba,Saxifraga.

Icon & defcriptio ex Manufcripto vetu-tißimo.

Quidam dicunt eam Scolopendriam, alij Scolioimos, alij Vitis canum, quidam vero Bru-cos. Itali Saxifragæm. Egyptij Peperem, alij Lamprocain eam nominant. Nafcitur enim in Cmontibus & locis faxofis.

Vna cura ipfius ad calculos expellendos.

Herbam iftam Saxifragam contufam calcu-lofo potum dabis in vino. Ipfe vero fi febrici-tauerit cum aqua calida, tam prefens effectum ab expertis traditum , vt eodem die perfectis eiectifque calculis ad fanitatem vfque produ-cit.

1   This firft little herbe,faith *Camera-rius*,hath been called *Saxifraga magna*, not from the greatneffe of his growth, but of his faculties: The ftalke is wooddie; writhen

writhen, and below ſometimes as thicke as ones little finger, from which grow many ſmall and hard branches,and thoſe ſlender ones, the leaues are little, long and ſharpe pointed:the floures are white and ſmall,and grow in cups,which are finely ſnipt at the top in manner of a coronet,wherein is contained a ſmall red ſeed:the roots grow ſo faſt impact in the Rockes, that they cannot by any means be got out. It grows vpon diuers rocks in Italy and Germany;and is the *Saxifraga magna* of *Matthiolus*, and the Italians.

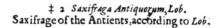

‡ 1 *Saxifraga magna Matthioli.*　　　　　　‡ 2 *Saxifraga Antiquorum, Lob.*
Matthiolus his great Saxifrage.　　　　　Saxifrage of the Antients,according to *Lob.*

2　*Pena* and *Lobel* ſay, this growes in great plenty in Italie, in Dolphonie in France, and England, hauing many ſmall ſlender branches a foot high,intricately wrapped within one another, where they are ſet with many graſſie joints:the root is ſmall and white,with ſome few fibres : the leaues ſtand by couples at the joints, being long and narrow;of the bigneſſe and ſimilitude of thoſe of the wild Pinkes, or Rocke Sauorie:vpon each wooddie,ſmall,capillarie, ſtraight, and creeping little branch,grows one little ſlour ſomwhat like a Pinke, being finely ſnipt about the edges:and in the head is contained a round ſmall reddiſh ſeed. The foreſaid Authors call this *Saxifragia,ſiue Saxifraga Antiquorum.*

¶ *The Vertues.*

1　*Matthiolus* ſaith,that *Calceolarius* of Verona mightily commended this plant to him,for the ſingular qualitie it had to expell or driue forth the ſtone of the kidneies, and that I might in verie deed beleeue it,he ſent me abundance of ſtones,whereof diuers exceeded the bigneſſe of a beane, which were voided by drinking of this plant by one onely Citizen of Verona, called *Hieronymo de Tortis*;but this made me moſt to wonder, for that there were ſome ſtones among them that ſeemed rather to come out of the Bladder,than forth of the Kidneies. **A**

2　This (ſay the Authors of the *Adverſ.*) as it is the lateſt receiued in vſe and name for Saxifrage, ſo is it the better and truer, eſpecially ſo thought by the Italians, both for the highly commended facultie, as alſo for the neere affinitie which it ſeems to haue with *Epithymum*, &c. ‡ **B**

Eee　　　　　　　　　　　CHAP.

Chap. 189.
Of Sneeſewoort.

¶ *The Deſcription.*

1   THe ſmall Sneeſe-woort hath many round and brittle branches, beſet with long & narrow leaues, hackt about the edges like a ſaw; at the tops of the ſtalks do grow ſmal ſingle floures like the wild field Daiſie. The root is tender and full of ſtrings, creeping far abroad in the earth, and in ſhort time occupieth very much ground: the whole plant is ſharp, biting the tongue and mouth like Pellitorie of Spaine, for which cauſe ſome haue called it wild Pellitorie. The ſmel of this plant procureth ſneeſing, whereof it tooke the name *Sternutamentoria*, that is, the herbe which procureth ſneeſing, or Neeſewoort.

2   Double floured Sneeſwoort, or *Ptarmica*, is like vnto the former in leaues, ſtalks, and roots, ſo that vnleſſe you behold the floure, you cannot diſcerne the one from the other, and it is exceeding white, and double like vnto double Fetherfew. This plant is of great beautie, and if it be cut downe in the time of his flouring there will come within a moneth after a ſupplie or crop of floures fairer than the reſt.

1 *Ptarmica.*
Sneeſewoort.

2 *Ptarmica duplici flore.*
Double floured Sneeſwort.

3   There is alſo another kind hereof, of exceeding great beauty, hauing long leaues ſomewhat narrow like thoſe of Oliue tree: the ſtalks are of a cubit high, on the top whercof grow very beautifull floures of the bigneſſe of a ſmall ſingle Marygold, conſiſting of fifteen or ſixteen large leaues, of a bright ſhining red colour tending to purple; ſet about a ball of thrummy ſubſtance, ſuch as is in the middle of the Daiſie, in manner of a pale; which floures ſtand in ſcalie knops like thoſe of Knapweed or Matfellon. The root is ſtraight, and thruſteth deepe into the ground.

‡ *Ptarmica Imperati* ; an *Ptarmicæ Auſtriacæ ſpecies Cluſ. Cur. poſt. p. 32 ?*

4   This riſeth vp with a ſmall hard tough cornered whitiſh woolly ſtalke, diuided into many
branches,

3 *Ptarmica Austriaca.*
Sneesewort of Austrich.

branches,& those againe diuided into other branches, like those of *Cyanus*, about two foot high,wheron grow long narrow whitish cottony leaues without order,of a bitter taft, whiter below than aboue, of the colour of the leaues of Wormewood, hauing but one rib or sinew, and that in the middle of the leafe,and commonly turne downewards : on the top of each slender branch groweth one small scaly head or knap like that of *Cyanus*, which bringeth forth a pale purple floure without smell,containing six,seuen, eight,or more small hard dry sharpe pointed leaues ; in the middle whereof groweth many stiffe chiues,their tops being of the colour of the floures : these floures fall not away till the whole herb perisheth,but change into a rustie colour:amongst those chiues grow long flat blackish seed, with a little beard at the top. The root is small, whitish, hard, and threddy,and perisheth when the feed is ripe, and soon springeth vp by the fall of the feed, and remaineth green all the Winter,and at the Spring sendeth forth a stalk as aforesaid. The herb touched or rubbed sendeth forth a pleasant aromaticall smell. Iuly 26. 1620. *Iohn Goodyer.* ‡

¶ *The Place.*
The first kind of Sneesewort growes wild in dry and barren pastures in many places, and in the three great fields next adioyning to a village neer London called Kentish towne, and in sundry fields in Kent about Southfleet.
† The rest grow only in gardens.

¶ *The Names.*
Sneesewort is called of some,*Ptarmica,*and *Pyrethrum syluestre,*and also *Draco syluestris,*or *Tarcon syluestris :* of most,*Sternutamentoria,* taken from his effect, because it procureth sneesing : of *Tragus* and *Tabern Tanacetum acutum album :* in English,wild Pellitorie, taking that name from his sharp and biting taste : but it is altogether vnlike in proportion to the true Pellitorie of Spain.

¶ *The Temperature.*
They are hot and dry in the third degree,

¶ *The Vertues.*
The juice mixed with vineger and holden in the mouth,easeth much the pain of the tooth-ach. **A**
The herb chewed and held in the mouth,bringeth mightily from the braine slimie flegme like **B** Pellitorie of Spain ; and therefore from time to time it hath bin taken for a wild kinde thereof.

CHAP. 190.
Of Hares Eares.

¶ *The Description.*

2   NArrow leaued Hares eares is called in Greek, ‌ ; and it is reputed of the late writers to be *Bupleurum Pliny,*from which name the figure disagreeth not:it hath the long
narrow and grassy leaues of *Lachrima Iob,*or *Gladiolus,*streaked or balked as it were with
sundry stiffe streaks or ribs running along euery leafe, as *Pliny* speaketh of his *Heptaplenrum.* The

ſtalks are a cubit and a halfe long, full of knots or knees, very rough or ſtiffe, ſpreading themſelues into many branches: at tops whereof grow yellow floures in round tufts or heads like Dill. The root is as big as a finger, and blacke like *Peucedanum*, whereunto it is like in taſte, ſmell, and reſemblance of ſeed, which doth the more perſuade me that it is the true *Bupleurum*, wherof I now ſpeak, and by the authoritie of *Nicander* and *Pliny* confirmed.

1 *Bupleurum angustifolium monspeliense.*
Narrow leaued Hares eare.

2 *Bupleurum latifolium monspeliense.*
Broad leaued Hares eare.

2   The ſecond kinde, called broad leaued Hares eares, in figure, tufts, and floures is the verie ſame with the former kinde, ſaue that the leaues are broader and ſtiffer, and more hollow in the midſt; which hath cauſed me to call it Hares eares, hauing in the middle of the leafe ſome hollow-neſſe reſembling the ſame. The root is greater, and of a wooddy ſubſtance.

¶ *The Place.*

They grow among Oken woods in ſtony and hard grounds in Narbon. I haue found them grow-ing naturally among the buſhes vpon Bieſton caſtle in Cheſhire.

¶ *The Time.*

They floure and bring forth their ſeed in Iuly and Auguſt.

¶ *The Names.*

Hares eare is called in Latine *Bupleurum*: in Greeke, Βούπλευρον: the Apothecaries of Montpelier in France do call it *Auricula Leporis*, and therefore I terme it in Engliſh Hares eare: *Valerius Cordus* nameth it *Iſophyllon*, but whence hs had that name it is not knowne.

¶ *The Temperature.*

They are temperate in heate and drineſſe.

¶ *The Vertues.*

A   *Hippocrates* hath commended it in meats; for ſallads and Pot-herbs: but by the authority of *Glau-con* and *Nicander*, it is effectuall in medicine, hauing the taſt and ſauor of *Hypericon*, ſeruing in the place thereof for wounds, and is taken by *Tragus* for *Panax Chironium*, who reckoneth it *inter Herbas vulnerarias*.

The

The leaues stamped with salt and wine, and applied, do consume and driue away the swelling of the necke called the Kings euill ; and are vsed against the stone and grauell.

Chap. 191.   Of Gromell.

¶ The Description.

1   THe great Gromel hath long slender and hairy stalks, beset with long browñ and hoarie leaues, among which grow certain bearded husks, bearing at the first smal blew flours; which being past, there succeedeth a gray stony seed somewhat shining. The root is hard, and of a wooddy substance.

2   The second kinde of Gromel hath straight round wooddy stalks full of branches, the leaues long, smal, and sharp, of a dark green colour, smaller than the leaues of great Gromel: among which come forth little white floures ; which being past, there followes such seed as the former hath, but smaller.

‡ 3   There is another kinde of Gromel which hath leaues and stalks like the small kind : the seed is not so white, neither so smooth and plain, but somewhat shriueled or wrinkled : the leaues are somewhat rough like vnto the common Gromel, but the floures are of a purple colour, and in shape like those of that wilde kinde of Buglosse called *Anchusa* ; for which cause it carrieth that additament *Anchusæ facie*.

. 4   There is also a degenerat kind hereof called *Anchusa degener*, being either a kind of wilde Buglosse or wilde Gromel, or else a kinde of neither of both, but a plant participating of both kindes. It hath the seeds and stalks of *Milium solis* or Gromel, the leaues and roots of *Anchusa*, which is Alkanet, and is altogether of a red colour like the same.

1 *Lithospermum majus.*
Great Gromel.

2 *Lithospermum minus.*
Small Gromel.

‡ 3 *Lithospermum Anchusa facie.*
Purple floured Gromell.

‡ 4 *Anchusa degener facie Milij solis.*
Baftard Gromell.

¶ *The Place.*

The two firft kindes do grow in vntoiled places, as by the highwayes fides and barren places, in the ftreet at Southfleet in Kent, as you go from the Church vnto a worfhipfull gentlemans houfe called Mr *William Swan,* and in fundry other places.

The two laft kindes grow vpon the fands and baich of the fea, in the Ifle of Thanet neere Reculuers, among the kindes of wilde Bugloffe there growing.

¶ *The Time.*

They floure in the Summer Solftice, or from the twelfth day of Iune euen vnto Autumne, and in the mean feafon the feed is ripe.

¶ *The Names.*

Gromel is called in Greeke Λιθόσπερμον, of the hardneffe of the feed : of diuers *Gorgonium :* of others, *Aegonychon, Leontion,* or *Diofporon,* or *Diofpyron,* as *Pliny* readeth it, and alfo *Heracleos :* of the Arabians, *Milium foler :* in fhops and among Italians, *Milium folis :* in Spanifh, *Mijo del fol :* in French, *Gremil,* and *Herbe aux perles :* in Englifh, Gromell : of fome, Pearle plant ; and of others, Lichwale.

¶ *The Temperature.*

The feed of Gromel is hot and dry in the fecond degree.

¶ *The Vertues.*

A   The feed of Gromel pound and drunke in white wine, breaketh, diffolueth, and driueth forth the ftone, and prouoketh vrine, and efpecially breaketh the ftone in the bladder.

CHAP. 192. *Of Chick-Weed.*

1   THe great Chickweed rifeth vp with ftalks a cubit high and fometimes higher, a great many from one root, long and round, flender, full of joints, with a couple of leaues growing

ing

ing out of euery knot or joint aboue an inch broad, and longer than the leaues of Pellitorie of the wall, whereunto they are very like in ſhape, but ſmooth without haires or downe, and of a light green colour: the ſtalks are ſomething cleare, and as it were tranſparent or thorow-ſhining, and about the joints they be oftentimes of a very light red colour, as be thoſe of Pellitorie of the wall: the floures be whitiſh on the top of the branches, like the floures of Stitchwort, but yet leſſer: in whoſe places ſucceed long knops, but not great, wherein the ſeed is contained. The root conſiſts of fine little ſtrings like haires.

　　2　The ſecond Chickweed for the moſt part lieth vpon the ground: the ſtalks are ſmall, ſlender, long, and round, and alſo jointed: from which ſlender branches doe ſpring leaues reſembling the precedent, but much leſſer, as is likewiſe the whole herbe, which in no reſpeſt attaineth to the greatneſſe of the ſame: the floures are in like ſort little and white: the knops or ſeed-heads are like the former: the root is alſo full of little ſtrings.

1 *Alſine major.*
Great Chickweed.

2 *Alſine minor, ſiue media.*
Middle or ſmall Chickweed.

　　3　The third is like the ſecond, but far leſſe: the ſtalks be moſt tender and fine, the leaues verie ſmall, the floures very little, the root maruellous ſlender.

　　4　Alſo there is a fourth kinde growing by the ſea, which is like the ſecond, but the ſtems are thicker, ſhorter, and fuller of juice: the leaues alſo be thicker, the knops or ſeed-heads be not long and round, but ſomewhat broad, in which are three or foure ſeeds contained.

　　5　The vpright Chickweed hath a very ſmall ſingle thready root, from which riſeth vp a ſlender ſtem diuiding it ſelfe into diuers branches euen from the bottome to the top: whereon grow ſmall leaues thicke and fat in reſpeſt of the others, in ſhape like thoſe of Rue or herb Grace. The floures grow at the top of the branches, conſiſting of foure ſmall leaues of a blew colour.

　　6　The ſtone Chickweed is one of the common Chickeweeds, hauing very thready branches couering the ground far abroad as it groweth: the leaues be ſet together by couples: the floures be ſmall and very white: the root is tough and very ſlender.

Speedwell

3 *Alſine minima.*
Fine Chickweed.

4 *Alſine marina.*
Sea Chickweed.

5 *Alſine recta.*
Right Chickweed.

6 *Alſine petraa.*
Stone Chickweed.

7. *Alſine folijs Veronicæ.*
Speed-well Chickweed.

8. *Alſine fontana.*
Fountaine Chickweed.

9. *Alſine fluviatilis.*
Riuer Chickweed.

10. *Alſine paluſtris.*
Mariſh Chickweed.

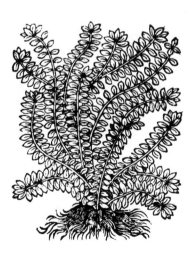

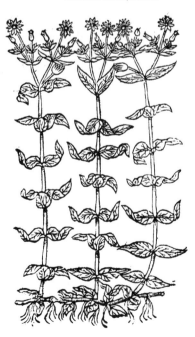

7   Speedwel Chickweed hath a little tender ſtalk, from which come diuers ſmal arms or branches as it were wings, ſet together by couples ; whereon grow leaues ſet likewiſe by couples; like thoſe of *Veronica* or herbe Fluellen, whereof it tooke his name. The floures grow along the branches, of a blew colour ; after which come little pouches wherein is the ſeed. The root is ſmall and likewiſe threddy. ‡ This in the *Hiſt. Lugd.* is called *Elatine polyſchides* and *Fabius Columna* iudgeth it to be the *Alyſſon* of *Dioſcorides.* ‡

8   There is a kinde of Chickweed growing in the brinks and borders of Wels, Fountaines, and ſhallow ſprings, hauing many threddy roots from which riſe vp diuers tender ſtalkes ; whereupon grow long narrow leaues, from the boſomes of which come forth diuers ſmaller leaues of a bright green colour. The floures grow at the top of the ſtalks, ſmall, and white of colour.

9   There is likewiſe another water Chickweed ſmaller than the laſt deſcribed, hauing for his root a thicke haſſock or tuft of threddy ſtrings ; from which riſe vp very many tender ſtems, ſtretching or trailing along the ſtream, whereupon grow long leaues ſet vpon a middle rib like thoſe of Lentils or wilde Fetch : the floures and ſeeds are like the precedent, but much ſmaller.

11   *Alſine rotundifolia, ſiue Portulaca aquatica.*
Water Purſlane.

13   *Alſine baccifera.*
Berry-bearing Chickweed.

‡   12   *Alſine paluſtris ſerpillifolia.*
Creeping water Chickweed.

10   There growes in the mariſh or waterie grounds another ſort of Chickweed, not much vnlike the reſt of the ſtock or kindred of the Chickweeds. It hath a long root of the bigneſſe of a wheat ſtraw, with diuers ſtrings hanging thereat, very like the root of Couch-graſſe : from which riſeth vp diuers vpright ſlender ſtalks, ſet with pretty large ſharpe pointed leaues ſtanding by couples at certain diſtances : on the top of the ſtalkes grow ſmall white floures like thoſe of Stitchwort, but leſſer, and of a white colour.

‡   11   To theſe water Chickweeds may fitly be added thoſe two which I mentioned and figured in my laſt Iournall : the former of which, that I haue there called *Alſine aquatica folijs rotundioribus, ſiue Portulaca aquatica* ; that is, round leaued Chickeweed, or water Purſlane, hath a ſmall ſtringy root which ſends forth diuers creeping ſquare branches, which here and there at the joints

put out ſmall fibres,and take root againe: the leaues grow at the joints by couples, ſomewhat lon-giſh,and round at the points, reſembling thoſe of Purſlane, but much ſmaller,and of a yellowiſh greene colour:at the boſomes of the leaues come forth little floures, which are ſucceeded by little round ſeed-veſſels containing a ſmall round ſeed. *Bauhine* hath ſet this forth by the name of *Alſine paluſtris minor folijs oblongis.*

12　The other water Chickweed, which *Iohn Bauhine* hath mentioned by the name of *Serpilli-folia*; and *Gaſper Bauhine* by the title of *Alſine paluſtris minor Serpillifolia*, hath alſo weake and tender creeping branches lying ſpred vpon the ground ; ſet with two narrow ſharp pointed leaues at each joynt,greene aboue, and of a whitiſh colour below:at the ſetting on of theſe leaues grow ſmall veſ-ſels parted as it were into two, with a little creſt on each ſide,and in theſe is contained a very ſmall ſeed. Both theſe may be found in waterie places in Iuly and Auguſt,as betweene Clapham heath and Touting,and betweene Kentiſh towne and Hampſtead.

13　This plant that *Cluſius* and others haue called *Alſine repens major*, and ſome haue thought the *Ciclaminus altera* of *Dioſcorides*; and *Cucubalus* of *Pliny*, may fitly be put in this ranke ; for it ſen-deth vp many long weake branches like the great Chickweed,ſet with two leaues at a joint,bigger than thoſe of the greateſt Chickweed, yet like them in ſhape and colour : at the tops of the bran-ches, out of pretty large cups come whitiſh greene floures, which are ſucceeded by berries as big as thoſe of Iuniper,at firſt green,but afterwards black:the ſeed is ſmal and ſmooth:the root white, very fibrous,long and wooddy , and it endures for many yeares. It floures moſt part of Summer, and growes wild in ſundry places of Spain and Germany, as alſo in Flanders and England, accor-ding to *Pena* and *Lobel:* yet I haue not ſeen it growing but in the garden of my friend Mr. *Pemble* at Marribone. The Authors laſt mentioned affirme the berries hereof to haue a poiſonous facultie like as thoſe of Dwale or deadly Nightſhade. ‡

¶ *The Place.*

Chickweeds, ſome grow among buſhes and briers,old walls , gutturs of houſes, and ſhadowie places.The places where the reſt grow are ſet forth in their ſeuerall deſcriptions.

¶ *The Time.*

The Chickweeds are greene in Winter, they floure and ſeed in the Spring.

¶ *The Names.*

Chickweed or Chickenweed is called in Greeke Ἀλσίνη: in Latine it retaineth the ſame name *Al-ſine:* of ſome of the Antients it is called *Hippia*. The reſt of the plants are diſtinguiſhed in their ſe-uerall titles,with proper names which likewiſe ſetteth forth the place of their growings.

¶ *The Temperature.*

Chickweed is cold and moiſt , and of a wateriſh ſubſtance; and therefore it cooleth without a-ſtriction or binding,as *Galen* ſaith.

¶ *The Vertues.*

The leaues of Chickweed boiled in water very ſoft,adding thereto ſome hogs greaſe,the pouder **A** of Fenugreek and Lineſeed,and a few roots of marſh Mallowes, and ſtamped to the form of a cata-plaſme or pulteſſe, take away the ſwellings of the legs or any other part , bring to ſuppuration or matter hot apoſtumes,diſſolue ſwellings that will not willingly yeeld to ſuppuration; eaſe mem-bers that are ſhrunk vp;comfort wounds in ſinewie parts;defend foule maligne and virulent lcers from inflammation during the cure:in a word,it comforteth,digeſteth,defendeth,and ſuppurateth very notably.

The leaues boiled in Vineger and ſalt are good againſt mangineſſe of the hands and legs,if they **B** be bathed therewith.

Little birds in cages (eſpecially Linnets)are refreſhed with the leſſer Chickeweed when they **C** loath their meat : whereupon it was called of ſome *Paſſerina.*

CHAP. 193.　*Of the baſtard Chickweeds.*

¶ *The Deſcription.*

1　GErmander Chickweed hath ſmal tender branches trailing vpon the ground,beſet with leaues like vnto thoſe of *Scordium* or Water Germander. Among which come forth little blew floures:which being faded,there appear ſmall flat huſks or pouches, wherin lieth the ſeed. The root is ſmall and threddie ; which being once gotten into a garden ground, is hard to be deſtroied,but naturally commeth vp from yeare to yeare as a noiſome weed.

2 *Cluſius*

1 *Alſine folijs Triſſaginis.*
Germander Chickweed.

2 *Alſine Corniculata Cluſtj.*
Horned Chickweed.

3 *Alſine Hederacea.*
Iuy Chickweed.

4 *Alſine Hederula altera.*
Great Henne-bit.

2　*Clusius*, a man singular in the knowledge of plants, hath set downe this herbe for one of the Chickweeds, which doth very well resemble the Storks bill, and might haue beene there inserted. But the matter being of small moment I let it passe; for doubtlesse it participateth of both, that is, the head or beake of Storkes bill, and the leaues of Chickweed, which are long and hairy, like those of Scorpions Mouse-eare. The floures are small, and of an herby colour; after which come long horned cods or seed vessels, like vnto those of the Storks bill. The root is small and single, with strings fastened thereto.

3　Iuie Chickweed or small Henbit, hath thin hairy leaues somewhat broad, with two cuts or gashes in the sides, after the manner of those of ground Iuie, whereof it tooke his name, resembling the backe of a Bee when she flieth. The stalks are small, tender, hairy, and lying flat vpon the ground. The floures are slender, and of a blew colour. The root is little and threddy.

4　The great Henbit hath feeble stalkes leaning toward the ground, whereupon do grow at certaine distances leaues like those of the dead Nettle; from the bosome whereof come forth slender blew floures tending to purple; in shape like those of the small dead Nettle. The root is tough, single, and a few strings hanging thereat.

¶ *The Place.*

These Chickweeds are found in gardens among pot-herbes, in darke shadowie places, and in the fields after the corne is reaped.

¶ *The Time.*

They flourish and are greene when the other Chickweeds are.

¶ *The Names.*

The first and third is called *Morsus Gallinæ*, Hens bit, *Alsine Hederula*, and *Hederacea*: Lobel also cals the fourth, *Morsus Galli, folio Hederulæ alter*: in high Dutch, **Hunerbiß**: in French, *Morgelin*, and *Morgeline*: in low Dutch, **Hoenderebeet**: in English, Henbit the greater and the lesser.

¶ *The Temperature and Vertues.*

These are thought also to be cold and moist, and like to the other Chickweeds in vertue and operation.

Chap. 185.　Of Pimpernell.

1　*Anagallis mas.*
Male Pimpernell.

2　*Anagallis fœmina.*
Female Pimpernell.

¶ *The Deſcription.*

1   Pimpernell is like vnto Chickweed; the ſtalkes are foure ſquare, trailing here and there vpon the ground, whereupon do grow broad leaues, and ſharpe pointed, ſet together by couples : from the boſomes whereof come forth ſlender tendrels, whereupon doe grow ſmall purple floures tending to redneſſe : which being paſt there ſucceed fine round bullets, like vnto the ſeed of Coriander, wherein is contained ſmall duſty ſeed. The root conſiſteth of ſlender ſtrings.

2   The female Pimpernell differeth not from the male in any one point, but in the colour of the floures; for like as the former hath reddiſh floures, this plant bringeth forth floures of a moſt perfect blew colour; wherein is the difference.

‡ 3   Of this there is another variety ſet forth by *Cluſius* by the name of *Anagallis tenuifolio Monelli*, becauſe he receiued the figure and Hiſtory thereof from *Iohn Monell* of Tourney in France; it differs thus from the laſt mentioned, the leaues are longer and narrower, ſomewhat like thoſe of *Gratiola*, and they now and then grow three at a joynt, and out of the boſomes of the leaues come commonly as many little foot-ſtalkes as there are leaues, which carry floures of a blew colour with the middle purpliſh, and theſe are ſomewhat larger than them of the former, otherwiſe like. ‡

‡ 3 *Anagallis tenuifolia.*
   Narrow leaued Pimpernell.

             4 *Anagallis lutea.*
              Yellow Pimpernell.

4   The yellow Pimpernell hath many weake and feeble branches trailing vpon the ground, beſet with leaues one againſt another like the great Chickweed, not vnlike to *Nummularia*, or Moneywoort; betweene which and the ſtalkes, come forth two ſingle and ſmall tender foot-ſtalkes, each bearing at their top one yellow floure and no more. The root is ſmall and threddy.

¶ *The Place.*

They grow in plowed fields neere path waies, in Gardens and Vineyards almoſt euery where. I found the female with blew floures in a chalkie corne field in the way from Mr. *William Swaines* houſe of Southfleet to Long field downes, but neuer any where elſe. ‡ I alſo being in Eſſex in the company of my kinde friend Mr. *Nathaniel Wright* found this among the corne at Wrightsbridge, being the ſeat of Mr. *Iohn Wright* his brother. ‡ The yellow Pimpernell groweſ in the woods betweene High gate and Hampſtead, and in many other woods.

¶ *The Time.*

They floure in Summer, and eſpecially in the moneth of Auguſt, at what time the husbandmen hauing occaſion to go vnto their harueſt worke, will firſt behold the floures of Pimpernell, whereby they know the weather that ſhall follow the next day after; as for example, if the floures be ſhut cloſe vp, it betokeneth raine and foule weather; contrariwiſe, if they be ſpread abroad, faire weather.

¶ *The Names.*

It is called in Greeke, Ἀναγαλλὶς: in Latine also *Anagallis*: of diuers, (as *Pliny* reporteth) *Corchorus*, but vntruly : of *Marcellus* an old Writer, *Macia* ; the word is extant in *Dioscorides* among the bastard names. That with the crimson floure, being the male, is named *Phœnicion*, and *Corallion* of this is made the composition or receit called *Diacorallion*, that is vsed against the gout; which composition *Paulus Ægineta* setteth downe in his seuenth booke. Amongst the bastard names it hath been called *Aëtitis*, *Ægitis*, and *Sauritis* : in English, red Pimpernell, and blew Pimpernell.

¶ *The Temperature.*

Both the sorts of Pimpernell are of a drying faculty without biting, and somewhat hot, with a certaine drawing quality, insomuch that it doth draw forth splinters and things fixed in the flesh, as *Galen* writeth.

¶ *The Vertues.*

Dioscorides writes, That they are of power to mitigate paine, to cure inflammations and hot swellings, to draw out of the body and flesh thornes, splinters, or shiuers of wood, and to helpe the Kings Euill.   **A**

The iuyce purgeth the head by gargarising or washing the throat therewith ; it cures the tooth-ach being snift vp into the nosethrils, especially into the contrary nosethrill.   **B**

It helpeth those that be dim sighted : the juyce mixed with hony cleanses the vlcers of the eye called in Latine *Argema*.   **C**

Moreouer he affirmeth, That it is good against the stinging of Vipers, and other venomous beasts.   **D**

It preuaileth against the infirmities of the liuer and kidnies, if the juyce be drunke with wine. He addeth further, how it is reported, That Pimpernel with the blew floure helpeth vp the fundament that is fallen downe, and that red Pimpernell applied, contrariwise bringeth it downe.   **E**

Chap. 184.   *Of Brooke-lime, or water Pimpernell.*

¶ *The Description.*

1   **B**Rooke-lime or Brooklem hath fat thicke stalkes, round, and parted into diuers branches : the leaues be thicke, smooth, broad, and of a deepe greene colour. The floures grow vpon small tender foot-stalkes, which thrust forth of the bosome of the leaues, of a perfect blew colour, not vnlike to the floures of land Pimpernell : the root is white, low creeping, with fine strings fastned thereto : out of the root spring many other stalkes, whereby it greatly encreaseth.

‡ There is a lesser variety of this, which our Author set forth in the fourth place, differing not from this but only that it is lesse in all the parts thereof ; wherefore I haue omitted the history and figure, to make roome for more conspicuous differences. ‡

2   The great water Pimpernell is like vnto the precedent, sauing that this plant hath sharper pointed or larger leaues, and the floures are of a more whitish or a paler blew colour, wherein consisteth the difference.

‡ There is also a lesser varietie of this, whose figure and description our Author gaue in the next place ; but because the difference is in nothing but the magnitude I haue made bold to omit it also.

3   Now that I haue briefely giuen you the history of the foure formerly described by our Author, I will acquaint you with two or three more plants which may fitly be here inserted : The first of these *Lobel* cals *Anagallis aquatica tertia* ; and therefore I haue thought fit to giue you it in the same place here. It hath a white and fibrous root ; from which ariseth a round smooth stalke a foot and more high, (yet I haue sometimes found it not aboue three or foure inches high :) vpon the stalkes grow leaues round, greene and shining, standing not by couples, but one aboue another on all sides of the stalkes. The leaues that lie on the ground are longer than the rest, and are in shape somewhat like those of the common Daisie, but that they are not snipped about the edges : the floures are white, consisting of one leafe diuided into fiue parts ; and they grow at the first as it were in an vmbel, but afterwards more spike fashioned. It floures in Iune and Iuly, and groweth in many watery places, as in the marithes of Dartford in Kent, also between Sandwich and Sandowne castle, and in the ditches on this side Sandwich. *Bauhine* saith, That *Guillandinus* called it sometimes *Alisma*, and otherwhiles *Cochlearia* : and others would haue it to be *Samolum* of *Pliny, lib. 25. cap. 11.* *Bauhine* himselfe fitly calls it *Anagallis aquatica folio rotundo non crenato.*

1 *Anagallis ſeu Becabunga.*
Brooke-lime.

2 *Anagallis aquatica major.*
Great long leaued Brook-lime.

‡ 3 *Anagallis aquatica rotundifolia.*
Round leaued water Pimpernell.

4 I conjecture this figure which we here giue you with the Authors title to be onely the leſſer variety of that which our Author deſcribes in the ſecond place ; but becauſe I haue no certainty hereof (for that *Lobel* hath giuen vs no deſcription thereof in any of his Latine Workes, and alſo *Bauhinus* hath diſtinguiſhed them) I am forced to giue you onely the figure thereof ; not intending to deceiue my Reader by giuing deſcriptions from my fancie and the figure, as our Author ſometimes made bold to do.

5 This which is ſet forth by moſt Writers for *Cepaa*, and which ſome may object to be more fit to be put next the Purſlanes, I will here giue you, hauing forgot to do it there ; and I thinke this place not vnfit, becauſe our Author in the Names in this Chapter takes occaſion in *Dodonæus* his words to make mention thereof. It hath a ſmall vnprofitable root, ſending vp a ſtalke ſome foot high, diuided into many weake branches, which are here and there ſet with thicke leaues like thoſe of Purſlane, but much leſſe, and narrower, and ſharper pointed : the floures which grow in good plenty vpon the tops of the branches are compoſed of fiue ſmall white leaues ; whereto ſucceeds ſmall heads, wherein is contained a ſeed like that of Orpine. This by *Matthiolus* and others is called *Cepaa*: but *Cluſius* doubts that it is not the true *Cepaa* of the Antients. ‡

¶ *The*

‡ 4 *Anagallis aquatica quarta, Lob.*
Lobels fourth water Pimpernel.

‡ 5 *Cepæa.*
Garden Brook-lime.

¶ *The Place.*

They grow by riuers sides, small running brookes, and watery ditches. The yellow Pimpernell I found growing in Hampsted wood neere London, and in many other woods and copses.

¶ *The Time.*

They bring forth their floures and seed in Iune, Iuly, and August.

¶ *The Names.*

Water Pimpernell is called *Anagallis aquatica*: of most, *Becabunga*, which is borrowed of the Germane word, Bachpunghen: in low Dutch, Beeckpunghen: in French, *Berle*; whereupon some do call it *Berula*: notwithstanding, *Marcellus* reporteth, That *Berula* is that which the Grecians call σιον. or rather Cresses: it is thought to be *Cepæa*, that is to say, of the garden; which *Dioscorides* writeth to be like vnto Purslane, whereunto this Brook-lime doth very well agree. But if it be therefore said to be σιον because it groweth either only or for the most part in Gardens, this Pimpernell or Brooklime shall not be like vnto it, which groweth no where lesse than in gardens, being altogether of his owne nature wilde, desiring to grow in watery places, and such as be continually ouerflowne: in English the first is called Brooklime, and the rest by no particular names; but we may call them water Pimpernels, or Brooklimes.

¶ *The Temperature.*

Brooklime is of temperature hot and dry like water Cresses, yet not so much.

¶ *The Vertues.*

Brooke-lime is eaten in sallads as Water-Cresses are, and is good against that ὀνθρωπινον malum of **A** such as dwell neere the Germane seas, which they call Seuerbuycke: or as we terme it, the Scuruie, or Skirby, being vsed after the same manner that Water-Cresses and Scuruy grasse is vsed, yet is it not of so great operation and vertue.

The herbe boyled maketh a good fomentation for swollen legs and the dropsie. **B**

The leaues boyled, strained, and stamped in a stone morter with the pouder of Fenugreeke, Line- **C** seeds, the root of marish Mallowes, and some hogs grease, vnto the forme of a cataplasme, or pultesse, take away any swelling in leg or arme; wounds also that are ready to fall into apostumation it mightily defendeth, that no humor or accident shall happen thereunto.

　　　　The

D   The leaues of Brooke-lime ſtamped, ſtrained, and giuen to drinke in wine, helpe the ſtrangurie, and griefes of the bladder.

E   The leaues of Brook-lime, and the tendrels of *Aſparagus*, eaten with oyle, vinegre, and pepper, helpe the ſtrangurie and ſtone.

CHAP. 196. Of ſtinking Ground-Pine.

¶ *The Kindes.*

‡   **D***ioſcorides* hath antiently mentioned two ſorts of *Anthyllis* : one with leaues like to the Lentill, and the other like to *Chamæpytis*. To the firſt, ſome late writers haue referred diuers plants, as the two firſt deſcribed in this Chapter ; The *Anthyllis Leguminoſa Belgarum* hereafter to be deſcribed ; the *Anthyllis Valentina Cluſij* formerly ſet forth Chap. 171. To the ſecond are referred the *Iua Moſcata Monſpeliaca*, deſcribed in the fourth place of the 150. Chap. of this booke ; the *Linaria adulterina* deſcribed formerly chap. 165. in the 14. place, and that which is here deſcribed in the third place of this chapter, by the name of *Anthyllis altera Italorum*. ‡

¶ *The Deſcription.*

1   **T**Here hath beene much adoe among Writers about the certaine knowledge of the true *Anthyllis* of *Dioſcorides* : I will therefore ſet downe that plant which of all others is found moſt agreeable thereunto. It hath many ſmall branches full of joynts, not aboue an handfull high, creeping ſundry waies, beſet with ſmall thicke leaues of a pale colour, reſembling *Lenticula*, or rather *Alſine minor*, the leſſer Chickeweed. The floures grow at the top of the ſtalke, ſtarre-faſhion, of an herby colour like boxe, or *Sedum minus* : it foſtereth his ſmall ſeeds in a three cornered huske. The root is ſomewhat long, ſlender, joynted, and deepely thruſt into the ground like *Soldanella* : all the whole plant is ſaltiſh, bitter in taſte, and ſomewhat heating.

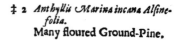

‡ 1 *Anthyllis lentifolia, ſiue Alſine cruciata marina.*
Sea Pimpernell.

‡ 2 *Anthyllis Marina incana Alſinefolia.*
Many floured Ground-Pine.

‡   This deſcription was taken out of the *Aduerſaria, pag.* 195. where it is called *Anthyllis prior lentifolia Peplios effigie maritima :* alſo *Cluſius* hath deſcribed it by the name of *Alſines genus pelagicum :* I haue called it in my laſt journall by the name of *Alſine cruciata marina*, becauſe the leaues which grow thicke together by couples croſſe each other, as it happens in moſt plants which haue ſquare ſtalkes with two leaues at each joynt. I haue Engliſhed it ſea Pimpernell, becauſe the leaues in ſhape are liker thoſe of Pimpernell then of any other Plant ; and alſo for that our Author hath called another plant by the name of Sea Chickweed. The figure of the *Aduerſaria* was not good, and *Cluſius* hath none ; which hath cauſed ſome to reckon this *Anthyllis* of *Lobel*, and *Alſine* of *Cluſius* for two ſeuerall plants, which indeed are not ſo. I haue giuen you a figure hereof which I tooke from the growing plant, and which well expreſſeth the growing hereof. ‡

2 There

3 *Anthyllis altera Italorum.*
Stinking ground Pine.

2   There is likewise another fort of *Anthyllis* or Sea Ground Pine, but in truth nothing els than a kinde of Sea Chickeweed, hauing small branches trailing vpon the ground of two hands high;whereupon do grow little leaues like thofe of Chickweed, not vnlike thofe of *Lenticula marina*, or Sea Lentils :on the top of the ftalkes ftand many fmall moffie floures of a white colour. The whole plant is of a bitter and faltifh tafte. ‡ This is the *Marina incana Anthyllis Alfine folia Narbonenfium* of Lobel : the *Paronychia altera* of *Matthiolus*. ‡

‡ 3   To this figure (which formerly was giuen for the firft of thefe by our Author) I will now giue you a briefe defcription. This in the branches, leaues, and whole face thereof is very like the French Herbe-Iure,or Ground Pine,but that it is much leffe in all the parts thereof, but chiefely in the leaues which alfo are not fnipt like thofe of the French Ground Pine,but fharp pointed : the tops of the branches are downy or woolly, and fet with little pale yellow floures. ‡

¶ *The Place.*

Thefe do grow in the South Ifles belonging to England,efpecially in Portland in the grauelly and fandy foords, which lie low and againft the fea;and likewife in the Ifle of Shepey neere the water fide. ‡ I haue only found the firft defcribed, and that both in Shepey, as alfo in Weft-gate bay by Margate in the Ifle of Thanet. ‡

¶ *The Time.*

They floure and flourifh in Iune and Iuly.

¶ *The Names.*

Their titles and defcriptions fufficiently fet forth their feuerall names.

¶ *The Temperature.*

Thefe fea herbes are of a temperate faculty betweene hot and cold.

¶ *The Vertues.*

Halfe an ounce of the dried leaues drunke, preuaileth greatly againft the hot piffe, the ftrangurie,or difficulty of making water,and purgeth the reines.   A

The fame taken with Oxymell or honied water is good for the falling fickeneffe, giuen firft at morning,and laft at night.   B

† There was formerly three defcriptions,yet but one figure in this chapter, and that was marked with the figure 1. and called *Anthyllis lentifolia*, but vnfitly: wherefore I haue giuen you the title which *Lobel* the firft Author thereof gaue it, vpon the fame defcription the two, that it may not ftand as a cipher,as it formerly did. That defcription which formerly held the fecond place was of the *Anthyllis Valentina* of *Clufius*,defcribed chap.171.and therefore I haue omitted it here.

CHAP. 197.   *Of Whiteblow, or Whitelow Graffe.*

¶ *The Defcription.*

1   THe firft is a very flender plant hauing a few fmall leaues like the leaft Chickeweed, growing in little tufts, from the midft whereof rifeth vp a fmall ftalke, three or foure inches long ; on whofe top do grow very little white floures ; which being paft there come in place fmall flat pouches compofed of three filmes ; which being ripe, the two outfides fall away,leauing the middle part ftanding long time after,which is like white Sattin,as is that of *Bolbonac*,which our women call white Sattin,but much leffer : the tafte is fomewhat fharpe.

2   This kinde of *Paronychia*, hath fmall thicke and fat leaues, cut into three or more diuifions, much refembling the leaues of Rue,but a great deale fmaller. The ftalkes are like the former,& the
leaues

leaues alſo ; but the caſes wherein the ſeed is contained, are like vnto the ſeed veſſels of *Myoſitis Scorpioides,* or mouſ-eare Scorpion graſſe. The floures are ſmall and white.

There is another ſort of Whitlow graſſe or Nailewoort, that is likewiſe a low or baſe herbe, hauing a ſmall tough root, with ſome threddy ſtrings annexed thereto: from which riſe vp diuerſe ſlender tough ſtalks, ſet with little narrow leaues confuſedly like thoſe of the ſmalleſt Chickweed, whereof doubtleſſe theſe be kindes: alongſt the ſtalks do grow very little white floures, after which come the ſeeds in ſmall buttons, of the bigneſſe of a pins head. ‡ Our Author ſeemes here to deſcribe the *Paronychia* 2. of *Tabern.* ‡

1 *Paronychia vulgaris.*
Common Whitlow graſſe.

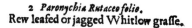

2 *Paronychia Rutaceo folio.*
Rew leafed or jagged Whitlow graſſe.

¶ *The Place.*
Theſe ſmall baſe and low herbs grow vpon bricke and ſtone walls, vpon old tiled houſes, which are growne to haue much moſſe vpon them, and vpon ſome ſhadowie and dry muddy wals. It groweth plentifully vpon the bricke wall in Chancery Lane, belonging to the Earle of Southampton, in the Suburbs of London, and ſundry other places.

¶ *The Time.*
Theſe floure many times in Ianuary and February, and when hot weather approacheth, they are no more to be ſeene all the yeare after.

¶ *The Names.*
The Græcians haue called theſe plants παρωνυχία : which *Cicero* calleth *Reduvia* : There be many kindes of plants, called by the ſame name of *Paronychia,* which hath cauſed many writers to doubt of the true kinde : but you may very boldly take theſe plants for the ſame, vntill time hath reueeled or raiſed vp ſome new plant, approching neerer vnto the truth: which I thinke will neuer be, ſo that we may call them in Engliſh, Naile-woort, and Whitlow graſſe.

¶ *The Temperature and Vertues.*

A   As touching the quality hereof, we haue nothing to ſet downe : only it hath beene taken to heale the diſeaſe of the nailes called a Whitlow, whereof it tooke his name.

‡ Our Author here gaue vs two figures, and as many deſcriptions of both theſe plants, wherefore I haue omitted two of the figures, and the more vnperfect deſcription.

CHAP.

CHAP. 198.   *Of the female Fluellen, or Speedwell.*

¶ *The Deſcription.*

1   THe firſt kinde of *Elatine*, being of *Fuchſius* and *Matthiolus*, called *Veronica fœmina*, or the female Fluellen, ſhooteth from a ſmall and fibrous root many flexible and tender branches, diſperſed flat vpon the ground, ramping and creeping with leaues like *Nummularia*, but that the leaues of *Elatine* are of an hoarie, hairie, and ouerworne greene colour; among which come forth many ſmall floures, of a yellow colour mixed with a little purple, like vnto the ſmall Snapdragon, hauing a certaine taile or Spurre faſtened vnto euery ſuch floure, like the herbe called Larkes Spurre. The lower jaw or chap of the floure is of a purple colour, and the vpper jaw of a faire yellow; which being paſt, there ſucceeds a ſmall blacke ſeed contained in round husks.

2   The ſecond kinde of *Elatine* hath ſtalkes, branches, floures, and roots, like the firſt: but the leaues are faſhioned like the former, but that they haue two little ears at the lower end, ſomewhat reſembling an arrow head, broad at the ſetting on: but the ſpur or taile of the floure is longer, and more purple mixed with the yellow in the floure.

1 *Veronica fœmina Fuchſij, ſiue Elatine.*
  The Female Fluellen.

2 *Elatine altera.*
  Sharpe pointed Fluellen.

¶ *The Place.*

Both theſe plants I haue found in ſundry places where corne hath growne, eſpecially barley, as in the fields about Southfleet in Kent, where within ſix miles compaſſe there is not a field wherein it doth not grow.

Alſo it groweth in a field next vnto the houſe ſometime belonging to that honourable Gentleman Sir *Francis Walſingham*, at Barn-elmes, and in ſundry places of Eſſex; and in the next field vnto the Churchyard at Chiſwicke neere London, toward the midſt of the field.

¶ *The Time.*

They floure in Auguſt and September.

¶ *The Names.*

Their seuerall titles set forth their names as well in Latine as English.

¶ *The Nature and Vertues.*

A   These plants are not onely of a singular astringent faculty, and thereby helpe them that be grieued with the Dysentery and hot swelling ; but of such singular efficacy to heale spreading and eating cankers, and corosiue vlcers, that their vertue in a manner passeth all credit in these fretting sores, vpon sure proofe done vnto sundry persons, and especially vpon a man whom *Pena* reporteth to haue his nose eaten most grieuously with a canker or eating sore, who sent for the Physitions and Chirurgions that were famously knowne to be the best, and they with one consent concluded to cut the said nose off, to preserue the rest of his face : among these Surgeons and Physitions came a poore sorie Barber, who had no more skill than he had learned by tradition, and yet vndertooke to cure the patient. This foresaid Barbar standing in the company and hearing their determination, desired that he might make triall of an herbe which he had seene his Mr. vse for the same purpose, which herbe *Elatine*, though he were ignorant of the name whereby it was called, yet he knew where to fetch it. To be short, this herbe he stamped, and gaue the juyce of it vnto the patient to drinke, and outwardly applied the same plaisterwise, and in very short space perfectly cured the man, and staied the rest of his body from further corruption, which was ready to fall into a leprosie, *Aduers.* pag. 197.

B   *Elatine* helpeth the inflammation of the eies, and defendeth humors flowing vnto them, being boiled and as a pultus applied thereto.

C   The leaues sodden in the broth of a hen, or Veale, stay the dysentery.

D   The new writers affirme, that the female Fluellen openeth the obstructions or stoppings of the liuer and spleene, prouoketh vrine, driueth forth stones, and clenseth the kidnies and bladder, according to *Paulus*.

E   The weight of a dram or of a French crowne, of the pouder of the herbe, with the like weight of treacle, is commended against pestilent Feuers.

CHAP. 199. *Of Fluellen the male, or Pauls Betonie.*

1 *Veronica vera & major.*
Fluellen, or Speedwell.

† 2 *Veronica recta mas.*
The male Speedwell.

¶ *The*

¶ *The Deſcription.*

1   THe firſt kind of *Veronica* is a ſmal herbe, and creepeth by the ground, with little reddiſh and hairy branches. The leafe is ſomething round and hairy, indented or ſnipped round about the edges. The floures are of a light blew colour, declining to purple: the ſeed is contained in little flat pouches: the root is fibrous and hairy.

†   2   The ſecond doth alſo creepe vpon the ground, hauing long ſlender ſtemmes, ſome foot high, and ſomewhat large leaues a little hairy, and pleaſantly ſoft. The floures be blew like as thoſe of the former, but ſomewhat bigger, and of a brighter colour; and they are alſo ſucceeded by round ſeed veſſels.

3   The third kinde of *Veronica*, creepeth with branches and leaues like vnto *Serpillum*, for which cauſe it hath been called *Veronica Serpillifolia.* The floures grow along the ſmall and tender branches, of a whitiſh colour declining to blewneſſe. The root is ſmall and threddy, taking hold vpon the vpper face of the earth, where it ſpreadeth. The ſeed is contained it ſmall pouches like the former.

4   The fourth hath a root ſomewhat wooddy, from the which riſe vp leaues like vnto the former. The ſmall vpright ſtalke is beſet with the like leaues, but leſſer; at the top whereof commeth forth a ſlender ſpike cloſely thruſt together, and full of blewiſh floures, which are ſucceeded by many horned ſeed veſſels.

‡   5   This hath many wooddy round ſmooth branches, ſome handfull and halfe high or better: the leaues are like thoſe of wilde Tyme, but longer, and of a blacker colour, ſometimes lightly ſnipt: at the tops of the branches grow floures of a whitiſh blew colour, conſiſting of foure, fiue, or elſe ſix little leaues a piece; which falling, there follow round ſeed veſſels, containing a round ſmall and blacke ſeed. It floures in Auguſt, and growes vpon cold and high mountaines, as the Alpes. *Pona* cals this *Veronica Alpina minima Serpilli folio:* and *Cluſius* hath it by the name of *Veronica 3. fruticans.* ‡

2 *Vetonica minor.*
Little Fluellen.

4 *Veronica recta minima.*
The ſmalleſt Fluellen.

‡ 5 *Veronica fruticans Serpillifolia.*
Shrubby Fluellen.

6 *Veronica assurgens, sive Spicata.*
Tree Fluellen.

† 7 *Veronica spicata latifolia.*
Vpright Fluellen.

‡ 8 *Veronica supina.*
Leaning Fluellen.

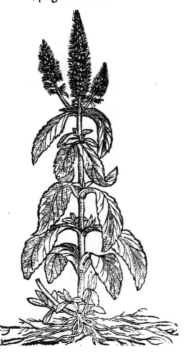

6　The ſixt kinde of *Veronica* hath many vpright branches a foot high and ſomtimes more, diuiding themſelues into ſundry other ſmall twigs; at the tops whereof grow faire ſpiky tuſts, bearing bright and ſhining blew floures. The leaues are ſomewhat long, indented about the edges like a ſaw: the root is compact of many threds, or ſtrings.

‡　7　This hath ſtalkes ſome cubit high and ſometimes more, and theſe not very full of branches, yet hauing diuers joints, at each whereof do grow forth two leaues, two or three inches long, and one broad, and theſe leaues are alſo thicke, ſmooth, and ſhining, lightly ſnipt or cut about the edges, and of a very aſtringent and drying taſt, and at laſt ſomwhat biting. At the top of the ſtalkes grow ſpekie tufts or blew floures like thoſe of the laſt mentioned, but of ſomwhat a lighter color, and they begin firſt to floure or ſhew themſelues below, and ſo go vpwards; the ſeed, which is ſmal and blacke, is contained in flat ſeed veſſells: the root is thicke with many fibres, euery yeare thruſting vp new ſhoots. There is a variety of this with the leaues not ſo black and ſhining, but hauing more branches; and another which hath a longer ſpike or tuft of floures. *Cluſius* calls this *Veronica erection latifolia.* ‡

8　The eighth hauing his ſtalks leaning vpon the ground, looketh with his face vpright, hauing ſundry flexible branches, ſet with leaues like vnto wilde Germander by couples, one right againſt another, deepely jagged about the edges, in reſpect of the other before mentioned. The floures are of a blew colour: the root is long, with ſome threds appendant thereto.

¶ *The Place.*

Veronica groweth vpon bankes, borders of fields, and graſſie mole-hils, in ſandy grounds, and in woods, almoſt euery where.

The fourth kinde, my good friend Mʳ. *Stephen Bredwell*, practitioner in phyſick, found and ſhewed it me in the cloſe next adjoining to the houſe of Mʳ. *Bele*, chiefe of the Clerkes of her Majeſties Counſell, dwelling at Barns neere London. The ſixth is a ſtranger in England, but I haue it growing in my garden.

¶ *The Time.*

Theſe floure from May to September.

¶ *The Names.*

†　Theſe plants are comprehended vnder this generall name *Veronica*; and *Dodonæus* would haue the firſt of them to be the *Betonica* or *Paulus Ægineta*; and *Turner* and *Geſner* the third: wee doe call them in Engliſh, Pauls Betonie, or Speedwell: in Welch it is called Fluellen, and the Welch people attribute great vertues to the ſame: in high Dutch, **Grovondheill**: in low Dutch, **Er eu priis**, that is to ſay, honor and praiſe.

¶ *The Temperature.*

Theſe are of a meane temperature, betweene heat and drineſſe.

¶ *The Vertues.*

The decoction of *Veroneca* drunke, ſodereth and healeth all freſh and old wounds, cleanſeth the bloud from all corruption, and is good to be drunke for the kidneies, and againſt ſcuruineſſe and foule ſpredding tettars, and conſuming and fretting ſores, the ſmall pox and meaſels.　　A

The water of *Veronica* diſtilled with wine, and re-diſtilled ſo often till the liquor wax of a reddiſh colour, preuaileth againſt the old cough, the drineſſe of the lungs, and all vlcers and inflammation of the ſame.　　B

† The ſecond and therd were both figures of that deſcribed in the third place: and thoſe that were formerly in the fifth and ſixth places, were alſo of the ſame plant, to wit that which is here deſcribed in the ſixth place and which was formerly in the fifth.

Chap. 199.　*Of Herbe Two-pence.*

¶ *The Deſcription.*

1　Herbe Two-pence hath a ſmal and tender root, ſpreding and diſperſing it ſelfe far within the ground, from which riſe vp many little, tender, flexible ſtalks trailing vpon the ground, ſet by couples at certaine ſpaces, with ſmooth greene leaues ſomewhat round, whereof it tooke his name: from the boſome of which leaues ſhoot forth ſmall tender foot-ſtalks, whereon do grow little yellow floures, like thoſe of Cinkefoile or Tormentill.

2　There is a kinde of Money-woort or herbe Two-pence, like the other of his kind in each reſpect, ſauing it is altogether leſſer, wherein they differ.

‡　3　There is another kind of Money-wort which hath many very ſlender creeping branches which here and there put forth fibres, and take root againe: the leaues are ſmall and round, ſtanding by couples one againſt another; and out of the boſomes come ſlender foot-ſtalkes bearing pretty

little whitish purple floures consisting of fiue little leaues standing together in manner of a little bell-floure,and seldome otherwise : the seed is small,and contained in round heads. This grows in many wet rotten grounds, and vpon bogges : I first found it *Anno* 1626,in the Bishopricke of Durham,and in two or three places of Yorkshire,and not thinking any had taken notice thereof,I drew a figure of it and called it *Nummularia pusilla flore ex albo purpurascente* ; but since I haue found that *Bauhine* had formerly set it forth in his *Prodromus* by the name of *Nummularia flore purpurascente*. It growes also on the bogs vpon the heath neer Burnt wood in Essex. It floures in Iuly and August. ‡

1 *Nummularia.*
Herbe Two-pence.

‡ 3 *Nummularia flore purpurascente.*
Purple floured Money-wort.

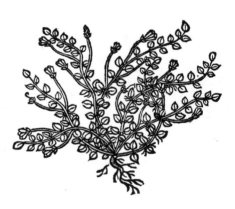

¶ *The Place.*

It groweth neere vnto ditches and streames,and other waterie places, and is somtimes found in moist woods : I found it vpon the banke of the riuer of Thames, right against the Queenes palace of White-hall;and almost in euery countrey where I haue trauelled.

¶ *The Time.*

It floureth from May till Summer be well spent.

¶ *The Names.*

Herb Two-pence is called in Latine *Nummularia* and *Centummorbia*:and of diuers *Serpentaria*.It is reported,that if serpents be hurt or wounded,they do heale themselues with this herb,wherupon came the name *Serpentaria*:it is thought to be called *Centummorbia*, of the wonderfull effect which it hath in curing diseases;and it is called *Nummularia* of the forme of money,whereunto the leaues are like:in Dutch, **Penninckcrupt** : in English,Money-woort, Herbe Two-pence, and Two-penny grasse.

¶ *The Temperature.*

That this herbe is dry,the binding tast thereof sheweth:it is also moderatly cold.

¶ *The Vertues.*

A   The floures and leaues stamped and laid vpon wounds and vlcers do cure them : but it worketh most effectually being stamped and boiled in oile oliue,with some rosin,wax,and turpentine added thereto.

The

The juice drunke in wine is good for the bloudy flix and all other issues of bloud in man or wo- B
man, the weaknesse and loosnesse of the belly and laske; it helpeth those that vomit bloud, & the
Whites in such as haue them.

Boiled with wine and hony it cureth the wounds of the inward parts, and vlcers of the lungs; & C
in a word, there is not a better wound herb, no not Tabaco it selfe, nor any other whatsoeuer.

The herb boiled in wine, with a little hony or mead, preuaileth much against the cough in chil- D
dren, called the Chin-cough.

CHAP. 200.
Of Bugle or middle Comfrey.

¶ The Description.

1 Bugula spreadeth and creepeth along the ground like Monywort: the leaues be long, fat,
and oleous, and of a browne colour for the most part. The floures grow about the stalks
in rundles, compassing the stalke, leauing between euery rundle bare or naked spaces, and
are of a faire blew colour, and often white. I found many plants of it in a moist ground vpon black-
Heath neere London, fast by a village called Charleton; but the leaues were green, and not browne
at all like the other.

1 *Bugula.*
Middle Consound.

2 *Bugula flore albo siue carneo.*
White or carnation floured Bugle.

2 Bugle with the white floure differs not from the precedent in roots, leaues, and stalkes; the
only difference is, that this plant bringeth forth faire milke white floures, and the other those that
are blew. ‡ It is also found with a flesh coloured floure, and the leaues are lesse snipt than those
of the former. *Bauhine* makes mention of one much lesse than those, with round snipt leaues and a
yellow floure, which he saith he had out of England, but I haue not as yet seene it, nor found any
other mention thereof. ‡

¶ *The Place.*

Bugula groweth almoſt in euery wood and copſe, and ſuch like ſhadowie and moiſt places, and is much planted in gardens: the other varieties are ſeldome to be met withall.

¶ *The Time.*

Bugula floureth in Aprill and May.

¶ *The Names.*

Bugle is reckoned among the Conſounds or wound-herbs; and it is called of ſome, *Conſolida media*, *Bugula*, and *Buglum*: in high Dutch, **Guntzel**: in low Dutch, **Senegroen**: of *Matthiolus, Herba Laurentina*: in Engliſh, browne Bugle: of ſome, Sicklewort, and herb Carpenter, but not truly.

¶ *The Temperature.*

Bugle is of a mean temperature between heate and dryneſſe.

¶ *The Vertues.*

A   It is commended againſt inward burſtings, and members torne, rent, and bruiſed: and therefore it is put into potions that ſerue for nodes, in which it is of ſuch vertue, that it can diſſolue & waſte away congealed and clotted bloud. *Ruellius* writeth that they commonly ſay in France, how hee needs neither Phyſition nor Surgeon, that hath Bugle and Sanicle; for it doth not onely cure rotten wounds, being inwardly taken, but alſo applied to them outwardly: it is good for the infirmities of the liuer, taking away the obſtruƈtions, and ſtrengthning it.

B   The decoƈtion of Bugle drunke diſſolueth clotted or congealed bloud within the body, healeth
C   and maketh ſound all wounds of the body both inward and outward.

   The ſame openeth the ſtoppings of the liuer and gall, and is good againſt the jaundice and feuers of long continuance.
D

E   The ſame decoƈtion cureth the rotten vlcers and ſores of the mouth and gums.

Bugula is excellent in curing wounds and ſcratches, and the juice cureth the wounds, vlcers, and ſores of the ſecret parts, or the herb bruiſed and laid theron.

Cʜᴀᴘ. 201. *Of Selfe-heale.*

1 *Prunella.*   2 *Prunella Lobelij.*
Selfe-heale.   The ſecond Selfe-heale.

3 *Prunella flore albo.*
White floured Selfe-heale.

¶ *The Deſcription.*

1. PRunell or Brunell hath ſquare hairie ſtalks of a foot high, beſet with long hairy and ſharp pointed leaues, and at the top of the ſtalkes grow floures thicke ſet together like an eare or ſpiky knap, of a brown colour, mixed with blew floures and ſometimes white; of which kinde I found ſome plants in Eſſex neere Heningham caſtle. The root is ſmall and very thready.

† 2. *Prunella altera*, or after *Lobel* and *Pena*, *Symphytum petraum*, hath leaues like the laſt deſcribed, but ſomewhat narrower, and the leaues that grow commonly toward the tops of the ſtalkes are deeply diuided or cut in after the manner of the leaues of the ſmall Valerian, and ſometimes the lower leaues are alſo diuided, but that is more ſeldome: the beads and floures are like thoſe of the former, and the colour of the floures is commonly purple, yet ſometimes it is found with fleſh coloured, and otherwhiles with white or aſh-coloured floures.

3. The third ſort of Selfe-heale is like vnto the laſt deſcribed in root, ſtalk, and leaues, and in euery other point, ſauing that the floures hereof are of a perfect white colour, and the others not ſo, which maketh the difference.

‡ The figure which our Author gaue in the third place, was of the *Prunella ſecunda* of *Tabern.* which I iudge to be all one with the *Prunella 1. non vulgaris* of *Cluſius*; & that becauſe the floures of that in *Tabernamontanus* are expreſſed *Ventre laxiore*, which *Cluſius* complaines his drawer did not obſerue; the orher parts alſo agree: now this of *Cluſius* hath much larger floures than the ordinarie, and thoſe commonly of a deeper purple colour, yet they are ſometimes whitiſh, and otherwhiles of an aſh-colour: the leaues alſo are ſomewhat more hairy, long and ſharper pointed than the ordinarie, and herein conſiſts the greateſt difference. ‡

¶ *The Place.*

The firſt kinde of Prunell or Brunel groweth very commonly in all our fields throughout England.

The ſecond Brunel or *Symphytum petraum* groweth naturally vpon rocks, ſtony mountaines, and grauelly grounds.

‡ The third for any thing I know is a ſtranger with vs; but the firſt common kind I haue found with white floures. ‡

¶ *The Time.*

Theſe plants floure for the moſt part all Summer long.

¶ *The Names.*

Brunell is called in Engliſh, Prunel, Carpenters herb, Selfe-heale, Hook heale, and Sicklewort. It is called of the later Herbariſts, *Brunella* and *Prunella*: of *Matthiolus*, *Conſolida minor*, and *Solidago minor*: but ſaith *Ruellius*, the Daiſie is the right *Conſolida minor* and alſo the *Solidago minor*.

¶ *The Nature.*

Theſe herbs are of the temperature of *Bugula*; that is to ſay, moderatly hot and dry, and ſomething binding.

¶ *The Vertues.*

The decoction of Prunell made with wine and water, doth ioine together and make whole and A ſound all wounds both inward and outward, euen as Bugle doth.

Prunel bruiſed with oile of roſes and vineger, and laid to the forepart of the head, ſwageth and B helpeth the pain and aking thereof.

To be ſhort, it ſerueth for the ſame that Bugle doth, and in the world there are not two better C wound herbs, as hath bin often proued.

D   It is commended against the infirmities of the mouth, and especially the ruggednesse, blacknes, and drinesse of the tongue, with a kinde of swelling in the same. It is an infirmitie among soldiers that lie in campe : the Germanes call it **de Braun**, which hapneth not without a continuall ague and frensie.   The remedie hereof is the decoction of Selfe-heale with common water, after bloud-letting out of the veins of the tongue ; and the mouth and tongue must be often washed with the same decoction, and sometimes a little vineger mixed therewith. This disease is thought to be vn-knowne to the old writers : but notwithstanding if it be conferred with that which *Paulus Ægine-ta* calleth *Eryſipelas cerebri*, an inflammation of the brain, then wil it not be thought to be much dif-fering, if it be not the very same.

CHAP. 202. *Of the great Daiſie, or Maudlin-wort.*

1 *Bellis major.*
The great Daiſie.

¶ *The Deſcription.*

1   THe great Daiſie hath very many broad leaues ſpred vpon the ground, ſomwhat indented about the edges, of a fingers bredth, not vnlike thoſe of Groundſwell : amongſt which riſe vp ſtalkes of the height of a cubit, ſet with the like leaues, but leſſer ; in the top whereof grow large white floures with yellow thrummes in the middle like thoſe of the ſingle field Daiſie or May-weed, without any ſmell at all. The root is ful of ſtrings.

¶ *The Place.*
It groweth in medowes, and in the borders of fields almoſt euery where.

¶ *The Time.*
It floureth and flouriſheth in May and Iune.

¶ *The Names.*
It is called (as we haue ſaid) *Bellis major*, and alſo *Conſolida media vulnerariorum*, to make a difference between it and *Bugula*, which is the true *Conſolida media*. Notwithſtanding this is holden of all to be *Conſolida medij generis*, or a kinde of middle Con-found : in high Dutch, as *Fuchſius* reports, **Gentz-blume :** in Engliſh, the great Daiſie, and Maudlin-wort.

¶ *The Nature.*
This great Daiſie is moiſt in the end of the ſe-cond degree, and cold in the beginning of the ſame

¶ *The Vertues.*

A   The leaues of the great Maudlin-wort are good againſt all burning vlcers and apoſtems, againſt the inflammation and running of the eies, being applied thereto.

B   The ſame made vp in an vnguent or ſalue, with Wax, oile, and Turpentine, is moſt excellent for wounds, eſpecially thoſe wherein is any inflammation, and will not come to digeſtion or maturati-on, as are thoſe weeping wounds made in the knees, elbowes, and other joints.

C   The juice, decoction, or diſtilled water is drunke to very good purpoſe againſt the rupture or any inward burſtings.

D   The herb is good to be put into vulnerarie drinks or potions, as one ſimple belonging thereto moſt neceſſarie, to the which effect, the beſt practiſed do vſe it, as a ſimple in ſuch caſes of great ef-fect.

E   It likewiſe aſſwageth the cruell torments of the gout, vſed with a few mallowes and butter boi-led and made to the forme of a pultis.

F   The ſame receit aforeſaid vſed in cliſters, profiteth much ngainſt the vehement heate in agues, and ceaſeth the torments or wringing of the guts or bowels.

CHAP.

CHAP. 203. *Of little Daisies.*

¶ *The Description.*

1   THe Daisie bringeth forth many leaues from a threddy root, smooth, fat, long, and somewhat round withall, very sleightly indented about the edges, for the most part lying vpon the ground: among which rise vp the floures, euerie one with his owne slender stem, almost like those of Camomill, but lesser, of a perfect white colour, and very double.

2   The double red Daisie is like vnto the precedent in euerie respect, sauing in the color of the floures; for this plant bringeth forth floures of a red colour; and the other white as aforesaid.

‡   These double Daisies are of two sorts, that is to say, either smaller or larger; and these again either white or red, or of both mixed together: wherefore I haue giuen you in the first place the figure of the small, and in the second that of the larger.

3   Furthermore, There is another pretty Daisie which differs from the first described onely in the floure, which at the sides thereof puts forth many footstalks, carrying also little double flours, being commonly of a red colour; so that each stalke carrieth as it were an old one and the brood thereof; whence they haue fitly termed it the childing Daisie. ‡

1 *Bellis minor multiplex flore albo vel rubro.*
The lesser double red or white Daisie.

2 *Bellis media multiplex flore albo vel rubro.*
The larger double white or red Daisie.

4   The wild field Daisie hath many leaues spred vpon the ground like those of the garden Daisie; among which rise vp slender stems; on the top whereof grow small single floures like those of Camomill, set about a bunch of yellow thrums, with a pale of white leaues, sometimes white, now and then red, and often of both mixed together: the root is threddy.

5   There doth likewise grow in the fields another sort of wilde Daisie, agreeing with the former in each respect, sauing that it is somewhat greater than the other, and the leaues are somewhat more cut in the edges, and larger,

6   The blew Italian Daisie hath many small threddie roots, from the which rise vp leaues like
those

‡ 3 *Bellis minor prolifera.*
Childing Daiſie.

4 *Bellis minor ſylueſtris.*
The ſmall wilde Daiſie.　　　　5 *Bellis media ſylueſtris.*
The middle wilde Daiſie.

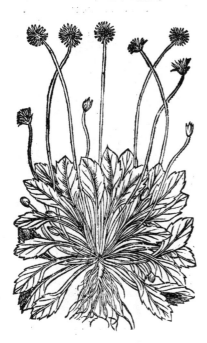

thoſe of the common Daiſie, of a darke green colour : among which comes vp a fat ſtem ſet round
abont with the like leaues, but leſſer. The floures grow at the top globe-faſhion, or round like a
ball, of a perfeſt blew colour, very like vnto the floures of mountain Scabious.

　　7. The French blew Daiſie is like vnto the other blew Daiſies in each reſpeſt, ſauing it is altogether leſſe, wherein conſiſteth the difference.

　　‡　There were formerly three figures and deſcriptions of this blew Daiſie, but one of them
might haue ſerued ; for they differ but in the talneſſe of their growth, and in the bredth and nar-
rowneſſe of their leaues.　　‡

¶ *The Place.*
The double Daiſies are planted in gardens : the others grow wilde euery where.

The

The blew Daiſies are ſtrangers in England ; their naturall place of abode is ſet forth in their ſeuerall titles.

6 *Bellis cœrulea ſiue Globularia Apula.*
The blew Italian Daiſie.

7 *Bellis cœrulea Monſpeliaca.*
Blew French Daiſies.

¶ *The Time.*

The Daiſies do floure moſt part of the Summer.

¶ *The Names.*

The Daiſie is called in high-Dutch, **Maßlieben** : in low-Dutch, **Margrieten** : in Latine, *Bellis minor,* and *Conſolida minor,* or the middle Confound : of *Tragus, Primula veris* ; but that name is
more proper vnto Primroſe : of ſome, *Herba Margaritá,* or Margarites herb : in French, *Marguerites,*
and *Caßaudes* : in Italian, *Fiori di prima veri gentili* : In Engliſh, Daiſies, and Bruiſewort.

The blew Daiſie is called *Bellis cœrulea* : of ſome, *Globularia,* of the round forme of the floure : it
is alſo called *Aphyllanthes,* and *Frondiflora* : in Italian, *Botanaria* : in Engliſh, Blew Daiſies,& Globe
Daiſie.

¶ *The Temperature.*

The leſſer Daiſes are cold and moiſt, being moiſt in the end of the ſecond degree, and cold in the
beginning of the ſame.

¶ *The Vertues.*

The Daiſies do mitigate all kinde of paines, but eſpecially in the joints, and gout proceeding A
from an hot and dry humor, if they be ſtamped with new butter vnſalted, and applied vpon the
pained place : but they worke more effectually if Mallowes be added thereto.

The leaues of Daiſies vſed among other pot-herbs, do make the belly ſoluble ; and they are al- B
ſo put into Clyſters with good ſucceſſe, in hot burning feuers, and againſt the inflammation of the
inteſtines.

The juice of the leaues and roots ſnift vp into the noſthrils, purgeth the head mightily of ſoule C
and filthy ſlimy humors, and helpeth the megrim.

The ſame giuen to little dogs with milke, keepeth them from growing great.   D

The leaues ſtamped take away bruiſes and ſwellings proceeding of ſome ſtroke, if they be ſtam- E
ped and laid thereon ; whereupon it was called in old time Bruiſewort.

The juice put into the eies cleareth them, and taketh away the watering of them.   F

The decoction of the field Daiſie (which is the beſt for phyſicks vſe) made in water and drunke, G
is good againſt agues, inflammation of the liuer and all other the inward parts.

CHAP. 204. Of Mouse-eare.

¶ The Description.

1   THe great Mouse-eare hath great and large leaues greater than our common *Pylosella*, or Mouse-eare, thicke, and full of substance : the stalkes and leaues bee hoary and white, with a silken mossinesse in handling like silke, pleasant and faire in view : it bears three or foure quadrangle stalkes somewhat knotty, a foot long: the roots are hard, wooddy, and full of strings; the floures come forth at the top of the stalk, like vnto the small Pisseabed or Dandelion, of a bright yellow colour.

2   The second kinde of *Pylosella* is that which we call *Auricula muris*, or Mous-eare, being a very common herbe, but few more worthy of consideration because of his good effect, and yet not remembred of the old writers. It is called *Pylosella*, of the rough hairy and whitish substance growing vpon the leaues, which are somewhat long like the little Daisie, but that they haue a small hollownesse in them resembling the eare of a mouse : vpon which consideration some haue called it *Myosotis* ; wherein they were greatly deceiued, for it is nothing like vnto the *Myosotis* of *Dioscorides*: his small stalks are likewise hairy, slender, and creeping vpon the ground ; his floures are double, and of a pale yellow colour, much like vnto *Sonchus*, or *Hieracium*, or Hawk-weed.

1 *Pylosella major.*	2 *Pylosella repens.*
Great Mouse-eare.	Creeping Mouse-eare.

3   The small Mouse-eare with broad leaues hath a small tough root, from which rise vp many hairy and hoary broad leaues spred vpon the ground, among which growes vp a slender stem, at the top whereof stand two or three small yellow floures, which being ripe turne into down that is caried away with the winde.

¶ The Place.
They grow vpon sandy banks and vntoiled places that lie open to the aire.

¶ The Time.
They floure in May and Iune.

¶ *The Names.*

Great Moufe-eare is called of the later Herbarifts, *Pylofella* : the fmaller likewife *Pylofella* , and *Auricula muris:* in Dutch, **Nageltruit,** and **Muyſoor:** *Lacuna* thinkes it *Holoftium:* in French, *Oreille de rit, ou ſouris :* in Italian, *Pelofella :* in English, Moufe-eare.

¶ *The Temperature.*

They are hot and dry of temperature, of an excellent aftringent facultie, with a certaine hottenuitie admixed.

¶ *The Vertues.*

The decoction of *Pylofella* drunke doth cure and heale all wounds, both inward and outward : it   A
cureth hernies, ruptures, or burftings.

The leaues dried and made into pouder, doe profit much in healing of wounds, beeing ftrewed   B
thereupon.

The decoction of the juice is of fuch excellencie, that if fteele-edged tooles red hot be dren-   C
ched, and cooled therein oftentimes, it maketh them fo hard, that they will cut ftone or iron, bee
they neuer fo hard, without turning the edge or waxing dull.

This herbe being vfed in gargarifmes cureth the loofeneffe of the Vvula.   D

Being taken in drinke it healeth the fluxes of the wombe, as alfo the difeafes called *Dyfenteria*   E
and *Enterocele:* it glueth and confoundeth wounds, ftaieth the fwelling of the fpleen, and the bloudy excrements procured thereby.

The Apothecaries of the Low-countries make a fyrrup of the juice of this herb, which they vfe   F
for the cough, confumption and ptifike.

¶   I haue in this chapter omitted two figures and one defcription : the firft of the two omitted figures, which fhould haue been the third, differs little from the firft, but in the fmallneffe of the ftalke, and fewneffe of the floures at the top thereof : the other, which was in the fourth place, was figured and defcribed by me formerly in the fourth place of the 54 chapter of this booke.

CHAP. 205.   *Of Cotton-weed or Cud-weed.*

1 *Gnaphalium Anglieum.*
English Cudweed.

2 *Gnaphalium vulgare.*
Common Cudweed.

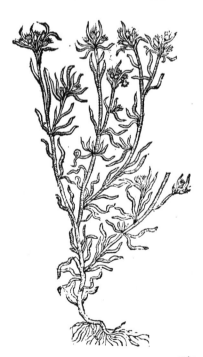

¶ *The Description.*

1   ENglish Cudweed hath sundry slender and vpright stalks diuided into many branches, and groweth as high as common Wormwood, whose colour and shape it much resembleth. The leaues shoot from the bottome of the turfe full of haires, in shape somwhat like a Willow leafe below, but aboue they be narrower, and like the leaues of *Psyllium* or Flea-wort: among which do grow small pale coloured floures like those of the smal *Coniza* or Flea-bane. The whole plant is of a bitter taste.

2   The second being our common *Gnaphalium* or Cudweed is a base or low herbe, nine or ten inches long, hauing many small stalks or tender branches, and little leaues couered all ouer with a certaine white cotton or fine wooll, and very thick : the floures be yellow, and grow like buttons at the top of the stalkes.

3   The third kind of Cudweed or Cotton-weed, being of the sea, is like vnto the other Cudweed last described, but is altogether smaller and lower, seldome growing much aboue a handfull high : the leaues grow thicke vpon the stalkes, and are short, flat, and very white, soft and woolly. The floures grow at the top of the stalkes in small round buttons, of colour and fashion like the other Cudweed.

4   The fourth being the Cotton-weed of the hils and stony mountaines is so exceeding white and hoary, that one would think it to be a plant made of wooll, which may very easily be known by his picture, without other description.

3 *Gnaphalium marinum.* ·
Sea Cudweed.

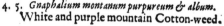
4. 5. *Gnaphalium montanum purpureum & album.*
White and purple mountain Cotton-weed.

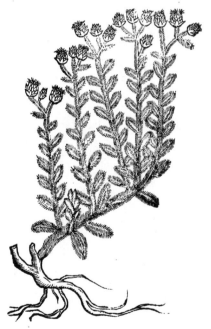

5   The fift kinde of Cotton-weed hath leaues and stalkes like the other of his kinde, and differeth in that, that this plant beareth a bush or tuft of purple floures, otherwise it is very like.

6   The sixth is like vnto the last recited, but greater : the leaues are of an exceeding bright red colour, and of an aromaticall sweet smell.

7   The seuenth kind of *Gnaphalium* or Cotton-weed of *Clusius* his description, growes nine or ten inches high, hauing little long leaues like the small Mouse-eare, woolly within, and of a hoarie colour on the outside : the stalkes in like manner are very woolly, at the top whereof commeth forth a faire floure and a strange, hauing such woolly leaues bordering the floure about, that a man would thinke it to be nothing else but wooll it selfe : and in the middest of the floure come forth
sundry

6 *Gnaphalium montanum suaue rubens.*
Bright red mountaine Cottonweed.

† 7 *Gnaphalium Alpinum.*
Rocke Cotton-weed.

‡ 8 *Gnaphalium Americanum.*
Liue for euer.

9 *Filago minor.*
Small Cud-weed.

ſundry ſmall heads of a pale yellow colour, like vnto the other of this kinde. The root is blacke and ſomewhat fibrous.

8　There is a kinde of Cotton-weed, being of greater beautie than the reſt, that hath ſtrait and vpright ſtalkes 3 foot high or more, couered with a moſt ſoft and fine wooll, and in ſuch plentifull manner, that a man may with his hands take it from the ſtalke in great quantitie : which ſtalke is beſet with many ſmall long and narrow leaues, greene vpon the inner ſide, and hoary on the other ſide, faſhioned ſomewhat like the leaues of Roſemary, but greater. The floures do grow at the top of the ſtalkes in bundles or tufts, conſiſting of many ſmall floures of a white colour, and very dou-ble, compact, or as it were conſiſting of little ſiluer ſcales thruſt cloſe together, which doe make the ſame very double. When the floure hath long flouriſhed, and is waxen old, then comes there in the middeſt of the floure a certaine browne yellow thrumme, ſuch as is in the middeſt of the Daiſie : which floure being gathered when it is young, may be kept in ſuch manner as it was gathered (I meane in ſuch freſhneſſe and well liking) by the ſpace of a whole yeare after, in your cheſt or elſe-where : wherefore our Engliſh women haue called it Liue-long, or Liue for euer, which name doth aptly anſwer his effects. ‡ Cluſius receiued this plant out of England, and firſt ſet it forth by the name of Gnaphalium Americanum, or Argyrocome. ‡

9　This plant hath three or foure ſmall grayiſh cottony or woolly ſtalkes, growing ſtrait from the root, and commonly diuided into many little branches : the leaues be long, narrow whitiſh, ſoft and woolly, like the other of his kinde : the floures be round like buttons, growing very many together at the top of the ſtalkes, but nothing ſo yellow as Mouſe-eare, which turne into downe, and are caried away with the winde.

10　Filago ſiue Herba impia.
Herbe impious, or wicked Cudweed.

11　Leontopodium, ſiue Pes Leoninus.
Lions Cudweed.

10　The tenth is like vnto the laſt before mentioned, in ſtalkes, leaues, and floures, but much larger, and for the moſt part thoſe floures which appeare firſt are the loweſt, and baſeſt, and they are ouertopt by other floures which come on younger branches, and grow higher, as children ſeeking to ouergrow or ouertop their parents, (as many wicked children do) for which cauſe it hath beene called Herba impia, that is, the wicked Herbe, or Herbe Impious.

11　This plant may be comprehended vnder the title of Gnaphalium, being without doubt a kinde thereof, as may appeare by the ſhape of his floures and ſtalkes, couered ouer with a ſoft wooll like vnto the other kindes of Cotton-weed it is an handfull high or thereabouts, beſet with leaues
like

† 12 *Leontopodium parvum.*
Small Lyons Cudweed.

‡ 13 *Gnaphalium oblongo folio.*
Long leaued Cudweed.

‡ 14 *Gnaphalium minus latiore folio.*
Small broad leaued Cudweed.

like *Gnaphalium Anglicum,* but somewhat
broader. At the top of the stalke groweth
a floure of a blackish brown violet colour,
beset about with rough and woolly hairie
leaues, which make the whole floure to resemble the rough haired foot of a Lyon, of
a Hare, or a Beare, or rather in mine opinion of a rough footed Doue. The heads of
these floures when they are spread abroad
carry a greater circumference than is re-
quired in so small a plant; and when the
floure is faded, the seed is wrapped in such
a deale of wooll that it is scarsely to be
found out.

12   This small kinde of *Leontopodium*
being likewise a kind of Cotton-weed, neither by *Dioscorides* or any other antient
writer once remembred, hath one single
stalke nine inches in height, and the leaues
of *Gnaphalium montanum;* which leaues and
stalkes are white, with a thick hoary woollinesse, bearing at the top pale yellow
floures like *Gnaphalium montanum:* the root
is slender and wooddy.

‡   13   This, which *Clusius* calls *Gnaphalium Plateau* 2. hath small stalkes some
handfull high or somewhat more, of which

some

ſome ſtand vpright,others lie along vpon the ground,being round,hairy,and vnorderly ſet with ſoft hoary leaues ingirting their ſtalkes at their ſetting on, and ſharpe pointed at their vpper ends. The tops of the ſtalkes carry many whitiſh heads full of a yellowiſh downe: the root is thicke and blac-kiſh,with ſome fibres.

14   This ſends vp one ſtalke parted into ſeuerall branches ſet here and there with broad ſoft and hoary leaues, and at the diuiſion of the branches and amongſt the leaues grow ſeuen or eight little heads thicke thruſt together, being of a grayiſh yellow colour, and full of much downe : the root is vnprofitable,and periſhes as ſoone as it hath perfected his ſeed. *Cluſius* calls this *Gnaphalium Plateau* 3. he hauing as it ſeemes receiued them both from his friend *Iaques Plateau.* ‡

¶ The Place.

The firſt groweth in the darke woods of Hampſted, and in the woods neere vnto Deptford by London. The ſecond groweth vpon dry ſandy banks. The third groweth at a place called Merezey, ſix miles from Colcheſter, neere vnto the ſea ſide. ‡ I alſo had it ſent me from my worſhipfull friend Mr *Thomas Glynn*,who gathered it vpon the ſea coaſt of Wales. ‡

The reſt grow vpon mountaines,hilly grounds,and barren paſtures.

The kinde of *Gnaphalium* newly ſet forth (to wit *Americanum*) groweth naturally neere vnto the Mediterranean ſea, from whence it hath beene brought and planted in our Engliſh gardens. ‡ If this be true which our Author here affirmes,it might haue had a fitter (at leaſt a neerer) denomina-tion than from America : yet *Bauhine* affirmes that it growes frequently in Braſill, and it is not im-probable that both their aſſertions be true. ‡

¶ The Time.

They floure for the moſt part from Iune to the end of Auguſt.

¶ The Names.

Cotton-weed is called in Greeke, *Gnaphalion*; and it is called *Gnaphalion*, becauſe men vſe the tender leaues of it in ſtead of bombaſte or Cotton,as *Paulus Ægineta* writeth. *Pliny* ſaith it is called *Chamaxylon*,as though he ſhould ſay Dwarfe Cotton ; for it hath a ſoft and white cotton like vnto bombaſte, whereupon alſo it was called of diuers *Tementitia*, and *Cotonaria* : of others, *Centunculus*, *Centuncularis*,and *Albinum* ; which word is found among the baſtard names : but the later word, by reaſon of the white colour,doth reaſonably well agree with it. It is alſo called *Bombax*, *Humilis filago*, and *Herba Impia*,becauſe the younger, or thoſe floures that ſpring vp later, are higher, and ouertop thoſe that come firſt, as many wicked children do vnto their parents,as is before touched in the de-ſcription : in Engliſh,Cottonweed,Cudweed,Chaffe-weed,and petty Cotton.

¶ The Temperature.

Theſe herbes be of an aſtringent or binding and drying quality.

¶ The Vertues.

A   *Gnaphalium* boyled in ſtrong lee cleanſeth the haire from nits and lice : alſo the herb being laid in ward-robes and preſſes keepeth apparell from moths.

B   The ſame boiled in wine and drunken, killeth wormes and bringeth them forth, and preuaileth againſt the bitings and ſtingings of venomous beaſts.

C   The fume or ſmoke of the herbe dried, and taken with a funnell, being burned therein,and recei-ued in ſuch manner as we vſe to take the fume of Tabaco, that is, with a crooked pipe made for the ſame purpoſe by the Potter, preuaileth againſt the cough of the lungs, the great ache or paine of the head, and cleanſeth the breaſt and inward parts.

† The figure that was formerly in the ſeuenth place ſhould haue beene in the eleuenth ; and that in the eleuenth in the ſeuenth.

CHAP. 206.

Of Golden Moth-wort, or Cudweed.

¶ The Deſcription.

1   GOlden Mothwort bringeth forth ſlender ſtalkes ſomewhat hard and wooddy, diuided in diuers ſmall branches ; whereupon doe grow leaues ſomewhat rough, and of a white colour, very much jagged like Southernwood. The floures ſtand on the tops of the ſtalkes,ioyned together in tufts,of a yellow colour glittering like gold,in forme reſembling the ſcaly floures of Tanſie,or the middle button of the floures of Camomill; which being gathered before they be ripe or withered, remaine beautifull long time after, as my ſelfe did ſee in the hands of Mr *Wade*, one of the Clerks of her Majeſties counſell, which were ſent him among other things

from

from Padua in Italy. For which caufe of long lafting, the images and carued gods were wont to weare garlands thereof: whereupon fome haue called it Gods floure. For which purpofe *Ptolomy* King of Ægypt did moft diligently obferue them, as *Pliny* writeth.

1 *Elyochryfon, fiue Coma aurea.*
Golden Moth-wort.

¶ *The Place.*

It growes in moft vntilled places of Italy and Spaine, in medowes where the foile is barren, and about the bankes of riuers; it is a ftranger in England.

¶ *The Time.*

It floures in Auguft and September: notwithftanding *Theophraftus* and *Pliny* reckon it among the floures of the Spring.

¶ *The Names.*

Golden Moth-wort is called of *Diofcorides Elichryfon*: *Pliny* and *Theophraftus* call it *Helichryfon*: *Gaza* tranflates it *Aurelia*: in Englifh, Gold-floure, Golden Moth-wort.

¶ *The Temperature.*

It is (faith *Galen*) of power to cut and make thinne.

¶ *The Vertues.*

Diofcorides teacheth, that the tops thereof A drunke in wine are good for them that can hardly make water; againft the ftingings of Serpents, paines of the huckle bones: and taken in fweet wine it diffolueth congealed bloud.

The branches and leaues laid amongft B cloathes keepeth them from mothes, whereupon it hath beene called of fome Moth-weed, or Moth-wort.

† Here formerly were two figures and defcriptions of the fame Plant.

CHAP. 207. *Of Golden Floure-Gentle.*

¶ *The Defcription.*

1 THis yellow Euerlafting or Floure-Gentle, called of the later Herbarifts Yellow Stœcas, is a plant that hath ftalkes of a fpan long, and flender, whereupon do grow narrow leaues white and downie, as are alfo the ftalkes. The floures ftand on the tops of the ftalkes, confifting of a fcattered or difordered fcaly tuft, of a reafonable good fmell, of a bright yellow colour; which being gathered before they be ripe, do keep their colour and beautie a long time without withering, as do moft of the Cottonweeds or Cudweeds, whereof this is a kinde. The root is blacke and flender. ‡ There is fome variety in the heads of this plant, for they are fometimes very large and longifh, as *Camerarius* notes in his Epitome of *Matthiolus*; otherwhiles they are very compact and round, and of the bigneffe of the ordinary.

2 This growes to fome foot or more high, and hath round downy leaues like the former, but broader: the floures are longer, but of the fame yellow colour and long continuance as thofe of the laft defcribed. This varies fomething in the bredth and length of the leaues, whence *Tabernamontanus* gaue three figures thereof, and therein was followed by our Author, as you fhall find more particularly fpecified at the end of the chapter. ‡

3 About Nemaufium and Montpelier there growes another kinde of *Chryfocome*, or as *Lobel* termes it, *Stœchas Citrina altera*, but that as this plant is in all points like, fo in all points it is leffer and flenderer, blacker, and not of fuch beautie as the former, growing more neere vnto an afh colour, confifting of many fmall twigs a foot long. The root is leffer, and hath fewer ftrings annexed thereto; and it is feldome found but in the cliffes and crags, among rubbifh, and on walls of cities. This plant is browne, without fent or fauor like the other: euery branch hath his owne bunch of floures comming forth of a fcaly or round head, but not a number heaped together, as in the firft kinde. It profpereth well in our London Gardens.

† 1 *Stæchas Citrina, ſiue Amaranthus luteus.*
Golden Scœchas, or Goldilockes.

† 2 *Amaranthus luteus latifolius.*
Broad leaued Goldilockes.

† 3 *Chryſocome capitulis conglobatis.*
Round headed Goldilockes.

† 4 *Amaranthus luteus flore oblongo.*
Golden Cudweed.

   4   There is kinde hereof being a very rare plant, and as rare to be found where it naturally groweth, which is in the woods among the Scarlet-Okes betweene Sommieres and Mountpellier. It is
a fine and beautifull plant, in ſhew paſſing the laſt deſcribed *Stæchas Citrina altera:* but the leaues of
this kinde are broad, and ſomewhat hoarie, as is all the reſt of the whole plant; the ſtalke a foot
long, and beareth the very floures of *Stæchas Citrina altera,* but bigger and longer, and ſomewhat like
the floures of *Lactuca agreſtis:* the root is like the former, without any manifeſt ſmell, little knowne,
hard to finde, whoſe faculties be yet vnknowne.

† 5 *Heliochryſon ſylueſtris.*
Wile Goldylockes.

† 5 This is a wilde kinde (which *Lobel* ſetteth forth) that here may be inſerted, called *Eliochryſos ſylueſtris.* The woolly or flockey leaſe of this plant reſembleth *Gnaphalium vulgare*, but that it is ſomewhat broader in the middle: the floures grow cluſtering together vpon the tops of the branches, of a yellow colour, and almoſt like thoſe of Maudline: the roots are blacke and wooddy.

¶ *The Place.*

The firſt mentioned growes in Italy, and other hot Countries: and the ſecond growes in rough and grauelly places almoſt euery where neere vnto the Rhene, eſpecially between Spires and Wormes.

¶ *The Time.*

They floure in Iune and Iuly.

¶ *The Names.*

Golden floure is called in Latine *Coma aurea*, of his Golden lockes or beautifull buſh, and alſo *Tineraria*: in ſhops *Stæchas citrina*, *Amaranthus luteus*, *Fuchſij*, *& Tragi*: of ſome, *Linaria aurea*, but not truely: in Greeke, *Chryſocome*: in Dutch, Repubio=emen, and Motten crupt: in Italian, *Amarantho Giallo*: in Engliſh, Gold-floure, Gods floure, Goldilockes, and Golden Stæchas.

¶ *The Temperature and Vertues.*

The floures of Golden Stœchados **A** boiled in wine and drunke, expell wormes out of the belly; and being boiled in Lee made of ſtrong aſhes doth kill lice and nits, if they bee bathed therewith. The other faculties are referred to the former plants mentioned in the laſt chapter.

. † There were formerly the ſame number of figures as are now in this Chapter, but no way agreeing with the deſcriptions; the firſt was of *Milleſolium Luteum*, being the *Heliebryſum Italicum* of *Matthiolus*: The ſecond was of the *Amaranthus primus* of *Tragus* which ſtill keeps the 2.place: and the 4.and 5. were only varieties of this, according to *Baubine*: but if they be not varieties, but made to expreſſe the 2. figures of the *Aduerſ.* which we here giue, as I conieċture they were, then ſhould the fourth haue been put in the third place, and the fifth in the fourth, and the third ſhould haue beene put in the fifth, as you may ſee now it is.

CHAP. 208. *Of Coſtmarie and Maudelein.*

¶ *The Deſcription.*

1　COſtmary groweth vp with round hard ſtalkes two foot high, bearing long broad leaues finely nicked in the edges, of an ouerworne whitiſh greene colour. The tuft or bundle is of a Golden colour, conſiſting of many little floures like cluſters, ioyned together in a rundle after the manner of golden Stœchados. The root is of a wooddy ſubſtance, by nature very durable, not without a multitude of little ſtrings hanging thereat. The whole plant is of a pleaſant ſmell, ſauour or taſte.

2　Maudeline is ſomewhat like to Coſtmary (whereof it is a kinde) in colour, ſmell, taſte, and in the golden floures, ſet vpon the tops of the ſtalkes in round cluſters. It bringeth forth a number of ſtalkes, ſlender, and round. The leaues are narrow, long, indented, and deepely cut about the edges. The cluſter of floures is leſſer than that of Coſtmarie, but of a better ſmell, and yellower colour. The roots are long laſting and many.

‡　3　There is another kinde of *Balſamita minor*, or *Ageratum*, which hath leaues leſſer and narrower than the former, and thoſe not ſnipt about the edges: the vmbel or tuft of floures is
yellow

1 *Balſamita mas.*
Coſtmaric.

2 *Balſamita fœmina, ſive Ageratum.*
Maudelein.

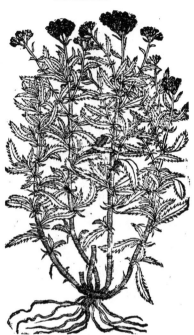

‡ 3 *Ageratum folijs non ſerratis.* 4. *Ageratum floribus albis.*
Maudelein with vncut leaues. White floured Maudlein.

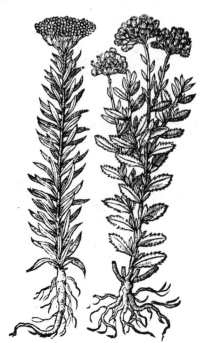

yellow like as the former, and you may call each of theſe laſt deſcribed at your pleaſure, either *Ageratum*, or *Balſamita*: the Grӕcians call it Ἀγήρατον, which is in Latine *Ageratum, vel non ſeneſcens,* called in ſhops (though vntruly) *Eupatorium Meſuæ.* The floures are of a beautifull and ſeemely ſhew, which will not loſe their excellency of grace in growing, vntill they be very old, and therefore called *Ageratum,* or *Non ſeneſcens,* as before, and are like in tuft to *Eliochryſon,* and this is thought to be the true and right *Ageratum* of *Dioſcorides,* although there hath beene great controuerſie which ſhould bethe true plant.

‡ 4 This differeth not from the common Maudelein, but in the colour of the floures, which are white, when as thoſe of the ordinary ſort are yellow. ‡

¶ *The Place.*
They grow euery where in gardens, and are cheriſhed for their ſweet floures and leaues.

¶ *The Time.*
They bring forth their tufts of yellow floures in the Summer moneths.

¶ *The*

¶ *The Names.*

Costmarie is called in Latine *Balsamita maior* or *mas* : of some, *Costus hortorum* : it is also called *Mentha Graca* : and *Saracenica Officinarum* : of *Tragus*, *Alisma* : of *Matthiolus*, *Herba Graca* : of others, *Saluia Romana*, and *Herba lassulata*: of some, *Herba D. Maria*: in English, Costmarie, and Ale-coast: in high Dutch, **Frauwenkraut**: in low Dutch, **Heydonisch windkraut**: in French, *Coq.*

Maudlein is without doubt a kinde of Costmarie, called of the Italians *Herba Giulia*: of *Valerius Cordus*, *Mentha Corymbifera minor*: and *Eupatorium Mesue*. It is iudged to be *Dioscorides* his *Ageratum*, and it is the *Costus minor hortensis* of *Gesner* : we call it in English, Maudelein.

¶ *The Nature.*

They are hot and dry in the second degree.

¶ *The Vertues.*

These plants are very effectuall, especially Maudlein, taken either inwardly or else outward- A
ly to prouoke vrine ; and the fume thereof doth the same, and mollifieth the hardnesse of the Matrix.

Costmarie is put into Ale to steepe, as also into the barrels and Stands amongst those herbes B
wherewith they do make Sage Ale ; which drinke is very profitable for the diseases before spoken of.

The leaues of Maudleine and A lders tongue stamped and boyled in Oile Oliue, adding there- C
to a little wax, rosin, and a little turpentine, make an excellent healing vnguent, or incarnatiue salue to raise or bring vp flesh from a deepe and hollow wound or vlcer, whereof I haue had long experience.

The Conserue made with the leaues of Costmarie and Sugar, doth warme and dry the braine, D
and openeth the stoppings of the same : stoppeth all Catarrhes, rheumes, and distillations, taken in the quantitie of a beane.

The leaues of Costmarie boyled in wine and drunken, cure the griping paine of the belly, the E
guts and bowels, and cureth the bloudy flix.

It is good for them that haue the greene sicknesse, or the dropsie, especially in the beginning; and F
it helpeth all that haue a weake and cold liuer.

The seed expelleth all manner of wormes out of the belly, as wormeseed doth. G

Chap. 209. *Of Tansie.*

¶ *The Description.*

1   Tansie groweth vp with many stalkes, bearing on the tops of them certaine clustered tufts, with floures like the round buttons of yellow Romane Cammomill, or Feuerfew (without any leaues paled about them) as yellow as gold. The leaues be long, made as it were of a great many set together vpon one stalke, like those of Agrimony, or rather wild Tansie, very like to the female Ferne, but softer and lesser, and euery one of them slashed in the edges as are the leaues of Ferne. The root is tough and of a wooddy substance. The whole plant is bitter in taste, and of a strong smell, but yet pleasant.

2   The double English Tansie hath leaues infinitly jagged and nicked, and curled withall, like vnto a plume of Feathers : it is altogether like vnto the other, both in smell and taste, as also in floures, but more pleasantly smelling by many degrees, wherein especially consisteth the difference.

3   The third kinde of Tansie hath leaues, roots, stalkes, and branches like the other, and differeth from them, in that this hath no smell or sauour at all, and the floures are like the common single Fetherfew.

‡ 4   *Clusius* hath described another bigger kind of vnsauorie Tansie, whose figure here we giue you; it grows some cubit and halfe high, with crested stalks, hauing leaues set vpon somwhat longer stalkes than those of the last described, otherwise much like them : the floures are much larger, being of the bignesse of the great Daisie, and of the same colour : the seed is long and blacke : The root is of the thicknesse of ones finger, running vpon the surface of the ground, and putting forth some fibres, and it lasts diuers yeares, so that the plant may be encreased thereby. This floures in May and Iune, and growes wilde vpon diuers hills in Hungary and Austria. ‡

5 The

1 *Tanacetum.*
Tanſie.

2 *Tanacetum criſpum Anglicum.*
Double Engliſh Tanſie.

3 *Tanacetum non odorum.*
Vnſauorie Tanſie.

‡ 4 *Tanacetum inodorum majus.*
Great vnſauorie Tanſie.

† 5 *Tanacetum minus album.*
Small white Tansie.

5   The fifth kinde of Tansie hath broad leaues, much jagged and wel cut, like the leaues of Fetherfew, but smaller, and more deeply cut. The stalke is small, a foot long, whereupon doe grow little tufts of little white floures, like the tuft of Milfoile or Yarrow. The herbe is in smell and sauour like the common Tansie, but not altogether so strong.

¶ *The Place.*

The first groweth wilde in fields as well as in gardens : the others grow in my garden.

¶ *The Time.*

They floure in Iuly and August.

¶ *The Names.*

The first is called Tansie ; the second, double Tansie; the third, vnsauory Tansie; the last, white Tansie : in Latine, *Tanacetum*, and *Athanasia*, as though it were immortall : because the floures do not speedily wither : of some, *Artemisia*, but vntruly.

¶ *The Nature.*

The Tansies which smel sweet are hot in the second degree, and dry in the third. That without smell is hot and dry, and of a meane temperature.

¶ *The Vertues.*

In the Spring time are made with the leaues **A** hereof newly sprung vp, and with egs, cakes or tansies, which be pleasant in taste, and good for the stomacke. For if any bad humours cleaue thereunto, it doth perfectly concoct them, and scowre them downewards. The root preserued with hony or sugar, is an especiall thing against the gout, if euery day for a certaine space, a reasonable quantity thereof be eaten fasting.

The seed of Tansie is a singular and approued medicine against Wormes, for in what sort soe- **B** uer it be taken, it killeth and driueth them forth.

The same pound, and mixed with oyle Oliue, is very good against the paine and shrinking of the **C** sinewes.

Also being drunke with wine, it is good against the paine in the bladder, and when a man cannot **D** pisse but by drops.

† The figure that was formerly in the fourth place was only the varietie of the ordinary Tansie, hauing a white floure, but that which agreed with the description was pag. 915. vnder the title of *Achillea, siue Millefolium nobile.*

CHAP. 210. *Of Fetherfew.*

¶ *The Description.*

1   FEuerfew bringeth forth many little round stalkes, diuided into certaine branches. The leaues are tender, diuersly torne and jagged, and nickt on the edges like the first and nethermost leaues of Coriander, but greater. The floures stand on the tops of the branches, with a small pale of white leaues, set round about a yellow ball or button, like the wild field Daisie. The root is hard and tough : the whole plant is of a light whitish greene colour, of a strong smell, and bitter taste.

2   The second kinde of Feuerfew, *Matricaria*, or *Parthenium*, differeth from the former, in that it hath double floures ; otherwise in smell, leaues, and branches, it is all one with the common Feuerfew.

3   There is a third sort called Mountaine Feuerfew, of *Carolus Clusius* his description, that hath

1 *Matricaria.*
Feuerſew.

2 *Matricaria duplici flore.*
Double Feuerſew.

‡ 3 *Matricaria Alpina Cluſij.*
Mountaine Feuerſew.

ſmall and fibrous roots; from which proceed
ſlender wooddie ſtalks, a foot high and ſomewhat more, beſet or garniſhed about with
leaues like Camomill, deepely jagged or cut,
of the ſauour or ſmell of Feuerſew, but not ſo
ſtrong, in taſte hot, but not vnpleaſant. At the
top of the ſtalks there come forth ſmall white
floures not like vnto the firſt, but rather like
vnto *Abſynthium album,* or White Wormewood.

   4   I haue growing in my Garden another
ſort, like vnto the firſt kinde, but of a moſt
pleaſant ſweet ſauour, in reſpect of any of the
reſt. ‡ This ſeemes to be the *Matricaria altera
ex Ilua,* mentioned by *Camerarius* in his *Hortus
medicus.* ‡

   ¶ *The Place.*

   The common ſingle Feuerſew groweth in
hedges, gardens, and about old wals, it joyeth
to grow among rubbiſh. There is oftentimes
found when it is digged vp a little cole vnder
the ſtrings of the root, and neuer without it,
whereof *Cardane* in his booke of Subtilties
ſetteth down diuers vaine and trifling things.

   ¶ *The Time.*

   They floure for the moſt part all the Summer long.

   ¶ *The Names.*

   Feuerſew is called in Greeke of *Dioſcorides*
παρθένιον: of *Galen,* and *Paulus* one of his ſect,
Ἀμάρακος: in Latine, *Parthenium, Matricaria,* and
Febrifu-

Febriſuga: of *Fuchſius,* *Artemeſia tenuifolia,* in Italian, *Amarella:* in Dutch 𝕸𝖔𝖊𝖉𝖊𝖗 𝖈𝖗𝖚𝖕𝖙 : in French, *Eſparagoute :* in Engliſh, Fedderſew and Feuerfew, taken from his force of driuing away Agues.

¶ *The Temperature.*

Feuerfew manifeſtly heateth, it is hot in the third degree, and dry in the ſecond, it clenſeth, purgeth, or ſcoureth, openeth, and fully performeth all that bitter things can do.

¶ *The Vertues.*

It is a great remedie againſt the diſeaſes of the matrix; it procureth womens ſicknes with ſpeed; A
it bringeth forth the after birth and the dead child, whether it be drunke in a decoction, or boiled
in a bath and the woman ſit ouer it ; or the herbes ſodden and applied to the priuie part, in manner
of a cataplaſme or pultis.

Dioſcorides alſo teacheth, that it is profitably applied to S. Anthonies fire, to all hot inflamma- B
tions, and hot ſwellings, if it be laid vnto, both leaues and floures.

The ſame Author affirmeth, that the pouder of Feuerfew drunk with Oxymell, or ſyrrup of Vi- C
neger, or wine for want of the others, draweth away flegme and melancholy, and is good for them
that are purſie, and haue their lungs ſtuffed with flegme; and is profitable likewiſe to be drunke a-
gainſt the ſtone, as the ſame Author ſaith.

Feuerfew dried and made into pouder, and two drams of it taken with hony or ſweet wine, pur- D
geth by ſiege melancholy and flegme ; wherefore it is very good for them that are giddie in the
head, or which haue the turning called *Vertigo,* that is, a ſwimming and turning in the head. Alſo it
is good for ſuch as be melancholike, ſad, penſiue, and without ſpeech.

The herbe is good againſt the ſuffocation of the mother, the hardnes and ſtopping of the ſame, E
being boiled in wine, and applied to the place.

The decoction of the ſame is good for women to ſit ouer, for the purpoſes aforeſaid. F

It is vſed both in drinks, and bound to the wreſts with bay ſalt, and the pouder of glaſſe ſtamped G
together, as a moſt ſingular experiment againſt the Ague.

CHAP. 211.   *Of Poley, or Pellamountaine.*

1 *Polium montanum album.*   2 *Polium montanum luteum.*
White Poley mountaine.   Yellow Poley mountaine.

¶ *The Deſcription.*

1   THe firſt kind of *Polium*, or in Engliſh Poley of the mountain, is a little tender and ſweet ſmelling herbe, very hoarie, whereupon it tooke his name : for it is not onely hoarie in part, but his hoarie flockineſſe poſſeſſeth the whole plant, tufts and all, being no leſſe hoarie than *Gnaphalium*, eſpecially where it groweth neer the ſea at the bending of the hils, or neer the ſandie ſhores of the Mediterranean ſea: from his wooddie and ſomewhat threddie root ſhoote forth ſtraight from the earth a number of ſmall round ſtalkes nine inches long, and by certaine diſtances from the ſtalk proceed ſomwhat long leaues like *Gnaphalium*, which haue light necks about the edges, that ſtand one againſt another, incloſing the ſtalke : in the top of the ſtalkes ſtand ſpokie tufts of floures, white of colour like *Serpillum*. This plant is ſtronger of ſent or ſauor than any of the reſt following, which ſent is ſomwhat ſharp, and affecting the noſe with his ſweetneſſe.

2   The tufts of the ſecond kind of *Polium* are longer than the tufts or floures of the laſt before mentioned, and they are of a yellow color; the leaues alſo are broader, otherwiſe they are very like.

3   From the wooddie roots of this third kind of *Polium*, proceed a great number of ſhoots like vnto the laſt rehearſed, lying flat vpright vpon the ground, whoſe ſlender branches take hold on the vpper part of the earth where they creep. The floures are like the other, but of a purple colour.

4   The laſt kinde of *Polium*, and of all the reſt the ſmalleſt, is of an indifferent good ſmell, in all points like vnto the common *Polium*, but that it is foure times leſſer, hauing the leaues not ſnipt, and the floures white.

‡   5   This ſends vp many branches from one root like to thoſe of the firſt deſcribed, but ſhorter and more ſhrubbie, lying partly vpon the ground, the leaues grow by couples at certain ſpaces, ſomwhat like, but leſſer than thoſe of Roſemarie or Lauander, greene aboue, and whitiſh beneath, not ſnipt about their edges; their taſt is bitter, and ſmell ſomewhat pleaſant: the floures grow plentifully vpon the tops of the branches, white of colour, and in ſhape like thoſe of the other Poleyes: they grow on a bunch together, and not Spike faſhion: the ſeed is blackiſh and contained in ſmall veſſels: the root is hard and wooddie with many fibres. *Cluſius* calls this *Polium 7. albo flore*. It is the *Polium alterum* of *Matthiolus*, and *Polium recentiorum fæmina Lavandulæ folio* of *Lobel*. I here giue you as (*Cluſius* alſo hath don) two figures to make one good one: the former ſhews the floures and their manner of growing; the other, the ſeed veſſels, and the leaues growing by couples together, with a little better expreſſion of the root. ‡

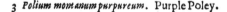

3 *Polium montanum purpureum*. Purple Poley.

¶ *The Place.*

Theſe plants do grow naturally vpon the mountaines of France, Italie, Spaine, and other hot regions. They are ſtrangers in England, notwithſtanding I haue plants of that Poley with yellow floures by the gift of *Lobel*.

¶ *The Time.*

They floure from the end of May, to the beginning of Auguſt.

¶ *The*

4 *Polium montanum minimum.*
Creeping Poley.

‡ 5 *Polium Lauandula folio, flore albo.*
Lauander leaued Poley.

Another figure of the Lauander leaued Poley.

¶ *The Names.*

Poley mountaine is called in Greeke, πώλιον, of his hoarinesse; and in Latine also *Polium*. Diuers suspect that *Polium* is *Leucas*; and that *Dioscarides* hath twice treated of that herbe, vnder diuers names: the kinds, the occasion of the names, and likewise the faculties do agree. There be two of the *Leucades*; one ἴμιν, that is, of the mountaine; the other, ἥμερος, which is that of the garden. It is called *Leucas* of the whitish colour, and *Polion*, of the hoarinesse, because it seemeth like to a mans hoary head; for whatsoeuer waxeth hoary is said to be white.

¶ *The Temperature.*

Poley is of temperature dry in the third degree, and hot in the end of the second.

¶ *The Vertues.*

Dioscorides saith it is a remedie for those that A haue the dropsie, the yellow jaundice, and that are troubled with the spleen.

It prouoketh vrine, and is put into Mithridate, B Treacle, and counterpoisons.

It profiteth much against the bitings of veno- C mous beasts, driuing them away from the place where it is strewed or burnt.

The same drunke with vineger is good for the D diseases of the milt and spleen; it troubleth the stomacke, afflicteth the head, and prouokes loosnesse of the belly.

Iii 2 CHAP.

Chap. 212. *Of Germander.*

¶ *The Kindes.*

THe old writers haue set downe no certain kindes of **Germander** ; yet we haue thought it good, and not without cause, to treat of more sorts than haue been obserued by all , diuiding those vnder the title of *Teucrium*, from *Chamædrys*,although they are both of one kinde,but yet differing very notably.

¶ *The Description.*

1   THe first Germander groweth low , with very many branches lying vpon the ground, tough,hard,and wooddy,spreading it selfe here and there ;whereupon are placed small leaues snipt about the edges like the teeth of a Saw,resembling the shape of an **Oken** leafe. The floures are of a purple colour, very small, standing close to the leaues toward the top of the branches : the seed is little and black : the root slender and full of strings,creeping,and alwaies spreading within the ground, whereby it greatly encreaseth. ‡ This is sometimes found with bigger leaues, otherwhiles with lesse ; also the floure is sometimes white and otherwhiles red in the same plant, whence *Tabernam.* gaue two figures,and our Author two figures and descriptions, whereof I haue omitted the later,and put the two titles into one. ‡

2   The second Germander riseth vp with a little straight stalke a span long and sometimes longer,wooddy and hard like vnto a little shrub : it is afterwards diuided into very many little small branches.The leaues are indented and nicked about the edges, lesser than the leaues of the former great creeping Germander. The floures likewise stand neere to the leaues, and on the vpper parts of the sprigs,of colour sometimes purple,and oftentimes tending to blewnesse:the root is diuersly dispersed with many strings.

1 *Chamædrys maior latifolia.*
Great broad leaued Germander.

2 *Chamædrys minor.*
Small Germander.

3 *Chamædrys syluestris.*
Wilde Germander.

3　Wilde Germander hath little stalks weake and feeble, edged or cornered, somwhat hairie, and set as it were with joints; about the which by certaine distances there come forth at each joint two leaues somthing broad, nicked in the edges, & somwhat greater than the leaues of creeping Germander, and softer. The floures be of a gallant blew colour, made of foure small leaues apiece, standing orderly on the tops of the tender spriggy spraies; after which come in place little husks or seedvessels. The root is small and threddy.

¶ *The Place.*

These plants grow in rocky and rough grounds, and in gardens they do easily prosper.

The wilde Germander groweth in many places about London in medowes and fertill fields, and in euerie place wheresoeuer I haue trauelled in England.

¶ *The Time.*

They floure and flourish from the end of May to the later end of August.

¶ *The Names.*

Garden Germander is called in Greeke, χαμαίδρυς *Chamædrys* : of some, *Trissago,* & *Trixago,* and likewise *Quercula minor;* notwithstanding most of these names do more properly belong to *Scordium* or water Germander: in Italian, *Quercinola* : in English, Germander, or English Treacle: in French, *Germandre.* Before creeping Germander was knowne, this wilde kinde bare the name of Germander amongst the Apothecaries, and was vsed for the right Germander in the compositions of medicines: but after the former were brought to light, this began to be named *Syluestris,* and *Spuria Camædrys* ; that is, wilde and bastard Germander : of some, *Teucrium pratense;* and without error, because al the sorts of plants comprehended vnder the title of *Teucrium,* are doubtlesse kindes of Germander: of some it hath bin thought to be the plant that *Dioscorides* called ιερβοτανη . *Hierabotane,* that is to say, the holy herb, if so be that the holy herb and *Verbenaca* or Veruain (which is called in Greek περιστερη) be sundry herbs. *Dioscorides* maketh them sundry herbs, describing them apart one after another : but other Authors, as *Paulus, Aëtius,* and *Oribasius,* make no mention of *Herba sacra,* the holy Herbe, but only of *Peristereon* ; and this is found to be likewise called *Hierabotane,* or the holy Herb ; and therefore it is euident, that it is one and the selfe same plant, called by diuers names : the which things considered, if they say so, and say truly, this wilde Germander cannot be *Hierabotane* at all, as diuers haue written and said it to be.

¶ *The Temperature.*

Garden Germander is of thin parts, and hath a cutting facultie, it is hot and dry almost in the third degree, euen as *Galen* writeth of *Teucrium* or wilde Germander.

The wilde Germander is likewise hot and dry, and is not altogether without force or power to open and clense : it may be counted among the number of them that do open the liuer and spleen.

¶ *The Vertues.*

Germander boiled in water and drunke, deliuereth the body from all obstructions or stoppings, A diuideth and cutteth tough and clammy humors : being receiued as aforesaid, it is good for them that haue the cough and shortnesse of breath, the strangurie or stopping of vrine, and helps those which are entring into a dropsie.

The leaues stamped with hony and strained, and a drop at sundry times put into the eies, taketh B away the web or haw in the same, or any dimnesse of sight.

It prouoketh the terms mightily, being drunke in wine, and the decoction drunke ; with a so- C mentation or bath made also thereof, and the secret parts bathed therewith.

Cʜᴀᴘ. 213. *Of Tree Germander.*

¶ *The Deſcription.*

1   THe firſt kind of Tree Germander riſeth vp with a little ſtraight ſtalk a cubit high, woo-die and hard like vnto a ſmall wooddy ſhrub : the ſtalke diuideth it ſelfe from the bottome to the top into diuers branches, whereon are ſet indented leaues nicked about the edges, in ſhape not much vnlike the leafe of the common Germander. The floures grow among the leaues, of a purple colour. The root is wooddy, as is all the reſt of the plant.

1 *Teucrium latifolium.*
Tree Germander with broad leaues.

2 *Teucrium Pannonicum.*
Hungarie Germander.

2   The tree Germander of Hungarie hath many, tough threddy roots, from which riſe vp diuers weake and feeble ſtalks reeling this way and that way ; whereupon are ſet together by couples long-leaues jagged in the edges, not vnlike thoſe of the vpright Fluellen : on the tops of the ſtalks ſtand the floures ſpike-faſhion, thicke thruſt together, of a purple colour tending towards blewneſſe.

‡ 3   This (which is the fourth of *Cluſius* deſcription) hath diuers ſtalkes ſome cubit high, foure ſquare, rough, and ſet at certaine ſpaces with leaues growing by couples like thoſe of Wilde Germander : the tops of the ſtalks are diuided into ſundry branches, carrying long ſpokes of blew floures conſiſting of foure leaues, wherof the vppermoſt leafe is the largeſt, and diſtinguiſhed with veins : after the floures are paſt, follow ſuch flat ſeed-veſſels as in Fluellen : the root is fibrous, and liues long, ſending forth euerie yeare new branches. ‡

4   This dwarfe Germander ſends vp ſtalks ſome handfull high, round, not branched: the leaues grow vpon theſe ſtalks by couples, thicke, ſhining, a little hairy and green on their vpper ſides, and

whitiſh

‡ 3 *Teucrium majus Pannonicum.*
Great Auſtrian Germander.

‡ 4 *Teucrium petræum pumilum.*
Dwarfe Rock Germander.

5 *Teucrium Bæticum.*
Spaniſh tree Germander.

6 *Teucrium Alpinum Ciſti flore.*
Rough headed tree Germander.

whitish below: the tops of the stalks carry spoky tufts of flours consisting of foure or fiue blewish leaues; which falling, there followes a seed-vessell as in the *Veronica's.* The root is knotty and fibrous, and growes so fast among the rocks that it cannot easily be got out. It floures in Iuly. *Clusius* describes this by the name of *Teucrium 6. pumilum :* and *Pona* sets it forth by the name of *Veronica petræa semper-virens.* ‡

5   This Spanish Germander riseth vp oft times to the height of a man, in manner of an hedge bush, with one stiffe stalke of the bignesse of a mans little finger, couered ouer with a whitish bark diuided sometimes into other branches, which are alwaies placed by couples one right against another, of an ouerworne hoary colour: and vpon them are placed leaues not much vnlike the common Germander; the vpper parts whereof are of a grayish hoary colour, and the lower of a deepe green, of a bitter taste, and somwhat crooked, turning and winding themselues after the manner of a welt. The flours come forth from the bosome of the leaues, standing vpon small tender footstalks of a white colour, without any helmet or hood on their tops, hauing in the middle many threddy strings : the whole plant keepeth green all the winter long.

6   Among the rest of the tree Germanders this is not of least beauty and account, hauing many weake and feeble branches trailing vpon the ground, of a darke reddish colour, hard and woody; at the bottom of which stalks come forth many long broad jagged leaues not vnlike the precedent, hoary vnderneath, and green aboue, of a binding and drying taste. The floures grow at the top of the stalks, not vnlike to those of *Cistus fæmina,* or Sage-rose, and are white of colour, consisting of eight or nine leaues, in the middle whereof do grow many threddy chiues without smel or sauor : which being past, there succeedeth a tuft of rough threddy or flocky matter, not vnlike to those of the great *Auens* or *Pulsatilla :* the root is wooddy, and set with some few hairy strings fastned to the same.

¶ *The Place.*

These plants do ioy in stony rough mountaines and dry places, such as lie open to Sun and aire, and prosper well in gardens : and of the second sort I haue receiued one plant for my garden of Mr *Garret* Apothecarie.

¶ *The Time.*

They floure, flourish, and seed when the other Germanders do.

¶ *The Names.*

Tree Germander is called in Greeke χαμαίδρυς ; retaining the name of the former *Chamædrys;* and διυνμιη , according to the authoritie of *Dioscorides* and *Pliny :* in Latine, *Teucrium :* in English, great Germander, vpright Germander, and tree Germander.

¶ *The Temperature and Vertues.*

Their temper and faculties are referred vnto the garden Germander, but they are not of such force in working, wherefore they be not so much vsed in physicke.

Chap. 214.
Of Water Germander, or Garlicke Germander.

¶ *The Description.*

1   *S Cordium* or water Germander hath square hairy stalks creeping by the ground, beset with soft whitish crumpled leaues, nickt and snipt round about the edges like a Saw : among which grow small purple floures like the floures of dead Nettle. The root is small and threddy, creeping in the ground very deeply. The whole plant being bruised smelleth like garlick, whereof it tooke that name *Scordium.* ‡ This by reason of goodnesse of soile varies in the largenesse thereof; whence *Tabernamontanus* and our Author made a bigger and a lesser thereof, but I haue omitted the later as superfluous. ‡

¶ *The Place.*

Water Germander groweth neere to Oxenford, by Ruley, on both sides of the water, and in a

medow

medow by Abington called Nierford, by the relation of a learned gentleman of S. *Iohns* in the said towne of Oxford, a diligent *************, my very good friend, called M.ʳ *Richard Slater*. Also it groweth in great plenty in the Ilse of Ely, and in a medow by Harwood in Lancashire, and in diuers other places.

Scordium.
Water Germander.

¶ The Time.

These floures appeare in Iune and Iuly : it is best to gather the herb in August : it perisheth not in Winter, but only loseth the stalks which come vp againe in Summer: the root remaineth fresh all the yeare.

¶ The Names.

The Grecians call it σκόρδιον : the Latines do also call it *Scordium* : the Apothecaries haue no other name : it is called of some, *Trixago palustris, Quercula,* and also *Mithridatium,* of *Mithridates* who first found it out. It tooke the name *Scordium* from the smell of Garlicke, by the Grecians called σκόροδον, and σκόρδον, of the rancknese of the smell : in high-Dutch, **Waster battenig** : in French, *Scordion* : in Itauan, *Chalamandrina palustre* : in English, Scordium, Water Germander, and garlicke Germander.

¶ The Nature.

Water Germander is hot and dry : it hath a certain bitter taste, harsh and sharpe, as *Galen* witnesseth.

¶ The Vertues.

Water Germander clenseth the intrals and likewise old vlcers, being mixed with honey according to art : it prouoketh vrine, and bringeth downe the monethly sicknesse : it draweth out of the chest thicke flegme and rotten matter : it is good for an old cough, paine in the sides comming of stopping and cold, and for burstings and inward ruptures. **A**

The decoction made in wine and drunke, is good against the bitings of serpents, and deadly poisons ; and is vsed in antidotes or counterpoysons with good successe. **B**

It is reported to mitigate the pain of the gout, being stamped and applied with a little vineger and water. **C**

Some affirm, That raw flesh being laid among the leaues of Scordium, may be preserued a long time from corruption. **D**

Being drunke with wine, it openeth the stoppings of the liuer, milt, kidnies, bladder, and matrix, prouoketh vrine, helpeth the strangurie, that is, when a man cannot pisse but by drops, and is a most singular cordial to comfort and make merry the heart. **E**

The pouder of Scordion taken to the quantitie of two drams in Mede or honied water, cureth and stoppeth the bloudy flix, and comforteth the stomacke. Of this Scordium is made a most singular medicine called *Diescordium*, which serueth very notably for all the purposes aforesaid. **F**

The same medicine made with Scordium is giuen with very good successe vnto children and aged people, that haue the small pocks, measles, or purples, or any other pestilent sicknesse whatsoeuer, euen the plague it selfe, giuen before the sicknesse haue vniuersally possessed the whole body. **G**

Chap. 215.
Of Wood Sage, or Garlicke Sage.

¶ *The Description.*

THat which is called wilde Sage hath stalkes foure square, somewhat hairy, about which are leaues like those of Sage, but shorter, broader, and softer: the floures grow vp all vpon one side of the stalke, open and forked as those of dead Nettle, but lesser, of a pale white colour. Then grow the seeds foure together in one huske: the root is full of strings. It is a plant that liueth but a yeare: it smelleth of Garlicke when it is bruised, being a kinde of Garlick Germander, as the said smell of garlick testifieth.

† *Scorodonia, siue Saluia agrestis.*
Wood Sage, or Garlicke Sage.

¶ *The Place.*

It groweth vpon heaths and barren places: it is also found in woods, and neere vnto hedge rowes about the borders of fields: it somewhat delighteth in a lean soile, and yet not altogether barren and dry.

¶ *The Time.*

It floureth and seedeth in Iune, Iuly, & August, and it is then to be gathered and laid vp.

¶ *The Names.*

It is called of the later Herbarists, *Saluia Agrestis*: of diuers also *Ambrosia*; but true *Ambrosia* which is Oke of Cappadocia, differs from this. *Valerius Cordus* nameth it *Scordonia*, or *Scorodonia*, and *Scordium alterum*. *Ruellius* saith it is called *Boscisaluia*, or *Saluia Bosci*: in high-Dutch, **waldt salbey**: in English, wilde Sage, wood Sage, and Garlicke Sage.

It seemeth to be *Theophrastus* his σφάκελος, *Sphacelus*, which is also taken for the small Sage, but not rightly.

¶ *The Temperature.*

Wilde Sage is of temperature hot and drie, yet lesse than common Sage, being hot and dry in the second degre.

¶ *The Vertues.*

A   It is commended against burstings, dry-beatings, and wounds: the decoction thereof is giuen to them that fall and are inwardly bruised: it also prouoketh vrine.

B   Some likewise giue the decoction hereof to drinke with good successe to them that are infected with the French pox: for it causeth sweat, drieth vp vlcers, digesteth humors, wasteth away & consumeth swellings, if it be taken thirty or forty dayes together, or put into the decoction of *Guaiacum* in stead of *Epithymum* and other adjutories belonging to the said decoction.

† The figure which was formerly here was of *Calamintha montana prastantior* of *Lobel*.

Chap. 216.   Of Eye-bright.

¶ *The Description.*

EVphrasia or Eye-bright is a small low herbe not aboue two handfulls high, full of branches, couered with little blackish leaues dented or snipt about the edges like a Saw. The floures are
small

ſmall and white, ſprinkled and poudered on the inner ſide, with yellow and purple ſpeckes mixed therewith. The root is ſmall and hairie.

Euphraſia.
Eye-bright.

¶ *The Place.*

This plant groweth in dry medowes, in greene and graſſie waies and paſtures ſtanding againſt the Sun.

¶ *The Time.*

Eye-bright beginneth to floure in Auguſt, and continueth vnto September, and muſt bee gathered while it floureth for phyſicks vſe.

¶ *The Names.*

It is commonly called *Euphraſia*, as alſo *Euphroſine*; notwithſtanding there is another *Euphroſine, viz.* Bugloſſe: it is called of ſome *Ocularis* and *Ophthalmica,* of the effect: in high-Dutch, **Augentroſt :** in low-Dutch, **Ooghentrooſt:** in Italian, Spaniſh, and French, *Eufraſia,* after the Latine name: in Engliſh, Eye-bright.

¶ *The Nature.*

This herbe is hot and dry, but yet more hot than dry.

¶ *The Vertues.*

It is very much commended for the eies. Being taken it ſelfe alone, or any way **A**
elſe, it preſerues the ſight, and being feeble & loſt it reſtores the ſame: it is giuen moſt fitly being beaten into pouder; oftentimes a like quantitie of Fennell ſeed is added thereto, and a little mace, to the which is put ſo much ſugar as the weight of them all commeth to.

Eye-bright ſtamped and laid vpon the eyes, or the juice thereof mixed with white Wine, and dropped into the eyes, or the diſtilled water, taketh away the darkneſſe and dimneſſe of the eyes, **B**
and cleareth the ſight.

Three parts of the pouder of Eye-bright, and one part of maces mixed therewith, taketh away all hurts from the eyes, comforteth the memorie, and cleareth the ſight, if halfe a ſpoonefull be taken **C**
euery morning faſting with a cup of white wine.

‡ That which was formerly here ſet forth in the ſecond place vnder the title of *Euphraſia Cærulea Tabor* was deſcribed by our Author amongſt the Scorpion graſſes, in the third place, Chap. 54. and the figure is pag. 338. vnder the title of *Myoſotis Scorpioides paluſtris.*

CHAP. 217. *Of Marjerome.*

¶ *The Deſcription.*

1   Sweet Marjerome is a low and ſhrubbie plant, of a whitiſh colour and maruellous ſweet ſmell, a foot or ſomwhat more high. The ſtalkes are ſlender, and parted into diuers branches, about which grow forth little leaues ſoft and hoarie: the floures grow at the top in ſcalie or chaffie ſpiked eares, of a white colour like vnto thoſe of Candy Organy. The root is compact of many ſmall threds. The whole plant and euerie part thereof is of a moſt pleaſant taſt and aromaticall ſmell, and periſheth at the firſt approch of Winter.

2   Pot Marjerome or Winter Marjerome hath many threddy tough roots, from which riſe immediatly diuers ſmall branches, whereon are placed ſuch leaues as the precedent, but not ſo hoary, nor yet ſo ſweet of ſmell, bearing at the top of the branches tufts of white floures tending to purple. The whole plant is of long continuance, and keepeth greene all the Winter; whereupon our Engliſh women haue called it, and that very properly, Winter Marjerome.

3   Marjerome gentle hath many branches, riſing from a threddy root, whereupon do grow ſoft and ſweet ſmelling leaues of an ouerworne ruſſet colour. The floures ſtand at the top of the ſtalks,

compact

1 *Marjorana major.*
Great ſweet Marjerome.

Marjorana major Anglica.
Pot Marjerome.

3 *Marjorana tenuifolia.*
Marjerome gentle.

compact of diuers ſmall chaffie ſcales, of a white colour tending to a bluſh. The whole plant is altogether like the great ſweet Marjerome, ſauing that it is altogether leſſer, and far ſweeter, wherein eſpecially conſiſteth the difference.

4 *Epimajorana* is likewiſe a kind of Marjerome, differing not from the laſt deſcribed, ſauing in that, that this plant hath in his naturall countrey of Candy, and not elſewhere, ſome laces or threds faſtned vnto his branches, ſuch, and after the ſame manner as thoſe are that doe grow vpon Sauorie, wherein is the difference.

¶ *The Place.*

Theſe plants do grow in Spain, Italy, Candy, and other Iſlands thereabout, wild, and in the fields; from whence wee haue the ſeeds for the gardens of our cold countries.

¶ *The Time.*

They are ſowne in May, and bring forth their ſcaly or chaffie husks or ears in Auguſt. They are to be watered in the middle of the day, when the Sun ſhineth hotteſt, euen as Baſill ſhould be, and not in the euening nor morning, as moſt plants are.

¶ *The Names.*

Marjerome is called *Marjarana*, and *Amaracus*, and alſo *Marum* and *Sampſychum* of others: in High-Dutch, **Mayoꝛan**: in Spaniſh, *Mayorana*, *Moradux*, and *Almoradux* : in French, *Mariolaine* : in Engliſh, Sweet Marjerome, Fine Marjerome, and Marjerome

rome gentle ; of the beſt ſort of Marjerane. The pot Marjerome is alſo called Winter Marjerome.
Some haue made a doubt whether *Majorana* and *Sampſycum* be all one ; which doubt, as I take it,
is becauſe that *Galen* maketh a difference betweene them, intreating of them apart, and attributeth
to either of them their operations. But *Amaracus Galeni* is *Parthenium*, or Feuerfew. *Dioſcorides* likewiſe witneſſeth, that ſome do call *Amaracus*, *Parthenium* ; and *Galen* in his booke of the faculties of
ſimple medicines, doth in no place make mention of *Parthenium*, but by the name of *Amaracus*.
Pliny in his 21. booke, chap. 2. witneſſeth, that *Diocles* the phyſition, and they of Cicily did call that
Amaracus, which the Ægyptians and the Syrians did call *Sampſycum*.

Virgil in the firſt booke of his *Æneidos* ſheweth, that *Amaracus* is a ſhrub bearing floures, writing thus :

> ——— *Vbi mollis Amaracus illum*
> *Floribus, & dulci aſpirans complectitur vmbra.*

Likewiſe *Catullus* in his *Epithalamium*, or mariage ſong of *Iulia* and *Mallius* ſaith.

> *Cinge tempora floribus*
> *Suaue olentis Amaraci.*

> Compaſſe the temples of the head with floures
> Of Amarac, affording ſweet ſauours.

Notwithſtanding it may not ſeeme ſtrange, that Majorane is vſed in ſtead of *Sampſycum*, ſeeing that
in *Galens* time alſo *Marum* was in the mixture of the ointment called *Amaracinum vnguentum*, in the
place of *Sampſycum*, as he himſelfe witneſſeth in his firſt booke of counterpoyſons.

¶ *The Temperature.*

They are hot and dry in the ſecond degree ; after ſome copies, hot and dry in the third.

¶ *The Vertues.*

Sweet Marjerome is a remedy againſt cold diſeaſes of the braine and head, being taken any way   A
to your beſt liking, put vp into the noſthrils it prouokes ſneeſing, and draweth forth much baggage
flegme: it eaſeth the tooth-ache being chewed in the mouth ; being drunke it prouoketh vrine and
draweth away wateriſh humors, and is vſed in medicines againſt poyſon.

The leaues boiled in water, and the decoction drunke, helpeth them that are entering into the   B
dropſie : it eaſeth them that are troubled with difficulty of making water, and ſuch as are giuen to
ouermuch ſighing, and eaſeth the paines of the belly.

The leaues dried and mingled with honey, and giuen, diſſolue congealed or clotted bloud, and   C
put away blacke and blew markes after ſtripes and bruſes, being applied thereto.

The leaues are excellent good to be put into all odoriferous ointments, waters, pouders, broths,   D
and meates.

The dried leaues poudered, and finely ſearched, are good to put into Cerotes, or Cere-clothes,   E
and ointments, profitable againſt cold ſwellings, and members out of joynt.

There is an excellent oyle to be drawne forth of theſe herbes, good againſt the ſhrinking of ſi-   F
newes, crampes, conuulſions, and all aches proceeding of a colde cauſe.

Cʜᴀᴘ. 218.   *Of wilde Marjerome.*

¶ *The Deſcription.*

1   Baſtard Marjerome groweth ſtraight vp with little round ſtalkes of a reddiſh colour, full
of branches, a foot high and ſometimes higher. The leaues be broad, more long than
round, of a whitiſh greene colour : on the top of the branches ſtand long ſpikie ſcaled
eares, out of which ſhoot forth little white floures like the flouring of wheat. The whole plant is of
a ſweet ſmell, and ſharpe biting taſte.

2   The white Organy, or baſtard Marjerome with white floures, differing little from the precedent, but in colour and ſtature. This plant hath whiter and broader leaues, and alſo much higher,
wherein conſiſteth the difference.

3   Baſtard Marjerome of Candy hath many thready roots ; from which riſe vp diuers weake
and feeble branches trailing vpon the ground, ſet with faire greene leaues, not vnlike thoſe of Penny Royall, but broader and ſhorter : at the top of thoſe branches ſtand ſcalie or chaffie eares of a
purple colour. The whole plant is of a moſt pleaſant ſweet ſmell. The root endured in my garden

1 *Origanum Heracleoticum.*
Baſtard Marjerome.

† 2 *Origanum album, Tabern.*
White baſtard Marjerome.

† 3 *Origanum Creticum.*
Wilde Marjerome of Candy.

4 *Origanum Anglicum.*
Engliſh wilde Marjerome.

and the leaues alſo greene all this Winter long, 1597. although it hath beene ſaid that it doth periſh at the firſt froſt, as ſweet Marjerome doth.

4 Engliſh wilde Marjerome is exceedingly well knowne to all, to haue long, ſtiffe, and hard ſtalkes of two cubits high, ſet with leaues like thoſe of ſweet Marjerome, but broader and greater, of a ruſſet greene colour, on the top of the branches ſtand tufts of purple floures, compoſed of many ſmall ones ſet together very cloſely vmbell faſhion. The root creepeth in the ground, and is long laſting.

¶ The Place.

Theſe plants do grow wilde in the Kingdome of Spaine, Italy, and other of thoſe hot regions. The laſt of the foure doth grow wilde in the borders of fields, and low copſes, in moſt places of England.

¶ The Time.

They floure and flouriſh in the Summer moneths, afterward the ſeed is perfected.

¶ The Names.

Baſtard Marjerome is called in Greeke, ἱρίγανος, and that which is ſurnamed *Heracleoticum*, ἱρίγανον ἡρακλεωτικόν of diuers it is called *Cunila* in ſhops, *Origanum Hiſpanicum*, Spaniſh Organy : our Engliſh wilde Marjerome is called in Greeke of *Dioſcorides*, *Galen*, and *Pliny*, *Onitis* : of ſome *Agriorioganum*, or *Sylueſtre Origanum* in Italian, *Origano* : in Spaniſh, *Oregano* : in French, *Mariolaine baſtarde* : in Engliſh, Organe, baſtard Marjerome : and that of ours, wilde Marjerome, and groue Marjerome.

¶ The Temperature.

All the Organies do cut, attenuate, or make thin, dry, and heate, and that in the third degree ; and *Galen* teacheth, that wilde Marjerome is more forceable and of greater ſtrength ; notwithſtanding Organy of Candy which is brought dry out of Spaine (wherof I haue a plant in my garden) is more biting than any of the reſt, and of greater heate.

¶ The Vertues.

Organy giuen in wine is a remedy againſt the bitings, and ſtingings of venomous beaſts, and cureth them that haue drunke *Opium*, or the juyce of blacke Poppy, or hemlockes, eſpecially if it be giuen with wine and raiſons of the ſunne.　**A**

The decoction of Organy prouoketh vrine, bringeth downe the monethly courſe, and is giuen with good ſucceſſe to thoſe that haue the dropſie.　**B**

It is profitably vſed in a looch, or a medicine to be licked, againſt the old cough and the ſtuffing of the lungs.　**C**

It healeth ſcabs, itches, and ſcuruineſſe, being vſed in bathes, and it taketh away the bad colour which commeth of the yellow jaundiſe.　**D**

The weight of a dram taken with meade or bonied water, draweth forth by ſtoole blacke and filthy humors, as *Dioſcorides* and *Pliny* write.　**E**

The juyce mixed with a little milke, being poured into the eares, mitigateth the paines thereof.　**F**

The ſame mixed with the oile of *Ireos*, or the roots of the white Florentine floure-de-luce, and drawne vp into the noſthrils, draweth downe water and flegme : the herbe ſtrowed vpon the ground driueth away ſerpents.　**G**

The decoction looſeth the belly, and voideth choler ; and drunke with vinegre helpeth the infirmities of the ſpleene, and drunke in wine helpeth againſt all mortall poyſons, and for that cauſe it is put into mithridate and treacles prepared for that purpoſe.　**H**

Theſe plants are eaſie to be taken in potions, and therefore to good purpoſe they may be vſed and miniſtred vnto ſuch as cannot brooke their meate, and to ſuch as haue a ſowre ſquamiſh and watery ſtomacke, as alſo againſt the ſwouning of the heart.　**I**

† The ſecond and third figures were formerly tranſpoſed.

Chap. 219.　*Of Goates Marjerome, or Organy.*

¶ The Deſcription.

1 THe ſtalkes of Goates Organy are ſlender, hard and wooddy, of a blackiſh colour; whereon are ſet long leaues, greater than thoſe of the wilde Time, ſweet of ſmell, rough, and ſomewhat hairy. The floures be ſmall, and grow out of little crownes or wharles round about the top of the ſtalkes, tending to a purple colour. The root is ſmall and threddy.

† 1 *Tragoriganum Dod.*   † *Tragoriganum. Lob.*
Goates Marjerome.

† 2 *Tragoriganum Cluſij.*
Cluſius his Goats Marjerome.

‡ 3 *Tragoriganum Cretenſe.*
Candy Goats Marjerome.

2   *Carolus Clusius* hath set forth in his Spanish Obseruations another sort of Goats Marjerome growing vp like a small shrub : the leaues are longer and more hoarie than wilde Marjerome, and also narrower, of a hot biting taste, but of a sweet smell, though not very pleasant. The floures doe stand at the top of the stalkes in spokie rundles, of a white colour. The root is thicke and wooddy.

‡ 3   This differs little in forme and magnitude from the last described, but the branches are of a blacke colour, with rougher and darker coloured leaues : the floures also are lesser, and of a purple colour. Both this and the last described continue alwaies greene, but this last is of a much more fragrant smell. This floures in March, and was found growing wilde by *Clusius* in the fields of Valentia : he cals it *Tragoriganum Hispanicum tertium. Pena* and *Lobel* call it *Tragoriganum Cretense apud Venetas* ; that is, the Candy Goats Marjerome of the Venetians. ‡

¶ *The Place.*

These plants grow wild in Spaine, Italy, and other hot countries. The first of these I found growing in diuers barren and Chalky fields and high-waies neere vnto Sittingburne and Rochester in Kent, and also neere vnto Cobham house and Southfleet in the same county.

‡   I doubt our Author was mistaken, for I haue not heard of this growing wild with vs. ‡

¶ *The Time.*

They floure in the moneth of August. I remember (saith *Dodonæus*) that I haue seene *Tragoriganum* in the Low-countries, in the gardens of those that apply their whole study to the knowledge of plants ; or as we may say, in the gardens of cunning Herbarists.

¶ *The Names.*

Goats Organie is called in Greeke τραγορίγανον : in Latine likewise *Tragoriganum* : in English, Goats Organie, and Goats Marjerome.

¶ *The Temperature.*

Goats Organies are hot and dry in the third degree : They are (saith *Galen*) of a binding qualitie.

¶ *The Vertues.*

Tragoriganum or Goats Marjerome is very good against the wambling of the stomacke, and the foure belchings of the same, and staieth the desire to vomit, especially at sea.   A

These bastard kindes of Organie or wild Marjeromes haue rhe same force and faculties that the other Organies haue for the diseases mentioned in the same chapter.   B

† There were formely two figures in this chapter; the first whereof was of that which is described in the second place; the second was of *Tragoriganum* of *Matthiolus*, whereof here it no mention made. The figure of the *Tragoriganum alterum* of *Lobel* (which, as I haue formerly said, *Bauhine* would haue all one with that of *Dodonæus*) was formerly vnder the name of *Thymum Creticum*, pag 459. of the former edition.

CHAP. 220.   *Of Herbe Masticke.*

¶ *The Description.*

1   THe English and French Herbarists at this day do in their vulgar tongues call this herb Masticke or Mastich, taking this name *Marum* of *Maro* King of Thrace ; though some rather suppose the name corruptly to be deriued from this word *Amaracus*, the one plant being so like the other, that many learned haue taken them to be one and the selfe same plant : others haue taken *Marum* for *Sampsucus*, which doubtlesse is a kinde of Marjerome. Some (as *Dodonæus*) haue called this our *Marum* by the name of *Clinopodium* : which name rather belongs to another plant than to Masticke. ‡ This growes some foot high, with little longish leaues set by couples : at the tops of the stalkes amongst white downy heads come forth little white floures : the whole plant is of a very sweet and pleasing smell. ‡

2   If any be desirous to search for the true *Marum*, let them be assured that the plant last mentioned is the same : but if any do doubt thereof, for nouelties sake here is presented vnto your view a plant of the same kinde (which cannot be rejected) for a speciall kind thereof, which hath a most pleasant sent or smell, and in shew resembleth Marjerome and *Origanum*, consisting of small twigs a foot and more long ; the heads tufted like the common Marjerome ; but the leaues are lesse, and like *Myrtus* : the root is of a wooddy substance, with many strings hanging thereat.

3   There is another kind hereof set forth by *Lobel*, which I haue not as yet seen, nor himselfe hath well described, which I leaue to a better consideration. ‡   Though our Author knew not how to describe this creeping *Marum* of *Lobel*, yet no question, if hee had knowne so much, hee would haue giuen vs the figure thereof as well in this place : as in the third place of the next chapter

for

1 *Marum.*
Herbe Maſticke.

2 *Marum Syriacum.*
Aſſyrian Maſticke.

† 3 *Marum ſupinum Lobelÿ.*
Creeping Maſticke.

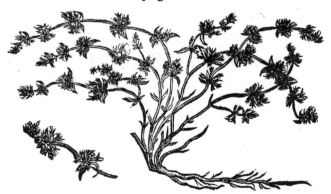

for a Penny-Royall; and might as well here as there, and much more fitly haue ventured at a deſcription. But that which is defectiue in him and *Lobel*, I will endeauour to ſupply out of *Caſalpinus*. This plant hath many creeping branches like to thoſe of wilde Time, but ſet with whiter and ſhorter leaues like to thoſe of the ſmaller Marjerome, but ſomewhat narrower : the floures grow in rundles amongſt the leaues, as in Calamint and are of a purple colour : the whole plant is of a ſtrong and ſweet ſmell, and of an hot and bitter taſte. *Caſalpinus* thinkes this to be the *Sampſuchum* of *Dioſcorides* : and ſo alſo doe the Authors of the *Aduerſaria*. *Tabernamontanus* calls it *Marum repens.* ‡

¶ *The Place.*

Theſe plants are ſet and ſowne in the gardens of England, and there maintained with great care and diligence from the injurie of our cold clymate.

¶ *The*

¶ *The Time.*

They floure about Auguſt, and ſomewhat later in cold Summers.

¶ *The Names.*

‡ Maſticke is called of the new writers *Marum :* and ſome, as *Lobel* and *Anguillara*, thinke it the *Helenium odorum* of *Theophraſtus.* *Dodonæus* iudges it to be the *Clinopodium* of *Dioſcorides. Cluſius* makes it his *Tragoriganum* 1. and ſaith he receiued the ſeeds thereof by the name of *Ambra dulcis.* ‡

¶ *The Nature.*

Theſe plants are hot and dry in the third degree.

¶ *The Vertues.*

Dioſcorides writeth, that the herbe is drunke, and likewiſe the decoction thereof, againſt the bitings of venomous beaſts, crampes and convulſions, burſtings and ſtrangury.   A

The decoction boiled in wine till the third part be conſumed, and drunke, ſtoppeth the laske in them that haue an ague, and vnto others in water.   B

† That we here giue you in the third place was formerly vnfitly figured in the third place of the enſuing Chapter, by the name of *Pulegium Anguſtifolium.*

CHAP. 221.   *Of Pennie Royall, or pudding graſſe.*

† 1 *Pulegium regium.*
Penny Royall.

† 2 *Pulegium mas.*
Vpright Penny Royall.

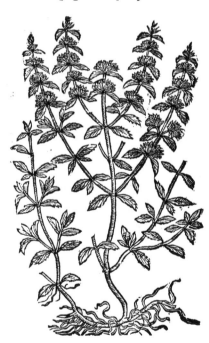

¶ *The Deſcription.*

1   Pᴠ*legium regium vulgatum* is ſo exceedingly well knowne to all our Engliſh Nation, that it needeth no deſcription, being our common Pennie Royall.

2   The ſecond being the male Penny Royall is like vnto the former, in leaues, floures and ſmell, and differeth not in that this male kinde groweth vpright of himſelfe without creeping, much like in ſhew vnto wilde Marjerome.

3 The

† 3 *Pulegium angustifolium.*
Narrow leafed Penny-Royall.

3   The third kinde of Pennie Royall growes like vnto Tyme, and is of a wooddy substance, somewhat like vnto the thinne leafed Hyssope, of the sauour of common Pennie-Royall, ‡ but much stronger and more pleasant: the longish narrow leaues stand vpon the stalkes by couples, with little leaues comming forth of their bosomes : and towards the tops of the branches grow rundles of small purple floures. This grows plentifully about Montpellier, & by the Authors of the *Aduersaria,* who first set it forth, it is stiled *Pulegium angustif. siue cervinum Monspeliensium.* ‡

¶ *The Place.*

The first and common Penny Royall groweth naturally wild in moist and ouerflown places, as in the Common neere London called Miles end, about the holes and ponds thereof in sundry places, from whence poore women bring plenty ro sell in London markets ; and it groweth in sundry other Commons neere London likewise.

The second groweth in my garden : the third I haue not as yet seene.

¶ *The Time.*

They floure from the beginning of Iune to the end of August.

¶ *The Names.*

Pennie Royall is called in Greeke γλήχων, and oftentimes βλήχων : in Latine, *Pulegium,* and *Pulegium regale,* for difference sake betweene it and wilde Time, which of some is called *Pulegium montanum :* in Italian, *Pulegio :* in Spanish, *Poleo :* in Dutch, **Poley:** in French, *Pouliot :* in English, Pennie Royall, Pudding grasse, Puliall Royall, and of some Organie.

¶ *The Temperature.*

Pennie Royall is hot and dry in the third degree, and of subtill parts, as *Galen* saith.

¶ *The Vertues.*

A   Pennie Royall boyled in wine and drunken, prouoketh the monethly termes, bringeth forth the secondine, the dead childe and vnnaturall birth : it prouoketh vrine, and breaketh the stone, especially of the kidnies.

B   Pennie Royall taken with hony clenseth the lungs, and cleareth the breast from all grosse and thicke humours.

C   The same taken with hony and Aloes, purgeth by stoole malancholy humours ; helpeth the crampe and drawing together of sinewes.

D   The same taken with water and vinegre asswageth the inordinate desire to vomit, and the paines of the stomacke.

E   If you haue when you are at the sea Penny Royall in great quantitie dry, and cast it into corrupt water, it helpeth it much, neither will it hurt them that drinke thereof.

F   A Garland of Pennie Royall made and worne about the head is of great force against the swimming in the head, and the paines and giddinesse thereof.

G   The decoction of Penny Royall is very good against ventositie, windines, or such like, & against the hardnesse and stopping of the mother being vsed in a bath or stew for the woman to sit ouer.

† It is apparant by the titles and descriptions that our Authour in this chapter followed *Isbel.* but the figures were not agreeable to the History, for the two first figures were of the *Pulegium Angustifolium* described in the third place ; and the third figure was of the *Marum supinum* described in the last place of the foregoing Chapter.

CHAP. 222. *Of Basill.*

¶ *The Description.*

1   GArden Basill is of two sorts, differing one from another in bignes. The first hath broad, thicke and fat leaues, of a pleasant sweet smell, and of which some one here and there are of a black reddish colour, somwhat snipped about the edges, not vnlike the leaues of French Mercury. The stalke groweth to the height of halfe a cubit, diuiding it selfe into diuers branches, whereupon do stand small and base floures and sometimes whitish, and often tending to a darke purple. The root is threddy, and dieth at the approach of Winter.

2 The

1 *Ocimum magnum.*
Great Basill.

2 *Ocimum medium citratum.*
Citron Basill.

3 *Ocimum minus Caryophyllatum.*
Bush Basill.

‡ 4 *Ocimum Indicum.*
Indian Basill.

2   The middle Basill is very like vnto the former, but it is altogether lesser. The whole plant is of a most odoriferous smell, not vnlike the smell of a Limon, or Citron, whereof it tooke his surname.

3   Bush Basill, or fine Basill, is a low and base plant, hauing a thready root, from which rise vp many small and tender stalkes, branched into diuers armes or boughes; whereupon are placed many little leaues, lesser than those of Penny Royall. The whole plant is of a most pleasing sweet smell.

‡ 4   This which some call *Ocimum Indicum*, or rather (as *Camerarius* saith) *Hispanicum*, sends vp a stalke a foot or more high, foure square, and of a purple color, set at each joynt with two leaues, and out of their bosomes come little branches : the largest leaues are some two inches broad, and some three long; growing vpon long stalkes, and deepely cut in about their edges, being also thicke, fat, and juicie, and either o. a darke purple colour, or else spotted with more or lesse such coloured spots. The tops of the branches end in spokie tufts of white floures with purple veines running a-longst them. The seed is contained in such seed vessels as that of the other Basils, and is round, blacke and large. The plant perishes euery yeare as soone as it hath perfected the seed. *Clusius* cals this *Ocimum Indicum*. ‡

¶ *The Place.*

Basil is sowne in gardens, and in earthen pots. It commeth vp quickly, and loueth little moisture except in the middle of the day; otherwise if it be sowne in rainie weather, the seed will putrifie, and grow into a jellie or slime, and come to nothing.

¶ *The Time.*

Basill floureth in Iune and Iuly, and that by little and little, whereby it is long a flouring, beginning first at the top.

¶ *The Names.*

Basill is called in Greeke, ωκιμον, and more commonly with α in the first syllable ακιμον: in Latine, *Ocimum.* It differeth from *Ocymum* which some haue called *Cereale* as we (saith *Dodonæus*) haue shewed in the Historie of Graine. The later Græcians haue called it βασιλικον: in shop, likewise *Basilicum*, and *Regium* : in Spanish, *Albahaca* · in French, *Basilic* : in English, Basill, garden Basill, the greater Basill Royall, the lesser Basill gentle, and bush Basill : of some, *Basilicum Gariophyllatum*, or Cloue Basill.

¶ *The Temperature.*

Basill, as *Galen* teacheth, is hot in the second degree, but it hath adjoyned with it a superfluous moisture, by reason whereof he doth not like that it should be taken inwardly; but being applied outwardly, it is good to digest or distribute, and to concoct.

¶ *The Vertues.*

A   *Dioscorides* saith that if Basill be much eaten, it dulleth the sight, it mollifieth the belly, breedeth winde, prouoketh vrine, drieth vp milke, and is of a hard digestion.

B   The juyce mixed with fine meale of parched barly, oyle of Roses, and Vinegre, is good against inflammations, and the stinging of venomous beasts.

C   The juyce drunke in wine of *Chios* or strong Sacke, is good against head-ache.

D   The juyce clenseth away the dimnesse of the eies, and drieth vp the humour that falleth into them.

E   The seed drunke is a remedy for melancholy people; for those that are short winded, and them that can hardly make water.

F   If the same be snift vp in the nose, it causeth often neesing: also the herbe it selfe doth the same.

G   There be that shunne Basill and will not eat thereof, because that if it be chewed and laid in the Sun, it engendreth wormes.

H   They of Africke do also affirme, that they who are stung of the Scorpion and haue eaten of it, shall feele no paine at all.

I   The later Writers, among whom *Simeon Zethy* is one, doe teach, that the smell of Basill is good for the heart and for the head. That the seed cureth the infirmities of the heart, taketh away sorrowfulnesse which commeth of melancholy, and maketh a man merry and glad.

C H A P.

CHAP. 223.   Of wilde Baſill.

¶ The Deſcription.

1   The wilde Baſil or *Acynos*, called of *Pena*, *Clinopodium vulgare*, hath ſquare hairie ſtems, beſet with little leaues like vnto the ſmall Baſil, but much ſmaller, and more hairy, ſharpe pointed, and a little ſnipt toward the end of the leaſe, with ſmall floures of a purple colour, faſhioned like vnto the garden Baſill. The root is full of hairie threds, and creepeth along the ground, and ſpringeth vp yearely anew of it ſelfe without ſowing. ‡ This is the *Clinopodium alterum* of *Matthiolus*. ‡

2   This kinde of wilde Baſill called amongſt the Grӕcians ἄκνος, which by interpretation is *Sine ſemine*, or *Sterilis*, hath cauſed ſundry opinions and great doubts concerning the words of *Pliny* and *Theophraſtus*, affirming that this herbe hath no floures nor ſeeds; which opinion I am ſure of mine owne knowledge to be without reaſon: but to omit controuerſies, this plant beareth purple floures, wharled about ſquare ſtalkes, rough leaues and hairy, very like in ſhape vnto Baſil: ‡ The ſtalkes are ſome cubit and more high, parted into few branches, and ſet at certaine ſpaces with leaues growing by couples. This is the *Clinopodium vulgare* of *Matthiolus*, and that of *Cordus*, *Geſner*, and others; it is the *Acinos* of *Lobel*. ‡

3   *Serapio* and others haue ſet forth another wilde Baſill vnder the title of *Molochia*; and *Lobel* after the minde of *Iohn Brancion*, calleth it *Corcoros*, which we haue Engliſhed, Fiſh Baſill, the ſeeds whereof the ſaid *Brancion* receiued from Spaine, ſaying that *Corcoros Plinij* hath the leaues of Baſill: the ſtalkes are two handfuls high, the floures yellow, growing cloſe to the ſtalkes, bearing his ſeed in ſmall long cods. The root is compact and made of an innumerable company of ſtrings, creeping far abroad like running Time. ‡ This figure of *Lobels* which here we giue you is (as *Camerarius* hath obſerued) vnperfect, for it expreſſes not the long cods wherein the ſeed is contained, neither the two little ſtrings or beards that come forth at the ſetting on of each leafe to the ſtalke. ‡

1   *Ocymum ſylueſtre.*
  Wilde Baſill.

2   *Acynos.*
  Stone Baſill.

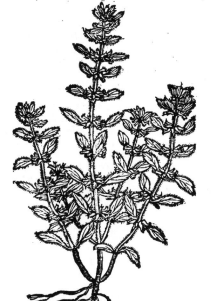 

‡ 3 *Corchoros.*
Iriſh Baſill.

‡ 5 *Clinopodium Auſtriacum.*
Auſtrian field Baſill.

‡ 6 *Clinopodium Alpinum.*
Wilde Baſill of the Alpes.

‡ 4 It may be our Authour would haue
deſcribed this in the firſt place, as I conjecture
by thoſe words which he vſed in mentioning the
place of their growing ; and [*Clinopodium vulgare*
groweth in great plentie vpon Long-field downs
in Kent ;] but to this neither figure nor deſcription did agree, wherefore I will giue you the Hiſtory therof. It ſends vp many little ſquare ſtalks
ſome handfull and halfe high, ſeldome diuided
into branches : at each joynt ſtand two ſmal greeniſh leaues, little, hairy, and not diuided or ſhipt
about the edges, and much like thoſe of the next
deſcribed, as you ſee them expreſt in the figure :
the little hollow and ſomewhat hooded floures
grow in roundles towards the tops of the ſtalkes,
as in the firſt deſcribed, and they are of a blewiſh
violet colour. The ſeeds I haue not yet obſerued :
the root is fibrous and wooddy, and laſts for many yeares. The whole plant hath a pretty pleaſing
but weake ſmell. It floures in Iuly and Auguſt. I
firſt obſerued it *Anno* 1626, a little on this ſide
Pomfret in Yorkſhire, and ſince by Datford in
Kent, and in the Iſle of Tenet. I haue ſometimes
ſeene it brought to Cheapſide market, where the
herbe women called it Poley mountaine, ſome it
may bee that haue taken it for *Polium montanum*
miſinforming them ; *Cluſius* firſt tooke notice of
this plant, and called it *Acinos Anglicum*, finding it
growing in Kent, *Anno*, 1581. and he thinkes it to
be

be the *Acinos* of *Dioscorides*: now the vertues attributed by *Dioscorides* to his *Acinos* are set downe at the end of the chapter, vnder the letter B.

5　This which *Clusius* hath also set forth by the name of *Clinopodium*, or *Acinos Austriacum*, differeth not much from the last described, for it hath tender square hard stalks like those of the last described, set also with two leaues at each joint, here and there a little snipt (which is omitted in the figure) the floures grow onely at the tops of the stalkes, and these pretty large and of a violet colour (yet they are somtimes found white:) they hang commonly forward, and as it were with their vpper parts turned downe. The seed vessels are like those of the first described, and containe each of them foure little black seeds: This floures in May, and the seed is ripe in Iune : It growes about the bathes of Badon, and in diuers places of Austria.

6　*Pena* also hath giuen vs knowledge of another, that from a fibrous root sends vp many quadrangular rough branches, of the height of the two former, set also with two leaues at each joint, and these rough and lightly snipt about the edges; the floures grow thick together at the tops of the stalks of a dark red colour, and in shape like those of the mountain Calamint. It floures in the beginning of Iuly, and growes vpon mount Baldus in Italy ; *Pena* sets it forth by the name of *Clinopodium Alpinum.*

7　To these I thinke fit to adde another, whose description was sent me by M[r]. *Goodyer*, and I question whether it may not be the plant which *Fabius Columna Phytobasani, pag.23.* sets forth by the name of *Acinos Dioscoridis*, for he makes his to be endured *odore fragrantissimo* : but to the purpose.

Acinos odoratissimum.

This herbe hath foure, fiue, or more, foure square hard wooddy stalks growing from one root, diuided into many branches, couered with a soft white hoarinesse, two or three foot long, or longer, not growing vpright, but trailing vpon the ground ; the leaues grow on little short footstalkes by couples of a light greene colour, somewhat like the leaues of Basill, very like the leaues of *Acinos Lobelij*, but smaller, about three quarters of an inch broad, and not fully an inch long, somwhat sharp pointed, lightly notched about the edges, also couered with a light soft hoary hairinesse, of a very sweet smell, little inferiour to garden Marjerome, of a hot biting tast: out of their bosomes grow other smaller leaues, or else branches ; the floures also grow forth of the bosomes of the leaues toward the tops of the stalkes and branches, not in whorles like the said *Acinos*, but hauing one little short footstalke growing forth of the bosome of each leafe, on which is placed three, foure, or more small floures, gaping open, and diuided into foure vnequall parts at the top, like the floures of Basil, and very neer of the likenes and bignes of the flours of garden Marjerome, but of a pale blewish colour tending towards a purple. The seed I neuer obserued by reason it floured late. This plant I first found growing in the garden of M[r]. *Wil. Yalden* in Sheete neer Petersfield in Hampshire, *Anno* 1620. among sweet Marjerome, and which by chance they bought with the seeds therof. It is to be considered whether the seeds of sweet Marjerome degenerate and send forth this herbe or not.

11. October, 1621. *Iohn Goodyer.* ‡

¶ *The Place.*

The wild kinds do grow vpon grauelly grounds by waters sides, and especially I found the three last in the barren plaine by an house in Kent, two miles from Dartford, called Saint Iones in a village called Sutton ; and *Clinopodium vulgare* groweth in great plentie vpon Longfield downes in Kent. ‡ One of the three last of our Authors description is omitted, as you may find noted at the end of the chapter, yet I canot be perswaded that euer he found any of the foure he described euer wilde in this kingdome, vnlesse the second, which growes plentifully in Autumne almost by euery hedge : also the fourth being of my description growes neere Dartford, and in many such dry barren places in sundry parts of this kingdome. ‡

¶ *The Time.*

These herbes floure in Iune and Iuly.

¶ *The Names.*

Vnprofitable Basil, or wild Basill is called by some *Clinopodium.*

¶ *The Temperature.*

The seed of these herbes are of complexion hot and dry.

¶ *The Vertues.*

Wild Basill pound with wine appeaseth the pain of the eies, and the juice mundifieth the same,　A and putteth away all obscurity and dimnesse, all catarrhes and flowing humors that fall into the eies, being often dropped into the same.

B　†　The ſtone Baſill howſoeuer it be taken ſtoppeth the laske, and courſes, and outwardly applied it helps hot tumors and inflammations.
　　‡　Theſe plants are good for all ſuch effects as require moderate heate and aſtriction. ‡

† The figure that was formerly in the third place of this chapter was of the *Calamentha Ocymoides of Tabernamontanus*, and it was deſcribed by our Author in
the fourth place of the next chapter ſaue one, and there you ſhall finde it: the deſcription ſeemes to be of the *Ocymoides repens Polygonifolia* of the *Aduerſaria*, formerly deſcribed by me in the fifth place of the 128. chapter of this booke; if that the place and floures in the omitted deſcription of our Author did not ſeeme to
vary; howeuer I iudge it the ſame, and therefore haue here excluded it.

Cʜᴀᴘ. 224. *Of Baſill Valerian.*

¶ *The Deſcription.*

1　THe firſt kinde of *Ocymaſtrum*, called of *Dodonæus*, *Valerianarubra*, bringeth forth long
　　and brittle ſtalks two cubits high, full of knots or joints, in which place is joined long
　　leaues much like vnto great Baſill, but greater, broader, and larger, or rather like the
leaues of the Woad, at the top of the ſtalks grow very pleaſant and long red floures, of the faſhion
of the floures of Valerian, which hath cauſed *Dodonæus* to call this plant red Valerian; which being
paſt, the ſeeds are carried away with the winde, being few in number, and little in quantity, ſo that
without great diligence the ſeed is not to be gathered or preſerued: for my ſelfe haue often indeauored to ſee it, and yet haue loſt my labour. The root is very thick, and of an excellent ſweet ſauor.

　　　1　*Valerianarubra Dodonai.*　　　　　2　*Behen album.*
　　　　　Red Valerian.　　　　　　　　　　Spatling poppy.

2　The ſecond is taken for *Spumeum papauer*, in reſpect of that kind of frothy ſpattle, or ſpume,
which we call Cuckow ſpittle, that more aboundeth in the boſomes of the leaues of theſe plants,
than in any other plant that is knowne : for which cauſe *Pena* calleth it *Papauer ſpumeum*, that
is, frothy, or ſpatling Poppy : his floure doth very little reſemble any kinde of Poppey, but onely
the ſeede and cod, or bowle wherein the ſeed is contained, otherwiſe it is like the other *Ocyma-*
　　ſtrum;

ſtrum : the floures grow at the top of the ſtalks, hanging downwards, of a white colour, and it is taken generally for *Behen album :* the root is white, plain, and long, and very tough and hard to break.

¶ *The Place.*

The firſt groweth plentifully in my garden, being a great ornament to the ſame, and not common in England.

The ſecond groweth almoſt in euery paſture.

¶ *The Time.*

Theſe plants floure from May to the end of Auguſt.

¶ *The Names.*

Red Valerian hath bin ſo called of the likeneſſe of the floures and ſpoked rundles with Valerian; by which name we had rather haue it called, than raſhly to lay vpon it an vnproper name. There are ſome alſo who would haue it to be a kinde of *Behen* of the later Herbariſts, naming the ſame *Behen rubram,* for difference between it and the other *Behen album,* that of ſome is called *Ocymaſtrum,* and *Papauer ſpumeum,* which I haue Engliſhed Spatling Poppy; and is in truth another plant much differing from *Behen* of the Arabians : it is alſo called *Valerianthon, Saponaria altera, Struthium Aldroandi,* and *Condurdum :* in Engliſh, red Valerian, and red Cow Baſill.

Spatling Poppy is called *Behen album, Ocymaſtrum alterum ;* of ſome, *Polemonium,* and *Papauer ſpumeum :* in Engliſh, Spatling Poppy, frothy Poppy, and white Ben.

¶ *The Temperature.*

Theſe plants are dry in the ſecond degree.

¶ *The Vertues.*

The root of *Behen album* drunke in wine is good againſt the bloudy flix; and being pound leaues **A** and flours, and laid to, cureth the ſtingings of Scorpions and ſuch like venomous beaſts; inſomuch that whoſo doth hold the ſame in his hand, can receiue no damage or hurt by any venomous beaſt.

The decoction of the root made in water and drunke, prouoketh vrin, helpeth the ſtrangury and **B** paines about the necke and huckle bone.

† That which was formerly here ſet forth in the third place by the name of *Ocymaſtrum multiflorum,* is nothing elſe but the *Lychnis ſylueſtris alba multiplex,* which I haue deſcribed amongſt the reſt of the ſame kinde in the 128 Chapter of this booke.

Chap. 225. *Of Mints.*

¶ *The Kindes.*

THere be diuers ſorts of Mints, ſome of the garden, others wilde or of the field; and alſo ſome of the water.

¶ *The Deſcription.*

1   THe firſt tame or garden Mint commeth vp with ſtalks foure ſquare, of an obſcure red colour ſomewhat hairy, which are couered with round leaues nicked in the edges like a Saw, of a deep green colour: the floures are little and red, and grow about the ſtalkes circle-wiſe as thoſe of Penny-Royall : the root creepeth aſlope in the ground, hauing ſome ſtrings on it, and now and then in ſundry places it buddeth out afreſh: the whole herb is of a pleaſant ſmel, and it rather lieth downe than ſtandeth vp.

2   The ſecond is like to the firſt in hairy ſtalks ſomthing round, in blackiſh leaues, in creeping roots, and alſo in ſmell, but the floures do not at all compaſſe the ſtalke about, but ſtand vp in the tops of the branches, being orderly placed in little eares, or rather catkins or aglets.

3   The leaues of Speare-Mint are long like thoſe of the Willow tree, but whiter, ſofter, and more hairy : the floures are orderly placed in the tops of the ſtalks, and in eares like thoſe of the ſecond. The root hereof doth alſo creepe no otherwiſe than doth that of the firſt, vnto which it is like.

4   There is another ſort of Mint which hath long leaues like to the third in ſtalks, yet in leaues and roots leſſer; but the flours hereof ſtand not in the tops of the branches, but compaſſe the ſtalks about circle-wiſe as do thoſe of the firſt, which be of a light purple colour.

‡ 5   This hath round leaues broader than the common Mint, rounder alſo, and as criſp or curled as thoſe deſcribed in the ſecond place (of which it ſeemes but a larger varietie:) the ſtalkes are

† 1 *Mentha ſativa rubra.*
Red garden Mints.

† 2 *Mentha cruciata, ſiue criſpa.*
Croſſe Mint or curled Mint.

† 3 *Mentha Romana.*
Speare Mint.

‡ 4 *Mentha Cardiaca.*
Heart Mint.

foure ſquare, and the floures grow in eares or ſpoky tufts like thoſe of the ſecond.   ‡

‡ 5 *Mentha ſpicata altera.*
Balſam Mint.

¶ *The Place.*

Moſt vſe to ſet Mints in gardens almoſt euery where.

¶ *The Time.*

Mints doe floure and flouriſh in Summer : in winter the roots only remain: being once ſet, they continue long, and remaine ſure and faſt in the ground.

¶ *The Names.*

Mint is called in Greeke ἡδύοσμος, and μίνθη : The ſweet ſmell, ſaith *Pliny, lib.* 19. *ca.* 8. hath changed the name among the Grecians, whenas otherwiſe it ſhould be called *Mintha*, from whence our old Writers haue deriued the name : for μίνν ſignifieth ſweet, and ὀσμή ſmell : the Apothecaries, Italians, and French men doe keepe the Latine name *Mentha* : the Spaniards do call it *Terua buena*, and *Ortelana* : in high-Dutch, 𝕸𝖚𝖓𝖙𝖟 : in low-Dutch, 𝕸𝖚𝖓𝖙𝖊 : in Engliſh, Mint.

The firſt Mint is called in high-Dutch, 𝕯𝖎𝖊𝖒𝖊𝖓𝖙 : in low-Dutch, 𝕭𝖗𝖚𝖞𝖓 𝖍𝖊𝖕𝖑𝖎𝖌𝖍𝖊 : he that would tranſlate it into Latine muſt call it *Sacra nigricans*, or the holy blackiſh Mint : in Engliſh, browne Mint, or red Mint.

The ſecond is alſo called in high Dutch, 𝕶𝖗𝖆𝖚𝖘 𝖇𝖎𝖊𝖒𝖊𝖓𝖙, 𝕶𝖗𝖆𝖚𝖘 𝖒𝖚𝖓𝖙𝖟, and 𝕶𝖗𝖆𝖚𝖘 𝕭𝖆𝖑𝖘𝖆𝖒 ; that is to ſay, *Mentha cruciata* : in French, *Beaume criſpu* : in Engliſh, Croſſe Mint, or curled Mint.

The third is called of diuers, *Mentha Sarracenica*, *Mentha Romana* : it is called in high-Dutch, 𝕭𝖆𝖑𝖘𝖆𝖒 𝖒𝖚𝖓𝖙𝖟, 𝕺𝖓𝖘𝖊𝖗 𝖋𝖗𝖆𝖜𝖊𝖓 𝖒𝖚𝖓𝖙𝖟, 𝕾𝖕𝖎𝖙𝖟𝖊𝖗 𝖒𝖚𝖓𝖙𝖟, 𝕾𝖕𝖎𝖙𝖟𝖊𝖗 𝖇𝖆𝖑𝖘𝖆𝖒 : it may be called *Mentha anguſtifolia*, that is to ſay, Mint with the narrow leaſe ; and in Engliſh, Speare-Mint, Common garden Mint, our Ladies Mint browne Mint, and Macrel Mint.

The fourth is called in high-Dutch, 𝕳𝖊𝖗𝖙𝖟𝖐𝖗𝖆𝖚𝖙, as though it were to be named *Cardiaca*, or *Cardiaca Mentha* : in Engliſh, Heart-wort, and Heart-Mint. ‡ This is the *Siſymbrium ſatiuum* of *Matthiolus*, and *Mentha hortenſis alter* of *Geſner* : the Italians call it *Siſembrio domeſtico*, and *Balſamita*; the Germanes, 𝕶𝖆𝖙𝖊𝖓𝖇𝖆𝖑𝖘𝖆𝖒. ‡

¶ *The Temperature.*

Mint is hot and dry in the third degree. It is, ſaith *Galen*, ſomwhat bitter and harſh, and is inferior to Calamint. The ſmell of Mint, ſaith *Pliny*, doth ſtir vp the minde, and the taſte to a greedy deſire of meat.

¶ *The Vertues.*

Mint is maruellous wholeſome for the ſtomacke, it ſtayeth the Hicket, parbreaking, vomiting, **A** and ſcouring in the Cholericke paſſion, if it be taken with the juice of a ſoure pomegranat.

It ſtoppeth the caſting vp of bloud, being giuen with water and vineger, as *Galen* teacheth. **B**

In broth, ſaith *Pliny*, it ſtayeth the flours, and is ſingular good againſt the Whites, namely that **C** Mint which is deſcribed in the firſt place : for it is found by experience, that many haue had this kinde of flux ſtayed by the continuall vſe of this only Mint. The ſame beeing applied to the forehead or temples, taketh away the head-ache, as *Pliny* teacheth.

It is good againſt watering eies, and all maner of breakings out in the head, as alſo for childrens **D** ſore heads, and againſt the infirmities of the fundament.

It is poured into the eares with honied water. It is taken inwardly againſt Scolopendres, Beare-**E** wormes, ſea Scorpions, and ſerpents.

It is applied with ſalt to the bitings of mad dogs. It will not ſuffer milke to cruddle in the ſto-**F** macke (*Pliny* addeth, to wax ſoure) therefore it is put in milke that is drunke, leſt thoſe that drinke thereof ſhould be ſtrangled.

It is thought, that by the ſame vertue it is an enemy to generation, by ouerthickning the ſeed. **G**

Dioſcorides teacheth, That being applied to the ſecret part of a woman before the act, it hindreth **H** conception.

I   Garden Mint taken in meat or drinke warmeth and ſtrengthneth the ſtomacke, and drieth vp all ſuperfluous humors gathered in the ſame, and cauſeth good digeſtion.

K   Mints mingled with the floure of parched barly conſume tumors and hard ſwellings.

L   The water Mint is of like operation in diuers medicines, it cureth the trenching and griping paines of the belly and bowels ; it appeaſeth head-ache, ſtayeth yexing and vomiting.

M   It is ſingular againſt the grauel and ſtone in the kidnies, and againſt the ſtrangurie, being boiled in wine and drunke.

N   It is laid to the ſtinging of waſps and bees with good ſucceſſe.

† The figures which were formerly in this Chapter were no way agreeable to the deſcription and names taken forth of *Dodonæus*. The firſt was of the *Calamintha montana vulgaris* of *Lobel* and *Tab*. The ſecond was of that which is deſcribed in the third place : the third was of the *Mentha Cattaria anguſtifolia* deſcribed in the third place of the next Chapter. The figure agreeing to the fourth deſcription was in the Chapter next ſaue one afore, by the title of *Ocymoides repens*.

Chap. 226. *Of Nep or Cat-Mint.*

¶ *The Deſcription.*

1   Cat-Mint or Nep growes high, it brings forth ſtalks aboue a cubit long, couered, chamfered, and full of branches : the leaues are broad, nickt in the edges like thoſe of Bawm or Hore-hound, but longer. The floures are of a whitiſh colour, they partly compaſſe about the vppermoſt ſprigs, and partly grow on the very top, ſet in manner of an eare or catkin: the root is diuerſly parted, and endureth a long time: the whole herb together with the leaues & ſtalks are ſoft, and couered with a white down, but leſſer than Horſe-mint : it is of a ſharp ſmel, and pierceth into the head: it hath a hot taſte with a certain bitterneſſe.

‡ 2   Our Author figured this, and deſcribed the next in the ſecond place of this Chap. This hath pretty large ſquare ſtalkes, ſet at each joint with two leaues like thoſe of Coſtmary, but of a gray or ouerworne colour : the floures grow at the tops of the ſtalks in long ſpoky tufts like thoſe of the laſt deſcribed, and of a whitiſh colour ; the ſmell is pleaſanter than that of the laſt deſcribed. ‡

1 *Mentha Felina, ſeu Cattaria.*   2 *Mentha Cattaria altera.*
Nep or Cat-mint.   Great Cat-mint.

3   There is also another kind hereof that hath a longer and narrower leafe, and not of so white a colour: the stalks hereof are foure square, the floures be more plentifull, of a red light purple colour inclining to blew, sprinkled with little fine purple specks: the smell hereof is stronger, but the taste is more biting. ‡ The figure of this was formerly in the third place of the last chapter. ‡

† 3 *Calentha Cattaria angustifolia.*
Small Cat-mint.

¶ *The Place.*

The first groweth about the borders of gardens and fields, neere to rough banks, ditches, and common wayes: it is delighted with moist and watery places, and is brought into gardens.

¶ *The Time.*

The Cat-mints flourish by and by after the Spring: they floure in Iuly and August.

¶ *The Names.*

The later Herbarists doe call it *Herba Cattaria*, & *Herba Catti*, because cats are very much delighted herewith; for the smell of it is so pleasant vnto them, that they rub themselues vpon it, & wallow or tumble in it, and also feed on the branches and leaues very greedily. It is named of the Apothecaries *Nepeta* (but *Nepeta* properly so called is a Calamint, hauing the smell of Penny-Royall:) in high-Dutch, **Katzen Muntz**: in Low-Dutch, **Catte cruijt**: in Italian, *Cattaria*, or *Herba Gatta*: in Spanish, *Terua Gatera*: in English, Cat-Mint and Nep.

¶ *The Nature.*

Nep is of temperature hot and dry, and hath the faculties of the Calamints.

¶ *The Vertues.*

It is commended against cold paines of the Å head, stomacke, and matrix, and those diseases that grow of flegme, raw humors, and winde. It is a present helpe for them that be bursten inwardly by means of some fall receiued from an high place, and that are very much bruised, if the juice be giuen with wine or meade.

It is vsed in baths and decoctions for women to sit ouer, to bring downe their sicknesse & make B them fruitfull.

‡ It is also good against those diseases for which the ordinarie Mints do serue and are vsed. ‡ C

Chap. 227.

Of Horse-Mint or Water-Mint.

¶ *The Description.*

1   WAter Mint is a kinde of wilde Mint like to the first garden Mint: the leaues thereof are round, the stalkes cornered, both the leaues and stalkes are of a darke red colour: the roots creep far abroad, but euery part is greater, and the herb it selfe is of a stronger smell: the floures in the tops of the branches are gathered together into a round eare, of a purple colour.

†   2   The second kinde of water-Mint in each respect is like the others, sauing that it hath a more odoriferous sauor being lightly touched with the hand; but being touched hard, it is ouer hot to smell vnto: it beareth his floures in sundry tufts or rundles ingirting the stalks in many places, being of a light purple colour; the leaues are also lesse than those of the former, and of an hoary gray colour.

‡   3   This common Horse-Mint hath creeping roots like as the other Mints; from which proceed stalkes partly leaning, and partly growing vpright: The leaues are pretty large, thicke, wrinkled,

† 1 *Mentha aquatica, ſiue Siſymbrium.*
Water Mint.

† 2 *Calamintha aquatica.*
Water Calamint.

‡ 3 *Mentaſtrum.*
Horſe-Mint.

‡ 4 *Mentaſtrum niveum Anglicum.*
Party-coloured Horſe-Mint.

‡ 5 *Mentaſtrum minus.*
Small Horſe-Mint.

‡ 6 *Mentaſtrum montanum* 1.*Cluſij.*
Mountain Horſe-Mint.

‡ 7 *Mentaſtrum tuberoſa radice Cluſij.*
Turnep rooted horſe-Mint.

wrinkled, hoary, and rough both aboue and below, and lightly ſnipt about the edges; the
floures grow in thick compact ears at the tops
of the ſtalkes, and are like thoſe of common
Mint. The whole plant is of a more vnpleaſant
ſent than any of the other Mints: It groweth in
diuers wet & moiſt grounds, and flours in Iune
and Iuly. This by moſt writers is called onely
*Mentaſtrum,*without any other attribute.

4   In ſome of our Engliſh gardens (as *Pena*
and *Lobel* obſerued) grows another Horſe Mint
much leſſe, and better ſmelling than the laſt
mentioned, hauing the leaues partly green, and
partly milke white; yet ſometimes the leaues
are ſome of them wholly white, but more and
more commonly all green : the ſtalkes, floures,
and other parts are like thoſe of the former, but
leſſe. This is the *Mentaſtrum niueum Anglicum*
of *Lobel;* and *Mentaſtrum alterum* of *Dodonæus.*

5   This growes in waterie places, hauing a
ſtalk of a cubit or cubit and halfe high, ſet with
longiſh hoary leaues like thoſe of horſe mint :
the flours grow in ſpoky tufts at the top of the
ſtalks, of a dusky purple colour, & in ſhape like
thoſe of the common Mint. The ſmell of this
comes neere to that of the water Mint. This is
the *Mentaſtrifolia aquatica hirſuta,ſiue Calamintha*
3.Dioſcoridis of *Lobel.* In the *Hiſt.Lugd.*it is called *Mentaſtrum minus ſpicatum.*

6 The

6   The stalke of this is some cubit and halfe high, square and full of pith: the leaues are like in shape to those of Cat-Mint, but not hoary, but rather green : the tops of the branches are set with rundles of such white floures as those of the Cats-mint : the smell of this plant is like to that of the horse-Mint ; whence *Clusius* calls it *Mentastrum montanum primum*. It floures in August, and growes in the mountainous places in Austria.

7   The same Author hath also set forth another, by the name of *Mentastrum tuberosa radice*. It hath roughish stalkes like the former, any longish crumpled leaues somewhat sniht about the edges like those of the last described : the floures grow in rundles alongst the tops of the branches, white of colour, and like those of Cat-Mint. The root of this (which, as also the leaues, is not well exprest in the figure) is like a Radish, and blackish on the outside, sending forth many suckers like to little turneps, and also diuers fibres ; these suckers taken from the main root will also take root and grow. It floures in Iune. *Clusius* receiued the seed of it from Spaine. ‡

¶ *The Place.*

They grow in moist and waterie places , as in medowes neere vnto ditches that haue water in them, and by riuers.

¶ *The Time.*

They floure when the other Mints do, and reuiue in the Spring.

¶ *The Names.*

It is called in Greeke ξισνμβριον : in Latine *Sisymbrium* ; in high-Dutch **Rofsmuntz**, and **wasser-muntz** : in French, *Menthe sauvage* : in English, Water-Mint, Fish-Mint, Brooke-Mint, and Horse-Mint.

¶ *The Temperature.*

Water Mint is hot and dry as is the garden Mint, and is of a stronger smell and operation.

¶ *The Vertues.*

A   It is commended to haue the like vertues that the garden Mint hath, & also to be good against the stinging of bees and wasps, if the place be rubbed therewith.

B   The sauor or smell of the water Mint reioyceth the heart of man , for which cause they vse to strew it in chambers and places of recreation, pleasure, and repose, and where feasts and banquets are made.

C   There is no vse hereof in physick whilest we haue with vs the garden Mint, which is sweeter and more agreeing to mans nature.

† The figure that was in the first place was of the horse Mint, and that in the second place should haue been in the first, as now it is.

CHAP. 228.   *Of Mountaine Mint or Calamint.*

¶ *The Description.*

1   Mountaine Calamint is a low herb seldome aboue a foot high, parted into many branches : the stalks are foure square, and full of joints as it were, out of euery one wherof grow forth leaues somewhat round, lesser than those of Basil, couered with a very thin hairy down, as are also the stalks, somewhat whitish, and of a sweet smell : the tops of the branches are notably deckt with floures somwhat of a purple colour ; then groweth the seed which is black : the roots are full of strings and continue.

2   This most excellent kinde of Calamint hath vpright stalks a cubit high, couered ouer with a woolly mossinesse, beset with rough leaues like a Nettle, somewhat notched about the edges : among the leaues come forth blewish or sky-coloured floures : the root is wooddy, and the whole plant is of a very good smell.

3   There is another kinde of Calamint which hath hard square stalkes, couered in like manner as the other: with a certaine hoary or fine cotton : the leaues be in shape like to Basil, but that they are rough ; and the floures grow in rundles toward the tops of the branches, somtimes three or four vpon a stem, of a purplish colour. The root is threddy and long lasting.

† 4   There is a kinde of strong smelling Calamint that hath also square stalks couered with soft cotton, and almost creeping by the ground, hauing euermore two leaues standing one against another, small and soft, not much vnlike the leaues of Penny-Royal, sauing that they are larger and whiter : the floures grow about the stalks like wharles or garlands, of a blewish purple colour. The root is small and threddy : the whole plant hath the smell of Penny-royal, whence it hath the addition of *Pulegij odore*.

1 *Calamintha montana vulgaris.*
Calamint,or mountaine Mint.

† 2 *Calamintha montana præſtantior.*
The more excellent Calaminth.

† 3 *Calamintha vulgaris Officinarum.*
Common Calamint.

† 4 *Calamintha odore Pulegij.*
Field Calamint.

¶ *The Place.*

It delighteth to grow in mountaines, and in the ſhadowy and grauelly ſides thereof: it is found in many places of Italy & France, and in other countries: it is brought into gardens, where it proſpereth maruellous wel, and very eaſily ſoweth it ſelfe. I haue found theſe plants growing vpon the chalkie grounds and highwaies leading from Graueſend to Canterbury, in moſt places, or almoſt euery where. ‡ I haue onely obſerued the third and fourth to grow wilde with vs in England. ‡

¶ *The Time.*

It flouriſheth in Summer, and almoſt all the yeare thorough : it bringeth forth floures and ſeed from Iune to Autumne.

¶ *The Names.*

It is called in Greeke Καλαμίνθη, as though you ſhould ſay, *Elegans aut vtilis Mentha*, a gallant or profitable Mint: the Latines keep the name *Calamintha: Apuleius* alſo nameth it amiſſe, *Mentaſtrum*, and confoundeth the names one with another : the Apothecaries call it *Montana Calamintha*, *Calamentum*, and ſomtime *Calamentum montanum* in French, *Calament*: in Engliſh, Mountaine Calamint. ‡ The fourth is certainly the ſecond Calamint of *Dioſcorides*, and the true *Nepeta* of the Antients. ‡

¶ *The Temperature.*

This Calamint which groweth in mountaines is of a feruent taſt, and biting, hot, and of a thinne ſubſtance, and dry after a ſort in the third degree, as *Galen* ſaith: it digeſteth or waſteth away thinne humors, it cutteth, and maketh thick humors thin.

¶ *The Vertues.*

A   Therefore being inwardly taken by it ſelfe, and alſo with mead or honied water, it manifeſtly heates, prouokes ſweat, and conſumes ſuperfluous humors of the body; it takes away the ſhiuering s of Agues that come by fits.

B   The ſame alſo is performed by the ſallet oile in which it is boiled, if the body be anointed and well rubbed and chafed therewith.

C   The decoction thereof drunke prouoketh vrine, bringeth downe the monethly ſickneſſe, and expelleth the childe, which alſo it doth being but only applied.

D   It helps thoſe that are bruiſed, ſuch as are troubled with cramps and convulſions, and that canot breathe vnleſſe they hold their necks vpright (that haue the wheeſing of the lungs, ſaith *Galen*) and it is a remedy ſaith *Dioſcorides* for the cholerick paſſion, otherwiſe called the Felony.

E   It is good for them that haue the yellow jaundiſe, for that it remoueth the ſtoppings of the liuer and gall, and withal clenſeth: being taken aforehand in wine, it keepes a man from being poiſoned: being inwardly taken, or outwardly applied it cureth them that are bitten of Serpents : being burned or ſtrewed it driues ſerpents away: it takes away black and blew ſpots that come by blowes or dry beatings, making the skinne faire and white ; but for ſuch things (ſaith *Galen*) it is better to be laid to greene than dry.

F   It killeth all manner of wormes of the belly, if it be drunke with ſalt and honey: the juice dropped into the eares doth in like manner kill the wormes thereof.

G   *Pliny* ſaith, that if the juice be conueyed vp into the noſthrils , it ſtancheth the bleeding at the noſe; and the root (which *Dioſcorides* writeth to be good for nothing) helpeth the Squincie, if it be gargariſed, or the throat waſhed therewith, being vſed in Cute, and Myrtle ſeed withall.

H   It is applied to thoſe that haue the Sciatica or ache in the huckle bone, for it drawes the humor from the very bottome, and bringeth a comfortable heat to the whole joint: *Paulus Ægineta* ſaith, that for the paine of the haunches or huckle bones it is to be vſed in Clyſters.

I   Being much eaten it is good for them that haue the leproſie, ſo that the patient drink whay after it, as *Dioſcorides* witneſſeth.

K   *Apuleius* affirmeth, that if the leaues be often eaten , they are a ſure and certaine remedy againſt the leproſie.

L   There is made of this an Antidote or compoſition, which *Galen* in his fourth booke of the Gouernment of health deſcribes by the name of *Diacalaminthos*, that doth not only notably digeſt or waſt away crudities , but alſo is maruellous good for young maidens that want their courſes , if their bodies be firſt well purged ; for in continuance of time it bringeth them downe very gently without force.

† The figure which formerly was in the ſecond place belonged to the fourth deſcription; and the figure that belonged thereto, was before falſly put for the *Scordonia* or Wood-ſage. As alſo that which ſhould haue been put in the fourth place was put in the firſt place of the laſt chapter ſaue two, for the Red Garden Mint.

Chap. 229. Of Bawme.

¶ *The Description.*

1 A Piaſtrum, or *Meliſſa*, is our common beſt knowne Balme or Bawme, hauing many ſquare ſtalkes and blackiſh leaues like to *Ballote*, or blacke Horehound, but larger, of a pleaſant ſmell, drawing neere in ſmell and ſauour vnto a Citron: the floures are of a Carnation colour; the root of a wooddy ſubſtance.

2 The ſecond kinde of Bawme was brought into my garden and others, by his ſeed from the parts of Turky, wherefore we haue called it Turky Balme: it excelleth the reſt of the kindes, if you reſpect the ſweet ſauour and goodly beauty thereof, and deſerueth a more liuely deſcription than my rude pen can deliuer. This rare plant hath ſundry ſmall weake and brittle ſquare ſtalkes and branches, mounting to the height of a cubit and ſomewhat more, beſet with leaues like to Germander or *Scordium*, indented or toothed very bluntly about the edges, but ſomewhat ſharpe pointed at the top. The floures grow in ſmall coronets of a purpliſh blew colour: the root is ſmall and threddy, and dieth at the firſt approch of Winter, and muſt be ſowne anew in the beginning of May, in good and fertill ground.

1 *Meliſſa.*
Bawme.

2 *Meliſſa Turcica.*
Turky Bawme.

3 *Fuchſius* ſetteth forth a kinde of Bawme hauing a ſquare ſtalke, with leaues like vnto common Bawme, but larger and blacker, and of an euill ſauour; the floures white, and much greater than thoſe of the common Bawme; the root hard, and of a wooddy ſubſtance. ‡ This varies with the leaues ſometimes broader, otherwhiles narrower: alſo the floures are commonly purple, yet ſometimes white, and otherwhiles of diuers colours: the leaues are alſo ſometimes broader, otherwhiles narrower: wherefore I haue giuen you one of the figures of *Iaſius*, and that of *Lobel*, that you may ſee the ſeuerall expreſſions of this plant. *Cluſius*, and after him *Bauhine*, referre it to the *Lamium*, or Arch-angell: and the former calls it *Lamium Pannonicum*: and the later, *Lamium montanum Meliſſæfolio*. ‡

4 There is a kinde of Bawme called *Herba Iudaica*, which *Lobel* calls *Tetrahit*, that hath many

‡ 3 *Meliſſa Fuchſij flore albo.*
Baſtard Bawme with white floures.

‡ 3 *Meliſſa Fuchſij flore purpureo.*
Baſtard Bawme with purple floures.

‡ 4 *Herba Iudaica Lobelij.*
Smiths Bawme, or Iewes All-heale.

weake and tender ſquare hairie branches; ſome leaning backward, and others turning inward, diuiding themſelues into ſundry other ſmall armes or twigs, which are beſet with long rough leaues dented about, and ſmaller than the leaues of Sage. And growing in another ſoile or clymat, you ſhall ſee the leaues like the oken leaf; in other places like *Marrubium Creticum*, very hoary, which cauſed *Dioſcorides* to deſcribe it with ſo many ſhapes, and alſo the floures, which are ſometimes blew and purple, and oftentimes white : the root is ſmall and crooked, with ſome hairie ſtrings faſtned thereto. All the whole plant draweth to the ſauor of Balme, called *Meliſſa.* ‡ This might much more fitly haue been put to the reſt of the *Sideritides*, but that our Author had thruſt it as by force into this Chapter. ‡

5   There be alſo two other plants comprehended vnder the kindes of Balme, the one very like vnto the other, although not knowne to many Herbariſts, and haue been of ſome called by the title of *Cardiaca* : the firſt kinde *Pena* calleth *Cardiaca Melica*, or *Molucca Syriaca*, ſo called for that it was firſt brought out of Syria : it groweth three cubits

cubits high, and yeelding many ſhoots from a wooddy root, full of many whitiſh ſtrings; the ſtalkes be round, ſomewhat thicke, and of a reddiſh colour, which are hollow within, with certaine obſcure prints or ſmall furrowes along the ſtalkes, with equall ſpaces halfe kneed or knotted, and at euery ſuch knee or joynt ſtand two leaues one againſt another, tufted like Meliſſa, but more rough and deeply indented, yet not ſo deeply as our common Cardiaca, called Mother-wort, nor ſo ſharpe pointed: about the knees there come forth ſmall little prickles, with ſix or eight ſmall open wide bells, hauing many corners thinne like parchment, and of the ſame colour, ſomewhat ſtiffe and long; and at the top of the edge of the bell it is cornered and pointed with ſharpe prickles; and out of the middle of this prickly bell riſeth a floure ſomewhat purple tending to whiteneſſe, not vnlike our Lamium or Cardiaca, which bringeth forth a cornered ſeed, the bottome flat, and ſmaller toward the top like a ſteeple: the ſauour of the plant draweth toward the ſent of Lamium.

6　　The other kinde of Melica, otherwiſe called Molucca aſperior (whereof Pena writeth) differeth from the laſt before mentioned, in that the cups or bells wherein the floures grow are more prickly than the firſt, and much ſharper, longer, and more in number: the ſtalke of this is foure ſquare, lightly hollowed or furrowed; the ſeed three cornered, ſharpe vpward like a wedge; the tunnels of the floures browniſh, and not ſo white as the firſt.

<table>
<tr><td>5　Meliſſa Molucca lævis.
Smooth Molucca Bawme.</td><td>6　Molucca ſpinoſa.
Thorny Molucca Bawme.</td></tr>
</table>

¶ The Place.

Bawme is much ſowen and ſet in Gardens, and oftentimes it groweth of it ſelfe in Woods and mountaines, and other wilde places: it is profitably planted in Gardens, as Pliny writeth, lib. 21. cap. 12. about places where Bees are kept, becauſe they are delighted with this herbe aboue others, whereupon it hath beene called Apiaſtrum: for, ſaith he, when they are ſtraied away, they doe finde their way home againe by it, as Virgil writeth in his Georgicks:

　　　　　——— Huc tu juſſos aſperge liquores,
　　Trita Meliphylla, & Cerinthe ignobile gramen.

　　——— Here liquors caſt in fitting ſort,
　　Of bruiſed Bawme and more baſe Honywort.

All theſe I haue in my garden from yeare to yeare.
　　　　　　　　　　　　　　　　　　　　　　　　　　¶ The

¶ *The Time.*

Bawme floureth in Iune, Iuly, and Auguſt : it withereth in the Winter ; but the root remaineth, which in the beginning of the Spring bringeth forth freſh leaues and ſtalkes.

The other ſorts do likewiſe flouriſh in Iune, Iuly, and Auguſt ; but they doe periſh when they haue perfected their ſeed.

¶ *The Names.*

Bawme is called in Greeke, μελισσόφυλλον : by *Pliny, Melitis* : in Latine, *Meliſſa, Apiaſtrum,* and *Citrago* : of ſome, *Meliſſophyllon,* and *Meliphyllon* : in Dutch, **Confille de greyn** ; in French, *Poucyrade, ou Meliſſe* : in Italian, *Cedronella,* and *Arantiata* : in Spaniſh, *Torongil* : in Engliſh, Balme, or Bawme.

¶ *The Temperature.*

Bawme is of temperature hot and dry in the ſecond degree, as *Auicen* ſaith : *Galen* ſaith it is like Horehound in faculty.

¶ *The Vertues.*

A   Bawme drunke in wine is good againſt the bitings of venomous beaſts, comforts the heart, and driueth away all melancholy and ſadneſſe.

B   Common Bawme is good for women which haue the ſtrangling of the mother, either being eaten or ſmelled vnto.

C   The juyce thereof glueth together greene wounds, being put into oyle, vnguent, or Balme, for that purpoſe, and maketh it of greater efficacy.

D   The herbe ſtamped, and infuſed in *Aqua vitæ,* may be vſed vnto the purpoſes aforeſaid (I meane the liquour and not the herbe) and is a moſt Cordiall liquor againſt all the diſeaſes before ſpoken of.

E   The hiues of Bees being rubbed with the leaues of Bawme, cauſeth the Bees to keep together, and cauſeth others to come vnto them.

F   The later age, together with the Arabians and Mauritanians, affirme Balme to be ſingular good for the heart, and to be a remedy againſt the infirmities thereof ; for *Auicen* in his booke written of the infirmities of the heart, teacheth that Bawme makes the heart merry and joyfull, and ſtrengtheneth the vitall ſpirits.

G   *Serapio* affirmeth it to be comfortable for a moiſt and cold ſtomacke, to ſtir vp concoction, to open the ſtopping of the braine, and to driue away ſorrow and care of the minde.

H   *Dioſcorides* writeth, That the leaues drunke with wine, or applied outwardly, are good againſt the ſtingings of venomous beaſts, and the biting of mad dogs : alſo it helpeth the tooth-ache, the mouth being waſhed with the decoction, and is likewiſe good for thoſe that cannot take breath vnleſſe they hold their neckes vpright.

I   The leaues being mixed with ſalt (ſaith the ſame Author) helpeth the Kings Euill, or any other hard ſwellings and kernels, and mitigateth the paine of the Gout.

K   Smiths Bawme or Carpenters Bawme is moſt ſingular to heale vp greene wounds that are cut with yron ; it cureth the rupture in ſhort time ; it ſtaieth the whites. *Dioſcorides* and *Pliny* haue attributed like vertues vnto this kinde of Bawme, which they call Iron-wort. The leaues (ſay they) being applied, cloſe vp wounds without any perill of inflammation. *Pliny* ſaith that it is of ſo great vertue, that though it be but tied to his ſword that hath giuen the wound, it ſtancheth the bloud.

Chap. 230.   *Of Horebound.*

¶ *The Deſcription.*

1   WHite Horehound bringeth forth very many ſtalkes foure ſquare, a cubit high, couered ouer with a thin whitiſh downineſſe : whereupon are placed by couples at certaine diſtances, thicke whitiſh leaues ſomwhat round, wrinkled and nicked on the edges, and couered ouer with the like downineſſe ; from the boſomes of which leaues come forth ſmall floures of a faint purpliſh color, ſet round about the ſtalke in round wharles, which turne into ſharpe prickly husks after the floures be paſt. The whole plant is of a ſtrong ſauor, but not vnpleaſant : the root is threddy.

2   The ſecond kinde of Horehound hath ſundry crooked ſlender ſtalkes, diuided into many ſmall branches couered ouer with a white hoarineſſe or cottony downe. The leaues are likewiſe hoary and cottony, longer and narrower than the precedent, lightly indented about the edges, and ſharply pointed like the Turky Bawme, and of the ſame bigneſſe, hauing ſmall wharles of white

floures

1 *Marrubium album.*
White Horehound.

2 *Marrubium candidum.*
Snow white Horehound.

3 *Marrubium Hiſpanicum.*
Spaniſh Horehound.

4 *Marrubium Creticum.*
Candy Horehound.

floures, and prickly rundles or feed-veffels fet about the ftalke by certaine diftances. The root is likewife threddy.

3   Spanifh Horehound hath a ftiffe hoarie and hairy ftalke, diuiding it felfe at the bottome into two or more armes, and likewife toward the top into two others; whereupon are placed by couples at certaine fpaces faire broad leaues, more round than any of the reft, and likewife more woolly and hairy. The floures grow at the top of the ftalkes, fpike fafhion, compofed of fmall gaping floures of a purple colour. The whole plant hath the fauour of Stœchados.

4   Candy Horehound hath a thicke and hard root, with many hairy threds faftned thereunto; from which rife vp immediately rough fquare ftalkes, fet confufedly with long leaues of a hoarie colour, of a moft pleafant ftrong fmell. The floures grow toward the top of the ftalkes in chaffie rundles, of a whitifh colour.

¶ *The Place.*

The firft of thefe Horehounds, being the common kinde, groweth plentifully in all places of England, neere vnto old walls, highwaies, and beaten pathes, in vntilled places. It groweth in all other countries likewife, where it altereth according to the fcituation and nature of the countries; for commonly that which growes in Candy and in Hungary is much whiter, and of a fweeter fmell, and the leaues oftentimes narrower and leffer than that which groweth in England and thefe Northerne Regions.

¶ *The Time.*

They floure in Iuly and Auguft, and that in the fecond yeare after the fowing of them.

¶ *The Names.*

Horehound is called in Greeke, σρίσων: in Latine, *Marrubium* : in fhops, *Praffum*, and alfo *Marrubium*. There be certaine baftard names found in *Apuleius*, as *Melittena*, *Labeonia*, and *Vlceraria* : in Italian, *Marrubio* : in Spanifh, *Marruuio* : in Dutch, 𝕸𝖆𝖑𝖗𝖔𝖚𝖊 : in French, *Marubin* : in Englifh, Horehound. ‡ *Clufius* calls the third *Ocimaftrum Valentinum.* ‡

¶ *The Temperature.*

Horehound (as *Galen* teacheth) is hot in the fecond degree, and dry in the third, and of a bitter tafte.

¶ *The Vertues.*

A   Common Horehound boyled in water and drunke, openeth the liuer and fpleene, cleanfeth the breft and lungs, and preuailes greatly againft an old cough, the paine of the fide, fpitting of bloud, the Ptyficke, and vlcerations of the lungs.

B   The fame boyled in wine and drunke, bringeth downe the termes, expelleth the fecondine, or afterbirth, and dead childe, and alfo eafeth thofe that haue fore and hard labour in childe-bearing.

C   Syrrup made of the greene frefh leaues and fugar, is a moft fingular remedy againft the cough and wheefing of the lungs.

D   The fame fyrrup doth wonderfully and aboue credit eafe fuch as haue lien long ficke of any confumption of the lungs, as hath beene often prooued by the learned Phyfitions of our London Colledge.

E   It is likewife good for them that haue drunke poyfon, or that haue beene bitten of Serpents. The leaues are applied with honey to cleanfe foule and filthy vlcers. It ftaieth and keepeth backe the pearle or web in the eies.

F   The juyce preffed forth of the leaues, and hardned in the Sun, is very good for the fame things, efpecially if it be mixed with a little wine and hony; and dropped into the eies, it helps them, and cleereth the fight.

G   Being drawne vp into the nofthrils it cleanfeth the yellowneffe of the eies, and ftaieth the running and watering of them.

Chap. 231.   *Of wilde Horehound.*

¶ *The Defcription.*

1   WIld Horehound is alfo like to common Horehound : there rifeth from the root hereof a great number of ftalkes high and joynted, and out of euery joynt a couple of leaues oppofite, or fet one againft another, fomewhat hard, a little longer than thofe of common Horehound, and whiter, as alfo the ftalkes are fet with foft haires, and of a fweet fmell : the floures do compaffe the ftalke about as thofe doe of common Horehound, but they are yellow, and the wharles be narrower : the root is wooddy and durable.

1 *Stachys.*
Wilde Horehound.

2 *Stathys Fuchsij.*
Wilde ftinking Horehound.

‡ 3 *Stachys ſpinoſa Cretica.*
Thorny Horehound.

‡ 4 *Stachys Luſitanica.*
Portugall Wilde Horehound.

‡ 5 *Sideritis Scordioides.*
Germander Ironwort.

‡ 6 *Sideritis Alpina Hyſſopifolia.*
Hyſſop leaued Iron-wort.

2 Beſides this there is alſo another deſcribed by *Fuchſius*: the ſtalkes hereof are thicke, foure ſquare, now and then two or three foot long: the leaues be broad, long, hoarie, nicked in the edges, hairie as are alſo the ſtalkes, and much broader than thoſe of the common Horehound: the floures in the whorles which compaſſe the ſtalke about, are of a purple colour; the ſeede is round and blackiſh: the root hard and ſomthing yellow.

‡ 3 This thorny *Stachys* hath leaues before it comes to ſend forth the ſtalke, like thoſe of the leſſer Sage, but more white and hairie, thoſe that grow vpon the ſtalkes are much narrower: the ſtalkes are ſquare ſome foot high: and at the parting of them into branches grow alwaies two leaues one oppoſit againſt another: the tops of the branches end in long ſharpe thornie prickles: the floures grow about the tops of the branches like thoſe of Sage, but of ſomewhat a lighter colour. This groweth naturally in Candy, about a Towne called Larda, where *Honorius Bellus* firſt obſerued it: there it is called *Guidarothymo*, or Aſſes Time, though it agree with Tyme in nothing but the place of growth. *Cluſius* ſets it forth by the name of *Stachys ſpinoſa.*

4 *Lobel* hath giuen vs the figure and firſt deſcription of this by the name of *Stachys Luſitanica*. It hath creeping and downy ſtalkes ſome handfull and halfe high, ſet with little leaues: amongſt which in rundles grow ſmall floures like thoſe of the other wilde Horehounds; the whole plant is of ſomewhat a gratefull ſmell. ‡

5 There is another wilde Horehound of Mountpelier, called *Sideritis Monſpelliaca Scordioides, ſiue Scordij folio*: being that kind of *Sideritis* or wilde Horehound which is like vnto *Scordium*, or water Germander, which groweth to the height of a handfull and a halfe, with many ſmall branches riſing vpright, of a wooddy ſubſtance, hauing the tops and ſpokie coronets of Hyſſop, but the leaues doe reſemble *Dioſcorides* his *Scordium*, ſaue that they be ſomewhat leſſer, ſtiffer, more wrinkled and curled and hairie, than *Tetrahit*, or the Iudaicall herbe: the floures do reſemble thoſe of the common Sauorie, in taſte bitter, and of an aromaticall ſmell.

6 Mountaine *Sideritis* being alſo of the kindes of Horehound, was firſt found by *Valerandus Donraz*, in the mountains of Sauoy, reſembling very well the laſt deſcribed, but the leaues are much narrower, and like thoſe of Hyſſope: the floures grow in ſmall rough rundles or tufts, pale of colour like *Marrubium* or *Tetrahit*; the root long and bending, of a wooddy ſubſtance, and purple colour, bitter in taſte, but not vnpleaſant, whoſe vertue is yet vnknowne.

¶ *The Place.*

Theſe herbes are forreiners, they grow in rough and barren places, notwithſtanding I haue them growing in my garden. ‡ My kinde friend Mr *Buckner* an Apothecary of London, the laſt yeere being

being 1632, found the ſecond of theſe growing wilde in Oxfordſhire in the field joyning to Witney Parke a mile from the Towne. ‡

¶ *The Time.*

They floure in the Summer moneths, and wither towards Winter: the root remaineth aliue a certaine time.

¶ *The Names.*

The former is taken for the right *Stachys*, which is called in Greeke, ʃταχυ: it is knowne in ſhoppes and euery where: we name it in Engliſh yellow Horehound, and wilde Horehound. ‡ *Lobel* calls it *Stachys Lychnites ſpuria flandrorum.* ‡

The other wilde Horehound, ſeeing it hath no name, is to be called *Stachys ſpuria*; for it is not the right, neither is it *Sphacelus* (as moſt haue ſuſpected) of which *Theophraſtus* hath made mention ; it is called in Engliſh, purple Horehound, baſtard wilde Horehound, & *Fuchſius* his wild Horehound. ‡ *Fabius Columna* proues the ſecond to be the *Sideritis Heraclia* of *Dioſcorides* and the Antients. ‡

¶ *The Temperature.*

Theſe herbes are of a biting and bitter taſte, and are hot in the third degree, according to *Galen.* ‡ The *Stachys Fuchſij* and *Sideritides* ſeeme to be hot and dry in the firſt degree. ‡

¶ *The Vertues.*

The decoction of the leaues drunke both draw downe the menſes and the ſecondine, as *Dioſcori-*   A
des teacheth.

‡ 2   This is of ſingular vſe (as moſt of the herbes of this kinde are) to keep wounds from in-   B
flammation, and ſpeedily to heale them vp, as alſo to ſtay all fluxes and defluctions, hauing a drying and moderate aſtrictiue faculty.

Aëtius and *Ægineta* commend the vſe of it in medicines vſed in the cure of the biting of a mad   C
Dog. ‡

---

‡ CHAP. 232.   *Of the Ironworts or All-heales.*

‡ 1 *Sideritis vulgaris.*
Ironwort or All-heale.

‡ 2 *Sideritis Anguſtifolia.*
Narrow leaued All-heale.

¶ *The Kindes.*

‡ THere are many plants that belong to this kindred of the *Sideritides*, or Ironwoorts, and
  ſome of them are already treated of, though in ſeuerall places, and that not very fitly by
our Author; and one of them is alſo ſet forth hereafter by the name of Clownes All-heale : theſe
that are formerly handled, and properly belong to this Chapter, are firſt the *Herba Iudaica Lobelij*,
being in the fourth place of the 226. Chapter. Secondly, the *Stachys Fuchſij* (being the firſt *Sideritis*
of *Dioſcorides*) deſcribed in the ſecond place of the laſt Chapter. Thirdly, the *Sideritis Scordioides*
ſet forth in the fifth place; and fourthly the *Sideritis Alpina Hyſſopifolia* ſet forth in the ſixth place of
the laſt Chapter. Now beſides all theſe, I will in this Chapter giue you the Deſcriptions of ſome
others like to them in face and Vertues, and all of them may be referred to the firſt *Sideritis* of *Dioſcorides* his deſcription.

¶ *The Deſcription.*

1 THis hath ſquare ſtalkes ſome cubit high, rough, and joynted with two leaues at each
  joynt which are wrinkled and hairie, of an indifferent bigneſſe, ſnipt about the edges,
  of a ſtrong ſmell, and of a bitteriſh and ſomewhat hottiſh taſte : almoſt forth of euery
joynt grow branches, ſet with leſſer leaues : the floures which in roundles incompaſſe the tops of
the ſtalkes end in a ſpike, being ſomewhat hooded, whitiſh, well ſmelling, and marked on the inſide
with ſanguine ſpots. The ſeed is rough and blacke, being contained in fiue cornered ſeed veſſels.
The root is hard and wooddy, ſending forth many ſtalkes. This is the *Sideritis prima* of *Fuchſius*,
Cordus, Cluſius, and others ; it hath a very great affinitie with the *Panax Coloni*, or Clownes All-heale
of our Author, and the difference betweene them certainly is very ſmall.

‡ 3 *Sideritis procumbens ramoſa.*   ‡ 4 *Sideritis procumbens non ramoſa.*
   Creeping branched Ironwort.   Not branched Creeping Ironwort.

   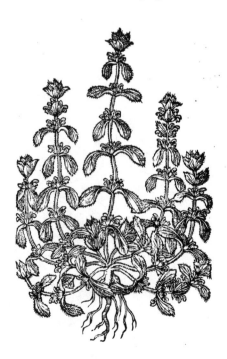

2 The foure ſquare ſtalke of this plant is not aboue a foot high, and it is preſently from the
root diuided into diuers branches ; the leaues are long and narrow with ſome nerues or veines running

‡ 6 *Sideritis latifolia glabra.*
Smooth broad leafed All-heale.

ning alongst them, being also very hairie, but not snipt about the edges : the floures grow alongst the branches, and vpon the main stalke in roundles like those of the first mentioned, but lesser, and of a darke colour, with a yellowish spot on their insides : the seed is also contained in fiue cornered vessels like as the former. It floures in Iune and Iuly, and growes amongst the corne in Hungary and Austria. This is onely set forth by *Clusius,* and that vnder the name of *Sideritis 6. Pannonica.*

3 This hath some branches lying along vpon the ground, slender, quadrangular & hairie, which at certaine spaces are set with leaues growing by couples, almost like those of the first, but much lesse, and snipt onely from the middle to the end : the floures grow after the manner of the former, and (as *Clusius* thinkes) are like them as is also the seed. *Clusius* hath this by the name of *Sideritis 4.*

4 The same Authour hath also giuen vs another, which from the top of the root sends forth many branches, partly lying spred on the ground, and partly standing vpright, being hairy, joynted, and square like those of the former, and such also are the leaues, but that they are lesse snipt about the edges : and in their bosomes from the bottome of the stalkes to the top grow rundles of whitish floures shaped like others of this kinde. *Clusius* calls this *Sideritis 5.* He had onely the figures of these elegantly drawne by the hand of *Iaques Plateau,* and so sent him.

5 This from a small wooddie root sends forth a square hairie stalke some halfe foot high, and sometimes higher, and this stalke most commonly sends forth some foure branches, which subdiuide themselues into smaller ones, all of them sometimes lying vpon the ground, and the stalke standing vpright ; the leaues grow by couples at each joynt, from a broader bottome, ending in an obtuse point, the lower leaues being some inch long, and not much lesse in breadth : the floures are whitish, or light purple, small and hooded, engirting the stalkes in roundles, which falling, foure longish blacke seeds are contained in fiue cornered vessels. I first found it August 1626 in floure and seed amongst the corne in a field joyning to a wood side not far from Greene-hiue in Kent, and I at that time, not finding it to be written of by any, called it *Sideritis humilis lato obtuso folio,* but since I finde that *Bauhine* hath set it forth in his *Prodromus* by the name of *Sideritis Alsine Trissaginis folio.*

6 This (which *Tabernamontanus* calls *Alyssum Germanicum,* and whose figure was formerly giuen with the same title by our Author in the 118 Chapter of the former Edition, with a Description no waies agreeing therewith) growes vp with square stalkes some cubit high, set with pretty large and greene smooth leaues snipt about the edges : the floures grow in roundles at the tops of the branches, being hooded, and of a pale yellow colour. This growes in the Corne fields in some places of Germany and Italy : and it is the *Sideritis 2. of Matthiolus* in *Bauhines* opinion, who calls it *Sideritis aruensis latifolia glabra.*

7 There is another plant that growes frequently in the Corne fields of Kent, and by Purfleet in Essex which may fitly be joyned to these, for *Camerarius* calls it *Sideritis aruensis flore rubro,* and in the *Historia Lugd.* it is named *Tetrahit angustifolium,* and thought to be *Ladanum segetum* of *Pliny,* mentioned *lib.29.cap.8.* and *lib.26.cap.11.* It hath a stalke some foot or better high, set with sharpe pointed longish leaues, hauing two or three nickes on their sides, and growing by couples ; at the top of the branches, and also the maine stalke it selfe, stand in one or two roundles faire red hooded floures : the root is small and fibrous, dying euery yeere when it hath perfected his seed. It floures in Iuly and August. This is also sometimes found with a white floure.

¶ *The Place, Time, &c.*
All these are sufficiently deliuered in the descriptions.

¶ *The Temperature and Vertues.*

A　Theſe plants are dry with little or no heat,and are endued with an aſtrictiue faculty. They conduce much to the healing of greene wounds being beaten and applied, or put in vnguents or plaiſters made for that purpoſe.

B　They are alſo good for thoſe things that are mentioned in the laſt chapter,in B, and C.

C　*Cluſius* ſaith,the firſt and ſecond are vſed in Stiria in fomentations,to bathe the head againſt the paines or aches thereof,as alſo againſt the ſtiffeneſſe and wearineſſe of the limbs or joynts.

D　And the ſame Author affirmes that he hath knowne the decoction vſed with very good ſucceſſe in curing the inflammations and vlcerations of the legs. ‡

CHAP. 233. *Of Water Horehound.*

‡ 1 *Marubium aquaticum.*
Water Horehound.

¶ *The Deſcription.*

1　WAter Horehound is very like to blacke and ſtinking Horehound in ſtalke and floured cups,which are rough,pricking,& compaſſing the ſtalks round about like garlands:the leaues thereof be alſo blacke,but longer,harder,more deeply gaſhed in the edges than thoſe of ſtinking Horehound,yet not hairy at all,but wrinkled:the floures be ſmall and whitiſh:the root is faſtened with many blacke ſtrings.

¶ *The Place.*

It growes in Brooks on the brinks of water ditches and neere vnto motes,for it requireth ſtore of water, and groweth not in dry places.

¶ *The Time.*

It flouriſhes and floures in the Summer moneths, in Iuly and Auguſt.

¶ *The Names.*

It is called *Aquatile*,and *Paluſtre Marrubium:* In Engliſh, water Horehound. *Matthiolus* taketh it to be *Species prima Sideritidis*; or a kinde of Ironwort, which *Dioſcorides* hath deſcribed in the firſt place ; but with this doth better agree that which is called *Herba Iudaica*, or Glidwort ; it much leſſe agreeth with *Sideritis ſecunda*,or the ſecond Ironwort,which opinion alſo hath his fauorers, for it is like in leaſe to none of the Fernes. Some alſo thinke good to call it *Herba Ægyptia*, becauſe they that feine themſelues Ægyptians (ſuch as many times wander like vagabonds from citie to citie in Germanie and other places) do vſe with this herbe to giue themſelues a ſwart colour,ſuch as the Ægyptians and the people of Affricke are of; for the juyce of this herbe doth die euery thing with this kind of colour,which alſo holdeth ſo faſt, as that it cannot be wiped or waſhed away : inſomuch as linnen cloth being died herewith, doth alwaies keepe that colour.

¶ *The Temperature.*

It ſeemeth to be cold,and withall very aſtringent or binding.

¶ *The Vertues.*

There is little vſe of the water Horehound in Phyſicke.

† The figure that heretofore was in the firſt place was of the *Marrubium nigrum* deſcribed in the next chapter ; and the figure and deſcription that were in the ſecond place by the name of *Marrubium aquaticum acutum*,were of theſe ſo much magnified *Panax Coloni* or Clownes Al-heale of our Author,and therefore here omitted to auoid *Tautologie*.

CHAP. 234.　*Of blacke or ſtinking Horehound.*

¶ *The Deſcription.*

1　BLacke Horehound is ſomewhat like the white kinde : the ſtalks be alſo ſquare and hairie ; the leaues ſomewhat larger, of a darke ſwart or blackiſh colour, ſomewhat like the leaues of Nettles, ſnipt about the edges, of an vnpleaſant and ſtinking ſauor : the floures grow about the ſtalks in certaine ſpaces, of a purple colour, in ſhape like thoſe of Arch-Angell or dead Nettle : the root is ſmall and threddy. ‡ I haue found this alſo with white floures.

‡ 2　To this may fitly be referred that plant which ſome haue called *Parietaria, Sideritis,* and *Herba venti,* with the additament of *Monſpelienſium* to each of theſe denominations : *Bauhine,* who I herein follow, calls it *Marrubium nigrum longifolium.* It is thus deſcribed : the root is thicke and very fibrous, ſending vp many ſquare rough ſtalks ſome cubit high, ſet at certain ſpaces with leaues longer and broader than Sage, rough alſo, and ſnipt about the edges, and out of their boſoms come floures hooded and purple of colour, engirting the ſtalkes as in other plants of this kinde. Some haue thought this to be the *Othonna* of the Antients, becauſe the leaues not falling off in Winter, are either eaten by the wormes, or waſted by the iniury of the weather to the very nerues or veines that run by them ; ſo that by this means they are all perforated and eaſily blowne thorow by each blaſt of winde : which cauſed ſome to giue it alſo the name of *Herba venti.* It growes in the Corne fields about Montpelier. ‡

† 1　*Marrubium nigrum.*　　　　　‡ 2　*Marrubium nigrum longifolium.*
Stinking Horehound.　　　　　　　Long leaued Horehound.

¶ *The Place.*
It is found in gardens amongſt pot herbs, and oft times among ſtones and rubbiſh in dry ſoiles.

　　　　　　　　　　　　　　The

¶ *The Time.*

It floureth and flouriſheth when the others do.

¶ *The Names.*

It is called in Greeke Βαλλωτή, and μέλαν πράσιον, as *Pliny* teſtifieth, *lib.27.cap.*8. of ſome, *Marrubia-ſtrum,*or *Marrubium ſpurium,*or baſtard Horehound : in ſhops, *Praſium fœtidum,*and *Ballote :* in Itali-an,*Marrubiaſtro :* in Spaniſh,*Marrauio negro :* in French,*Marrubin noir & putant :* in Engliſh,ſtinking Horehound.

¶ *The Temperature.*

Stinking Horehound is hot and dry,as *Paulus Ægineta* teacheth,of a ſharp and clenſing faculty.

¶ *The Vertues.*

A   Being ſtamped with ſalt and applied, it cureth the biting of a mad dog, againſt which it is of great efficacie,as *Dioſcorides* writeth.

B   The leaues roſted in hot embers do waſte or conſume away hard lumps or knots in or about the fundament.

C   It alſo clenſeth foule and filthy vlcers,as the ſame Author teacheth.

† The figure was of *Lamium album,*or Archangell with the white floure ; and the figure that ſhould haue been here was in the former Chapter.

Cʜᴀᴘ. 235.   Of *Archangell or dead Nettle.*

† 1 *Lamium album.*   2 *Lamium luteum.*
  White Archangell.   Yellow Archangell.

¶ *The Deſcription.*

1   WHite Archangell hath foure ſquare ſtalkes a cubit high, leaning this way and that way,by reaſon of the great weight of his ponderous leaues, which are in ſhape like vnto thoſe of Nettles,nicked round about the edges,yet not ſtinging at all,
                                                                      but

but ſoft and as it were downy : the floures compaſſe the ſtalkes round about at certaine diſta ces, euen as thoſe of Horehound do,whereof doubtleſſe this is a kinde, and not of Nettles, as hath bin generally holden ; which floures are white of colour, faſhioned like to little gaping hoods or helmets : the root is very threddy. ‡ There is alſo a varietie of this hauing red or purple floures. ‡

2   Yellow Archangell hath ſquare ſtalks riſing from a threddy root,ſet with leaues by couples very much cut or hackt about the edges,and ſharp pointed, the vppermoſt whereof are oftentimes of a faire purple colour : the flours grow among the ſaid leaues,of a gold yellow colour,faſhioned like thoſe of the white Archangell,but greater,and wider gaping open.

3   Red Archangell,being called *Vrtica non mordax,*or dead Nettle,hath many leaues ſpred vpon the ground ; among which riſe vp ſtalkes hollow and ſquare, whereupon grow rough leaues of an ouerworne colour, among which come forth purple floures ſet about in round wharles or rundles. The root is ſmall,and periſheth at the firſt approch of winter.

‡ 4 *Lamium Pannonicum, ſiue Galeopſis.*
    Hungary dead Nettle.

4   Dead Nettle of Hungarie hath many large rough leaues very much curled or crumpled like thoſe of the ſtinging Nettle,of a darke green colour,ſnipt about the edges like the teeth of a Saw, ſet vpon a foure ſquare ſtalke by couples ; from the boſome of which leaues come forth the flours cloſe to the ſtalks,of a perfect purple colour,in ſhape like thoſe of the white Archangell , gaping like a dragons mouth,the lower chap whereof is of a bright purple ſpotted with white;which being paſt there followes ſeed incloſed in little round husks,with fine ſharp points ſticking out:the root is thicke,tough,conſiſting of many threds and long ſtrings.

‡   5   To this of *Cluſius* we may fitly refer two others plant;the firſt of which *Tragus* & others call *Vrtica Heraclea* or *Herculea,* and *Cluſius* iudges it to be the true *Galeopſis* of *Dioſcorides,* as *Tragus* alſo thought before him. The root hereof is fibrous and creeping,ſending forth many foure ſquare ſtalks,vpon which at each ioint grow two leaues vpon long ſtalkes very like thoſe of Nettles,but more ſoft and hairy,not ſtinging : the tops of the branches end as it were in a ſpike made of ſeuerall roundles of floures like thoſe of Archangell,but leſſe,and of a purple colour ſpotted with white on their inſides : the ſeeds are contained foure in a veſſell,and are black when they come to be ripe.It growes about hedges in very many places,and floures in Iune and Iuly.

6 This

‡ 6   This hath roots like thoſe of the laſt deſcribed, ſending vp alſo ſquare ſtalks a foot high, ſet at each joint with leaues growing vpon long ſtalkes like thoſe of the ſmall dead Nettle, or rather like thoſe of Ale-hooſe: out of the boſoms of thoſe come three or foure ſtalks carying floures like thoſe of Alehooſe, gaping, without a hood, but with a lip turned vp, which is variegated with blew, white, and purple. This hiſtory *Cluſius* (who did not ſee the plant, but an exact figure thereof in colours) giues vs, and he names it as you finde expreſt in the title. ‡

<table>
<tr><td>‡ 5 Galeopſis vera.
Hedge Nettle.</td><td>6 Lamium Pannonicum 3.Cluſij.
Hungary Nettle with the variegated floure.</td></tr>
</table>

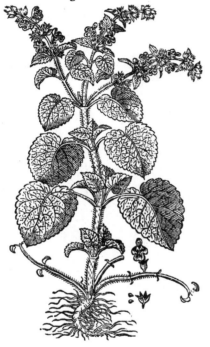 

¶ *The Place.*

Theſe plants are found vnder hedges, old walls, common waies, among rubbiſh, in the borders of fields, and in eatable grounds, oftentimes in gardens ill husbanded.

That with the yellow floure groweth not ſo common as the others. I haue found it vnder the hedge on the left hand as you go from the village of Hampſted neer London to the Church, & in the wood therby, as alſo in many other copſes about Lee in Eſſex, neer Watford & Buſhy in Middleſex, and in the woods belonging to the Lord Cobham in Kent.

¶ *The Time.*

They floure for the moſt part all Summer long, but chiefely in the beginning of May.

¶ *The Names.*

Archangell is called of ſome *Vrtica iners*, and *Mortua* : of ſome, *Lamium* : in Engliſh, Archangell, blinde Nettle, and dead Nettle.

¶ *The Temperature.*

They are hotter and drier than Nettle, approching to the temperature of Horehound.

¶ *The Vertues.*

A   Archangel [or rather the hedge Nettle] ſtamped with vineger, and applied in manner of a pultis, taketh away wens and hard ſwellings, the Kings euill, inflammation of the kernels vnder the eares and jawes, and alſo hot fiery inflammations of the kernels of the neck, arm-holes, and flanks.

B   It is good to bathe thoſe parts with the decoction of it, as *Dioſcorides* and *Pliny* ſay.

C   The later Phyſitions think, that the white floures of Archangel ſtay the whites, and for the ſame purpoſe diuers do make of them a Conſerue, as they call it, of the floures and ſugar, which they appoint to be taken for certaine daies together.

The

The floures are baked with ſugar as Roſes are, which is called Sugar roſet : as alſo the diſtilled **D** water of them, which is vſed to make the heart merry, to make a good colour in the face, and to re-freſh the vitall ſpirits.

† The firſt figure that was formerly in this Chapter was of the *Galeopſis* 1. of *Tabern.* being a kinde of dead Nettle, that hath the leaues ſpotted with white, and ſomewhat ſmaller than the ordinarie one : the figure that ſhould haue been here was in the laſt Chapter ; the third was the ſame with the firſt (that ſhould haue been) differing only in colour of floures, and that which ſhould haue been in the third place was in the fourth.

Chap. 236.   Of *Motherwort*.

Cardiaca.
Mother-wort.

¶ *The Deſcription.*

MOtherwort brings forth ſtalkes foure ſquare thick, hard, two cubits high, of an obſcure or ouerworne red colour : the leaues are ſome-what blacke like thoſe of Nettles, but greater and broader than the leaues of Horehound, deepely in-dented or cut in the edges. The husks are hard and pricking which do compaſſe the ſtalkes about like wharles or little coronets ; out of which doe grow purpliſh flours not vnlike thoſe of dead Nettle, but leſſer : the root is compact of many ſmall ſtrings ; the whole plant is of a very ranke ſmell and bitter taſte.

¶ *The Place.*

It ioyeth among rubbiſh in ſtony and other bar-ren and rough places, eſpecially about Oxford : it profiteth well in gardens.

¶ *The Time.*

It flouriſheth, floureth, and ſeedeth from Iune to September : the leaues and ſtalks periſh in winter, but the root endureth.

¶ *The Names.*

It is called in our age *Cardiaca* : in high-Dutch, **Hertzgeſpor̄t** ; in low-Dutch, **Hertegeſpan** ; in French *Agripaulme* : in Engliſh, Motherwort. Some there be that make it a kinde of Bawme : it ſeemes that it may alſo be referred to *Sideritis Herculana,* or Hercules Ironwort.

¶ *The Temperature.*

Motherwort is hot and dry in the ſecond degree, by reaſon of the clenſing and binding qualitie that it hath.

¶ *The Vertues.*

Diuers commend it againſt the infirmities of the heart : it is iudged to be ſo forcible, that it is **A** thought it tooke his name *Cardiaca* of the effect.

It is alſo reported to cure convulſions, cramps, and palſies, to open the obſtructions or ſtoppings **B** of the intrals, and to kill all kindes of wormes in the belly.

The pouder of the herbe giuen in wine, prouoketh not only vrine and the monethly courſes, but **C** alſo is good for them that are in hard trauell with childe.

Moreouer, the ſame is commended for green wounds ; it is alſo a remedie againſt certain diſea- **D** ſes in cattell, as the cough and murren ; and for that cauſe diuers husbandmen oftentimes much deſire it.

CHAP. 237.   Of stinging Nettle.

¶ The Description.

1 THe stalks of the first be now and then halfe a yard high, round and hollow within : the leaues are broad, sharp pointed, cut round about like a saw, rough on both sides, and co-uered with a stinging down, which with a light touch only causeth a great burning, and raiseth hard knots in the skin like blisters, somtimes making it red. The seed comes from the roots of the leaues in round pellets bigger than pease : it is slipperie, glittering like Lineseed, but yet lesser and rounder : the root is set with strings.

1 *Vrtica Romana.*   2 *Vrtica vrens.*
Romane Nettle.   Common stinging Nettle.

2   The second Nettle being our common Nettle is like to the former in leaues and stalks, but yet now and then higher and more full of branches ; it is also couered with a downe that stingeth and burneth as well as the other : the seed hereof is small, and groweth not in round bullets, but on long slender strings as it were in clusters, as those of the female Mercurie, which grow along the stalks and branches aboue the leaues, very many. The root is full of strings, of colour somthing yel-low, and creepeth all about.   ‡ This hath the stalks and roots somtimes a little yellow, whence *Tabernamontanus* and our Author gaue another figure thereof by the name of *Vrtica rubra,* Red Nettle. ‡

3   The third is like to the second in stalks, leaues, & seed growing by clusters close to the stalks, but lesse, and commonly fuller of branches of a light green, more burning and stinging : the root is small, and not without strings.

3 *Vrtica minor.*
Small Nettle.

¶ *The Place.*

Nettles grow in vntilled places, & the first in thicke woods, and is a stranger in England, notwithstanding it groweth in my garden.

The second is more common, and groweth of it selfe neere hedges, bushes, brambles, and old walls almost euery where.

The third also commeth vp in the same places, which notwithstanding groweth in gardens and most earable grounds.

¶ *The Time.*

They all flourish in Summer: the second suffereth the winters cold : the seed is ripe & may be gathered in Iuly and August.

¶ *The Names.*

It is called in Greeke Ακαλήφη : in Latine, *Vrtica, ab vrendo,* of his burning and stinging qualitie; whereupon *Macer* saith,

—— *nec immerito nomen sumpsisse videtur,*
Tacta quod exurat digitos vrtica tenentis.

Neither without desert his name he seemes to git,
As that which quickly burnes the fingers touching it.

And of diuers also ἀκιὴ ; because it stings with his hurtfull downe : in high-Dutch 𝕹𝖊𝖘𝖘𝖊𝖑 : in Italian, *Ortica :* in Spanish, *Hortiga :* in French, *Ortie :* in English, Nettle. The first is called in low-Dutch 𝕽𝖔𝖔𝖒𝖘𝖈𝖍𝖊 𝕹𝖊𝖙𝖊𝖑𝖊𝖓, that is, *Romana vrtica,* or Roman Nettle : and likewise in high-Dutch, 𝖂𝖆𝖑𝖘𝖈𝖍𝖊 𝕹𝖊𝖘𝖘𝖊𝖑𝖊𝖓, that is, *Italica vrtica,* Italian Nettle, because it is rare, and groweth but in few places, and the seed is sent from other Countries, and sowne in gardens for his vertues : it is also called of diuers *Vrtica mas* ; and of *Dioscorides, Vrtica syluestris,* or wilde Nettle, which he saith is more rough, with broader and longer leaues, and with the seed of flax, but lesser. *Pliny* maketh the wilde Nettle the male, and *Lib. 22. cap. 15.* saith, that it is milder and gentler. It is called in English, Romane Nettle, Greeke Nettle, male Nettle. The second is called *Vrtica fœmina,* and oftentimes *Vrtica maior,* that it may differ from the third Nettle : in English, Female Nettle, great Nettle, or common Nettle. The third is named in high-Dutch, 𝕳𝖊𝖞𝖙𝖊𝖗 𝕹𝖊𝖘𝖘𝖊𝖑 , in the Brabanders speech, 𝕳𝖊𝖎𝖙𝖊 𝕹𝖊𝖙𝖊𝖑𝖊𝖓, so called of the stinging qualitie : in English, Small Nettle, Small burning Nettle : but whether this be that or no which *Pliny* calleth *Cania,* or rather the first, let the Studious consider. There is in the wild Nettle a more stinging qualitie, which, saith he, is called *Cania,* with a stalke more stinging, hauing nicked leaues.

¶ *The Temperature.*

Nettle is of temperature dry, a little hot, scarse in the first degree : it is of thin and subtil parts; for it doth not therefore burne and sting by reason it is extreme hot, but because the downe of it is stiffe and hard, piercing like fine little prickles or stings, and entring into the skin : for if it be withered or boiled it stingeth not at all, by reason that the stiffenesse of the down is fallen away.

¶ *The Vertues.*

Being eaten, as *Dioscorides* saith, boiled with Periwinkles, it maketh the body soluble, doing it **A** by a kinde of clensing facultie : it also prouoketh vrine, and expelleth stones out of the kidnies: being boiled with barly cream it is thought to bring vp tough humors that sticke in the chest.

Being stamped, and the juice put vp into the nosthrils, it stoppeth the bleeding of the nose : the **B** juice is good against the inflammation of the Vvula.

The seed of Nettle stirreth vp lust, especially drunke with Cute : for (saith *Galen*) it hath in it a **C** certaine windinesse.

It concocteth and draweth out of the chest draw humors. **D**

It is good for them that cannot breathe vnlesse they hold their necks vpright, and for those that **E** haue the pleurisie, and for such as be sick of the inflammation of the lungs, if it be taken in a looch

or

ot licking medicine,and alſo againſt the troubleſome cough that children haue, called the Chin-
cough.

F   *Nicander* affirmeth,that it is a remedie againſt the venomous qualitie of Hemlocke, Muſhroms,
and Quickſiluer.

G   And *Apollodorus* ſaith that it is a counterpoiſon for Henbane,Serpents,and Scorpions.

H   *Pliny* ſaith,the ſame Author writeth, that the oile of it takes away the ſting that the Nettle it
ſelfe maketh.

I   The ſame groſſely powned and drunke in White wine, is a moſt ſingular medicine againſt the
ſtone either in the bladder or reins,as hath bin often proued,to the great eaſe and comfort of thoſe
that haue been grieuouſly tormented with that maladie.

K   It expelleth grauell,and prouoketh vrine.

L   The leaues or ſeeds of any kinde of Nettle do work the like effect,but not with that good ſpeed
and ſo aſſured as the Roman Nettle.

Cʜᴀᴘ. 238.   *Of Hempe.*

1 *Cannabis mas.*	‡ 2 *Cannabis fœmina.*
Male or ſteele Hempe.	Femeline or female Hempe.

¶ *The Deſcription.*

1   HEmpe bringeth forth round ſtalks,ſtraight,hollow, fiue or ſix foot high, full of branches when it groweth wilde of it ſelfe ; but when it is ſowne in fields it hath very few
or no branches at all. The leaues thereof be hard,tough,ſomewhat blacke,and if they
be bruiſed they be of a rank ſmell,made vp of diuers little leaues joined together,euery particular
leaſe whereof is narrow, long,ſharpe pointed, and nicked in the edges : the ſeeds come forth from
the bottoms of the wings and leaues,being round,ſomwhat hard, full of white ſubſtance:the roots
haue many ſtrings.

2   There is another,beeing the female Hempe, yet barren and without ſeed,contrarie vnto the
nature

nature of that fex; which is very like to the other being the male, and one muſt be gathered before the other be ripe, elſe it will wither away and come to no good purpoſe.

¶ *The Place.*

Hempe, as *Columella* writeth, delighteth to grow in a fat dunged and watery ſoile, or plaine and moiſt, and deeply digged.

¶ *The Time.*

Hemp is ſowne in March and Aprill ; the firſt is ripe in the end of Auguſt, the other in Iuly.

¶ *The Names.*

This is named of the Grecians κάνναβις : alſo of the Latines *Cannabis* : the Apothecaries keep that name : in high-Dutch **Zamer hanff** : of the Italians *Canape* : of the Spaniards *Canamo* : in French *Chanure* : of the Brabanders **Kemp** : in Engliſh Hemp. The male is called Charle Hemp, and winter Hemp : the female, barren Hemp, and Summer Hemp.

¶ *The Temperature and Vertues.*

The ſeed of Hemp, as *Galen* writeth in his booke of the faculties of ſimple medicines, is hard of **A** digeſtion, hurtful to the ſtomack and head, and containeth in it an ill juice: notwithſtanding ſome do vſe to eat the ſame parched, *cum alijs tragematis*, with other junkets.

The ſame Author in his ſaid booke addeth, that it conſumes winde, and is ſo great a drier, that **B** it drieth vp the ſeed if too much be eaten of it.

Dioſcorides ſaith, That the juice of the herb dropped into the ears aſſwage their pain, proceeding **C** (as *Galen* addeth) of obſtruction or ſtopping.

The inner ſubſtance or pulp of the ſeed preſſed out in ſome kind of liquor, is giuen to thoſe that **D** haue the yellow jaundice, when the diſeaſe firſt appeares, and oftentimes with good ſucceſſe, if the diſeaſe come of obſtruction without an ague ; for it openeth the paſſage of the gall, and diſperſeth and concocteth the choler through the whole body.

Matthiolus ſaith, that the ſeed giuen to hens cauſeth them to lay eggs more plentifully. **E**

Chap. 239. *Of wilde Hempe.*

1 *Cannabis ſpuria.*
Wilde Hempe.

‡ 2 *Cannabis ſpuria altera.*
Baſtard Hemp.

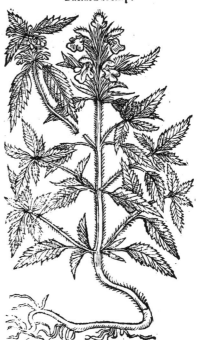

¶ *The*

‡ 3 *Cannabis spuria tertia.*
Small baſtard Hemp.

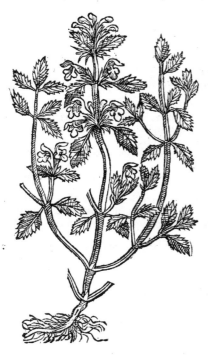

¶ *The Deſcription.*

1   THe wilde Hemp, called *Cannabis ſpu-ria*, or baſtard Hemp, hath ſmal ſlen-der hoary & hairy ſtalks a foot high, beſet at euery joint with two leaues, ſmally in-dented about the edges ſomewhat like a Nettle. The floures grow in rundles about the ſtalkes, of a purple colour, and ſometimes alſo white: The root is little and threddy.

2   There is likewiſe another kinde of wilde Hempe which hath hairy ſtalkes and leaues like the former, but the floures are greater, gaping wide open like the flours of *Lamium* or dead net-tle. The floures are of a cleare and light Carna-tion colour declining to purple.

3   There is alſo another kinde of wild Hemp like vnto the laſt before mentioned, ſauing that it is ſmaller in each reſpect, and not ſo hairy: the leafe is ſomwhat rounder, the root ſmal & thred-die; the floure is larger, beeing purple or white, with a yellow ſpot in the inſide.

¶ *The Place.*

Theſe kindes of wilde or baſtard Hemp grow vpon barren hills and mountaines, eſpecially in earable land, as I haue often ſeene in the corne fields of Kent, as about Graueſend, Southfleet, and in all the tract from thence to Canterburie, and in many places about London.

¶ *The Time.*
Theſe herbs floure from Iuly to the end of Auguſt.

¶ *The Names.*
What hath been ſaid ſhall ſuffice for the Latine names: in Engliſh, wilde Hemp, Nettle hemp, and baſtard Hemp.

¶ *The Nature and Vertues.*
The temperature and faculties are referred to the manured Hemp, notwithſtanding they are not vſed in phyſicke where the other may be had.

CHAP. 240. *Of Water Hempe.*

¶ *The Deſcription.*

1   WAter Hemp or water Agrimonie is ſeldome found in hot regions, for which cauſe it is called *Eupatorium Cannibinum fæmina Septentrionalium*, and groweth in the cold Northern countries in moiſt places, and in the midſt of ponds, ſlow running riuers and ditches. The root continues long, hauing long and ſlender ſtrings after the nature of Water herbs: the ſtalks grow a cubit and a halfe high, of a darke purple colour, with many branches ſtan-ding by diſtances one from another. The leaues are more indented and leſſe hairy than the male kind: the floures grow at the top, of a brown yellow colour, ſpotted with black ſpots like *After At-ticus*; which conſiſts of ſuch a ſubſtance as is in the midſt of the Daiſie or Tanſie floure, & is ſet about with ſmall and ſharpe leaues ſuch as are about the Roſe, which cauſeth the whole floure to reſemble a ſtar, & it ſauoureth like Gum *Elemni, Roſine*, or Cedar wood when it is burned. The ſeed is long like *Pyrethrum*, cloſely thruſt together, and lightly cleaueth to any woollen garment that it toucheth

toucheth by reaſon of his roughneſſe. ‡ This is found with the leaues whole, and alſo with them parted into three parts : the firſt varietie was expreſt by our Authors figure, and the ſecond is expreſt by this we giue you in the place thereof. ‡

2   There is another wilde Hempe growing in the water, whereof there be two ſorts more, delighting to grow in the like ground, in ſhew differing very little. This ſprings vp with long round ſtalks, and ſomewhat reddiſh, about two cubits high, or ſomthing hi her : they are beſet with long green leaues indented about the edges, wherof you ſhal commonly ſee fiue or ſeuen of thoſe leaues hanging vpon one ſtem like the leaues of Hemp, but yet ſofter. The floures are little, of a pale reddiſh colour, conſiſting of ſoft round tufts, and ſtand perting vpon the top of the ſprigges, which at length vaniſh away into downe : the root vnderneath is full of thready ſtrings of a mean bigneſſe.

1 *Eupatorium Cannabinum fœmina.*
  Water Hemp or water Agrimonie.

‡ 2 *Eupatorium Cannabinum mas.*
  Common Dutch Agrimony.

¶ *The Place.*
They grow about the brinks of ditches, running waters, and ſtanding pooles, and in waterie places almoſt euerie where.

¶ *The Time.*
They floure and flouriſh in Iuly and Auguſt, the root continues, but the ſtalks and leaues wither away in Winter.

¶ *The Names.*
The baſtard or wild Hemps, eſpecially thoſe of the water, are commonly called *Hepatorium Cannabinum* : of diuers alſo *Eupatorium*. *Fuchſius* nameth it *Eupatorium adulterinum* : of moſt, *Cannabina*, of the likeneſſe it hath with the leaues of *Cannabis*, Hemp, and *Eupatorium Auicennæ*. It is thought alſo to be that which *Baptiſta Sardus* termeth *TerZola* : in high Dutch, **S. Kunigund kraut** That is to ſay in Latine, *Sanĉta Cunigunda herba*, S Cunigunds herb : in low-Dutch, **Beltzens kruit** : in Engliſh, water Hemp, baſtard and water Agrimonie. It is called *Hepatorium* for that it is good for the liuer.

‡   I haue named the ſecond common Dutch Agrimonie, becauſe it is commonly vſed for Agrimonie in the ſhops of that country. ‡

The

¶ *The Temperature.*

The leaues and roots of theſe herbs are bitter, alſo hot and dry in the ſecond degree : they haue vertue to ſcoure and open, to attenuat or make thin thicke and groſſe humors, and to expel or driue them forth by vrin : they clenſe and purifie the bloud.

¶ *The Vertues, which chiefely belong to the laſt deſcribed.*

A   The decoction hereof is profitably giuen to thoſe that be ſcabbed and haue filthy ſkinnes, and likewiſe to ſuch as haue their ſpleen and liuer ſtopped or ſwolne ; for it takes away the ſtopping of both thoſe intrals, and alſo of the gall : wherefore it is good for them that haue the jaundice, eſpecially ſomewhat after the beginning.

B   The herb boiled in wine or water is ſingular good againſt tertian feuers.

C   The decoction drunke, and the leaues outwardly applied, do heale all wounds both inward and outward.

D   ‡ *Fuchſius* ſaith that the ſecond is very effectuall againſt poiſon. And *Geſner* in his Epiſtles affirmeth, that he boiled about a *pugil* of the fibres of the root of this plant in Wine, and drunke it, which an houre after gaue him one ſtoole, and afterwards twelue vomits, whereby he caſt vp much flegme : ſo that it works (ſaith he) like white Hellebor, but much more eaſily and ſafely, and it did me very much good. ‡

Chap. 241. *Of Egrimonie.*

Agrimonia.
Agrimonie.

¶ *The Deſcription.*

THe leaues of Agrimonie are long and hairie, green aboue, and ſomewhat grayiſh vnderneath, parted into diuers other ſmal leaues ſnipt round about the edges, almoſt like the leaues of Hemp : the ſtalke is two foot and a halfe long, rough and hairie ; whereupon grow many ſmal yellow flours one aboue another vpwards toward the top : after the flours come the ſeeds ſomwhat long and rough, like to ſmall burs hanging downewards ; which when they be ripe doe catch hold vpon peoples garments that paſſe by it. The root is great, long, and blacke.

¶ *The Place.*

It grows in barren places by highwayes, incloſures of medowes, and of corne fields, almoſt euery where, and oftentimes in woods and copſes.

¶ *The Time.*

It floureth in Iune and ſomewhat later, and ſeedeth a great part of Summer after that.

¶ *The Names.*

The Grecians call it ευπατοριον : and the Latines alſo *Eupatorium : Pliny. Eupatoria :* yet there is another *Eupatorium* in *Apuleius*, and that is *Marrubium*, Horehound : in like manner the Apothecaries of Germany haue another *Hepatorium* that is there commonly vſed, being deſcribed in the laſt chapter, and may be named *Hepatorium adulterinum*. Agrimonie is named *Lappa inuerſa*: being ſo called, becauſe the ſeeds which are rough like burres, do hang downewards : of ſome, *Philanthropos*, of the cleauing qualitie of the ſeeds hanging to mens garments : the Italians and Spaniards call it *Agramonia* : in high-Dutch, Odermeng, Bructwurtz : in low-Dutch, in French, and in Engliſh, Agrimonie, and Egrimonie : *Eupatorium* taketh the name of *Eupator* the finder of it out : and (ſaith *Pliny*) it hath a royal and princely authoritie.

¶ *The Temperature.*

It is hot, and doth moderatly binde, and is of a temperat drineſſe. *Galen* ſaith, that Agrimony is of fine and ſubtill parts, cutting and ſcouring : therefore, ſaith he, it remoues obſtructions or ſtoppings of the liuer, and doth likewiſe ſtrengthen it by reaſon of it's binding qualitie.

The

¶ *The Vertues.*

The decoction of the leaues of Egrimony is good for them that haue naughty liuers, and for A
such as pisse bloud vpon the diseases of the kidnies.

The seed being drunke in wine (as *Pliny* affirmeth) doth helpe the bloudy flix. B

Dioscorides addeth, that it is a remedy for them that haue bad liuers, and for such as are bitten C
with serpents.

The leaues being stamped with old swines grease, and applied, close vp vlcers that be hardly hea- D
led, as *Dioscorides* saith.

‡ Agrimony boiled in wine and drunke, helpes inueterate hepaticke fluxes in old people. ‡ E

Chap. 242. *Of Saw-wort.*

1. 2. *Serratula purpurea, sive alba.*
Saw-wort with purple, or white floures.

¶ *The Description.*

1 THe plant which the new Writers
haue called *Serratula* differeth from
Betonica, although the Antients
haue so called Betony; It hath large leaues
somewhat snipt about the edges like a saw
(whereof it took his name) rising immediately
from the root: among which come vp stalkes of
a cubit high, beset with leaues very deeply cut
or jagged euen to the middle of the rib, not
much vnlike the male Scabious. The stalkes
towards the top diuide themselues into other
small branches, at the top whereof they beare
floures somewhat scaly, like the Knapweed, but
not so great nor hard: at the top of the knap
commeth forth a bushie or thrummy floure, of
a purple colour. The root is threddy, and thereby increaseth and becommeth of a great quantitie.

2 Saw-wort with white floures differeth
not from the precedent, but in the colour of
the floures: for as the other bringeth forth a
bush of purple floures; in a manner this plant
bringeth forth floures of the same fashion, but
of a snow white colour, wherein consisteth the
difference.

‡ Our Authour out of *Tabernamontanus*
gaue three figures, with as many descriptions
of this plant, yet made it onely to vary in the
colour of the floures, being either purple, white,
or red; but he did not touch the difference which *Tabernamontanus* by his figures exprest, which
was, the first had all the leaues whole, being onely snipt about the edges; the lower leaues of the second were most of them whole, and those vpon the stalkes deepely cut in, or diuided, and the third
had the leaues both below and aboue all cut in or deeply diuided. The figure which we here giue
you expresses the first and third varieties: and if you please, the one may be with white, and the
other with red or purple floures. ‡

¶ *The Place.*

Sawewort groweth in woods and shadowie places, and sometimes in medowes. They grow in
Hampsted wood: likewise I haue seene it growing in great abundance in the wood adjoyning to
Islington, within halfe a mile from the further end of the towne, and in sundry places of Essex and
Suffolke.

¶ *The Time.*

They floure in Iuly and August.

¶ *The Names.*

The late Writers cal this *Serratula*, and *Serratula tinctoria*, it differeth as we haue said from Betony, which is also called *Serratula*: it is called in English Saw-wort. ‡ *Cæsalpinus* calls it *Cerretta*, and *Serretta*; and *Thalius*, *Centauroides*, or *Centaurium majus syluestre Germanicum*. ‡

¶ *The Temperature and Vertues.*

A   *Serratula* is wonderfully commended to be most singular for wounds, ruptures burstings, and such like: and is referred vnto the temperature of Sanicle.

CHAP. 243.   *Of Betony.*

¶ *The Description.*

1   BEtony groweth vp with long leaues and broad, of a darke greene colour, slightly indented about the edges like a saw. The stalke is slender, foure square, somewhat rough, a foot high more or lesse. It beareth eared floures, of a purplish colour, and sometimes reddish; after the floures, commeth in place long cornered seed. The root consisteth of many strings.

1 *Betonica.*
Betony.

2   Betony with white floures is like the precedent in each respect, sauing that the flours of this plant are white, and of greater beautie, and the others purple or red as aforesaid.

¶ *The Place.*

Betony loues shadowie woods, hedge-rowes, and copses, the borders of pastures, and such like places.

Betony with white floures is seldome seene. I found it in a wood by a Village called Hampstead, neere vnto a Worshipfull Gentlemans house, one of the Clerkes of the Queenes counsell called M^r.*Wade*, from whence I brought plants for my Garden, where they flourish as in their naturall place of growing.

¶ *The Time.*

They floure and flourish for the most part in Iune and Iuly.

¶ *The Names.*

Betony is called in Greeke, κεςον: in Latine, *Betonica*: of diuers, *Vetonica*: but vnproperly. There is likewise another *Betonica*, which *Paulus Ægineta* described; and *Galen* in his first booke of the gouernment of health sheweth that it is called κεςον. that is to say, *Betonica*, Betonie, and also *Sarxiphagon*: *Dioscorides* notwithstanding doth describe another *Sarxiphagon*.

¶ *The Temperature.*

Betony is hot and dry in the second degree: it hath force to cut, as *Galen* saith.

¶ *The Vertues.*

A   Betony is good for them that be subject to the falling sickenesse, and for those also that haue ill heads vpon a cold cause.

B   It cleanseth the lungs and chest, it taketh away obstructions or stoppings of the liuer, milt, and gall: it is good against the yellow jaundise.

C   It maketh a man to haue a good stomacke and appetite to his meate: it preuaileth against sower belchings:

belchings : it maketh a man piſſe well : it mitigateth paine in the kidnies and bladder : it breaketh ſtones in the kidnies,and driueth them forth.

It is alſo good for ruptures, cramps, and convulſions : it is a remedy againſt the biting of mad **D** dogs and venomous ſerpents,being drunk,and alſo applied to the hurts,and is moſt ſingular againſt poyſon.

It is commended againſt the paine of the Sciatica,or ache of the huckle bone.　　　**E**

There is a conſerue made of the floures and ſugar good for many things, and eſpecially for the **F** head-ache. A dram weight of the root of Betony dried,and taken with meade or honied water, procureth vomit,and bringeth forth groſſe and tough humors,as diuers of our age do report.

The pouder of the dried leaues drunke in wine is good for them that ſpit or piſſe bloud, and cu- **G** reth all inward wounds eſpecially the greene leaues boyled in wine and giuen.

The pouder taken with meate looſeth the belly very gently,and helpeth them that haue the fal- **H** ling ſickeneſſe with madneſſe and head-ache.

It is ſingular againſt all paines of the head : it killeth wormes in the belly ; helpeth the ague : it **I** cleanſeth the mother, and hath great vertue to heale the body, being hurt within by bruiſing or ſuch like.

Chap. 244. *Of Water-Betony.*

¶ *The Deſcription.*

WAter Betony hath great ſquare hollow and brown ſtalks,whereon are ſet very broad leaues notched about the edges like vnto thoſe of Nettles, of a ſwart greene colour, growing for the moſt part by two and two as it were from one joynt, oppoſite or ſtanding one right againſt another. The floures grow at the top of the branches, of a darke purple colour,in ſhape like to little helmets. The ſeed is ſmall,contained in round bullets or buttons. The root is compact of many and infinite ſtrings.

Betonica aquatica.
Water Betony.

¶ *The Place.*

It groweth by brookes and running waters, by ditch ſides,and by the brinks of riuers,and is ſeldome found in dry places.

¶ *The Time.*

It floureth in Iuly and Auguſt, and from that time the ſeed waxeth ripe.

¶ *The Names.*

Water Betony is called in Latine *Betonica aquatica.* ſome haue thought it *Dioſcorides* his *Clymenum* · others, his *Galeopſis* : it is *Scrophularia altera* of *Dedonæus* : of *Turner, Clymenon* : of ſome, *Seſamoides minus,* but not properly:of others, *Serpentaria* : in Dutch, **S. Antonies cruyt** : in Engliſh, Water Betony : and by ſome, Browne-wort : in Yorke-ſhire, Biſhops leaues.

¶ *The Temperature.*

Water Betony is hot and dry.

¶ *The Vertues.*

The leaues of Water Betony are of a ſcou- **A** ring or cleanſing quality,and are very good to mundifie foule and ſtinking vlcers, eſpecially the juyce boyled with hony.

It is reported, if the face be waſhed with **B** the juyce thereof, it taketh away the redneſſe and deformity of it.

Chap. 245.
Of Great Figge-wort, or Brownewoort.

¶ The Description.

1   THe great Fig-wort springeth vp with stalkes foure square, two cubits high, of a darke
purple colour, and hollow within: the leaues grow alwaies by couples, as it were from
one ioynt, opposite or standing one right against another, broad, sharpe pointed, snipped round about the edges like the leaues of the greater Nettle, but bigger, blacker, and nothing at
all stinging when they be touched: the floures in the tops of the branches are of a darke purple colour, very like in forme to little helmets: then commeth vp little small seed in pretty round buttons, but sharpe at the end: the root is whitish, beset with little knobs and bunches as it were knots
and kernels.

2   There is another Figge-wort called *Scrophularia Indica*, that hath many and great branches
trailing here and there vpon the ground, full of leaues, in fashion like the wilde or common Thistle, but altogether without prickes: among the leaues appeare the floures in fashion like a hood,
on the outside of a feint colour, and within intermixt with purple; which being fallen and withered, there come in place small knops very hard to breake, and sharpe at the point as a bodkin: which
containeth a small seed like vnto Time. The whole plant perisheth at the first approach of Winter,
and must be sowen againe in Aprill, in good and fertile ground. ‡ This is the *Scrophularia Cretica*
1. of Clusius. ‡

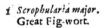

1 *Scrophularia major.*
Great Fig-wort.

‡ 2 *Scrophularia Indica.*
Indian Fig-wort.

‡ 3   The stalke of this is also square, and some yard high, set with leaues like those of the
hedge Nettle, but somewhat larger and thicker, and a little deeper cut in: out of the bosomes of
these leaues come little rough foot-stalkes some inch or two long, carrying some foure or fiue hollow round floures of a greenish yellow colour, with some threds in them, being open at the top,
and cut in with fiue little gashes: the seeds are blacke, and contained in vessels like those of the
first

first described: the root is like that of the Nettle, and liues many yeares : it floures in May, and the seeds are ripe in Iune. I haue not found nor heard of this wilde with vs, but seen it flourishing in the garden of my kinde friend Mr *Iohn Parkinson*. *Clusius* calls it *Lamium* 2. *Pannonicum exoticum :* and *Bauhine* hath set it forth by the name of *Scrophularia flore luteo :* whom in this I follow. ‡

‡ 3 *Scrophularia flore luteo.*
Yellow floured Fig-wort.

¶ *The Place.*

The great *Scrophularia* groweth plentifully in shadowie Woods, and sometimes in moist medowes, especially in greatest abundance in a wood as you goe from London to Herneley, and also in Stow wood and Shotouer neere Oxford.

The strange Indian fig-woort was sent mee from Paris by *Iohn Robin* the Kings Herbarist, and it now groweth in my Garden.

¶ *The Time.*

They floure in Iune and Iuly.

¶ *The Names.*

Fig-wort or Kernel-wort is called in Latine *Scrophularia major*, that it might differ from the lesser Celandine, which is likewise called *Scrophularia*, with this addition *minor*, the lesser : it is called of some *Millemorbia*, and *Castrangula :* in English, great Fig-wort, or Kernel-wort, but most vsually Brown-wort.

¶ *The Vertues.*

Fig-woort is good against the hard kernells A
which the Græcians call κχοινς : the Latines, *Strumas*, and commonly *Scrophulas*, that is, the Kings Euill : and it is reported to be a remedy against those diseases whereof it tooke his name, as also the painefull piles and swelling of the hæmorrhoides.

Diuers do rashly teach, that if it be hanged B
about the necke, or else carried about one, it keepeth a man in health.

Some do stampe the root with butter, and set it in a moist shadowie place fifteene daies toge- C
ther : then they do boyle it, straine it, and keepe it, wherewith they anoint the hard kernels, and the hæmorroide veines, or the piles which are in the fundament, and that with good successe.

CHAP. 246. *Of Veruaine.*

¶ *The Description.*

1   THe stalke of vpright Veruaine riseth from the root single, cornered, a foot high, seldome aboue a cubit, and afterwards diuided into many branches. The leaues are long, greater than those of the Oke, but with bigger cuts and deeper : the floures along the sprigs are little, blew, or white, orderly placed : the root is long, with strings growing on it.

2   Creeping Veruaine sendeth forth stalkes like vnto the former, now and then a cubit long, cornered, more slender, for the most part lying vpon the ground. The leaues are like the former, but with deeper cuts, and more in number. The floures at the tops of the sprigs are blew, and purple withall, very small as those of the last described, and placed after the same manner and order. The root groweth straight downe, being slender and long, as is also the root of the former.

¶ *The*

1 *Verbena communis.*
Common Veruaine.

2 *Verbena ſacra.*
Holy Veruaine

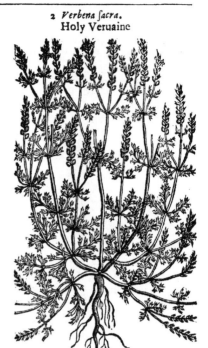

¶ *The Place.*

Both of them grow in vntilled places neere vnto hedges, high-waies, and commonly by ditches almoſt euery where. ‡ I haue not ſeene the ſecond, and doubt it it is not to be found wilde in England. ‡

¶ *The Time.*

The Veruaines floure in Iuly and Auguſt.

¶ *The Names.*

Veruaine is called in Greeke, ꝑερεριον : in Latine, *Verbena,* and *Verbenaca, Herculania, Ferraria,* and *Exupera :* of ſome, *Matricalis,* and *Hiera botane :* of others, *Veruena,* and *Sacra herba : Verbena* are any manner of herbes that were taken from the Altar, or from ſome holy place, which becauſe the Conſull or Pretor did cut vp, they were likewiſe called *Sagmina,* which oftentimes are mentioned in *Liuy* to be graſſie herbes cut vp in the Capitoll. *Pliny* alſo in his two and twentieth booke, and eleuenth Chapter witneſſeth, That *Verbena* and *Sagmina* be all one : and this is manifeſt by that which we reade in *Andræa* in *Terence : Ex ara verbenas hinc ſume :* Take herbes here from the Altar : in which place *Terence* did not meane Veruain to be taken from the Altar, but ſome certain herbes : for in *Menander,* out of whom this Comedy was tranſlated, is read μυρσιν, or Myrtle, as *Donatus* ſaith. In Spaniſh it is called *Vrgebaom :* in Italian, *Verminacula :* in Dutch, **Iſer cruijt :** in French, *Veruaine :* in Engliſh, Iuno's teares, Mercuries moiſt bloud, Holy-herbe ; and of ſome, Pigeons graſſe, or Columbine, becauſe pigeons are delighted to be amongſt it, as alſo to eat thereof, as *Apuleius* writeth.

¶ *The Temperature.*

Both the Veruaines are of temperature very dry, and do meanly binde and coole.

¶ *The Vertues.*

A   The leaues of Veruaine pound with oyle of Roſes or hogs greaſe, doe mitigate and appeaſe the paines of the mother, being applied thereto.

B   The leaues of Veruaine and Roſes ſtamped with a little new hogs greaſe, and emplaiſtered after the manner of a pulteſſe, doe ceaſe the inflammation and grieuous paines of wounds, and ſuffereth them not to come to corruption : and the greene leaues ſtamped with hogs greaſe takes away the ſwelling and paine of hot impoſtumes and tumors, and clenſe corrupt and rotten vlcers.

C   It is reported to be of ſingular force againſt the Tertian and Quartaine Feuers : but you muſt
obſerue.

obſerue mother *Bombiss* rules, to take iuſt ſo many knots or ſprigs, and no more, leſt it fall out ſo that it do you no good, if you catch no harme by it. Many odde old wiues fables are written of Veruaine tending to witchcraft and ſorcery, which you may reade elſewhere, for I am not willing to trouble your eares with reporting ſuch trifles, as honeſt eares abhorre to heare

Archigenes maketh a garland of Veruaine for the head-ache, when the cauſe of the infirmitie proceedeth of heat.   **D**

The herbe ſtamped with oyle of Roſes and Vinegre, or the decoction of it made in oile of roſes, keepeth the haires from falling, being bathed or annointed therewith.   **E**

It is a remedy againſt putrified vlcers, it healeth vp wounds, and perfectly cureth Fiſtulaes, it waſteth away old ſwellings, and taketh away the heat of inflammations.   **F**

The decoction of the roots and leaues ſwageth the tooth-ache, and faſteneth them, and healeth the vlcers of the mouth.   **G**

They report, ſaith *Pliny*, that if the dining roome be ſprinkled with water in which the herbe hath beene ſteeped, the gueſts will be the merrier, which alſo *Dioſcorides* mentioneth.   **H**

Moſt of the later Phyſitions do giue the juice or decoction hereof to them that haue the plague: but theſe men are deceiued, not only in that they looke for ſome truth from the father of falſhood and leaſings, but alſo becauſe in ſtead of a good and ſure remedy they miniſter no remedy at all; for it is reported, that the Diuell did reueale it as a ſecret and diuine medicine.   **I**

CHAP. 247. *Of Scabious.*

† 1 *Scabioſa major vulgaris.*
Common Scabious.

† 2 *Scabioſa minor, ſive Columbaria.*
The ſmall common Scabious.

¶ *The Deſcription.*

1   THe firſt kinde of **Scabious** being the moſt common and beſt knowne, hath leaues long and broad, of a grayiſh, hoary, and hairy colour, ſpred abroad vpon the ground, among which riſe vp round and rough ſtems, beſet with hairy jagged leaues, in faſhion like

great

great Valerian,which we call Setwall. At the top of the ſtalkes grow blew floures in thicke tuſts or buttons. The root is white and ſingle.

   2   The ſecond is like vnto the former, ſauing that his leaues are much cut or jagged, and the whole plant is altogether leſſer, ſcarcely growing to the height of a foot.

   3   The third kinde of Scabious is in all things like vnto the ſecond, ſauing that the knap or head doth not dilate himſelfe ſo abroad,and is not ſo thicke or cloſely thruſt together,and the loweſt leaues are not ſo deepely cut or jagged, but the vpper are much ſmaller, and alſo the more diuided.

   † 4   The fourth groweth with large ſtalkes, hauing two leaues one ſet right againſt another, very much jagged,almoſt like vnto common Ferne,or rather Aſh : and at the top of the ſtalks there grow larger floures, like vnto the firſt, but greater, and the root is alſo like it, and it differs no way from the firſt deſcribed,but onely by reaſon of the ſoile.

 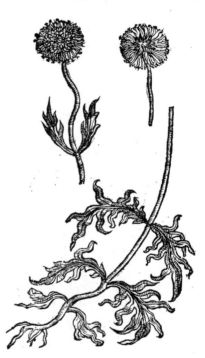

   † 3 *Scabioſa media.*   4 *Scabioſa campeſtris, ſiue ſegetum.*
        Middle Scabious.   Corne Scabious.

   5   Purple floured Scabious hath a rough hairie ſtalke, whereon doe grow broad leaues deepely cut in the edges, in forme like thoſe of Sowthiſtle, rough likewiſe and hairie ; the floures grow at the top of the ſtalkes, compoſed of an innumerable ſort of purple thrums : after which come ſcaly knaps like thoſe of *Iacea*,or Knapweed,wherein is the ſeed. The root is ſmall and threddy.

   ‡ 6   The ſixth ſort of Scabious hath ſtalkes ſome cubit high, round, and ſet with leaues not cut and jagged almoſt to the middle rib, as in the former, yet ſomewhat rough and hairy, ſnipt about the edges,and of a light greene colour;amongſt which riſe vp rough ſtalkes, on the top whereof doe grow faire red floures conſiſting of a bundle of thrummes. The root is long, tough, and fibrous. ‡

   7   The ſeuenth kinde of Scabious hath ſundry great rough and round ſtemmes, as high as a tall man, beſet with leaues like the firſt Scabious, but far greater. The floures grow at the top of the ſtalkes like vnto the others, but of a faint yellow colour, which fall as ſoone as it is touched with the hand,whereby it mightily increaſeth, notwithſtanding the root endureth for many years, and groweth to be wonderfull great : and in my Garden it did grow to the bigneſſe of a mans body.

                                                           ‡ 8 The

5 *Scabiofa flore purpureo.*
Purple floured Scabious.

† 6 *Scabiofa rubra Auftriaca.*
Red Scabious of Auftrich.

† 7 *Scabiofa montana maxima.*
Mountaine Scabious.

‡ 8 *Scabiofa montana alba.*
White mountaine Scabious.

9 *Scabioſa major Hiſpanica.*
Spaniſh Scabious.

10 *Scabioſa peregrina.*
Strange Scabious.

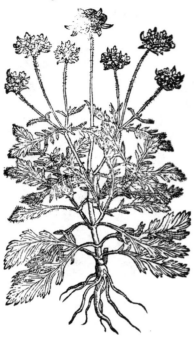

† 11 *Scabioſa omnium minima.*
Sheepes Scabious.

‡ 8 The white mountaine Scabious hath broad leaues ſpred vpon the ground, like thoſe of the field Primroſe, but greater. Amongſt which riſeth vp a great ſtiffe ſtalke ſmooth and plain, garniſhed with leaues not like thoſe next the ground, but leſſer, much more diuided, and of a greener colour & harder. The floures are like thoſe of the common Scabious, but white of colour: the root of this periſhes euery yeare after the perfecting of the ſeed. ‡

9 The ninth kinde of Scabious is like vnto the mountaine Scabious, but lower and ſmaller, hauing ſurdry large and broad leaues next the ground, ſui₁t confuſedly and out of order at the edges like the Oken leaſe; among which riſeth vp a ſtem two cubits high, diuiding it ſelfe into ſundry other branches. The floures are ſet at the top of the naked ſtalkes, of a whitiſh colour; which being paſt, the ſeed appeareth like a tuft of ſmall bucklers, round, and ſomewhat hollow within, and made as it were of parchment, very ſtrange to behold : and within the bucklers there are ſundry ſmall croſſes of blacke faſtened to the bottome, as it were the needle in a diall, running vpon the point of a needle. The plant dieth at the beginning of Winter, and muſt be ſowne in Aprill in good and fertile ground.

10 The tenth is like vnto the laſt before mentioned, in ſtalkes, root, and floures, and differeth that this plant hath leaues altogether without any cuts or jagges about the edges, but is ſmooth and plaine like the leaues of Marigolds, or Diuels bit, and the floures are like vnto thoſe of the laſt deſcribed.

11 Sheeps Scabious hath ſmall and tender branches trailing vpon the ground, whereupon doe grow ſmall leaues very finely jagged or minced euen almoſt to the middle ribbe, of an ouerworne colour. The floures grow at the top of a blewiſh colour, conſiſting of much thrummie matter, hard thruſt together like a button : the root is ſmall, and creepeth in the ground.

12 The

12 *Scabiosa minima hirsuta.*
Hairie Sheepes Scabious.

‡ 13 *Scabiosa minima Bellidis folio.*
Daisie leaued Scabious.

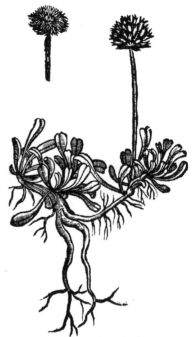

‡ 14 *Scabiosa flore pallido.*
Yellow Scabious.

‡ 15 *Scabiosa prolifera.*
Childing Scabious.

‡ 12   The other Sheeps Scabious of our Author(according to the figure) is greater than the last described, growing some foot or better high, with slender rough branches set with leaues not so much diuided, but onely nicked about the edges: the floures are in colour and shape like those of the last described, or of the blew Daisie; the root is single, and like that of a Rampion, whence *Fabius Columna* (the seed and milkie juyce inducing him) hath refer'd this to the Rampions, calling it *Rapuntium montanum capitatum leptophyllon. Lobel* calls it *Scabiosa media:* and *Dodonæus, Scabiosa minor.*

13   To these little plants we may fitly adde another small one refer'd by *Clusius* to this Classis, and called *Scabiosa.* 10.*siue repens:* yet *Bauhine* refers it to the Daisies, and termes it *Bellis cærulea montana frutescens*; but it matters not to which we referre it: the description is thus; The root is hard, blacke, and creeping, so that it spreds much vpon the surface of the ground, sending forth many thicke, smooth, greene leaues, like those of the blew Daisie, not sharpe pointed, but ending as we vulgarly figure an heart, hauing a certaine grassie but not vnpleasant smell, and somewhat a bitter and hot taste: out of the middest of these leaues grow slender naked stalks some hand high, hauing round floures on their tops, like those of Diuells bit, and of the same colour, yet sometimes of a lighter blew. It growes in the mountaines of Hungary and Austria. It floures in Aprill and May, and ripens the seed in Iuly and August.

‡ 16 *Scabiosa rubra Indica.*
Red Indian Scabious.

‡ 17 *Scabiosa æstivalis Clusij.*
Summer Scabious.

14   This (which is the seuenth Scabious of *Clusius*, and which hee termes ‡χεωιωδις. of the whitish yellow colour of the floure) hath round, slender, stiffe, and greene stalkes set at each joynt with two large and much diuided leaues of a whitish greene colour: those leaues that come from the root before the stalke grow vp are broader, and lesse diuided; vpon the tops of the branches and stalkes grow floures like those of the common Scabious, being white or rather (before they be throughly open) of a whitish yellow colour; which fading, there follow seeds like as in the ordinary kinde. This floures in Iune and Iuly, and growes very plentifully in all the hilly grounds and dry Meades of Austria and Morauia.

15   There is also a kinde of Scabious hauing the leaues much cut and diuided, and the stalkes and floures like to the common sort, of a blewish purple colour, but differing in this, that at
the

the sides of the floure it puts forth little stalkes, bearing smaller floures, as is seene in some other plants, as in Daisies and Marigolds, which therefore are fitly tearmed in Latine *Prolifera* or Childing. This growes only in Gardens, and floures at the same time with the former.

16 The stalks of the red Scabious grow some cubit or more in height, and are diuided into many very slender branches, which at the tops carry floures composed after the manner of the other sorts of Scabious, that is, of many little floures diuided into fiue parts at the top, and these are of a perfect red colour, and haue small threds with pendants at them comming forth of the middle of each of these little floures, which are of a whitish colour, and make a pretty show. The leaues are green, and very much diuided or cut in. The starry seeds grow in long round hairy heads handsomly set together. This is an annuall, and perishes as soone as it hath perfected the seed. *Clusius* maketh it his sixt Scabious, and calls it *Scabiosa Indica*. It floures in Iuly, and growes in the gardens of our prime Herbarists.

17 The same Author hath also giuen vs the figure and description of another Scabious, which sends vp a stalke some three cubits or more high, set at certaine spaces with leaues large, and snipt about their edges, and a little cut in neer their stalks. The stalks are diuided into others, which at their tops carry blewish floures in long scaly heads, which are succeeded by long whitish seed: the root is whitish and fibrous, and dyes euery yeare. This is the *Scabiosa 9. siue æstinalis* of *Clusius*. ‡

¶ *The Place.*

These kindes of Scabious do grow in pastures, medowes, corn fields, and barren sandy grounds almost euery where.

The strange sorts do grow in my garden, yet are they strangers in England.

¶ *The Time.*

They floure and flourish in the Summer moneths.

¶ *The Names.*

Scabious is commonly called *Scabiosa*, diuers think it is named ψωρα, which signifieth a scab, and a certaine herbe so called by *Aëtius* : I do not know, saith *Hermolaus Barbarus*, whether it be Scabious which *Aëtius* doth call *Psora*, the smoke of which being burnt killeth cankers or little wormes. The Author of the Pandects doth interpret *Scabiosa* to be *Dioscorides* his *Stæbe* : *Dioscorides* describeth *Stæbe* by no markes at all, being commonly known in his time ; and *Galen, lib. 1.* of Antidotes saith thus : There is found amongst vs a certaine shrubbie herbe, hot, very sharpe and biting, hauing a little kinde of aromaticall or spicy smell, which the inhabitants call *Colymbada*, and *Stæbe*, singular good to keepe and preserue wine : but it seemeth that this *Stæbe* doth differ from that of which he hath made mention in his book of the faculties of medicines, which agrees with that of *Dioscorides*: for he writeth that this is of a binding quality without biting, so that it cannot be very sharpe.

¶ *The Temperature.*

Scabious is hot and dry in the later end of the second degree, or neere hand in the third, and of thin and subtile parts : it cutteth, attenuateth, or maketh thin, and throughly concocteth tough and grosse humors.

¶ *The Vertues.*

Scabious scoureth the chest and lungs ; it is good against an old cough, shortnesse of breath, **A** paine in the sides, and such like infirmities of the chest.

The same prouoketh vrine, and purgeth now and then rotten matter by the bladder, which hap- **B** peneth when an impostume hath somewhere lien within the body.

It is reported that it cureth scabs, if the decoction thereof be drunk certaine daies, and the juice **C** vsed in ointments.

The later Herbarists do also affirm, that it is a remedy against the bitings of Serpents and stin- **D** gings of venomous beasts, being outwardly applied or inwardly taken.

The juice beeing drunke procureth sweat, especially with Treacle ; and it speedily consumeth **E** plague sores, if it be giuen in time, and forthwith at the beginning : but it must be vsed often.

It is thought to be forceable against all pestilent feuers. **F**

‡ Formerly the 1. 2. 3. 11. figures were all nothing else than the varieties of one plant, being of the 1. 2. 3. 4. *Scabiosa minor* of *Tabern.* they differ only in the more or lesse cutting or diuiding of the leaues : I haue of these only reserued the third, and in other places put such figures as are agreeable to the titles. The figure that was in the sixt place was of the ordinary first described Scabious ; and the figure that should haue been there was in the eighth place ; and that which was in the seuenth place belongs to the plant described by me in the fourteenth place.

CHAP. 248.   Of Diuels bit.

Morſus Diaboli.
Diuels bit.

¶ *The Deſcription.*

Diuels bit hath ſmall vpright round ſtalkes of a cubite high, beſet with long leaues ſomwhat broad, very little or nothing ſnipt about the edges, ſomwhat hairie and euen. The floures alſo are of a dark purple colour, faſhioned like the floures of Scabious: the ſeeds are ſmal and downy, which being ripe are carried away with the winde. The root is blacke, thick, hard and ſhort, with many threddie ſtrings faſtned thereto. The great part of the root ſeemeth to be bitten away: old fantaſticke charmers report, that the diuel did bite it for enuie, becauſe it is an herbe that hath ſo many good vertues, and is ſo beneficiall to mankinde.

¶ *The Place.*

Diuels bit groweth in dry medows and woods, & about waies ſides. I haue found great ſtore of it growing in Hampſtead wood neer London, at Lee in Eſſex and at Raleigh in Eſſex, in a wood called Hammerell, and ſundrie other places.

¶ *The Time.*

It floureth in Auguſt, and is hard to be knowne from Scabious, ſauing when it floureth.

¶ *The Names.*

It is commonly called *Morſus Diaboli*, or Diuels bit, of the root (as it ſeems) that is bitten off : for the ſuperſtitious people hold opinion, that the diuell for enuy that he beareth to mankinde, bit it off, becauſe it would be otherwiſe good for many vſes: it is called of *Fuchſius Succiſa :* in high-Dutch, **Teuffels abbiſʒ:** in low-Dutch, **Duyuelles beet:** in French, *Mors du diable :* in Engliſh, Diuels bit, and Forebit. ‡ *Fabius Columna* iudges it to be the *Pycnocomon* of *Dioſcorides*, deſcribed by him *lib.4.cap.176.* ‡

¶ *The Temperature.*

Diuels bit is ſomthing bitter, and of a hot and drie temperature, and that in the later end of the ſecond degree.

¶ *The Vertues.*

There is no better thing againſt old ſwellings of the Almonds, and vpper parts of the throat that be hardly ripened.

B   It clenſeth away ſlimie flegme that ſticketh in the jawes, it digeſteth and conſumeth it : and it quickely taketh away the ſwellings in thoſe parts, if the decoction thereof bee often held in the mouth and gargarized, eſpecially if a little quantitie of *Mel Roſarum,* or honie of Roſes be put into it.

C   It is reported to be good for the infirmities that Scabious ſerueth for, and to be of no leſſe force againſt the ſtingings of venomous beaſts, poiſons, and peſtilent diſeaſes, and to conſume and waſt away plague ſores, being ſtamped and laid vpon them. ·

D   And alſo to mitigate the paines of the matrix or mother, and to driue forth winde, if the decoction thereof be drunke.

C H A P.

Chap. 249.

Of Matfellon or Knapweed.

¶ *The Deſcription.*

1 MAtfellon or blacke Knapweed is doubtleſſe a kinde of Scabious, as all the others are intituled with the name of *Iacea* ; yet for diſtinction I haue thought good to ſet them in a ſeuerall Chapter, beginning with that kind which is called in Engliſh Knapweed and Matfellon, or *Materſilon* : it hath long and narrow leaues of a blackiſh greene colour, in ſhape like Diuels bit, but longer, ſet vpon ſtalks two cubits high, ſomewhat bluntly cut or ſnipt about the edges : the floures grow at the top of the ſtalkes, beeing firſt ſmall ſcaly knops like to the knops of Corn-floure or Blew-Bottles, but greater : out of the midſt thereof groweth a purple thrummy or threddy floure. The root is thicke and ſhort.

2 The great Knapweed is very like vnto the former, but that the whole plant is much greater, the leaues bigger, and more deeply cut euen to the middle rib : the floures come forth of ſuch like ſcaly heads, of an excellent faire purple colour, and much greater.

3 The third kinde of Matfellon or Knapweed is very like vnto the former great Knapweed laſt before mentioned, ſauing that the floures of this plant are of an excellent faire yellow colour, proceeding forth of a ſcaly head or knop, beſet with moſt ſharpe prickes not to bee touched without hurt : the floure is of a pleaſing ſmell, and very ſweet ; the root is long and laſting, and creepes far abroad, by meanes whereof it greatly encreaſeth.

1 *Iacea nigra.*
Blacke Matfellon.

† 2 *Iacea major.*
Great Matfellon.

‡ 4 The mountain Knapweed of Narbone in France hath a ſtrong ſtem of 2 cubits high, and is very plentifull about Couentry among the hedges and buſhes : the leaues are very much jagged in forme of *Lonchitis* or Spleene-wort : the floures are like the reſt of the Knapweeds, of a purple colour.

3 *Iacea major lutea.*
Yellow Knapweed.

4 *Iacea montana.*
Mountain Knapweed.

5 *Iacea flore albo.*
White floured Knapweed.

6 *Iacea tuberoſa.*
Knobbed Knapweed.

‡ 7. *Iacea Austriaca villosa.*
Rough headed Knapweed.

‡ 5   The white floured Knapweed hath creeping roots, which send vp pretty large whitish greene leaues much diuided or cut in almost to the middle rib; from the middest of which rises vp a stalke some two foot high, set also with the like diuided leaues, but lesser: the floures are like those of the common sort, but of a pleasant white colour. I first found this growing wild in a field neere Martin Abbey in Surry, and since in the Isle of Tenet. ‡

6   The tuberous or knobby Knapweed being set forth by *Tabern.* and which is a stranger in these parts, hath many leaues spread vpon the ground, rough, deeply gasht or hackt about the edges like those of Sow-thistle. Among which riseth vp a straight stalke diuiding it self into other branches, wheron grow the like leaues, but smaller: the knappy flours stand on the top of the branches, of a bright red colour, in shape like the other Knapweeds The root is great, thicke, and tuberous, consisting of many cloggy parcels, like those of the Asphodill.

‡ 7   This (saith *Clusius*) is a comely plant, hauing broad and long leaues, white, soft, and lightly snipt about the edges: the tast is gummy, and not a little bitter: it sends vp many crested stalks from one root, some cubit high or more: at the tops of them grow the heads some two or three together, consisting of many scales, whose ends are hairy, and they are set so orderly, that by this meanes the heads seem as they were inclosed in little nets. The floures are purple, and like those of the first described: the seed is small and long, and of an ash colour. This *Clusius* calls *Iacea 4 Austriaca villoso capite.*

Iacea capitulis hirsutis Boelij.

8   This hath many smal cornered straked hairy trailing branches growing from the root, & those again diuided into many other branches, trailing or spreading vpon the ground three or foure foot long, imploying or couering a good plot of ground, whereon grow hairy leaues diuided or jagged into many parts, like the leaues of *Iacea maior*, or Rocket, of a very bitter taste: at the top of each branch groweth one scaly head, each scale ending with fiue, six, or seuen little weak prickles growing orderly like halfe the rowell of a spurre, but far lesse: the floures grow forth of the heads, of a light purple colour, consisting of many small floures, like those of the common *Iacea*, the bordering floures, being bigger and larger than those of the middle of the floure, each small floure being diuided into fiue small parts or leaues, not much vnlike those of *Cyanus*: the seed is small, and inclosed in down. The root perisheth when the seed is ripe.

This plant hath not been hitherto written of, that I can find. Seeds of it I receiued from Mr *William Coys*, with whom also I obserued the plant, Octob. 10. 1621. he receiued it from *Boelius* a low-Country man. *Iohn Goodyer.* ‡

¶ *The Place.*
The two first grow commonly in euery fertile pasture. The rest grow in my garden.

¶ *The Time.*
They floure in Iune and Iuly.

¶ *The Names.*
The later age calls it *Iacea nigra*, putting *nigra* for a difference betweene it and the Hearts-ease or Pancie, which is likewise called *Iacea*: it is called also *Materfillon* and *Matrefillen*: in English, Matsellon, Bulweed, and Knapweed.

¶ *The*

¶ *The Temperature and Vertues.*

A   Theſe plants are of the nature of Scabious, whereof they be kindes, therefore their faculties are like, although not ſo proper to phyſicks vſe.

B   They be commended againſt the ſwellings of the Vvula, as is Diuels bit, but of leſſe force and vertue.

† The figure that was formerly in the ſecond place was of the *faces tertia* of *Tabern.* which differs from that our Author meant and deſcribed, whoſe figure we haue giuen you in the place thereof.

Chap. 250.   *Of Siluer Knapweed.*

¶ *The Deſcription.*

1   THe great ſiluer Knapweed hath at his firſt comming vp diuers leaues ſpred vpon the ground, of a deep green colour, cut and jagged as are the other Knapweeds, ſtraked here and there with ſome ſiluer lines down the ſame, whereof it took his ſyrname *Argentea*: among which leaues riſeth vp a ſtraight ſtalk of the height of two or three cubits, ſomwhat rough and brittle, diuiding it ſelfe toward the top into other twiggy branches; on the tops whereof doe grow floures ſet in ſcaly heads or knaps like the other Matfellons, of a gallant purple colour, conſiſting of a number of threds or thrums thicke thruſt together: after which the ſeeds appeare, ſlipperie, ſmooth at one end, and bearded with blacke haires at the other end, which makes it to leape and skip away when a man doth but lightly touch it. The root is ſmall, ſingle, and periſheth when the ſeed is ripe. ‡ This is not ſtreaked with any lines, as our Author imagined, nor called *Argentea* by any but himſelfe, and that very vnfitly. ‡

† 1 *Stæbe argentea maior.*
Great ſiluer Knapweed.

† 2 *Stæbe argentea minor.*
Little ſiluer Knapweed.

2   The ſecond agreeth with the firſt in each reſpect, ſauing that the leaues hereof are more jagged, and the ſiluer lines or ſtrakes are greater, and more in number, wherin conſiſteth the difference.

‡ The

‡ 4 *Stæbe Rosmarini folio.*
Narrow leafed Knapweed.

‡ 5 *Stæbe ex Codice Cæsareo.*
Thorny Knapweed.

‡ The leaues of this are very much diuided and hoary, the stalkes some two cubits high, set also with much diuided leaues that end in soft harmlesse prickles: at the tops of the branches stand the heads composed as it were of siluer scales (whence *Lobel* and others haue called this plant *Stæbe argentea*) and out of these siluer heads come floures like those of the Blew-bottles, but of a light purple colour : the seed is small, blackish, and hairy at the tops.

3　There is another like this in each respect, but that the heads haue not so white a shining siluer colour : and this I haue also seen growing with M' *Iohn Tradescant* at South Lambeth.

4　To these may be added that plant which *Pona* hath set forth by the name of *Stæbe capitata Rosmarini folio*. It hath a whitish wooddy root, from whence arise diuers branches set with long narrow leaues somewhat like those of Rosemary, but liker those of the Pine, of a greenish colour aboue, & whitish below : at the tops of the branches grow such heads as in the first described *Stæbe*, with floures of somewhat a deeper purple colour : the seed is like that of *Carthamus*, but blackish. The root is not annual, but lasts many yeares.

5　Though these plants haue of late bin vulgarly set forth by the name of *Stæbe's*, yet are they not iudged to be the true *Stæbe* of *Dioscorides* and the Antients, but rather of another, whose figure which we here giue was by *Dodonæus* taken forth of a Manuscript in the Emperors Library; and he faith, *Paludanus* brought home some of the same out of Cyprus and Morea, as he returned from his journey out of Syria : the bottom leaues are said to be much diuided, those on the stalks long, and only snipt about the edges, and white : the floures white, contained in scaly heads like the Blewbottles, and the tops of the branches end in sharpe prickles. ‡

¶ *The Place.*

.·. These grow of themselues in fields neere common high-wayes and in vntilled places ; but are strangers in England, neuerthelesse I haue them in my garden.

¶ *The*

¶ *The Time.*

They spring vp in Aprill, they floure in August, and the seed is ripe in September.

¶ *The Names.*

Siluer Knapweed is called of *Lobel, Stœbe Salamantica:* of *Dodonæus, Aphyllanthes,* that is, without leaues; for the floures consist only of a number of threds, without any leaues at all : in English, siluer Knapweed, or siluer Scabious, whereof doubtlesse it is a kinde.

¶ *The Nature and Vertues.*

The faculties of these Matfellons are not as yet found out, neither are they vsed for meat or medicine.

† *The Faculties of* Stœbe *out of* Dioscorides.

A   The seed and leaues are astringent, wherefore the decoction of them is cast vp in Dysenteries, and into purulent eares, and the leaues applied in manner of a pultis are good to hinder the blackenesse of the eyes occasioned by a blow, and stop the flowing of bloud.

† The figures were formerly transposed.

CHAP. 251.
Of the Blew-Bottle or Corne-Floure.

1 *Cyanus maior.*
Great Blew-Bottle.

2 *Cyanus vulgaris.*
Common Blew-Bottle.

¶ *The Description.*

1   THe great Blew-Bottle hath long leaues smooth, soft, downy, and sharp pointed: among the leaues rise vp crooked and pretty thicke branches, chamfered, furrowed, and garnished with such leaues as are next the ground: on the tops wherof stand faire blew flours tending to purple, consisting of diuers little flours, set in a scaly huske or knap like those of Knapweed : the seed is rough or bearded at one end, smooth at the other and shining : the root is tough and long lasting (contrary to the rest of the Corne-floures) and groweth yearly into new shoots whereby it greatly encreaseth.

? The

Cyanus cæruleus multiflorus.
Double Blew-Bottles.

Cyanus purpureus multiflorus.
Double purple Bottles.

‡ 9 *Cyanus repens latifolius.*
Broad leafed creeping Blew-Bottle.

‡ 10 *Cyanus repens anguſtifolius.*
Small creeping Blew-Bottle.

2   The common Corn-floure hath leaues ſpred vpon the ground, of a whitiſh green colour, ſomwhat hackt or cut in the edges like thoſe of corne Scabious : among which riſeth vp a ſtalke diuided into diuers ſmall branches, whereon do grow long leaues of an ouerworne green colour, with few cuts or none at all. The floures grow at the top of the ſtalks, of a blew colour, conſiſting of many ſmal floures ſet in a ſcaly or chaffie head like thoſe of the Knapweeds: the seed is ſmooth, bright ſhining, and wrapped in a woolly or flocky matter. The root is ſmall and ſingle, and periſheth when it hath perfected his ſeed.

3   This Bottle is like to the laſt deſcribed in each reſpect, ſauing in the colour of the floures, which are purple, wherein conſiſteth the difference.

4   The fourth Bottle is alſo like the precedent, not differing in any point but in the floures; for as the laſt before mentioned are of a purple colour, contrariwiſe theſe are milke white, which ſets forth the difference.

5   The violet coloured Bottle or Corne-floure is like the precedent in ſtalkes, leaues, ſeeds, and roots : the only difference is, that this bringeth floures of a Violet colour, and the others not ſo.

6   Variable Corn-floure is ſo like the others in ſtalks, leaues, and proportion, that it cannot be diſtinguiſhed with words; only the floures hereof are of two colours, purple and white mixt together, wherein it differeth from the reſt.

7   There is no difference to be found in the leaues, ſtalks, ſeed, or roots of this Corn-floure, from the other, but only that the floures hereof are of a faire blew colour, and very double.

8   The eighth Corne-floure is like vnto the precedent without any difference at all, ſauing in the colour of the floures, which are of a bright purple, that ſetteth forth the difference.

‡   9   This from a ſmall root ſends vp diuers creeping branches ſome foot long, ſet with long hoary narrow leaues : at the tops of the ſtalkes ſtand the floures in ſcaly heads, like as the other Blew-Bottles, but of a darke purple colour. The whole plant is very bitter and vngratefull to the taſte. *Lobel* calls this *Cyanus repens*.

10   This is like the laſt deſcribed, but that the leaues are much ſmaller or narrower, alſo the ſcaly heads of this are of a finer white ſiluer colour, and this plant is not poſſeſſed with ſuch bitterneſſe as the former. *Lobel* calls this *Cyanus minimus repens*. ‡

¶ *The Place.*

The firſt groweth in my garden, and in the gardens of Herbariſts, but not wilde that I know of. The others grow in corn fields among Wheat, Rie, Barley, and other graine : it is ſowne in gardens, and by cunning looking to doth oft times become of other colours, and ſome alſo double, as hath beene touched in their ſeuerall deſcriptions. ‡ The two laſt grow wilde about Montpellier in France. ‡

¶ *The Time.*

They bring forth their floures from the beginning of May to the end of Harueſt.

¶ *The Names.*

The old Herbariſts call it *Cyanus flos*, of the blew colour which it naturally hath : moſt of the later ſort following the common Germane name, call it *Flos frumentorum* : for the Germanes name it 𝕮𝖔𝖟𝖓 𝖇𝖑𝖚𝖒𝖊𝖓 : in low-Dutch, 𝕮𝖔𝖟𝖓 𝖇𝖑𝖔𝖊𝖒𝖊𝖚 : in French, *Blaueole*, and *Bluet* : in Italian, *Fior campeſe*, and *Bladiſeris*, i. *Ceris bladi*, and *Battiſecula*, or *Baptiſecula*, as though it ſhould be called *Blaptiſecula*, becauſe it hindereth and annoyeth the Reapers, by dulling and turning the edges of their ſicles in reaping of corne : in Engliſh it is called Blew-Bottle, Blew-Blow, Corne-floure, and hurt-Sicle. ‡ *Fabius Columna* would haue it to be the *Papauer ſpumeum* or *Heracleum* of the Antients. ‡

¶ *The Temperature and Vertues.*

A   The faculties of theſe floures are not yet ſufficiently known. Sith there is no vſe of them in phyſicke, we will leaue the reſt that might be ſaid to a further conſideration : notwithſtanding ſome haue thought the common Blew-Bottle to be of temperature ſomething cold, and therefore good againſt the inflammation of the eyes, as ſome thinke.

C H A P.

CHAP. 252.   *Of Goats Beard, or Go to bed at noone.*

¶ *The Deſcription.*

1   Goats-beard, or Go to bed at noone hath hollow ſtalks, ſmooth, and of a whitiſh green colour, whereupon do grow long leaues creſted downe the middle with a ſwelling rib, ſharp pointed, yeelding a milkie juice when it is broken, in ſhape like thoſe of Garlick: from the boſome of which leaues thruſt forth ſmal tender ſtalks, ſet with the like leaues, but leſſer: the floures grow at the top of the ſtalks, conſiſting of a number of purple leaues, daſht ouer as it were with a little yellow duſt, ſet about with nine or ten ſharp pointed green leaues: the whole floure reſembles a Star when it is ſpred abroad ; for it ſhutteth it ſelfe at twelue of the clock, and ſheweth not his face open vntill the next daies Sunne doth make it floure anew, whereupon it was called Go to bed at noone : when theſe floures be come to their full maturitie and ripeneſſe, they grow into a downy Blow-ball like thoſe of Dandelion, which is carried away with the winde. The ſeed is long, hauing at the end one peece of that downy matter hanging at it. The root is long and ſingle, with ſome few threds thereto annexed, which periſhes when it hath perfeſted his ſeed, yeelding much quantitie of a milky juice when it is cut or broken, as doth all the reſt of the plant.

2   The yellow Goats beard hath the like leaues, ſtalks, root, ſeed, and downie blow-balls that the other hath, and alſo yeeldeth the like quantitie of milke, inſomuch that if the pilling while it is greene be pulled from the ſtalks, the milky juice followeth : but when it hath there remained a little while it waxeth yellow. The floures hereof are of a gold yellow colour, and haue not ſuch long green leaues to garniſh it withall, wherein conſiſteth the difference.

1 *Tragopogon purpureum.*	2 *Tragopogon luteum.*
Purple Goats-beard.	Yellow Goats-beard.

3   There is another ſmall ſort of Goats-beard or Go to bed at noone, which hath a thicke root full of a milky ſap, from which riſe vp many leaues ſpred vpon the ground, very long, narrow, thin, and like vnto thoſe of graſſe, but thicker and groſſer : among which riſe vp tender ſtalkes , on the tops whereof do ſtand faire double yellow floures like the precedent, but leſſer. The whole plant
yeeldeth

yeeldeth a milkie ſap or juice as the others do : it periſhes like as the other when it hath perfected his ſeed. This may be called *Tragopogon minus anguſtifolium*, Little narrow leaued Goats-beard.

¶ *The Place.*

The firſt growes not wild in England that I could euer ſee or heare of, except in Lancaſhire on the banks of the riuer Chalder, neere to my Lady *Heskiths* houſe, two miles from Whawley : it is ſown in gardens for the beauty of the floures almoſt euery where. The others grow in medows and fertil paſtures in moſt places of England. It growes plentifully in moſt of the fields about London, as at Iſlington, in the medows by Redriffe, Deptford, and Putney, and in diuers other places.

¶ *The Time.*

They floure and flouriſh from the beginning of Iune to the end of Auguſt.

¶ *The Names.*

Goats-beard is called in Greeke, τραγοπώγων : in Latine, *Barba hirci*, and alſo *Coma* : in high-Dutch, **Bocrbaert** : in low-Dutch, **Joſephs bloemen** : in French, *Barbe de bouc*, and *Saſſify* : in Italian, *Saſſefrica* : in Spaniſh, *Barba Cabruna* : in Engliſh, Goats beard, Ioſephs floure, Star of Ieruſalem, Noon-tide, and Go to bed at noone.

¶ *The Temperature.*

Theſe herbes are temperate between heate and moiſture.

¶ *The Vertues.*

A   The roots of Goats-beard boiled in wine and drunk, aſſwageth the pain and pricking ſtitches of the ſides.

B   The ſame boiled in water vntill they be tender, and buttered as parſneps and carrots, are a moſt pleaſant and wholeſome meate, in delicate taſte far ſurpaſſing either Parſenep or Carrot : which meat procures appetite, warmeth the ſtomacke, preuaileth greatly in conſumptions, and ſtrengthneth thoſe that haue been ſicke of a long lingring diſeaſe.

Chap. 253. *Of Vipers-graſſe.*

1 *Viperaria, ſiue Scorzonera Hiſpanica.*
Common Vipers graſſe.

2 *Viperaria humilis.*
Dwarfe Vipers graſſe.

† 3 *Viperaria Pannonica.*
Auftrian Vipers graffe.

‡ 4 *Viperaria anguftifolia elatior.*
Hungary Vipers graffe.

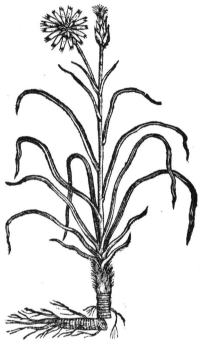

5 *Viperaria Pannonica anguftifolia.*
Narrow leaued Vipers graffe.

¶ *The Defcription.*

1 THe firft of the Viper graffes hath long
broad leaues, fat, or ful bodied, vneuen
about the edges, fharpe pointed, with a high
fwolne ribbe downe the middle, and of an ouerworne colour, tending to the colour of Woade :
among which rifeth vp a ftiffe ftalke, fmooth
and plaine, of two cubits high, whereon do grow
fuch leaues as thofe next the ground. The flours
ftand on the top of the ftalkes, confifting of many fmall yellow leaues thicke thruft together,
very double, as are thofe of Goats-beard, whereof it is a kinde, as are all the reft that doe follow
in this prefent chapter : the root is long, thicke,
very brittle, continuing many yeeres, yeelding
great increafe of roots, blacke without, white
within, and yeelding a milkie iuyce, as doe the
leaues alfo, like vnto the Goates-beard.

2 The dwarfe Vipers graffe differeth not
from the precedent, fauing that it is altogether
leffer, wherein efpecially confifteth the difference.

† 3 The broad leaued Auftrian Vipers-graffe
hath broad leaues fharpe pointed, vneuen about
the edges, of a blewifh greene colour: the ftalke
rifeth vp to the height of a foot or better on the
top whereof doe ftand faire yellow floures, very
double, greater and broader than any of the reft

Qqq

of

of a reasonable good smell. The seed followeth, long and sharpe, like vnto those of Goates-beard. The root is thicke, long, and full of milkie juyce, as are the leaues also.

4   The narrow leaued Hungary Vipers-grasse hath long leaues like to those of Goates-beard, but longer and narrower, among which riseth vp a slender hollow stalke, stiffe and smooth, on the top whereof do stand faire double floures of a faire blew colour tending to purple, in shape like the other of his kinde, of a pleasant sweet smell, like the smell of sweet balls made of *Benzoin*. The seed is conteined in small cups like those of Goates-beard, wrapped in a downy matter, that is carried away with the winde. The root is not so thicke nor long as the others, very single, bearded at the top, with certaine hairy thrums yeelding a milky juyce of a resinous taste, and somewhat sharpe withall. It endureth the Winter euen as the others do.

1   5   This (whose figure was by our Author put to the last description) hath leaues like those of Goates-beard, but stiffer and shorter, amongst which there growes vp a short hollow stalke some handfull high, set with a few short leaues, bearing a yellow floure at the top, almost like that of the last saue one, but lesse : the seed is conteined in such cups as the common Vipers-grasse, and being ripe is carried away with the least winde. The root is blacke with a wrinkled barke, and full of milke, hauing the head hairy, as also the last described hath. This by *Clusius* is called *Scorsonera humilis angustifolia Pannonica.* ‡

¶ The Place and Time.

Most of these are strangers in England. The two first described do grow in my Garden. The rest are touched in their seuerall titles.

They floure and flourish from May to the end of Iuly.

¶ The Names.

Vipers grasse is called of the Spaniards, *Scorzonera*, which soundeth in Latine *Viperaria*, or *Viperina*, or *Serpentaria*, so called because it is accounted to be of force and efficacy against the poysons of Vipers and serpents, for *Vipera* or a viper is called in Spanish *Scurzo* : it hath no name either in the high or low Dutch, nor in any other, more than hath been said, that I can reade : in English we may call it Scorzoner, after the Spanish name, or Vipers-grasse.

¶ The Temperature.

They are hot and moist as are the Goats-beards.

¶ The Vertues.

A   It is reported by those of great judgement, that Vipers-grasse is most excellent against the infections of the plague, and all poysons of venomous beasts, and especially to cure the bitings of vipers, (of which there be very many in Spaine and other hot countries, yet haue I heard that they haue beene seene in England) if the juyce or herbe be drunke.

B   It helpeth the infirmities of the heart, and such as vse to swoune much : it cureth also them that haue the falling sickenesse, and such as are troubled with giddinesse in the head.

C   The root being eaten, either rosted in embers, sodden, or raw, doth make a man merry, and remoueth all sorrow.

D   The root condited with sugar, as are the roots of *Eringos* and such like, worke the like effects : but more familiarly, being thus dressed.

† Formerly there were six figures in this chapter whereof the first and fourth were both of one plant and the fifth which was of the *Scorsonera Botanica* of *Matthiolus* did not much differ from them; if it differ at all. In the title and history of the third there should haue been put *Pannonica;* as now it is.

CHAP. 254.   Of Marigolds.

¶ The Description.

1   THe greatest double Marigold hath many large, fat, broad leaues, springing immediatly from a fibrous or threddy root ; the vpper sides of the leaues are of a deepe greene, and the lower side of a more light and shining greene : among which rise vp stalkes somewhat hairie, and also somewhat joynted, and full of a spungeous pith. The floures in the top are beautifull, round, very large and double, something sweet, with a certaine strong smell, of a light saffron colour, or like pure gold: from the which follow a number of long crooked seeds, especially the outmost, or those that stand about the edges of the floure; which being sowne commonly bring forth single floures, whereas contrariwise those seeds in the middle are lesser, and for the most part bring forth such floures as that was from whence it was taken.

2   The common double Marigold hath many fat, thicke, crumpled leaues set vpon a grosse and spungeous stalke : whereupon do grow faire double yellow floures, hauing for the most part in the middle a bunch of threddes thicke thrust together : which being past there succeed such crooked seeds as the first described. The root is thicke and hard, with some threds annexed thereto,

3 The

1. 2. *Calendula major polyanthos.*
The great double Marigold.

4 *Calendula multiflora orbiculata.*
Double globe Marigold.

6 *Calendula simplici flore.*
Single Marigold.

7 *Calendula prolifera.*
Fruitfull Marigold.

3   The ſmaller or finer leaſed double Marigold groweth vpright, hauing for the moſt part one ſtem or fat ſpongeous ſtalke, garniſhed with ſmooth and flat leaues confuſedly. The floures grow at the top of the ſmall branches, very double, but leſſer than the other, conſiſting of more fine jag-gedneſſe, and of a faire yellow gold colour. The root is like the precedent.

4   The Globe-flouring Marigold hath many large broad leaues riſing immediately forth of the ground, among which riſeth vp a ſtalke of the height of a cubit, diuiding it ſelfe toward the top into other ſmaller branches, ſet or garniſhed with the like leaues, but confuſedly, or without order. The floures grow at the tops of the ſtalkes, very double; the ſmall leaues whereof are ſet in comely order by certaine rankes or rowes, as ſundry lines are in a Globe, trauerſing the whole com-paſſe of the ſame; whereupon it tooke the name *Orbiculata*.

5   The fifth ſort of double Marigold differeth not from the laſt deſcribed, ſauing in the colour of the floures; for this plant bringeth forth floures of a ſtraw or light yellow colour, and the others not ſo, wherein conſiſteth the difference.

‡   All theſe fiue here deſcribed, and which formerly had ſo many figures, differ nothing but in the bigneſſe and littleneſſe of the plants and floures, and in the intenſeneſſe and remiſneſſe of their colour, which is either orange, yellow, or of a ſtraw colour. ‡

6   The Marigold with ſingle floures differeth not from thoſe with double floures, but in that it conſiſteth of fewer leaues, which we therefore terme ſingle, in compariſon of the reſt, and that maketh the difference.

7   This fruitfull or much bearing Marigold is likewiſe called of the vulgar ſort of women, Iacke-an-apes on horſe backe: it hath leaues, ſtalkes, and roots like the common ſort of Marigold, differing in the ſhape of his flours, for this plant doth bring forth at the top of the ſtalke one floure like the other Marigolds; from the which ſtart forth ſundry other ſmall floures, yellow likewiſe, and of the ſame faſhion as the firſt, which if I be not deceiued commeth to paſſe *per accidens*, or by chance, as Nature oftentimes liketh to play with other floures, or as children are borne with two thumbes on one hand, and ſuch like, which liuing to be men, do get children like vnto others; euen ſo is the ſeed of this Marigold, which if it be ſowen, it brings forth not one floure in a thouſand like the plant from whence it was taken.

8   The other fruitfull Marigold is doubtleſſe a degenerate kind, comming by chance from the ſeed of the double Marigold, whereas for the moſt part the other commeth of the ſeed of the ſingle floures, wherein conſiſteth the difference. ‡ The floure of this (wherein the only difference con-ſiſts) you ſhall finde expreſt at the bottome of the fourth figure. ‡

9 *Calendula Alpina.*
Mountaine Marigold.

9   The Alpiſh or mountaine Marigold, which *Lobelius* ſetteth downe for *Nardus Celtica*, or *Plantago Alpina*, is called by *Tabernamontanus, Caltha*, or *Calendula Alpina:* and be-cauſe I ſee it rather reſembles a Marigold, than any other plant, I haue not thought it a-miſſe to inſert it in this place, leauing the conſideration thereof vnto the friendly Rea-der, or to a further conſideration, becauſe it is a plant that I am not well acquainted with-all; yet I do reade that it hath a thicke root, growing aſlope vnder the vpper cruſt of the earth, of an aromaticall or ſpicie taſte, and ſomewhat biting, with many threddy ſtrings annexed thereto: from which riſe vp broad thicke and rough leaues of an ouerworn green colour, not vnlike to thoſe of Plantaine: a-mong which there riſeth vp a rough and ten-der ſtalke ſet with the like leaues: on the top wherof commeth forth a ſingle yellow floure, paled about the edges with ſmall leaues of a light yellow, tending to a ſtraw colour; the middle of the floure is compoſed of a bundle of threds, thicke thruſt together, ſuch as is in the middle of the field Daiſie, of a deepe yel-low colour.

‡   This plant is all one with the two de-ſcribed in the next Chapter: they vary onely thus;

thus ; the ſtalkes and leaues are ſometimes hairy,otherwhiles ſmooth ; the floure is yellow, or elſe blew. I hauing three figures ready cut, thinke it not amiſſe to giue you one to expreſſe each va-rietie. ‡

10 The wilde Marigold is like vnto the ſingle garden Marigold,but altogether leſſer, and the whole plant periſheth at the firſt approach of Winter, and recouereth it ſelfe againe by falling of the ſeed.

¶ *The Place.*

Theſe Marigolds, with double floures eſpecially, are ſet and ſowne in Gardens : the reſt,their titles do ſet forth their naturall being.

¶ *The Time.*

The Marigold floureth from Aprill or May euen vntill Winter, and in Winter alſo, if it bee warme.

¶ *The Names.*

The Marigold is called *Calendula :* it is to be ſeene in floure in the Calends almoſt of euery moneth : it is alſo called *Chryſanthemum,* of his golden colour : of ſome,*Caltha,*and *Caltha Poetarum :* whereof *Columella* and *Virgil* doe write, ſaying, That *Caltha* is a floure of a yellow colour : whereof *Virgil* in his Bucolickes,the ſecond Ecloge,writeth thus :

Tum Caſia atque alijs intexens ſuauibus herbis
Mollia Luteola pingit vaccinia Caltha.

And then ſhee'l Spike and ſuch ſweet herbes infold,
And paint the Iacinth with the Marigold.

Columella alſo in his tenth booke of Gardens hath theſe words ;

Candida Leucoia & flauentia Lumina Caltha.

Stock-Gillofloures exceeding white,
And Marigolds moſt yellow bright.

It is thought to be *Gromphena Plinij :* in low Dutch it is called, **Goudt bloemen:**in high Dutch, **Ringleblumen :** in French, *Souſij & Goude :* in Italian, *Fior d'ogni meſe :* iu Engliſh, Marigolds and Ruddes.

¶ *The Temperature and Vertues.*

The floure of the Marigold is of temperature hot, almoſt in the ſecond degree, eſpecially when **A** it is dry : it is thought to ſtrengthen and comfort the heart very much, and alſo to withſtand poy-ſon, as alſo to be good againſt peſtilent Agues,being taken any way. *Fuchſius* hath written, That being drunke with wine it bringeth downe the termes, and that the fume thereof expelleth the ſe-condine or after-birth.

But the leaues of the herbe are hotter ; for there is in them a certaine biting,but by reaſon of the **B** moiſture joyned with it, it doth not by and by ſhew it ſelfe ; by meanes of which moiſture they mollifie the belly,and procure ſolubleneſſe if it be vſed as a pot-herbe.

Fuchſius writeth, That if the mouth be waſhed with the juyce it helpeth the tooth-ache. **C**

The floures and leaues of Marigolds being diſtilled, and the water dropped into red and watery **D** eies,ceaſeth the inflammation, and taketh away the paine.

Conſerue made of the floures and ſugar taken in the morning faſting, cureth the trembling of **E** the heart,and is alſo giuen in time of plague or peſtilence,or corruption of the aire.

The yellow leaues of the floures are dried and kept throughout Dutchland againſt Winter, to **F** put into broths, in Phyſicall potions, and for diuers other purpoſes,in ſuch quantity, that in ſome Grocers or Spice-ſellers houſes are to be found barrels filled with them, and retailed by the penny more or leſſe,inſomuch that no broths are well made without dried Marigolds.

CHAP. 255.   *Of Germane Marigolds.*

¶ *The Deſcription.*

GOlden Marigold with the broad leafe doth forthwith bring from the root long leaues ſpred vpon the ground,broad,greene, ſomething rough in the vpper part, vnderneath ſmooth, and of a light greene colour : among which ſpring vp ſlender ſtalkes a cubit high,

high,ſomething hoarie, hauing three or foure joynts, out of euery one whereof grow two leaues ſet one right againſt another, and oftentimes little ſlender ſtems; on the tops wherof ſtand broad round floures like thoſe of Ox-eie, or the corne Marigold, hauing a round ball in the middle (ſuch as is in the middle of thoſe of Camomill) bordered about with a pale of white yellow leaues. The whole floure turneth into downe that is carried away with the winde; among which downe is found long blackiſh ſeed. The root conſiſteth of threddy ſtrings.

† 2　The leſſer ſort hath foure or fiue leaues ſpred vpon the ground like vnto thoſe of the laſt deſcribed, but altogether leſſer and ſhorter: among which riſeth vp a ſlender ſtalke two hands high; on the top whereof ſtand ſuch floures as the precedent, but not ſo large, and of a blew colour.

‡　Theſe two here deſcribed, and that deſcribed in the ninth place of the foregoing chapter, are all but the varieties of one & the ſame plant, differing as I haue ſhewed in the foregoing chapter. ‡

1　*Chryſanthemum latifolium.*
Golden Marigold with the broad leafe.

2　*Chryſanthemum latifolium minus.*
The leſſer Dutch Marigold.

¶ *The Place.*
They be fonnd euery where in vntilled places of Germany, and in woods, but are ſtrangers in England.

¶ *The Time.*
They are to be ſeene with their floures in Iune, and Iuly, in the Gardens of the Low-countries.

¶ *The Names.*
Golden Marigold is called in high-Dutch, **Goldtblume.** There are that would haue it to be *A-liſma Dioſcoridis*; which is alſo called *Damaſonium*, but vnproperly; therefore we muſt rather call it *Chryſanthemum latifolium*, than raſhly attribute vnto it the name of *Aliſma.* ‡ This plant indeed is a *Doronicum*, and the figure in the precedent chapter by *Cluſius* is ſet forth by the name of *Doronicum 6. Pannonicum*: *Matthiolus* calls this plant *Aliſma*: *Geſner, Caltha Alpina*: *Dodonæus, Chryſanthemum latifolium*: *Pena* and *Lobel, Nardus Celtica altera.* Now in the *Hiſtoria Lugd.* it is ſet forth in foure ſeuerall places by three of the former names; and pag. 1169. by the name of *Ptarmica montana Daleſchampij.* ‡

¶ *The Temperature.*
It is hot and dry in the ſecond degree being greene, but in the third being dry.

¶ *The*

¶ *The Vertues.*

The women that liue about the Alps wonderfully commend the root of this plant againſt the ſuffocation of the mother, the ſtoppings of the courſes, the greene ſickeneſſe and ſuch like affeēts in maids. *Hiſtor. Lugd.* ‡

CHAP. 256. *Of Corne-Marigold.*

¶ *The Deſcription.*

1 COrne Marigold or golden Corne floure hath a ſoft ſtalke, hollow, and of a greene colour, whereupon do grow great leaues, much haēt and cut into diuers ſeētions, and placed confuſedly or out of order: vpon the top of the branches ſtand faire ſtar-like flours, yellow in the middle, and ſuch likewiſe is the pale or border of leaues that compaſſeth the ſoft ball in the middle, like that in the middle of Camomill floures, of a reaſonable pleaſant ſmel. The roots are full of ſtrings.

† 1 *Chryſanthemum ſegetum.*
Corne Marigold.

2 *Chryſanthemum Valentinum.*
Corne Marigold of Valentia.

2 The golden floure of Valentia hath a thick fat ſtalk, rough, vneuen, and ſomewhat crooked, whereupon do grow long leaues, conſiſting of a long middle rib, with diuers little fether-like leaues ſet thereon without order. The floures grow at the top of the ſtalks, compoſed of a yellow thrummie matter, ſuch as in the middle of the Camomill floures, and is altogether like the Corne Marigold laſt deſcribed, ſauing it doth want that border or pale of little leaues that doe compaſſe the ball or head: the root is thicke, tough, and diſperſeth it ſelfe far abroad.

‡ 3 To theſe may be added diuers other, as the *Chryſanthema Alpina,* of *Cluſius,* and his *Chryſanthemum Creticum,* & others. The firſt of theſe ſmal mountaine Marigolds of *Cluſius* his deſcription hath leaues like thoſe of white Wormewood, but greener and thicker: the ſtalkes grow ſome handfull high, ſet with few and much diuided leaues; and at the tops, as in an vmbell, they carry ſome dozen floures

floures more or leſſe, not much vnlike in ſhape, colour, and ſmell, to thoſe of the common *Iacobæa*, or Ragwort. The root is ſomewhat thicke, and puts forth many long white fibres. It floures in Iuly and Auguſt, and growes vpon the Alpes of Stiria. *Cluſius* calls it *Chryſanthemum Alpinum*. 1.

4   The ſecond of his deſcription hath many leaues at the root, like to the leaues of the male Sothernwood, but of a lighter and brighter greene, and of no vnpleaſant ſmell, though the taſte be bitteriſh and vngratefull : in the middeſt of the leaues grow vp ſtalkes ſome foot high, diuided at their tops into ſundry branches, which carry each of them two or three floures bigger than, yet like thoſe of the common Camomill, but without ſmell, and wholly yellow : the root is fibrous, blackiſh and much ſpreading. It floures in Auguſt, and growes in the like places as the former. *Bauhine* iudges this to be the *Achillæa montana Artemiſiæ tenuifoliæ facie* of the *Aduerſ.* and the *Ageratum ferulaceum* in the *Hiſt. Lugd*. But I cannot be of that opinion; yet I iudge the *Achillæa montana*, and *Ageratum ferulaceum* to be but of the ſame plant. But different from this, and that chiefly in that it hath many more, and thoſe much leſſe floures than thoſe of the plant here figured and deſcribed.

5   Now ſhould I haue giuen you the Hiſtorie of the *Chryſanthemum Creticum* of the ſame Authour, but that my friend Mr. *Goodyer* hath ſaued me the labour, by ſending an exact deſcription thereof, together with one or two others of this kinde, which I thinke fit here to giue you.

‡ 3 *Chryſanthemum Alpinum* 1. *Cluſ.*   ‡ 4 *Chryſanthemum Alpinum* 2. *Cluſ.*
Small mountaine Marigold.   The other Alpine Marigold.

Chryſanthemum Creticum primum Cluſij, pag. 334.

The ſtalkes are round, ſtraked, branched, hard, of a whitiſh greene, with a very little pith within ; neere three foot high : the leaues grow out of order, diuided into many parts, and thoſe againe ſnipt or diuided, of the colour of the ſtalkes : at the tops of the ſtalkes and branches grow great floures, bigger than any of the reſt of the Corne-floures, forth of ſcaly heads, conſiſting of twelue or more broad leaues apiece, notched at the top, of a ſhining golden colour at the firſt, which after turne to a pale, whitiſh, or very light yellow, and grow round about a large yellow ball, of ſmell ſomewhat ſweet. The floures paſt, there commeth abundance of ſeed cloſely compact or thruſt together, and it is ſhort, blunt at both ends, ſtraked, of a ſalue colour, ſomwhat flat, and of a reaſonable bignes. The
root

‡ 5 *Chryſanthemum Creticum.*
Candy Corne Marigold.

root is whitiſh, neere a fingers bigneſſe, ſhort, with many threds hanging thereat, and periſheth when the ſeed is ripe ; and at the Spring groweth vp againe by the falling of the ſeed.

Chryſanthemum Bæticum Boelij, inſcriptum.

The ſtalks are round, ſtraked, reddiſh brown, diuided into branches, containing a ſpungious white pith within, a cubit high ı the leaues grow out of order, without foot-ſtalkes, about three inches long, and an inch broad, notched about the edges, not at all diuided, of a darke greene colour : the floures grow at the tops of the ſtalkes and branches, forth of great ſcaly heads, containing twentie leaues apiece or more, notched at the top, of a ſhining yellow colour, growing about a round yellow ball, of a reaſonable good ſmell, very like thoſe of the common *Chryſanthemum ſegetum :* the ſeed groweth like the other, and is very ſmall, long, round, crooked and whitiſh : the root is ſmall, whitiſh, thready, and periſheth alſo when the ſeed is ripe.

Chryſanthemum tenuifolium Bæticum Boelij.

The ſtalks are round, ſmall, ſtraked, reddiſh, ſomewhat hairie, branched, a cubit high, or higher: the leaues are ſmal, much diuided, jagged, and very like the leaues of *Cotula fætida :* the floures are yellow, ſhining like gold, compoſed of thirteene or fourteene leaues a piece, notched at the top, ſet about a yellow ball, alſo like the common *Chryſanthemum ſegetum :* the ſeed groweth amongſt white flattiſh ſcales, which are cloſely compacted in a round head together, and are ſmall, flat, grayiſh, and broad at the top : the root is ſmall, whitiſh, with a few threds, and dyeth when the ſeed is ripe. Iuly 28. 1621. *Iohn Goodyer.* ‡

¶ *The Place.*

The firſt groweth among corne, and where corne hath been growing ı it is found in ſome places with leaues more jagged, and in others leſſe.

The ſecond is a ſtranger in England.

¶ *The Time.*

They floure in Iuly and Auguſt.

¶ *The Names.*

Theſe plants are called by one name in Greeke, of the golden glittering colour, χρυσάνθεμον: in High Dutch, **Sant Johans blum** : in Low Dutch, **Uokelaer** : in Engliſh, Corne Marigold, yellow Corne floure, and golden Corne floure.

There be diuers other floures called *Chryſanthemum* alſo, as *Batrachion,* a kinde of yellow Crowfoot, *Heliochryſon,* but theſe golden floures differ from them.

¶ *The Temperature.*

They are thought to be of a meane temperature betweene heat and moiſture.

¶ *The Vertues.*

The ſtalkes and leaues of Corne Marigold, as *Dioſcorides* ſaith, are eaten as other pot-herbes are. A

The floures mixed with wax, oile, roſine, and frankinſence, and made vp into a ſeare-cloth, waſte B away cold and hard ſwellings.

The herbe it ſelfe drunke, after the comming forth of the bath, of them that haue the yellow C jaundiſe, doth in ſhort time make them well coloured.

† The figure that was in the firſt place was of the *Chryſanthemum* of *Matthiolus,* which is a ſtranger with vs, and the leaues of it are much like thoſe of Feuerfew, or Mugwort, the floure is ſomewhat like, but larger than that of Feuerfew, and wholly yellow.

C H A P. 257.   Of Oxe-Eie.

¶ *The Description.*

1   THe plant which we haue called *Buphthalmum*, or Oxe-eie, hath slender stalkes growing from the roots, three, foure, or more, a foot high, or higher, about which be green leaues finely jagged like to the leaues of Fenell, but much lesser : the floures in the tops of the stalks are great, much like to Marigolds, of a light yellow colour, with yellow threds in the middle, after which commeth vp a little head or knap like to that of red Mathes before described, called *Adonis*, consisting of many seeds set together. The roots are slender, and nothing but strings, like to the roots of blacke Ellebor, whereof it hath beene taken to be a kinde.

2   The Oxe-eie which is generally holden to be the true *Buphthalmum*, hath many leaues spred vpon the ground, of a light greene colour, laied far abroad like wings, consisting of very many fine jags, set vpon a tender middle rib : among which spring vp diuers stalks, stiffe and brittle, vpon the top whereof do grow faire yellow leaues, set about a head or ball of thrummie matter, such as is in the middle of Cammomill, like a border or pale. The root is tough and thicke, with certain strings fastned thereto.

3   The white Oxe-eie hath small vpright stalkes of a foot high, whereon do grow long leaues, composed of diuers small leaues, and those snipt about the edges like the teeth of a saw. The flours grow on the tops of the stalks, in shape like those of the other Oxe-eie ; the middle part whereof is likewise made of a yellow substance, but the pale or border of little leaues are exceeding white, like those of great Dasie, called *Consolida media vulnerariorum*. The root is long, creeping alongst vnder the vpper crust of the earth, whereby it greatly increaseth. ‡ This by the common consent of all writers that haue deliuered the history thereof, hath not the pale or out-leaues of the floure white, as our Author affirmes, but of a bright and perfect yellow colour. And this is the *Buphthalmum*, of *Tragus*, *Matthiolus*, *Lobel*, *Clusius* and others. ‡

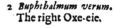

1 *Buphthalmum siue Helleborus niger ferulaceus.*   2 *Buphthalmum verum.*
     Oxe-eie.   The right Oxe-eie.

3 *Buphthalmum vulgare.*
White Oxe-eie.

¶ *The Place.*

The two first grow of themselues in Germany, Bohemia, and in the Gardens of the Low-countries; of the first I haue a plant in my Garden. The last groweth in barren pastures and fields almost euery where.

‡ The last is also a stranger with vs, for any thing that I know or can learne; neither can I conjecture what our Authour meant here: first in that he said the floures of this were white, and secondly in that it grew in barren pastures and fields almost euery where. ‡

¶ *The Time.*

They floure in May and Iune. The last in August.

¶ *The Names.*

Touching the naming of the first of those plants the later writers are of diuers opinions: some would haue it to be a kind of *Veratrum nigrum*, black Hellebor: othersome *Consiligo*; and againe, others, *Sesamoides*; and some, *Elleborastrum*: But there be found two kinds of blacke Ellebor among the old writers, one with a leafe like vnto Laurel, with the fruit of *Sesamum*: the other with a leafe like that of the Plane tree, with the seed of *Carthamus*, or Bastard Saffron. But it is most euident, that this *Buphthalmum*, in English, Oxe-eie, which in this Chapter we in the first place haue described, doth agree with neither of these: what form *Consiligo* is of, we find not among the old writers. *Pliny, 26. cap. 7.* saith, That in his time it was found amongst the *Marsi*, & was a present remedy for the infirmity of the lungs of swine, and of all kind of cattel, though it were but drawne through the eare. *Columella* in his 6. booke, chap. 5. doth also say, that in the mountaines called Marsi there is very great store thereof, and that it is very helpfull to all kind of cattell, and he telleth how and in what manner it must be put into the eare, the roots also of our Oxe-eie are said to cure certaine infirmities of cattell, if they be put into the slit or bored eare: but it followeth not that for the same reason it should be *Consiligo*; and it is an ordinary thing to finde out plants that are of a like force and quality: for *Pliny* doth testifie in his 25. booke, 5. chapter, That the roots also of blacke Ellebor can do the same: it cureth (saith he) the cough in cattell, if it be drawne through the eare, and taken out again the next day at the same houre: which is likewise most certaine by experiments of the country men of our age; who do cure the diseases of their cattell with the roots of common black Ellebor. The roots of white Ellebor also do the like, as *Absyrtus*, and after him *Hierocles* doth write: who notwithstanding do not thrust the roots of white Ellebor into the eare, but vnder the skin of the brest called the dew lap: after which manner also *Vegetius Renatus* doth vse *Consiligo*, in his first booke of the curing of cattell, chapter 12. intituled, Of the cure of the infirmities vnder the skin: although in his 3. booke 2 chapter, *de Malleo*, he writeth, that they also must be fastned thorow the eare: which things do sufficiently declare, that sundry plants haue oftentimes like faculties: and that it doth not at all follow by the same reason, that our Oxe-eie is *Consiligo*, because it doth cure diseases in cattell as well as *Consiligo* doth. But if we must conjecture by the faculties, *Consiligo* then should be White Ellebor: for *Vegetius* vseth *Consiligo* in the very same manner that *Absyrtus* and *Hierocles* do vse White Ellebor. This suspition is made the greater, because it is thought that *Vegetius* hath taken this manner of curing from the Græcians; for which cause also most doe take *Consiligo* to be nothing else but white Ellebor: the which if it be so, then shall this present Oxe-eie much differ from *Consiligo*; for it is nothing at all like to White Ellebor.

And that the same is not *Sesanoides*, either the first or the second, it is better knowne, than needfull to be confuted.

This same also is vnproperly called *Helleboraſtrum*; for that may aptly be called *Helleboraſtrum* which hath the forme and likeneſſe of Hellebor : and this Oxe-eie is nothing at all like to Ellebor. For all which cauſes it ſeemeth that none of theſe names agree with this plant, but only the name *Buphthalmum*, with whoſe deſcription which is extant in *Dioſcorides*, this plant doth moſt aptly agree. We take it to be the right Ox e-eie ; for Oxe-eie bringeth forth ſlender ſoft ſtalkes, and hath leaues of the likeneſſe or ſimilitude of Fennell leaues : the floure is yellow, bigger than that of Cammomill, euen ſuch an one is this preſent plant, which doth ſo exquiſitly expreſſe that forme or likeneſſe of Fennell leaues, both in ſlenderneſſe and manifold jaggedneſſe of the leaues, as no other little leafed herbe can doe better ; ſo that without all doubt this plant ſeemeth to be the true and right Oxe-eie. Oxe-eie is called *Cachla*, or rather *Caltha*; but *Caltha* is *Calendula*, or Marigold, which we ſaid that our Oxe-eie in floure did neereſt repreſent. There are ſome that would haue *Buphthalmum* or Oxe-eie to be *Chryſanthemum*, and ſay that *Dioſcorides* hath in ſundry places, and by diuers names intreated of this herbe; but if thoſe men had ſomewhat more diligently weighed *Dioſcorides* his words, they would haue been of another minde : for although deſcriptions of either of them do in many things agree, yet there is no property wanting that may ſhew the plants to differ. The leaues of *Chryſanthemum* are ſaid to be diuided and cut into many fine jags : and the leaues of *Buphthalmum* to be like the leaues of Fennell : for all things that be finely jagged and cut into many parts haue not the likeneſ of the leaues of Fennell. Moreouer, *Dioſcorides* ſaith, that *Chryſanthemum* doth bring forth a floure much glittering, but he telleth not that the floure of *Buphthalmum*, or Oxe-eie is much glittering, neither doth the floure of that which we haue ſet downe glitter, ſo that it can or ought to bee ſaid to glitter much. Doe not theſe things declare a manifeſt difference betweene *Buphthalmum* and *Chryſanthemum*, and confirme that which we haue ſet downe to be the true and right Oxe-eie ? we are of that minde, let others thinke as they will : and they that would haue *Chryſanthemum* to be *Buphthalmum*, let them ſeeke out another, if they deny this to be Oxe-eie : for that which we and others haue deſcribed for *Chryſanthemum* cannot be the true *Buphthalmum* or Oxe-eie ; for the leaues of it are not like Fennell, ſuch as thoſe of the true *Buphthalmum* ought to be.

¶ The Temperature.

But concerning the faculties, *Matthiolus* ſaith, that all the Phyſitions and Apothecaries in Bohemia, vſe the roots of this Oxe-eie in ſtead of thoſe of blacke Ellebor, namely for diſeaſes in cattell : but he doth not affirme that the roots hereof in medicines are ſubſtitutes, or *quid pro quo* ; for, ſaith he, I do remember that I once ſaw the roots hereof in a ſufficient big quantitie put by certain Phyſitions into decoctions which were made to purge by ſiege, but they purged no more than if they had not bin put in at all : which thing maketh it moſt plaine, that it cannot be any of the Ellebors, although it hath been vſed to be faſtned through the eares of cattell for certaine diſeaſes, and doth cure them as Ellebor doth. The roots of *Gentian* do mightily open the orifices of Fiſtulaes, which be too narrow, ſo do the roots of *Ariſtolochia*, or Birthwort, or Briony, or pieces of ſpunges, which notwithſtanding do much differ one from another in other operations : wherefore though the roots of Oxe-eie can do ſomething like vnto blacke Ellebor, yet for all that they canot perform all thoſe things that the ſame can. We know that thornes, ſtings, ſplinters of wood, and ſuch like, bring pain, cauſe inflammations, draw vnto them humors from the parts neere adjoyning, if they be faſtned in any part of the body; no part of the body is hurt without paine; the which is encreaſed if any thing be thruſt through, or put into the wound : peraduenture alſo if any other thing beſide be put into the ſlit or bored eare, the ſame effect would follow which hapned by the root of this plant thruſt in; notwithſtanding we here affirme nothing, we onely make way for curious men to make more diligent ſearch touching the operations hereof. ‡ *Cluſius* affirmes, that when he came to Vienna in Auſtria, this was vulgarly bought, ſold, and vſed for the true blacke Ellebor, the ignorance of the Phyſitions and Apothecaries in the knowledge of ſimples was ſuch to make vſe of this ſo far different plant, when as they had the true black Hellebor growing plentifully wilde within ſeuen miles of the citie, the which afterward vpon his admonition, they made vſe of. ‡

¶ The Vertues.

A   *Dioſcorides* ſaith, that the floures of Oxe-eie made vp in a ſeare-cloth doe aſſwage and waſte away cold hard ſwellings ; and it is reported that if they be drunke by and by after bathing, they make them in ſhort time well coloured that haue beene troubled with the yellow jaundiſe.

CHAP.

CHAP. 258.   *Of French Marigold, or African Marigold.*

¶ *The Description.*

1   THe great double African Marigold hath a great long browne reddish stalke, crested, furrowed and somewhat knobby, diuiding it selfe toward the top into other branches; whereupon grow leaues composed of many small leaues set vpon a middle rib by couples, much like vnto the leaues of wilde Valerian, bearing at the top very faire and beautifull double yellow floures, greater and more double than the greatest Damask Rose, of a strong smel, but not vnpleasant. The floures being past, there succeedeth long black flat seed: the whole plant perishes at the first approch of Winter.

2   There is little difference betweene this and the precedent, or last described, sauing that this plant is much lesser, and bringeth forth more store of floures, which maketh the difference. ‡ And we may therefore call it *Flos Aphricanus minor multiflorus,* The small double African Marigold. ‡

1 *Flos Aphricanus major Polyanthos.*
  The great African double Marigold.

3 *Flos Aphricanus major simplici flore.*
  The great single French Marigold.

3   The single great Africane Marigold, hath a thicke root, with some fibres annexed thereto; from which riseth vp a thicke stalke chamfered and furrowed, of the height of two cubits, diuided into other small branches; whereupon are set long leaues, compact or composed of many little leaues like those of the Ash tree, of a strong smell, yet not very vnpleasant: on the top of the branches do grow yellow single floures, composed in the middle of a bundle of yellow thrummes hard thrust together, paled about the edges with a border of yellow leaues; after which commeth long blacke seed. The whole plant perisheth with the first frost, and must be sowne yerely as the other sorts must be.

4   The common Africane, or as they vulgarly terme it French Marigold, hath small weake and tender branches trailing vpon the ground, reeling and leaning this way and that way, beset with leaues consisting of many particular leaues, indented about the edges, which being held vp against the sunne, or to the light, are seene to be full of holes like a sieue, euen as those of Saint Iohns

R r r   woort:

woort : the floures ſtand at the top of the ſpringy branches forth of long cups or husks, conſiſting of eight or ten ſmall leaues, yellow vnderneath, on the vpper ſide of a deeper yellow tending to the colour of a darke crimſon veluet, as alſo ſoft in handling : but to deſcribe the colour in words, it is not poſſible, but this way ; lay vpon paper with a penſill a yellow colour called Maſticot, which being dry, lay the ſame ouer with a little ſaffron ſteeped in water or wine, which ſetteth forth moſt liuely the colour. The whole plant is of a moſt ranke and vnwholeſome ſmell, and periſheth at the firſt froſt.

4 *Flos Aphricanus minor ſimplici flore.*
The ſmall French Marigold.

¶ *The Place.*

They are cheriſhed and ſowne in gardens e-uery yere : they grow euery where almoſt in A-fricke of themſelues, from whence wee firſt had them, and that was when *Charles* the fifth, Em-peror of Rome made a famous conqueſt of Tu-nis; whereupon it was called *Flos Aphricanus*, or *Flos Tunetanus*.

¶ *The Time.*

They are to be ſown in the beginning of A-pril, if the ſeaſon fall out to be warm, otherwiſe they muſt be ſown in a bed of dung, as ſhall be ſhewed in the chapter of Cucumbers. They bring forth their pleaſant floures very late, and therefore there is the more diligence to be vſed to ſow them very early, becauſe they ſhall not be ouertaken with the froſt before their ſeed be ripe.

¶ *The Names.*

The Africane or French Marigold is called in Dutch, 𝕮𝖍𝖚𝖓𝖎𝖘 𝖇𝖑𝖔𝖊𝖒𝖊𝖓 : in high-Dutch, 𝕴𝖓𝖉𝖎𝖆𝖓𝖎𝖘𝖈𝖍 𝖓𝖊𝖌𝖊𝖑𝖎𝖓 : that is, the floure, or Gil-loſloure of India : in Latine, *Caryophyllus Indicus*; whereupon the French men call it *Oeilletz d'In-de. Cordus* calls it *Tanacetum Peruvianum*, of the likeneſſe the leaues haue with Tanſie, and of Peru a Prouince of America, from whence hee thought it may be, it was firſt brought into Eu-rop. *Geſner* calleth it *Caltha Aphricana*, and ſaith that it is called in the Carthaginian tongue, *Pedua*: ſome would haue it to be *Petilius flos Pliniy*, but not properly : for *Petilius flos* is an Autumne floure growing among briers and brambles. *Andreas Lacuna* calleth it *Othonna*, which is a certain herbe of the Troglodytes, growing in that part of Arabia which lieth toward Ægypt, hauing leaues full of holes as though they were eaten with mothes. *Galen* in his firſt booke of the faculties of Simple medicines maketh mention of an herbe called *Lycoperſicum*, the juice whereof a certain Centurion did carrie out of Barbarie all Ægypt ouer with ſo ranke a ſmell, and ſo lothſome, as *Galen* himſelfe durſt not ſo much as taſt of it, but coniectured it to be deadly; yet that Centurion did vſe it againſt the extreme pains of the joints, and it ſeemeth to the patients themſelues, to be of a very cold tem-perature; but doubtleſſe of a poiſonſome quality, very neere to that of Hemlockes.

¶ *The Nature and Vertues.*

A   The vnpleaſant ſmel, eſpecially of that common ſort with ſingle floures (that ſtuffeth the head like to that of Hemlocke, ſuch as the juice of *Lycoperſium* had) doth ſhew that it is of a poiſonſome and cooling qualitie ; and alſo the ſame is manifeſted by diuers experiments: for I remember, ſaith *Dodonæus*, that I did ſee a boy whoſe lippes and mouth when hee began to chew the floures did ſwell extreamely ; as it hath often happened vnto them, that playing or piping with quils or kexes of Hemlockes, do hold them a while betweene their lippes : likewiſe he ſaith, we gaue to a cat the floures with their cups, tempered with freſh cheeſe, ſhee forthwith mightily ſwelled, and a little while after died : alſo mice that haue eaten of the ſeed thereof haue been found dead. All which things doe declare that this herbe is of a venomous and poyſonſome facultie ; and that they are

not

not to be heark'ned vnto,that ſuppoſe this herb to be a harmleſſe plant:ſo to conclude,theſe plants are moſt venomous and full of poiſon,and therefore not to be touched or ſmelled vnto,much leſſe vſed in meat or medicine.

Cʜᴀᴘ. 259.

Of the Floure of the Sun, or the Marigold of Peru.

¶ *The Deſcription.*

1   THe Indian Sun,or the golden floure of Peru,is a plant of ſuch ſtature and talneſſe,that in one ſummer, beeing ſowne of a ſeed in Aprill, it hath riſen vp to the height of fourteene foot in my garden, where one floure was in weight three pound and two ounces, & croſſe ouerthwart the floure by meaſure ſixteen inches broad. The ſtalks are vpright & ſtraight, of the bigneſſe of a ſtrong mans arme, beſet with large leaues euen to the top, like vnto the great Clot bur : at the top of the ſtalk commeth forth for the moſt part one floure, yet many times there ſpring out ſucking buds which come to no perfection : this great floure is in ſhape like to the Camomil floure,beſet round about with a pale or border of goodly yellow leaues, in ſhape like the leaues of the floures of white Lillies : the middle part whereof is made as it were of vnſhorn veluet,or ſome curious cloath wrought with the needle : which braue worke,if you do thorowly view and marke well,it ſeemeth to be an innumerable ſort of ſmall floures,reſembling the noſe or noſle of a candleſtick broken from the foot thereof; from which ſmall noſle ſweats forth excellent fine and cleare turpentine,in ſight,ſubſtance,ſauor,and taſt.The whole plant in like manner being broken ſmelleth of turpentine : when the plant groweth to maturitie the floures fall away, in place whereof appeareth the ſeed,black and large,much like the ſeed of Gourds,ſet as though a cunning workman had of purpoſe placed them in very good order,much like the hony-combs of Bees : the root is white,compact of many ſtrings,which periſh at the firſt approch of Winter,and muſt bee ſet in moſt perfect dunged ground : the manner how ſhall be ſhewed when vpon the like occaſion I ſhall ſpeake of Cucumbers and Melons.

1 *Flos Solis maior.*
The greater Sun-floure.

2 *Flos Solis minor.*
The leſſer Sun-floure.

2   The other golden floure of Peru is like the former, sauing that it is altogether lower, and the leaues more jagged, and very few in number.

3   The male floure of the Sun of the smaller sort hath a thicke root, hard, and of a woody substance, with many threddy strings anexed thereto, from which riseth vp a gray or russet stalk to the height of fiue or six cubits, of the bignesse of ones arme, whereupon are set great broad leaues with long footstalks, very fragill or easie to breake, of an ouerworne green colour, sharp pointed, & somwhat cut or hackt about the edges like a saw : the floure groweth at the top of the stalks, bordered about with a pale of yellow leaues : the thrummed middle part is blacker than that of the last described : the whole floure is compassed about likewise with diuers such russet leaues as they are that grow lower vpon the stalks, but lesser and narrower. The plant and euery part thereof smells of turpentine, and the floure yeeldeth forth most cleare turpentine, as my selfe haue noted diuers yeares. The seed is also long and blacke, with certain lines or strakes of white running alongst the same : the root and euery part thereof perisheth when it hath perfected his seed.

4   The female or Marigold Sun-floure hath a thicke and wooddy root, from which riseth vp a straight stem diuiding it selfe into one or more branches, set with smooth leaues sharpe pointed, sleightly indented about the edges. The floures grow at the top of the branches, of a feint yellow colour, the middle part is of a deeper yellow tending to blacknesse, of the forme and shape of a single Marigold, whereupon I haue named it the Sun Marigold. The seed as yet I haue not obserued.

¶ *The Place.*

These plants grow of themselues without setting or sowing, in Peru, and in diuers other prouinces of America, from whence the seeds haue beene brought into these parts of Europ. There hath bin seen in Spain and other hot regions a plant sowne and nourished vp from seed, to attaine to the height of 24 foot in one yeare.

¶ *The Time.*

The seed must be set or sowne in the beginning of April, if the weather be temperat, in the most fertill ground that may be, and where the Sun hath most power the whole day.

¶ *The Names.*

The flour of the Sun is called in Latine *Flos Solis*, for that some haue reported it to turn with the Sun, which I could neuer obserue, although I haue indeauored to finde out the truth of it : but I rather thinke it was so called because it resembles the radiant beams of the Sunne, whereupon some haue called it *Corona Solis*, and *Sol Indianus*, the Indian Sunne-floure : others, *Chrysanthemum Peruuianum*, or the Golden floure of Peru : in English, the floure of the Sun, or the Sun-floure.

¶ *The Temperature.*

They are thought to be hot and dry of complexion.

¶ *The Vertues.*

A   There hath not any thing bin set down either of the antient or later writers, concerning the vertues of these plants, notwithstanding we haue found by triall, that the buds before they be floured boiled and eaten with butter, vineger, and pepper, after the manner of Artichokes, are exceeding pleasant meat, surpassing the Artichoke far in procuring bodily lust.

B   The same buds with the stalks neere vnto the top (the hairinesse being taken away) broiled vpon a gridiron, and after eaten with oile, vineger, and pepper, haue the like propertie.

C H A P. 260.

Of Ierusalem Artichoke.

ONe may wel by the English name of this plant perceiue, that those that vulgarly giue names to plants, haue little either iudgement or knowledge of them : for this plant hath no similitude in leafe, stalke, root, or manner of growing, with an Artichoke, but only a little likenesse of taste in the dressed root ; neither came it from Ierusalem, or out of Asia, but out of America : whence *Fabius Columna* one of the first setters of it forth, fitly names it *After Peruuianus tuberosus*, and *Flos Solis Farnesianus*, because it so much resembles the *Flos Solis*, and for that he first obserued it growing in the garden of Cardinal *Farnesius*, who had procured roots thereof from the West Indies. *Pelliterius* calleth this *Heliotropium Indicum tuberosum* ; and *Bauhinus* in his *Prodromus* sets it forth by the name of *Chrysanthemum latifolium Brasilianum* ; but in his *Pinax* he hath it by the name of of *Helianthemum Indicum tuberosum*. Also out countryman Mr *Parkinson* hath exactly deliuered the historie of this by the name of *Battatas de Canada*, Englishing it Potatoes of Canada: now al these that haue written and mentioned it bring it from America, but from far different places, as from Peru, Brasill, and Canada: but this is not much materiall, seeing it now growes so well and plentifully in so many places of England. I will therefore deliuer you the historie as I haue receiued it from my oft mentioned friend Mr *Goodyer*, who, as you may see by the date, tooke it presently vpon the first arriuall into England.

‡ *Flos solis Pyramidalis.*
Ierusalem Artichoke.

¶ *The Description.*

Flos solis Pyramidalis, paruo flore, tuberosa radice Heliotropium Indicum quorundam.

THis wonderfull encreasing plant hath growing vp from one root, one, somtimes two, three, or more round greene rough hairy straked stalks, commonly about twelue foot high, somtimes sixteene foot or higher, as big as a childes arme, full of white spongeous pith within. The leaues grow all alongst the stalkes out of order, of a light greene colour, rough, sharpe pointed, about 8 inches broad, and ten or eleuen inches long, deeply notched or indented about the edges, very like the leaues of the common *Flos solis Peruanus*, but nothing crumpled, and not so broad: the stalks diuide themselues into many long branches euen from the roots to their very tops, bearing leaues smaller and smaller toward the tops, making the herbe appear like a little tree, narrower and slenderer toward the top, in fashion of a steeple or pyramide. The floures with vs grow onely at the tops of the stalks and branches, like those of the said *Flos solis*, but no bigger than our common single Marigold, consisting of twelue or thirteen straked sharpe pointed bright yellow bordering leaues, growing forth of a scaly small hairy head, with a small yellow thrummy matter within. These floures by reason of their late flouring, which is commonly two or three weekes after Michaelmasse, neuer bring their seed to perfection; and it maketh shew of abundance of smal heads neere the tops of the stalkes and branches, forth of the bosoms of the leaues, which neuer open and floure with vs, by reason they are destroyed with the frosts, which otherwise it seemeth would be a goodly spectacle. The stalke sends forth many small creeping roots, whereby it is fed or nourished, full of hairy threds euen from the vpper part of the earth, spreading far abroad: amongst which from the main root grow forth many tuberous roots, clustring together, somtimes fastned to the great root it selfe, sometimes growing on long strings a foot or more from the root, raising or heauing vp the earth aboue them, and sometimes appearing aboue the earth, producing from the increase of one root, thirty, forty, or fifty in number, or more, making in all vsually aboue a pecke, many times neere halfe a bushell, if the soile be good. These tuberous roots are of a reddish colour without, of a soft white substance within, bunched or bumped out many waies, sometimes as big as a mans fist, or not so big, with white noses or peakes where they wil sprout or grow the next yeare. The stalkes bowed downe, and some part of them couered ouer with earth, send forth small creeping threddy roots, and also tuberous roots like the former, which I haue found by experience. These tuberous roots will abide aliue in the earth all Winter, though the stalkes and roots by which they were nourished vtterly rot and perish away; and will begin to spring vp again at the beginning of May, seldome sooner.

¶ *The Place.*

Where this plant groweth naturally I know not. In *An.* 1617 I receiued two small roots there-of from M.ʳ *Franqueuill* of London, no bigger than hens egges; the one I planted, and the other I gaue to a friend: myne brought me a pecke of roots,wherewith I stored Hampshire.

¶ *The Vertues.*

These roots are dressed diuers wayes, some boile them in water, and after stew them with sacke and butter,adding a little ginger. Others bake them in pies,putting Marrow,Dates,Ginger,Raisons of the sun,Sacke,&c. Others some other way as they are led by their skill in Cookerie. But in my judgement,which way soeuer they be drest and eaten,they stirre and cause a filthy loathsom stinking winde within the body, thereby causing the belly to be pained and tormented; and are a meat more fit for swine, than men: yet some say they haue vsually eaten them, and haue found no such windy qualitie in them. 17 Octob. 1621.   *Iohn Goodyer.* ‡

C H A P. 261.   *Of Cammomill.*

1 *Chamæmelum.*
Cammomill.

2 *Chamæmelum nudum odoratum.*
Sweet naked Cammomill.

¶ *The Description.*

1   TO distinguish the kindes of Cammomils with sundry descriptions, would bee but to inlarge the Volume,and smal profit would thereby redound to the Reader,considering they are so well knowne to all: notwithstanding it shal not be amisse to say somthing of them, to keep the order and method of the book hitherto obserued. The common Cammomill
hath

hath many weak and feeble branches trailing vpon the ground, taking hold on the top of the earth as it runneth, whereby it greatly increaſeth. The leaues are very fine, and much jagged or deepely cut, of a ſtrong ſweet ſmell: among which come forth the floures, like vnto the field Daiſie, bordered about the edge with a pale of white leaues: the middle part is yellow, compoſed of the like thrums cloſe thruſt together as is that of the Daiſie. The root is very ſmall and threddy.

 2 The ſecond kinde of Cammomill hath leaues, roots, ſtalks, and creeping branches like the precedent: the floures grow at the tops of ſmal tender ſtems, which are nothing elſe but ſuch yellow thrummy matter as is in the midſt of the reſt of the Cammomils, without any pale or border of white floures, as the others haue: the whole plant is of a pleaſing ſweet ſmell, whereupon ſome haue giuen it this addition, *Odoratum*.

 3 This third Cammomill differeth not from the former, ſauing that the leaues herof are verie much doubled with white leaues, inſomuch that the yellow thrum in the middle is little ſeen, and the other very ſingle, wherein conſiſteth the difference.

3 *Chamæmelum Anglicum flore multiplici.*
 Double floured Cammomill.

4 *Chamæmelum Romanum.*
 Romane Cammomill.

 4 Roman Cammomil hath many ſlender ſtalks, yet ſtiffer and ſtronger than any of the others by reaſon whereof it ſtandeth more vpright, and doth not creep vpon the earth as the reſt do. The leaues are of a more whitiſh colour, tending to the colour of the leaues of Woad. The floures be likewiſe yellow in the middle, and placed about with a border of ſmall white floures.

¶ *The Place.*
Theſe plants are ſet in gardens both for pleaſure and alſo profit.

¶ *The Time.*
They floure moſt part of all the Summer.

¶ *The*

¶ *The Names.*

Cammomill is called *Chamæmelum* : of ſome, *Anthemis*, and *Leucanthemis*, and alſo *Leucanthemon*,eſpecially that double floured Cammomill ; which Greek name is takenfrom the whiteneſſe of his floure : in Engliſh, Cammomill ; ſo called becauſe the floures haue the ſmell of μῆλον, an apple,which is plainly perceiued in common Cammomill.

¶ *The Temperature.*

Cammomill,ſaith *Galen*,is hot and dry in the firſt degree,and of thin parts : it is of force to digeſt,ſlacken,and rarifie ; alſo it is thought to be like the Roſe in thinneſſe of parts,comming to the operation of oile in heate,which are to man familiar and temperat : wherefore it is a ſpecial helpe againſt weariſomneſſe : it eaſeth and mitigateth pain,it mollifieth and ſuppleth,and all theſe operations are in our vulgar Cammomill,as common experience teacheth, for it heateth moderately, and drieth little.

¶ *The Vertues.*

A   Cammomil is good againſt the Collicke and ſtone, it prouoketh vrine, and is moſt ſingular in Clyſters made againſt the foreſaid diſeaſes.

B   Oile of Cammomill is exceeding good againſt all manner of ache and paine, bruiſings, ſhrinking of ſinues,hardneſſe,and cold ſwellings.

C   The decoction of Cammomill made in wine and drunke, is good againſt coldneſſe in the ſtomack,ſoure belchings,voideth winde,and mightily bringeth downe the monethly courſes.

D   The Egyptians haue vſed it for a remedie againſt all cold agues,and they did therfore conſecrat it(as *Galen* ſaith)to their Deities.

E   The decoction made in white wine and drunke expelleth the dead child, and ſecondine or afterbirth ſpeedily,and clenſeth thoſe parts.

F   The herb boiled in poſſet ale and giuen to drink,eaſeth the pain of the cheſt comming of wind, expelleth tough and clammy flegme,and helpeth children of the Ague.

G   The herb vſed in baths prouoketh ſweat,rarifieth the skin,and openeth the pores : briefly,it mitigateth gripings and gnawings of the belly,allayeth the paine of the ſides,mollifieth hard ſwellings,and waſteth away raw and vndigeſted humors.

H   The oile compounded of the flours performeth the ſame,and is a remedie againſt all weariſomneſſe,being with good ſucceſſe mixed with all thoſe things that are applied to mitigate paine.

Chap.262.
Of May-weed or wilde Cammomill.

¶ *The Kindes.*

THere be three kindes of wilde Cammomil, which are generally called in Latine *Cotula* ; one ſtinking, and two other not ſtinking ; the one hath his floure all white throughout the compaſſe,and alſo in the middle, and the other yellow. Beſides theſe there is another with verie faire double floures void of ſmell,which Mr *Bartholmew Lane* a Kentiſh gentleman found growing wild in a field in the Iſland of Tenet,neere to a houſe called Queakes,ſomtime belonging to Sir *Henrie Criſpe*. Likewiſe Mr *Hesketh* before remembred,found it in the garden of his Inne at Barnet, if my memorie faile me not,at the ſigne of the red Lyon, or neere vnto it, and in a poore womans garden as he was riding into Lancaſhire.

‡ The double floured May-weed, in the yeare 1632, I (being in company with Mr *Wil. Broad*, Mr *Iames Clark*, and ſome other London Apothecaries in the Iſle of Tenet) found it growing wild vpon the cliffe ſide cloſe by the towne of Margate,and in ſome other places of the Iſland. ‡

¶ *The Deſcription.*

1   MAy-weed bringeth forth round ſtalks,green,brittle,and full of juice, parted into many branches thicker and higher than thoſe of Cammomil ; the leaues in like maner are broader,and of a blackiſh green colour. The floures are like in form and color,yet commonly larger,of a ranke and naughty ſmell. The root is wooddy,and periſheth when the ſeed is ripe. The whole plant ſtinketh,and giueth a ranke ſmell.

‡ This

‡ This herb varies, in that it is found somtimes with narrower, and otherwhiles with broader leaues ; as also with a strong vnpleasant smell, or without any smell at all : the flours also are single or else (which is seldome found) very double. ‡

2   The yellow May-weed hath a small and tender root, from which riseth vp a feeble stalke diuiding it selfe into many other branches, whereupon grow leaues not vnlike to Camomil, but thinner and fewer in number : the floures grow at the top of the stalks, of a gold yellow colour. ‡ This I take to be no other than the *Buphthalmum verum* of our Author, formerly described in the second place, Chap. 257.

3   The mountain Cammomill hath leaues somewhat deepely cut in almost to the middle rib, thicke also and juicie, of a bitterish taste, and of no vnpleasant smell : the stalks are weak, and some foot high, carrying at their tops single floures, bigger, yet like those of Cammomill, yellow in the middle, with a border of twenty or more long white leaues incompassing it. It increaseth much, as Cammomil doth, and hath creeping roots. It is found vpon the Stirian Alps, and floureth in Iuly and August. *Clusius* hath set this forth by the name of *Leucanthemum Alpinum*. ‡

1 *Cotula fœtida.*
May-weed.

‡ 3 *Leucanthemum Alpinum Clusij.*
Wilde Mountain Cammomil.

¶ *The Place.*

They grow in Corne fields neere vnto path waies, and in the borders of fields.

¶ *The Time.*

They floure in Iuly and August.

¶ *The Names.*

May-weed is called in shops *Cotula fœtida*: of *Fuchsius, Parthenium*, and *Virginea*, but not truly: of others, Χαμαίμηλον in high-Dutch, **Krotendill ;** in low-Dutch, **Paddebloemen** ; in French, *Espargoutte:* in English, May-weed, wilde Cammomill, and stinking Mathes.

¶ *The Temperature and Vertues.*

May-weed is not vsed for meat nor medicine, and therefore the faculties are vnknowne ; yet all **A** of them are thought to be hot and dry, & like after a sort in operation to Cammomil, but nothing at all agreeing with mans nature : notwithstanding it is commended against the infirmities of the mother, seeing all stinking things are good against those diseases.

It is

It is an vnprofitable weed among corne, raiſing bliſters vpon the hands of weeders and reapers.

C H A P. 263. *Of Pellitorie of Spaine.*

¶ *The Deſcription.*

1   PYrethrum, in Engliſh Pellitorie of Spain (by the name whereof ſome do vnproperly cal another plant, which is indeed the true *Imperatoria* or Maſterwort, & not Pellitory) hath great and fat leaues like vnto Fennell, trailing vpon the ground : amongſt which immediatly from the root riſeth vp a fat great ſtem, bearing at the top a goodly floure faſhioned like the great ſingle white Daiſie, whoſe bunch or knob in the midſt is yellow like that of the Daiſie, and bordered about with a pale of ſmall leaues exceeding white on the vpper ſide, and vnder of a faire purple colour : the root is long, of the bigneſſe of a finger, very hot, and of a burning taſte.

2   The wilde Pellitorie groweth vp like vnto wild Cheruile, reſembling the leaues of *Caucalis*, of a quicke and nipping taſte like the leaues of Dittander or Pepperwort : the floures grow at the top of the ſlender ſtalks, in ſmall tufts or ſpoky vmbels, of a white colour : the root is tough, of the bigneſſe of a little finger, with ſome threds thereto belonging, and of a quick biting taſte.

1 *Pyrethrum officinarum.*
Pellitorie of Spaine.

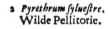

2 *Pyrethrum ſylueſtre.*
Wilde Pellitorie.

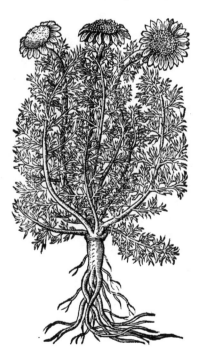 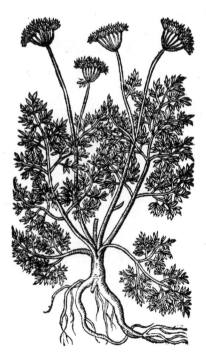

¶ *The Place.*
It groweth in my garden very plentifully.

¶ *The Time.*
It floureth and ſeedeth in Iuly and Auguſt.

¶ *The Names.*
Pellitorie of Spaine is called in Greeke πύρεθρον, by reaſon of his hot and fiery taſte : in Shops alſo *Pyrethrum* : in Latine, *Saliuaris* : in Italian, *Pyrethro* : in Spaniſh, *Pelitre* : in French, *Pied d'*
Alexandre,

Alexandre, that is to ſay, *Pes Alexandrinus*, or Alexanders foot : in high and low Dutch, Bertram: in Engliſh, Pellitorie of Spaine ; and of ſome, Bertram, after the Dutch name : and this is the right *Pyrethrum*, or Pellitorie of Spaine ; for that which diuers here in England take to bee the right, is not ſo, as I haue before noted.

¶ *The Temperature and Vertues.*

The root of Pellitorie of Spaine is very hot and burning, by reaſon whereof it taketh away the A cold ſhiuering of Agues, that haue been of long continuance, and is good for thoſe that are taken with a dead palſie, as *Dioſcorides* writeth.

The ſame is with good ſucceſſe mixed with Antidotes or counterpoyſons which ſerue againſt B the megrim or continuall paine of the head, the dizzineſſe called *Vertigo*, the apoplexy, the falling ſickneſſe, the trembling of the ſinewes, and palſies, for it is a ſingular good and effectuall remedy for all cold and continuall infirmities of the head and ſinewes.

Pyrethrum taken with hony is good againſt all cold diſeaſes of the braine. C

The root chewed in the mouth draweth forth great ſtore of rheume, ſlime, and filthy wateriſh D humors, and eaſeth the paine of the teeth, eſpecially if it be ſtamped with a little Stauef-acre, and tied in a ſmall bag, and put into the mouth, and there ſuffered to remaine a certaine ſpace.

If it be boiled in Vineger, and kept warme in the mouth it hath the ſame effect. E

The oyle wherein Pellitorie hath been boyled is good to anoint the body to procure ſweating, F and is excellent good to anoint any part that is bruiſed and blacke, although the member be declining to mortification : it is good alſo for ſuch as are ſtricken with the palſie.

It is moſt ſingular for the Surgeons of the Hoſpitals to put into their vnctions *contra Neapolita-* G *num morbum*, and ſuch other diſeaſes that be couſin germanes thereunto.

Chap. 264.   *Of Leopards bane.*

† 1 *Doronicum minus Officinarum.*
Small Leopards bane.

† 2 *Doronicum majus Officinarum.*
Great Leopards bane.

¶ *The Deſcription.*

1   OF this plant **Doronicum**, there be ſundry kindes, whereof I will onely touch foure. *Do-donæus* vnproperly **calleth** it *Aconitum Pardalianches*, which hath hapned through the negligence

negligence of *Dioſcorides* and *Theophraſtus*, who in deſcribing *Doronicum*, haue not onely omitted the floures thereof, but haue committed that negligence in many and diuers other plants, leauing out in many plants which they haue deſcribed, the ſpeciall accidents; which hath not a little troubled the ſtudy and determination of the beſt Herbariſts of late yeares, not knowing certainly what to determine and ſet downe in ſo ambiguous a matter, ſome taking it one way, and ſome another, and ſome eſteeming it to be *Aconitum*. But for the better vnderſtanding hereof, know that this word *Aconitum*, as it is a name attributed to diuers plants, ſo it is to be conſidered, that all plants called by this name are malignant and venomous, as with the juyce and root whereof ſuch as hunted after wilde and noyſome beaſts, were woont to embrue and dip their arrowes, the ſooner and more ſurely to diſpatch and ſlay the beaſt in chaſe. But for the proofe of the goodneſſe of this *Doronicum*, and the reſt of his kinde, know alſo, That *Lobel* writeth of one called *Iohn de Vroëde*, who ate very many of the roots at ſundry times, and found them very pleaſant in taſt, and very comfortable. But to leaue controuerſies, circumſtances, and obiections which here might be brought in and alledged, aſſure your ſelues that this plant *Doronicum minus Officinarum* (whoſe roots *Pena* reporteth he found plentifully growing vpon the Pede-mountaine hills and certaine high places in France) hath many leaues ſpred vpon the ground, ſomwhat like Plantaine : among which riſe vp many tender hairy ſtalks ſome handfull and an halfe high, bearing at the top certaine ſingle yellow floures, which when they ſade change into downe, and are carried away with the wind. The roots are thick and many, very crookedly croſſing and tangling one within another, reſembling a Scorpion, and in ſome yeares do grow in our Engliſh gardens into infinite numbers.

‡ 3 *Doronicum radice repente.*
Cray-fiſh Wolfes bane.

‡ 4 *Doronicum brachiata radice.*
Winged Wolfes bane.

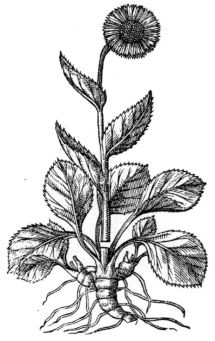

2　　The ſecond kinde of *Doronicum* hath larger leaues than the former, but round, and broader, almoſt like the ſmall leaues of the Clot or Bur : among which riſeth vp a ſtalk ſcarſe a cubit high : the floures are like the former: the root is longer and bigger than the former, barred ouer with many ſcaly barkes, in colour white, and ſhining like white marble, hauing on each ſide one arme or finne, not vnlike to the ſea Shrimpe called *Squilla marina*, or rather like the ribs or ſcales of a Scorpions body, and is ſweet in taſte.

3　　The third kinde of *Doronicum*, growing naturally in great abundance in the mountaines
of

of France,is also brought into and acquainted with our English grounds, bearing very large leaues of a light yellowish greene,and hairy like *Pilosella*,or *Cucumis agrestis*. The stalkes are a cubit high, hauing at the top yellow floures like *Buphthalmum*, or *Consolida media vulnerariorum*; all the root is barred and welted ouer with scales like the taile of a Scorpion, white of colour, and in taste sweet, with some bitternesse,yeelding forth much clamminesse,which is very astringent.

4　The fourth kinde hereof is found in the wooddy mountaines about Turin and Sauoy, very like vnto the former,sauing that the leaues are somewhat rougher, the floures greater,and the stalks higher. But to be short,each of these kindes are so like one another, that in shew, taste, smell, and manner of growing they seeme to be as it were all one : therefore it were superfluous to stand vpon their variety of names, *Pardalianches, Myoctonum, Thelyphonum, Camorum*, & such like,of *Theophrastus, Dioscorides, Pliny*, or any of the new Writers, which names they haue giuen vnto *Doronicum* ; for by the opinion of the most skilfull in plants, they are but Synonimaes of one kinde of plant. And though these old writers speake of the hurtfull qualities of these plants ; yet experience teacheth vs,that they haue written what they haue heard and read, and not what they haue knowne and proued ; for it is apparant that *Doronicum* (by the consent of the old and new writers) is vsed as an antidote or certaine treacle,as well in the confections *de Gemmis Mesua*, as in *Electuario Aramatum*. And though *Matthiolus* disclaimeth against the vse thereof, and calleth it *Pardalianches*, that is, Wolfes bane ; yet let the learned know, that *quantitas, non qualitas, nocet* : for though Saffron be comfortable to the heart,yet if you giue thereof,or of Muske,or any such cordial thing, too great a quantitie,it killeth the party which receiueth it.

‡ 5　*Doronicum angustifolium Austriacum.*
Narrow leafed Wolfes bane.

‡ 6　*Doronicum Stiriacum flore amplo.*
Large floured Wolfes bane.

‡　5　To these foure formerly intended by our Author, may we fitly adde some others out of *Clusius*. The first of these hath a stalke some foot high, soft, rough, and crested : the leaues are few, thicke, narrow, long, very greene and shining, yet hairie on their vpper sides, but smooth on the lower sides, and of a lighter greene ; yet those that adorne the stalkes are narrower : there groweth commonly at the top of the stalke one single floure of the shape and bignesse of the common *Doronicum* described in the second place, but of a brighter yellow : the seed is little and blackish, and is carried away with the winde : the root is small, blackish, and joynted, hauing somewhat thicke

　　　　white

white fibres, and an aromaticke tafte. This floures in Iuly and Auguft, and growes in rockie places vpon the higheft Alpes. *Clufius* (the firft and onely defcriber thereof) calls it *Doronicum 2. fiue Auftriacum 1.*

6   This growes fomewhat higher than the laft defcribed, and hath much broader and rounder leaues, and thofe full of veines, and fnipt about the edges. The knots and off-fets of the roots defcend not downe, but run on the furface of the ground, and fo fends forth fibres at each fide, to faften them and attract nourifhment. The floure is like that of the former, but much larger. This groweth in the high mountainous places of Stiria, and floures at the fame time as the former. *Clufius* calls this *Doronicum 4. Stiriacum.*

7   This is the largeft of all the reft, and hath a ftalke two cubits or more high, of the thickneffe of ones little finger, crefted, rough, and toward the top diuided into fundry branches. The leaues next to the root are round, wrinkled, hairy, and faftned to a long ftalke: thofe towards the top of the ftalke are longer and narrower, and ingirt the ftalke at their fetting on. The floures are large and yellow, like to the other plants of this kinde: the feed alfo is carried away with the winde, and is longifh, and of a greenifh colour: the root is knotty or joynted like to a little Shrimpe, and of a whitifh greene colour. This floures in Iune or Iuly, and growes vpon the like places as the former. *Clufius* calls this *Doronicum 7. Auftriacum 3.* ‡

‡ 7 *Doronicum maximum.*
The greateft Wolfe-bane.

¶ *The Place.*

The place is fufficiently fet forth in the defcription; yet you fhall vnderftand, that I haue the two firft in my Garden; the fecond hath beene found and gathered in the cold mountaines of Northumberland, by D^r. *Penny* lately of London deceafed, a man of much experience and knowledge in Simples, whofe death my felfe and many others do greatly bewaile.

¶ *The Time.*
They floure in the months of Iune and Iuly.

¶ *The Names.*
Concerning their names I haue already fpoken; yet fith I would be glad that our Englifh women may know how to call it, they may terme *Doronicum* by this name, Cray-fifh Piffe-a-bed, becaufe the floure is like Dandelion, which is called Piffe-a-bed.

‡ Our Author certainly at the beginning of this Chapter did not well vnderftand what he faid, when he affirmes, That the reafon of the not wel knowing the *Doronicum* of the Antients was, [through the negligence of *Diofcorides* and *Theophraftus*, who in defcribing *Doronicum*, &c.] Now it is manifeft, that neither of thefe Authors, nor any of the Antient Greekes euer fo much as named *Doronicum*: but that which he fhould haue faid, was, That the want of exact defcribing the *Aconitum thelyphonon* in *Theophraftus*, and *Aconitum Pardalianches* in *Diofcorides* (which are iudged to be the fame plant and all one with our *Doronicum*) hath beene the caufe, that the controuerfie which *Matthiolus* and others haue of late raifed cannot be fully determined, which is, Whether that vulgar *Doronicum*, vfed in fhops, and defcribed in this Chapter, be the *Aconitum Pardalianches*? *Matthiolus* affirmes it is, and much and vehemently exclaimes againft the vfe thereof in cordiall Electuaries, as that which is of a moft pernitious and deadly qualitie, becaufe that (as he affirmes) it will kill dogs: now *Dodonæus* alfo feemes to incline to his opinion: but others (and not without good reafon) deny it, as *Gefner* in his Epiftles, who made often triall of it vpon himfelfe: part of his words are fet downe hereafter by our Author (being tranflated out of *Dodonæus*) and fome part alfo you fhall finde added in the end of the vertues: and thefe are other fome; *Plura alia nunc omitto, quibus oftendere liquido poffem, nec Doronicum noftrum nec*

Aconitum

Aconitum in illo modo esse venenatum homini. Canibus autem letiferum esse scio, non solum si Drachmarum 4. sed etiam si vnius pondere sumant. And before he said, *quasi non alia multa canibus sint venena, quæ homini salubria sunt; vt de asparago fertur.* Of the same opinion with *Gesner* is *Pena* and *Lobel*, who, *Aduers. pag.290,& 291,* do largely handle this matter, and exceedingly deride and scoffe at *Matthiolus* for his vehement declaiming against the vse thereof. Now briefely my opinion is this, That the *Doronicum* here mentioned is not that mentioned and written of by *Serapio* and the Arabians : neither is it the *Aconitum Pardalianches* of *Dioscorides*, nor of so malignant a quality as *Matthiolus* would haue it, for I my selfe also haue often eaten of it, and that in a pretty quantity, without the least offence. ‡

¶ *The Nature and Vertues.*

I haue sufficiently spoken of that for which I haue warrant to write, both touching their natures **A** and vertues ; for the matter hath continued so ambiguous and so doubtfull, yea, and so full of controuersies, that I dare not commit that to the world which I haue read : these few lines therefore shall suffice for this present ; the rest which might be said I refer to the great and learned Doctors, and to your owne consideration.

These herbes are mixed with compound medicines that mitigate the paine of the eies, and by **B** reason of his cold quality, being fresh and greene, it helpeth the inflammation or fiery heate of the eyes.

It is reported and affirmed, that it killeth Panthers, Swine, Wolues, and all kinds of wilde beasts, **C** being giuing them with flesh, *Theophrastus* saith, That it killeth Cattell, Sheep, Oxen, and all foure-footed beasts, within the compasse of one day, not by taken it inwardly only, but if the herb or root be tied vnto their priuie parts. Yet hee writeth further, That the root being drunke is a remedy against the stinging of Scorpions ; which sheweth, that this herbe or the root thereof is not deadly to man, but to diuers beasts only : which thing also is found out by triall and manifest experience ; for *Conrade Gesner* (a man in our time singularly learned, and a most diligent searcher of many things) in a certaine Epistle written two *Adolphus Occo,* sheweth, That he himselfe hath oftentimes inwardly taken the root hereof greene, dry, whole, preserued with hony, and also beaten to pouder ; and that euen the very same day in which he wrote these things, he had drunke with warme water two drams of the roots made into fine pouder, neither felt he any hurt thereby : and that he often-times also had giuen the same to his sicke Patients, both by it selfe, and also mixed with other things, and that very luckily. Moreouer, the Apothecaries in stead of *Doronicum* doe vse (though amisse) the roots thereof without any manifest danger.

That this *Aconite* killeth dogs, it is very certaine, and found out by triall : which thing *Matthio-* **D** *lus* could hardly beleeue, but that at length he found it out to be true by a manifest example, as he confesseth in his Commentaries.

‡ I haue (saith *Gesner*) oft with very good successe prescribed it to my Patients, both alone, as **E** also mixed with other medicines, especially in the *Vertigo* and falling sicknesse : sometimes also I mix therewith Gentian, the pouder of Misle-toe, and *Astrantia* : thus it workes admirable effects in the Epilepsie, if the vse thereof be continued for some time. ‡

† Formerly the figure that was in the first place should haue beene in the second, and the first and second were confounded in the description.

CHAP. 265.   *Of Sage.*

¶ *The Description*

1   THe great Sage is very full of stalkes, foure square, of a wooddy substance, parted into branches, about the which grow broad leaues, long, wrinkled, rough, whitish, very like to the leaues of wilde Mullein, but rougher, and not so white, like in roughnesse to woollen cloath thread-bare : the floures stand forked in the tops of the branches like those of dead Nettle, or of Clarie, of a purple blew colour, in the place of which doth grow little blackish seeds, in small huskes. The root is hard and wooddy, sending forth a number of little strings.

2   The lesser Sage is also a shrubby plant, spred into branches like to the former, but lesser : the stalkes hereof are tenderer : the leaues be long, lesser, narrower, but not lesse rough : to which there do grow in the place wherein they are fixed to the stalke, two little leaues standing on either side one right against another, somewhat after the manner of finnes or little eares : the floures are

eared

eared blew like thoſe of the former : the root alſo is wooddy, both of them ar e of a certaine ſtrong ſmell, but nothing at all offenſiue ; and that which is the leſſer is the better.

3   This Indian Sage hath diuers branches of a wooddy ſubſtance, wheron do grow ſmall leaues, long, rough, and narrow, of an ouerworne colour, and of a moſt ſweet and fragrant ſmell. The floures grow alongſt the tops of the branches, of a white colour, in forme like the precedent. The root is tough and wooddy.

4   The Mountaine Sage hath an vpright ſtalke ſmooth and plaine, whereupon doe grow broad rough and rugged leaues, ſleightly nicked, and vneuenly indented about the edges, of an hoary colour, ſharpe pointed, and of a ranke ſmell : the floures grow alongſt the top of the ſtalke, in ſhape like thoſe of Roſemary, of a whitiſh colour. The root is likewiſe wooddy.

1 *Saluia major.*
Great Sage.

2 *Saluia minor.*
Small Sage.

5   We haue in our gardens a kinde of Sage, the leaues whereof are reddiſh ; part of thoſe red leaues are ſtripped with white, others mixed with white, greene, and red, euen as Nature liſt to play with ſuch plants. This is an elegant variety, and is called *Salvia variegata elegans*, Variegated or painted Sage.

6   We haue alſo another, the leaues whereof are for the moſt part white, ſomewhat mixed with greene, often one leafe white, and another greene, euen as Nature liſt, as we haue ſaid. This is not ſo rare as the former, nor neere ſo beautifull, wherefore it may be termed *Salvia variegata vulgaris*, Common painted Sage.

‡ 7   There is kept in ſome of our chiefe gardens a fine Sage, which in ſhape and manner of growing reſembles the ſmaller Sage, but in ſmell and taſte hath ſome affinity with Wormewood ; whence it may be termed *Saluia Abſinthites*, or Wormewood Sage. *Bauhine* onely hath mentioned this, and that in the fourth place in his *Pinax pag .* 237. by the name of *Saluia minor altera*, and hee addes, *Hæc odore & ſapore eſt Abſinthij, floreque rubente*: that is, This hath the ſmell and taſte of Wormewood, and a red floure : but ours (if my memory faile me not) hath a whitiſh floure : it is a tender plant, and muſt be carefully preſerued from the extremity of Winter. I firſt ſaw this Sage with M^r. *Cannon*, and by him it was communicated to ſome others.

3 *Salvia Indica.*
Indian Sage.

4 *Salvia Alpina.*
Mountaine Sage.

‡ 8 *Salvia Cretica pomifera.*
Apple-bearing Sage of Candy.

‡ 8 *Salvia Cretica non pomifera.*
Candy Sage without Apples.

8　This which we here giue you hath pretty large leaues, and thoſe alſo very hairy on the vnder ſide, but rough on the vpper ſide like as the ordinary Sage. The ſtalkes are rough and hairy, foure ſquare below, and round at their tops. The floures in their growing and ſhape are like thoſe of the ordinary, but of a whitiſh purple colour; and fading, they are each of them ſucceeded by three or foure ſeeds, which are larger than in other Sages, and ſo fill their ſeed-veſſels, that they ſhew like berries. The ſmell of the whole plant is ſomewhat more vehement than that of the ordinarie: the leaues alſo haue ſometimes little eares or appendices, as in the ſmaller or Pig-Sage: and in Candy (the naturall place of the growth) it beares excreſcences, or Apples (if we may ſo terme them) of the bigneſſe of large Galls, or Oke-Apples: whence *Cluſius* hath giuen you two figures by the ſame titles as I here preſent the ſame to your view. *Matthiolus, Dodonæus,* and others alſo haue made mention hereof. ‡

¶ *The Place.*

Theſe kindes of Sage grow not wilde in England: I haue them all in my garden: moſt of them are very common.

‡　The fine or elegant painted Sage was firſt found in a country Garden, by Mr. *Iohn Tradeſcant*, and by him imparted to other louers of plants. ‡

¶ *The Time.*

Theſe Sages floure in Iune and Iuly, or later: they are fitly remoued and planted in March.

¶ *The Names.*

Sage is called in Greeke, ιλελιϛφακος: the Apothecaries, the Italians, and the Spaniards keepe the Latine name *Salvia:* in high Dutch, **Salben:** in French, *Sauge:* in low Dutch, **Sauie:** in Engliſh, Sage.

¶ *The Temperature.*

Sage is manifeſtly hot and dry in the beginning of the third degre, or in the later end of the ſecond; it hath adjoyned no little aſtriction or binding.

¶ *The Vertues.*

A　*Agrippa* and likewiſe *Aëtius* haue called it the Holy-herbe, becauſe women with childe if they be like to come before their time, and are troubled with abortments, do eate thereof to their great good; for it cloſeth the matrix, and maketh them fruitfull, it retaineth the birth, and giueth it life, and if the woman about the fourth day of her going abroad after her childing, ſhall drinke nine ounces of the juyce of Sage with a little ſalt, and then vſe the company of her husband, ſhe ſhall without doubt conceiue and bring forth ſtore of children, which are the bleſſing of God. Thus far *Agrippa.*

B　Sage is ſingular good for the head and braine; it quickneth the ſences and memory, ſtrengthneth the ſinewes, reſtoreth health to thoſe that haue the palſie vpon a moiſt cauſe, takes away ſhaking or trembling of the members; and being put vp into the noſthrils, it draweth thin flegme out of the head.

C　It is likewiſe commended againſt the ſpitting of bloud, the cough, and paines of the ſides, and bitings of Serpents.

D　The juyce of Sage drunke with hony is good for thoſe that ſpit and vomit bloud, and ſtoppeth the flux thereof incontinently, expelleth winde, drieth the dropſie, helpeth the palſie, ſtrengthneth the ſinewes, and cleanſeth the bloud.

E　The leaues ſodden in water, with Wood-binde leaues, Plantaine, Roſemary, Hony, Allome, and ſome white wine, make an excellent water to waſh the ſecret parts of man or woman, and for cankers or other ſoreneſſe in the mouth, eſpecially if you boyle in the ſame a faire bright ſhining Seacole, which maketh it of greater efficacie.

F　No man needs to doubt of the wholeſomneſſe of Sage Ale, being brewed as it ſhould be, with Sage, Scabious, Betony, Spikenard, Squinanth, and Fennell ſeeds.

G　The leaues of red Sage put into a woodden diſh, wherein is put very quicke coles, with ſome aſhes in the bottome of the diſh to keepe the ſame from burning, and a little vinegre ſprinkled vpon the leaues lying vpon the coles, and ſo wrapped in linnen cloath, and holden very hot vnto the ſide of thoſe that are troubled with a grieuous ſtitch, taketh away the paine preſently: The ſame helpeth greatly the extremity of the pleuriſie.

CHAP.

CHAP. 166.   Of French Sage or wooddy Mullein.

1 *Verbascum Matthioli.*
French Sage.

‡ 2 *Verbascum angustis salvia foijs.*
The lesser French Sage.

† 3 *Phlomos Lychnites Syriaca.*
Syrian Sage-leaued Mullein.

¶ *The Description.*

1 WIld Mullein, wooddy Mullein, *Mat-thiolus* his Mullein, or French Sage groweth vp like a small wooddie shrub, hauing many wooddy branches of a woollie and hoarie colour, soft and downy: whereupon are placed thicke hoarie leaues, of a strong pontick fauour, in shape like the leaues of Sage, whereupon the vulgar people call it French Sage: toward the top of the branches are placed roundles or crownets of yellow gaping floures like those of dead nettle, but much greater. The root is thicke, tough, and of a wooddy substance, as is all the rest of the plant.

† 2 There is another sort hereof that is very like the other, sauing that the leaues & euery other part of this plant, hath a most sweet and pleasant smell, and the other more strong and offensiue: the leaues also are much lesser and narrower, somwhat resembling those of the lesser Sage.

‡ 3 I thinke it not amisse here to insert this no lesse rare than beautifull plant, which differs from the last described in the maner of growing & shape of the floures, which resemble those of the *Lychnis Chalcedonica,* or None-such, but are of a yellow colour. The leaues are hairy, narrow, and sharp pointed; the stalkes square, and root wooddy. *Lobel*

(to

(to whom we are beholden for this figure and deſcription) calls this, *Phlomos Lychnites altera Syriaca*. ‡

¶ *The Place.*

Theſe wilde Mulleins do grow wilde in diuers Prouinces of Spaine, and alſo in Languedoc, vpon dry bankes and ſtony places: I haue them both in my Garden, and many others likewiſe.

¶ *The Time.*

They floure in Iune and Iuly.

¶ *The Names.*

They are called of the learned men of our time, *Verbaſca Sylueſtria*: the firſt is called of the Grecians φλόμος, or ϲλογμος in Latine, *Elychnium*, or after others, *Elychinium*, becauſe of the Cottonie ſubſtance thereof, matches or weeks were made to keep light in lamps: *Verbaſcum Lychnitis*, as *Dioſcorides* himſelfe teſtifieth, is named alſo *Thryallis* or Roſe Campion; but the floure of *Thryallis* is red of colour, as *Nicander* in his Counterpoyſons doth ſhew, but the floures of theſe are yellow: therefore they are neither *Thryallis* nor *Lychnitis*, but *Sylueſtre Verbaſcum*, or wilde Mullein, as we haue already taught in the Chapter of Roſe Campion, that *Thryallis* is *Lychnitis ſatiua*, or Roſe Campion. There is nothing to the contrary, but that there may be many plants with ſoft downy leaues fit to make Candle weeke of: in Engliſh it is generally called French Sage: we may call it Sage Mullein.

¶ *The Temperature.*

As theſe be like in vertues to the others going before, ſo they be likewiſe dry in temperature.

¶ *The Vertues.*

Dioſcorides ſaith, that the leaues are ſtamped and laied in manner of a pultis vpon burnings and ſcaldings.

Chap. 267. *Of Clarie.*

1 *Gallitricum, ſiue Horminum.*
Common Clarie.

2 *Gallitricum alterum.*
Small Clarie.

¶ *The*

‡ 3. *Horminum sylatstre Fuchsij.*
Fuchsius his wilde Clarie.

¶ *The Description.*

1 THe first kinde of Clarie which is the right, bringeth forth thick stalks foure square, two foot long, diuided into branches: it hath many leaues growing both from the roots, and along the stalkes and branches by distances, one against another by two and two, great, a handfull broad or broader, somewhat rough, vnequall, whitish and hairie, as be also the stalkes. The floures are like those of Sage, or of dead Nettle, of colour white, out of a light blew: after which grow vp long toothed huskes in stead of cods, in which is blacke seed. The root is full of strings: the whole herbe yeeldeth forth a rank and strong smell that stuffeth the head: it perisheth after the seed is ripe, which is in the second yeare after it is sowne.

2 The second kinde of Clarie hath likewise stalkes foure square, a foot and a halfe high: the leaues also be rough and rugged, lesser, and not so white. The floures be alike, of colour purple or blew: the roots bee as those of the former are. This hath not so strong a sent by a great deale.

3 There is a kinde of Clarie, which *Fuchsius* pictureth for wilde Clarie, that hath shorter stalkes, hairie, and also foure square: the leaues lesser, long, deeper indented: the floures blew of colour, sweet of smell, but not so sweet as those of

† 4 *Colus Iouis.*   Iupiters Distaffe.

the right Clarie: the husks or cods when they are ripe bend downewards: the seed is blackish, the roots in like manner are blacke and full of strings.

4   The fourth kind of *Horminum*, called *Iovis Colus*, reprefenteth in the higheft top of the ftalk a diftaffe, wrapped about with yellow flax, whereof it tooke his name, hauing knobby roots, with certaine ftrings annexed thereto like *Galeopfis*, or like vnto the roots of Clary, which do yeeld forth fundry foure fquare rough ftalkes, two cubits high, whereon do grow leaues like thofe of the Nettle, rough, fharpe pointed, and of an ouerworne greene colour: the floures do grow alongft the top of the ftalkes, by certaine fpaces, fet round about in fmall coronets, or wharles, like thofe of Sage in forme, but of a yellow colour.

¶ *The Place.*

Thefe doe grow wilde in fome places, notwithftanding they are anured and planted in Gardens, almoft euery where, except Iupiters diftaffe, being a kinde thereof, which I haue in my Garden.

¶ *The Time.*

They floure in Iune, Iuly, and Auguft.

¶ *The Names.*

Clarie is called of the Apothecaries, *Gallitricum*; it is likewife named *Oruala:* of fome, *Tota bona*, but not properly: of others, *Scarlea, Sclarea, Centrum Galli*, and *Matrifalvia:* in Italian, *Sciaria:* in French, *Oruale:* in High Dutch, **Scharlach** : in Low Dutch, **Scharleye** : in Englifh, Clarie, or Cleere-eie.

Iupiters diftaffe is called *Colus Iovis:* of fome, *Galeopfis lutea*, but not properly: of diuers, *Horminum luteum*, or yellow Clarie, and *Horminum Tradentinum*, or Clarie of Trent.

¶ *The Temperature.*

Clarie is hot and dry in the third degree.

¶ *The Vertues.*

A   The feed of Clarie poudered, finely fcarced and mixed with hony, taketh away the dimneffe of the eies, and cleereth the fight.

B   The fame ftamped, infufed, or laied to fteepe in warme water, the muffilag or flimie fubftance taken and applied plaifterwife draweth fplinters of wood, thornes, or any other thing fixed in the bodie: it alfo fcattereth and diffolueth all kindes of fwellings, efpecially in the joynts.

C   The feed poudered and drunke with wine, ftirreth vp bodily luft.

D   The leaues of Clarie taken any manner of way, helpeth the weakeneffe of the backe proceeding of the ouermuch flowing of the whites, but moft effectually if they be fried with egges in maner of a Tanfie, either the leaues whole or ftamped.

†   The figure which formerly was vnder the title of *Colus Iovis*, was of the *Horminum filveftre* of *Fuchfius*, which is defcribed immediatly before it.

C H A P. 268.   *Of wilde Clarie, or* Oculus Chrifti.

¶ *The Defcription.*

1   **O**culus Chrifti is alfo a kinde of Clarie, but leffer: the ftalkes are many, a cubit high, fquared and fomewhat hairie: the leaues be broad, rough, and of a blackifh greene colour. The floures grow alongft the ftalkes, of a blewifh colour. The feed is round and blackifh, the root is thicke and tough, with fome threds annexed thereto. ‡ This is *Hormini fylveftris* 4. *quinta fpecies* of *Clufius*. ‡

2   The purple Clarie hath leaues fomewhat round, layed ouer with a hoary cottony fubftance, not much vnlike Horehound: among which rife vp fmall hairy fquare ftalkes, fet toward the top with little leaues of a purple colour, which appeare at the firft view to be flours, and yet are nothing elfe but leaues, turned into an excellent purple colour: and among thefe beautifull leaues come forth fmal floures of a blewifh or watched colour, in fafhion like to the floures of Rofemarie; which being withered, the husks wherein they do grow containe certaine blacke feed, that falleth forth vpon the ground very quickely, becaufe that euery fuch huske doth turne and hang downe his head toward the ground. The root dieth at the firft approch of Winter.

‡ 3   Broad leaued Clarie hath a fquare ftalke fome cubit high, hairy, firme, and joynted; the leaues are large, rough, and fharpe pointed, fnipt about the edges, wrinkled, and ftanding by couples at each joynt: vpon the branches in roundles grow purple floures, leffe than thofe of Clarie, and fcarce any bigger than thofe of Lauander: the feed is fmall and blacke: the root is large, hard, blacke,

† 1 *Horminum syluestre.*
Wilde Clarie, or *Oculus Christi.*

2 *Horminum syluestre folijs purpureis.*
Clarie with purple leaues.

‡ 3 *Horminum syluestre latifolium.*
Broad leaued wilde Clarie.

‡ 4 *Horminum syluestre flore albo.*
White floured wilde Clarie.

‡ 5 *Horminum ſylueſtre flore rubro.*
Red floured wilde Clarie.

blacke, and liues many yeares. It floures in Iune and Iuly, and growes wilde in many mountainous places of Germany. *Cluſius* calls it *Horminum ſylueſtre tertium.*

4 This hath long leaues next vnto the ground, growing vpon pretty long ſtalkes, broad at their ſetting on, and ſo ending by little and little in ſharpe points, they are not deeply cut in, but onely lightly ſnipt about the edges : they are alſo wrinkled on the vpper ſide, and whitiſh, but hairie on the vnder ſide. The ſquare ſtalkes are ſome cubit high, joynted, and ſet with two leaues at each joynt. The floures grow alongſt the tops of the branches, and are of a ſnow white colour. There is a variety of this with the leaues greener, and the floures of an elegant deepe purple colour. This is the *Hormini ſylueſtris quarti ſpecies prima* of *Cluſius*, and the variety with the white floures is his *Hormini ſylueſtris quarti ſpecies prima :* and the figure that our Authour gaue in the firſt place was of theſe.

5 There is another variety of the laſt deſcribed, which alſo hath ſquare ſtalks ſet with rough ſnipt leaues, which end in ſharpe points, but are narrower at the lower end than the former, and they are greene of colour : vpon the tops of the ſtalkes grow red hooded floures, and thoſe not very large : the ſeed is ſmall and blacke, and the root liues many yeares. This floures in Iuly. *Cluſius* makes this his *Hormini ſylueſtris quarti ſpecies quarta.* ‡

¶ *The Place.*

The firſt groweth wilde in diuers barren places, almoſt in euery country, eſpecially in the fields of Holborne necre vnto Grayes Inne, in the high way by the end of a bricke wall : at the end of Chelſey next to London, in the high way as you go from the Queenes pallace of Richmond to the waters ſide, and in diuers other places.

The other is a ſtanger in England : it groweth in my Garden.

¶ *The Time.*

They floure and flouriſh from Iune to the end of Auguſt.

¶ *The Names.*

Wilde Clarie is called after the Latine name *Oculus Chriſti,* of his effect in helping the diſeaſes of the eies : in Greeke, *ʜʊʍ :* and likewiſe in Latine, *Horminum :* of ſome, *Geminalis :* in Engliſh, wild Clarie, and *Oculus Chriſti.*

The ſecond is thought of ſome to be the right Clarie, and they haue called it *Horminum verum,* but with greater errour : it may be called in Latine *Horminum ſylueſtre folijs & floribus purpureis,* Clarie with leaues and floures of a purple colour.

‡ Our Author ſhould haue ſhowne his reaſons why this is not the *Horminum verum,* to haue conuicted the errour of *Anguillara, Matthiolus, Geſner, Dodonæus Lobel,* and others, who haue accounted it ſo, as I my ſelfe muſt needs do, vntill ſome reaſon be ſhewed to the contrary, the which I thinke cannot be done.

¶ *The Temperature and Vertues.*

A   The temperature and faculties are referred vnto the garden Claries : yet *Paulus Ægineta* ſaith it is hot and moderately dry, and it alſo clenſeth.

B   The ſeed of wilde Clarie, as *Dioſcorides* writeth, being drunke with wine, ſtirreth vp luſt, it clenſeth the eies from filmes and other imperfections, being mixed with hony.

C   The ſeed put whole into the eies, clenſeth and purgeth them exceedingly from wateriſh humors, redneſſe, inflammation, and diuers other maladies, or all that happen vnto the eies, and takes away the paine and ſmarting thereof, eſpecially being put into the eies one ſeed at one time, and

no

no more,which is a generall medicine in Cheſhire and other countries thereabout, knowne of all, and vſed with good ſucceſſe.

The leaues are good to be put into pottage or broths among other pot-herbes , for they ſcatter congealed bloud,warme the ſtomacke,and help the dimneſſe of the eies.

⸸ The figure that formerly was in the firſt place, was of that which you may here finde figured and deſcribed in the fourth place.

CHAP. 269.   Of Mullein.

¶ The Deſcription.

1   THe male Mullein or Higtaper hath broad leaues,very ſoft,whitiſh and downy ; in the midſt of which riſeth vp a ſtalk,ſtraight,ſingle,and the ſame alſo whitiſh all ouer,with a hoary down,and couered with the like leaues,but leſſer and leſſer euen to the top ; among which taperwiſe are ſet a multitude of yellow floures conſiſting of fiue leaues apiece:in the places wherof come vp little round veſſels,in which is contained very ſmall ſeed.The root is long, a finger thicke,blacke without,and full of ſtrings.

1 *Tapſus barbatus*.
Mullein or Higtaper.

2 *Tapſus barbatus flore albo*.
White floured Mullein.

2   The female Mullein hath likewiſe many white woolly leaues,ſet vpon an hoary cottony vpright ſtalke of the height of foure or fiue cubits : the top of the ſtalke reſembleth a torch decked with infinite white floures,which is the ſpeciall marke to know it from the male kinde,being like in euery other reſpeَ.

¶ The Place.

Theſe plants grow of themſelues neere the borders of paſtures, plowed fields, or cauſies & dry ſandy ditch banks,and in other vntilled places. They grow in great plenty neere vnto a lyme-kiln vpon the end of Blacke heath next to London,as alſo about the Queenes houſe at Eltham neere to Dartford in Kent ; in the highwayes about Highgate neere London,and in moſt countries of England that are of a ſandy ſoile.

¶ *The Time.*

They are found with their floure from Iuly to September, and bring forth their ſeed the ſecond yeare after it is ſowne.

¶ *The Names.*

Mullein is called in Greeke φλόμος: in ſhops, *Tapſus Barbatus*: of diuers, *Candela Regia, Candelaria,* and *Lanaria*: *Dioſcorides, Pliny,* and *Galen* call it *Verbaſcum*: in Italian, *Verbaſco,* and *Taſſo Barbaſſo*: in Spaniſh, *Gordolobo*: in high-Dutch, **Wullkraut**: in French, *Bouillon*: in Engliſh, Mullein, or rather Woollen, Higtaper, Torches, Longwort, and Bullocks Longwort; and of ſome, Hares beard.

¶ *The Temperature.*

Mullein is of temperature dry: the leaues are alſo of a digeſting and clenſing qualitie, as *Galen* affirmeth.

¶ *The Vertues.*

A   The leaues of Mullein boiled in water, and laid vpon hard ſwellings and inflammations of the eies, cure and eaſe the paine.

B   The root boiled in red wine and drunke, ſtoppeth the laske and bloudy flix.

C   The ſame boiled in water and drunke, is good for them that are broken and hurt inwardly, and preuaileth much againſt the old cough.

D   A little fine treacle ſpred vpon a leafe of Mullein, and layd to the piles or hemorrhoids, cureth the ſame: an ointment alſo made of the leaues thereof and old hogs greaſe worketh the ſame effect.

E   The leaues worne vnder the feet day and night in manner of a ſhoo ſole or ſock, bring downe in yong maidens their deſired ſickeneſſe, being ſo kept vnder their feet that they fall not away.

F   The country people, eſpecially the husbandmen in Kent, do giue their cattel the leaues to drink againſt the cough of the lungs, being an excellent approued medicine for the ſame, wherupon they call it Bullocks Lungwort.

G   Frankinceſe and Maſtick burned in a chafing diſh of coles, and ſet within a cloſe ſtoole, and the fume thereof taken vnderneath, doth perfectly cure the piles, hemofrhoids, and al diſeaſes hapning in thoſe lower parts, if there be alſo at euery ſuch fuming (which muſt be twice euery day) a leafe of the herb bound to the place, and there kept vntill the next dreſſing.

H   There be ſome who think that this herb being but carried about one, doth help the falling ſickneſſe, eſpecially the leaues of that plant which hath not as yet born floures, and gathered when the Sun is in Virgo, and the Moon in Aries; which thing notwithſtanding is vaine and ſuperſtitious.

I   The later Phyſitions commend the yellow floures, being ſteeped in oile and ſet in warme dung vntill they be waſhed into the oile and conſumed away, to be a remedie againſt the piles.

K   The report goeth (ſaith *Pliny*) that figs do not putrifie at all that are wrapped in the leaues of Mullein: which *Dioſcorides* alſo mentioneth.

Chap. 270. *Of baſe Mullein.*

¶ *The Deſcription.*

1   The baſe white Mullein hath a thicke wooddy root, from which riſeth vp a ſtiffe & hairy ſtalke of the height of foure cubits, garniſhed with faire grayiſh leaues like thoſe of Elecampane, but leſſer: the floures grow round about the ſtalks taper or torch faſhion, of a white colour, with certain golden thrums in the middle: the ſeed followeth, ſmall, and of the colour of duſt.

2   Blacke Mullein hath long leaues not downy at all, large and ſharp pointed, of an ouerworne blackiſh green colour, ſomewhat rough, and ſtrongly ſmelling: the floures grow at the top of the ſtalks, of a golden yellow colour, with certain threds in the middle thereof. The root differeth not from the precedent.

3   Candleweek Mullein hath large broad and woolly leaues, like vnto thoſe of the common Mullein: among which riſeth vp a ſtalke couered with the like leaues, euen to the branches wheron the floures do grow, but leſſer and leſſer by degrees. The ſtalke diuideth it ſelfe toward the top into diuers branches, whereon is ſet round about many yellow floures, which oftentimes doe change into white, varying according vnto the ſoile and clymat. The root is thicke and wooddy.

4 The

1 *Verbascum album.*
Base white Mullein.

2 *Verbascum nigrum.*
Base blacke Mullein.

3 *Verbascum Lychnite Matthioli.*
Candleweeke Mullein.

4 *Verbascum Lychnite minus.*
Small Candleweek Mullein.

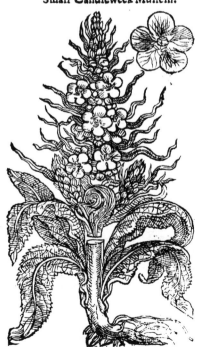

4   The ſmall Candleweek Mullein differs little from the laſt rehearſed, ſauing that the whole plant of this is of a better ſauor, wherein eſpecially conſiſteth the difference. ‡ The floure alſo is much larger, and of a ſtraw or pale yellow colour. ‡

¶ *The Place.*

Theſe plants grow where the other Mulleins do, and in the like ſoile.

¶ *The Time.*

The time likewiſe anſwereth their flouring and ſeeding.

¶ *The Names.*

Their capitall names expreſſed in the titles ſhal ſerue for theſe baſe Mulleins, conſidering they are all andeuery of them kindes of Mulleins.

¶ *The Temperature.*

Theſe Mulleins are dry without any manifeſt heate, yet doubtleſſe hotter and drier than the common Mullein or Hygtaper.

¶ *The Vertues.*

A   The blacke Mullein with his pleaſant yellow flours boiled in water or wine and drunk, is good againſt the diſeaſes of the breſt and lungs, and againſt all ſpitting of corrupt rotten matter.

B   The leaues boiled in water, ſtamped and applied pultis wiſe vpon cold ſwellings called *Oedemata*, and alſo vpon the vlcers and inflammations of the eyes, cure the ſame.

C   The floures of blacke Mullein are put into lie, which cauſeth the haire of the head to wax yellow, if it be waſhed or kembed therewith.

D   The leaues are put into cold ointments with good ſucceſſe, againſt ſcaldings and burnings with fire or water.

E   *Apuleius* reporteth a tale of *Vlyſſes, Mercury*, and the Inchantreſſe *Circe*, and their vſe of theſe herbs in their incantations and witchcrafts.

CHAP. 271. *Of Moth Mullein.*

1 *Blattaria Pliny.*
Plinies Moth Mullein.

2 *Blattaria flore purpureo.*
Purple moth Mullein.

¶ *The Deſcription.*

1   PLiny hath ſet forth a kind of *Blattaria* which hath long and ſmooth leaues ſomwhat jagged or ſnipt about the edges : the ſtalk riſeth vp to the height of three cubits, diuiding
it ſelfe toward the top into ſundry armes or branches, beſet with yellow leaues like vnto
blacke Mullein.

2   *Blattaria* with purple floures hath broad black leaues, without any manifeſt ſnips or notches
by the ſides, growing flat vpon the ground ; among which riſeth vp a ſtalke two cubits high, garniſhed with floures like vnto the common *Blattaria*, but that they are of a purple colour, and thoſe
few threds or chiues in the middle of a golden colour : the root is as thicke as a mans thumb, with
ſome threds hanging thereat, and it endures from yeare to yeare.

3   There is another kinde like vnto the blacke Mullein in ſtalks, roots, and leaues, and other reſpects, ſauing that his ſmall floures are of a green colour.

4   There is another like vnto the laſt before written, ſauing that his leaues are not ſo deepely
cut about the edges, and that the ſmall floures haue ſome purple colour mixed with the greeneſſe.

‡ 3 *Blattaria flore viridi.*
Greene Moth Mullein.

‡ *Blattaria flore ex viridi purpuraſcente.*
Moth Mullein with the greeniſh purple
coloured floure.

‡ 5   This is ſomewhat like the firſt deſcribed in leaues and ſtalks, but much leſſe, the floures
alſo are of a whittiſh or grayiſh colour, wherein conſiſts the greateſt difference.

6   There is alſo another varietie of this kinde, which hath very faire and large floures, & theſe
either of a bright yellow, or elſe of a purple colour.

7   This hath long narrow leaues like thoſe of the ſecond, ſnipt about the edges, and of a darke
green colour: the ſtalks grow ſome 2 cubits high, and ſeldom ſend forth any branches ; the floures
are large and yellow, with rough threds in their middles tipt with red, and theſe grow in ſuch an
order that they ſomwhat reſemble a fly : the ſeed is ſmall, and contained in round buttons. This is
an Annuall, and periſheth when the ſeed is ripe. ‡

¶ *The*

‡ 5 *Blattaria flore albo.*
White floured Moth Mullein.

‡ 6 *Blattaria flore amplo.*
Moth Mullein with the great floure.

‡ 7 *Blattaria flore luteo.*
Yellow Moth Mullein.

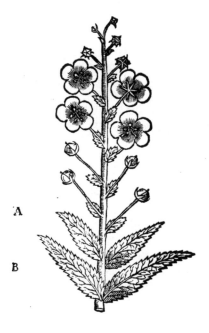

A

B

¶ *The Place.*

† The firſt and fift of theſe grow wild in ſundry places, and the reſt only in gardens with vs.

¶ *The Time.*

They floure in Iuly and Auguſt.

¶ *The Names.*

The later herbariſts cal Moth Mullein *Blattaria*, and do truely take it to be that deſcribed by *Pliny, lib. 22. cap. 9.* in theſe words: [There is an herb like Mullein or *Verbaſcum nigrum*, which oftentimes deceiueth, being taken for the ſame, with leaues not ſo white, moe ſtalks, & with yellow floures (as we haue written) which do agree with black Mullein; but we haue not as yet learned by obſeruation, that they gather Moths and flies vnto them, as we haue ſaid.] *Valerius Cordus* names it *Verbaſcum Leptophyllon*, or narrow leafed Mullein: their ſeuerall titles ſet forth their Engliſh names.

¶ *The Temperature and Vertues.*

Concerning the plants intituled *Blattaria*, or Moth Mulleins, I find nothing written, but that Moths and Butter-flies, and all other ſmall flies and bats, do reſort to the place where theſe herbs are layd or ſtrewed.

‡ The decoction of the floures or leaues of the firſt deſcribed opens the obſtructions of the bowels and meſeraick veines, as *Camerar.* affirmes. ‡

CHAP. 272. *Of Mullein of Ethiopia.*

Æ herb.
Ethyopian Mullein.

¶ *The Description.*

MVllein of Ethiopia hath many very broad hoary leaues spred vpon the ground, verie soft and downy, or rather woolly like those of Hygtaper, but farre whiter, softer, thicker, and fuller of woollinesse, which wooll is so long, that one may with his fingers pull the same from the leaues, euen as wool is pulled from a sheeps kin: amongst which leaues riseth vp a foure square downy stalk set with the like leaues, but smaller; which stalk is diuided at the top into other branches, set about and orderly placed by certain distances, hauing many floures like those of Archangell, of a white colour tending to blewnesse: which being past, there succedeth a three square brown seed: the root is black, hard, and of a woody substance.

¶ *The Place.*

It groweth naturally in Ethyopia, and in Ida a hill by Troy, and in Messenia a prouince of Morea, as *Pliny* sheweth, *lib. 27. cap. 4.* it also growes in Meroë an Island in the riuer Nilus; and also in my garden.

¶ *The Time.*

It floureth and flourisheth in Iune, and perfecteth his seed in August.

¶ *The Names.*

It is called in Greeke Αιθιοπις and in Latine *Æthiopis,* of the countrey; and for that cause it is likewise called *Meroides,* of *Meroë,* as *Pliny* writeth: of some, because the Greeke word Αιθος signifies in Latine *Fauilla adusta,* or *Cinere aspersa,* or couered with ashes: in English wee may call it Mullein of Ethyopia, or woolly Mullein.

¶ *The Temperature.*

Æthiopis is dry, without any manifest heate.

¶ *The Vertues.*

Æthiopis is good against the Pleurisie, and for those that haue their brests charged with corrupt A and rotten matter, and for the asperitie and roughnesse of the throat, and against the Sciatica, if one drinke the decoction of the root thereof.

For the diseases of the brest and lungs it is good to licke oftentimes of a confection made with B the root hereof and hony, and so are the roots condited with sugar, in such manner as they condite the roots of Eringos.

CHAP. 273. *Of Cowslips.*

¶ *The Description.*

1   THose herbs which at this day are called Primroses, Cowslips, and Oxlips, are reckoned among the kindes of Mulleins; notwithstanding for distinctions sake I haue marshalled them in a chapter, comming in the rereward as next neighbors to the Mulleins, for that the Antients haue named them *Verbasculi,* that is to say, small Mulleins. The first, which is called in English the field Cowslip, is as common as the rest, therefore I shal not need to spend much time about the description.

2   Thesecond is likewise well knowne by the name of Oxlip, and differeth not from the other
saue

ſaue that the floures are not ſo thicke thruſt together,and they are fairer,and not ſo many in number,and do not ſmell ſo pleaſant as the other : of which kind we haue one lately come into our gardens,whoſe floures are curled and wrinkled after a moſt ſtrange maner,which our women haue named Iack-an-apes on horſebacke.

1 *Primula veris maior.*
Field Cowſlips.

2 *Primula pratenſis inodorata lutea.*
Field Oxlips.

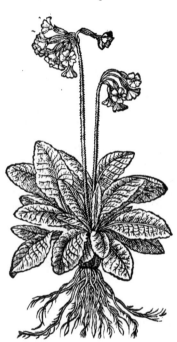 

   3   Double Paigle,called of *Pena, Primula hortenſis Anglica omnium maxima, & ſerotina floribus plenis* ; that is,The greateſt Engliſh garden Cowſlip with double yellow floures, is ſo commonly knowne that it needeth no deſcription.

   4   The fourth is likewiſe known by the name of double **Cowſlips,**hauing but one floure within another,which maketh the ſame once double, where the other is many times double,called by *Pena,Geminata,*for the likeneſſe of the floures,which are brought forth as things againſt nature, or twinnes.

   5   The fifth being the common whitiſh yellow field **Primroſe,**needeth no deſcription.

   6   The ſixth, which is our garden double Primroſe, of all the reſt is of greateſt beauty,the deſcription whereof I refer vnto your owne conſideration.

   7   The ſeuenth is alſo very well known,being a Primroſe with greeniſh floures ſomwhat welted about the edges ; for which cauſe *Pena* hath called it *Siluarum primula,floribus obſcure verentibus fimbriatis.*

   There is a ſtrange Primroſe found in a wood in Yorkſhire growing wilde,by the trauel and induſtry of a learned gentleman of Lancaſhire called M[r] *Tho. Hesketh,* a diligent ſearcher of Simples, who hath not only brought to light this amiable and pleaſant Primroſe, but many others alſo, neuer before his time remembred or found out. This kind of Primroſe hath leaues and roots like the wilde field Primroſe in each reſpect : it bringeth forth among the leaues a naked ſtalk of a grayiſh or ouerworne greeniſh colour,at the top whereof doth grow in the Winter time one floure and no more,like vnto that ſingle one of the field : but in the Summer time it bringeth forth a ſoft ruſſet huske or hoſe,wherein are contained many ſmal floures, ſomtimes foure or fiue,many times more, very thicke thruſt together,which make one entire floure,ſeeming to be one of the common double Primroſes,whereas indeed it is one double floure,made of a number of ſmall ſingle floures, neuer ceaſing to beare floures winter nor ſummer,as before is ſpecified.

                                                       ‡ Beſides

3 *Primula hortensis Anglica.*
Double Paigles.

4 *Primula veris flore geminato.*
Cowslips two in a hose.

5 *Primula veris minor.*
Field Primrose.

6 *Primula veris flore pleno.*
Double white Primrose.

‡ Besides these, there are kept in our gardens and set forth by Mr *Parkinson* (to whose Worke I refer the curious Reader) two or three more varieties, one a double Cowslip hose in hose, naked, without any huske: the other two beare many greene leaues on the tops of the stalkes, the one of them hauing yellowish floures among the leaues, and the other only longish narrow green leaues. The first bee calleth *Paralysis inodora flore geminato*, Double Oxlips hose in hose. The second, *Paralysis fatua*, the Foolish Cowslip: and the last, *Paralysis flore viridi roseo calamistrato*, the double greene feathered Cowslip. ‡

7 *Primula flore viridi.*
Green Primroſe.

‡ 8 *Primula veris Heskethi.*
Mʳ *Heskeths* Primroſe.

¶ *The Place.*

Cowſlips and Primroſes ioy in moiſt and dankiſh places, but not altogether couered with water: they are found in woods and the borders of fields: Mʳ *Heskeths* Primroſe growes in a wood called Clapdale, three miles from a towne in Yorkſhire called Settle.

¶ *The Time.*

They flouriſh from Aprill to the end of May, and ſome one or other of them do floure all Winterlong.

¶ *The Names.*

They are commonly called *Primula veris,* becauſe they are the firſt among thoſe plants that doe floure in the Spring, or becauſe they floure with the firſt. They are alſo named *Arthritica,* and *Herba paralyſis,* for they are thought to be good againſt the paines of the joints and ſinues. They are called in Italian, *Brache cuculi:* in Engliſh, Petty Mulleins or Palſie-worts: of moſt, Cowſlips.

The greater ſort, called for the moſt part Oxlips or Paigles, are named of diuers *Herba S. Petri:* in Engliſh, Oxlip, and Paigle.

The common Primroſe is vſually called *Primula veris:* moſt herbariſts doe refer the Primroſes to the Φλώμοι called in Latine *Verbaſcula,* or petty Mulleins: but ſeeing the leaues be neither woolly nor round, they are hardly drawne vnto them: for *Phlomides* are deſcribed by leaues, as *Pliny* hath interpreted it, *Hirſutis & rotundis,* hairy and round; tranſlating it thus, *lib. 25. cap. 10. Sunt & Phlomides duæ hirſutæ, rotundis folijs, humiles:* which is as much as to ſay in Engliſh, There be alſo two pretty Mulleins, hairy, round leafed, low or ſhort. ‡ *Fabius Columna* refers theſe to the *Aliſma* of *Dioſcorides,* and calls the Cowſlip *Aliſma pratorum:* and the Primroſe, *Aliſma ſyluarum.* ‡

¶ *The Temperature.*

The Cowſlips and Primroſes are in temperature dry and a little hot.

¶ *The Vertues.*

A　Cowſlips are commended againſt the pain of the joints called the gout, and ſlackneſſe of the ſinues, which is the palſie. The decoction of the roots is thought to be profitably giuen againſt the ſtone in the kidnies and bladder; and the juice of the leaues for members that are looſe and out of joints, or inward parts that are hurt, rent, or broken.

B　A dramme and a halfe of the pouder of the dried roots of field Primroſe gathered in Autumne,

giuen

giuen to drinke in ale or wine purgeth by vomit very forcibly (but ſafely) wateriſh humors, choler, and flegme, in ſuch manner as *Aʒarum* doth, experimented by a learned and skilfull Apothecary of Colcheſter Mr *Buckſtone*, a man ſingular in the knowledge of Simples.

A conſerue made with the flours of Cowſlips and ſugar preuaileth wonderfully againſt the pal- **C** ſie, convulſions, cramps, and all diſeaſes of the ſinues.

Cowſlips or Paigles do greatly reſtraine or ſtop the belly in time of a great laske or bloudy flix, **D** if the decoction thereof be drunke warme.

A practitioner of London who was famous for curing the phrenſie, after he had performed his **E** cure by the due obſeruation of phyſick, accuſtomed euery yeare in the moneth of May to dyet his Patients after this manner: Take the leaues and floures of Primroſe, boile them a little in foun-taine water, and in ſome roſe and Betony waters, adding thereto ſugar, pepper, ſalt, and butter, which being ſtrained, he gaue them to drinke thereof firſt and laſt.

The roots of Primroſe ſtampéd and ſtrained, and the iuice ſniffed into the noſe with a quill or **F** ſuch like, purgeth the brain, and qualifieth the pain of the megrim.

An vnguent made with the iuice of Cowſlips and oile of Linſeed, cureth all ſcaldings or bur- **G** nings with fire, water, or otherwiſe.

The floures of Primroſes ſodden in vineger and applied, heale the Kings euill, and the almonds **H** of the throat and vuula, if you gargariſe the part with the decoction thereof.

The leaues and floures of Primroſes boiled in wine and drunke, are good againſt all diſeaſes of **I** the breſt and lungs, and draweth forth of the fleſh any thorne or ſplinter, or bone fixed therein.

CHAP. 274. *Of Birds Eyne.*

1 *Primula veris flore rubro.*
Red Bird-eyne.

2 *Primula veris flore albo.*
White Bird-eyne.

¶ *The Deſcription.*

1　SOme Herbariſts call this plant by the name of *Sanicula anguſtifolia*, making thereof two kinds, and diſtinguiſhing them by theſe termes, *major*, & *minor ſiue media:* others cal them *Paralytica alpina*, which without controuerſie are kindes of Cowſlips, agreeing with them as well in ſhape, as in their nature and vertues, hauing leaues much like vnto Cowſlips, but ſmaller,

growing

growing flat vpon the ground, of a feint greenifh colour on the vpper fide, & vnderneath of a white or mealy colour : among which rife vp fmall and tender ftalks of a foot high, hauing at the top of euery ftalke a bufh of fmall floures in fhape like the common Oxlip, fauing that they are of a faire ftammel colour tending to purple : in the middle of euery fmall floure appeareth a little yellow fpot refembling the eye of a bird ; which hath moued the people of the North parts (where it a-boundeth) to call it Birds eyne. The feed is fmall like duft ; the root white and thready.

2   The fecond is like the firft, fauing the whole plant is greater in each refpect, and the floures of a whitifh colour.

¶ The Place.

Thefe plants grow very plentifully in moift and fqually grounds in the North parts of England, as in Harwood nere to Blackburn in Lancafhire, and ten miles from Prefton in Aundernefſe; alfo at Crosby, Rauenswaith, and Crag-Clofe in Weftmerland.

They likewife grow in the medowes belonging to a village in Lancafhire neere Maudfley, cal-led Harwood, and at Hesketh not far from thence, and in many other places of Lancafhire, but not on this fide Trent, that I could euer certainly know. *Lobel* reporteth, That Dr *Penny* (a famous phy-fition of our London Colledge) did finde them in thefe Southern parts.

¶ The Time.

They floure and flourifh from Aprill to the end of May.

¶ The Names.

The firft is called Primrofe with the red floure : the fecond, Primrofe with the white floure, and Birds eyne.

¶ The Temperature and Vertues.

The nature and vertues of thefe red and white Primrofes muft be fought out among thofe aboue named.

CHAP. 275.   Of Beares eares, or mountaine Cowflips.

1 *Auricula vrfiflore luteo.*
Yellow Beares eare.

2 *Auricula vrfiflore purpureo.*
Purple Beares eare.

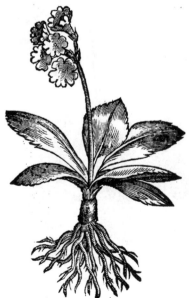

¶ *The Kindes.*

THere be diuers sorts of Mountaine Cowslips, or Beares-eares, differing especially in the colour of their floures, as shall be declared, notwithstanding it may appeare to the curious, that there is great difference in the roots also, considering some of them haue knobbed roots, and others thredddy: notwithstanding there is no difference in the roots at all.

‡ There are diuers varieties of these floures, and the chiefe differences arise, either from the leaues or floures; from their leaues, which are either smooth and greene, or else gray and hoary, againe they are smooth about the edges, or snipt more or lesse; The floures some are fairer than othersome, and their colours are so various, that it is hard to finde words to expresse them, but they may be referd to whites, reds, yellowes, and purples; for of all the varieties and mixtures of these they chiefly consist. The Gardens of Mr. *Tradescant* and Mr. *Tuggie* are at this present furnished with very great varieties of these floures. ‡

3 *Auricula Vrsi ij. Clusij.*
Red Beares-eare.

4 *Auricula Vrsi iiij. Clusij.*
Scarlet Beares-eare.

¶ *The Description.*

1   AVricula Vrsi was called of *Matthiolus*, *Pena*, and other Herbarists, *Sanicula Alpina*, by reason of his singular faculty in healing of wounds, both inward and outward. They doe all call it *Paralytica*, because of his vertues in curing the palsies, cramps, and convulsions, and is numbred among the kindes of Cowslips, whereof no doubt they are kindes as others are which do hereafter follow vnder the same title, although there be some difference in the colour of the floures. This beautifull and braue plant hath thicke, greene, and fat leaues, somewhat finely snipt about the edges, not altogether vnlike those of Cowslips, but smoother, greener, and nothing rough or crumpled: among which riseth vp a slender round stem a handfull high, bearing a tuft of floures at the top, of a faire yellow colour, not much vnlike to the floures of Oxe-lips, but more open and consisting of one only leafe like Cotiledon: the root is very thredddy, and like vnto the Oxe-lip

2   The leaues of this kinde which beareth the purple floures are not so much snipt about the edges: these said purple floures haue also some yellownesse in the middle, but the floures are not so much laid open as the former, otherwise in all respects they are like.

3   *Carolus Cluſius* ſetteth forth in the booke of his Pannonicke trauels two kindes more, which he hath found in his trauell ouer the Alpes and other mountaines of Germany and Heluetia, being the third in number, according to my computation : it hath leaues like the former, but longer, ſmaller, and narrower toward the bottome, greene aboue, and of a pale colour vnderneath. The floures are in faſhion like to the former, but of a moſt ſhining red colour within, and on the outſide of the colour of a mulberry : the middle or eie of the floure is of a whitiſh pale colour : the root is like the former.

4   The fourth is a ſmaller plant than any of the foreſaid, whoſe leaues are thicke and fat, nothing at all ſnipt about the edges, greene aboue, and grayiſh vnderneath. The floures are like the former, ſhining about the edges, of an ouerworne colour toward the middle, and in the middle commeth a forke couered with an hairineſſe : the root is blacke and threddy.

5 *Auricula Vrſi erubeſcens.*
Bluſh coloured Beares eare.

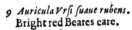
9 *Auricula Vrſi ſuaue rubens.*
Bright red Beares eare.

7 *Auricula Vrſi minima.*
Stamell Beares eare.

5   The bluſh-coloured Beares eare hath diuers thicke fat leaues ſpred vpon the ground, of a whitiſh greene colour, ſleightly or not at all indented in the edges : among which riſeth vp a naked ſtalke likewiſe hairy or whitiſh, on the top whereof ſtand very faire floures, in ſhape like thoſe of the common Cowſlip, but of a whitiſh colour tending to purple, which wee terme bluſh-colour. The root is tough and threddy, as are all the reſt.

6   The bright ſhining red Beares eare of *Matthiolus* deſcription ſeemes to late Herbariſts to be rather a figure made by conceit or imagination, than by the ſight of the plant it ſelfe ; for doubtleſſe we are perſuaded that there is no ſuch plant, but onely a figure foiſted for oſtentation ſake, the deſcription whereof we leaue to a further conſideration, becauſe we haue not ſeene any ſuch plant, neither do we beleeue there is any ſuch. ‡ Our Author is here without cauſe iniurious to *Matthiolus* ; for he figures and deſcribes onely the common firſt deſcribed yellow Beares eare : yet if he had ſaid the floures were of a light ſhining red, he had not erred ; for I haue ſeene theſe floures of all the reds both bright and darke that one may imagine. ‡

7 *Pena*

7 *Pena* ſetteth forth a kinde of Beares eare vnder the name of *Sanicula alpina*, hauing his vpper-moſt leaues an inch long, ſomewhat jagged and hem'd at the ends, and broad before like a ſhouel; the lower leaues next the ground are ſomewhat ſhorter, but of the ſame forme; among which riſeth a ſmall ſlender foot-ſtalke of an inch long, whereon doth ſtand a ſmall floure, conſiſting of fiue little leaues of a bright red or ſtammell colour.

8 The ſnow white Beares-eare differeth not from the laſt deſcribed but in the colour of the floure, for as the others are red, contrary theſe are very white, and the whole plant is leſſer, wherein conſiſteth the difference. The root is long, tough, with ſome fibres thereto belonging. Neither of theſe two laſt deſcribed will be content to grow in Gardens.

¶ *The Place.*

They grow naturally vpon the Alpiſh and Heluetian mountaines: moſt of them do grow in our London Gardens.

¶ *The Time.*

Theſe herbes do floure in Aprill and May.

¶ *The Names.*

Either the antient writers knew not theſe plants, or elſe the names of them were not by them or their ſucceſſors diligently committed vnto poſterity. *Matthiolus* and other later writers haue giuen names according to the ſimilitude, or of the ſhape that they beare vnto other plants, according to the likeneſſe of the qualities and operations: you may call it in Engliſh, Beares-eare: they that dwell about the Alpes doe call it **Diaſtkrawt**, and **Schwindlekrawt**, by reaſon of the effects thereof; for the root is amongſt them in great requeſt for the ſtrengthning of the head, that when they are on the tops of places that are high, giddineſſe and the ſwimming of the braine may not afflict them: it is there called the Rocke-Roſe, for that it groweth vpon the rockes, and reſembleth the braue colour of the roſe. ‡ *Fabius Columna* proues this to be the *Aliſma* or *Damaſonium* of *Dioſcorides* and the Antients. ‡

¶ *The Temperature.*

Theſe herbes are dry and very aſtringent.

¶ *The Vertues.*

It healeth all outward and inward wounds of the breſt, and the enterocele alſo, if for ſome reaſo- A
nable ſpace of time it be put in drinkes, or boyled by it ſelfe.

Theſe plants are of the nature and temperature of *Primula veris*, and are reckoned amongſt the B
Sanicles by reaſon of their vertue.

Thoſe that hunt in the Alps and high mountaines after Goats and bucks, do as highly eſteeme C
hereof as of *Doronicum*, by reaſon of the ſingular effects that it hath, but (as I ſaid before) one eſpe-cially, euen in that it preuenteth the loſſe of their beſt joynts (I meane their neckes) if they take the roots hereof before they aſcend the rocks or other high places.

‡ The root of *Damaſonium* (according to *Dioſcorides*) taken in the weight of one or two drams, D
helpeth ſuch as haue deuoured the *Lepus marinus* or ſea Hare, or haue beene bitten by a Toad, or ta-ken too great a quantity of *Opium*.

It is alſo profitably drunke, either by it ſelfe, or with the like quantity of *Daucus* ſeeds, againſt E
gripings in the belly, and the bloudy flux.

Alſo it is good againſt convulſions and the affects of the wombe. F

The herbe ſtaies the flux of the belly, moues the courſes, and applied in forme of a pultis aſſwa- G
geth œdematous tumors. ‡

Chap. 256. *Of Mountaine Sanicle.*

¶ *The Kindes.*

THere be ſundry ſorts of herbes contained vnder the name of Sanicle, and yet not one of them agreeing with our common Sanicle, called *Diapenſia*, in any one reſpect, except in the vertues, wherofno doubt they tooke that name, which number doth dayly increaſe, by reaſon that the later writers haue put downe more new plants, not written of before by the Antients, which ſhall be di-ſtinguiſhed in this chapter by ſeuerall titles.

¶ *The Deſcription.*

1 SPotted Sanicle of the mountaine hath ſmall fat & round leaues, bluntly indented about the edges, and faſhioned like vnto the leaues of *Saxifragia aurea*, or rather *Cyclamen folio hederæ*, of a darke greene colour, and ſomewhat hairy vnderneath: amongſt which riſe

1 *Sanicula guttata.*
Spotted Sanicle.

2 *Pinguicula ſiue Sanicula Eboracenſis.*
Butterwort, or Yorkſhire Sanicle.

3 *Sanicula Alpina Cluſij, ſiue* Cortuſa Matthioli.
Beares-eare Sanicle.

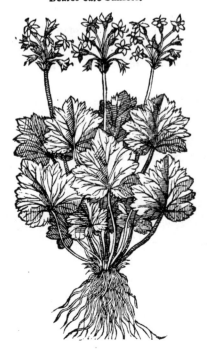

vp ſundry ſtalkes, beſet with like leaues, but
ſmaller, and of a cubit high, diuiding themſelues into many ſmall armes or branches, bearing diuers little white floures, ſpotted moſt
curiouſly with bloudy ſpecks or prickles, inſomuch that if you marke the admirable worke-
manſhip of the ſame wrought in ſuch glorious
manner, it muſt needs put euery creature in
minde of his Creator: the floures are in ſmell
like the May floures or Hawthorne: the ſeed
is ſmall and blacke, contained in ſmall pointals like vnto white Saxifrage: the root is ſcaly
and full of ſtrings.

    2   The ſecond kind of Sanicle, which *Clu-
ſius* calleth *Pinguicula,* not before his time remembred, hath ſmall thicke leaues, fat and full
of juyce, being broad towards the root, and
ſharpe towards the point, of a faint greene colour, and bitter in taſte: out of the middeſt
whereof ſprouteth or ſhooteth vp a naked ſlender ſtalke, nine inches long, euery ſtalke bearing one floure and no more, ſometimes white
but commonly of a blewiſh purple colour, faſhioned like vnto the common *Conſolida regalis,* hauing the like ſpur or Larks heele annexed
thereto.

    3   The third kinde of mountaine Sanicle
ſome

ſome Herbariſts haue called *Sanicula alpina floro rubro* : the leaues ſhoot forth in the beginning of the Spring,very thicke and fat, and are like a purſe or round lumpe at their firſt comming out of the ground ; and when it is ſpread abroad,the vpper part thereof is full of veines or ſinewes,and houen vp or curled like *Ranunculus Luſitanicus*, or like the crumpling of a cabbage leaſe ; and are not onely indented about the edges, but each leaſe is diuided into ſix or more jagges or cuts, deepely hacked,greeniſh aboue,and of an ouerworne greene colour vnderneath,hot in taſte;from the middle whereof ſhooteth forth a bar or naked ſtalke,ſix inches long,ſomewhat purple in colour, bearing at the top a tuſt of ſmall hollow floures,looking or hanging downewards like little bells,not vnlike in forme to the common Cowſlips, but of a fine deepe red colour tending to purple, hauing in the middle a certaine ring or circle of white, and alſo certaine pointals or ſtrings, which turne into an head wherein is contained ſeed. The whole plant is couered as it were with a rough woollineſſe : the root is fibrous and threddy.

¶ The Place.

Theſe plants are ſtrangers in England ; their naturall country is the Alpiſh mountaines of Heluetia : they grow in my Garden, where they flouriſh exceedingly, except Butterwort, which groweth in our Engliſh ſqually wet grounds, and will not yeeld to any culturing or tranſplanting : it groweth eſpecially in a field called Crag-Cloſe, and at Crosby, Rauenſwaith, in Weſtmerland, vpon Ingleborow fels twelue miles from Lancaſter, and in Harwood in the ſame county neere to Blackburne,ten miles from Preſton in Aunderneſſe vpon the bogs, and mariſh grounds, and in the boggie medowes about Biſhops Hatfield ; and alſo in the fens in the way to Wittles meare from London,in Huntingdonſhire. ‡ It groweth alſo in Hampſhire, and abundantly in many places of Wales. ‡

¶ The Time.

They floure and flouriſh from May to the end of Iuly.

¶ The Names.

The firſt is called *Sanicula guttata*, taken from the ſpots wherewith the floures are marked:of *Lobel,Geum Alpinum*,making it a kinde of Auens : in Engliſh,ſpotted Sanicle :of our London dames, Pratling Parnell.

The ſecond is called *Pinguicula*,of the fatneſſe or fulneſſe of the leafe, or of fatning : in Yorkeſhire,where it doth eſpecially grow,and in greateſt abundance,it is called Butterworts,Butter-root, and white root : but the laſt name belongeth more properly to Solomons Seale.

¶ The Temperature.

They are hot and dry in the third degree.

¶ The Vertues.

The husbandmens wiues of Yorkſhire do vſe to annoint the dugs of their kine with the fat and oilous juyce of the herbe Butterwort, when they are bitten with any venomous worme,or chapped, rifted and hurt by any other meanes.　　　　　　　　　　　　　　　　　　　　　　　　　　　A

They ſay it rots their ſheepe,when for want of other food they eat thereof.　　　　　　　　　　　　B

CHAP.277.　Of Fox-Gloues.

¶ The Deſcription.

FOx-gloue with the purple floure is moſt common;the leaues whereof are long,nicked in the edges,of a light greene,in manner like thoſe of Mullein,but leſſer,and not ſo downy: the ſtalke is ſtraight, from the middle whereof to the top ſtand the floures,ſet in a courſe one by another vpon one ſide of the ſtalke,hanging downwards with the bottome vpward,in forme long, like almoſt to finger ſtalkes, whereof it tooke his name *Digitalis*, of a red purple colour, with certaine white ſpots daſht within the floure ; after which come vp round heads, in which lies the ſeed ſomewhat browne, and as ſmall as that of Time. The roots are many ſlender ſtrings.

2　　The Fox-Gloue with white floures differs not from the precedent but in the colour of the floures ; for as the other were purple,theſe contrariwiſe are of a milke-white colour.

3　　We haue in our Gardens another ſort hereof, which bringeth forth moſt pleaſant yellow floures,and ſomewhat leſſer than the common kinde, wherein they differ. ‡ This alſo differs from the common kind in that the leaues are much ſmoother,narrower,and greener,hauing the nerues or veines running alongſt it,neither are the nerues ſnipt,nor ſinuated on their edges. ‡

1 *Digitalis purpurea.*
Purple Fox-gloues.

2 *Digitalis alba.*
White Fox-gloues.

‡ 3 *Digitalis lutea.*
Yellow Fox-gloues.

‡ 4 *Digitalis ferruginea.*
Dusky Fox-gloues.

4   We haue also another sort, which we call *Digitalis ferruginea*, whose floures are of the colour of rusty yron; whereof it tooke his name, and likewise maketh the difference. ‡ Of this sort there is a bigger and a lesser; the bigger hath the lower leaues some foot long, of a darke green colour, with veines running along them; the stalks are some yard and halfe high, the floures large and ending in a sharpe tarned vp end as you see in the figure, and they are of a rusty colour, mixed of a yellow and red.

5   The lesser duskie Fox-gloue hath much lesse leaues and those narrow, smooth, and exceeding greene: amongst which comes vp a stalke some foot high, hauing small floures of the colour of the last described. This I obserued in the yeare 1632. in floure with Mr. *Iohn Tradescant* in the middle of Iuly. It may fitly be called *Digitalis ferruginea minor*, small duskie Fox-gloues. ‡

¶ *The Place.*

Fox-gloue groweth in barren sandy grounds, and vnder hedges almost euery where.

Those with white floures do grow naturally in Landesdale, and Crauen, in a field called Cragge close, in the North of England: likewise by Colchester in Essex; neere Excester in the West parts, and in some few other places. The other two are strangers in England, neuerthelesse they doe grow with the other in my Garden.

¶ *The Time.*

They floure and flourish in Iune and Iuly.

¶ *The Names.*

Fox-gloues some call in Greeke, Ἀραβὶς, and make it to be *Verbasci speciem*, or a kinde of Mullein: in Latine, *Digitalis*: in High Dutch, **Fingerhut**, and **Finger kraut**: in Low Dutch **Vinger hoet**: in French, *Gantes nostre dame*: in English, Fox-gloues. ‡ *Fabius Columna* thinkes it to be that *Ephemerum* of *Dioscorides* described in his fourth booke, and *cap.* 75. ‡

¶ *The Temperature.*

The Fox-gloues in that they are bitter, are hot and dry, with a certaine kinde of clensing qualitie joyned therewith; yet are they of no vse, neither haue they any place amongst medicines, according to the Antients.

¶ *The Vertues.*

Fox-gloue boiled in water or wine, and drunken, doth cut and consume the thicke toughnesse of **A** grosse and slimie flegme and naughty humours; it openeth also the stopping of the liuer, spleene, and milt, and of other inward parts.

The same taken in like manner, or boiled with honied water or sugar, doth scoure and clense the **B** brest, ripeneth and bringeth forth tough and clammie flegme.

They serue for the same purposes whereunto Gentian doth tend, and hath been vsed in stead **C** thereof, as *Galen* saith.

‡   Where or by what name *Galen* either mentions, or affirmes this which our Authour cites **D** him for, I must confesse I am ignorant. But I probably conjecture that our Authour would haue said *Fuchsius*: for I onely finde him to haue these words set downe by our Author, in the end of his Chapter of *Digitalis*. ‡

CHAP. 278.   *Of Baccharis out of* Dioscorides.

¶ *The Description.*

1   ABout this plant *Baccharis* there hath beene great contention amongst the new Writers; *Matthiolus* and *Dodonaeus* haue mistaken this plant, for *Coniza major*, or *Coniza Helenitis Cordi*; *Virgil* and *Athenaeus* haue confounded *Baccharis*, and *Azarum* together: but following the antient Writers, it hath many blackish rough leaues, somwhat bigger than the leaues of Primrose: amongst which riseth vp a stalke two cubits high, bearing at the top little chaffie or scalie floures in small bunches, of a darke yellowish or purple colour, which turne into downe, and are carried away with the winde, like vnto the kindes of Thistles: the root is thicke, grosse, and fat, spreading about in the earth, full of strings: the fragrant smell that the root of this plant yeeldeth, may well be compared vnto the sauour of Cinnamon, *Helenium*, or *Enula Campana*, being a plant knowne vnto very many or most sort of people, I meane in most parts of England.

¶ *The Place.*

Baccharis delighteth to grow in rough and craggy places, and in a leane soile where no moisture is

Baccharis Monſpelienſium.
Plowmans Spikenard.

it : it groweth very plentifully about Mont-pellier in France, and diuers places in the Weſt parts of England.

¶ *The Time.*

It ſpringeth vp in April, it floureth in Iune, and perfecteth his ſeed in Auguſt.

¶ *The Names.*

The learned Herbariſts of Montpellier haue called this plant *Baccharis* : the Græci-ans, βάκχαρις : or after others, σάκαρις, by reaſon of that ſweet and aromaticall ſauour which his root containeth and yeeldeth : in Engliſh it may be called the Cinamom root, or Plow-mans Spiknard : *Virgil* in his ſeuenth Ecloge of his Bucolicks maketh mention of *Baccha-ris*, and doth not onely ſhew that it is a Gar-land plant, but alſo ſuch a one as preuaileth againſt inchantments, ſaying,

——— *Bacchare frontem*
Cingite, ne vati noceat mala lingua futuro.

With Plowmans Nard my forehead girt,
Leſt euill tongue thy Poët hurt.

Baccharis is likewiſe an ointment in *Athe-næus*, in his 15. booke, which may take his name of the ſweet herbe *Baccharis* : for as *Pliny* writeth, *Ariſtophanes* of old, being an antient comical Poët witneſſeth, that ointments were wont to bee made of the root thereof : to bee briefe, *Cratevas*, his *Aſarum* is the ſame that *Dioſcorides* his *Baccharis* is. ‡ This plant here deſcribed is the *Coniza major* of *Matthiolus, Tragus,* and others. ‡

¶ *The Temperature.*

Baccharis or Plowmans Spiknard is of temperature very aſtringent or binding.

¶ *The Vertues.*

A   *Baccharis* or the decoction of the root, as *Paulus Ægineta* briefely ſetteth downe, doth open the pipes and paſſages that are ſtopped, prouoketh vrine, and bringeth downe the deſired ſickeneſſe : the leaues thereof for that they are aſtringent or binding, ſtop the courſe of fluxes and rheumes.

B   *Baccharis* is a ſingular remedy to heale inflammations and Saint Anthonies fire, called *Ignis ſacer*, and the ſmell thereof prouoketh ſleepe.

C   The decoction of the roots of *Baccharis* helpeth ruptures and convulſions, thoſe alſo that haue falne from an high place, and thoſe that are troubled with the ſhortneſſe of breath.

D   It helpeth alſo the old cough, and difficulty to make water.

E   When it is boiled in wine, it is giuen with great profit againſt the biting of Scorpions, or any venomous beaſt, being implaiſtered and applied thereto.

F   A bath made thereof and put into a cloſe ſtoole, and receiued hot, mightily voideth the birth, and furthereth thoſe that haue extreame labour in their childing, cauſing them to haue eaſie deli-uerance.

C H A P. 279.   *Of Elecampane.*

¶ *The Deſcription.*

ELecampane bringeth forth preſently from the root great white leaues, ſharpe pointed, almoſt like thoſe of great Comfrey, but ſoft, and couered with a hairy downe, of a whitiſh greene co-lour,

Helenium.
Elecampane.

lour, and are more white vnderneath, fleightly nicked in the edges : the ftalke is a yard and a halfe long, about a finger thicke, not without downe, diuided at the top into diuers branches, vpon the top of euery fprig ftand great floures broad and round, of which not only the long fmal leaues that compaffe round about are yellow, but alfo the middle ball or circle, which is filled vp with an infinite number of threds, and at length is turned into fine downe; vnder which is flender and long feed : the root is vneuen, thicke, and as much as a man may gripe, not long, oftentimes blackifh without, white within, and full of fub-ftance, fweet of fmell, and bitter of tafte.

¶ *The Place.*

It groweth in medowes that are fat and fruit-full : it is alfo oftentimes found vpon mountains, fhadowie places, that be not altogether dry : it groweth plentifully in the fields on the left hand as you go from Dunftable to Puddlehill, alfo in an orchard as you go from Colbrooke to Ditton ferry, which is the way to Windfor, and in fundry other places, as at Lidde, and Folkeftone, neere to Douer by the fea fide.

¶ *The Time.*

The floures are in their brauery in Iune and Iuly : the roots be gathered in Autumne, and of-tentimes in Aprill and May.

¶ *The Names.*

That which the Græcians name *λἰνον,* the La-tines call *Inula* and *Enula :* in fhops, *Enula campa-na :* in high Dutch, **Alantwurtz :** in low Dutch, **Alandt wortele :** in Italian, *Enoa,* and *Enola :* in Spanifh, *Raiʒ del alla :* in French, *Enula Campane :* in Englifh, Elecampane and Scab.woort, and Horfe-heale : fome report that this plant tooke the name *Helenium* of *Helena* wife to *Menalaus,* who had her hands full of it when *Paris* ftole her away into Phrygia.

¶ *The Temperature.*

The root of this Elecampane is maruellous good for many things, being of nature hot and dry in the third degree, efpecially when it is dry : for being greene and as yet full of iuyce, it is full of fuperfluous moifture, which fomewhat abateth the hot and dry quality thereof.

¶ *The Vertues.*

It is good for fhortneffe of breath, and an old cough, and for fuch as cannot breathe vnleffe they A hold their neckes vpright.

It is of great vertue both giuen in a looch, which is a medicine to be licked on, and likewife pre- B ferued, as alfo otherwife giuen to purge and void out thicke, tough, and clammy humours, which fticke in the cheft and lungs.

The root preferued is good and wholefome for the ftomacke : being taken after fupper it doth C not onely helpe digeftion, but alfo keepeth the belly foluble.

The iuyce of the fame boyled, driueth forth all kinde of wormes of the belly, as *Pliny* teacheth : D who alfo writeth in his twentieth booke, and fifth chapter, the fame being chewed fafting, doth fa-ften the teeth.

The root of Elecampane is with good fucceffe mixed with counterpoifons: it is a remedy againft E the bitings of ferpents, it refifteth poifon : it is good for them that are burften and troubled with cramps and convulfions.

Some affirme alfo, that the decoction thereof, and likewife the fame beaten into pouder and F mixed with hony in manner of an ointment, doth clenfe and heale vp old vlcers.

Galen faith, that herewith the parts are to be made red, which be vexed with long and cold griefs: G as are diuers paffions of the huckle bones, called the Sciatica, and little and continuall bunnies and loofeneffe of certaine joynts, by reafon of ouermuch moifture.

The

H   The decoction of *Enula* drunken prouoketh vrine, and is good for them that are grieued with inward burſtings, or haue any member out of joynt.

I   The root taken with hony or ſugar, made in an electuary, clenſeth the breaſt, ripeneth tough flegme, and maketh it eaſie to be ſpit forth, and preuaileth mightily againſt the cough and ſhortneſſe of breath, comforteth the ſtomacke alſo, and helpeth digeſtion.

K   The roots condited after the manner of *Eringos* ſerue for the purpoſes aforeſaid.

L   The root of *Enula* boiled very ſoft, and mixed in a morter with freſh butter and the pouder of Ginger, maketh an excellent ointment againſt the itch, ſcabs, mangineſſe, and ſuch like.

M   The roots are to be gathered in the end of September, and kept for ſundry vſes, but it is eſpecially preſerued by thoſe that make Succade and ſuch like.

Chap. 264. *Of Sauce alone, or Iacke by the hedge.*

Alliaria.
Sauce alone.

SAuce alone hath affinity with Garlicke in name, not becauſe it is like it in forme, but in ſmell: for if it be bruiſed or ſtamped it ſmelleth altogether like Garlicke: the leaues hereof are broad, of a light greene colour, nicked round about, and ſharpe pointed: the ſtalke is ſlender, about a cubit high, about the branches whereof grow little white floures; after which come vp ſlender ſmal and long cods, & in theſe black ſeed: the root is long, ſlender, and ſomething hard.

¶ *The Place.*

It groweth of it ſelfe by garden hedges, by old wals, by highway ſides, or oftentimes in the borders of fields.

¶ *The Time.*

It floureth chiefely in Iune and Iuly, the ſeed waxeth ripe in the meane ſeaſon. The leaues are vſed for a ſauce in March or Aprill.

¶ *The Names.*

The later writers call it *Alliaria*, and *Alliaris*: of ſome, *Rima Maria*: it is not *Scordium*, or water Germander, which the apothecaries in times paſt miſtooke for this herbe: neither is it *Scordij ſpecies*, or a kinde of water Germander, whereof we haue written: it is named of ſome, *Pes Aſininus*: it is called in High Dutch, **Knoblauch kraut Leuchel**, and **Seſzkraut**: and in Low Dutch, **Loock ſonder Loock**: you may name it in Latine, *Allium non bulboſnm*: in French, *Alliayre*: in Engliſh, Sauce alone, and Iacke by the hedge.

¶ *The Temperature.*

Iacke of the hedge is hot and dry, but much leſſe than Garlicke, that is to ſay, in the end of the ſecond degree, or in the beginning of the third.

¶ *The Vertues.*

A   We know not what vſe it hath in medicine: diuers eat the ſtamped leaues hereof with Salt-fiſh for a ſauce, as they do thoſe of Ramſons.

B   Some alſo boile the leaues in cliſters which are vſed againſt the paine of the collicke and ſtone, in which not only winde is notably waſted, but the pain alſo of the ſtone mitigated and very much eaſed.

Chap.

CHAP. 281. *Of Dittany.*

¶ *The Deſcription.*

1   **D**Ittanie of Crete now called Candie (as *Dioſcorides* ſaith) is a hot and ſharpe hearbe, much like vnto Penni-Royall, ſauing that his leaues be greater and ſomewhat hoary, couered ouer with a ſoft downe or white woollie cotton : at the top of the branches grow ſmall ſpikie eares or ſcaly aglets, hanging by little ſmall ſtemmes, reſembling the ſpiky tufts of Marjerome, of a white colour : amongſt which ſcales there do come forth ſmall floures like the flouring of Wheat, of a red purple colour ; which being paſt, the knop is found full of ſmall ſeed, contrary to the ſaying of *Dioſcorides*, who ſaith, it neither beareth floure nor ſeed, but my ſelfe haue ſeene it beare both in my Garden : the whole plant periſhed in the next Winter following.

1 *Dictamnum Creticum.*
Dittany of Candy.

2 *Pſeudodictamnum.*
Baſtard Dittany.

1 *Dictamnum Creticum.*
Dittany of Candy.

2   The ſecond kind called *Pſeudodictamnum*, that is, Baſtard Dittany, is much like vnto the firſt, ſauing that it is not ſweet of ſmell, neither doth it bite the tongue, hauing round ſoft woolly ſtalkes with knots and joynts, and at euery knot two leaues ſomewhat round, ſoft, woolly, and ſomewhat bitter : the floures be of a light purple colour, compaſſing the ſtalks by certain ſpaces like garlands or wharles, and like floures of Penny-Roiall. The root is of a wooddy ſubſtance : the whole plant groweth to the height of a cubit and a halfe, and laſteth long.

¶ *The Place.*

The firſt Dittany commeth from Crete, an Iſland which we call Candie, where it growes natu-rally : I haue ſeene it in my garden, where it hath floured and borne ſeed ; but it periſhed by reaſon of the injury of our extraordinary cold Winter that then happened : neuertheleſſe *Dioſcorides* writeth

writeth againſt all truth,that it neither beareth floures nor ſeed : after *Theophraſtus, Virgil* wineſſeth that it doth beare floures in the twelfth of his Æneidos.

Dictamnum genetrix Cretæa carpit ab Ida,
Puberibus caulem folijs, & flore comantem
Purpureo.

In Engliſh thus :

His mother from the Cretæan Ida crops
Dictamnus hauing ſoft and tender leaues,
And purple floures vpon the bending tops, &c.

¶ *The Time.*

They floure and flouriſh in Summer Moneths, their ſeed is ripe in September.

¶ *The Names.*

It is called in Greeke δίκταμνος : in Latine, *Dictamnus* and *Dictamnum* : of ſome, *Pulegium ſylueſtre,* or wilde Pennie-roiall : the Apothecaries of Germany for *Dictamnum* with *e,* in the firſt ſyllable, doe reade *Diptamnum* with *p :* but (ſaith *Dodonæus*) this errour might haue beene of ſmall importance, if in ſtead of the leaues of Dittanie, they doe not vſe the roots of *Fraxinella* for Dittany, which they falſely call *Dictamnum :* in Engliſh, Dittany, and Dittany of Candie.

The other is called *Pſeudodictamnum,* or baſtard Dittany, of the likeneſſe it hath with Dittany, skilleth not, though the ſhops know it not : the reaſon why let the Reader gueſſe.

¶ *The Nature.*

Theſe plants are hot and dry of Nature.

¶ *The Vertues.*

A   Dittany being taken in drinke, or put vp in a peſſary, or vſed in a fume, bringeth away dead children : it procureth the monethly termes, and driueth forth the ſecondine or afterbirth.

B   The juyce taken with wine is a remedy againſt the ſtinging of ſerpents.

C   The ſame is thought to be of ſo ſtrong an operation, that with the very ſmell alſo it driues away venomous beaſts, and doth aſtoniſh them.

D   It is reported likewiſe that the wilde Goats or Deere in Candy when they be wounded with arrowes, do ſhake them out by eating of this plant, and heale their wounds.

E   It preuaileth much againſt all wounds, and eſpecially thoſe made with inuenomed weapons, arrowes ſhot out of guns, or ſuch like, and is very profitable for Chirurgians that vſe the ſea and land wars, to carry with them and haue in readineſſe : it draweth forth alſo ſplinters of wood, bones, or ſuch like.

F   The baſtard Dittany or *Pſeudodictamnum,* is ſomewhat like in vertues to the firſt, but not of ſo great force, yet it ſerueth exceeding well for the purpoſes aforeſaid.

Chap. 282. *Of Borage.*

¶ *The Deſcription.*

1   BOrage hath broad leaues, rough, lying flat vpon the ground, of a blacke or ſwart green colour : among which riſeth vp a ſtalke two cubits high, diuided into diuers branches, wherupon do grow gallant blew floures, compoſed of fiue leaues apiece; out of the middle of which grow forth blacke threds joined in the top, and pointed like a broch or pyramide : the root is threddy, and cannot away with the cold of Winter.

2   Borage with white floures is like vnto the precedent, but differeth in the floures, for thoſe of this plant are white, and other of a perfect blew colour, wherein is the difference.

† 3   Neuer dying Borage hath many very broad leaues, rough and hairy, of a blacke darke greene colour : among which riſe vp ſtiffe hairy ſtalkes, whereupon doe grow faire blew floures, ſomewhat rounder pointed than the former : the root is blacke and laſting, hauing leaues both Winter and Summer, and hereupon is was called *Semper virens,* and that very properly, to diſtinguiſh it from the reſt of this kinde, which are but annuall. ‡

                                                4 There

1 *Borago hortenſis.*
Garden Borage.

2 *Rorago flore albo.*
White floured Borage.

3 *Borago ſemper virens.*
Neuer-dying Borage.

4 There is a fourth ſort of Borage that hath leaues like the precedent, but thinner and leſſer, rough and hairy, diuiding it ſelfe into branches at the bottome of the plant, whereupon are placed faire red floures, wherein is the chiefeſt difference between this and the laſt deſcribed. ‡ The figure which belonged to this deſcription was put hereafter for *Lycopſis Anglica.* ‡

¶ *The Place.*

Theſe grow in my garden and in others alſo.

¶ *The Time.*

Borage floures and flouriſhes moſt part of all Summer, and till Autumne be far ſpent.

¶ *The Names.*

Borage is called in ſhops *Borago* : of the old writers, Βούγλωσσον, which is called in Latine *Lingua bubula* : *Pliny* calleth it *Euphroſinum*, becauſe it makes a man merry and ioyfull : which thing alſo the old verſe concerning Borage doth teſtifie :

Ego Borago gaudia ſemper ago.

I Borage bring alwaies courage.

It is called in high-Dutch, **Burretſch** : in Italian, *Boragine* : in Spaniſh, *Boraces:* in low-Dutch, **Beruagie** : in Engliſh, Borage.

¶ *The Temperature.*

It is euidently moiſt, and not in like ſort hot, but ſeems to be in a mean betwixt hot and cold.

¶ *The Vertues.*

Thoſe of our time do vſe the floures in ſallads, to exhilerate and make the minde glad. There be **A** alſo many things made of them, vſed for the comfort of the heart, to driue away ſorrow, & increaſe the ioy of the minde.

The

B   The leaues boiled among other pot-herbs much preuaile in making the belly ſoluble; & being boiled in honied water they are alſo good againſt the roughneſſe and hoarſeneſſe of the throat, as *Galen* teacheth.

C   The leaues and floures of Borrage put into wine make men and women glad and merry, driuing away all ſadneſſe, dulneſſe, and melancholy, as *Dioſcorides* and *Pliny* affirme.

D   Syrrup made of the floures of Borrage comforteth the heart, purgeth melancholy, and quieteth the phrenticke or lunaticke perſon.

E   The floures of Borrage made vp with ſugar, do all the aforeſaid with greater force and effect.

F   Syrrup made of the juice of Borrage with ſugar, adding thereto pouder of the bone of a Stags heart, is good againſt ſwouning, the cardiacke paſſion of the heart, againſt melancholy and the falling ſickneſſe.

G   The root is not vſed in medicine: the leaues eaten raw ingender good bloud, eſpecially in thoſe that haue bin lately ſicke.

Chap. 283. *Of Bugloſſe.*

¶ *The Kindes.*

Like as there be diuers ſorts of Borage, ſo are there ſundry of the Bugloſſes; yet after *Dioſcorides*, Borage is the true Bugloſſe: many are of opinion, & that rightly, that they may be both refetred to one kinde, yet will we diuide them according to the cuſtome of our time, and their vſuall denominations.

1 *Bugloſſa vulgaris.*
Common Bugloſſe, or garden Bugloſſe.

2 *Bugloſſum luteum.*
Lang de beefe.

¶ *The Deſcription.*

1   THat which the Apothecaries call Bugloſſe bringeth forth leaues longer than thoſe of Borage, ſharpe pointed, longer than the leaues of Beets, rough and hairy. The ſtalke groweth vp to the height of two cubits, parted aboue into ſundry branches, whereon are orderly placed blewiſh floures, tending to a purple colour before they be opened, and afterward more blew: the root is long, thicke, groſſe, and of long continuance.

‡ 3 *Bugloſſa ſylueſtris minor.*
Small wilde Bugloſſe.

2　Lang de beef is a kinde hereof, altogether leſſe, but the leaues herof are rougher
like the rough tongue of an Oxe or Cowe,
whereof it tooke his name. ‡ The leaues of
Lang de beef are very rough, the ſtalke ſome
cubit and halfe high, commonly red of colour: the tops of the branches carry floures
in ſcaly rough heads: theſe floures are compoſed of many ſmall yellow leaues in maner
of thoſe of Dandelion, and fly away in down
like as they do: the floures are of a very bitter taſt, whence *Lobel* calls it *Bugloſſum echioides luteum Hieracio cognatum.* *Tabernamontanus* hath fitly called it *Hieracium echioides.*
3　There is another wild Bugloſſe which
Dodonæus hath by name of *Bugloſſa ſylueſtris:*
it hath a ſmall white root, from which ariſes
a ſlender ſtalke ſome foot and halfe high, ſet
with ſmal rough leaues ſinuated or cut in on
the edges: the ſtalkes at the top are diuided
into three, or foure ſlender branches, bearing
little blew floures in rough husks. ‡

¶ *The Place.*

Theſe grow in gardens euery where. ‡ The
Lang de beefe growes wilde in many places,
as betweene Redriffe and Deptford by the
waterie ditch ſides. The little wild Bugloſſe
growes vpon the dry ditch banks about Pic-
kadilla, and almoſt euery where. ‡

¶ *The Time.*

They floure from May or Iune, euen to the end of Summer. The leaues periſh in Winter, and
new come vp in the Spring.

¶ *The Names.*

Garden Bugloſſe is called of the later Herbariſts, *Bugloſſa,* and *Bugloſſa domeſtica,* or garden Bugloſſe.

Lang de beef is called in Latine, *Lingua bouis,* and *Bugloſſum luteum Hieracio cognatum,* and alſo
Bugloſſa ſylueſtris, or wilde Bugloſſe.

‡　Small wild Bugloſſe is called *Borago ſylueſtris,* by *Tragus ; Echium Germanicum ſpinoſum* by
Fuchſius : and *Bugloſſa ſylueſtris* by *Dodonæus.* ‡

¶ *The Temperature and Vertues.*

The root, ſaith *Dioſcorides,* mixed with oile, cureth green wounds, and adding therto a little bar-　A
ley meale it is a remedie againſt S. *Anthonies* fire.

It cauſeth ſweat in agues, as *Pliny* ſaith, if the juice be mixed with a little aqua vitæ, and the bo-　B
dy rubbed therewith.

The Phyſitions of later time vſe the leaues, floures, and roots in ſtead of Borage, and put them in-　C
to all kindes of medicine indifferently, which are of force and vertue to driue away ſorrow & pen-
ſiueneſſe of the minde, and to comfort and ſtrengthen the heart. The leaues are of like operation
as thoſe of Borage, and are vſed as pot-herbes for the purpoſes aforeſaid, as well Bugloſſe as Lang
de beefe, and alſo to keepe the belly ſoluble.

CHAP. 284.　*Of Alkanet or wilde Bugloſſe.*

¶ *The Deſcription.*

THeſe herbes comprehended vnder the name of *Anchuſa,* were ſo called of the Greeke word
ἀγχοῦν; i. *Illinere ſucco, vel pigmentis,* that is, to colour or paint any thing: Whereupon thoſe

† 1 *Anchusa Alcibiadion.*
Red Alkanet.

† 2 *Anchusa lutea.*
Yellow Alkanet.

‡ 3 *Anchusa minor.*
Small Alkanet.

plants were called *Anchusa*, of that flourishing and bright red colour which is in the root, euen as red as pure and cleare bloud; for that is the onely marke or note whereby to diftinguifh thefe herbs from thofe which be called *Echium, Lycopfis*, and *Buglossa*, wherto they haue a great refemblance; I haue therefore expreffed foure differences of this plant *Anchusa* or Alkanet, from the other kindes, by the leaues, flours, and bigneffe.

1   The firft kinde of Alkanet hath many leaues like *Echium* or fmall Bugloffe, couered ouer with a pricky hoarineffe, hauing commonly but one ftalke, which is round, rough, and a cubit high. The cups of the floures are of a sky colour tending to purple, not vnlike the floures of *Echium*: the feed is fmall, fomwhat long, and of a pale colour: the root is a finger thicke, the pith or inner part thereof is of a wooddy fubftance, dying the hands or whatfoeuer toucheth the fame, of a bloudy colour, or of the colour of Sanders.

2   The fecond kinde of *Anchusa* or Alkanet is of greater beauty and eftimation than the firft, the branches are leffe and more bufhy in the top; it hath alfo greater plenty of leaues, and thofe more woolly or hairy: the ftalk groweth to the height of two cubits: at the top grow floures of a yellow colour, far different from the other: the root is more fhining, of an excellent delicat purple colour, and more ful of juice than the firft.

3 There

3 There is a small kinde of Alkanet, whose root is greater and more ful of juyce and substance than the roots of the other kindes : in all other respects it is lesse, for the leaues are narrower, smaller, tenderer, and in number more, very greene like vnto Borage, yeelding forth many little tender stalkes : the floures are lesse than of the small Buglosse, and red of colour : the seed is of an ash colour, somewhat long and slender, hauing the taste of Buglosse.

4 There is also another kinde of Alkanet, which is as the others before mentioned, a kinde of wilde Buglosse, notwithstanding for distinctions sake I haue separated and seuered them. This last *Anchusa* hath narrow leaues, much like vnto our common Summer Sauory. The stalkes are two handfuls high, bearing very small floures, and of a blewish or skie colour : the root is of a dark brownish red colour, dying the hands little or nothing at all, and of a wooddy substance.

¶ *The Place.*

These plants do grow in the fields of Narbone, and about Montpellier, and many other parts of France : I found these plants growing in the Isle of Thanet neere ,vnto the sea, betwixt the house sometime belonging to Sir *Henry Crispe*, and Margate; where I found some in their naturall ripenes, yet scarcely any that were come to that beautifull color of Alkanet : but such as is sold for very good in our Apothecaries shops I found there in great plenty,

‡ I doubt whether our Author found any of these in the place here set down, for I haue sought it but failed of finding ; yet if he found any it was only the first described, for 1 thinke the other three are strangers. ‡   ¶ *The Time.*

The Alkanets floure and flourish in the Summer moneths : the roots doe yeeld their bloudy juyce in haruest time, as *Dioscorides* writeth.

¶ *The Names.*

Alkanet is called in Greeke ἄγχουσα : in Latine also *Anchusa* : of diuers, *Fucus herba*, and *Onocleia*, *Buglossa Hispanica*, or Spanish Buglosse : in Spanish, *Soagem* : in French, *Orchanet* : and in English likewise, Orchanet and Alkanet.

¶ *The Temperature.*

The roots of Alkanet are cold and dry, as *Galen* writeth, and binding, and because it is bitter it clenseth away cholericke humours : the leaues be not so forceable, yet doe they likewise binde and drie.   ¶ *The Vertues.*

▬ *Dioscorides* saith, that the root being made vp in a cerote, or searecloth with oyle, is very good for   A old vlcers ; that with parched barley meale it is good for the leprey, and for tetters and ring-worms.

That being vsed as a pessarie it bringeth forth the dead birth.   B

The decoction being inwardly taken with Mead or honied water, cureth the yellow jaundise,   C diseases of the kidnies, the spleene and agues.

It is vsed in ointments for womens paintings : and the leaues drunke in wine is good against the   D laske.

Diuers of the later Physitions do boile with the root of Alkanet and wine, sweet butter, such as   E hath in it no salt at all, vntill such time as it becommeth red, which they call red butter, and giue it not only to those that haue falne from some high place, but also report it to be good to driue forth the measels and small pox, if it be drunke in the beginning with hot beere.

The roots of these are vsed to color sirrups, waters, gellies, & such like infections as Turnsole is.   F

Iohn of *Ardern* hath set down a composition called *Sanguis Veneris*, which is most singular in deep   G punctures or wounds made with thrusts, as follows : take of oile oliue a pint, the root of Alkanet two ounces, earth worms purged, in number twenty, boile them together & keep it to the vse aforesaid.

The Gentlewomen of France do paint their faces with these roots, as it is said.

† The two figures that were formerly here were both of the ordinary Buglosse, whereof the first might well enough serue, but the 2 was much different from that it should haue been.

Chap. 285.   *Of Wall and Vipers Buglosse.*

¶ *The Description.*

1 Lycopsis Anglica, or wilde Buglosse, so called for that it doth not grow so commonly elsewhere, hath rough and hairy leaues, somewhat lesser than the garden Buglosse : the floures grow for the most part vpon the side of the slender stalke, in fashion hollow like a little bell, whereof some be blew, and others of a purple colour.

2 There is another kinde of *Echium* that hath rough and hairy leaues likewise, much like vnto the former ; the stalke is rough, charged full of little branches, which are laden on euery side with diuers small narrow leaues, sharp pointed, and of a brown colour : among which leaues grow floures, each floure being composed of one leafe diuided into fiue parts at the top, lesse, and not so wide open as that of *Lycopsis* ; yet of a sad blew or purple colour at the first, but when they are open they shew to be of an azure colour, long and hollow, hauing certaine smal blew threds in the middle : the seed is small and black, fashioned like the head of a snake or viper : the root is long, and red without.

† 1 *Lycopſis Anglica.*
Wall Bugloſſe.

‡ 2 *Echium vulgare.*
Vipers Bugloſſe.

‡ 3 *Echium pullo flore.*
Rough Vipers Bugloſſe.

‡ 4 *Echium rubro flore.*
Red floured Vipers Bugloſſe.

‡ 3 This hath a crested very rough and hairy stalke some foot high ; the leaues are like those of Vipers Buglosse, and couered ouer with a soft downinesse, and grow disorderly vpon the stalke, which toward the top is parted into sundry branches, which are diuided into diuers foot-stalkes carrying small hollow floures diuided by fiue little gashes at their tops ; and they are of a darke purple colour, and contained in rough cups lying hid vnder the leaues. The seed, as in other plants of this kinde, resembles a Vipers head : the root is long, as thicke as ones little finger, of a dusky color on the outside, and it liues diuers yeares. This floures in May, and growes in the dry medowes and hilly grounds of Austria. *Clusius* calls it *Echium pullo flore.*

4 This other being also of *Clusius* his description hath long and narrow leaues like those of the common Vipers Buglosse, yet a little broader: the stalkes rise vp some cubit high, firme, crested, and hairy ; vpon which grow abundance of leaues, shorter and narrower than those below, and amongst these towards the top grow many floures vpon short foot-stalkes, which twine themselues round like a Scorpions taile : these floures are of an elegant red colour, and in shape somewhat like those of the common kinde ; and such also is the seed, but somewhat lesse : the root is lasting, long also, hard, wooddy, and blacke on the outside, and it sometimes sends vp many, but most vsually but one stalke. It floures in May, and was found in Hungary by *Clusius*, who first set it forth by the name of *Echium rubro flore.* ‡

¶ *The Place.*

Lycopsis groweth vpon stone walls, and vpon dry barren stony grounds,
Echium groweth where Alkanet doth grow, in great abundance.

¶ *The Time.*

They flourish when the other kindes of Buglosses do floure.

¶ *The Names.*

It is called in Greeke, *Echium*, and Ἀλκιβιάδιον, of *Alcibiades* the finder of the vertues thereof : of some it is thought to be *Anchusa species*, or a kinde of Alkanet : in high Dutch, wilde **Ochsenzungen**: in Spanish, *Yerua de la Biuora*, or *Chupamel* : in Italian, *Buglossa saluatica* : in French, *Buglosse sauuage* : in English, Vipers Buglosse, Snakes Buglosse ; and of some, Vipers herbe, and wilde Buglosse the lesser.

¶ *The Temperature.*

These herbes are cold and dry of complexion.

¶ *The Vertues.*

The root drunke with wine is good for those that be bitten with Serpents, and it keepeth such　A from being stung as haue drunk of it before : the leaues and seeds do the same, as *Dioscorides* writes. *Nicander* in his booke of Treacles makes Vipers Buglosse to be one of those plants which cure the biting of Serpents, and especially of the Viper, and that driue serpents away.

If it be drunke in wine or otherwise it causeth plenty of milke in womens brests.　B

The herbe chewed, and the juyce swallowed downe, is a most singular remedy against poyson　C and the biting of any venomous beast ; and the root so chewed, and laid vpon the sore, workes the same effect.

† That figure which formerly stood in the second place, vnder the title of *Onosma*, and whereof there was no more mention made by our Athor, neither in description name, nor otherwise, I take to be nothing else than the *Lycopsis* which lies with long leaues spred vpon the ground before it comes to send vp the stalke ; as you may see it exprest apart by it selfe in the figure we giue you ; which is the true figure of that plant our Author described and meant ; for the figure which he gaue was nothing but of the common Borage with narrower leaues, which he described in the fourth place of the chapter of Borage, as I haue formerly noted.

Chap. 286. *Of Hounds-tongue.*

¶ *The Description.*

1 THe common Hounds-tongue hath long leaues much like the Garden Buglosse, but broader, and not rough at all, yet hauing some fine hoarinesse or softnesse like veluer. These leaues stinke very filthily, much like to the pisse of dogs ; wherefore the Dutch men haue called it **Hounds pisse**, and not Hounds tongue. The stalkes are rough, hard, two cubits high, and of a browne colour, bearing at the top many floures of a darke purple colour : the seed is rough, cleauing to garments like Agrimony seed : the root is blacke and thicke. ‡ These plants for one yeere after they come vp of seed bring forth onely leaues, and those pretty large ; and the second yeare they send vp their stalkes, bearing both floures and seed, and then vsually the root perisheth. I haue therefore presented you with the figures of it, both when it floures, and when it sendeth forth onely leaues. ‡

2 We

1 *Cynoglossum maius vulgare sine flore.*
Hounds-tongue without the floure.

1 *Cynoglossum maius cnm flore & semine.*
Hounds-tongue with the floure and seed.

‡ 2 *Cynoglossum Creticum 1.*
The first Candy Dogs-tongue.

‡ 2 *Cynoglossum Creticum alterum.*
The other Candy Dogs-tongue.

2 We haue receiued another ſort hereof from the parts of Italy, hauing leaues like Woade, ſomewhat rough, and without any manifeſt ſmell, wherein it differeth from the common kinde; the ſeed hereof came vnder the title *Cynogloſſum Creticum*, Hounds tongue of Candy. ‡ The flours are leſſer, and of a lighter colour than thoſe of the former ; the ſeeds alſo are rough, and grow foure together, with a point comming out of the middle of them, as in the common kinde, but yet leſſe: the root is long and whitiſh. *Cluſius* hath this by the name of *Cynogloſſum Creticum* 1.

3 This ſecond *Cynogloſſum Creticum* of *Cluſius* hath leaues ſome handfull high, and ſome inch or better broad : among which the next yeare after the ſowing, comes vp a ſtalke ſome cubit or more high, creſted, ſtiffe, ſtraight, and ſomewhat downy, as are alſo the leaues which grow vpon the ſame, being ſomwhat broad at their ſetting on, and of a yellowiſh green colour: the top of the ſtalk is diuided into ſundry branches, which twine or turne on their tops like as the ſcorpion graſſe, and carry ſhorter yet larger floures than the ordinarie kinde, and thoſe of a whitiſh colour at the firſt, with many ſmall purpliſh veins, which after a few daies become blew. The ſeeds are like the former in their growing, ſhape, and roughneſſe. ‡

4 We haue another ſort of Hounds tongue like vnto the common kinde, ſauing it is altogether leſſe : the leaues are of a ſhining green colour.

‡ 4 *Cynogloſſum minus folio virente.*
Small greene leaued Hounds tongue.

¶ *The Place.*

The great Hounds tongue growes almoſt euery where by highwayes and vntoiled ground : the ſmall Hounds tongue groweth very plentifully by the wayes ſide as you ride Colcheſter highway from Londonward, between Eſterford and Wittam in Eſſex.

¶ *The Time.*
They floure in Iune and Iuly.

¶ *The Names.*
Houndstongue is called in Greek, Κυνόγλωσσον : in Latine, *Lingua canis* : of Pliny, *Cynogloſſos* : and he ſheweth two kinds thereof: in Engliſh, Hounds tongue, or Dogges tongue, but rather Hounds piſſe, for in the world there is not any thing that ſmelleth ſo like Dogs piſſe as the leaues of this plant doe.

¶ *The Nature.*
Hounds tongue, but eſpecially his root, is cold and dry.

¶ *The Vertues.*

A The roots of Hounds tongue roſted in the embers and laid to the fundament, heal the hemorrhoids, and the diſeaſe called *Ignis ſacer*, or wilde fire.

B The juice boiled with hony of roſes and turpentine to the forme of an vnguent, is moſt ſingular in wounds and deep vlcers.

C *Dioſcorides* ſaith, That the leaues boiled in wine and drunk, doth mollifie the belly; and being ſtamped with old ſwines greaſe, are good againſt the falling away of the haire, proceeding of hot humors.

D Likewiſe they are a remedy againſt ſcaldings or burnings, and againſt the biting of dogs, as the ſame Author addeth.

CHAP. 287.
Of Comfrey or great Conſound.

¶ *The Deſcription.*

1 THe ſtalke of this Comfrey is cornered, thicke, and hollow like that of Sow-thiſtle. It groweth two cubits or a yard high : the leaues that ſpring from the root, and thoſe that grow

1 *Conſolida maior flore purpureo.*
Comfrey with purple floures.

‡ 3 *Symphytum tuberoſum.*
Comfrey with the knobbed root.

‡ 4 *Symphytum parvum Boraginis facie.*
Borage-floured Comfrey.

grow vpon the ſtalkes are long, broad, rough, and pricking withall, ſomething hairy, and beeing handled make the hands itch, very like in colour and roughneſſe to thoſe of Borage, but longer, and ſharp pointed as bee the leaues of Elecampane: from out the wings of the ſtalkes appeare the floures orderly placed, long, hollow within, of a light red colour: after them growes the ſeed which is black: the root is long and thick, black without, white within, hauing in it a clammy juice, in which root conſiſteth the vertue.

2 The great Comfrey hath rough hairy ſtalks and long rough leaues much like the garden Bugloſſe, but greater and blacker: the flours be round and hollow like little bells, of a white colour: the root is black without, white within, and very ſlimy. ‡ This differs no way from the former but only in the colour of the flour, which is yellowiſh or white, when as the other is reddiſh or purple. ‡

3 There is another kind of Comfrey which hath leaues like the former, ſauing that they bee leſſer: the ſtalks are rough and tender: the flours be like the former, but that they be of an ouerworn yellow colour: the roots are thicke, ſhort, blacke without, and tuberous, ‡ which in the figure are not expreſſed ſo large and knobby as they ought to haue bin. ‡

4 T his

‡ 4. This pretty plant hath fibrous and blackiſh roots, from which riſe vp many leaues like thoſe of Borage or Comfrey, but much ſmaller and greener, the ſtalks are ſome eight incheshigh, and on their tops carry pretty floures like thoſe of Borage, not ſo ſharp pointed, but of a more pleaſing blew colour. This flours in the Spring, and is kept in ſome choice gardens: *Lobel* calls it *Symphytum pumilum repens Borraginis facie, ſiue Borrago minima Herbariorum.*

¶ *The Place.*

Comfrey ioyeth in waterie ditches, in fat and fruitfull medowes: they grow all in my garden.

¶ *The Time.*

They floure in Iune and Iuly.

¶ *The Names.*

It is called in Greeke χύφυτον: in Latine, *Symphytum*, and *Solidago*: in ſhops, *Conſolida major*, and *Symphytum majus*: of *Scribonius Largus*, *Inula ruſtica*, and *Alus Gallica*: of others, *Oſteocollon*: in high Dutch, **Walwurtʒ**: in low-Dutch, **Waelwortele**: in Italian, *Conſolida maggiore*: in Spaniſh, *Suelda majore*, and *Conſuelda maior*: in French, *Couſire*, and *Oreille d'aſne*: in Engliſh, Comfrey, Comfrey Conſound: of ſome, Knit-backe, and Blackwort.

¶ *The Temperature.*

The root of Comfrey hath a cold qualitie, but yet not much: it is alſo of a clammy and gluing moiſture, it cauſeth no itch at all, neither is it of a ſharpe or biting taſte, but vnſauorie or without taſte; ſo farre is the tough and gluing moiſture from the ſharpe clamminneſſe of the ſea Onion, as that there is no compariſon betweene them. The leaues may cauſe itching not through heate or ſharpeneſſe, but through their ruggedneſſe, as we haue already written, yet leſſe than thoſe of the Nettle.

¶ *The Vertues.*

The roots of Comfrey ſtamped, and the iuyce drunke with wine, helpeth thoſe that ſpit bloud, **A** and healeth all inward wounds and burſtings.

The ſame bruiſed and layd to in manner of a plaiſter, doth beale all freſh and green wounds, and **B** are ſo glutinatiue, that it wil ſoder or glew together meat that is chopt in pieces, ſeething in a pot, and make it in one lumpe.

The roots boiled and drunke, do clenſe the breſt from flegme, and cure the griefes of the lungs, **C** eſpecially if they be confectd with ſugar and ſyrrup: it preuaileth much againſt ruptures or burſtings.

The ſlimy ſubſtance of the root made in a poſſet of ale, and giuen to drinke againſt the paine in **D** the backe gotten by any violent motion, as wreſtling, or ouermuch vſe of women, doth in foure or fiue days preſently cure the ſame, although the involuntarie flowing of the ſeed in man be gotten thereby.

The roots of Comfry in number foure, Knot-graſſe and the leaues of Clary of each an handfull, **E** ſtamped all together, ſtrained, and a quart of Muſcadel put thereto, the yelks of three eggs, and the pouder of three nutmegs drunke firſt and laſt, is a moſt excellent medicine againſt a Gonorrhœa or running of the reins, and all paines and conſumptions of the backe.

There is likewiſe a ſyrrup made hereof to be vſed in this caſe, which ſtayeth voiding of bloud, **F** tempereth the heate of agues, allayeth the ſharpneſſe of flowing humors, healeth vp vlcers of the lungs, and helpeth the cough. The receit whereof is this: Take two ounces of the roots of great Comfrey, one ounce of Liquorice, two handfulls of Folefoot roots and all, one ounce and halfe of Pine-apple kernels, twenty Iuiubes, two drams or a quarter of an ounce of Mallow ſeed, one dram of the heads of Poppy; boile all in a ſufficient quantitie of water till one pint remain, ſtrain it, and adde to the liquor ſix ounces of very white ſugar, and as much of the beſt hony, and make thereof a ſyrrup that muſt be throughly boiled.

The ſame ſyrrup cureth the vlcers of the kidnies, though they haue bin of long continuance, and **G** ſtoppeth the bloud that commeth from thence.

Moreouer, it ſtayeth the ouermuch flowing of the monethly ſickneſſe, taken conſtantly for cer- **H** tain daies together.

It is highly commended for wounds or hurts of all the intrales and inward parts, and for bur- **I** ſtings or ruptures.

The root ſtamped and applied, taketh away the inflammation of the fundament, and ouermuch **K** flowing of the hemorrhoids.

CHAP.

CHAP. 288. *Of Cowslips of Ierusalem.*

1 *Pulmonaria maculosa.*
Spotted Cowslips of Ierusalem.

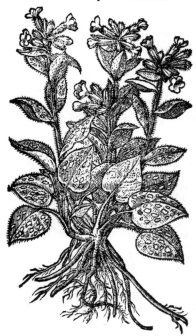

2 *Pulmonaria folijs Echij.*
Buglosse Cowslips.

3 *Pulmonaria angustifolia 2 Clusij.*
Narrow leafed Cowslips of Ierusalem.

¶ *The Description.*

1 COwslip of Ierusalem, or the true and
right Lung-woort hath rough hairie
and large leaues of a brown green color, confusedly spotted with diuers spots or drops of
white : among which spring vp certain stalks
a spanne long, bearing at the top many fine
flours, growing together in bunches like the
flours of Cowslips, sauing that they be at the
first red or purple, somtimes blew, and oftentimes all these colours at once. The floures
being fallen, there come small buttons ful of
seed. The root is black and threddy. ‡ This
is somtimes found with white floures. ‡

2 The second kinde of Lungwort is like
vnto the former, but greater in each respect ;
the leaues bigger than the former, resembling wilde Buglosse, yet spotted with white
spots like the former: the floures are like the
other, but of an exceeding shining red color.

3 *Clusius* setteth forth a third kinde of
Lungwort, which hath rough & hairy leaues
like vnto wild Buglosse, but narrower: among
which rises vp a stalke a foot high, bearing at
the top a bundle of blew floures, in fashion
like vnto those of Buglosse or the last described.

¶ *The*

¶ *The Place.*

Theſe plants do grow in moiſt ſhadowie woods, and are planted almoſt euery where in Gardens. ‡ Mr. *Goodyer* found the *Pulmonaria folijs Echij*, being the ſecond, May 25. *Anno* 1620. flouring in a wood by Holbury houſe in the New Forreſt in Hampſhire. ‡

¶ *The Time.*

They ſloure for the moſt part in March and Aprill.

¶ *The Names.*

Cowſlips of Ieruſalem, or Sage of Ieruſalem, is called of the Herbariſts of our time, *Pulmonaria*, and *Pulmonalis*; of *Cordus*, *Symphytum ſylueſtre*, or wilde Comfrey: but ſeeing the other is alſo of nature wilde, it may aptly be called *Symphytum maculoſum*, or *Maculatum*: in high Dutch, **Lungenkraut**: in low Dutch, **Onſer bzouwen melcruijt**: in Engliſh, ſpotted Comfrey, Sage of Ieruſalem, Cowſlip of Ieruſalem, Sage of Bethlem, and of ſome Lungwort; notwithſtanding there is another Lungwort, of which we will intreat among the kindes of Moſſes.

¶ *The Temperature.*

Pulmonaria ſhould be of like temperature with the great Comfrey, if the root of this were clammie: but ſeeing that it is hard and wooddy, it is of a more drying quality and more binding.

¶ *The Vertues.*

The leaues are vſed among pot-herbes. The roots are alſo thought to be good againſt the infirmities and vlcers of the lungs, and to be of like force with the great Comfrey.

† The figure which formerly was in the fourth place of this chapter, was onely of the firſt deſcribed with white ſloures. But the Title *Pulmonaria Gallorum*, and the deſcription ſited to it (though little to the purpoſe, and therefore omitted) were intended for the *Pulmonaria Gallorum ſiue aurea*, whereof I haue in the due place largely intreated, as you may ſee in this booke, *pag. 304, chap. 36.*

CHAP. 289.   *Of Clote Burre, or Burre Docke.*

1 *Bardana major.*
The great Burre-Docke.

2 *Bardana minor.*
The leſſer Burre Docke.

¶ *The Deſcription.*

1   CLot-Burre bringeth forth broad leaues and hairy, far bigger than the leaues of Gourds, and of greater compaſſe, thicker alſo, and blacker, which on the vpper ſide are of a darke greene colour, and on the neither ſide ſomewhat white : the ſtalke is cornered, thicke, beſet with like leaues, but farre leſſe, diuided into very many wings and branches, bringing forth great Burres round like bullets or balls, which are rough all ouer, and full of ſharpe crooked prickles, taking hold on mens garments as they paſſe by ; out of the tops whereof groweth a floure thrummed, or all of threds, of colour purple : the ſeed is perfected within the round ball or bullet, and this ſeed when the burres open, and the wind bloweth, is carried away with the winde : the root is long, white within, and blacke without.

‡   There is another kinde hereof which hath leſſer and ſofter heads, with weaker prickles ; theſe heads are alſo hairy or downy, and the leaues and whole plant ſomewhat leſſe, yet otherwiſe like the fore deſcribed ; *Lobel* calls this *Arction montanum*, and *Lappa minor Galeni* : it is alſo the *Lappa minor altera* of *Matthiolus*. *Lobel* found this growing in Somerſetſhire three miles from Bath, neere the houſe of one Mr. *Iohn Colt*.‡

2   The leſſer Burre hath leaues farre ſmall than the former, of a grayiſh ouerworne colour like to thoſe of Orach, nicked round about the edges : the ſtalke is a foot and halfe high, full of little blacke ſpots, diuiding it ſelfe into many branches: the flours before the Burres come forth do compaſſe the ſmall ſtalkes round about ; they are but little, and quickely vade away : then follow the Burres or the fruit out of the boſome of the leaues, in forme long, on the tops of the branches, as big as an Oliue or a Cornell berry, rough like the balls of the Plane tree, and being touched cleaue faſt vnto mens garments : they do not open at all, but being kept cloſe ſhut bring forth long ſeeds. The root is faſtned with very many ſtrings, and groweth not deepe.

¶ *The Place.*

The firſt groweth euery where : the ſecond I found in the high way leading from Draiton to Iuer, two miles from Colbrooke, ſince which time I haue found it in the high way between Stanes and Egham. ‡   It alſo groweth plentifully in Southwicke ſheet in Hampſhire, as I haue beene enformed by Mr. *Goodyer*. ‡

¶ *The Time.*

Their ſeaſon is in Iuly and Auguſt.

¶ *The Names.*

The great Burre is called in Greeke, ἀράιον: in Latine, *Perſonata, perſonatia,* and *Arcium* : in ſhops, *Bardana,* and *Lappa maior* : in high Dutch, **Grofskletten** : in low Dutch, **Groot cliſſen** : in French, *Glouteron* : in Engliſh, Great Burre, Burre Docke, or Clot Burre : *Apuleius* beſides theſe doth alſo ſet downe certaine other names belonging to Clot Burre, as *Dardana, Bacchion, Elephantoſis, Nephelion, Manifolium.*

The leſſer Burre Docke is called of the Græcians, ξάνθιον: in Latine, *Xanthium* : in ſhops, *Lappa minor, Lappa inuerſa,* and of diuers, *Strumaria* : *Galen* ſaith it is alſo called, *Phaſganion,* and *Phaſganon,* or herbe victory, being but baſtard names, and therefore not properly ſo called : in Engliſh, Louſe Burre, Ditch Burre, and leſſer Burre Docke : it ſeemeth to be called *Xanthium* of the effect, for the Burre or fruit before it be fully withered, being ſtamped and put into an earthen veſſell, and afterward when need requireth the weight of two ounces thereof and ſomewhat more, being ſteeped in warme water and rubbed on, maketh the haires of the head red : yet the head is firſt to be dreſſed or rubbed with niter, as *Dioſcorides* writeth.

¶ *The Temperature.*

The leaues of Clot Burre are of temperature moderately dry and waſting ; the root is ſomething hot.

The ſeed of the leſſer Burre, as *Galen* ſaith, hath power to digeſt, therefore it is hot and dry.

¶ *The Vertues.*

A   The roots being taken with the kernels of Pine Apples, as *Dioſcorides* witneſſeth, are good for them that ſpit bloud and corrupt matter.

B   *Apuleius* ſaith that the ſame being ſtamped with a little ſalt, and applied to the biting of a mad dog, cureth the ſame, and ſo ſpeedily ſetteth free the ſicke man.

C   He alſo teacheth that the juyce of the leaues giuen to drinke with hony, procureth vrine, and taketh away the paines of the bladder ; and that the ſame drunke with old wine doth wonderfully helpe againſt the bitings of ſerpents.

D   *Columella* declareth, that the herbe beaten with ſalt and laid vpon the ſcarifying, which is made with the launcet or raſer, draweth out the poyſon of the viper : and that alſo the root being ſtamped is more auaileable againſt ſerpents, and that the root in like manner is good againſt the Kings euill

The

The stalke of Clot-Burre before the burres come forth, the rinde pilled off, being eaten raw with **E**
salt and pepper, or boyled in the broth of fat meate, is pleasant to be eaten: being taken in that
manner it increaseth seed and stirreth vp lust.

Also it is a good nourishment, especially boyled: if the kernell of the Pine Apple be likewise **F**
added it is the better, and is no lesse auailable against the vlcers of the lungs, and spitting of bloud,
than the root is.

The root stamped and strained with a good draught of Ale is a most approued medicine for a **G**
windie or cold stomacke.

Treacle of Andromachus, and the whites of egges, of each a like quantitie, laboured in a leaden **H**
mortar, and spred vpon the Burre leafe, and so applied to the gout, haue beene proued many times
most miraculously to appeale the paine thereof.

Dioscorides commendeth the decoction of the root of *Arcion*, together with the seed, against the **I**
tooth-ach, if it be holden a while in the mouth: also that it is good to foment therewith both burnings and kibed heeles; and affirmeth that it may be drunke in wine against the strangury, and paine
in the hip.

Dioscorides reporteth that the fruit is very good to be laid vnto hard swellings. **K**

The root cleane picked, washed, stamped, and strained with Malmesey, helpeth the running of **L**
the reines, the whites in women, and strengtheneth the backe, if there be added thereto the yelks of
egges, the pouder of acornes and nutmegs brued or mixed together, and drunke first and last.

Chap. 290. *Of Colts-foot, or Horse-foot.*

1 *Tussilago florens.*
Colts-foot in floure.

1 *Tussilaginis folia.*
The leaues of Colts-foot.

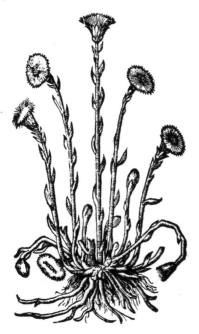 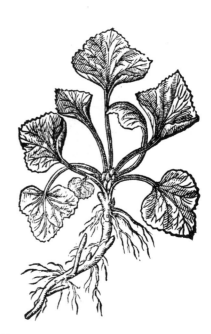

¶ *The Description.*

1   **T**ussilago or Fole-foot hath many white and long creeping roots, somewhat fat; from
which rise vp naked stalkes (in the beginning of March and Aprill) about a spanne
long, bearing at the top yellow floures, which change into downe and are caried away
with the winde: when the stalke and seed is perished, there appeare springing of out the earth many
broad

broad leaues,greene aboue,and next the ground of a white hoarie or grayiſh colour, faſhioned like an Horſe foot ; for which cauſe it was called Fole-foot, and Horſe-hoofe : ſeldome or neuer ſhall you find leaues and floures at once, but the flours are paſt before the leaues come out of the ground; as may appeare by the firſt picture, which ſetteth forth the naked ſtalkes and floures ; and by the ſecond,which pourtraireth the leaues onely.

‡ 2 Beſides the commonly growing and deſcribed Colts-foot, there are other two ſmall mountaine Colts feet deſcribed by *Cluſius* ; the firſt whereof I will here preſent you with, but the ſecond you ſhall finde hereafter in the chapter of *Aſarum,* by the name of *Aſarina Matthioli.* This here delineated hath fiue or ſix leaues not much vnlike thoſe of Alehoofe,of a darke ſhining greene colour aboue,and very white and downy below : the ſtalke is naked,ſome handfull high,hollow and downy,bearing one floure at the top compoſed of purpliſh threds, and flying away in downe : after which the ſtalke falls away, and ſo the leaues onely remaine during the reſt of the yeare : the root is ſmall and creeping. It growes on the tops of the Auſtrian and Stirian mountaines, where it floures in Iune or Auguſt. Brought into Gardens it floures in Aprill. *Cluſius* calls it *Tuſſilago Alpina,* 1. and he hath giuen two figures thereof,both which I here giue you by the ſame titles as he hath them. ‡

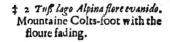

‡ 2 *Tuſſilago Alpina flore aperto.*
Mountaine Colts-foot full in floure

‡ 2 *Tuſſilago Alpina flore euanido.*
Mountaine Colts-foot with the floure fading.

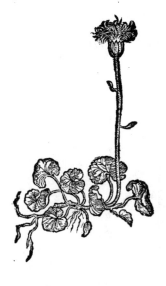

¶ *The Place.*

This groweth of it ſelfe neere vnto Springs, and on the brinkes of brookes and riuers,in wet furrowes,by ditches ſides,and in other moiſt and watery places neere vnto the ſea, almoſt euery where.

¶ *The Time.*

The floures which quickly fade, are to be ſeene in the end of March, and about the Calends of Aprill, which ſpeedily wither together with the ſtems : after them grow forth the leaues, which remaine greene all Summer long : and hereupon it came that Colts-foot was thought to be without floures ; which thing alſo *Pliny* hath mentioned in his ſix and twentieth booke,*cap. 6.*

¶ *The Names.*

Fole-foot is called in Greeke, Βήχιον : of the Latines likewiſe *Bechion,* and *Tuſſilago :* in ſhops,*Farfara,*and *Vngula Caballina :* of diuers,*Pata equina ·* in Italian, *Vnghia di Cauallo :* in Spaniſh, *Vnha d' aſno :* in French, *Pas d'aſne :* in Engliſh, Fole-foot, Colts-foot, Horſe-hoofe, and Bull-foot. The ſame is alſo *Chamæleuce,*which *Pliny* in his twenty eighth booke, and fifteenth chapter reporteth to be likewiſe called *Farfugium,*and *Farranum,*if there be not an errour in the copy : which thing alſo *Aëtius* in his firſt booke affirmeth, pretermitting the name of *Bechium,* and attributing vnto it all the verrues and faculties of *Bechium* or Colts-foot. Whoſe opinion *Orobaſius* ſeemeth to be of, in his fifteenth booke of his medicinable Collections, making mention of *Chamæleuce,* onely *Pliny*
also

also agreeth with them; shewing that some thinke, that *Bechium* is called by another name *Chamæleuce*, in his twenty sixth booke, *cap.6.* and it may bee that *Dioscorides* hath written of one and the selfe same herbe in sundry places, and by diuers names. *Bechium* and *Tussilago*, which may also be Englished Coughwort, so called of the effect, and *Farfara*, of the white Poplar tree, to whose leaues it is like; which was named of the Antients *Farfarus*, as *Plautus* writeth in his Comedie called *Pænulus* :

―――*viscum legioni dedi.*
fundasque eos prosternebam vt folia Farfari.

To the company I gaue both lime bush and sling.
That to the ground as Poplar leaues I might them fling.

‡ *Dodonæus* (from whom our Author tooke this) sets downe this place in *Plautus* as you finde it here, but not well; for the last verse should be *Fundasque, eo præsternebant folia Farfari.* Thus it is in most editions of *Plautus*, and that rightly, as the ensuing words in that place declare. ‡

The white Poplar tree is called in Greeke, Λευκη, and hereupon *Bechion* or Colts-foot was also called *Chamæleuce.*

¶ *The Temperature and Vertues.*

The leaues of Colts-foot being fresh and greene are somthing cold, and haue withall a drying quality; they are good for vlcers and inflammations: but the dried leaues are hot and dry, and somewhat biting.　　　A

A decoction made of the greene leaues and roots, or else a syrrup thereof, is good for the cough that proceedeth of a thin rheume.　　　B

The green leaues of Fole-foot pound with hony, do cure and heale the hot inflammation called Saint Anthonies fire, and all other inflammations.　　　C

The fume of the dried leaues taken through a funnell or tunnell, burned vpon coles, effectually helpeth those that are troubled with the shortnesse of breath, and fetch their winde thicke and often, and breaketh without perill the impostumes of the brest.　　　D

Being taken in manner as they take Tobaco, it mightily preuaileth against the diseases aforesaid.　　　E

Chap. 291. *Of Butter-Burre.*

¶ *The Description.*

1　BVtter-Burre doth in like manner bring forth floures before the leaues, as doth Coltsfoot, but they are small, mossie, tending to a purple colour; which being made vp into a big eare as it were, do quickely (together with the stem, which is thicke, full of substance, and brittle) wither and fall away : the leaues are very great like to a round cap or hat, called in Latine *Petasus*, of such a widenesse, as that of it selfe it is big and large enough to keepe a mans head from raine, and from the heate of the Sunne : and therefore they be greater than the leaues of the Clot-Burre, of colour somewhat white, yet whiter vnderneath : euery stem beareth his leafe; the stem is oftentimes a cubit long, thicke, full of substance; vpon which standeth the leafe in the centre or middlemost part of the circumference, or very neere, like to one of the greatest Mushromes, but that it hath a cleft that standeth about the stem, especially when they are in perishing and withering away : at the first the vpper superficiall or outside of the Mushromes standeth out, and when they are in withering standeth more in; and euen so the leafe of Butter-Bur hath on the outside a certaine shallow hollownesse : the root is thicke, long, blacke without, white within, of taste somewhat bitter, and is oftentimes worme-eaten.

¶ *The Place.*

This groweth in moist places neere vnto riuers sides, and vpon the brinks and banks of lakes and ponds, almost euery where.

¶ *The Time.*

The eare with the floures flourish in Aprill or sooner : then come vp the leaues, which continue till Winter, with new ones still growing vp.

　　　¶ The

1 *Petaſites florens.*
Butter-Burre in floure.

2 *Petaſitis folia.*
The leaues of Butter Burre.

¶ The Names.

Butter-Bur is called in Greeke, πετασῖτις : of the hugeneſſe of the leafe that is like to πέτασος, or a hat : the Latines call it, *Petaſites* : in high-Dutch, **Peſtilentzwurtz** : in low-Dutch, **Dockebladeren** : in Engliſh it is named, Butter-Burre : it is very manifeſt that this is like to Colts-foot, and of the ſame kinde.

¶ The Temperature.

Butter-Burre is hot and dry in the ſecond degree, and of thinne parts.

¶ The Vertues.

A　The roots of Butter-Burre ſtamped with ale, and giuen to drinke in peſtilent and burning Feuers, mightily coole and abate the heate thereof.

B　The roots dried and beaten to pouder, and drunke in wine, are a ſoueraigne medicine againſt the the plague and peſtilent ſeuers, becauſe they prouoke ſweat, and driue from the heart all venome and ill heate : it killeth wormes, and is of great force againſt the ſuffocation of the mother.

C　The ſame cureth all naughty filthy vlcers, if the pouder be ſtrewed therein.

D　The ſame kills wormes in the belly : it prouokes vrine, and brings downe the monethly termes.

‡ CHAP. 267.　Of Mountaine Horſe-foot.

¶ The Deſcription.

1　THis plant (which the moderne Writers haue referred to the *Cacalia* of the antients, and to the kindes of Colts-foot) I haue thought good to name in Engliſh, Horſe-foot, for that the leaues exceed Colts-foot in bigneſſe, yet are like them in ſhape : and of this plant *Cluſius* (whom I here chiefely follow) hath deſcribed two ſorts : the firſt of theſe haue many leaues almoſt like vnto thoſe of Colts-foot, but larger, very round, and ſnipt about the edges, of a light greene colour aboue, and hoarie vnderneath, hauing alſo many veines or nerues running vp and downe them ; and theſe leaues are of an vngratefull taſte, and grow vpon long purpliſh creſted ſtalkes : The ſtemme is ſome two cubits high, creſted likewiſe, and of a purpliſh colour, ſet alſo at certaine ſpaces with leaues very like vnto the other, but leſſer than thoſe next the ground, and more
cornered

‡ 1 *Cacalia incano folio.*
Hoarie leaued Horſefoot.

‡ 2 *Cacalia folio glabro.*
Smooth leaued Horſe-foot.

cornered and ſharper pointed, the tops of the ſtalkes and branches carry bunches of purple floures, as in an vmbell : and commonly in each bunch there are three little flours conſiſting of foure leaues a peece, and a forked peſtle, and theſe are of a purple colour, and a weake, but not vnpleaſant ſmell, and they at length turne into downe, amongſt which lies hid a longiſh ſeed : the root, if old, ſends forth diuers heads, as alſo ſtore of long whitiſh fibres.

2　　The leaues of this are more thin, tough and hard, and of a deeper greene on the vpper ſides, neither are they whitiſh below, nor come ſo round or cloſe whereas they are faſtened to their ſtalks (which are not creſted as thoſe of the other, but round and ſmooth) they are alſo full of veines and nickt about the edges, and of ſomewhat an vngratefull hot and bitter taſte. The ſtalkes are alſo ſmoother, and the floure of a lighter colour.

¶ *The Place.*
Both theſe grow in the Auſtrian and Stirian Alpes vnder the ſides of woods, among buſhes and ſuch ſhadowie places : but not in England, that I haue yet heard of.

¶ *The Time.*
I find it not ſet downe when theſe floure and ſeed, but iudge it about the ſame time that Coltſfoot doth.

¶ *The Names.*
This by *Cluſius*, *Lobel* and others, hath beene called *Cacalia*, and referred to that deſcribed by *Dioſcorides, lib. 4. cap.* 123. which is thought to be that ſet forth by *Galen* by the name of *Cancanus*. In the *Hiſtoria Lugd. pag.* 1052. The later of theſe two here deſcribed is figured by the name of *Tuſſilago Alpina ſiue montana*, and the former is there, *pag.* 1398. by the name of *Cacalia*, but the floures are not rightly expreſt : and if my iudgement faile me not, the figure which is in the ſeuenteenth page of the *Appendix* of the ſame Author, by the title of *Aconitum Pardalianches primum*, is of no other than this very plant. But becauſe I haue not as yet ſeene the plant, I will not poſitiuely affirme it : but referre this my opinion to thoſe that are iudicious and curious, to know the plant that raiſed ſuch controuerſie betweene *Matthiolus* and *Geſner*, and whereof neither *Camerarius* nor *Bauhine*, who hath ſet forth *Matthiolus* his Commentaries, haue giuen vs any certain or probable knowledge.

¶ *The*

¶ *The Temperature and Vertues, out of the Antients.*

The root of *Cacalia* is void of any biting qualitie, and moderately dries, and it is of a groſſe and emplaiſticke ſubſtance; wherefore ſteeped in wine and ſo taken it helpes the cough, the roughneſſe of the Arterie or hoarſnes, like as *Tragacanth :* neither if you chew it and ſwallow downe the juyce doth it leſſe auaile againſt thoſe effects than the juyce of Liquorice. ‡

CHAP. 219. *Of ſmall Celandine or Pilewort.*

¶ *The Kindes.*

THere be two kindes of Celandine, according to the old writers, much differing in forme and figure : the one greater, the other leſſer, which I intend to diuide into two diſtinct chapters, marſhailing them as neere as may be with their like, in forme and figure, and firſt of the ſmall Celandine.

Chelidonium minus.
Pilewort.

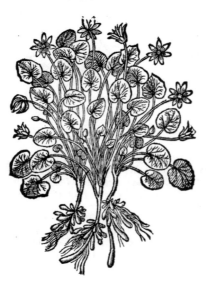

¶ *The Deſcription.*

THe leſſer Celandine hath greene round leaues, ſmooth, ſlipperie, and ſhining, leſſe than the leaues of the Iuie : the ſtalks are ſlender, ſhort, and for the moſt part creeping vpon the ground : they bring forth little yellow floures like thoſe of Crowfoot; and after the floures there ſpringeth vp a little fine knop or head full of ſeed : the root conſiſteth of ſlender ſtrings, on which doe hang as it were certain graines, of the bignes of Wheat cornes, or bigger.

¶ *The Place.*

It groweth in medows, by common waies, by ditches and trenches, and it is common euery where, in moiſt and dankiſh places.

¶ *The Time.*

It commeth forth about the Calends of March, and floureth a little after : it beginneth to fade away in Aprill, it is quite gone in May, afterwards it is hard to be found, yea ſcarcely the root.

¶ *The Names.*

Is is called in Greeke, χελιδόνιον of the Latines *Chelidonium minus,* and *Hirundinaria minor:* of diuers, *Scrophularia minor, Ficaria minor:* of *Serapio, Memiren :* in Italian, *Faueſcello :* in High Dutch, **Feigwurtzenkrant :** in French, *Eſclere,* and *Petit Baſſinet :* in Engliſh, little Celandine, Figwort and Pile-wort.

¶ *The Temperature.*

It is hot and dry, alſo more biting and hotter than the greater : it commeth neereſt in faculty to the Crowfoot.

‡ This which is here, and by moſt Authors ſet forth for *Chelidonium minus,* hath no ſuch great heat and Acrimony as *Dioſcorides* and *Galen* affirme to be in theirs ; making it hot in the fourth degree, when as this of ours ſcarce exceeds the firſt, as far as we may conjecture by the taſte. ‡

¶ *The Vertues.*

A   It preſently, as *Galen* and *Dioſcorides* affirme, exulcerateth or bliſtereth the skin : it maketh rough and corrupt nailes to fall away.

B   The juyce of the roots mixed with hony, and drawne vp into the noſthrils, purgeth the head of foule and filthy humors.

The

The later age vſe the roots and graines for the piles, which being often bathed with the jayce C mixed with wine, or with the ſickmans vrine, are drawne together and dried vp, and the paine quite taken away.

There be alſo who thinke, that if the herbe be but carried about one that hath the piles, the paine forthwith ceaſeth.

CHAP. 294.   Of Marſh Marigold.

¶ *The Deſcription.*

1   MArſh Marigold hath great broad leaues ſomewhat round, ſmooth, of a gallant greene colour, ſleightly indented or purld about the edges: among which riſe vp thicke fat ſtalkes, likewiſe greene; whereupon doe grow goodly yellow floures, glittering like gold, and like to thoſe of Crow-foot, but greater: the root is ſmall, compoſed of very many ſtrings.

1 *Caltha paluſtris maior.*
The great Marſh Marigold.

2 *Caltha paluſtris minor.*
The ſmall Marſh Marigold.

 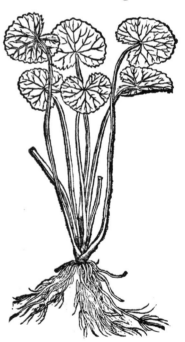

2   The ſmaller Marſh Marigold hath many round leaues ſpred vpon the ground, of a darke greene colour: amongſt which riſe vp diuers branches, charged with the like leaues: the floures grow at the toppes of the branches, of a moſt ſhining yellow colour: the root is alſo like the former.

3   The great Marſh Marigold with double floures is a ſtranger in England, his natiue Country ſhould ſeeme to bee in the furtheſt part of Germany, by the relation of a man of thoſe Countries that I haue had conference withall, the which he thus deſcribed: it hath (ſaith he) leaues, roots, and ſtalkes like thoſe of our common ſort, and hath double floures like thoſe of the garden Marigold, wherein conſiſteth the difference.

‡   *Camerarius* writes iuſt contrary to that which our Author here affirmes: for hee ſaith, *In Anglia ſua ſponte non ſolum plenis, ſed odoratus etiam floribus paſsim ſe ſe offert.* But I feare that both our
Author

3 *Caltha paluſtris multiplex.*
Double floured Marſh Marigold.

Author and *Camerarius* were deceiued by tru-
ſting the report of ſome lying, or elſe ignorant
perſons, for I could neuer find it growing wilde
with double floures here, nor *Camerarius* there:
yet I do not deny but by chance ſome one with
double floures may be found both here and
there, but this is not euery where. ‡

¶ *The Place.*

They joy in moiſt and mariſh grounds, and in
watery medowes. ‡ I haue not found the double one wilde, but ſeene it preſerued in diuers
Gardens for the beauty of the floure. ‡

¶ *The Time.*

They floure in the Spring when the Crowfoots doe, and oftentimes in Summer: the
leaues keepe their greeneneſſe all the Winter
long.

¶ *The Names.*

Marſh Marigold is called of *Valerius Cordus,*
Caltha paluſtris: of *Tabernamontanus, Populago:*
but not properly: in Engliſh ; March Marigolds: in Cheſhire and thoſe parts it is called
Bootes.

¶ *The Temperature and Vertues.*

Touching the faculties of theſe plants, wee
haue nothing to ſay, either out of other mens
writings, or our owne experience.

CHAP. 295.   *Of Frogge-bit.*

Morſus Rana.
Frogge-bit.

¶ *The Deſcription.*

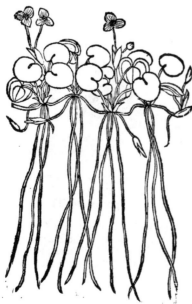

THere floteth or ſwimmeth vpon the vpper
parts of the water a ſmall plant, which we
vſually call Frog-bit, hauing little round
leaues, thicke and full of juyce, very like to the
leaues of wall Peniwort: the floures grow vpon
long ſtems among the leaues, of a white colour,
with a certain yellow thrum in the middle conſiſting of three leaues : in ſtead of roots it hath
ſlender ſtrings, which grow out of a ſhort and
ſmall head, as it were, from whence the leaues
ſpring, in the bottom of the water : from which
head alſo come forth ſlopewiſe certain ſtrings,
by which growing forth it multiplieth it ſelfe.

¶ *The Place.*

It is found ſwimming or floting almoſt in e-
uery ditch, pond, poole, or ſtanding water, in all
the ditches about Saint George his fields, and
in the ditches by the Thames ſide neere to
Lambeth Marſh, where any that is diſpoſed
may ſee it.

¶ *The Time.*

It flouriſheth and floureth moſt part of all
the yeare

¶ *The Names.*

It is called of ſome *Rana morſus,* and *Morſus*
Rana, and *Nymphæa parua.*

¶ *The*

¶ *The Temperature and Vertues.*

It is thought to be a kinde of Pond-weed (or rather of Water Lillie) and to haue the same fa- A
culties that belong vnto it.

Chap. 296. *Of Water Lillie.*

¶ *The Description.*

1   THe white water Lillie of *Nenuphar* hath great round leaues, in shape of a buckler, thick,
fat, and full of juyce, standing vpon long round and smooth foot-stalkes, ful of a spungi-
ous substance; which leaues do swim or flote vpon the top of the water: vpon the end
of each stalk groweth one floure only, of colour white, consisting of many little long sharpe pointed
leaues in the middest whereof be many yellow threds: after the floure it bringeth forth a round
head, in which lieth blackish glittering seed. The roots be thicke, full of knots, blacke without,
white and spungy within, out of which groweth a multitude of strings, by which it is fastned in the
bottome.

1 *Nymphæa alba.*
White Water Lillie.

2 *Nymphæa lutea.*
Yellow Water Lillie.

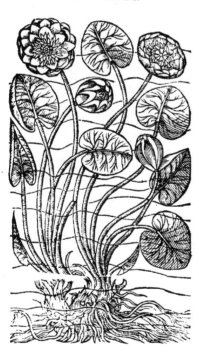
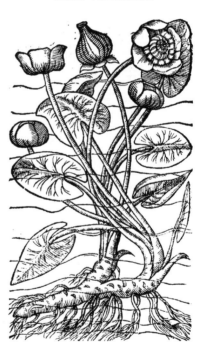

2   The leaues of the yellow water Lilly be like to the other, yet are they a little longer. The
stalkes of the floures and leaues be like: the floures be yellow, consisting onely of fiue little short
leaues something round; in the middest of which groweth a small round head, or button, sharpe towards the point, compassed about with many yellow threds, in which, when it is ripe, lie also glittering seeds, greater than those of the other, and lesser than wheat cornes. The roots be thicke, long,
set with certaine dents, as it were white both within and without, of a spungeous substance.

❦ 3   The small white water Lillie floteth likewise vpon the water, hauing a single root, with some
few fibres fastened thereto: from which riseth vp many long, round, smooth, and soft foot stalkes,
some of which doe bring forth at the end faire broad round buckler leaues like vnto the precedent,
but

but leſſer : on the other foot-ſtalks ſtand pretty white floures, conſiſting of fiue ſmall leaues apeece, hauing a little yellow in the middle thereof.

4   The ſmall yellow water Lillie hath a little thready root, creeping in the bottome of the water and diſperſing it ſelfe far abroad : from which riſe ſmall tender ſtalkes, ſmooth and ſoft, whereon doe grow little buckler leaues like the laſt deſcribed : likewiſe on the other ſmall ſtalke ſtandeth a tuft of many floures likewiſe floting vpon the water as the others do. ‡ This hath the floures larger than thoſe of the next deſcribed, wherefore it may be fitly named *Nymphæa lutea minor flore amplo.* ‡

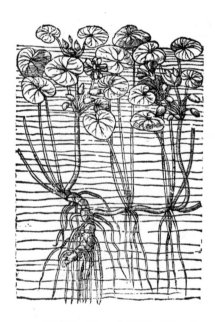

3   *Nymphæa alba minor.*
The ſmall white Water Lillie.

5   *Nymphæa lutea minima.*
Dwarfe Water Lillie.

5   This dwarfe water Lillie differeth not from the other ſmall yellow water Lillie, ſauing that, that this kinde hath ſharper pointed leaues, and the whole plant is altogether leſſer, wherein lieth the difference. ‡ This hath the floures much leſſe than thoſe of the laſt deſcribed, wherefore it is fitly for diſtinction ſake named *Nymphæa lutea minor flore paruo.* ‡

¶ *The Place.*

Theſe herbes do grow in fennes, ſtanding waters, broad ditches, and in brookes that run ſlowly, and ſometimes in great riuers.

¶ *The Time.*

They floure and flouriſh moſt of the Summer moneths.

¶ *The Names.*

Water Lillie is called in Greeke, Νυμφαια : and in Latine alſo *Nymphæa,* ſo named becauſe it loues to grow in watery places, as *Dioſcorides* ſaith : the Apothecaries call it *Nenuphar :* of *Apuleius, Mater Herculania, Alga paluſtris, Papauer paluſtre, Clauus Veneris,* and *Digitus Veneris : Marcellus* a very old writer reporteth, that it is called in Latine, *Claua Herculis :* in french, *Baditin :* in high Dutch, ſeaſſer Mahem : in low Dutch, Plompen : in Engliſh, Water Lillie, Water Roſe.

¶ *The Temperature.*

Both the root and ſeed of water Lillie haue a drying force without biting.

¶ *The Vertues.*

A   Water Lillie with yellow floures ſtoppeth laskes, the ouerflowing of ſeed which commeth away by dreames or otherwiſe, and is good for them that haue the bloudy flix.

But

But water Lilly which hath the white flours is of greater force, inſomuch as it ſtaies the whites: **B** but both this and the other that hath the blacke root muſt be drunke in red wine : they haue alſo a ſcouring qualitie, and clenſe away the morphew, being good alſo againſt the pilling away of the haire of the head : againſt the morphew they are ſteeped in water, and for the pilling away of the haire in Tarre : but for theſe things that is fitter which hath the blacke root, and for the other, that which hath the white root.

Theophraſtus ſaith, That being ſtamped and laid vpon the wound it is reported to ſtay the blee-**C** ding.

The Phyſitions of our age do commend the floures of white *Nymphæa* againſt the infirmities of **D** the head comming of an hot cauſe ; and do certainly affirme, that the root of the yellow cures hot diſeaſes of the kidnies and bladder, and is ſingular good againſt the running of the reines.

The root and ſeed of great water Lilly is very good againſt venerie or fleſhly deſire, if one drink **E** the decoction of it, or vſe the ſeed or root in pouder in his meats ; for it drieth vp the ſeed of gene-ration, and ſo cauſeth a man to be chaſte, eſpecially vſed in broth with fleſh.

The conſerue of the floures is good for the diſeaſes aforeſaid, as alſo againſt hot burning feuers. **F**

The floures made into oile as yee make oile of roſes, coole and refrigerate, cauſing ſweat and **G** quiet ſleepe, and put away all venereous dreames : the temples of the head, the palms of the hands and feet, and the breſt being anointed for the one, the genitoirs vpon and about them for the other.

The green leaues of the great water Lilly either the white or the yellow, laid vpon the region of **H** the backe in the ſmall, mightily ceaſe the involuntary flowing of the ſeed called *Gonorrhæa* or run-ning of the reins, being two or three times a day remoued, and freſh applied thereto.

CHAP. 297.

Of Pond-weed, or water Spike.

1 *Potamogeiton latifolium.*
Broad leafed Pond-weed.

2 *Potamogeiton anguſtifolium.*
Narrow leafed Pond-weed.

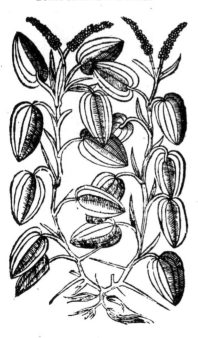

¶ *The Deſcription.*

1   POnd-weed hath little ſtalkes, ſlender, ſpreading like thoſe of the Vine, and jointed: the leaues be long, ſmaller than the leaues of Plantain, and harder, with manifeſt veins running alongſt them as in Plantains, which ſtanding vpon ſlender and long ſtems or footſtalks, ſhew themſelues aboue the water, and lie flat along vpon the ſuperficial or vpper part thereof, as do the leaues of water Lilly: the floures grow in ſhort eares, and are of a light red purple colour like thoſe of Red-ſhanks or Biſtort: the ſeed is hard.

‡ 2   This (whoſe figure was formerly vnfitly put by our Author to the following deſcription) hath longer, narrower, and ſharper pointed leaues than thoſe of the laſt deſcribed, hauing the veins running from the middle rib to the ſides of the leaues, as in a willow leaſe, which they ſomwhat reſemble: at the tops of the ſtalkes grow reddiſh ſpikes or eares like thoſe of the laſt deſcribed: the root is long, jointed, and fibrous. ‡

‡ 3 *Potamogeiton 3. Dodonæi.*
Small Pond-weed.

‡ 4 *Potamogeiton longis acutis folijs.*
Long ſharp leaued Pond-weed.

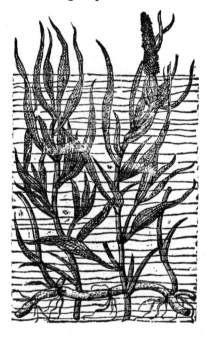

3   There is another Pond-weed deſcribed thus; it ſhooteth forth into many ſlender & round ſtems, which are diſtributed into ſundry branches: his leaues are broad, long, and ſharp pointed, yet much leſſe than the firſt kinde: out of the boſomes of the branches and leaues there ſpring certain little ſtalks which beare ſundry ſmall white moſſie floures, which turn into plaine and round ſeeds like the common Tare or Vetch: the root is fibrous, throughly faſtned in the ground.

‡ 4   There is alſo another Pondweed, which hath whitiſh and jointed roots creeping in the bottom of the water, with ſome fibres, but ſending vp ſlender jointed and long ſtalkes, ſmall below, and bigger aboue, hauing long narrow and very ſtiffe ſharpe pointed leaues. The floures grow in a reddiſh ſpike like thoſe of the firſt deſcribed. This is the *Potamogeiton altera* of *Dodonæus.* ‡

¶ *The Place.*

Theſe herbs grow in ſtanding waters, pooles, ponds, and ditches almoſt euery where.

¶ *The Time.*

They floure in Iune and Iuly.

¶ *The Names.*

It is called of the Greeks, ποταμογείτων: in Latine, *Fontalis,* and *Spicata:* in high Dutch, **Zamkraut:** in low-Dutch, **ſontepnerupt:** in French, *Eſpi d'eaue:* in Engliſh, Pondweed, and water Spike.

¶ *The*

¶ *The Temperature.*

Pondweed,ſaith *Galen*,doth binde and coole,like as doth Knot-graſſe,but his eſſence is thicker than that of Knot-graſſe.

¶ *The Vertues.*

It is good againſt the itch and conſuming and eating vlcers,as *Dioſcorides* writeth.

Alſo it is good being applied to the inflammation of the leggs,wherein *Ignis ſacer* hath got the ſuperioritie.

A
B

Chap. 298.
Of Water Saligot, water Caltrops, or water Nuts.

¶ *The Deſcription.*

1 WAter Caltrops haue long ſlender ſtalks growing vp and riſing from the bottom of the water,and mounting aboue the ſame : the root is long,hauing here & there vnder the water certaine taſſels full of ſmall ſtrings or thready baires : the ſtem towards the top of the water is very great in reſpect of that which is lower ; the leaues are large and ſomewhat round,not vnlike thoſe of the Poplar or Elme tree leaues,a little creuiſed or notched about the edges : amongſt or vnder the leaues grow the fruit,which is triangled,hard,ſharp pointed and prickly,in ſhape like thoſe hurtfull engins in the wars,caſt in the paſſage of the enemy to annoy the feet of their horſes,called Caltrops, whereof this tooke it's name : within theſe heads or Nuts is contained a white kernell in taſte almoſt like the Cheſnut,which is reported to bee eaten green,and being dried and ground to ſerue in ſtead of bread.

‡ There are two other plants which are found growing in many ponds & ditches of this kingdome both about London and elſewhere , and I will here giue you their figures out of *Lobel* and *Cluſius*,and their deſcriptions as they were ſent me by Mr *Goodyer*, who hath ſaued me the labor of deſcribing them.

Tribulus aquaticus minor quercus floribus, Cluſ. p.252.
Puſillum fontila pathum, Lobelij.

2 This water herb bringeth forth from the root,thin flat knotty ſtalks of a reddiſh colour,two or three cubits long,or longer,according to the depth of the water (which when they are dry are pliant or bowing) diuided towards the top into many parts or branches, bearing but one leafe at euery joint,ſomtimes two inches long,and halfe an inch broad,thin,and as it were ſhining,ſo wrinkled and crumpled by the ſides that it ſeemeth to be torn, of a reddiſh green colour:the foot ſtalks are ſomwhat long and thicke,and riſe vp from amongſt thoſe leaues,which alwayes grow two one oppoſit againſt another,in a contrary manner to thoſe that grow below on the ſtalk : neere the top of which foot-ſtalke groweth ſmall grape-like husks,out of which ſpring very ſmal reddiſh flours like thoſe of the Oke,euery floure hauing foure very ſmall round topped leaues : after euery floure commeth commonly foure ſharp pointed graines growing together,containing within them a little white kernell. The lower part of the ſtalke hath at euery joint ſmall white thready roots,ſomwhat long,whereby it taketh hold in the mud,and draweth nouriſhment vnto it. The whole plant is commonly couered ouer with water : it floures in Iune and the beginning of Iuly. I found it in the ſtanding pooles or fiſh-ponds adioyning to a diſſolued Abby called Durford,which ponds diuide Hampſhire and Suſſex,and in other ſtanding waters elſewhere. This deſcription was made vpon ſight of the plant,Iune 2, 1622.

Tribulus aquaticus minor, muſcatella floribus.

3 This hath not flat ſtalks like the other,but round, kneed, and alwaies bearing two leaues at euery joint,one oppoſite againſt another,greener, ſhorter, and leſſer than the other, ſharp pointed, not much wrinkled and crumpled by the edges : *Cluſius* ſaith that they are not at all crumpled. I neuer obſerued any without crumples and wrinkles : the floures grow on ſhort ſmall foot ſtalks of a whitiſh green colour like thoſe of *Muſcatella Cordi*,called by *Gerard, Radix cana minima virid flore: viz.*two floures at the top of euery foot-ſtalke,one oppoſit againſt another,euery floure containing foure ſmall leaues : which two floures being paſt,there come vp eight ſmall huskes,making ſix ſe-

ueral

1 *Tribulus aquaticus.*
Water Caltrops.

‡ 3 *Tribulus aquaticus minor, Muscatella floribus.*
Small Frogs Lettuce.

‡ 2 *Tribulus aquaticus minor quercus floribus.*
Small water Caltrops, or Frogs Lettuce.

uerall wayes a square of floures. The roots
are like the former. This groweth aboundantly in the riuer by Droxford in Hampshire. It floures in Iune and Iuly when the
other doth, and continueth couered ouer
with water, green both winter & sommer.
Iohn Goodyer. ‡

¶ *The Place.*

Cordus saith that it groweth in Germanie in myrie lakes, and in city ditches that
haue mud in them : in Brabant and other
places of the Lowcountries it is found oftentimes in standing waters and springs.
Mathiolus writeth, that it growes not only
in lakes of sweet water, but also in certain
ditches by the sea neere vnto Venice.

¶ *The Time.*
It flourisheth in Iune, Iuly, & August.

¶ *The Names.*

The Grecians cal it *τρίβολος ἔνυδρος:* the Latines, *Tribulus aquatilis* and *aquaticus,* and
Tribulus lacustris the Apothecaries, *Tribulus marinus:* in high Dutch, ꝏ aſſer nutʒ :
the Brabanders, ꝏ ater noten : and of the
likenesse of yron nailes , Minckiſſers:
the

the French men, *Macres* : in English it is named water Caltrops, Saligot, and water Nuts : most do call the fruit of this Caltrops, *Castanea aquatiles*, or water Chesnuts.

¶ *The Temperature.*

Water Caltrop is of a cold nature, it consisteth of a moist essence, which in this is more waterie than in the land Caltrops, wherein an earthy cold is predominant, as *Galen* saith.

¶ *The Vertues.*

The herb vsed in manner of a pultis, as *Dioscorides* teacheth, is good against all inflammations or　**A** hot swellings: boiled with hony and water it perfectly cureth cankers of the mouth, sore gums, and the almonds of the throat.

The Thracians, saith *Pliny*, that dwel in Strymona, fatten their horses with the leaues of Saligot,　**B** and they themselues feed of the kernels, making very sweet bread thereof, which bindes the belly.

The green nuts or fruit of *Tribulus aquaticus*, or Saligot, drunke in wine, are good for them that　**C** are troubled with the stone and grauell.

The same drunke in like manner, or laid outwardly to the place, helpeth those that are bitten　**D** with any venomous beast, and resisteth all venome and poison.

The leaues of Saligot be giuen against all inflammations and vlcers of the mouth, the putrifa-　**E** ction and corruption of the jawes, and against the Kings euill.

A pouder made of the nuts is giuen to such as pisse bloud and are troubled with grauell: it bin-　**F** deth the belly very much.

‡ The two lesser water Caltrops here described are in my opinion much agreeable in temper　**G** to the great one, and are much fitter *Succidanea* for it than Aron, which some in the composition of *Vnguentum Agrippa* haue appointed for it. ‡

<h3 style="text-align:center">Chap. 299.</h3>

<h2 style="text-align:center">Of Water Sengreen, or Fresh-water Souldier.</h2>

Militaris Aizoides.
Fresh-water Souldier.

¶ *The Description.*

FResh-water Souldier or Water Housleeke hath leaues like those of the Herb Aloë or *Sempervivum*, but shorter and lesser, set round about the edges with certaine stiffe and short prickles : amongst which commeth forth diuers cases or husks very like to crabs clawes, out of which when they open grow white floures, consisting of three leaues altogether like those of Frogs-bit, hauing in the middle little yellowish threads : in stead of roots there be long strings, round, white, very like to great harp strings or long worms, which falling down from a short head that brought forth the leaues, go to the bottome of the water, and yet be they seldome there fastned : there grow also from the same other strings aslope, by which the plant is multiplied after the manner of Frogs-bit.

¶ *The Place.*

‡ I found this growing plentifully in the ditches about Rotsey a small village in Holdernesse: and my friend Mr *William Broad* obserued it in the fennes in Lincolnshire. ‡ The leaues and flours grow vpon the top of the water, and the roots are sent downe through the water to the mud.

¶ *The Time.*

It floures in Iune, and sometimes in August.

¶ *The Names.*

It may be called *Sedum aquatile,* or water Sengreen, that is to ſay, of the likeneſſe of Herb Aloë, which is alſo called in Latine *Sedum :* of ſome, *Cancri chela,* or *Cancri forficula :* in Engliſh, Water Houſleeke, Knights Pondwort, and of ſome, Knights water Sengreen, Freſh-water Soldier, or wading Pondweed : it ſeems to be *Stratiotes aquatilis,* or *Stratiotes potamios,* or Knights water Woundwort, which may alſo be named in Latine, *Militaris aquatica,* and *Militaris Aizoides,* or ſoldiers Yarrow ; for it groweth in the water, and floteth vpon it, and if thoſe ſtrings which it ſendeth to the bottom of the water be no roots, it alſo liueth without roots.

¶ *The Temperature.*

This herb is of a cooling nature and remperament.

¶ *The Vertues.*

This Houſleeke ſtaieth the bloud which commeth from the kidnies, it keepeth greene wounds from being inflamed, and is good againſt S. *Anthonies* fire and hot ſwellings, beeing applied vnto them : and is equall in vertues with the former aforeſaid.

CHAP. 300.   *Of water Yarrow or water Gilloſloure.*

1 *Viola paluſtris.*   ‡ 2 *Viola paluſtris tenuifolia.*
Water Violet.   The ſmaller leaued water Violet.

¶ *The Deſcription.*

1   WAter Violet hath long and great jagged leaues very finely cut or rent like Yarrow, but ſmaller : among which come vp ſmall ſtalks a cubit and a halfe high, bearing at the top ſmall white floures like vnto ſtock Gilloſloures, with ſome yellowneſſe in the middle. The roots are long and ſmall like black threds, and at the end whereby they are faſtned to the ground they are white.

‡ There is another variety of this plant, which differs from it only in that the leaues are much ſmaller, as you may ſee them expreſt in the figure. ‡

2　Water Milfoile or water Yarrow hath long and large leaues deepely cut with many diuiſions like Fennell, but finelier jagged, ſwimming vpon the water. The root is ſingle, long, and round,
which brings vp a right ſtraight and ſlender ſtalke, ſet in ſundry places with the like leaues, but
ſmaller. The leaues grow at the top of the ſtalk, tuft-faſhion, and like vnto the land Yarrow.

3　This water Milfoile differeth from all the kindes aforeſaid, hauing a root in the bottom of
the water, made of many hairy ſtrings, which yeeldeth vp a naked ſlender ſtalke within the water,
and the reſt of the ſtalk which floteth vpon the water diuideth it ſelfe into ſundry other branches
or wings, which are bedaſht with fine ſmall jagged leaues like vnto Cammomill, or rather reſembling hairy taſſels or fringe, than leaues. From the boſomes whereof come forth ſmall and tender
branches, euery branch bearing one floure like vnto water Crow-foot, white of colour, with a little
yellow in the midſt: the whole plant reſembleth water Crow-foot in all things ſaue in the broad
leaues.

† 4　There is another kinde of water Violet very like the former, ſauing that his leaues are
much longer, ſomewhat reſembling the leaues of fennell, faſhioned like vnto wings, and the flours
are ſomewhat ſmaller, yet white, with ſome yellowneſſe in their middles, and ſhaped like thoſe of
the laſt deſcribed: the ſeed alſo growes like vnto that of the water *Ranunculus* laſt deſcribed.

5　There is alſo another kinde of water Milfoile which hath leaues very like vnto water Violet, ſmaller, and not ſo many in number: the ſtalk is ſmall and tender, bearing yellow gaping flours
faſhioned like a hood or the ſmall Snapdragon; which cauſed *Pena* to put vnto his name this additament *Galericulatum*, that is, hooded. The roots are ſmall and thready, with ſome few knobs
hanging thereat like the ſounds of fiſh.

2　*Millefolium aquaticum.*
Water Yarrow.

3　*Millefolium, ſiue Maratriphyllion, flore &*
ſemine Ranunculi aquatice, Hepatica facie.
Crow-foot, or water Milfoile.

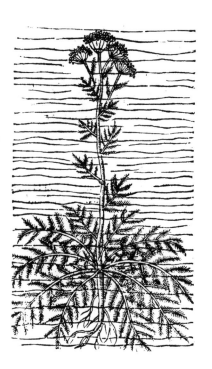

‡ 6　To theſe may we adde a ſmall water Milfoile ſet forth by *Cluſius*: it hath round greene
ſtalks ſet with many joints, whereout come at their lower ends many hairy fibres whereby it takes
hold of the mud: the tops of theſe ſtems ſtand ſome handfull aboue the water, and at each joint
ſend fiue long finely winged leaues, very greene, and ſome inch long, which wax leſſe and leſſe, as
they

they stand higher or neerer the top of the stalke; and at each of these leaues about the top of the stem growes one small white floure consisting of six little leaues ioined together, and not opening themselues, which at length turne into little knobs, with foure little pointals standing out of them. *Clusius* calls this *Myriophyllon aquaticum minus*. ‡

‡ 4 *Millefolium tenuifolium.*
Fennel-leaued water Milfoile.

‡ 5 *Millefolium palustre galericulatum.*
Hooded water Milfoile.

¶ *The Place.*

They be found in lakes and standing waters, or in waters that run slowly: I haue not found such plenty of it in any one place as in the water ditches adioyning to S. *George* his field neere London.

¶ *The Time.*

They floure for the most part in May and Iune.

¶ *The Names.*

The first is called in Dutch, **water Violetian**, that is to say, *Viola aquatilis :* in English, Water Gillofloure, or water Violet: in French, *Gyrofflees d'eaue. Matthiolus* makes this to be also *Myriophylli species*, or a kinde of Yarrow, although it doth not agree with the description thereof; for neither hath it one stalke only, nor one single root, as *Myriophyllon* or Yarrow is described to haue; for the roots are full of strings, and it bringeth forth many stalks.

The second is called in Greeke, Μυριοφυλλον : in Latine, *Millefolium*, and *Myriophyllon*, and also *Supercilium Veneris* : in shops it is vnknowne. This Yarrow differs from that of the land: the rest are sufficiently spoken of in their titles.

¶ *The Temperature and Vertues.*

A   Water Yarrow, as *Dioscorides* saith, is of a dry facultie, and by reason that it takes away hot inflammations and swellings, it seemeth to be of a cold nature; for *Dioscorides* affirmeth, That water Yarrow is a remedie against inflammation in green wounds, if with vineger it be applied greene or dry: and it is giuen inwardly with vineger and salt to those that haue fallen from an high place.

B   Water Gillofloure or water Violet is thought to be cold and dry, yet hath it no vse in physick.

C H A P.